PROCESS DYNAMICS AND CONTROL

Wiley Series in Chemical Engineering

Process Dynamics and Control

Dale E. Seborg
University of California, Santa Barbara

Thomas F. Edgar
University of Texas at Austin

Duncan A. Mellichamp
University of California, Santa Barbara

WILEY
John Wiley & Sons
New York   Chichester   Brisbane   Toronto   Singapore

ISBN 0-471-86389-0

Printed in the United States of America

10 9 8 7 6 5 4 3 2 1

About the Authors

Dale E. Seborg is a Professor of Chemical Engineering at the University of California, Santa Barbara. He received his B.S. degree from the University of Wisconsin and his Ph.D. degree from Princeton University. Before joining UCSB in 1977, he taught at the University of Alberta for nine years. Dr. Seborg has published numerous articles on process control and related topics. He is a co-author of *Multivariable Computer Control—A Case Study* and coeditor of *Chemical Process Control 2*. He has received awards from the Joint Automatic Control Conference and the AIChE Southern California Section. He is an active industrial consultant and has served as a director of three organizations: the American Automatic Control Council, the AIChE Computing and Systems Technology Division, and the ASEE ChE Division. Dr. Seborg serves on the editorial boards of the *IEEE Transactions on Automatic Control*, and *Adaptive Control and Signal Processing*.

Thomas F. Edgar is Professor/Chairman of Chemical Engineering at the University of Texas, Austin. He received his B.S.Ch.E. from the University of Kansas and the Ph.D. in Chemical Engineering at Princeton University. Dr. Edgar has been President of the CACHE Corporation, Chairman of the Computing and Systems Technology Division of AIChE, and Vice-President of the American Automatic Control Council. He has won the AIChE Colburn Award and the ASEE Westinghouse Award. He has served on the editorial boards for *Chemical Engineering Reviews, Computers and Chemical Engineering,* and the *AIChE Journal*. Dr. Edgar has published extensively in the fields of process control, optimization and mathematical modeling and is author of *Coal Processing and Pollution Control* and co-author of *Optimization of Chemical Processes*.

Duncan A. Mellichamp is a Professor of Chemical Engineering at the University of California, Santa Barbara. He received his B.S. degree from Georgia Institute of Technology and his Ph.D. degree from Purdue University. He has taught at UCSB since 1966, having started up the process control program when the department was founded. Prior to that, he was a Research Engineer with the Textile Fibers Department of Du Pont for four years. Dr. Mellichamp was a pioneer in real-time computing and has edited a book, *Real-Time Computing with Applications to Data Acquisition and Control*. He served in a variety of capacities with CACHE, including Trustee for fifteen years and President. He is the author of numerous research papers and editor or author of several monographs in the areas of process modeling, large-scale systems analysis, and computer control.

Preface

Process control has undergone significant changes since the 1970s when the availability of inexpensive digital technology began a radical change in instrumentation technology. Pressures associated with increased competition, rapidly changing economic conditions, more stringent environmental regulations, and the need for more flexible yet more complex processes have given process control engineers an expanded role in the design and operation of processing plants. High-performance measurement and control systems, most often based on the use of digital instrumentation, play a critical role in making modern industrial plants economically competitive.

In the foreseeable future there will be an expanded use of microprocessor-based instrumentation and networks of digital computers. More sophisticated control strategies—including feedforward, supervisory, multivariable, and adaptive control features as well as sophisticated digital logic—are easily justified to maintain plant operation closer to the economic optimum. On the other hand, conventional analog control systems continue to be used in many existing plants.

A modern undergraduate course in chemical process control should reflect this rather diverse milieu of theory and applications. It should incorporate process dynamics, computer simulation, feedback control, a discussion of measurement and control hardware (both analog and digital), advanced control strategies, and digital control techniques. This textbook allows the instructor to cover the basic material while providing the flexibility to pursue selected topics in the process control field. It can provide the basis of 20 to 30 weeks of instruction, covering a single course or a sequence of courses at the undergraduate or first year graduate levels. We have divided the text into reasonably short chapters to make the book more readable and to enhance its use in a modular fashion. This organization allows the student to skip some chapters without a loss of continuity. For example, the subject of digital control could be omitted or included readily in a particular course.

The mathematical level of the book is oriented toward the "typical" chemical engineering student (who has taken at least one course in differential equations) and the engineer in industry. The additional mathematical tools required for analysis of control systems are introduced as needed in the text. We have emphasized the control techniques that have been found useful in control system design practice and have provided detailed mathematical developments only when they help in understanding the material.

The text material has evolved at the Universities of California (Santa Barbara), Alberta, and Texas over the past 20 years. As part of our effort to rethink the teaching of process control, we have omitted outdated topics that are not being used today by process control engineers to design control systems. However, we have included material on frequency response, even though these methods are not often used in the process industries, because they provide valuable insight for the control engineer (indeed, they provide the source of much of the descriptive vocabulary used by control engineers). Many people believe that they should be a part of a first course in process control. At the same time, the book has been structured so that an instructor with a preference for transient response design techniques can teach an entire course with only passing reference to the frequency response material.

The book is divided into seven parts. Part One presents introductory concepts of process control along with an overview of mathematical modeling based on material and energy balances together with basic principles of chemistry and physics. In Chapter 1 a stirred-tank heater system is used as the vehicle for introducing the student to the concepts of feedback and feedforward control. The development of phenomenological models for representative unit operations is considered in Chapter 2, including liquid storage systems, continuous stirred-tank reactors, staged absorbers, and heat exchangers.

Part Two (Chapters 3 through 7) is concerned with dynamic behavior of processes. In Chapter 3 the Laplace transform is introduced as a general means for solving differential equations and obtaining dynamic responses. When the Laplace transform is used, the concept of a transfer function follows directly (Chapter 4). Once this transform tool is mastered, it can be used to determine transient responses of first- and second-order systems (Chapter 5) and higher-order and distributed-parameter systems (Chapter 6). Chapter 7 presents an alternative approach for obtaining process dynamics, one in which basic phenomenological models are ignored in favor of empirical models. Here we specifically consider methods for obtaining dynamic models from step response data.

Part Three is devoted to the subject of feedback control. Chapters 8 and 9 deal with hardware aspects of measurement and control components. These are "bridging" chapters to the topic of controller design, since we wish to emphasize that a control system contains not only a controller, but also other components— sensors, transmitters, and final control elements—that modify the dynamics of the control loop. Chapters 10 through 13 deal with the analysis and design of feedback control systems, including closed-loop analysis (Chapters 10 and 11), transient response design methods (Chapter 12), and field tuning and troubleshooting of control loops (Chapter 13).

Frequency response methods are the focus of Part Four. Frequency response analysis of dynamic systems is covered in Chapter 14, followed by the development of empirical methods to obtain process models using frequency response methods (Chapter 15). Chapter 16 presents the fundamentals of controller design based on frequency response methods, including the important Bode and Nyquist stability criteria. An integrated computer-aided design procedure that makes use of Bode plots, the Nichols chart, and time-domain response is discussed.

Chapters 1 through 16 plus Chapter 28 can serve as the core material for most undergraduate courses. However, an undergraduate course should also expose students to advanced control concepts in Part Five. Chapters 17 through 19 include well-known topics such as feedforward, ratio, cascade, and multivariable control, plus time-delay compensation. In Chapter 18 we briefly introduce three promising techniques: adaptive control, statistical quality control, and expert systems. The final chapter in this group (Chapter 20) is concerned with steady-state optimization in process control, which is often referred to as "supervisory control." Much of the material in Part Five has not appeared in previous process control textbooks and hence, will be of interest to the industrial practitioner as well.

Part Six (Chapters 21 through 27) is devoted to digital computer control. In many ways this material parallels that for analog control techniques covered earlier in the text. Chapter 21 focuses on digital instrumentation, real-time computing, programmable logic controllers and batch sequencing and control. Chapter 22 deals with sampling operations and the filtering of noisy data.

Chapters 23 and 24 deal with the dynamics of sampled-data systems. Chapter 23 employs a difference equation approach to represent the discrete model. Chapter 24 presents the classical z-transform approach for calculating dynamic responses (in analogy to Chapters 3, 4, and 5).

Chapters 25 through 27 deal with stability criteria and design methods for digital control systems, in analogy to Chapters 11, 12, and 16. Several alternative ways to design controllers are given in Chapter 26, with the emphasis on model-based design methods. Several recent improvements in the standard digital design approaches are outlined, especially for handling disturbances. Chapter 27 presents a powerful new approach, predictive control, in which a process model is explicitly incorporated into the control strategy to predict future process behavior.

In Chapter 28 we emphasize that process control is as much an art as it is a science. We consider general issues such as the influence of process design on process control and the selection of controlled and manipulated variables. We conclude with an industrial case study. Chapter 28 should be used at the end of a typical undergraduate course.

This book has been classroom-tested for several years at the University of California, Santa Barbara and at the University of Texas at Austin, where we have received invaluable "feedback" from our undergraduate and graduate students. Engineering personnel from many companies who utilized portions of the manuscript in various short courses have provided useful suggestions. David Cardner (Du Pont) provided useful background information for the industrial case study in Chapter 28. We also gratefully acknowledge the very helpful classroom evaluations by Dominique Bonvin (ETH, Zurich), Sandra Harris (Clarkson University), Manfred Morari (Caltech), Jim Rawlings (University of Texas), Sirish Shah (University of Alberta), and Alan Schneider (University of California, San Diego), who used earlier versions of the book in their classes. Mukul Agarwal provided numerous helpful suggestions while preparing the Solution Manual for the book. Finally, we gratefully commend Carina Billigmeier, Barbara Merlo, and Bee Hanson for their skill and patience in typing the numerous versions of the text.

Dale E. Seborg
Thomas F. Edgar
Duncan A. Mellichamp

Contents

PART FIVE
ADVANCED CONTROL TECHNIQUES

PART SIX
DIGITAL CONTROL TECHNIQUES

PART ONE

INTRODUCTORY CONCEPTS

— CHAPTER 1 —

Introduction to Process Control

In recent years the performance requirements for process plants have become increasingly difficult to satisfy. Stronger competition, tougher environmental and safety regulations, and rapidly changing economic conditions have been key factors in the tightening of plant product quality specifications. A further complication is that modern processes have become more difficult to operate because of the trend toward larger, more highly integrated plants with smaller surge capacities between the various processing units. Such plants give the operators little opportunity to prevent upsets from propagating from one unit to other interconnected units. In view of the increased emphasis placed on safe, efficient plant operation, it is only natural that the subject of *process control* has become increasingly important in recent years. In fact, without process control it would not be possible to operate most modern processes safely and profitably, while satisfying plant quality standards.

1.1 ILLUSTRATIVE EXAMPLE

As an introduction to process control, consider the continuous stirred-tank heater shown in Fig. 1.1. The inlet liquid stream has a mass flow rate w and a temperature T_i. The tank contents are well agitated and heated by an electrical heater that provides Q watts. It is assumed that the inlet and outlet flow rates are identical and that the liquid density ρ remains constant, that is, the temperature variations are small enough that the temperature dependence of ρ can be neglected. Under these conditions the volume V of liquid in the tank remains constant.

The control objective for the stirred-tank heater is to keep the exit temperature T at a constant reference value T_R. The reference value is referred to as a *set point* in control terminology. Next we consider two questions.

Question 1. How much heat must be supplied to the stirred-tank heater to heat the liquid from an inlet temperature T_i to an exit temperature T_R?

To determine the required heat input for the design operating conditions, we need to write a steady-state energy balance for the liquid in the tank. In writing this balance, it is assumed that the tank is perfectly mixed and that heat losses are negligible. Under these conditions there are no temperature gradients within the tank contents and consequently, the exit temperature is equal to the temperature

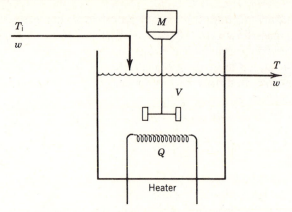

Figure 1.1. Continuous stirred-tank heater.

of the liquid in the tank. A steady-state energy balance for the tank indicates that the heat added is equal to the change in enthalpy between the inlet and exit streams:

$$\overline{Q} = \overline{w}C(\overline{T} - \overline{T}_i) \tag{1-1}$$

where $\overline{T}_i$, $\overline{T}$, $\overline{w}$, and $\overline{Q}$ denote the nominal steady-state design values of T_i, T, w, and Q, respectively, and C is the specific heat of the liquid. We assume that C is constant. At the design conditions, $\overline{T} = T_R$ (the set point). Making this substitution in Eq. 1-1 gives an expression for the nominal heat input $\overline{Q}$:

$$\overline{Q} = \overline{w}C(T_R - \overline{T}_i) \tag{1-2}$$

Equation 1-2 is the design equation for the heater. If our assumptions are correct and if the inlet flow rate and inlet temperature are equal to their nominal values, then the heat input given by Eq. 1-2 will keep the exit temperature at the desired value, T_R. But what if conditions change? This brings us to the second question:

Question 2. Suppose that inlet temperature T_i changes with time. How can we ensure that T remains at or near the set point T_R?

As a specific example, assume that T_i increases to a new value greater than $\overline{T}_i$. If Q is held constant at the nominal value of $\overline{Q}$, we know that the exit temperature will increase so that $T > T_R$. (cf. Eq. 1-1).

To deal with this situation, there are a number of possible strategies for controlling exit temperature T.

Method 1. *Measure T and adjust Q.*   One way of controlling T despite disturbances in T_i is to adjust Q based on measurements of T. Intuitively, if T is too high, we should reduce Q; if T is too low, we should increase Q. This control strategy will tend to move T toward the set point T_R and could be implemented in a number of different ways. For example, a plant operator could observe the measured temperature and compare the measured value to T_R. The operator would then change Q in an appropriate manner. This would be an application of *manual control*. However, it would probably be more convenient and economical to have this simple control task performed automatically by an electronic device rather than a person, that is, to utilize *automatic control*.

Method 2. *Measure T_i, adjust Q.*   As an alternative to Method 1, we could measure disturbance variable T_i and adjust Q accordingly. Thus, if T_i is greater than $\overline{T}_i$, we would decrease Q; for $T_i < \overline{T}_i$ we would set $Q > \overline{Q}$.

Method 3. *Measure T, adjust w.*   Instead of adjusting Q, we could choose to manipulate mass flow rate w. Thus, if T is too high we would increase w to reduce the energy input rate in the stirred tank relative to the mass flow rate and thereby reduce the exit temperature.

Method 4. *Measure T_i, adjust w.*   In analogy with Method 3, if T_i is too high, w should be increased.

Method 5. *Measure T_i and T, adjust Q.*   This approach is a combination of Methods 1 and 2.

Method 6. *Measure T_i and T, adjust w.*   This approach is a combination of Methods 3 and 4.

Method 7. *Place a heat exchanger on the inlet stream.*   The heat exchanger is intended to reduce the disturbances in T_i and consequently reduce the variations in T. This approach is sometimes called "hog-tieing" an input.

Method 8. *Use a larger tank.*   If a larger tank is used, fluctuations in T_i will tend to be damped out due to the larger thermal capacitance of the tank contents. However, increased volume of tankage would be an expensive solution for an industrial plant due to the increased capital costs of the larger tank. Note that this approach is analogous to the use of water baths in chemistry laboratories where the large thermal capacitance of the bath serves as a heat sink and thus provides an isothermal environment for a small-scale research apparatus.

1.2 CLASSIFICATION OF CONTROL STRATEGIES

Next, we will classify the eight control strategies of the previous section and discuss their relative advantages and disadvantages. Methods 1 and 3 are examples of *feedback control* strategies. In feedback control, the process variable to be controlled is measured and the measurement is used to adjust another process variable which can be manipulated. Thus, for Method 1, the measured variable is T and the manipulated variable is Q. For Method 3, the measured variable is still T but the manipulated variable is now w. Note that in feedback control the disturbance variable T_i is not measured.

It is important to make a distinction between *negative feedback* and *positive feedback*. Negative feedback refers to the desirable situation where the corrective action taken by the controller tends to move the controlled variable toward the set point. In contrast, when positive feedback exists, the controller tends to make things worse by forcing the controlled variable farther away from the set point. Thus, for the stirred-tank heater, if T is too high we would decrease Q (negative feedback) rather than increase Q (positive feedback).[1]

Methods 2 and 4 are *feedforward control* strategies. Here, the disturbance variable T_i is measured and used to manipulate either Q (Method 2) or w (Method 4). Note that in feedforward control, the controlled variable T is *not* measured. Method 5 is a feedforward–feedback control strategy since it is a combination of Methods 1 and 2. Similarly, Method 6 is also a feedforward–feedback control strategy since it is a combination of Methods 3 and 4. Methods 7 and 8 consist of equipment design changes and thus are not really control strategies. Note that Method 7 is somewhat inappropriate since it involves adding a heat exchanger to

---

[1]Note that social scientists use the terms, negative feedback and positive feedback, in a very different way. For example, they would say that school teachers provide "positive feedback" when they compliment students who correctly do assignments. Criticism of a poor performance would be an example of negative feedback.

the inlet line of the stirred-tank heater which in itself was designed to function as a heat exchanger! The control strategies for the stirred-tank heater are summarized in Table 1.1.

So far we have considered only one source of process disturbances, fluctuations in T_i. We should also consider the possibility of disturbances in other process variables such as the ambient temperature, which would affect heat losses from the tank. Recall that heat losses were assumed to be negligible earlier. Changes in process equipment are another possible source of disturbances. For example, the heater characteristics could change with time due to scaling by the liquid. It is informative to examine the effects of these various types of disturbances on the feedforward and feedback control strategies discussed above.

First, consider the feedforward control strategy of Method 2 where the disturbances in T_i are measured and the measurements are used to adjust the manipulated variable Q. From a theoretical point of view, this control scheme is capable of keeping the controlled variable T exactly at set point T_R despite disturbances in T_i. Ideally, if accurate measurements of T_i were available and if the adjustments in Q were made in an appropriate manner, then the corrective action taken by the heater would cancel out the effects of the disturbances before T is affected. Thus, in principle, feedforward control is capable of providing *perfect control* in the sense that the controlled variable would be maintained at the set point.

But how will this feedforward control strategy perform if disturbances occur in other process variables? In particular, suppose that the flow rate w cannot be held constant but, instead, varies over time. In this situation, w would be considered a disturbance variable. If w increases, then the exit temperature T will decrease unless the heater supplies more heat. However, in the control strategy of Method 2 the heat input Q is maintained constant as long as T_i is constant. Thus *no* corrective action would be taken for unmeasured flow disturbances. In principle, we could deal with this situation by measuring *both* T_i and w and then adjusting Q to compensate for both of these disturbances. However, as a practical matter it is generally uneconomical to attempt to measure all potential disturbances. It would be more practical to use a combined feedforward–feedback control system, since feedback control provides corrective action for unmeasured disturbances, as discussed below. Consequently, in industrial applications feedforward control is normally used in combination with feedback control.

Next, we will consider how the feedback control strategy of Method 1 would perform in the presence of disturbances in T_i or w. If Method 1 were used, no

Table 1.1   Temperature Control Strategies for the Stirred-Tank Heater

Method	Measured Variable	Manipulated Variable	Category[a]
1	T	Q	FB
2	T_i	Q	FF
3	T	w	FB
4	T_i	w	FF
5	T_i and T	Q	FF/FB
6	T_i and T	w	FF/FB
7	—	—	Design change
8	—	—	Design change

[a]FB, feedback control; FF, feedforward control; FF/FB, feedforward control and feedback control.

corrective action would occur until after the disturbance had upset the process, that is, until after T differed from T_R. Thus, by its inherent nature, feedback control is not capable of perfect control since the controlled variable must deviate from the set point before corrective action is taken. However, an extremely important advantage of feedback control is that corrective action is taken regardless of the *source* of the disturbance. Thus, in Method 1, corrective action would be taken (by adjusting Q) after a disturbance in T_i or w caused T to deviate from the set point. The ability to handle unmeasured disturbances of unknown origin is a major reason why feedback controllers have been so widely used for process control.

The previous discussion has cited some of the major advantages and disadvantages of feedforward and feedback control. A more detailed analysis of feedforward control is presented in Chapter 17 while the basic concepts of feedback control are covered in Chapters 8, 10 through 13, and 16.

1.3 PROCESS CONTROL AND BLOCK DIAGRAMS

In Section 1.2 we discussed the rationale for various control strategies using intuitive arguments rather than a quantitative analysis. We have not yet addressed the question of hardware for implementing control strategies.

Again consider the stirred-tank heater system in Fig. 1.1. Suppose that we want to measure the exit temperature T and adjust the electrical heater so that the exit temperature returns to the desired value T_R. One simple type of feedback control strategy is the proportional control law,

$$Q(t) = \overline{Q} + K_c[T_R - T(t)] \tag{1-3}$$

where K_c is called the controller gain and $Q(t)$ and $T(t)$ denote variables that vary with time t. If $K_c > 0$, then this control law provides *negative feedback,* since the heat input moves in a direction opposite to the exit temperature. This control law is referred to as a *proportional control law* since the change in the heat input, $Q(t) - \overline{Q}$, is proportional to the deviation from the set point, $T_R - T(t)$. Thus, a large deviation produces a large corrective action.

Having specified the form of the control law (or control algorithm), we now discuss how this control strategy could be implemented. A schematic diagram for the stirred-tank heater and feedback control system is shown in Fig. 1.2. The hardware components in Fig. 1.2 can be represented in a block diagram, which provides a convenient starting point for analyzing process control problems. A block diagram of the temperature control system is shown in Fig. 1.3. Note that the schematic diagram shows the *physical connections* between the components of the control system, while the block diagram shows the *flow of information* within the control system. The block labeled "heater" has $p(t)$ as its input signal and $Q(t)$ as its output signal, which illustrates that the signals on a block diagram can represent either a physical variable such as Q or an instrument signal such as p. Physical units for each input or output signal are also shown in Fig. 1.3. In general, the blocks in a block diagram do not necessarily have a one-to-one correspondence with process equipment and instrumentation, although the correspondence will be close.

The operation of the temperature control system can be summarized as follows:

1. The tank exit temperature is measured with a thermocouple which generates a corresponding millivolt-level signal. This time-varying signal must be amplified to a dc voltage-level signal before being sent to the controller as the controller input $V(t)$.

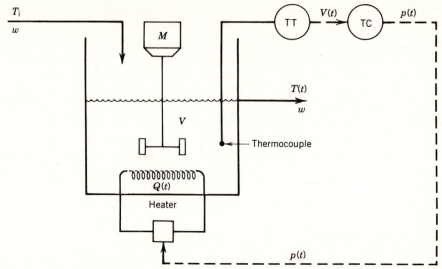

Figure 1.2. Schematic diagram of a temperature feedback control system for a stirred-tank heater. –––, Electrical instrument line; TT, temperature transmitter; TC, temperature controller.

2. The controller performs three distinct calculations. First, it converts the set point T_R that is input by the operator into an internal controller voltage V_R. The block labeled "thermocouple calibration" indicates this operation. Second, it calculates an error signal $e(t)$ by subtracting the controller input $V(t)$ from the set-point (reference) voltage V_R. Thus, $e(t) = V_R - V(t)$. In Fig. 1.3 the *comparator* represents this subtraction operation. Third, heat duty $Q(t)$ is calculated from the proportional control law in Eq. 1-3. The dashed line around the two blocks and the comparator is included to emphasize that these calculations are physically performed in the controller.

3. The controller output $p(t)$ is a dc voltage signal that is sent to the electrical heater to generate $Q(t)$. It is assumed that the heater contains a silicon-controlled rectifier (SCR) to convert the dc voltage signal to an alternating current that is compatible with the heating element.

4. In response to its input signal $p(t)$ the heater generates a heat input $Q(t)$ to the stirred tank. $Q(t)$, the heating rate, is proportional to $p(t)$.

Each block in Fig. 1.3 represents a dynamic or static process element whose behavior can be described by a differential or algebraic equation. One of the tasks facing a control engineer is to develop suitable mathematical descriptions for each block; the development and analysis of dynamic mathematical models are considered in Chapters 2–7 of this book. In Chapter 10 we demonstrate how block diagrams such as Fig. 1.3 are analyzed and how the dynamic response of the controlled system can be predicted.

1.4 CONTROL AND MODELING PHILOSOPHIES

In the previous section we considered a number of alternative temperature control strategies for a simple stirred-tank heating system. Now we briefly consider how control systems are actually designed.

In the process industries control systems are usually designed using one of two general approaches:

1. **Traditional Approach.** The control strategy and control system hardware are selected based on knowledge of the process, experience, and insight. After the

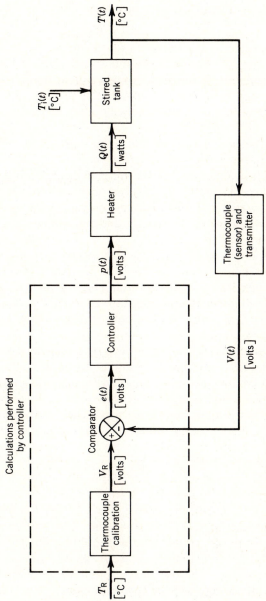

Figure 1.3. Block diagram for temperature feedback control system in Fig. 1.2.

control system is installed in the plant, the controller settings (such as K_c in Eq. 1-3) are adjusted, that is, the controller is *tuned*.

2. **Model-Based Approach.**   A process model is developed that can be helpful in at least three ways: (i) It can be used as the basis for classical controller design methods. (ii) It can be incorporated directly in the control law, an approach that now is the starting point for many advanced control techniques. (iii) It can be used to develop a computer simulation of the process to allow exploration of alternative control strategies and to calculate preliminary values of controller settings.

The model-based approach is becoming more advantageous for several reasons. First, modern processing plants are highly integrated with respect to the flow of both material and energy. This integration makes plant operation more difficult. Second, there are economic incentives for operating plants closer to limiting constraints to maximize profitability while satisfying safety and environmental restrictions.

In this book we subscribe to the philosophy that, for complex processes, a mathematical model of the process should be developed so that the control system can be properly designed. Of course, there are many simple process control problems where controller specification is relatively straightforward and does not require a detailed analysis or an explicit model. But for complex processes, a process model is invaluable both for control system design and for gaining an improved understanding of the process.

The major steps involved in designing and installing a control system using the model-based approach are shown in Fig. 1.4. The first step, formulation of the control objectives, is a critical decision. For example, in a distillation column control problem, the objective might be to regulate a key component in the distillate stream, the bottoms stream, or key components in both streams. An alternative would be to minimize energy consumption (e.g., heat input to the reboiler) while meeting product quality specifications on one (or both) product streams. In formulating the control objectives, process constraints must also be considered. For example, flooding conditions should be avoided in distillation columns, while, for safety reasons, temperature and pressure limits must not be exceeded in many processes.

After the control objectives have been formulated, a dynamic model of the process is developed. The process model can have a theoretical basis, for example, physical and chemical principles such as conservation laws and rates of reactions, or the model can be developed empirically from experimental data. The model development usually involves computer simulation. Alternative modeling strategies are described in Chapters 2, 7, and 15.

The next step in the control system design is to devise an appropriate control strategy that will meet the control objectives while satisfying process constraints. Computer simulation of the controlled process is used to screen alternative control strategies and to determine preliminary estimates of suitable controller settings.

Finally, the control system hardware is selected, ordered, and installed in the plant. Then the controllers are tuned in the plant using the preliminary estimates from the design step as a starting point. Controller tuning usually involves trial and error procedures, as described in Chapters 12 and 13.

1.5   ANALOG OR DIGITAL CONTROL?

Up to this point we have discussed *analog* controllers, namely controllers that have continuous input and output signals. Conventional feedback controllers are analog

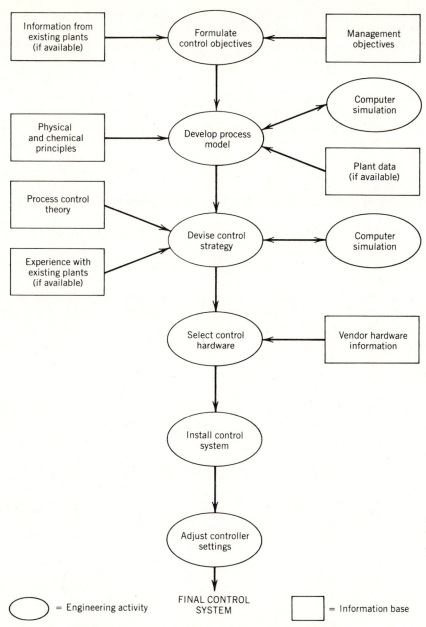

Figure 1.4. Major steps in control system development.

devices that utilize either pneumatic or electronic input/output signals. An alternative approach is to use a *digital* controller, which inherently involves input and output signals that change only at discrete instants in time, the so-called sampling instants.

Recent advances in the performance and cost of digital equipment—minicomputers, microcomputers, and corresponding digital interface elements—have made digital control systems generally preferred over conventional analog controllers. The advantages of digital control include increased flexibility and accuracy, and improved monitoring of the plant through data acquisition, storage, and analysis. With the availability of dedicated microcomputer-based instrumentation, it is no longer necessary to justify computer control for the total plant. It is now possible to select specific locations in a plant for application of digital controllers and realize an attractive return on investment. The design and analysis of digital controllers is the subject of Chapters 21–27.

1.6 ECONOMIC JUSTIFICATION OF PROCESS CONTROL

The justification of process instrumentation and control hardware (including a computer) is sometimes solely based on the need to operate the plant safely while satisfying environmental constraints. In fact, the application of process control methodology is essential to operate most plants. In addition, there are a number of potential economic benefits to be derived from the effective use of process control techniques. For example, the resulting increased production levels, reduced raw material costs, or enhanced product quality provide tangible improvements in profitability. Control systems also provide important auxiliary benefits, such as coordinated alarm systems and extended equipment life.

One of the simplest ways of identifying the potential benefits of improved process control is to examine the decreased product variability resulting from the application of process control. Figure 1.5 shows a typical pattern for product variability. The limit represents a hard constraint, in other words, one that should not be violated, such as a product specification or an unacceptable impurity level. Suppose that in a refinery we can substitute ethane (a less valuable product) for propane in the overhead product stream of the distillation column referred to as the depropanizer. If, through improved control, the depropanizer could be operated more consistently near the upper limit on ethane in the overhead product (such as shown in Fig. 1.5), then the resulting cost savings would be directly proportional to the average amount of ethane substituted, or $A_2 - A_1$ in Fig. 1.5.

Figure 1.6 shows the corresponding frequency distribution or histogram for variations in product quality before and after the improved process control system is installed. This type of figure is an important analytical tool in statistical quality control, which is discussed in more detail in Chapter 18. As is evident from the two product quality distribution curves in Fig. 1.6, improved control allows the operating point (controller set point) to be changed from A_1 to A_2 and thus to be moved closer to the constraint on product quality. This change generates significant economic benefits.

Process control computers can also be used to maximize plant profitability by performing on-line optimization to determine the most economical plant operating conditions. Thus, the computer analyzes process data and then recalculates the optimum operating point on a regular basis, for example, at one-hour intervals. This strategy, which is also referred to as optimizing control or supervisory control, is the subject of Chapter 20. Supervisory control systems that optimize plant op-

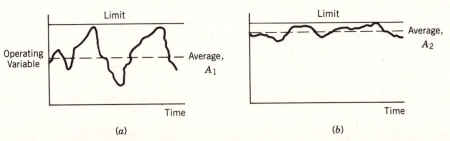

Figure 1.5.   Product variability over time: (*a*) before improved control; (*b*) after. The operating variable is % ethane.

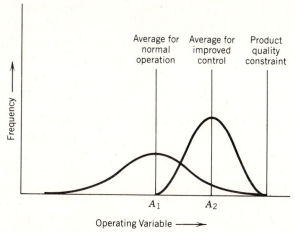

Figure 1.6. Product quality distribution curves showing justification for improved control.

eration are currently of great interest because of increased competition in the process industries. Since energy costs can be a significant factor in evaluating overall profitability and determining the optimum operating conditions for a given process, it has become necessary to evaluate all of the trade-offs among product quality, energy consumption, and the cost of process improvements. Supervisory control alone may be sufficient reason to justify an extensive investment in process control hardware for a particular process.

SUMMARY

In this chapter we introduced the basic principles of feedback and feedforward control and indicated where each approach is likely to be applicable. Although most controllers are used simply to obtain operable processes, we have indicated how control systems can be justified economically based on their ability to hold the process closer to an operating constraint. In many cases the potential economic rewards from closer (tighter) control justify the application of sophisticated digital techniques that go far beyond the feedback and feedforward loops one normally encounters. The remainder of this book deals with a full development of the theory and design techniques required for successful process control applications, covering the spectrum from simple to complex. The underlying idea in all of these developments is to use a model-based approach for analysis and design wherever possible. Thus we present in Chapter 2 a discussion of process modeling.

EXERCISES

1.1. Consider a home heating system consisting of a natural gas-fired furnace and a thermostat. In this case the process consists of the interior space to be heated. The thermostat contains both the measuring element and the controller. The furnace is either on (heating) or off. Draw a schematic diagram for this control system. On your diagram, identify the process outputs and all inputs, including disturbance variables.

1.2. The distillation column shown in the drawing is used to distill a binary mixture. Symbols x, y, and z denote mole fractions of the more volatile component while B, D, R, and F represent molar flow rates. It is desired to control distillate composition y despite disturbances in feed flow rate F. All flow rates can be measured and manipulated with the exception of F which can only be measured. A composition analyzer provides measurements of y.

(a) Propose a feedback control method and sketch the schematic diagram.
(b) Suggest a feedforward control method and sketch the schematic diagram.

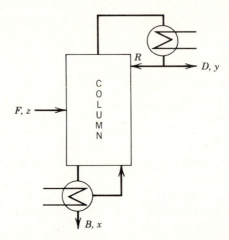

1.3. Two flow control loops are shown in the drawing. Indicate whether each system is either a feedback or a feedforward control system. Justify your answer. It can be assumed that the distance between the flow transmitter (FT) and the control valve is quite small in each system.

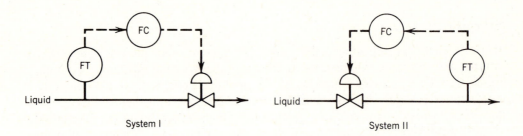

1.4. I.M. Appelpolscher, supervisor of the process control group of the Ideal Gas Company, has installed a $25 \times 40 \times 5$-ft swimming pool in his backyard. The pool contains level and temperature sensors used with feedback controllers to maintain the pool level and temperature at desired values. Appelpolscher is satisfied with the level control system, but he feels that the addition of one or more feedforward controllers would help maintain the pool temperature more nearly constant. As a new member of the control group, you have been selected to check Appelpolscher's own mathematical analysis and to give your advice. The following information may or may not be pertinent to your analysis:

   i. Appelpolscher is particular about cleanliness and hence has a high capacity pump that continually recirculates the water through an activated charcoal filter.
  ii. The pool is equipped with a natural gas-fired heater that adds heat to the pool at a rate $Q(t)$ that is directly proportional to the output signal from the controller $p(t)$.
 iii. There is a leak in the pool which Appelpolscher has determined is constant and equal to F (volumetric flow rate). The liquid-level control system adds water from the city supply system to maintain the level in the pool exactly at the specified level. The temperature of the water in the city system is T_w, a variable.

iv. A significant amount of heat is lost by conduction to the surrounding ground which has a constant, year-round temperature T_G. Experimental tests by Appelpolscher showed that essentially all of the temperature drop between the pool and the ground occurred across the homogeneous layer of gravel that surrounded his pool. The gravel thickness is Δx, and the overall thermal conductivity is k_G.

v. The main challenge to Appelpolscher's modeling ability was the heat loss term accounting for convection, conduction, radiation, and evaporation to the atmosphere. He determined that the heat losses per unit area of open water could be represented by

$$\text{losses} = U(T_p - T_a)$$

where T_p = temperature of pool

T_a = temperature of the air, a variable

U = overall heat transfer coefficient

Appelpolscher's detailed model included radiation losses and heat generation due to added chemicals, but he determined that these terms are negligible.

(a) Draw a schematic diagram for the pool and all control equipment. Show all inputs and outputs, including all disturbance variables.

(b) What additional variable(s) would have to be measured to add feedforward control to the existing pool temperature feedback controller?

(c) Write a steady-state energy balance. How can you determine which of the disturbance variables you listed in part (a) are most/least likely to be important?

(d) What recommendations concerning the prospects of adding feedforward control would you make to Appelpolscher?

CHAPTER 2

Mathematical Modeling of Chemical Processes

2.1 THE RATIONALE FOR MATHEMATICAL MODELING

A model is nothing more than a mathematical abstraction of a real process. The equation or set of equations that comprise the model are at best an approximation to the true process. Hence, the model cannot incorporate all of the features, both macroscopic and microscopic, of the real process. The engineer normally must seek a compromise involving the cost of obtaining the model, that is, the time and effort required to obtain and verify it. These considerations are related to the level of physical and chemical detail in the model and the expected benefits to be derived from its use. The model accuracy necessarily is intertwined in this compromise and the ultimate use of the model influences how accurate it needs to be. Mathematical models can be helpful in process analysis and control in the following ways:

1. *To improve understanding of the process.* Process models can be analyzed or used in a computer simulation of the process to investigate process behavior without the expense and, perhaps, without the unexpected hazards of operating the real process. This approach is necessary when it is not feasible to perform dynamic experiments in the plant or before the plant is actually constructed.

2. *To train plant operating personnel.* Plant operators can be trained to operate a complex process and to deal with emergency situations by use of a process simulator. By interfacing a process simulator to standard process control equipment, a realistic environment can be created for operator training without the costs or exposure to dangerous conditions that might exist in a real plant situation.

3. *To design the control strategy for a new process.* A process model allows alternative control strategies to be evaluated, for example, the selection of the variables that are to be measured (controlled) and those that are to be manipulated. In Chapter 1 we discussed some of the considerations that enter into the design for a simple temperature control system. With more complicated processes, or with new processes for which we have little operating experience, the design of an appropriate control strategy seldom is straightforward.

4. *To select controller settings.* A dynamic model of the process may be used to develop appropriate controller settings, either via computer simulation or by direct analysis of the dynamic model. Prior to start-up of a new process it is desirable to have reasonable estimates of the controller settings. For some

operating processes it may not be feasible to perform experiments that would lead to better controller settings.

5. *To design the control law.* Modern control techniques often incorporate a process model into the control law. Such techniques are called model-predictive or model-based control.

6. *To optimize process operating conditions.* In most processing plants there is an incentive to adjust operating conditions periodically so that the plant maximizes profits or minimizes costs. For example, blending operations for production of gasoline, fuel oil, and jet fuel in a refinery need to be modified in response to changes in the physical properties of the crude oil feedstock, market conditions, and product inventory capacity. A steady-state model of the process and appropriate economic information can be used to determine the most profitable process conditions, as in supervisory control.

For many of the examples cited above—particularly where new, hazardous, or expensive-to-operate processes are involved—development of a mathematical model may be crucial to success. Models can be considered in three different classifications, depending on how they are derived:

a. Theoretical models developed using the principles of chemistry and physics.
b. Empirical models obtained from a mathematical (statistical) analysis of process operating data.
c. Semiempirical models that are a compromise between (a) and (b), with one or more parameters to be evaluated from plant data.

In the last classification, certain theoretical model parameters such as reaction rate coefficients, heat transfer coefficients, and similar fundamental relations usually must be evaluated from physical experiments or from process operating data. Such semiempirical models do have several inherent advantages. They often can be extrapolated over a wider range of operating conditions than purely empirical models which are usually accurate over a very limited range. Semiempirical models also provide the capability to infer how unmeasured or unmeasurable process variables vary as the process operating conditions change.

The remainder of this chapter deals with models based on application of physical and chemical principles (classifications a and c, above). First, we review general techniques that are used in dynamic modeling. Then steady-state and dynamic models are developed for several important processes. Finally, we discuss how the differential equation models might be solved numerically using digital computer simulation. Typical dynamic responses are shown in later chapters. Methods for obtaining empirical models (classification b) are presented in Chapters 7 and 14.

2.2 DYNAMIC VERSUS STEADY-STATE MODELS

Recall that in Chapter 1 we developed a simple model, Eq. 1-2, for a stirred-tank heating process through use of an energy balance. This equation was obtained assuming that the process was at steady state. A dynamic (unsteady-state) model of the stirred-tank process differs from the steady-state model in that an accumulation term is required. For the stirred-tank heater the dynamic energy balance is

$$\text{rate of energy accumulation} = wC(T_i - T_{\text{ref}}) - wC(T - T_{\text{ref}}) + Q$$

$$= wC(T_i - T) + Q \tag{2-1}$$

since the thermodynamic reference temperature T_{ref} used in expressing the inlet and exit enthalpies cancels out. In this case, because potential and kinetic energy terms are unimportant, an appropriate expression for the rate at which energy is accumulated in the tank would be $Cd[V\rho(T - T_{ref})]/dt$, or $V\rho C \, dT/dt$ for the case where the liquid volume V and the density may be assumed constant. Consequently, the dynamic energy balance can be written as

$$V\rho C \frac{dT}{dt} = wC(T_i - T) + Q \tag{2-2}$$

which represents a simple dynamic model for this process.

Note that at this point we have made no assumptions concerning the variables on the right side of (2-2): whether the model inputs (w, T_i, and Q), vary and, if so, how. The dynamic model gives a relation for determining the output variable T as a function of time for arbitrary variations in the inputs. If w can be assumed to be constant, the dynamic model in Eq. 2-2 is a *linear* differential equation model because the inputs T_i and Q and the single output variable T and its derivative are multiplied only by constants. Hence, the solution $T(t)$ can be found analytically in cases where $T_i(t)$ and $Q(t)$ can be specified as simple functions of time. If w is not constant, model linearization (Section 4.4) or numerical integration (Section 2.6) can be employed to obtain a solution.

One benefit of starting with an unsteady-state model is that a steady-state model can always be determined from it directly. For example, the steady-state model in Eq. 1-2 is obtained immediately from Eq. 2-2 by noting that T is constant at steady state; hence dT/dt can be set to zero.

2.3   GENERAL MODELING PRINCIPLES

In this section we first review some general principles, emphasizing the importance of the mass and energy conservation laws. Because the coverage of mathematical modeling is limited here to one chapter, the interested student may wish to refer to the books by Franks [1], Denn [2], Douglas [3], and Luyben [4] for additional information.

Dynamic models of chemical processes invariably consist of one or more differential equations—ordinary differential equations (o.d.e.) and/or partial differential equations (p.d.e.)—often combined with one or more algebraic relations. For the most part we will restrict our discussion to o.d.e. models, although one p.d.e. model is considered in Section 2.5.

For process control problems, a dynamic model can be obtained from the application of unsteady-state conservation relations, usually material and energy balances. Force–momentum balances are employed less often. For processes with momentum effects that cannot be neglected, for example, some fluid and solid transport systems, such balances should be considered. Algebraic equations in the process model can arise from thermodynamic and transport relations; for example, a viscosity may vary as a function of temperature or a heat transfer coefficient may be a function of fluid velocity (flow rate).

Next we consider the stirred-tank heater example but with a slight modification: The amount of liquid contained in the tank (the holdup) is not constant. The modified process is shown in Fig. 2.1; note that the inlet and outlet flow rates may differ.

In this case neither the temperature nor the mass (volume) of the tank contents is constant. *Both* unsteady-state material and energy balances will be required to

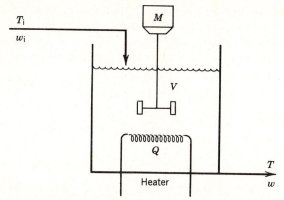

Figure 2.1. A stirred-tank heating system with variable holdup and temperature. (Both the inlet and outlet flow rates and the heat input are set externally.)

model the changing mass and temperature, respectively, of the tank contents. We first utilize the fundamental law for mass conservation,

$$\left\{ \begin{array}{c} \text{rate of mass} \\ \text{accumulation} \end{array} \right\} = \left\{ \begin{array}{c} \text{rate of mass} \\ \text{in} \end{array} \right\} - \left\{ \begin{array}{c} \text{rate of mass} \\ \text{out} \end{array} \right\} \qquad (2\text{-}3)$$

The mass balance is then

$$\frac{d(V\rho)}{dt} = w_i - w \qquad (2\text{-}4)$$

The general law for energy conservation is

$$\left\{ \begin{array}{c} \text{rate of energy} \\ \text{accumulation} \end{array} \right\} = \left\{ \begin{array}{c} \text{rate of energy in} \\ \text{by flow or convection} \end{array} \right\} - \left\{ \begin{array}{c} \text{rate of energy out} \\ \text{by flow or convection} \end{array} \right\}$$

$$+ \left\{ \begin{array}{c} \text{net rate of heat addition to} \\ \text{system from surroundings} \end{array} \right\} \qquad (2\text{-}5)$$

Using Eq. 2-5, the energy balance is

$$C \frac{d[V\rho(T - T_{\text{ref}})]}{dt} = w_i C(T_i - T_{\text{ref}}) - wC(T - T_{\text{ref}}) + Q \qquad (2\text{-}6)$$

where w_i and w are the inlet and outlet mass flow rates and Q is the heat input to the contents of the tank from the electrical heating element. Later we will consider the case where condensing steam inside a heating coil is used for energy transfer. Note that in Eq. 2-6 the input and output energy terms due to convective flow have been written using T_{ref} as the reference or base temperature for calculation of enthalpy. In Eq. 2-2, the inlet and outlet flow rates were assumed to be equal at all times and the specific heat was not a function of temperature. Thus the T_{ref} terms canceled out directly. Here they do not.

Equations 2-4 and 2-6 constitute a dynamic model for the stirred-tank process that can be solved by numerical or analytical integration. First, we must specify the initial conditions (for V and T), the inputs (w_i, w, Q, and T_i) as functions of time, the relations for all algebraic functions (ρ, C), and the constant (T_{ref}). Several simplifications can be made in Eqs. 2-4 and 2-6 for the accumulation terms if ρ and C are assumed constant:

$$\frac{d(V\rho)}{dt} = \rho \frac{dV}{dt} = w_i - w \qquad (2\text{-}7)$$

The rule for differentiation of a product of two functions yields

$$C \frac{d[V\rho(T - T_{ref})]}{dt} = \rho C(T - T_{ref}) \frac{dV}{dt} + V\rho C \frac{d(T - T_{ref})}{dt} \qquad (2\text{-}8)$$

The right side of Eq. 2-8 can be simplified by substituting the material balance relation (2-7) into (2-8) and noting that $dT_{ref}/dt = 0$,

$$C \frac{d[V\rho(T - T_{ref})]}{dt} = C(T - T_{ref})(w_i - w) + V\rho C \frac{dT}{dt} \qquad (2\text{-}9)$$

Next we equate the right sides of Eqs. 2-9 and 2-6 to obtain

$$C(T - T_{ref})(w_i - w) + V\rho C \frac{dT}{dt} = w_i C(T_i - T_{ref}) - wC(T - T_{ref}) + Q$$

$$(2\text{-}10)$$

After we cancel common terms and rearrange, a considerably simpler dynamic model for the process results:

$$\frac{dV}{dt} = \frac{1}{\rho}(w_i - w) \qquad (2\text{-}11)$$

$$\frac{dT}{dt} = \frac{w_i}{V\rho}(T_i - T) + \frac{Q}{V\rho C} \qquad (2\text{-}12)$$

A number of circumstances may lead to a constant volume situation in which $w = w_i$ and $dV/dt = 0$, for example,

1. When an overflow weir is used in the tank.
2. When the tank is closed and filled to capacity.
3. When a liquid-level controller is used to maintain V constant by adjusting w.

In all of these cases, Eq. 2-12 reduces to Eq. 2-2 not because w_i and w are constant, but because w_i is equal to w at all times. Three points are worth mentioning here:

1. Many dynamic models involve unsteady-state component mass balances and/or energy balances with nonconstant holdup. The overall material balance usually can be substituted, as in Eq. 2-9, to simplify the component and energy balances.
2. The algebraic manipulations involved in model simplification may be tedious. If the model equations are to be solved using numerical techniques, it usually is easier to work with the unsimplified model, such as Eqs. 2-4 and 2-6. Although more computer time may be required, the results will be comparable.
3. Equations 2-11 and 2-12 contain physical property parameters (ρ, C) that may be available in handbooks. However, for a mixture of several components, these property parameters may not be available. In that case, it may be necessary to estimate values of ρ and C from response data. Refer to Chapter 7 for more details on parameter estimation procedures for semiempirical models.

In general, modeling is very much an art. The modeler must bring a significant level of creativity to the task, namely to make a set of simplifying assumptions that result in a realistic model. A realistic model incorporates all of the important dynamic effects, is no more complicated than necessary, and keeps the number of equations and parameters at a reasonable level. The failure to choose an appropriate set of simplifying assumptions invariably leads to either (1) rigorous but overly complicated models, or (2) models that are overly simplistic. Both extremes should be avoided. Fortunately, modeling is also a science to the extent that the principles can be learned and predictions from alternative models can be compared both qualitatively and quantitatively. In the next section we briefly present some techniques that are useful in model development for complicated processes.

2.4 DEGREES OF FREEDOM IN MODELING

To use a mathematical model for process simulation we must ensure that the model equations (differential and algebraic) provide a unique relation among all inputs and outputs. This requirement is analogous to the requirement for a set of linear algebraic equations to have a unique solution, which is that the number of variables must equal the number of independent equations. It is not easy to make a similar evaluation for a large, complicated steady-state or dynamic model. However, for such a system of equations to have a unique solution, the number of unknown variables must equal the number of independent model equations. An equivalent way of stating this condition is to require that the *degrees of freedom* [5] be zero, that is,

$$N_F = N_V - N_E = 0 \qquad (2\text{-}13)$$

where N_F is the degrees of freedom, N_V is the total number of variables (unspecified inputs plus outputs), and N_E is the number of independent equations (both differential and algebraic). Hence, a degrees of freedom analysis separates modeling problems into three categories:

1. $N_F = 0$: *exactly determined* (*exactly specified*) *process*. If $N_F = 0$, then the number of equations is equal to the number of process variables and the set of equations has a unique solution.
2. $N_F > 0$: *underdetermined* (*underspecified*) *process*. If $N_F > 0$, then $N_V > N_E$ so there are more process variables than equations. Consequently, the N_E equations have an infinite number of solutions since N_F process variables can be specified arbitrarily. The process model is said to be *underdetermined* or *underspecified*.
3. $N_F < 0$: *overdetermined* (*overspecified*) *process*. For $N_F < 0$, there are fewer process variables than equations and consequently the set of equations has no solution. The process model is said to be *overdetermined* or *overspecified*.

Note that $N_F = 0$ is the only satisfactory case. If $N_F > 0$, then sufficient inputs have not been identified. If $N_F < 0$, then additional independent model equations must be developed.

A structured approach to modeling would involve the following steps: First, note which quantities in the model are *known* constants or parameters that can be fixed on the basis of equipment dimensions, constant physical properties, and so on. Second, identify the N_E output variables, those that will be obtained through solution of the model differential equations (integration using specified boundary conditions) and the algebraic equations. Third, identify the variables that are specified functions of time, the inputs in the model. These will be fixed (determined) by the process environment; for example, the process feed flow rate may be the output flow rate of an upstream process unit. Or they will be specified by the control system designer to serve as manipulated variables in a control strategy (Chapter 28). Note that time t is not one of the N_V process variables because it is neither a process input nor an output.

Findley [5] has reviewed several examples where degrees of freedom must be carefully checked. Below we examine some simple examples of this type of analysis, then return to development of additional process models.

EXAMPLE 2.1

Analyze the degrees of freedom of the nonlinear model represented by Eq. (2-2).

Solution

First, we note that there are

> 3 parameters: V, ρ, C
> 4 variables ($N_V = 4$): T, w, T_i, Q
> 1 equation ($N_E = 1$): Eq. 2-2

The degrees of freedom are calculated as $N_F = 4 - 1 = 3$. Hence, we must identify three variables as inputs which can be specified as functions of time for the equation to have a unique solution. T, the dependent variable in the equation, is an obvious choice for the output in this simple example. Consequently, we have:

> 3 inputs: w, T_i, Q
> 1 output: T

Specifying the three inputs as functions of time uses all three degrees of freedom. Therefore, the single equation is exactly determined and can be solved.

EXAMPLE 2.2

Analyze the degrees of freedom of the model represented by Eqs. 2-11 and 2-12. Is this set of equations linear or nonlinear according to our working definition?

Solution

In this case, volume is no longer one of the known, fixed parameters. Consequently, we have

> 2 parameters: ρ, C

Analyzing the degrees of freedom, we have

> 6 variables ($N_V = 6$): V, T, w_i, w, T_i, Q
> 2 equations: ($N_E = 2$): Eqs. 2-11 and 2-12

so that $N_F = 6 - 2 = 4$. The variables on the left side of the two differential equations can be selected as model outputs. The remaining four variables must be chosen as inputs. Thus we have

> 4 inputs: w_i, w, T_i, Q
> 2 outputs: V, T

Since the outputs are the only variables to be determined in solving the system of two equations, no degrees of freedom are left. The system of equations is exactly specified and hence solvable.

Note that allowing one parameter, V, from the previous model (2-2) to vary has added

> 1 output (V)
> 1 input (w_i)
> 1 independent model equation (2-11)

Also, note that w now is classified as a *mathematical* input, that is, a function of time that must be known or specified, even though it is a *physical* output flow rate.

Equation 2-11 is a linear o.d.e.; Eq. 2-12 is a nonlinear o.d.e. because of the presence of products and quotients of the output and input variables (T, w_i, V, T_i, and Q).

Electrically Heated Stirred Tank

Let us return to the stirred-tank heating system. Assume that the input and output flow rates are equal (constant tank holdup) but relax the assumption that energy is transferred instantaneously from the heating element to the contents of the tank as in Eq. 2-2. Suppose that the metal heating element has a significant thermal capacitance and that the electrical heating rate Q directly affects the temperature of the heating element rather than the liquid contents. Unsteady-state energy balances for the tank and the heating element then become, after simplification

$$mC \frac{dT}{dt} = wC(T_i - T) + h_e A_e(T_e - T) \tag{2-14}$$

$$m_e C_e \frac{dT_e}{dt} = Q - h_e A_e(T_e - T) \tag{2-15}$$

Here $mC = V\rho C$ and $m_e C_e$, the product of the mass of metal in the heating element and its specific heat, represent the thermal capacitances; T_e is the average element temperature; and $h_e A_e$ is the product of the heat transfer coefficient and area available for heat transfer into the tank. In this case, Q, the thermal equivalent of the instantaneous electrical power dissipation in the heating element, is assumed to be a known function of time.

Is the model given by Eqs. 2-14 and 2-15 in suitable form for calculation of the unknown output variables T_e and T? The number of output variables (2) that can be calculated is equal to the the total number of differential equations. All of the other quantities must be system parameters (constants) or inputs (known functions of time). For a specific process, m, C, m_e, C_e, h_e, and A_e are known parameters related to the design of the process, its materials of construction, and its operating conditions. Then w, T_i, and Q must be specified (or measured) functions of time for the model to be completely determined, that is, for there to be no remaining degrees of freedom. The model can then be solved for T and T_e as functions of time by integration if initial conditions for both T and T_e are specified.

If w is constant, the system of two first-order differential equations (2-14 and 2-15) can be converted into a single second-order differential equation model. First solve Eq. 2-14 for T_e and then differentiate to find dT_e/dt. Substituting the expressions for T_e and dT_e/dt into Eq. 2-15, yields

$$\frac{m\, m_e\, C_e}{w h_e A_e} \frac{d^2 T}{dt^2} + \left(\frac{m_e\, C_e}{h_e A_e} + \frac{m_e\, C_e}{wC} + \frac{m}{w} \right) \frac{dT}{dt} + T$$

$$= \frac{m_e\, C_e}{h_e A_e} \frac{dT_i}{dt} + T_i + \frac{1}{wC} Q \tag{2-16}$$

The reader should verify that the dimensions of each term in the equation are consistent and have units of temperature. The model (2-16) can be simplified when $m_e C_e$, the thermal capacitance of the heating element, is very small. When $m_e C_e = 0$, Eq. 2-16 reverts to the first-order model, Eq. 2-2, which was derived for the case where there is no thermal lag in the heating element. In addition, consider the steady-state version of (2-16), when $dT/dt = d^2T/dt^2 = 0$. The resulting steady-state equation is the same as that for (2-2), which is to be expected. Analyzing limiting cases is one way to check the consistency of a more complicated model.

It is important to note that the model in Eq. 2-16 has only a single output variable, T, corresponding to the single equation. The intermediate variable, T_e, which is less important than T, has been eliminated from the earlier model (Eqs. 2-14 and 2-15). In either case, the two models we have developed are exactly determined, that is, they have no remaining degrees of freedom. To integrate Eq. 2-16, we require two initial conditions on T, namely T and dT/dt at $t = 0$, since it is a second-order differential equation. The initial condition on dT/dt can be found by evaluating the right side of Eq. 2-14 when $t = 0$ using the values of $T_e(0)$ and $T(0)$. For both models the inputs (w, T_i, Q) must be specified as functions of time.

Steam-Heated Stirred Tank

Steam (or some other heating medium) can be condensed within a coil to heat liquid in the stirred tank, and the inlet steam pressure varied by adjusting a control valve. The condensation pressure P_s then fixes the steam temperature T_s through an appropriate thermodynamic relation or from tabular information (steam tables):

$$T_s = f(P_s) \tag{2-17}$$

Energy balances corresponding to those of Eqs. 2-14 and 2-15 are given by

$$mC\frac{dT}{dt} = wC(T_i - T) + h_p A_p(T_w - T) \tag{2-18}$$

$$m_w C_w \frac{dT_w}{dt} = h_s A_s(T_s - T_w) - h_p A_p(T_w - T) \tag{2-19}$$

where the subscripts w, s, and p refer respectively to the wall of the heating coil, and to its steam and process sides.

There are three output variables (T_s, T, and T_w) and three equations: an algebraic equation with T_s related to P_s (a specified function of time or a constant) and two differential equations. Thus Eqs. 2-17 through 2-19 constitute an exactly determined model in terms of steam pressure, inlet temperature, and flow rate as inputs. Several important features should be noted at this point.

1. The thermal capacitance of the condensate is neglected in this model. This assumption would be reasonable for the case where a steam trap is used to remove liquid condensate from the coil as it is produced.
2. Usually $h_s A_s \gg h_p A_p$, because the resistance to heat transfer on the steam side of the coil is much lower than on the process side,
3. The change from electrical heating to steam as the heating medium increases the complexity of the model (three equations instead of two) but does not increase the model order (number of first-order differential equations).
4. As models become more complicated, the input and output variables may be coupled through certain parameters. For example, h_p may be a function of w or of the mechanical agitation rate while h_s may vary with steam condensation

rate. Sometimes algebraic equations cannot be solved explicitly for a key variable. In this case, short of simplifying the model through alternative assumptions, numerical solution techniques have to be used if the model equations are to be integrated. Usually, implicit algebraic equations must be solved by iterative methods at each time step in the numerical integration.

In closing this section we present a set of procedural steps to follow in obtaining dynamic models:

Step 1. Draw a schematic diagram of the process and label all process variables.

Step 2. List all assumptions to be used in developing the model. Try for parsimony: The model should be no more complicated than necessary to meet the modeling objectives.

Step 3. Determine whether independent variables other than time are required. (If spatial variations are important, a p.d.e. model will result.)

Step 4. Write appropriate dynamic balances (overall mass, component, energy, etc.).

Step 5. Introduce equilibrium and other algebraic relations (from thermodynamics, reaction stoichiometry, equipment geometry, etc.).

Step 6. Identify the system parameters (constants).

Step 7. Identify the model variables.

Step 8. Calculate the degrees of freedom.

Step 9. Specify N_F inputs to utilize the available degrees of freedom. This leaves N_E outputs (dependent variables). If this step is feasible, the process model is exactly determined and can be solved. If Step 9 is not feasible, go back to Step 2.

Step 10. Simplify the model equations if possible. For example, at this point it often is possible to arrange the equations so that the dependent variables (outputs) appear on the left side of each equation and the inputs appear on the right side to facilitate numerical solution.

2.5 MODELS OF SEVERAL REPRESENTATIVE PROCESSES

We begin this section with some simple models for liquid storage systems utilizing a single tank. In the event that two or more tanks are connected in series (cascaded), the single-tank models developed here are easily extended, as discussed in Chapter 5.

Liquid Storage Systems

A typical liquid storage process is shown in Fig. 2.2; q_i and q are volumetric flow rates. A mass balance yields

$$\frac{d(V\rho)}{dt} = q_i\rho - q\rho \tag{2-20}$$

Since $V = Ah$ for a tank that is cylindrical with cross-sectional area A, and if the liquid density is assumed to be constant, (2-20) becomes

$$A \frac{dh}{dt} = q_i - q \tag{2-21}$$

Note that Eq. 2-21 appears to be a *volume balance*. However, in general, volume

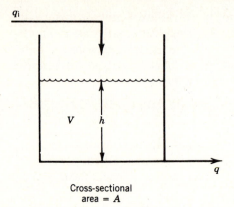

Cross-sectional
area = A

Figure 2.2.   A liquid-level storage system.

is *not* conserved. Consequently, this result only follows from the constant density assumption.

There are three important variations of the liquid storage process:

1. The inlet or outlet flow rates might be constant; for example, outflow q might be maintained by a constant-speed, fixed-volume (metering) pump. An important consequence of this configuration is that the outflow rate is then completely independent of head (height of liquid) in the tank over a wide range, that is, $q = \bar{q}$, (the overbar denotes a steady-state value of flow) and the tank operates essentially as a flow *integrator*. We will return to this case in Section 5.3.

2. The tank exit line may function simply as a resistance to flow from the tank (distributed along the entire line), or it may contain a valve that provides significant resistance to flow at a single point. In the simplest case, the flow may be assumed to be linearly related to the driving force, the liquid head, in analogy to Ohm's law for electrical circuits (E = IR)

$$h = qR_v \tag{2-22}$$

where R_v is the resistance of the line. Rearranging (2-22) gives,

$$q = \frac{1}{R_v} h \tag{2-23}$$

The choice of Eq. 2-23 as an outflow rate relation in conjunction with (2-21) yields a first-order linear differential equation:

$$A \frac{dh}{dt} = q_i - \frac{1}{R_v} h \tag{2-24}$$

This equation exhibits dynamic behavior similar to that of the heating process described by Eq. 2-2.

3. A more realistic expression for flow rate q can be obtained when a valve has been placed in the exit line and turbulent flow can be assumed. The pressure difference driving flow through the valve is

$$\Delta P = P - P_a \tag{2-25a}$$

Because ΔP is proportional to q^2 from the Bernoulli relation,

$$q = C_v \sqrt{P - P_a} \tag{2-25b}$$

where C_v, called the valve coefficient, depends on the particular valve used and its flow rating. C_v must be obtained empirically or from manufacturer's information (see Chapter 9 for more details). Here we assume that the flow discharges at ambient pressure P_a and that the upstream pressure P is the pressure at the bottom of the tank

$$P = P_a + \frac{\rho g}{g_c} h \tag{2-26}$$

The acceleration of gravity g and conversion factor g_c are constants. The model for this process is nonlinear since combining Eqs. 2-21, 2-25b, and 2-26 gives

$$A \frac{dh}{dt} = q_i - C_v \sqrt{\rho \frac{g}{g_c} h} \tag{2-27}$$

and we note that the square root function includes the dependent (output) variable h. Since a model is nonlinear if any dependent variable appears to a power other than one or if dependent variables and/or inputs appear in combination—as products, quotients, and so on—the square root of h is the source of the nonlinearity here.

The liquid storage processes discussed above could be operated by controlling the level in the tank or by allowing the level to fluctuate without attempting to control it. In the latter case (a true surge tank), it may be of interest to predict whether the tank would overflow or run dry for particular variations in the inlet and outlet flow rates. Hence the dynamics of the process may be important even when automatic control is not utilized.

The Continuous Stirred-Tank Reactor (CSTR)

Although other types of reactors are more widely used in industry, the CSTR still is an important process because it embodies many of the features of more commonly encountered reaction systems. At the same time CSTR models tend to be simpler than models for other types of continuous reactors.

Consider a simple first-order, irreversible chemical reaction where chemical species A reacts to form species B:

$$A \longrightarrow B$$

For the CSTR shown in Fig. 2.3, assume perfect mixing. The rate of reaction per unit volume is

$$r = kc_A \tag{2-28}$$

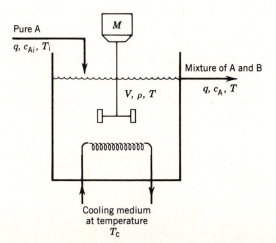

Figure 2.3. A nonisothermal continuous stirred-tank reactor.

where k is the reaction rate coefficient of species A with units of time^{-1}, and c_A is the molar concentration of A. Typically, the rate coefficient k is a strong function of reaction temperature and can be described by the Arrhenius relation [6]:

$$k = k_0 \exp(-E/RT) \tag{2-29}$$

where k_0 is the frequency factor, E is the activation energy, and R is the gas constant.

Assume that the mass densities of the feed and product streams are equal. If the reactor operates with constant mass holdup, the input and output mass flow rates are equal (w). Under conditions of perfect mixing in the reactor, the unsteady-state component balance for species A becomes

$$V \frac{dc_A}{dt} = q\,(c_{Ai} - c_A) - Vkc_A \tag{2-30}$$

where the molecular weight of A appearing in each term has been canceled, and q is the volumetric flow rate w/ρ.

The reactor shown in Fig. 2.3 contains a cooling coil to maintain the reaction mixture at the desired operating temperature by removing heat that is liberated in the exothermic reaction. The energy balance for this system, neglecting shaft work, can be written as [2,5]

$$V\rho C \frac{dT}{dt} = wC(T_i - T) + (-\Delta H)Vkc_A + UA(T_c - T) \tag{2-31}$$

where ΔH is the heat of reaction, U is the overall heat transfer coefficient, and the other variables and constants have been defined previously. In comparing Eq. 2-31 with Eq. 2-18, note the presence of the heat generation term and the use of the overall heat transfer coefficient U. We have neglected the thermal capacitance of the coil wall and of the cooling medium. If a more accurate heat transfer model is desired, then an equation for the wall similar to Eq. 2-19 could be included, or the slightly more rigorous treatment employed by Douglas [3, pp. 59–61] could be used. Since heat transfer coefficients or reaction rate parameters generally are not accurately known (and should be determined from experimental data whenever possible), a more sophisticated model probably is not justified.[1] The stirred-tank reactor represents a good example of a system that usually does not justify a complicated model; a simpler model, containing several empirically determined parameters, likely will be satisfactory for dynamic simulation and control system design.

The coupled equations (2-30) and (2-31) constitute the basis for the unsteady-state reactor model. The model is nonlinear because of the exponential dependence of k on T in Eq. 2-29; normally such a model can be solved only by numerical integration techniques unless the rate expression is linearized (Section 4.4). Notice that the CSTR model becomes considerably more complex if:

1. More complicated rate expressions are appropriate, for example, for a second-order reaction

$$2A \longrightarrow B \tag{2-32}$$

a mass action kinetics model (elementary second-order) would be given by

$$r = k_2\, c_A^2 \tag{2-33}$$

---

[1]Stephanopoulos [7, pp. 59–64] derives the energy balance (2–31) rigorously, assuming that the liquid enthalpies are functions of the concentrations and temperature. In this case, the complex expressions involving partial molar enthalpies ultimately reduce to Eq. 2–31.

2. Additional species or chemical reactions are involved. If the reaction mechanism involved production of an intermediate species

$$2A \longrightarrow B^* \longrightarrow B \tag{2-34}$$

then unsteady material balances for both A and B^* would be necessary (to calculate c_A and c_B^*) or balances for both A and B could be written (to calculate c_A and c_B). Information concerning the reaction mechanisms would also be required.

Reactions involving multiple species are described by high-order, highly coupled, nonlinear reaction models because several component balances must be written [6]. However, though the modeling task becomes much more complex, the same principles illustrated above can be extended and applied. We will return to the simple CSTR model again in Chapter 4.

Staged Systems (A Three-Stage Absorber)

Chemical processes, particularly separation processes, often consist of a sequence of stages. In each stage materials are brought into intimate contact to obtain (or approach) equilibrium between the individual phases. The most important examples of staged processes include distillation, absorption, and extraction. The stages are usually arranged as a cascade with immiscible or partially miscible materials (the separate phases) flowing either cocurrently or countercurrently. Countercurrent contacting, shown in Fig. 2.4, usually permits the highest degree of separation to be attained in a fixed number of stages and is treated here.

The feeds to staged systems may be introduced at each end of the process, as in absorption units, or a single feed may be introduced at a middle stage, as is usually the case with distillation. The stages may be physically connected in either a vertical or horizontal configuration, depending on how the materials are transported, that is, whether pumps are used between stages, and so forth. Below we consider a gas–liquid absorption process since its dynamics are somewhat simpler to develop than those of distillation and extraction processes. At the same time it illustrates the characteristics of more complicated countercurrent staged processes [8].

For the three-stage absorption unit shown in Fig. 2.5, a gas phase is introduced at the bottom (molar flow rate G) and a single component is to be absorbed into a liquid phase introduced at the top (molar flow rate L, flowing countercurrently). A practical example of such a process is the removal of sulfur dioxide (SO_2) from combustion gas by use of a liquid absorbent. The gas passes up through the perforated (sieve) trays and contacts the liquid cascading down through them. A series of weirs and downcomers typically would be used to retain a significant holdup of liquid on each stage while forcing the gas to flow upward through the perforations. Because of intimate mixing, we can assume that the component to be absorbed will be in equilibrium between the gas and liquid streams leaving each stage i. For

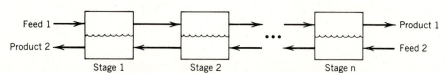

Figure 2.4. A countercurrent-flow staged process.

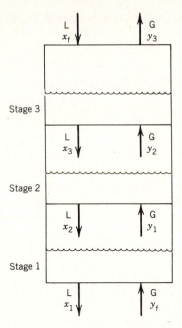

Figure 2.5. A three-stage absorption unit.

example, a simple linear relation is often assumed. For stage i:

$$y_i = ax_i + b \tag{2-35}$$

where y_i and x_i denote gas and liquid concentration of the absorbed component. Assuming constant liquid holdup H, perfect mixing on each stage, and neglecting the holdup of gas, the component material balance for any stage i is

$$H \frac{dx_i}{dt} = G(y_{i-1} - y_i) + L(x_{i+1} - x_i) \tag{2-36}$$

In Eq. 2-36 we also assume that molar liquid and gas flow rates L and G are unaffected by the absorption of SO_2—since changes in concentration of the absorbed component are small, L and G are approximately constant. Substituting Eq. 2-35 into Eq. 2-36 yields

$$H \frac{dx_i}{dt} = aGx_{i-1} - (L + aG)x_i + Lx_{i+1} \tag{2-37}$$

Dividing through by L and substituting $\tau = H/L$ (the stage liquid residence time), $S = aG/L$ (the so-called stripping factor), and $K = G/L$ (the gas-to-liquid ratio), the following model is obtained for the three-stage absorber:

$$\tau \frac{dx_1}{dt} = K(y_f - b) - (1 + S)\, x_1 + x_2 \tag{2-38a}$$

$$\tau \frac{dx_2}{dt} = Sx_1 - (1 + S)x_2 + x_3 \tag{2-38b}$$

$$\tau \frac{dx_3}{dt} = Sx_2 - (1 + S)x_3 + x_f \tag{2-38c}$$

In the model of (2-38) note that the individual equations are linear but also coupled, meaning that each output variable—x_1, x_2, x_3—appears in more than one equation.

This feature can make it difficult to convert these three equations into a single higher-order equation in one of the outputs, as done in Eq. 2-15. However, we will demonstrate how this can be done in Chapter 6.

Distributed Parameter Systems (The Double-Pipe Heat Exchanger)

All of the process models discussed up to this point have been of the *lumped parameter* type, meaning that any dependent variable can be assumed to be a function only of time and not of spatial position. For the stirred-tank systems discussed earlier, we assumed that any spatial variations of the liquid temperature or concentration within the tank could be neglected. Perfect mixing in each stage was also assumed for the absorber. Even when perfect mixing cannot be assumed, a lumped or average temperature may be taken as representative of the tank contents to simplify the process model.

While lumped parameter models are normally used to describe processes, many important process units are inherently *distributed parameter,* that is, the output variables are functions of both time and position. Hence, their process models contain one or more partial differential equations. Pertinent examples include shell-and-tube heat exchangers, packed-bed reactors, packed columns, and long pipelines carrying compressible gases. In each of these cases the output variables should be modeled (if done rigorously) as a function of distance down the tube (pipe), height in the bed (column), or some other measure of location. In some cases, two or even three spatial variables may be considered; for example, concentration and temperature in a tubular reactor may depend on both axial and radial positions, as well as time.

Figure 2.6 illustrates a double-pipe heat exchanger where a fluid flowing through the inside tube with velocity v is heated by steam condensing in the outer tube. If the fluid is assumed to be in plug flow, the temperature of the liquid is expressed as $T_L(z, t)$ where z denotes distance from the fluid inlet. The fluid heating process is truly distributed parameter; at any instant in time there is a temperature profile along the inside tube. The steam condensation, on the other hand, might justifiably be treated as a lumped process, since the steam temperature $T_s(t)$ can be assumed to be a function of the condensation pressure, itself presumably a function only of time and not a function of position. Hence it is a lumped variable.

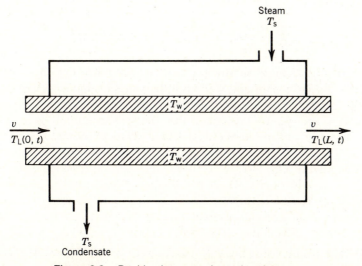

Figure 2.6. Double-pipe steam-heated exchanger.

We also assume that the wall temperature $T_w(z, t)$ is different from T_L and T_s due to the thermal capacitance and resistances.

In developing a model for this process, assume that the liquid enters at temperature $T_L(0, t)$, that is, at $z = 0$. Heat transfer coefficients (steam to wall h_s and wall to liquid h_L) can be used to approximate the energy transfer processes. We neglect the effects of axial energy conduction, the resistance to heat transfer within the metal wall, and the thermal capacitance of the steam condensate.[2] A distributed parameter model for the heat exchanger can be obtained by applying Eq. 2-5 over a differential tube length Δz of the exchanger. In such a *shell* energy balance, the partial differential equation is obtained by taking the limit as $\Delta z \to 0$ [9]. Using the conservation law, Eq. 2-5, the following p.d.e. results [10, p. 354]

$$\rho_L C_L S_L \frac{\partial T_L}{\partial t} = -\rho_L C_L S_L v \frac{\partial T_L}{\partial z} + h_L A_L (T_w - T_L) \qquad (2\text{-}39)$$

where the following parameters are constant: ρ_L = liquid density, C_L = liquid heat capacity, S_L = cross-sectional area for liquid flow, h_L = liquid heat transfer coefficient, and A_L = wall heat transfer area of the liquid. This relation can be rearranged to yield

$$\frac{\partial T_L}{\partial t} = -v \frac{\partial T_L}{\partial z} + \frac{1}{\tau_{HL}} (T_w - T_L) \qquad (2\text{-}40)$$

where $\tau_{HL} = \rho_L C_L S_l / h_L A_L$ has units of time and might be called a characteristic time for heating of the liquid. An energy balance for the wall gives

$$\rho_w C_w S_w \frac{\partial T_w}{\partial t} = h_s A_s (T_s - T_w) - h_L A_L (T_w - T_L) \qquad (2\text{-}41)$$

where the parameters associated with the wall are denoted by subscript w and the steam-side transport parameters are denoted by subscript s. Because T_w depends on T_L, it is also a function of time and position, $T_w(z, t)$. T_s is a function only of time, as noted above. Upon rearrangement, this relation gives

$$\frac{\partial T_w}{\partial t} = \frac{1}{\tau_{sw}} (T_s - T_w) - \frac{1}{\tau_{wL}} (T_w - T_L) \qquad (2\text{-}42)$$

where

$$\tau_{sw} = \frac{\rho_w C_w S_w}{h_s A_s} \qquad \text{and} \qquad \tau_{wL} = \frac{\rho_w C_w S_w}{h_L A_L}$$

are also characteristic time quantities, here representing the thermal transport processes between the steam and wall and the wall and liquid, respectively.

To be able to solve Eqs. 2-40 and 2-42, boundary conditions for both T_L and T_w at time $t = 0$ are required. Assume that the system initially is at steady state ($\partial T_L/\partial t = \partial T_w/\partial t = 0$; the initial steam temperature is known). The steady-state profile, $T_L(z, 0)$, can be obtained by integrating Eq. 2-40 with respect to z simultaneously with solving Eq. 2-42, which is an algebraic expression. Note that (2-40) is an o.d.e. in z, with $T_L(0, 0)$ as the boundary condition. $T_w(z, 0)$ is found algebraically from T_s and $T_L(z, 0)$.

With the initial conditions completely determined, the variations in $T_L(z, t)$ and $T_w(z, t)$ resulting from a change in the inputs, $T_s(t)$ or $T_L(0, t)$, now can be

---

[2]The condensate temperature can be chosen as the reference temperature for energy balances.

found by solving Eqs. 2-40 and 2-42 simultaneously using an analytical approach [9] or a numerical procedure [11]. Because analytical methods can be used only in special cases, we illustrate the use of one type of numerical procedure here. A numerical approach invariably requires that either z, t, or both z and t be *discretized*. Here we use a finite difference approximation to convert the p.d.e.'s to o.d.e.'s. While numerically less efficient than other techniques such as those based on weighted residuals [12], finite difference methods yield more physical insight into both the method and the result of physical lumping.

To obtain ordinary differential equation models with time as the independent variable, the dependence on z is eliminated by discretization. In Fig. 2.7 the double-pipe heat exchanger has been redrawn with a set of grid lines to indicate points at which the liquid and wall temperatures will be evaluated. We now rewrite Eqs. 2-40 and 2-42 in terms of the liquid and wall temperatures $T_L(0)$, $T_L(1)$, . . . , $T_L(N)$ and $T_w(0)$, $T_w(1)$, . . . , $T_w(N)$. Utilizing the backward difference approximation [11] for the derivative $\partial T_L / \partial z$ yields

$$\frac{\partial T_L}{\partial z} \approx \frac{T_L(j) - T_L(j-1)}{\Delta z} \tag{2-43}$$

where $T_L(j)$ is the liquid temperature at the jth node (discretization point). Substituting Eq. 2-43 into Eq. 2-40, the equation for the jth node is

$$\frac{dT_L(j)}{dt} = -v \frac{T_L(j) - T_L(j-1)}{\Delta z} + \frac{1}{\tau_{HL}} [T_w(j) - T_L(j)] \quad (j = 1, \ldots, N) \tag{2-44}$$

The boundary condition at $z = 0$ becomes

$$T_L(0, t) = T_F(t) \tag{2-45}$$

where $T_F(t)$ is some forcing (input) function. Rearranging Eq. 2-44 yields

$$\frac{dT_L(j)}{dt} = \frac{v}{\Delta z} T_L(j-1) - \left(\frac{v}{\Delta z} + \frac{1}{\tau_{HL}}\right) T_L(j) + \frac{1}{\tau_{HL}} T_w(j) \quad (j = 1, \ldots, N) \tag{2-46}$$

Similarly, for the wall equation,

$$\frac{dT_w(j)}{dt} = -\left(\frac{1}{\tau_{sw}} + \frac{1}{\tau_{wL}}\right) T_w(j) + \frac{1}{\tau_{wL}} T_L(j) + \frac{1}{\tau_{sw}} T_s(t) \quad (j = 1, \ldots, N) \tag{2-47}$$

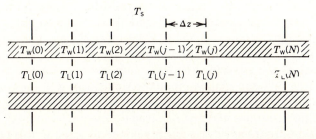

Figure 2.7. Finite-difference approximations for double-pipe heat exchanger.

Note that Eqs. 2-46 and 2-47 represent $2N$ separate linear ordinary differential equations for N liquid and N wall temperatures. There are a number of anomalies associated with this simplified approach compared to the original p.d.e.'s. For example, it is clear that heat transfer from steam to wall to liquid is not accounted for at the zeroth node (the entrance), but is accounted for at all other nodes. Also, a detailed analysis of the discrete model will show that the steady-state relations between $T_L(j)$ and either input, T_s or T_F, are a function of the number of grid points and thus the grid spacing Δz. The discrepancy can be minimized by making N large, that is, Δz small. The lowest-order model for this system that retains some distributed nature would be for $N = 2$. In this case, four equations result:

$$\frac{dT_{L1}}{dt} = \frac{v}{\Delta z} T_F(t) - \left(\frac{v}{\Delta z} + \frac{1}{\tau_{HL}}\right) T_{L1} + \frac{1}{\tau_{HL}} T_{w1} \tag{2-48a}$$

$$\frac{dT_{L2}}{dt} = \frac{v}{\Delta z} T_{L1} - \left(\frac{v}{\Delta z} + \frac{1}{\tau_{HL}}\right) T_{L2} + \frac{1}{\tau_{HL}} T_{w2} \tag{2-48b}$$

$$\frac{dT_{w1}}{dt} = -\left(\frac{1}{\tau_{sw}} + \frac{1}{\tau_{wL}}\right) T_{w1} + \frac{1}{\tau_{wL}} T_{L1} + \frac{1}{\tau_{sw}} T_s(t) \tag{2-48c}$$

$$\frac{dT_{w2}}{dt} = -\left(\frac{1}{\tau_{sw}} + \frac{1}{\tau_{wL}}\right) T_{w2} + \frac{1}{\tau_{wL}} T_{L2} + \frac{1}{\tau_{sw}} T_s(t) \tag{2-48d}$$

where the node number has been denoted by the second subscript on the output variables to simplify the notation.

Equations 2-48a to d are coupled, linear, ordinary differential equations. In Chapter 6 we will consider some of the response properties of such a lumped parameter model for a distributed parameter process.

2.6 SOLUTION OF DYNAMIC MODELS AND THE USE OF DIGITAL SIMULATORS

Once a dynamic model has been developed, methods are available to solve it numerically. By solve, we mean that the transient responses of the dependent variables can be found to some degree of accuracy by numerically integrating the differential equations, given that appropriate initial values for the dependent variables have been specified and that the inputs (forcing functions) also have been specified as functions of time.

Over the years applied mathematicians have developed a large number of numerical integration techniques ranging from the simple (e.g., the Euler method) to the complicated (e.g., the Runge-Kutta method) [11]. All of these techniques represent some compromise between computational effort (computing time) and accuracy. While a dynamic model can always be solved, there may be difficulties in obtaining useful numerical solutions in some cases. With large models, for example, a very short integration interval may be required to obtain an accurate solution; but this may require so much computation time that the solution is impractical to obtain. Collections of computer programs for integrating ordinary differential equations are available on most computer systems; the IMSL Package [13], for most large or even small computer systems, is a well-known example.

Obtaining dynamic solutions of models with large numbers of equations using standard integration routines may not be straightforward. A number of equation-oriented simulation programs have been developed to assist in this task. The user supplies the set of algebraic and ordinary differential equations to the simulation program along with conditions on the total integration period, error limitations, variables to be printed out or plotted, and so on. The simulation program then assumes responsibility for

1. Testing to see if the set of equations is exactly determined (if not, an error message is printed).
2. Sorting the equations into an appropriate sequence for iterative solution.
3. Integrating the equations.
4. Generating tables of printed results and/or plots.

An example of an equation-oriented simulator is ACSL [14]. Catalogs of simulation methods have been given in [15] and [16]. While the equation-oriented digital simulator is easy to use, one disadvantage is the much larger expenditure of computer time and, hence, greater expense.

A second disadvantage of equation-oriented packages is the amount of time required to develop all of the equations for a large process or plant. An alternative approach is to use so-called modular simulation programs. The modular approach uses prewritten subroutines to represent an entire process unit, such as a distillation tower or a reactor. Thus, it has the significant advantage that plant-scale simulations require the programmer only to include subroutine calls to the appropriate modules and to supply the numerical values of specific parameters. The simulator itself is responsible for all aspects of the solution and includes sophisticated numerical integration procedures. Consequently, this type of simulator bears a direct physical relation to the plant or process flowsheet. Since each module is rather general in form, the user can simulate any of a number of possible configurations for a complex process, for example, one or more distillation towers, extraction columns, tubular and stirred tank reactors, heat exchangers, condensers, reboilers, precipitators, evaporators, and automatic controllers. Often the user is allowed to program and attach special-purpose modules for unusual or specific applications.

Modular dynamic simulators have been available since the early 1970s; one of the first was DYFLOW [17]. A recently developed "hybrid simulator" called SPEED-UP [18] supplies a set of modular programs and a physical properties package. Equation-oriented capabilities are directly available to the user. Such simulators are achieving a high degree of acceptance in process engineering and control studies because they allow plant dynamics, alternative control configurations, and resulting operability of a plant to be evaluated prior to construction.

SUMMARY

In this chapter we have first mentioned the importance of process modeling, particularly in determining correct operating procedures and in designing control systems for today's complex processes. Next, a set of principles and procedures has been presented that can make process modeling much easier and more rewarding. The representative models derived and discussed here were intentionally kept simple to illustrate general ideas; however, even very simple models can be used effectively if key model parameters are fit to actual process operating data. Such semiempirical models are potentially of great utility in industrial environments where the time and expense to construct detailed fundamental models and to verify them cannot be justified. Finally, we note that the solution of dynamic models has been made considerably faster and easier by recent developments in computer hardware and simulation packages. Beginning with Chapter 3 we show how analytical solutions for linear models can be routinely obtained. Approximate solutions, obtained for nonlinear models through linearization techniques, also are discussed.

REFERENCES

1. Franks, R. G. E., *Mathematical Modeling in Chemical Engineering,* Wiley, New York, 1967.
2. Denn, M. M., *Process Modeling,* Longman, New York, 1986.
3. Douglas, J. M., *Process Dynamics and Control: Analysis of Dynamic Systems,* Vol. 1, Prentice-Hall, Englewood Cliffs, NJ, 1972.
4. Luyben, W. L., *Process Modeling, Simulation, and Control for Chemical Engineers,* McGraw-Hill, New York, 1973.
5. Findley, M. E., Selection of control measurements, *AIChEMI Modular Instruction, Series A,* Vol. 4, T. F. Edgar (Ed.), AIChE, New York: 1983, p. 55.
6. Fogler, H. S., *Elements of Chemical Reaction Engineering,* Prentice-Hall, Englewood Cliffs, NJ, 1986.
7. Stephanopoulos, G., *Chemical Process Control: An Introduction to Theory and Practice,* Prentice-Hall, Englewood Cliffs, NJ, 1984.
8. King, C. J., *Separation Processes, 2d ed.,* McGraw-Hill, New York, 1980.
9. Bird, R. B., W. E. Stewart, and E. N. Lightfoot, *Transport Phenomena,* Wiley, New York, 1960.
10. Coughanowr, D. R., and L. B. Koppel, *Process Systems Analysis and Control,* McGraw-Hill, New York, 1965.
11. Carnahan, B., and J. O. Wilkes, *Digital Computing and Numerical Methods,* Wiley, New York, 1973.
12. Finlayson, B. A., *Nonlinear Analysis in Chemical Engineering,* McGraw-Hill, New York, 1980.
13. IMSL Mathematics and Statistics Library, IMSL, Inc., Houston, TX, 1988.
14. *Advanced Continuous Simulation Language(ACSL),* Mitchell and Gauthier Associates, Inc., Concord, MA, 1988.
15. Catalog of simulation software, *Simulation* **41,** 156 (1983).
16. Additions to the catalog of simulation software, *Simulation* **42,** 88 (1984).
17. Franks, R. G. E., *Modelling and Simulation in Chemical Engineering,* Wiley, New York, 1972.
18. Perkins, J. D. and R. W. H. Sargent, SPEED-UP: A computer program for steady-state and dynamic simulation and design of chemical processes, in R. S. H. Mah and G. V. Reklaitis (Eds.), *Selected Topics in Computer-Aided Process Design and Analysis,* AIChE Symposium Series, Vol. 78, 1982, p. 1.

EXERCISES

2.1. Temperature sensors are often protected from rough treatment, corrosion, and erosion by placing them in thermowells. Suppose that a filled-bulb sensor containing a liquid is placed in a thermowell as shown in the figure. Develop an unsteady-state model describing how the bulb temperature T_b is affected by the ambient fluid temperature T_a. In your derivation the following assumptions can be made:

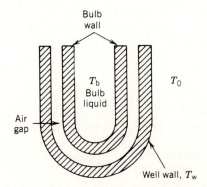

i. Temperature gradients in the well wall and in the bulb wall are negligible.
ii. The temperature difference between the liquid in the bulb and the bulb wall is negligible.

Be sure to define all variables.

2.2. Consider the stirred-tank heater example with steam heating. If the thermal capacitance of the coil wall is negligible, the dynamic model in Eqs. 2-17–2-19 can be simplified. In particular, show that the model can be reduced to a single differential equation which relates temperature T to steam pressure P_s. Interpret the heat transfer coefficient you obtain in terms of the usual overall coefficient between the steam and process liquid.

2.3. A jacketed vessel is used to cool a process stream as shown in the figure. The following information is available:

i. The volume of liquid in the tank V and the volume of coolant in the jacket V_J remain constant. Volumetric flow rate q_F is constant but q_J varies with time.
ii. Heat losses from the jacketed vessel are negligible.
iii. Both the tank contents and the jacket contents are well-mixed and have significant thermal capacitances.
iv. The thermal capacitances of the tank wall and the jacket wall are negligible.
v. The overall heat transfer coefficient for transfer between the tank liquid and the coolant varies with coolant flow rate:

$$U = Kq_J^{0.8}$$

where $U\;[=]$ Btu/h ft^2 °F

$\qquad q_J\;[=]$ ft^3/h

$\qquad K = $ constant

Derive a dynamic model for this system. (State any additional assumptions that you make.)

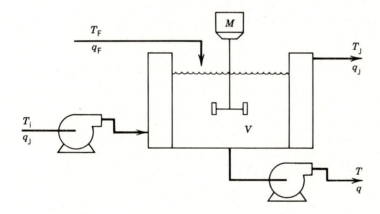

2.4. A jacketed vessel similar to the one in Exercise 2.3 is used to *heat* a liquid by means of condensing steam. The following information is available:

i. The volume of liquid within the tank may vary.
ii. Heat losses are negligible.
iii. The tank contents are well mixed. Steam condensate is removed from the jacket by a steam trap as soon as it has formed.
iv. Thermal capacitances of the tank and jacket walls are negligible.
v. The steam condensation pressure P_s is set by a control valve and is not necessarily constant.
vi. The overall heat transfer coefficient U for this system is constant.
vii. Flow rates q_F and q are independently set by external valves and may vary.

Derive a dynamic model for this process. The model should be simplified as much as possible. State any additional assumptions that you make.

2.5. Consider a liquid flow system consisting of a sealed tank with noncondensible gas above the liquid as shown in the drawing. Derive an unsteady-state model relating the liquid level h to the input flow rate q_i. Is operation of this system independent of the ambient presure P_a? What about for a system open to the atmosphere?

You may make the following assumptions:

i. The gas obeys the ideal gas law. A constant amount of m_g/M moles of gas are present in the tank.

ii. The operation is isothermal.

iii. A square root relation holds for flow through the valve.

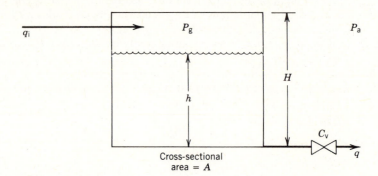

Cross-sectional
area = A

2.6. The unsteady-state flow network in the drawing is to be used for surge damping. P_i is the applied pressure (q_i is not specified, i.e., is unknown). The volumetric flow rate in any pipe leg is given by the pressure drop across the leg divided by the resistance of the leg. The resistances R_{bc}, R_{be}, R_{de} are known.

(a) Find expressions for all flow rates, pressures, and the liquid level in the tank in terms of the pressures at points b and e. (The pressure at e can be taken equal to P_a, the ambient pressure.)

(b) Obtain expressions for q_4 and for h in terms of P_i and P_e.

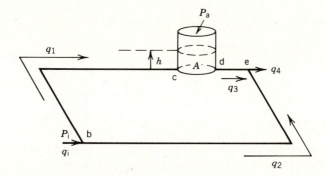

2.7. Two surge tanks are used to dampen pressure fluctuations caused by erratic operation of a large air compressor (see drawing).

(a) If the discharge pressure of the compressor is $P_d(t)$ and the operating pressure of the furnace is P_f (constant), develop a dynamic model for the pressures in the two surge tanks as well as for the air mass flows at points a, b, and c. You may assume that the valve resistances are constant, that the valve flow characteristics are linear, that the surge processes operate isothermally, and that the ideal gas law holds.

(b) How would you modify your model if the surge tanks operated adiabatically? If the ideal gas law were not a good approximation?

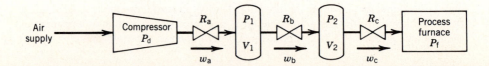

2.8. The Ideal Gas Company has built a stirred-tank chemical reactor with a paddle-type agitator. Dr. A. Quirk, head of the Chemical Kinetics and Reactor Design Group, is attempting to determine why his group is unable to duplicate results obtained earlier with a bench-scale reactor. Some mixing tests have been performed with a chemical tracer which show that circulation is set up in both the top and bottom parts of the reactor and that there is relatively little exchange of material between the two sections, as shown in part *a* of the figure.

I. M. Appelpolscher, head of the Systems Group, feels that the reactor may be inadequately baffled. He asks you to write out a set of equations (model) that describes transient changes in the outflow tracer concentration as a function of inflow tracer concentration (constant volume and no reaction), assuming that the reactor can be modeled as two separate tanks, each perfectly stirred, as in part (b) of the figure. State all additional assumptions, identify all variables, and report to Appelpolscher which parameters in your model will have to be obtained from physical experiments on the full-scale reactor.

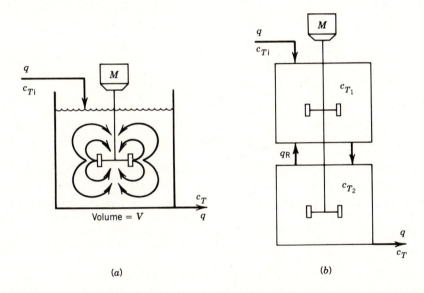

(a)   (b)

2.9. Water is blended with a slurry to give the slurry the proper consistency. They are mixed in a blending tank that has a constant volume V. The slurry solids mass fraction in the inlet is x_s with volumetric flow rate q_s. Since x_s and q_s vary somewhat, the water makeup mass flow rate w is changed to compensate for these variations. Write an unsteady-state model for this blender that can be used to predict the dynamic behavior of the mass fraction of solids in the exit stream x_e for changes in x_s, q_s, or w.

2.10. The chemical reaction sequence

$$A \longrightarrow B$$
$$A + B \longrightarrow C$$

takes place isothermally in a continuous, stirred-tank reactor. Batch kinetic studies have indicated that the first reaction is second order with respect to c_A while the reaction rate for the second reaction is first order with respect to both c_A and c_B:

$$\left. \begin{array}{l} r_1 = k_1 c_A^2 \\[2mm] r_2 = k_2 c_A c_B \end{array} \right\} \quad r_1, r_2 \; [=] \; \frac{\text{mol}}{(\text{ft}^3)(\text{h})}$$

It can be assumed that the reactor has a constant volume V and constant feed rate q, and that the feed contains traces of B but no C. Derive an unsteady-state model that will yield the concentrations of A, B, and C for variations in the concentration of B in the feed.

2.11. Irreversible consecutive reactions $A \rightarrow B \rightarrow C$ occur in a jacketed, stirred-tank reactor as shown in the figure. Derive a dynamic model based on the following assumptions:

  i. The contents of the tank and cooling jacket are well mixed. The volumes of material in the jacket and in the tank do not vary with time.

  ii. The reaction rates are given by

$$r_1 = k_1 e^{-E_1/RT} c_A \ [=] \ \text{mol } A/\text{h} \cdot \text{L}$$

$$r_2 = k_2 e^{-E_2/RT} c_B \ [=] \ \text{mol } B/\text{h} \cdot \text{L}$$

  iii. The thermal capacitances of the tank contents and the jacket contents are significant relative to the thermal capacitances of the jacket and tank walls, which can be neglected.

  iv. Constant physical properties and heat transfer coefficients can be assumed.

Note: All flow rates are volumetric flow rates in L/h. The concentrations have units of mol/L. The heats of reaction are ΔH_1 and ΔH_2.

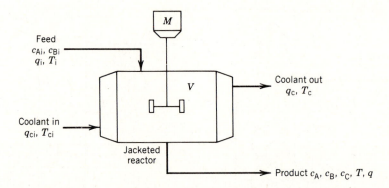

2.12. A binary mixture of A and C are heated just to the saturation point and then fed to a single-stage flash unit where additional heat is added at a known rate (see drawing). The feed flow rate and concentration are known functions of time. Assume that A is the more volatile component, that the molar heat of vaporization λ is the same for both components, and that the molar liquid holdup in the stage H is constant. Flow rates are in moles per unit time. Concentrations are in mole fractions.

  (a) Derive expressions for the vapor and liquid outflow rates and concentrations.

  (b) What can you say about the stage temperature (which you were not asked to deal with specifically as part of the model in (a)), that is, how might you estimate or calculate it?

  (c) Rederive your model for the case where a semi-batch distillation is carried out, that is, there is no bottoms stream and the feed F is supplied to the unit only intermittently during the cycle. Note that in this case you cannot assume the liquid holdup H is constant. Be sure to state all additional assumptions explicitly.

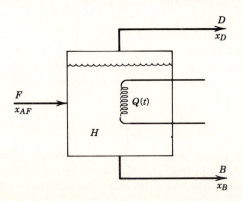

PART TWO

TRANSIENT BEHAVIOR OF PROCESSES

— CHAPTER 3 —

Laplace Transforms

In Chapter 2 we developed a number of mathematical models that describe the dynamic operation of selected processes. Solving such models, that is, finding the output variables as functions of time for some change in the input variable(s), requires either analytical or numerical integration of the differential equations. Sometimes considerable effort is involved in obtaining the solutions. One important class of models includes systems described by linear differential equations. Such linear systems represent the starting point for many analytical techniques in process control.

In this chapter we introduce a mathematical tool, the *Laplace transform*, that can significantly reduce the effort required to solve linear differential equation models analytically. A major benefit is that this transformation converts differential equations to algebraic equations, which can simplify the mathematical manipulations required to obtain a solution.

First we define the Laplace transform and show how it can be used to convert the elements of any linear differential equation—both the time derivatives and the inputs, which are functions of time—into a standard set of *transforms* that can be placed in a compact table. How a simple differential equation can be Laplace transformed by referring to this table is then illustrated. Since many practical problems involve functions that are not found in the standard tables, we next demonstrate how any transform can be expanded into functions that *are* found in the table. This technique, called partial fraction expansion, is the key part of a general technique that can be used to solve virtually any linear ordinary differential equation problem. Some important general properties of Laplace transforms are then presented, and, finally, we illustrate the use of these techniques for a practical modeling situation.

3.1 THE LAPLACE TRANSFORM OF REPRESENTATIVE FUNCTIONS

The Laplace transform of a function $f(t)$ is defined as

$$F(s) = \mathcal{L}[f(t)] = \int_0^\infty f(t)e^{-st}\, dt \tag{3-1}$$

where $F(s)$ is the symbol for the Laplace transform, $f(t)$ is some function of time, and $\mathcal{L}$ is an operator, defined by the integral. The function $f(t)$ must satisfy mild conditions [1] which include being piecewise continuous for $0 < t < \infty$; this requirement almost always holds for functions that are useful in modeling and control. When the integration is performed, the transform becomes a function of the Laplace transform variable s which is a complex variable. The *inverse Laplace transform* ($\mathcal{L}^{-1}$) operates on the function $F(s)$ and converts it to $f(t)$. Notice that $F(s)$ contains no information about $f(t)$ for $t < 0$. Hence $f(t) = \mathcal{L}^{-1}\{F(s)\}$ is not defined for $t < 0$.

One of the important properties of the Laplace transform and the inverse Laplace transform is that they are linear operators; a linear operator satisfies the general superposition principle:

$$\mathcal{F}(ax(t) + by(t)) = a\mathcal{F}(x(t)) + b\mathcal{F}(y(t)) \tag{3-2}$$

where $\mathcal{F}$ denotes a particular operation to be performed, such as differentiation or integration with respect to time. If $\mathcal{F} \equiv \mathcal{L}$, then Eq. 3-2 would be

$$\mathcal{L}(ax(t) + by(t)) = aX(s) + bY(s) \tag{3-3}$$

Therefore, the Laplace transform of a sum of functions is the sum of the individual Laplace transforms; in addition, multiplicative constants can be factored out of the operator, as shown in (3-3).

In this book we are more concerned with operational aspects of Laplace transforms, that is, using them to obtain solutions of *linear* differential equations. For more details on mathematical aspects of the Laplace transform, the text by Churchill [1] is recommended.

Before we proceed to solution techniques, the application of Eq. 3-1 should be discussed. The Laplace transform can be calculated easily for most simple functions, as shown below.

Constant Function.   For $f(t) = a$ (a constant),

$$\mathcal{L}(a) = \int_0^\infty ae^{-st}\,dt = -\frac{a}{s}e^{-st}\Big|_0^\infty$$

$$= 0 - \left(-\frac{a}{s}\right) = \frac{a}{s} \tag{3-4}$$

Step Function.   The unit step function, defined as

$$\mathbf{S}(t) = \begin{cases} 0 & t < 0 \\ 1 & t \geq 0 \end{cases} \tag{3-5}$$

is an important input that is used frequently in process dynamics and control applications. The Laplace transform of the unit step function is similar to that obtained for the constant a above.

$$\mathcal{L}[\mathbf{S}(t)] = \frac{1}{s} \tag{3-6}$$

If the step magnitude is a, the Laplace transform is a/s. The step function incorporates the idea of *initial time, zero time, or time zero;* all of these terms are ways to refer to the time at which this function changes from 0 to 1. To avoid any ambiguity concerning the value of the step function at $t = 0$ (it has two values), we hereafter consider $f(t = 0)$ to be the value of the function approached from the positive side, $t = 0^+$.

Derivatives. The transform of a first derivative is important because such derivatives appear in linear differential equations. This transform is

$$\mathscr{L}(df/dt) = \int_0^\infty (df/dt)e^{-st}\, dt \tag{3-7}$$

Integrating by parts,

$$\mathscr{L}(df/dt) = \int_0^\infty f(t)e^{-st}s\, dt + fe^{-st}\Big|_0^\infty \tag{3-8}$$

$$= s\mathscr{L}(f) - f(0) = sF(s) - f(0) \tag{3-9}$$

where $F(s)$ is the symbol for the Laplace transform of $f(t)$. Generally, the point at which we *start keeping time* for a solution is arbitrary. Model solutions are ordinarily obtained assuming that time *starts* (i.e., $t = 0$) at the moment the process model is first perturbed. For example, if the process initially is assumed to be at steady state and an input changes by a step function, *time zero* is taken to be the moment at which the input changes in magnitude. In many process modeling applications, functions are often defined so that they are zero at initial time, that is, at $t = 0$, $f(t) = 0$, or $f(0) = 0$. In these cases, (3-9) reduces to $\mathscr{L}(df/dt) = sF(s)$.

The Laplace transform for higher-order derivatives can be found using Eq. 3-9. To calculate $\mathscr{L}[f''(t)]$, we define a new variable ($\phi = df/dt$) such that

$$\mathscr{L}\left(\frac{d^2f}{dt^2}\right) = \mathscr{L}\left(\frac{d\phi}{dt}\right) = s\phi(s) - \phi(0) \tag{3-10}$$

From the definition of ϕ ($= df/dt$),

$$\phi(s) = sF(s) - f(0) \tag{3-11}$$

Substituting into Eq. 3-10,

$$\mathscr{L}\left(\frac{d^2f}{dt^2}\right) = s[sF(s) - f(0)] - \phi(0) \tag{3-12}$$

$$= s^2F(s) - sf(0) - f'(0) \tag{3-13}$$

where $f'(0)$ denotes the value of df/dt at $t = 0$. The Laplace transform for derivatives higher than second order can be found by the same procedure. An nth-order derivative, when transformed, yields a series of $(n + 1)$ terms:

$$\mathscr{L}\left(\frac{d^nf}{dt^n}\right) = s^nF(s) - s^{n-1}f(0) - s^{n-2}f^{(1)}(0) - \cdots$$
$$- sf^{(n-2)}(0) - f^{(n-1)}(0) \tag{3-14}$$

where $f^i(0)$ is the ith derivative evaluated at $t = 0$. If $n = 2$, Eq. 3-13 is obtained.

Exponential Functions. The Laplace transform of an exponential function is important because exponential functions appear in the solution to all linear differential equations. For a negative exponential, e^{-bt}, with $b > 0$

$$\mathscr{L}(e^{-bt}) = \int_0^\infty e^{-bt}e^{-st}\, dt = \int_0^\infty e^{-(b+s)t}\, dt \tag{3-15}$$

$$= \frac{1}{b+s}[-e^{-(b+s)t}]\Big|_0^\infty = \frac{1}{s+b} \tag{3-16}$$

The Laplace transform for $b < 0$ is unbounded if $s < b$; therefore, the real part of s must be restricted to be larger than $-b$ for the integral to be finite. This condition is satisfied for all problems we consider in this book.

Trigonometric Functions. In modeling processes and in studying control systems, there are many other important time functions, such as the trigonometric functions, $\cos \omega t$ and $\sin \omega t$, where ω is the frequency in radians per unit time. The Laplace transform of $\cos \omega t$ or $\sin \omega t$ can be calculated using integration by parts. An alternative method is to use the Euler identity

$$\cos \omega t = \frac{e^{j\omega t} + e^{-j\omega t}}{2}, \qquad j \triangleq \sqrt{-1} \tag{3-17}$$

and to apply (3-1). Since the Laplace transform of a sum of two functions is the sum of the Laplace transforms,

$$\mathcal{L}(\cos \omega t) = \tfrac{1}{2}\mathcal{L}(e^{j\omega t}) + \tfrac{1}{2}\mathcal{L}(e^{-j\omega t}) \tag{3-18}$$

Using Eqs. 3-15 and 3-16 gives

$$\mathcal{L}(\cos \omega t) = \frac{1}{2}\left(\frac{1}{s - j\omega} + \frac{1}{s + j\omega}\right)$$
$$= \frac{1}{2}\left(\frac{s + j\omega + s - j\omega}{s^2 + \omega^2}\right) = \frac{s}{s^2 + \omega^2} \tag{3-19}$$

Note that the use of imaginary variables above was merely a device to avoid integration by parts; imaginary numbers do not appear in the final result. To find $\mathcal{L}(\sin \omega t)$, we can use a similar approach.

Table 3.1 lists some important Laplace transform pairs that occur in the solution of linear differential equations. For a more extensive list of transforms, see Ref. 2.

Note that in all the transform cases derived above, $F(s)$ is a ratio of polynomials in s, a so-called rational form. There are some important cases when nonpolynomial (nonrational) forms appear. One such case is discussed next.

The Rectangular Pulse Function. An illustration of the rectangular pulse is shown in Fig. 3.1. The pulse has height h and width t_w. This type of signal might be used to depict the opening and closing of a valve regulating flow into a tank. The flow rate would be held at h for a duration of t_w units of time. The area under the curve in Fig. 3.1 could be interpreted as the amount of material delivered to the tank $(= ht_w)$. Mathematically, the function $f(t)$ is defined as follows:

$$f(t) = \begin{cases} 0 & t < 0 \\ h & 0 \leq t < t_w \\ 0 & t \geq t_w \end{cases} \tag{3-20}$$

The Laplace transform of the rectangular pulse can be found by evaluating the integral (3-1) between $t = 0$ and $t = t_w$ since $f(t)$ is zero everywhere else:

$$F(s) = \int_0^\infty f(t)e^{-st}\, dt = \int_0^{t_w} he^{-st}\, dt \tag{3-21}$$

$$F(s) = \frac{h}{s} e^{-st} \Big|_0^{t_w} = \frac{h}{s}(1 - e^{-t_w s}) \tag{3-22}$$

Table 3.1   Laplace Transforms for Various Time-Domain Functions[a]

$f(t)$	$F(s)$
1. $\delta(t)$ (unit impulse)	1
2. $\mathbf{S}(t)$ (unit step)	$\dfrac{1}{s}$
3. t (ramp)	$\dfrac{1}{s^2}$
4. t^{n-1}	$\dfrac{(n-1)!}{s^n}$
5. e^{-bt}	$\dfrac{1}{s+b}$
6. $\dfrac{1}{\tau} e^{-t/\tau}$	$\dfrac{1}{\tau s + 1}$
7. $\dfrac{t^{n-1}e^{-bt}}{(n-1)!}$ $(n>0)$	$\dfrac{1}{(s+b)^n}$
8. $\dfrac{1}{\tau^n(n-1)!}\, t^{n-1}e^{-t/\tau}$	$\dfrac{1}{(\tau s + 1)^n}$
9. $\dfrac{1}{b_1 - b_2}(e^{-b_2 t} - e^{-b_1 t})$	$\dfrac{1}{(s+b_1)(s+b_2)}$
10. $\dfrac{1}{\tau_1 - \tau_2}(e^{-t/\tau_1} - e^{-t/\tau_2})$	$\dfrac{1}{(\tau_1 s + 1)(\tau_2 s + 1)}$
11. $\dfrac{b_3 - b_1}{b_2 - b_1} e^{-b_1 t} + \dfrac{b_3 - b_2}{b_1 - b_2} e^{-b_2 t}$	$\dfrac{s + b_3}{(s+b_1)(s+b_2)}$
12. $\dfrac{1}{\tau_1}\dfrac{\tau_1 - \tau_3}{\tau_1 - \tau_2} e^{-t/\tau_1} + \dfrac{1}{\tau_2}\dfrac{\tau_2 - \tau_3}{\tau_2 - \tau_1} e^{-t/\tau_2}$	$\dfrac{\tau_3 s + 1}{(\tau_1 s + 1)(\tau_2 s + 1)}$
13. $1 - e^{-t/\tau}$	$\dfrac{1}{s(\tau s + 1)}$
14. $\sin \omega t$	$\dfrac{\omega}{s^2 + \omega^2}$
15. $\cos \omega t$	$\dfrac{s}{s^2 + \omega^2}$
16. $\sin(\omega t + \phi)$	$\dfrac{\omega \cos \phi + s \sin \phi}{s^2 + \omega^2}$
17. $e^{-bt}\sin \omega t$ $\left.\begin{array}{c} \\ \\ \end{array}\right\}$ b, ω real $\left\{\begin{array}{c} \\ \\ \end{array}\right.$	$\dfrac{\omega}{(s+b)^2 + \omega^2}$
18. $e^{-bt}\cos \omega t$	$\dfrac{s + b}{(s+b)^2 + \omega^2}$
19. $\dfrac{1}{\tau\sqrt{1 - \zeta^2}} e^{-\zeta t/\tau} \sin(\sqrt{1 - \zeta^2}\, t/\tau)$ $(0 \le |\zeta| < 1)$	$\dfrac{1}{\tau^2 s^2 + 2\zeta\tau s + 1}$
20. $1 + \dfrac{1}{\tau_2 - \tau_1}(\tau_1 e^{-t/\tau_1} - \tau_2 e^{-t/\tau_2})$ $(\tau_1 \ne \tau_2)$	$\dfrac{1}{s(\tau_1 s + 1)(\tau_2 s + 1)}$

(continued on page 48)

Table 3.1 (*Continued*)

$f(t)$	$F(s)$
21. $1 - \dfrac{1}{\sqrt{1 - \zeta^2}} e^{-\zeta t/\tau} \sin[\sqrt{1 - \zeta^2}\, t/\tau + \psi]$ $\psi = \tan^{-1} \dfrac{\sqrt{1 - \zeta^2}}{\zeta}$ $(0 \le |\zeta| < 1)$	$\dfrac{1}{s(\tau^2 s^2 + 2\zeta\tau s + 1)}$
22. $1 - e^{-\zeta t/\tau} [\cos(\sqrt{1 - \zeta^2}\, t/\tau)$ $+ \dfrac{\zeta}{1 - \zeta^2} \sin(\sqrt{1 - \zeta^2}\, t/\tau)]$ $(0 \le |\zeta| < 1)$	$\dfrac{1}{s(\tau^2 s^2 + 2\zeta\tau s + 1)}$
23. $1 + \dfrac{\tau_3 - \tau_1}{\tau_1 - \tau_2} e^{-t/\tau_1} + \dfrac{\tau_3 - \tau_2}{\tau_2 - \tau_1} e^{-t/\tau_2}$ $(\tau_1 \ne \tau_2)$	$\dfrac{\tau_3 s + 1}{s(\tau_1 s + 1)(\tau_2 s + 1)}$
24. $\dfrac{df}{dt}$	$sF(s) - f(0)$
25. $\dfrac{d^n f}{dt^n}$	$s^n F(s) - s^{n-1}f(0) - s^{n-2}f^{(1)}(0) - \cdots$ $\quad - sf^{(n-2)}(0) - f^{(n-1)}(0)$

[a]Note that $f(t)$ and $F(s)$ are defined for $t \ge 0$ only.

Note that an exponential term in $F(s)$ results. A special case of (3-22) is the unit rectangular pulse, where $h = 1/t_w$. In this case, the area under the pulse is unity.

Impulse Function. A limiting case of the unit rectangular pulse is the impulse, or Dirac delta function, which has the symbol $\delta(t)$. This function is obtained when $t_w \to 0$ while keeping the area under the pulse constant. A pulse of infinite height and infinitesimal width but with a finite area of unity is obtained. Mathematically, this can be accomplished by substituting $h = 1/t_w$ into (3-22); the Laplace transform of $\delta(t)$ is

$$\mathscr{L}(\delta(t)) = \lim_{t_w \to 0} \frac{1}{t_w s} (1 - e^{-t_w s}) \tag{3-23}$$

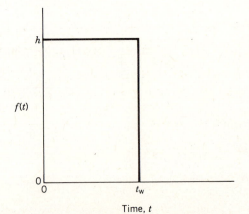

Time, t

Figure 3.1. The rectangular pulse function.

Equation 3-23 is an indeterminate form that can be evaluated by application of L'Hospital's rule. (Also spelled L'Hôpital) Taking derivatives with respect to t_w of both numerator and denominator,

$$\mathcal{L}(\delta(t)) = \lim_{t_w \to 0} \frac{se^{-t_w s}}{s} = 1 \tag{3-24}$$

If the impulse magnitude (i.e., area $t_w h$) is taken to be equal to a rather than unity, then

$$\mathcal{L}(a\delta(t)) = a \tag{3-25}$$

as given in Table 3.1. The unit impulse function may also be interpreted as the time derivative of the unit step function $S(t)$.

A physical example of an impulse function might be given by rapid injection of dye or tracer into a fluid stream, where $f(t)$ would correspond to the concentration or the flow rate of the tracer. This type of signal is sometimes used in process testing, for example, to obtain the residence time distribution of a piece of equipment, as illustrated in Section 3.5. The response of a process to a unit impulse is called its *impulse response*.

3.2 SOLUTION OF DIFFERENTIAL EQUATIONS BY LAPLACE TRANSFORM TECHNIQUES

In the previous section we developed the techniques required to obtain the Laplace transform of each element in a linear ordinary differential equation. Table 3.1 lists important functions of time, including derivatives, and their Laplace transform equivalents. Since the Laplace transform converts any function $f(t)$ to $F(s)$ and the inverse Laplace transform converts $F(s)$ back to $f(t)$, the table provides a simple way to carry out these transformations.

The procedure that we will use to solve a differential equation is quite simple: Laplace transform both sides of the differential equation, substituting values for the initial conditions in the derivative transforms. Rearrange the resulting algebraic equation, finding the transform of the dependent (output) variable. Find the inverse of the transformed output variable. The method is best illustrated by means of several examples.

EXAMPLE 3.1

Solve the differential equation,

$$5\frac{dy}{dt} + 4y = 2 \qquad y(0) = 1 \tag{3-26}$$

using Laplace transforms.

Solution

First take the Laplace transform of both sides of Eq. 3-26:

$$\mathcal{L}\left(5\frac{dy}{dt} + 4y\right) = \mathcal{L}(2) \tag{3-27}$$

Using the principle of superposition, each term can be transformed individually:

$$\mathcal{L}\left(5\frac{dy}{dt}\right) + \mathcal{L}(4y) = \mathcal{L}(2) \tag{3-28}$$

$$\mathcal{L}\left(5\frac{dy}{dt}\right) = 5\mathcal{L}\left(\frac{dy}{dt}\right) = 5(sY(s) - 1) = 5sY(s) - 5 \qquad (3\text{-}29)$$

$$\mathcal{L}(4y) = 4\mathcal{L}(y) = 4Y(s) \qquad (3\text{-}30)$$

$$\mathcal{L}(2) = \frac{2}{s} \qquad (3\text{-}31)$$

Substitute the individual terms:

$$5sY(s) - 5 + 4Y(s) = \frac{2}{s} \qquad (3\text{-}32)$$

Rearrange (3-32) and factor out $Y(s)$:

$$Y(s)(5s + 4) = 5 + \frac{2}{s} \qquad (3\text{-}33)$$

or

$$Y(s) = \frac{5s + 2}{s(5s + 4)} \qquad (3\text{-}34)$$

Having obtained an explicit relation for $Y(s)$, now take the inverse Laplace transform of Eq. 3-34:

$$\mathcal{L}^{-1}[Y(s)] = \mathcal{L}^{-1}\left[\frac{5s + 2}{s(5s + 4)}\right] \qquad (3\text{-}35)$$

The inverse Laplace transform of the right side of (3-35) can be found by using Table 3.1. First divide the numerator and denominator by 5 to put all factors in the $s + b$ form corresponding to the table entries:

$$y(t) = \mathcal{L}^{-1}\left[\frac{s + 0.4}{s(s + 0.8)}\right] \qquad (3\text{-}36)$$

Noting that the table expression $(s + b_3)/[(s + b_1)(s + b_2)]$ matches (3-36) with $b_1 = 0.8$, $b_2 = 0$, and $b_3 = 0.4$, the solution can be written immediately,

$$y(t) = 0.5 + 0.5e^{-0.8t} \qquad (3\text{-}37)$$

Note that in solving (3-26) both the forcing function (the constant 2 on the right side) and the initial condition have been incorporated easily and directly. As with any differential equation solution, (3-37) should be checked to make sure it satisfies the initial condition and that it satisfies the original differential equation.[1]

Next we apply the Laplace transform solution to a higher-order differential equation.

---

[1]In some books, Eq. 3-37 would be written as

$$y(t) = 0.5 + 0.5e^{-0.8t}\mathbf{S}(t)$$

to emphasize the fact that the time function $e^{-0.8t}$ is defined only for $t \geq 0$, that is, when $\mathbf{S}(t) = 1$. However, we have already noted that $Y(s)$ is defined only for $t \geq 0$; hence the solution $y(t) = \mathcal{L}^{-1}[Y(s)]$ clearly has meaning only for $t > 0$. Inclusion of the unit step function is redundant.

EXAMPLE 3.2

Solve the ordinary differential equation

$$\frac{d^3y}{dt^3} + 6\frac{d^2y}{dt^2} + 11\frac{dy}{dt} + 6y = 1 \qquad (3\text{-}38)$$

with initial conditions $y(0) = y'(0) = y''(0) = 0$.

Solution

Take Laplace transforms, term by term, using Table 3.1:

$$\mathcal{L}\left(\frac{d^3y}{dt^3}\right) = s^3 Y(s)$$

$$\mathcal{L}\left(6\frac{d^2y}{dt^2}\right) = 6s^2 Y(s)$$

$$\mathcal{L}\left(11\frac{dy}{dt}\right) = 11s Y(s)$$

$$\mathcal{L}(6y) = 6Y(s)$$

$$\mathcal{L}(1) = \frac{1}{s}$$

Rearranging and factoring out $Y(s)$, we obtain

$$Y(s)(s^3 + 6s^2 + 11s + 6) = \frac{1}{s} \qquad (3\text{-}39)$$

$$Y(s) = \frac{1}{s(s^3 + 6s^2 + 11s + 6)} \qquad (3\text{-}40)$$

To invert (3-40) to find $y(t)$, we must find a similar expression in Table 3.1. Unfortunately no formula in the table has a fourth-order polynomial in the denominator. This example will be continued later, after we develop the techniques necessary to generalize the method.

We conclude from this example that a general transform may not exactly match any of the entries in Table 3.1. In the case of higher-order differential equations, this problem will always arise because the order of the denominator polynomial (characteristic polynomial) of the transform is equal to the order of the original differential equation and no table entries are higher than third order in the denominator. Hence, the method used in Examples 3.1 and 3.2 must be made more general. It is simply not practical to expand the number of entries in the table ad infinitum. Rather, we choose to develop a method based on elementary transform building blocks. This procedure, called *partial fraction expansion*, is discussed in the next section.

3.3 PARTIAL FRACTION EXPANSION

When a high-order denominator polynomial arises in a Laplace transform solution, we first factor it. Note that the denominator polynomial consists of terms arising from the differential equation (its *characteristic polynomial*) plus terms contributed by the inputs. The factors of the characteristic polynomial may have to be obtained by finding the roots of the *characteristic equation* which is obtained by setting the characteristic polynomial equal to zero. The input factors usually are quite simple. Once the factors are obtained, the Laplace transform is then expanded into *partial fractions*. As an example, consider

$$Y(s) = \frac{s + 5}{(s + 1)(s + 4)} \tag{3-41}$$

which can be expanded into the sum of two partial fractions:

$$\frac{s + 5}{(s + 1)(s + 4)} = \frac{\alpha_1}{s + 1} + \frac{\alpha_2}{s + 4} \tag{3-42}$$

where α_1 and α_2 are unspecified coefficients that must satisfy Eq. 3-42. The expansion in (3-42) indicates that the original denominator polynomial has been factored into a product of first-order terms. In general, for every partial fraction expansion, there will be a unique set of the α_i.

There are several methods for calculating the appropriate values of α_1 and α_2:

Method 1.   Multiply both sides of (3-42) by $(s + 1)(s + 4)$

$$s + 5 = \alpha_1(s + 4) + \alpha_2(s + 1) \tag{3-43}$$

Equating coefficients of each power of s, we have

$$s^1: \alpha_1 + \alpha_2 = 1 \tag{3-44a}$$
$$s^0: 4\alpha_1 + \alpha_2 = 5 \tag{3-44b}$$

Solving for α_1 and α_2 simultaneously yields $\alpha_1 = \frac{4}{3}$, $\alpha_2 = -\frac{1}{3}$.

Method 2.   Since Eq. 3-42 should hold for all values of s, we can specify two values of s and solve for the two constants:

$$s = -5: \quad 0 = -\tfrac{1}{4}\alpha_1 - \alpha_2 \tag{3-45a}$$
$$s = -3: -\tfrac{2}{2} = -\tfrac{1}{2}\alpha_1 + \alpha_2 \tag{3-45b}$$

Solving, $\alpha_1 = \frac{4}{3}$, $\alpha_2 = -\frac{1}{3}$.

Method 3.   The fastest and most popular method is called the *Heaviside expansion*. In this method we multiply both sides of the equation by one of the denominator terms $(s + b_i)$ and then set $s = -b_i$, which causes all terms except one to be multiplied by zero. Multiplying Eq. 3-42 by $s + 1$ and then letting $s = -1$, gives

$$\alpha_1 = \frac{s + 5}{s + 4}\bigg|_{s = -1} = \frac{4}{3}$$

Similarly, after multiplying by $(s + 4)$ and letting $s = -4$, the expansion gives

$$\alpha_2 = \frac{s + 5}{s + 1}\bigg|_{s = -4} = -\frac{1}{3}$$

Thus, the coefficients essentially can be found by simple calculations.

For a more general transform, where the factors are real and distinct (no complex or repeated factors appear), the following expansion formula can be used:

$$Y(s) = \frac{N(s)}{D(s)} = \frac{N(s)}{\displaystyle\prod_{i=1}^{n}(s + b_i)} = \sum_{i=1}^{n}\frac{\alpha_i}{s + b_i} \tag{3-46}$$

where $D(s)$, an nth-order polynomial, is the denominator of the transform. $D(s)$ is the characteristic polynomial. $N(s)$, the numerator, has a maximum order of $n - 1$. The ith coefficient can be calculated using the Heaviside expansion

$$\alpha_i = (s + b_i) \frac{N(s)}{D(s)} \bigg|_{s = -b_i} \tag{3-47}$$

Alternatively, an expansion for real, distinct factors may be written as

$$Y(s) = \frac{N'(s)}{D'(s)} = \frac{N'(s)}{\prod\limits_{i=1}^{n} (\tau_i s + 1)} = \sum_{i=1}^{n} \frac{\alpha_i'}{\tau_i s + 1} \tag{3-48}$$

In this case the coefficients are calculated by

$$\alpha_i' = (\tau_i s + 1) \frac{N'(s)}{D'(s)} \bigg|_{s = -\frac{1}{\tau_i}} \tag{3-49}$$

Note that several entries in Table 3.1 follow the $\tau s + 1$ format.

We now are in position to use the methods of partial fraction expansion to complete the solution of Example 3.2.

EXAMPLE 3.2 (CONTINUED)

First factor the denominator of Eq. 3-40 into a product of first-order terms ($n = 4$ in Eq. 3-46). Simple factors, as in this case, rarely occur:

$$s(s^3 + 6s^2 + 11s + 6) = s(s + 1)(s + 2)(s + 3) \tag{3-50}$$

This result determines the four terms that will appear in the partial fraction expansion, namely

$$Y(s) = \frac{1}{s(s + 1)(s + 2)(s + 3)} = \frac{\alpha_1}{s} + \frac{\alpha_2}{s + 1} + \frac{\alpha_3}{s + 2} + \frac{\alpha_4}{s + 3} \tag{3-51}$$

The Heaviside expansion method gives $\alpha_1 = \frac{1}{6}$, $\alpha_2 = -\frac{1}{2}$, $\alpha_3 = \frac{1}{2}$, $\alpha_4 = -\frac{1}{6}$.

Once the transform has been expanded into a sum of first-order terms, merely invert each term individually using Table 3.1:

$$y(t) = \mathcal{L}^{-1}[Y(s)] = \mathcal{L}^{-1}\left(\frac{1/6}{s} - \frac{1/2}{s + 1} + \frac{1/2}{s + 2} - \frac{1/6}{s + 3}\right)$$

$$= \frac{1}{6} \mathcal{L}^{-1}\left(\frac{1}{s}\right) - \frac{1}{2} \mathcal{L}^{-1}\left(\frac{1}{s + 1}\right) + \frac{1}{2} \mathcal{L}^{-1}\left(\frac{1}{s + 2}\right) - \frac{1}{6} \mathcal{L}^{-1}\left(\frac{1}{s + 3}\right)$$

$$= \frac{1}{6} - \frac{1}{2} e^{-t} + \frac{1}{2} e^{-2t} - \frac{1}{6} e^{-3t} \tag{3-52}$$

Equation 3-52 is thus the solution $y(t)$ to the differential equation (3-38). The α_i's are simply the coefficients of the solution. Equation 3-52 also satisfies the three initial conditions of the differential equation. The reader should verify this result.

General Solution Procedure

Having solved Example 3.2, we now can state a general procedure to solve ordinary differential equations using Laplace transforms. The procedure consists of four steps:

Step 1.   Take the Laplace transform of both sides of the differential equation.
Step 2.   Solve for $Y(s)$.
If the expression for $Y(s)$ does not appear in Table 3.1,
Step 3a.   Factor the characteristic equation polynomial.
Step 3b.   Perform the partial fraction expansion.
Step 4.   Use the inverse Laplace transform relations to find $y(t)$.

Figure 3.2 illustrates the four steps for solving an ordinary differential equation; note that solution of the differential equation involves use of Laplace transforms as an intermediate step. Step 3 can be bypassed if the transform found in Step 2 matches an entry in Table 3.1.

In carrying out Step 3, other cases that we have not covered can arise. Both repeated factors and complex factors require a modified partial fraction expansion procedure.

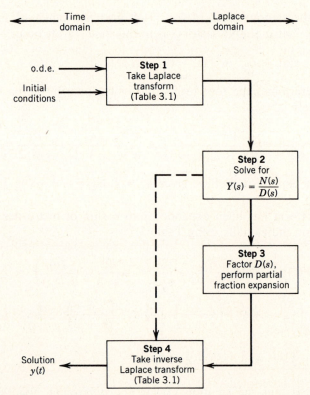

Figure 3.2.   The general procedure for solving an ordinary differential equation using Laplace transforms.

Repeated Factors

If a denominator term $s + b$ occurs r times in the denominator, r terms must be included in the expansion that incorporate the $s + b$ factor

$$Y(s) = \frac{\alpha_1}{s + b} + \frac{\alpha_2}{(s + b)^2} + \cdots + \frac{\alpha_r}{(s + b)^r} + \cdots \qquad (3\text{-}53)$$

in addition to the other factors. Repeated factors arise very infrequently in models of real systems. They are most often encountered in hypothetical models, for example, in the model of a cascade of identical stages. Because real system models exhibit slight physical differences from one stage to the next, the factors in real system cascade models will be only approximately equal, not truly repeated in the mathematical sense.

EXAMPLE 3.3

For

$$Y(s) = \frac{s + 1}{s(s^2 + 4s + 4)} = \frac{\alpha_1}{s + 2} + \frac{\alpha_2}{(s + 2)^2} + \frac{\alpha_3}{s} \qquad (3\text{-}54)$$

set up the partial fractions and evaluate their coefficients.

Solution

To find α_1 in (3-54), the Heaviside rule cannot be used for multiplication by $(s + 2)$, since $s = -2$ causes the second term on the right side to be unbounded, rather than 0 as desired. We therefore employ the Heaviside expansion method for the other two coefficients (α_2 and α_3) that can be evaluated normally, then solve for α_1 by arbitrarily selecting some other value of s. Multiplying (3-54) by $(s + 2)^2$ and letting $s = -2$ yields

$$\alpha_2 = \left.\frac{s + 1}{s}\right|_{s=-2} = \frac{1}{2} \qquad (3\text{-}55)$$

Multiplying (3-54) by s and letting $s = 0$ yields

$$\alpha_3 = \left.\frac{s + 1}{s^2 + 4s + 4}\right|_{s=0} = \frac{1}{4} \qquad (3\text{-}56)$$

Substituting the value $s = -1$ in (3-54) yields

$$0 = \alpha_1 + \alpha_2 - \alpha_3 \qquad (3\text{-}57)$$

$$\alpha_1 = -\frac{1}{4} \qquad (3\text{-}58)$$

An alternative approach to find α_1 is to use differentiation of the transform. Equation 3-54 is multiplied by $s(s + 2)^2$,

$$s + 1 = \alpha_1(s + 2)s + \alpha_2 s + \alpha_3(s + 2)^2 \qquad (3\text{-}59)$$

Then (3-59) is differentiated twice with respect to s,

$$0 = 2\alpha_1 + 2\alpha_3; \qquad \text{so that } \alpha_1 = -\alpha_3 = -\tfrac{1}{4} \qquad (3\text{-}60)$$

Differentiation in this case is tantamount to equating powers of s, as demonstrated earlier.

The differentiation approach illustrated above can be used as the basis of a more general method to evaluate the coefficients of repeated factors. If the denominator polynomial $D(s)$ contains the repeated factor $(s + b)^r$, first form the quantity

$$Q(s) = \frac{N(s)}{D(s)} (s + b)^r = (s + b)^{r-1}\alpha_1 + (s + b)^{r-2}\alpha_2 + \cdots$$
$$+ \alpha_r + (s + b)^r[\text{other partial fractions}]$$

(3-61)

Setting $s = -b$ will generate α_r directly. Differentiating $Q(s)$ with respect to s and letting $s = -b$ generates α_{r-1}. Successive differentiation a total of $r - 1$ times will generate all α_i, $i = 1, 2, \ldots, r$ from which we obtain the general expression

$$\alpha_{r-i} = \frac{1}{i!} \frac{d^{(i)}Q(s)}{ds^{(i)}}\bigg|_{s=-b} \qquad i = 0, \ldots, r - 1$$

(3-62)

For $i = 0$ in (3-62), 0! is defined to be 1 and the zeroth derivative of $Q(s)$ is defined to be simply $Q(s)$ itself.

Returning to the problem in Example 3.3,

$$Q(s) = \frac{s + 1}{s}$$

(3-63)

from which

$$i = 0: \alpha_2 = \frac{s + 1}{s}\bigg|_{s=-2} = \frac{1}{2}$$

(3-64a)

$$i = 1: \alpha_1 = \frac{d\left(\dfrac{s + 1}{s}\right)}{ds}\bigg|_{s=-2} = \frac{-1}{s^2}\bigg|_{s=-2} = -\frac{1}{4}$$

(3-64b)

Complex Factors

Another important case arises when the factored characteristic polynomial yields terms of the form

$$\frac{c_1 s + c_0}{s^2 + d_1 s + d_0} \qquad \text{where} \qquad \frac{d_1^2}{4} < d_0$$

In this case the denominator cannot be written as the product of two real factors. However, we can put it into factorable form by completing the square

$$s^2 + d_1 s + d_0 = \left(s^2 + d_1 s + \frac{d_1^2}{4}\right) + \left(d_0 - \frac{d_1^2}{4}\right)$$

$$= \left(s + \frac{d_1}{2}\right)^2 + \left(d_0 - \frac{d_1^2}{4}\right)$$

$$= \left[\left(s + \frac{d_1}{2}\right) + j\left(d_0 - \frac{d_1^2}{4}\right)^{1/2}\right]\left[\left(s + \frac{d_1}{2}\right) - j\left(d_0 - \frac{d_1^2}{4}\right)^{1/2}\right]$$

(3-65)

Hence, it is possible to rewrite the denominator in the form of two complex factors

$$s^2 + d_1 s + d_0 = (s + b + j\omega)(s + b - j\omega) \tag{3-66}$$

where $b = \dfrac{d_1}{2}$ and $\omega = +\sqrt{d_0 - \dfrac{d_1^2}{4}}$

Note that the quantity under the radical is positive, according to the earlier assumption.

The complex factors in Eq. 3-66 lead to a pair of complex terms in the partial fraction equation, as follows:

$$Y(s) = \frac{\alpha_1 + j\beta_1}{s + b + j\omega} + \frac{\alpha_2 + j\beta_2}{s + b - j\omega} + \text{other terms} \tag{3-67}$$

Note that one factor is the complex conjugate of the other. Appearance of these terms implies an oscillatory behavior: Terms of the form $e^{-bt} \sin \omega t$ and $e^{-bt} \cos \omega t$ arise after combining the inverse transforms $e^{-(b+j\omega)t}$ and $e^{-(b-j\omega)t}$. The sine and cosine terms yield periodic, oscillatory behavior which ultimately damps to zero $(e^{-bt} \to 0)$ if b is positive. Although dealing with complex factors is more tedious than dealing with real factors, the Heaviside expansion (3-47) can still be employed.

First we examine inherent requirements on the α_i, β_i, b, and ω. if the two complex terms in (3-67) are combined into a single fraction,

$$Y(s) = \frac{[(\alpha_1 + \alpha_2) + j(\beta_1 + \beta_2)]s}{(s + b)^2 + \omega^2} + \frac{\alpha_1 b + \beta_1 \omega + \alpha_2 b - \beta_2 \omega}{(s + b)^2 + \omega^2}$$
$$+ \frac{j(\beta_1 b - \alpha_1 \omega + \alpha_2 \omega + \beta_2 b)}{(s + b)^2 + \omega^2} + \cdots \tag{3-68}$$

Since $Y(s)$ is a transform of real quantities, the right side must also contain only real quantities. Hence $\beta_1 = -\beta_2$ and $\alpha_1 = \alpha_2$. Equation 3-68 then simplifies to

$$Y(s) = \frac{\alpha_1 + j\beta_1}{s + b + j\omega} + \frac{\alpha_1 - j\beta_1}{s + b - j\omega} + \cdots \tag{3-69}$$

The two complex numbers now combine to yield

$$Y(s) = \frac{2\alpha_1(s + b) + 2\beta_1 \omega}{(s + b)^2 + \omega^2} + \cdots \tag{3-70}$$

When the individual terms in (3-69) are inverted,

$$y(t) = \alpha_1 e^{(-b-j\omega)t} + j\beta_1 e^{(-b-j\omega)t}$$
$$+ \alpha_1 e^{(-b+j\omega)t} - j\beta_1 e^{(-b+j\omega)t} + \cdots \tag{3-71}$$

Rearranging gives

$$y(t) = \alpha_1 e^{-bt}(e^{-j\omega t} + e^{j\omega t}) + j\beta_1 e^{-bt}(e^{-j\omega t} - e^{j\omega t}) + \cdots \tag{3-72}$$

With the use of the identities

$$\cos \omega t = \frac{e^{-j\omega t} + e^{j\omega t}}{2} \quad \text{and} \quad \sin \omega t = \frac{j(e^{-j\omega t} - e^{j\omega t})}{2} \tag{3-73}$$

Eq. 3-72 becomes

$$y(t) = 2\alpha_1 e^{-bt} \cos \omega t + 2\beta_1 e^{-bt} \sin \omega t + \cdots \tag{3-74}$$

Therefore, once α_1 and β_2 are evaluated using partial fraction decomposition, the inverse transform can be written out immediately.

An alternative partial fraction form that avoids complex algebra is as follows: Let

$$Y(s) = \frac{a_1(s + b) + a_2}{(s + b)^2 + \omega^2} + \cdots \tag{3-75}$$

Using Table 3.1, this fraction form inverts to

$$y(t) = a_1 e^{-bt} \cos \omega t + \frac{a_2}{\omega} e^{-bt} \sin \omega t + \cdots \tag{3-76}$$

However, the coefficients a_1 and a_2 must be found by simultaneous equations rather than by the Heaviside expansion, as is shown in Example 3.4.

EXAMPLE 3.4

Find the inverse Laplace transform of

$$Y(s) = \frac{s + 1}{s^2(s^2 + 4s + 5)} \tag{3-77}$$

Solution

Using the quadratic formula, the factors of $s^2 + 4s + 5$ are found to be $(s + 2 + j)$ and $(s + 2 - j)$, so that

$$Y(s) = \frac{s + 1}{s^2(s^2 + 4s + 5)} = \frac{s + 1}{s^2(s + 2 + j)(s + 2 - j)} \tag{3-78}$$

The partial fraction expansion is, therefore,

$$Y(s) = \frac{s + 1}{s^2(s^2 + 4s + 5)} = \frac{\alpha_1}{s} + \frac{\alpha_2}{s^2} + \frac{\alpha_3 + j\beta_3}{s + 2 + j} + \frac{\alpha_3 - j\beta_3}{s + 2 - j} \tag{3-79}$$

First, use the Heaviside expansion to evaluate α_3 and β_3; multiply by $s + 2 + j$ and let $s = -2 - j$:

$$\alpha_3 + j\beta_3 = \frac{s + 1}{s^2(s + 2 - j)}\bigg|_{s = -2-j}$$

$$= \frac{-2 - j + 1}{(-2 - j)^2(-2j)} = \frac{-1 - j}{8 - 6j} \tag{3-80}$$

Rationalize the complex number, multiply (3-80) by $(8 + 6j)/(8 + 6j)$, to obtain

$$\alpha_3 + j\beta_3 = \frac{-2 - 14j}{100} = -0.02 - 0.14j \tag{3-81}$$

or $\alpha_3 = -0.02$ and $\beta_3 = -0.14$.

Now evaluate the repeated root using the formula (3-62):

$$Q(s) = \frac{s + 1}{s^2 + 4s + 5} \tag{3-82}$$

$$\alpha_2 = Q(s)\bigg|_{s=0} = \frac{s + 1}{s^2 + 4s + 5}\bigg|_{s=0} \tag{3-83}$$

$$= 0.2$$

$$\alpha_1 = \frac{dQ(s)}{ds}\bigg|_{s=0} = \frac{s^2 + 4s + 5 - (s + 1)(2s + 4)}{(s^2 + 4s + 5)^2}\bigg|_{s=0} \tag{3-84}$$

$$= 0.04$$

Use Eq. 3-79 and Table 3.1 to obtain the corresponding time-domain expression:

$$y(t) = 0.04 + 0.2t - 0.04e^{-2t}\cos t - 0.28e^{-2t}\sin t \tag{3-85}$$

The alternative partial fraction form for (3-77), the form that completely avoids using complex factors, is

$$Y(s) = \frac{s + 1}{s^2(s^2 + 4s + 5)} = \frac{\alpha_1}{s} + \frac{\alpha_2}{s^2} + \frac{\alpha_5 s + \alpha_6}{s^2 + 4s + 5} \tag{3-86}$$

Multiply both sides of Eq. 3-86 by $s^2(s^2 + 4s + 5)$ and collect terms:

$$s + 1 = (\alpha_1 + \alpha_5)s^3 + (4\alpha_1 + \alpha_2 + \alpha_6)s^2 + (5\alpha_1 + 4\alpha_2)s + 5\alpha_2 \tag{3-87}$$

Equate coefficients of like powers of s:

$$s^3: \quad \alpha_1 + \alpha_5 = 0 \tag{3-88a}$$
$$s^2: \quad 4\alpha_1 + \alpha_2 + \alpha_6 = 0 \tag{3-88b}$$
$$s^1: \quad 5\alpha_1 + 4\alpha_2 = 1 \tag{3-88c}$$
$$s^0: \quad 5\alpha_2 = 1 \tag{3-88d}$$

Solving simultaneously, we obtain $\alpha_1 = 0.04$, $\alpha_2 = 0.2$, $\alpha_5 = -0.04$, $\alpha_6 = -0.36$. The inverse Laplace transform of $Y(s)$ is

$$y(t) = \mathcal{L}^{-1}\left(\frac{0.04}{s}\right) + \mathcal{L}^{-1}\left(\frac{0.2}{s^2}\right) + \mathcal{L}^{-1}\left(\frac{-0.04s - 0.36}{s^2 + 4s + 5}\right) \tag{3-89}$$

Before using Table 3.1, the denominator term $(s^2 + 4s + 5)$ must be converted by completing the square to $(s + 2)^2 + 1^2$; the numerator is $-0.04(s + 9)$. Note that, to match the expressions in Table 3.1, the argument of the last term in (3-89) must be written as

$$\frac{-0.04s - 0.36}{(s + 2)^2 + 1} = \frac{-0.04(s + 2)}{(s + 2)^2 + 1} + \frac{-0.28}{(s + 2)^2 + 1} \tag{3-90}$$

This procedure yields the same results as given in Eq. 3-85.

3.4 OTHER LAPLACE TRANSFORM PROPERTIES

Final Value Theorem

The asymptotic value of $y(t)$ for large values of time $y(\infty)$ can be found from (3-91) as long as $\lim_{t \to \infty} y(t)$ exists, that is providing the limit exists for all $Re(s) \geq 0$:

$$y(\infty) = \lim_{s \to 0}[sY(s)] \tag{3-91}$$

Equation 3-91 can be proved using the relation for the Laplace transform of a derivative (Eq. 3-9):

$$\int_0^\infty \frac{dy}{dt} e^{-st}\, dt = sY(s) - y(0) \tag{3-92}$$

Taking the limit as $s \to 0$ and assuming that dy/dt is continuous and $sY(s)$ has a limit for all $Re(s) \geq 0$, we find

$$\int_0^\infty \frac{dy}{dt} \, dt = \lim_{s \to 0}[sY(s)] - y(0) \tag{3-93}$$

Integrating the left side and simplifying yields

$$y(\infty) = \lim_{s \to 0} [sY(s)] \tag{3-94}$$

If the transform corresponds to a function of time that is unbounded for $t \to \infty$, Eq. 3-94 will give erroneous results. For example, if $Y(s) = 1/(s - 5)$, Eq. 3-94 predicts $y(\infty) = 0$. Note that Eq. 3-9, which is the basis of (3-92), requires that $\lim y(t \to \infty)$ exist. In this case $y(t) = e^{5t}$, which is unbounded for $t \to \infty$. However, we can see that (3-92) does not apply here without having to evaluate $y(t)$, by noting that $sY(s) = s/(s - 5)$ does not have a limit for some real value of $s \geq 0$, in particular, for $s = 5$.

Initial Value Theorem

In analogy to the final value theorem, the initial value theorem can be stated as

$$y(0) = \lim_{s \to \infty}[sY(s)] \tag{3-95}$$

The proof of this theorem is similar to the development in (3-92) through (3-94). It also requires continuous functions; however, the limit is taken as $s \to \infty$. The proof is left to the reader as an exercise.

EXAMPLE 3.5

Apply the initial and final value theorems to the transform derived in Example 3.1:

$$y(s) = \frac{5s + 2}{s(5s + 4)} \tag{3-34}$$

Solution

Initial value

$$y(0) = \lim_{s \to \infty} [sY(s)] = \lim_{s \to \infty} \frac{5s + 2}{5s + 4} = 1 \tag{3-96}$$

Final value

$$y(\infty) = \lim_{s \to 0} [sY(s)] = \lim_{s \to 0} \frac{5s + 2}{5s + 4} = 0.5 \tag{3-97}$$

The initial value of 1 corresponds to the initial condition given in Eq. 3-26. The final value of 0.5 agrees with the time-domain solution in Eq. 3-37. Both theorems are useful for checking mathematical errors that may occur in the course of obtaining Laplace transform solutions.

Transform of an Integral

Occasionally, it is necessary to find the Laplace transform of a function that is integrated with respect to time. In general, by application of the definition (Eq. 3-1),

$$\mathcal{L}\left\{\int_0^t f(t^*)\, dt^*\right\} = \int_0^\infty \left\{\int_0^t f(t^*)dt^*\right\} e^{-st}\, dt \tag{3-98}$$

$$= -\frac{1}{s}\left[e^{-st}\int_0^t f(t^*)dt^*\right]_0^\infty + \frac{1}{s}\int_0^\infty e^{-st}f(t)\, dt \tag{3-99}$$

after integration by parts. The first term in (3-99) yields 0 when evaluated at both the upper and lower limits, as long as $f(t^*)$ possesses a transform (is bounded). The integral in the second term obviously represents the definition of the Laplace transform of $f(t)$. Hence,

$$\mathcal{L}\left\{\int_0^t f(t^*)dt^*\right\} = \frac{1}{s}\, F(s) \tag{3-100}$$

Note that Laplace transformation of an integral function of time leads to division of the transformed function by s. We have already seen that transformation of time derivatives leads to the reverse procedure, that is, multiplication of the transform by s.

Time Delay (Translation in Time)

Functions that exhibit time delay represent an important case in process modeling and control. Time delays are commonly encountered in process control problems because of the transport time required for a fluid to flow through piping. Consider the stirred-tank heater example presented in Chapter 1. Suppose one thermocouple is located at the outflow point of the stirred tank, and a second thermocouple is

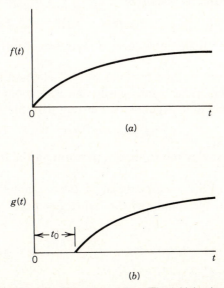

Figure 3.3. A time function with and without time delay. (The initial value has been subtracted from both functions.) (a) Original function (no delay); (b) function with delay.

immersed in the fluid a short distance (10 m) downstream. The heater is off initially and, at time zero, it is turned on. If there is no fluid mixing in the pipe (the fluid is in plug flow) and if no heat losses occur from the pipe between the two thermocouples, the shape of the temperature response of the two sensors should be identical. However, the second sensor response will be translated in time, that is, it will exhibit a time delay. If the fluid velocity is 1 m/s, the time delay $(t_0 = L/v)$ is 10 s. If we denote $f(t)$ as the transient temperature response at the first sensor and $g(t)$ as the temperature response at the second sensor, Fig. 3.3 shows typical behavior of the functions $f(t)$ (no delay) and $g(t)$ (added delay of t_0 units). In Fig. 3.3 we have plotted only the response; in other words, we have subtracted the initial sensor output value. The function $g(t)$ is 0 for $t < t_0$. Therefore, we can state that

$$g(t) = f(t - t_0)\, \mathbf{S}(t - t_0) \tag{3-101}$$

which is the function $f(t)$ delayed by t_0 time units. Here we include the unit step function $\mathbf{S}(t - t_0)$ to denote explicitly that $g(t) = 0$ for all values of $t < t_0$. If $\mathcal{L}\,(f(t)) = F(s)$, then

$$\mathcal{L}\,(g(t)) = \mathcal{L}\,(f(t - t_0)\mathbf{S}(t - t_0)) = \int_0^\infty f(t - t_0)\mathbf{S}(t - t_0)e^{-st}\, dt$$

$$= \int_0^{t_0} f(t - t_0)(0)e^{-st}\, dt + \int_{t_0}^\infty f(t - t_0)e^{-st}\, dt$$

$$= \int_{t_0}^\infty f(t - t_0)e^{-s(t-t_0)}e^{-st_0}d(t - t_0) \tag{3-102}$$

Since $(t - t_0)$ is now the artificial variable of integration, which can be replaced by t^*

$$\mathcal{L}\,(g(t)) = e^{-st_0}\int_0^\infty f(t^*)e^{-st^*}\, dt^* \tag{3-103}$$

yielding the *Real Translation Theorem*

$$G(s) = \mathcal{L}\,(f(t - t_0)\mathbf{S}(t - t_0)) = e^{-st_0}\, F(s) \tag{3-104}$$

Consequently, the time delay introduces a nonrational (nonpolynomial) element into the transform.

In inverting a transform that contains an e^{-st_0} element (time-delay term), the following procedure will yield results easily and also avoid the pitfalls of dealing with translated (shifted) time arguments. Starting with the Laplace transform

$$Y(s) = e^{-st_0}F(s) \tag{3-105}$$

1. Invert $F(s)$ in the usual manner, that is, perform partial fraction expansion, and so forth, to find $f(t)$.
2. Find $y(t) = f(t - t_0)\, \mathbf{S}(t - t_0)$ by replacing the t argument, wherever it appears in $f(t)$, by $(t - t_0)$; then multiply the entire function by the shifted unit step function.

EXAMPLE 3.6

Find the inverse transform of

$$Y(s) = \frac{1 + e^{-2s}}{(4s + 1)(3s + 1)} \tag{3-106}$$

Solution

Equation 3-106 first can be split into two terms

$$Y(s) = Y_1(s) + Y_2(s) \tag{3-107}$$

$$= \frac{1}{(4s + 1)(3s + 1)} + \frac{e^{-2s}}{(4s + 1)(3s + 1)} \tag{3-108}$$

From Table 3.1, the inverse transform of $Y_1(s)$ can be obtained directly:

$$y_1(t) = e^{-t/4} - e^{-t/3} \tag{3-109}$$

Since $Y_2(s) = e^{-2s} Y_1(s)$, its inverse transform can be written out immediately by replacing t by $(t - 2)$ in (3-109), then multiplying by the shifted step function:

$$y_2(t) = [e^{-(t-2)/4} - e^{-(t-2)/3}]S(t - 2) \tag{3-110}$$

Hence, the complete inverse transform is

$$y(t) = e^{-t/4} - e^{-t/3} + [e^{-(t-2)/4} - e^{-(t-2)/3}]S(t - 2) \tag{3-111}$$

Equation 3-111 can be numerically evaluated without difficulty for particular values of t, noting that the bracketed terms are multiplied by 0 (the value of the unit step function) when $t < 2$, and by 1 when $t \geq 2$. An equivalent and simpler method is to evaluate the contributions from the bracketed time functions only when the time arguments are nonnegative. An alternative way of writing Eq. 3-111 is as two relations, each one applicable over a particular interval of time:

$$y(t) = e^{-t/4} - e^{-t/3} \qquad 0 \leq t < 2 \tag{3-112}$$

and

$$y(t) = e^{-t/4} - e^{-t/3} + [e^{-(t-2)/4} - e^{-(t-2)/3}]$$

$$= e^{-t/4}(1 + e^{2/4}) - e^{-t/3}(1 + e^{2/3})$$

$$= 2.6487e^{-t/4} - 2.9477e^{-t/3} \qquad t \geq 2 \tag{3-113}$$

Note that (3-112) and (3-113) give equivalent results for $t = 2$ since, in *this* case, $y(t)$ is continuous at $t = 2$.

3.5 A TRANSIENT RESPONSE EXAMPLE

In Chapter 4 we will develop a standardized approach for the use of Laplace transforms in solving transient problems. That approach will unify the way process models are manipulated after transforming them, and it will further simplify the way initial conditions and inputs (forcing functions) are handled. However, we already have the tools to analyze an example of a transient response situation in some detail. The example below illustrates many features of Laplace transform methods in investigating the dynamic characteristics of a physical process.

EXAMPLE 3.7

The Ideal Gas Company has two fixed-volume stirred-tank reactors connected in series as shown in Fig. 3.4. The three IGC engineers who are responsible for reactor operations—Larry, Moe, and Curly—are concerned about the adequacy of mixing in the two tanks and want to run a tracer test on the system to determine if dead zones and/or channeling exist in the reactors.

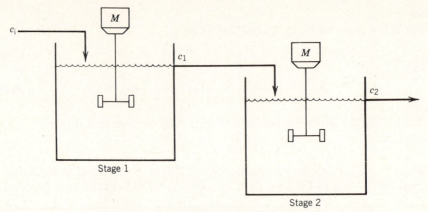

Figure 3.4.   Two-stage stirred-tank reactor system.

Their idea is to operate the reactors at a temperature low enough that reaction will not occur and to apply a rectangular pulse in the reactant concentration to the first stage for test purposes. In this way, available instrumentation on the second-stage outflow line can be used without modification to measure reactant (tracer) concentration.

Before performing the test, the engineers would like to have a good idea of the results that should be expected if perfect mixing actually is accomplished in the reactors. A rectangular pulse input for the change in reactant concentration will be used with the restriction that the resulting output concentration changes must be large enough to be measured precisely.

The process data and operating conditions required to model the reactor tracer test are given in Table 3.2. Figure 3.5 shows the proposed pulse change of 0.25 min duration that can be made while maintaining the total reactor input flow rate constant. As part of the theoretical solution, Larry, Moe, and Curly would like to know how closely the rectangular pulse response might be approximated by the system response to an impulse of equivalent magnitude. Based on all of these considerations, they need to obtain the following information:

(a) The magnitude of an impulse input equivalent to the rectangular pulse of Fig. 3.5.
(b) The impulse and pulse responses of the reactant concentration leaving the first stage.
(c) The impulse and pulse responses of the reactant concentration leaving the second stage.

Solution
The reactor model for a single-stage CSTR was given in Chapter 2 as

$$V \frac{dc}{dt} = q(c_i - c) - Vkc \qquad (2\text{-}34)$$

Table 3.2   Two-Stage Stirred-Tank Reactor Process and Operating Data

Volume of Stage 1	=	4 m³
Volume of Stage 2	=	3 m³
Total flow rate	=	2 m³/min
Nominal reactant concentration in input	=	1 kg mol/m³

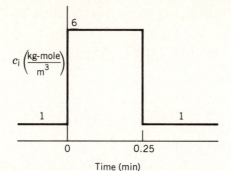

Figure 3.5. Proposed input pulse in reactant concentration.

where c is the reactant concentration. Since the reaction term can be neglected in this example ($k = 0$), the stages serve simply as continuous-flow mixers. Two simple material balance equations are required to model the two stages:

$$4 \frac{dc_1}{dt} + 2c_1 = 2c_i \tag{3-114}$$

$$3 \frac{dc_2}{dt} + 2c_2 = 2c_1 \tag{3-115}$$

If the system initially is at steady state

$$c_2(0) = c_1(0) = c_i(0) = 1 \text{ kg mol/m}^3 \tag{3-116}$$

(a) The pulse input is described by

$$c_i^p = \begin{cases} 1 & t < 0 \\ 6 & 0 \le t < 0.25 \\ 1 & t \ge 0.25 \end{cases} \tag{3-117}$$

A convenient way to interpret (3-117) is as a constant input of 1 added to a rectangular pulse of height = 5:

$$c_i^p = 1 + 5(\text{rectangular pulse with height} = 1 \text{ and width} = 0.25) \tag{3-118}$$

The magnitude of an impulse input that is equivalent to the time-varying portion of (3-118) is

$$M = 5 \frac{\text{kg mol}}{\text{m}^3} \times 0.25 \text{ min} = 1.25 \frac{\text{kg mol} \cdot \text{min}}{\text{m}^3}$$

Note that M is simply the integral of the rectangular pulse. Therefore, the impulse input in concentration $\mathbf{c}_i^\delta$ would be

$$c_i^\delta(t) = 1 + 1.25\delta(t) \tag{3-119}$$

Although the units of M may have little physical meaning, we note that Eq. 3-114 is written with units of kg mol/min. In evaluating its right side, we see that

$$qM = 2 \frac{\text{m}^3}{\text{min}} \times 1.25 \frac{\text{kg mol} \cdot \text{min}}{\text{m}^3} = 2.5 \text{ kg mol}$$

can be interpreted as the amount of additional reactant fed into the reactor in either the rectangular pulse or the impulse.

(b) The impulse response of Stage 1 is obtained by Laplace transforming (3-114), using $c_1(0) = 1$:

$$4sC_1(s) - 4(1) + 2C_1(s) = 2C_i(s) \qquad (3\text{-}120)$$

Rearranging we obtain $C_1(s)$:

$$C_1(s) = \frac{4}{4s + 2} + \frac{2}{4s + 2} C_i(s) \qquad (3\text{-}121)$$

The transform of the impulse input in (3-119) is

$$C_i^\delta(s) = \frac{1}{s} + 1.25 \qquad (3\text{-}122)$$

Substituting (3-122) into (3-121), we have

$$C_1^\delta(s) = \frac{2}{s(4s + 2)} + \frac{6.5}{4s + 2} \qquad (3\text{-}123)$$

Equation 3-123 does not correspond exactly to any entries in Table 3.1. However, putting the denominator in $\tau s + 1$ form yields

$$C_1^\delta(s) = \frac{1}{s(2s + 1)} + \frac{3.25}{2s + 1} \qquad (3\text{-}124)$$

which can be directly inverted using the table to yield

$$c_1^\delta(t) = 1 - e^{-t/2} + 1.625e^{-t/2} = 1 + 0.625e^{-t/2} \qquad (3\text{-}125)$$

The rectangular pulse response is obtained in the same way. The transform of the input pulse (3-117) is given by (3-22) so that

$$C_i^p(s) = \frac{1}{s} + \frac{5(1 - e^{-0.25s})}{s} \qquad (3\text{-}126)$$

Substituting (3-126) into (3-121) and solving for $C_1^p(s)$ yields

$$C_1^p(s) = \frac{4}{4s + 2} + \frac{12}{s(4s + 2)} - \frac{10e^{-0.25s}}{s(4s + 2)} \qquad (3\text{-}127)$$

Again, we have to put (3-127) into a form suitable for inversion

$$C_1^p(s) = \frac{2}{2s + 1} + \frac{6}{s(2s + 1)} - \frac{5e^{-0.25s}}{s(2s + 1)} \qquad (3\text{-}128)$$

Before inverting (3-128), note that the term containing $e^{-0.25s}$ will involve a translation in time. Utilizing the procedure discussed above, the following inverse transform is obtained

$$c_1^p(t) = e^{-t/2} + 6(1 - e^{-t/2}) - 5[1 - e^{-(t-0.25)/2}]S(t - 0.25) \qquad (3\text{-}129)$$

Note that there are two solutions; for $t < 0.25$ min (or t_w) the right-most term, including the time delay, is zero in the time solution. Hence

$$c_1^p(t) = e^{-t/2} + 6(1 - e^{-t/2}) = 6 - 5e^{-t/2} \qquad t < 0.25 \text{ min} \qquad (3\text{-}130)$$

$$c_1^p(t) = e^{-t/2} + 6(1 - e^{-t/2}) - 5(1 - e^{-(t-0.25)/2})$$

$$= 1 - 5e^{-t/2} + 5e^{-t/2}e^{+0.25/2}$$

$$= 1 + 0.6657e^{-t/2} \qquad t \geq 0.25 \text{ min} \qquad (3\text{-}131)$$

Plots of (3-125), (3-130), and (3-131) are given in Fig. 3.6. Note that the rectangular pulse response approximates the impulse response fairly well for $t > 0.25$ min. Obviously, the approximation cannot be very good before $t = 0.25$ because the full effect of the rectangular pulse is not felt until that time, whereas the full effect of the hypothetical impulse is felt immediately at $t = 0$.

(c) For the impulse response of Stage 2, Laplace transform (3-115), using $c_2(0) = 1$

$$3sC_2(s) - 3(1) + 2C_2(s) = 2C_1(s) \tag{3-132}$$

Rearrange to obtain $C_2(s)$:

$$C_2(s) = \frac{3}{3s + 2} + \frac{2}{3s + 2} C_1(s) \tag{3-133}$$

For the input to (3-133), substitute the Laplace transform of the output from Stage 1, namely (3-124)

$$C_2^\delta(s) = \frac{3}{3s + 2} + \frac{2}{3s + 2} \left[\frac{1}{s(2s + 1)} + \frac{3.25}{2s + 1} \right] \tag{3-134}$$

which can be rearranged to a form suitable for inversion

$$C_2^\delta(s) = \frac{1.5}{1.5s + 1} + \frac{1}{s(1.5s + 1)(2s + 1)} + \frac{3.25}{(1.5s + 1)(2s + 1)} \tag{3-135}$$

Since each term in (3-135) appears as an entry in Table 3.1, partial fraction expansion is not required.

$$c_2^\delta(t) = e^{-t/1.5} + \left[1 + \frac{1}{0.5} (1.5e^{-t/1.5} - 2e^{-t/2}) \right] + \frac{3.25}{0.5} e^{-t/2} - e^{-t/1.5}$$

$$= 1 - 2.5e^{-t/1.5} + 2.5e^{-t/2} \tag{3-136}$$

For the rectangular pulse response of Stage 2, substitute the Laplace transform of

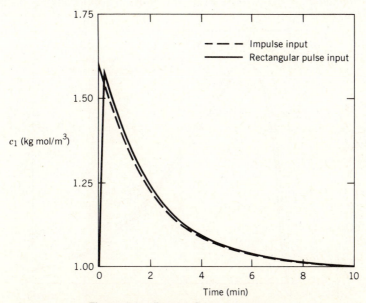

Figure 3.6.   Reactor Stage 1 response.

the appropriate stage output, Eq. 3-128, into Eq. 3-133 to obtain

$$C_2^p(s) = \frac{1.5}{1.5s + 1} + \frac{2}{(1.5s + 1)(2s + 1)}$$

$$+ \frac{6}{s(1.5s + 1)(2s + 1)} - \frac{5e^{-0.25s}}{s(1.5s + 1)(2s + 1)} \qquad (3\text{-}137)$$

Again, the right-most term in (3-137) must be excluded from the inverted result or included, depending on whether $t < 0.25$ min or not. The calculation of the inverse transform of (3-137) gives:

$$c_2^p(t) = 6 + 15e^{-t/1.5} - 20e^{-t/2} \qquad t < 0.25 \qquad (3\text{-}138)$$

$$c_2^p(t) = 1 - 2.7204e^{-t/1.5} + 2.663e^{-t/2} \qquad t \geq 0.25 \qquad (3\text{-}139)$$

Plots of Eqs. 3-136, 3-138, and 3-139 are given in Fig. 3.7. The rectangular pulse response is virtually indistinguishable from the impulse response for t larger than about 1 min. Hence, Larry, Moe, and Curly can use the simpler impulse response solution to compare with real data obtained from the reactor when forced by a rectangular pulse, as long as $t \geq 1$ min. The maximum expected value of $c_2(t)$ is approximately 1.25 kg mol/m^3. This value should be compared with the nominal concentration before and after the test ($\bar{c}_2 = 1.0$ kg mol/m^3) to determine if the instrumentation is precise enough to record the change in concentration. If the change in concentration is too small, then the pulse amplitude or the pulse width or both must be increased.

Since this system is linear, multiplying the pulse magnitude (h) by a factor of four would yield a maximum concentration of reactant in the second stage of about 2.0 (since the difference between initial and maximum concentration will be four times as large). On the other hand, the solutions obtained above strictly apply only for $t_w = 0.25$ min. Hence, the effect of a fourfold increase in t_w can be predicted only by resolving the model response for $t_w = 1$ min. Qualitatively, we know that the maximum value of c_2 will increase with an increase in t_w. Since the impulse response model is a reasonably good approximation with $t_w = 0.25$ min, we might expect that *small changes* in the pulse width will yield an approximately proportional

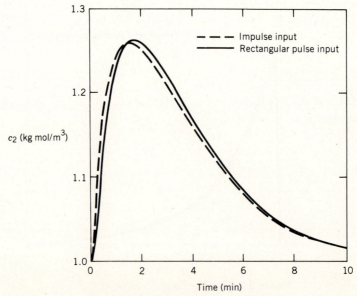

Figure 3.7. Reactor Stage 2 response.

effect on the maximum concentration change. This argument is based on a proportional increase in the approximately equivalent impulse input. A quantitative verification is left as an exercise.

SUMMARY

In this chapter we have considered the application of Laplace transform techniques to solve linear differential equations. Although this material may be a review for some readers, an attempt has been made to concentrate on the important properties of the Laplace transform and its inverse, and to point out the techniques that make manipulation of transforms easier and less prone to error. The final example illustrates the use of a number of these techniques in a practical setting.

The use of Laplace transforms to obtain solutions can be extended to models consisting of multiple simultaneous differential and algebraic equations. However, before discussing further complications, it makes sense to discuss the definition and use of transfer functions. The conversion of differential equation models into transfer function models, covered in the next chapter, represents an important simplification in the methodology, one that can be exploited considerably in process modeling and control system design.

REFERENCES

1. Churchill, R. V., *Operational Mathematics*, McGraw-Hill, New York, 1972.
2. Nixon, F. E., *Handbook of Laplace Transformation—Fundamental Applications, Tables and Examples,* Prentice-Hall, Englewood Cliffs, NJ, 1965.

EXERCISES

3.1. Use Eq. 3-1 to show that the Laplace transform of

(a) $e^{-bt} \sin \omega t$ is $\dfrac{\omega}{(s + b)^2 + \omega^2}$

(b) $e^{-bt} \cos \omega t$ is $\dfrac{s + b}{(s + b)^2 + \omega^2}$

3.2. Find the Laplace transform of $x(t) = \cosh(at) \sin(\omega t)$ by substituting for the hyperbolic function and then using Table 3.1.

3.3. An input function that is sometimes used to test the dynamics of processes is the so-called half-sine-wave pulse shown in the drawing. Here

$$x(t) = \begin{cases} 0 & t < 0 \\ A \sin \omega t & 0 \le t < \dfrac{\pi}{\omega} \\ 0 & \dfrac{\pi}{\omega} \le t \end{cases}$$

Find $X(s)$.

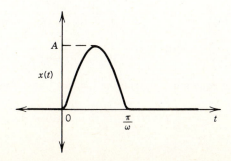

3.4. Calculate the Laplace transforms of the graphical input signals in the accompanying figure.

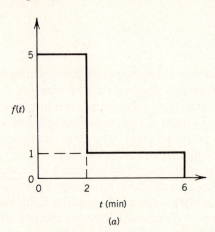

(a)

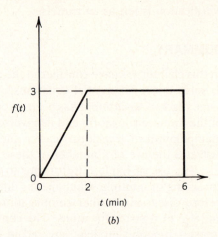

(b)

3.5. The start-up procedure for a batch reactor includes a heating step where the reactor temperature is gradually heated to the nominal operating temperature of 75°C. The desired temperature profile $T(t)$ is shown in the drawing. What is $T(s)$?

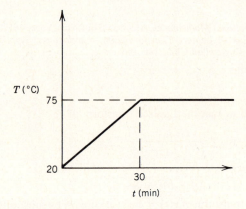

3.6. Using partial fraction expansion where required, find $x(t)$ for

(a) $X(s) = \dfrac{s(s + 1)}{(s + 2)(s + 3)(s + 4)}$

(c) $X(s) = \dfrac{s + 4}{(s + 1)^2}$

(b) $X(s) = \dfrac{s + 1}{(s + 2)(s + 3)(s^2 + 4)}$

(d) $X(s) = \dfrac{1}{s^2 + s + 1}$

3.7. Expand each of the following s-domain functions into partial fractions:

(a) $Y(s) = \dfrac{6(s + 1)}{s^2(s + 1)}$

(c) $Y(s) = \dfrac{(s + 2)(s + 3)}{(s + 4)(s + 5)(s + 6)}$

(b) $Y(s) = \dfrac{12(s + 2)}{s(s^2 + 9)}$

(d) $Y(s) = \dfrac{1}{[(s + 1)^2 + 1]^2 (s + 2)}$

3.8. Solve the following equation for $y(t)$ using Laplace transforms:

$$\int_0^t y(\tau)d\tau = \frac{dy(t)}{dt} \qquad y(0) = 1$$

3.9. For each of the following functions $X(s)$, what can you say about $x(t)$ ($0 \le t \le \infty$) without solving for $x(t)$? In other words, what are $x(0)$ and $x(\infty)$? Is $x(t)$ converging or diverging? Is $x(t)$ smooth or oscillatory?

(a) $X(s) = \dfrac{6(s + 2)}{(s^2 + 9s + 20)(s + 4)}$

(b) $X(s) = \dfrac{10s^2 - 3}{(s^2 - 6s + 10)(s + 2)}$

(c) $X(s) = \dfrac{16s + 5}{s^2 + 9}$

3.10. Find the mathematical *form* of the solutions to the following equations, that is, determine the form of the time solution but do not numerically evaluate the coefficients. Be sure to obtain correct arguments of all exponential and periodic functions. Use the Final Value Theorem to obtain the solution for large values of time. In all cases $x(0) = \dot{x}(0) = 0$ and the dots denote differentiation with respect to time.

(a) $\ddot{x} + 4\dot{x} + 8x = 10$ (e) $\ddot{x} + 4\dot{x} = 10$

(b) $\ddot{x} + 4\dot{x} + 4x = 10$ (f) $\ddot{x} + 4\dot{x} - 5x = 10$

(c) $\ddot{x} + 4\dot{x} + 3x = 10$ (g) $\ddot{x} - 4\dot{x} + 3x = 10$

(d) $\ddot{x} + 4\dot{x} + x = 10$ (h) $\ddot{x} - 4\dot{x} + 4x = 10$

3.11. Which solutions of the following equations will exhibit convergent behavior? Which oscillatory?

(a) $\dfrac{d^3x}{dt^3} + 2\dfrac{d^2x}{dt^2} + 2\dfrac{dx}{dt} + x = 3$ (c) $\dfrac{d^3x}{dt^3} + x = \sin t$

(b) $\dfrac{d^2x}{dt^2} - x = 2e^t$ (d) $\dfrac{d^2x}{dt^2} + \dfrac{dx}{dt} = 4$

Note: All of the above differential equations have one common factor in their characteristic equations.

3.12. For a system described by the following equations

$$\frac{dx}{dt} + x + y = e^{-3t} \qquad x(0) = 1$$

$$\frac{dy}{dt} + y + 4x = 0 \qquad y(0) = 0$$

(a) Find $X(s)$. Be sure to eliminate $Y(s)$ from this expression.

(b) Determine the final value of $x(t)$ as $t \to \infty$.

(c) Find $y(t)$

3.13. Find the complete time-domain solutions for the following differential equations using Laplace transforms:

(a) $\dfrac{d^3x}{dt^3} + 4x = e^t$ with $x(0) = 0$, $\dfrac{dx(0)}{dt} = 0$, $\dfrac{d^2x(0)}{dt^2} = 0$

(b) $\dfrac{dx}{dt} - 12x = \sin 3t$ $x(0) = 0$

(c) $\dfrac{d^2x}{dt^2} + 6\dfrac{dx}{dt} + 25x = e^{-t}$ $x(0) = 0$, $\dfrac{dx(0)}{dt} = 0$

(d) $\left. \begin{array}{l} \dfrac{dx_1}{dt} + 2x_1 - x_2 = t \\[2mm] \dfrac{dx_2}{dt} - x_1 + 2x_2 = \sin t \end{array} \right\}$ $x_1(0) = 0$, $x_2(0) = 0$

3.14. Find the solution to the following differential equation:

$$\ddot{x} + \dot{x} + x = \sin t$$
$$x(0) = \dot{x}(0) = 0$$

Use the final value theorem to determine $x(t)$ as $t \to \infty$ from $X(s)$. What can you say about this result? Note: Dots denote differentiation with respect to time.

3.15. Find the solution of

$$\frac{dx}{dt} + 4x = f(t)$$

where $f(t) = \begin{cases} 0 & t < 0 \\ h & 0 \le t < 1/h \\ 0 & t \ge 1/h \end{cases}$

$$x(0) = 0$$

Plot the solution for values of $h = 1, 10, 100$, and the limiting solution ($h \to \infty$) from $t = 0$ to $t = 2$. Put all plots on the same graph.

3.16. Three stirred tanks in series are used in a reactor train (see drawing). The flow rate into the system of some inert species is maintained constant while tracer tests are conducted. Assuming that mixing in each tank is perfect and volumes are constant:

(a) Derive model expressions for the concentration of tracer leaving each tank. c_i is the concentration of tracer entering the first tank.

(b) If c_i has been constant and equal to zero for a long time and an operator suddenly injects a large amount of tracer material in the inlet to tank 1, what will be the form of $c_3(t)$ (i.e., what kind of time functions will be involved) if

1. $V_1 = V_2 = V_3$
2. $V_1 \ne V_2 \ne V_3$.

(c) If the amount of tracer injected is unknown, is it possible to back-calculate the amount from experimental data? How?

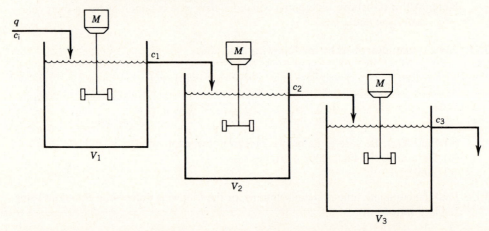

3.17. The system pictured is used to dilute a concentrated caustic solution. It is started up with pure water in the mix tank.

(a) Derive a mathematical model to describe the concentration of caustic in the mix tank effluent c. You may assume that the caustic solution has approximately the same density as the water and that volume in the tank is constant.

(b) Simplify the model for conditions which give rise to $c_c \gg c$. When will this occur?

(c) Find the solution $c(t)$ for both $q_w(t)$ and $q_c(t)$ constant with the assumption $c_c \gg c$.

Volumetric rate of caustic: $q_c(t)$

Volumetric rate of water: $q_w(t)$

Volume of stirred tank: V

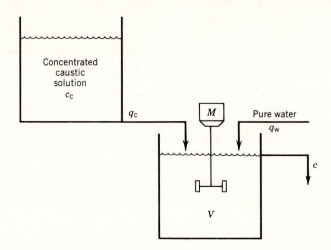

3.18. An electrical heating element is being tested in a large mass of water at temperature T_w. Determine how the element temperature T_e varies in time if the power dissipation rate in the element Q changes suddenly from 0 to $\overline{Q}$. The element initially is at the water temperature T_w. You may assume that the entire element is at the same temperature at any instant in time and that the convective heat transfer coefficient is constant. What is the relation between the unsteady-state model for this case and that of the stirred-tank heater described by Eqs. 2-10 and 2-11?

3.19. I. M. Appelpolscher has just returned home from a hard day at work and is looking forward to a cool, refreshing drink of his favorite beer, Old Froth and Slosh. Unfortunately, the only can available has been left sitting on the kitchen counter. Since sunlight from the kitchen window has been warming the beer all afternoon, it is hardly fit to drink. He is interested in cooling the beer to the desired temperature of 60 °F as quickly as possible.

(a) Appelpolscher immediately springs into action by placing the can in the freezer section of the refrigerator, which has a temperature of −6 °F. If the initial temperature of the beer is 85 °F, estimate how long it will take to cool it to 60 °F. (You may assume that a can of beer weighs 0.8 lb and that the heat capacity is ~1 Btu/lb °F).

(b) Is there a faster method of cooling the beer, using only the resources that normally would be found in Appelpolscher's house? Justify your answer.

3.20. A stirred-tank reactor is operated with a feed mixture containing reactant A at a mass concentration c_{Ai}. The feed flow rate is w_i, as shown below in the drawing. Under certain conditions the system operates according to the model

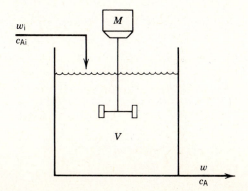

$$\frac{d(\rho V)}{dt} = w_i - w$$

$$\frac{d(\rho V c_A)}{dt} = w_i c_{Ai} - w c_A - \rho V k c_A$$

(a) For cases where the feed flow rate and feed concentration may vary and the volume is not fixed, simplify the model to one or more equations that do not contain product derivatives. The density may be assumed to be constant. Is the model in a satisfactory form for Laplace transform operations? Why or why not?

(b) For the case where the feed flow rate has been steady at $\overline{w}_i$ for some time, determine how c_A changes with time if a step change in c_{Ai} is made from c_{A1} to c_{A2}. List all assumptions necessary to solve the problem using Laplace transform techniques.

—— CHAPTER 4 ——

The Transfer Function

In Chapter 3 we discussed how Laplace transform techniques can be used to determine transient responses from ordinary differential equation models. One disadvantage of this approach is that the full procedure must be applied for each model solution. Any change in the initial conditions or in the type of forcing function requires that the complete solution be rederived. In this chapter we present a modified approach, once the differential equation model has been transformed, based on the concept of the *transfer function*. The transfer function is an algebraic expression for the dynamic relation between the input and the output of the process model. It is defined so as to be independent of the initial conditions and of the particular choice of forcing function. If the transfer function of the process is written in standard form, the fundamental dynamic properties of the process itself become apparent. Hence, in this chapter, we first show how the transfer function is derived and then demonstrate how alternative transfer function models can be compared, once they are put in standard form.

A transfer function can be derived only for a linear differential equation model because Laplace transforms can be applied only to linear equations. If the model is nonlinear, then it must be linearized first, as will be demonstrated in Section 4.3.

In this chapter, and in Chapters 5 and 6, the primary value of the transfer function will be in expressing process models compactly. The transfer function form is easy to interpret and use in calculating output responses for particular input changes. In later chapters, we show how transfer functions can be used to simplify the analysis and synthesis (design) of control systems.

4.1 DEVELOPMENT OF TRANSFER FUNCTIONS

Consider the simple first-order differential equation (2-2) derived earlier for the stirred-tank heating system:

$$V\rho C \frac{dT}{dt} = wC(T_i - T) + Q \tag{2-2}$$

Assuming constant liquid holdup and constant inflow (w is constant), a linear model results. If the process is at steady state initially, $T(0) = \overline{T}$, $T_i(0) = \overline{T}_i$, and

75

$Q(0) = \overline{Q}$. The output $\overline{T}$ is related to the inputs $\overline{T}_i$ and $\overline{Q}$ by the steady-state energy balance,

$$0 = wC(\overline{T}_i - \overline{T}) + \overline{Q} \tag{4-1}$$

In analyzing process dynamics we wish to make the model as general as possible. One way to eliminate the explicit dependence of the model on the original steady-state conditions is to subtract the steady-state relation (4-1) from the differential equation model (2-2)

$$V\rho C\frac{dT}{dt} = wC[(T_i - \overline{T}_i) - (T - \overline{T})] + (Q - \overline{Q}) \tag{4-2}$$

Now divide both sides of (4-2) by wC and note that $dT/dt = d(T - \overline{T})/dt$ because $\overline{T}$ is a constant

$$\frac{V\rho}{w}\frac{d(T - \overline{T})}{dt} = (T_i - \overline{T}_i) - (T - \overline{T}) + \frac{1}{wC}(Q - \overline{Q}) \tag{4-3}$$

The use of parentheses to group variables in Eq. 4-3 motivates the definition of some important new variables

$$T' \stackrel{\Delta}{=} T - \overline{T}$$
$$T_i' \stackrel{\Delta}{=} T_i - \overline{T}_i \tag{4-4}$$
$$Q' \stackrel{\Delta}{=} Q - \overline{Q}$$

which, when substituted, simplify (4-3) considerably:

$$\frac{V\rho}{w}\frac{dT'}{dt} = T_i' - T' + \frac{1}{wC}Q' \tag{4-5}$$

In (4-5) the original input and output variables have been replaced by *deviation variables*, that is, in terms of deviations measured from the original steady-state values as defined in (4-4). Deviation variables are sometimes referred to as perturbation variables, and the words "deviation" and "perturbation" are used interchangeably in this context.

The particular form of Eq. 4-5 has dimensional significance. The term $V\rho/w$ has units of time and is called the system time constant τ for this first-order system (first-order differential equation model). The time constant is indicative of the speed of response of the process. Large values of τ mean a slow process response; small values of τ indicate a fast response. The term $1/wC$ is called the steady-state gain K. It relates the input Q' to the output T' *at steady state*. This feature can be seen by rewriting (4-5) with $dT'/dt = 0$,

$$T' = T_i' + \frac{1}{wC}Q' \tag{4-6}$$

Suppose $T_i' = 0$. Then any change in Q' (deviation in Q from $\overline{Q}$) will result in a change in T' (deviation in T from $\overline{T}$) that is $1/wC$ times as large. We also note that the process gain relating changes in T_i' (deviation in T_i from $\overline{T}_i$) to T' is 1, meaning that each degree increase or decrease in T_i' will be matched by a similar change in T' *at steady state* if $Q' = 0$. An important concept regarding process gains is that they relate steady-state changes in the process output resulting from sustained changes in one or both inputs (step-type input changes).

Transfer Functions

Suppose we now apply Laplace transforms to the situation where both inputs T_i' and Q' are general functions of time. Here Eq. 4-5 is employed after substitution of K and τ:

$$\tau \frac{dT'}{dt} = T_i' - T' + KQ' \tag{4-7}$$

At time zero the system is at its steady state; hence $T'(0) = 0$. Taking the Laplace transform of both sides of the equation, we have

$$\tau \mathcal{L}\left(\frac{dT'}{dt}\right) = \mathcal{L}(T_i' - T' + KQ') \tag{4-8}$$

or

$$\tau \mathcal{L}\left(\frac{dT'}{dt}\right) = \mathcal{L}(T_i') - \mathcal{L}(T') + K\mathcal{L}(Q') \tag{4-9}$$

The constant has been factored out of the transform. Since $T'(t)$, $T_i'(t)$, and $Q'(t)$ are unspecified, their transforms can be expressed in a general manner:

$$\tau s T'(s) = T_i'(s) - T'(s) + KQ'(s) \tag{4-10}$$

Note that by using deviation variables based on the initial steady state, the initial condition does not appear in the transformed relation (4-10). Rearranging (4-10) gives

$$(\tau s + 1) T'(s) = KQ'(s) + T_i'(s) \tag{4-11}$$

$$T'(s) = \left(\frac{K}{\tau s + 1}\right) Q'(s) + \left(\frac{1}{\tau s + 1}\right) T_i'(s) \tag{4-12}$$

$$T'(s) = G_1(s)Q'(s) + G_2(s)T_i'(s) \tag{4-13}$$

where both $G_1(s)$ and $G_2(s)$ are called *transfer functions*. $G_1(s)$ relates the input $Q'(s)$ to the output $T'(s)$; $G_2(s)$ has a similar role for input $T_i'(s)$. Note that the parameters in the transfer functions (K and τ) depend on the operating conditions of the process and that the transfer function does not contain the initial condition explicitly.

Briefly consider how Eq. 4-13 can be used. Assume the inlet temperature is held constant ($T_i = \overline{T}_i$). Then $T_i'(s) = 0$. For any input $Q'(t)$ we can determine the corresponding Laplace transform $Q'(s)$ and then find $T'(s)$ directly from (4-13) because $T'(s) = G_1(s)\,Q'(s)$. Suppose the heat input is changed by a step input at $t = 0$ from its value of $\overline{Q}$ to a new value, $\overline{Q} + \Delta Q$. Therefore, $Q' = \Delta Q$ for $t \geq 0$. Use Table 3.1 to obtain $Q'(s) = \Delta Q/s$ and substitute into (4-12):

$$T'(s) = \frac{K}{\tau s + 1} \frac{\Delta Q}{s} \tag{4-14}$$

Let $A = K\Delta Q$ and observe from Table 3.1 that $T'(s)$ in (4-14) corresponds to the time domain function

$$T'(t) = A(1 - e^{-t/\tau}) = K\Delta Q\,(1 - e^{-t/\tau}) \tag{4-15}$$

Equation 4-15 indicates that the response of the stirred-tank heater output temperature to a step change in heater input follows an exponential approach to a final value. From (4-15), $T'(t = 0) = 0$ and $T'(t \to \infty) = K\Delta Q$. This latter value corresponds to the steady-state response of the system (see Eq. 4-6).

We demonstrated above that the transfer function obtained from a linear process model does not depend explicitly on the initial conditions if suitable deviation variables are defined. Consequently, transfer functions are almost always expressed in terms of deviation variables. Equation 4-13 also implies that the transfer functions G_1 and G_2 are independent of the specific form of the associated input forcing function:

$$\frac{T'(s)}{Q'(s)} = G_1 = \frac{K}{\tau s + 1} \qquad (T_i'(s) = 0) \tag{4-16}$$

$$\frac{T'(s)}{T_i'(s)} = G_2 = \frac{1}{\tau s + 1} \qquad (Q'(s) = 0) \tag{4-17}$$

Consequently, for any initial condition and for any choice of input forcing, the output temperature change can be found by multiplying the appropriate transfer function ((4-16) or (4-17)) by the Laplace transform of the input, converting back to the time domain, and substituting the actual process variables. Note that each transfer function indicates the dynamic relation between a single input variable (Q' or T_i') and the output variable T'. The situation where simultaneous changes in both inputs (Q' and T_i') occur can be analyzed using (4-13). The effects of simultaneous input changes are additive because of the Principle of Superposition for linear systems.

We illustrate this procedure in several examples below.

EXAMPLE 4.1

A stirred-tank heating process described by Eqs. 2-2 and 4-12 is operating with a flow rate of 200 lb/min of liquid with specific heat and density equal to 0.32 Btu/lb°F and 62.4 lb/ft^3, respectively. The tank volume is constant at 1.60 ft^3.

a. A heater input of 1920 Btu/min and inlet temperature of 70 °F have been maintained constant long enough for the system to come to steady state (several minutes). Calculate the system response for a sudden change in inlet temperature to 90 °F.

b. A heater input of 2880 Btu/min and inlet temperature of 85 °F yield a new steady state. Calculate the system response for a sudden decrease (a negative step change) in inlet temperature to 75 °F.

Solution
In both cases the appropriate transfer function is G_2. We begin each solution by noting that $\tau = (1.60)(62.4)/200 = 0.5$ min, and hence

$$G_2 = \frac{1}{0.5s + 1} \tag{4-18}$$

applies in each case. For case (a), first find the basis for deviation variables. Using Eq. 4-1 yields the steady-state temperature value

$$\overline{T} = \overline{T}_i + \overline{Q}/\overline{w}C$$

$$\overline{T} = 70 + \frac{1}{(200)(0.32)} (1920) = 100 \text{ °F}$$

We then note that the input change from 70 to 90 °F in a stepwise fashion implies that

$$T_i'(s) = \frac{90 - 70}{s} = \frac{20}{s}$$

Multiplying the transfer function by the transformed input yields

$$T'(s) = \frac{1}{0.5s + 1} \frac{20}{s}$$

which corresponds to the time-domain expression

$$T'(t) = 20(1 - e^{-2t})$$

and since $T = \overline{T} + T'$,

$$T(t) = 100 + 20(1 - e^{-2t}) \tag{4-19}$$

For case (b) the above procedure is repeated, first by finding the new steady-state temperature which is the basis for a new set of deviation variables

$$\overline{T} = 85 \text{ °F} + \frac{1}{(200)(0.32)}(2880) = 130 \text{ °F}$$

In this case the input change is:

$$T_i'(s) = \frac{75 - 85}{s} = \frac{-10}{s}$$

The transformed output deviation is

$$T'(s) = \frac{1}{0.5s + 1}\left(\frac{-10}{s}\right)$$

from which the time-domain solution is

$$T'(t) = -10(1 - e^{-2t})$$

or

$$T(t) = 130 - 10(1 - e^{-2t}) \tag{4-20}$$

In this example we have shown that solutions corresponding to different initial conditions and input changes can be found without having to rederive the process transfer function G_2. Since the process transfer function is applicable to any heating process of the type described by Eq. 2-2, a different liquid or a different flow rate can be handled simply by changing τ and K in Eq. 4-12.

Any process that can be put into the form of Eq. 4-12, that is, any process described by a first-order linear differential equation, can be characterized in terms of its values of τ and K. These characteristic parameters are independent of the initial condition and the nature of the input.

Transfer Functions for Complicated Models

In the next example we extend the concept of a transfer function based on a single differential equation model to a model consisting of several equations. A more complicated transfer function results, but the approach remains the same.

EXAMPLE 4.2

Consider the model of the stirred-tank heater given in Chapter 2, namely the system with heating element dynamics that cannot be neglected

$$mC\frac{dT}{dt} = wC(T_i - T) + h_eA_e(T_e - T) \tag{2-13}$$

$$m_eC_e\frac{dT_e}{dt} = Q - h_eA_e(T_e - T) \tag{2-14}$$

Subscript e refers to the heating element. Find transfer functions relating changes in temperature of the outlet liquid T to changes in the two input variables: (a) heater input Q assuming no change in inlet temperature, and (b) inlet temperature T_i with no change in heater input.

Solution

First write the steady-state equations:

$$0 = wC(\overline{T}_i - \overline{T}) + h_eA_e(\overline{T}_e - \overline{T}) \tag{4-21}$$

$$0 = \overline{Q} - h_eA_e(\overline{T}_e - \overline{T}) \tag{4-22}$$

Next subtract (4-21) from (2-13) and (4-22) from (2-14):

$$mC\frac{dT}{dt} = wC[(T_i - \overline{T}_i) - (T - \overline{T})] + h_eA_e[(T_e - \overline{T}_e) - (T - \overline{T})] \tag{4-23}$$

$$m_eC_e\frac{dT_e}{dt} = (Q - \overline{Q}) - h_eA_e(T_e - \overline{T}_e - (T - \overline{T})) \tag{4-24}$$

Note that $dT/dt = dT'/dt$ and $dT_e/dt = dT'_e/dt$. Substitute deviation variables, then multiply (4-23) by $1/wC$ and (4-24) by $1/h_eA_e$:

$$\frac{m}{w}\frac{dT'}{dt} = -(T' - T'_i) + \frac{h_eA_e}{wC}(T'_e - T') \tag{4-25}$$

$$\frac{m_eC_e}{h_eA_e}\frac{dT'_e}{dt} = \frac{Q'}{h_eA_e} - (T'_e - T') \tag{4-26}$$

The Laplace transform of each equation, after rearrangement, is

$$\left(\frac{m}{w}s + 1 + \frac{h_eA_e}{wC}\right)T'(s) = T'_i(s) + \frac{h_eA_e}{wC}T'_e(s) \tag{4-27}$$

$$\left(\frac{m_eC_e}{h_eA_e}s + 1\right)T'_e(s) = \frac{Q'(s)}{h_eA_e} + T'(s) \tag{4-28}$$

We can eliminate one of the output variables $T'(s)$ or $T'_e(s)$ by solving either (4-27) or (4-28). Since $T'_e(s)$ is the intermediate variable, remove it from Eq. 4-27 by using Eq. 4-28. Multiply (4-27) by $[(m_eC_e/h_eA_e)s + 1]$ and substitute to obtain

$$\left[\frac{mm_eC_e}{wh_eA_e}s^2 + \left(\frac{m_eC_e}{h_eA_e} + \frac{m_eC_e}{wC} + \frac{m}{w}\right)s + 1\right]T'(s)$$

$$= \left(\frac{m_eC_e}{h_eA_e}s + 1\right)T'_i(s) + \frac{1}{wC}Q'(s) \tag{4-29}$$

By inspection it is clear that Eq. 2-15, obtained by time-domain manipulation, is equivalent to (4-29). However, the above use of Laplace transforms makes this derivation more direct, consisting largely of algebraic manipulation.

In Eq. 4-29 we have included both input variables (Q' and T_i') on the right side, while the output variable (T') is on the left side of the equation. Both inputs influence the dynamic behavior of T'; thus, it is necessary to develop two transfer functions for the model. The effect of Q' on T' is obtained when T_i' is assumed to be zero:

$$\frac{T'(s)}{Q'(s)} = \frac{1/wC}{b_2 s^2 + b_1 s + 1} \qquad (T_i'(s) = 0) \qquad (4\text{-}30)$$

Similarly, the effect of T_i' on T' is obtained when Q' is assumed not to vary:

$$\frac{T'(s)}{T_i'(s)} = \frac{\dfrac{m_e C_e}{h_e A_e} s + 1}{b_2 s^2 + b_1 s + 1} \qquad (Q'(s) = 0) \qquad (4\text{-}31)$$

where

$$b_1 = \frac{m_e C_e}{h_e A_e} + \frac{m_e C_e}{wC} + \frac{m}{w}$$

$$b_2 = \frac{m m_e C_e}{w h_e A_e}$$

In deriving (4-30) and (4-31) we have successively set each input equal to zero to identify individual transfer functions. Equations 4-30 and 4-31 contain the same second-order polynomial in s in the denominator. Hence, each of these transfer functions represents second-order dynamic behavior of the stirred-tank heater; that is, a second-order differential equation (cf. Eq. 2-15) governs the dynamic behavior. Any set of n first-order differential equations can be combined to yield a single nth-order differential equation in terms of one of the original output variables. In this example we eliminated the output variable T_e' and retained T'. We could obtain transfer functions relating T_e' to Q' and to T_i' in a similar manner by eliminating output variable T' and retaining T_e'.

In the case of the first-order transfer function (4-16), the steady-state (process) gain is $K = 1/wC$. The second-order model of the stirred-tank heater, Eq. 4-30, also has a gain $K = 1/wC$. In other words, the second-order and first-order models have the same steady-state gain between Q' and T'. The gain for the inlet temperature input is unity for both models. This means that a 20 °F increase in T_i' ultimately brings about a 20 °F change in T', which is intuitively correct. For the second-order model, there will be two time constants (τ_1, τ_2), which can be obtained by factoring the characteristic polynomial, $b_2 s^2 + b_1 s + 1$, as $(\tau_1 s + 1)(\tau_2 s + 1)$. These time constants provide measures of the speed of response, similar to the single time constant associated with the first-order system. In the next section, these characteristics of transfer functions are generalized.

4.2 PROPERTIES OF TRANSFER FUNCTIONS

One important property of the transfer function is that the steady-state output change for a sustained change in input can be calculated directly. Very simply, setting $s = 0$ in $G(s)$ gives the steady-state gain of a process if the gain exists.[1] This feature is a consequence of the final value theorem presented in Chapter 3. If a unit step change in input is assumed, the corresponding output change for $t \rightarrow \infty$ is $\lim G(s)$ as $s \rightarrow 0$.

The steady-state gain is the ratio of the output variable change to an input variable change when the input is adjusted to a new value and held there, thus allowing the process to reach a new steady state. Stated another way, the steady-state gain K of a process corresponds to the following expression

$$K = \frac{\bar{y}_2 - \bar{y}_1}{\bar{x}_2 - \bar{x}_1} \tag{4-32}$$

where 1 and 2 indicate different steady states and $(\bar{y}, \bar{x})$ denote the corresponding steady-state values of the output and input variables.

Another important property of the transfer function is that the order of the denominator polynomial (in s) is the same as the order of the equivalent differential equation. A general nth-order differential equation has the form

$$a_n \frac{d^n y}{dt^n} + a_{n-1} \frac{d^{n-1} y}{dt^{n-1}} + \cdots + a_1 \frac{dy}{dt} + a_0 y$$

$$= b_m \frac{d^m x}{dt^m} + b_{m-1} \frac{d^{m-1} x}{dt^{m-1}} + \cdots + b_1 \frac{dx}{dt} + b_0 x \tag{4-33}$$

where x and y are input and output deviation variables, respectively. The transfer function obtained by Laplace transformation of (4-33) with all initial conditions on y, its derivatives, and the derivatives of x set equal to zero is

$$G(s) = \frac{Y(s)}{X(s)} = \frac{\sum\limits_{i=0}^{m} b_i s^i}{\sum\limits_{i=0}^{n} a_i s^i} = \frac{b_m s^m + b_{m-1} s^{m-1} + \cdots + b_o}{a_n s^n + a_{n-1} s^{n-1} + \cdots + a_o} \tag{4-34}$$

Note that the numerator and denominator polynomials of the transfer function have the same orders (m and n, respectively) as the differential equation.

The steady-state gain of $G(s)$ in (4-34) is b_o/a_o, obtained by setting $s = 0$ in $G(s)$. If both the numerator and denominator of (4-34) are divided by a_o, the characteristic (denominator) polynomial can be factored into the product form $\Pi_i(\tau_i s + 1)$. In this form, the so-called gain and time constant form, inspection of the individual time constants (the τ_i) yields information about the speed and qualitative features of the system response. This important point is discussed in detail in Chapter 6, after some additional mathematical tools have been developed.

The orders of the numerator and denominator polynomials in Eq. 4-34 are restricted by physical reasons so that $n \geq m$. Suppose that a real process could be modeled by

---

[1]Some processes do not exhibit a steady-state gain, for example, integrating elements discussed in Chapter 5.

$$a_0 y = b_1 \frac{dx}{dt} + b_0 x \tag{4-35}$$

that is, $n = 0$ and $m = 1$ in (4-34). This system will respond to a step change in $x(t)$ with an impulse at time zero because dx/dt is infinite at the time the step change occurs. The ability to respond infinitely fast to a sudden change in input is impossible to achieve with any real (physical) process, although it is approximated in some instances, for example, an explosion. Therefore, we refer to the restriction $n \geq m$ as a *physical realizability* condition. It provides a diagnostic check on transfer functions derived from a high-order differential equation or from a set of first-order differential equations. Those transfer functions where $m > 0$, such as (4-31), are said to exhibit *numerator* or *input dynamics*. There are, however, many important cases where m is zero.

We have already illustrated the important *additive* property of transfer functions in deriving Eq. 4-13. A general form of that equation is

$$X_3(s) = G_1(s) X_1(s) + G_2(s) X_2(s) \tag{4-36}$$

In Fig. 4.1 observe that a single process output variable (X_3) may be influenced by more than one input (X_1 and X_2) acting singly or together. In such a case the total output change is calculated by summing the individual input contributions in the s-domain before inverting to the time domain. In this case, $X_3(s)$ is the composite output response that results from both input dynamic effects, $X_1(s)$ and $X_2(s)$.

EXAMPLE 4.3

The stirred-tank heating process described in Example 4.1 is operating at steady state with an inlet temperature of 70 °F and a heater input of 1920 Btu/min. At the same instant, the inlet temperature is changed to 90 °F *and* the heater input is changed to 1600 Btu/min. Calculate the output temperature response.

Solution
From Example 4.1, the steady-state output temperature is 100 °F for an inlet temperature of 70 °F and a heater input of 1920 Btu/min. The input changes are

$$T_i'(s) = \frac{90 - 70}{s} = \frac{20}{s}$$

$$Q'(s) = \frac{1600 - 1920}{s} = -\frac{320}{s}$$

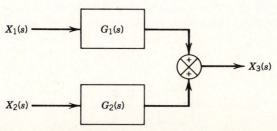

Figure 4.1. Block diagram of additive transfer function model.

Substituting in Eq. 4-12 yields

$$T'(s) = \frac{K}{\tau s + 1} \left(-\frac{320}{s} \right) + \frac{1}{\tau s + 1} \left(\frac{20}{s} \right) \tag{4-37}$$

with

$$\tau = 0.5 \text{ min}$$

$$K = \frac{1}{(200)(0.32)} = 1.56 \times 10^{-2} \frac{°F}{Btu/min}$$

After simplification of (4-37) we obtain

$$T'(s) = \frac{-5}{s(0.5s + 1)} + \frac{20}{s(0.5s + 1)} = \frac{15}{s(0.5s + 1)} \tag{4-38}$$

which yields the total (composite) solution

$$T(t) = 100 + 15(1 - e^{-2t}) \tag{4-39}$$

Note that, in evaluating Eq. 4-37, the units on K appropriately yield units of °F for the product $KQ'(s)$.

Transfer functions also exhibit a *multiplicative* property for sequential processes or process elements. Suppose two processes with transfer functions G_1 and G_2 are placed in series. The input $X_1(s)$ to G_1 yields an output $X_2(s)$, which is the input to G_2. The output from G_2 is $X_3(s)$. In equation form,

$$X_2(s) = G_1(s)X_1(s) \tag{4-40a}$$

$$X_3(s) = G_2(s)X_2(s) = G_2(s)G_1(s)X_1(s) \tag{4-40b}$$

In other words, the transfer function between the original input X_1 and the output X_3 can be found by multiplying G_2 by G_1. Figure 4.2 shows the block diagrams for Eqs. 4-40a and 4-40b.

EXAMPLE 4.4

Suppose that two liquid surge tanks are placed in series so the outflow from the first tank is the inflow to the second tank, as shown in Fig. 4.3. If the outflow from each tank is linearly related to the height of the liquid (head) in that tank, find the transfer function relating changes in outflow from the second tank, $Q_2'(s)$, to changes in inflow to the first tank, $Q_i'(s)$. Show how this transfer function is related to the individual transfer functions, $H_1'(s)/Q_i'(s)$, $Q_1'(s)/H_1'(s)$, $H_2'(s)/Q_1'(s)$, and $Q_2'(s)/H_2'(s)$. $H_1'(s)$ and $H_2'(s)$ denote the deviations in Tank 1 and Tank 2 levels, respectively. Assume that the two tanks have different cross-sectional areas A_1 and A_2, and that the outflow valve resistances are fixed at R_1 and R_2.

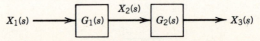

Figure 4.2. Block diagram of multiplicative (series) transfer function model.

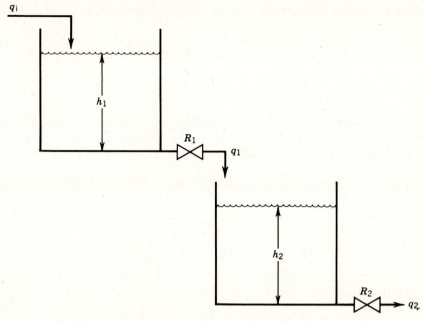

Figure 4.3. Schematic diagram of two liquid surge tanks in series.

Solution
Equations 2-20 and 2-22 are valid for each tank; for Tank 1,

$$A_1 \frac{dh_1}{dt} = q_i - q_1 \tag{4-41}$$

$$q_1 = \frac{1}{R_1} h_1 \tag{4-42}$$

Substituting (4-42) into (4-41) eliminates q_1:

$$A_1 \frac{dh_1}{dt} = q_i - \frac{1}{R_1} h_1 \tag{4-43}$$

Putting (4-43) and (4-42) into deviation variable form gives

$$A_1 \frac{dh_1'}{dt} = q_i' - \frac{1}{R_1} h_1' \tag{4-44}$$

$$q_1' = \frac{1}{R_1} h_1' \tag{4-45}$$

The transfer function relating $H_1'(s)$ to $Q_{1i}'(s)$ is found by transforming (4-44) and rearranging to obtain

$$\frac{H_1'(s)}{Q_i'(s)} = \frac{R_1}{A_1 R_1 s + 1} = \frac{K_1}{\tau_1 s + 1} \tag{4-46}$$

where $K_1 = R_1$ and $\tau_1 = A_1 R_1$. Similarly, the transfer function relating $Q_1'(s)$ to $H_1'(s)$ is obtained by transforming (4-45)

$$\frac{Q_1'(s)}{H_1'(s)} = \frac{1}{R_1} = \frac{1}{K_1} \tag{4-47}$$

The same procedure leads to the corresponding transfer functions for Tank 2

$$\frac{H_2'(s)}{Q_1'(s)} = \frac{R_2}{A_2 R_2 s + 1} = \frac{K_2}{\tau_2 s + 1} \tag{4-48}$$

$$\frac{Q_2'(s)}{H_2'(s)} = \frac{1}{R_2} = \frac{1}{K_2} \tag{4-49}$$

where $K_2 = R_2$ and $\tau_2 = A_2 R_2$. Finally, note that the desired transfer function relating the outflow from Tank 2 to the inflow to Tank 1 can be derived by forming the product of (4-46) through (4-49)

$$\frac{Q_2'(s)}{Q_i'(s)} = \frac{Q_2'(s)}{H_2'(s)} \frac{H_2'(s)}{Q_1'(s)} \frac{Q_1'(s)}{H_1'(s)} \frac{H_1'(s)}{Q_i'(s)} \tag{4-50}$$

or

$$\frac{Q_2'(s)}{Q_i'(s)} = \frac{1}{K_2} \frac{K_2}{\tau_2 s + 1} \frac{1}{K_1} \frac{K_1}{\tau_1 s + 1} \tag{4-51}$$

which can be simplified to yield

$$\frac{Q_2'(s)}{Q_i'(s)} = \frac{1}{(\tau_1 s + 1)(\tau_2 s + 1)} \tag{4-52}$$

a second-order transfer function (does unity gain make sense on physical grounds?). Figure 4.4 is a block diagram showing *information flow* for this system.

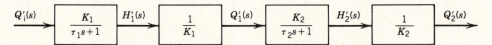

Figure 4.4.   Input–output model for two liquid surge tanks in series.

The multiplicative property of transfer functions proves to be quite valuable in designing process control systems because of the series manner in which process units and instrument systems generally are connected.

As a final point on the properties of transfer functions, it is worth noting that *for linear system models* the transfer function can often be written directly from the original differential equation model. For example, for the stirred-tank heating system model, note the similarity between Eq. 2-2 and Eq. 4-5. A shortcut approach of adding primes to the variables in (2-2) yields the same results as the rigorous approach that leads to Eq. 4-5. From (4-5), the appropriate transfer function can be obtained by inspection. Such shortcut methods should be used with caution, however, and *never* applied to nonlinear process models. The methods developed in the next section apply for nonlinear cases.

4.3 LINEARIZATION OF NONLINEAR MODELS

In the previous sections we have limited the discussion to those processes that can be accurately described by linear ordinary differential equations. There is a wide variety of processes for which the dynamic behavior depends on the process variables in a nonlinear fashion. Prominent examples include the exponential dependence of reaction rate on temperature in a chemical reactor (such as considered in Chapter 2), the nonlinear behavior of pH with flow rate of acid or base, and the asymmetric responses of distillate and bottoms compositions in a distillation column to changes in feed flow. Classical linear process control theory has been developed for linear process systems, and its use, therefore, is restricted to linear approximations of the actual nonlinear processes. A linear approximation of a nonlinear steady-state model is most accurate near the point of linearization. The same is true for dynamic process models. Wide changes in operating conditions for a nonlinear process cannot be approximated satisfactorily by linear expressions. Other approaches, such as adaptive control methods (Chapter 18), must be implemented in this case.

However, there are many instances where nonlinear processes remain in the vicinity of a particular operating state. Under such conditions a linearized model of the process may be sufficiently accurate. Suppose a conservation law (material, energy, or momentum balance) is obtained for some process, yielding a nonlinear unsteady-state equation of the form

$$\frac{dy}{dt} = f(y, x) \tag{4-53}$$

where y is the output and x is the input. A linear approximation of this equation can be obtained by using a Taylor series expansion and truncating with first-order terms. The reference point for linearization is the normal steady-state operating point $(\bar{y}, \bar{x})$.

$$f(y, x) \cong f(\bar{y}, \bar{x}) + \left.\frac{\partial f}{\partial y}\right|_{\bar{y}, \bar{x}} (y - \bar{y}) + \left.\frac{\partial f}{\partial x}\right|_{\bar{y}, \bar{x}} (x - \bar{x}) \tag{4-54}$$

By definition, the steady-state condition corresponds to $f(\bar{y}, \bar{x}) = 0$. In addition, note that deviation variables (from the steady state) arise naturally out of the Taylor series expansion (4-54), namely, $y' = y - \bar{y}$ and $x' = x - \bar{x}$. Hence, the linearized differential equation in terms of y' and x' is (after substituting $dy'/dt = dy/dt$)

$$\frac{dy'}{dt} = \left.\frac{\partial f}{\partial y}\right|_{s} y' + \left.\frac{\partial f}{\partial x}\right|_{s} x' \tag{4-55}$$

where $(\partial f/\partial y)|_s$ is used in place of $(\partial f/\partial y)|_{\bar{y}, \bar{x}}$. If another input variable, z, is contained nonlinearly in the physical model, then Eq. 4-55 must be generalized further:

$$\frac{dy'}{dt} = \left.\frac{\partial f}{\partial y}\right|_{s} y' + \left.\frac{\partial f}{\partial x}\right|_{s} x' + \left.\frac{\partial f}{\partial z}\right|_{s} z' \tag{4-56}$$

where $z' = z - \bar{z}$.

EXAMPLE 4.5

Let us return to the stirred-tank heater presented in Section 4.1. However, assume that the inlet and outlet flow rates are equal ($w = w_i$) but are allowed to vary in time, with the volume of liquid in the heating tank remaining constant. Find all transfer functions relating input variables to the output, T.

Solution

In this system there now are three input variables that can affect the process output T: the tank flow rate w, the inlet temperature T_i, and the heat input Q. The nonlinearities in Eq. 4-2 are found in the product terms wT and wT_i. After linearizing the right side of (4-2) about the steady-state values $(\overline{T}, \overline{Q}, \overline{T}_i, \overline{w})$, the resulting linear process model is

$$\frac{dT}{dt} = \frac{dT'}{dt} = \left(\frac{\partial f}{\partial T}\right)_s (T - \overline{T}) + \left(\frac{\partial f}{\partial Q}\right)_s (Q - \overline{Q})$$

$$+ \left(\frac{\partial f}{\partial T_i}\right)_s (T_i - \overline{T}_i) + \left(\frac{\partial f}{\partial w}\right)_s (w - \overline{w}) \qquad (4\text{-}57)$$

where $f(T, Q, T_i, w)$ denotes the right side of (4-2).

The partial derivatives can be obtained from (4-2):

$$\left(\frac{\partial f}{\partial T}\right)_s = \frac{-\overline{w}}{V\rho} \qquad (4\text{-}58a)$$

$$\left(\frac{\partial f}{\partial Q}\right)_s = \frac{1}{V\rho C} \qquad (4\text{-}58b)$$

$$\left(\frac{\partial f}{\partial T_i}\right)_s = \frac{\overline{w}}{V\rho} \qquad (4\text{-}58c)$$

$$\left(\frac{\partial f}{\partial w}\right)_s = \frac{\overline{T}_i - \overline{T}}{V\rho} \qquad (4\text{-}58d)$$

The coefficients determined above are functions of the steady state selected. Note that in (4-57) there are four variables: one output (T) and three inputs, (Q, T_i, w). In terms of deviation variables (T', Q', T_i', w'),

$$\frac{dT'}{dt} = \frac{-\overline{w}}{V\rho} T' + \frac{1}{V\rho C} Q' + \frac{\overline{w}}{V\rho} T_i' + \frac{\overline{T}_i - \overline{T}}{V\rho} w' \qquad (4\text{-}59)$$

The above equation is general in that it applies to any specified operating point.

Next take the Laplace transform of Eq. 4-59 with the initial steady-state condition $T'(0) = 0$:

$$sT'(s) = \frac{-\overline{w}}{V\rho} T'(s) + \frac{1}{V\rho C} Q'(s) + \frac{\overline{w}}{V\rho} T_i'(s) + \frac{\overline{T}_i - \overline{T}}{V\rho} W'(s) \qquad (4\text{-}60)$$

where $W'(s) = \mathcal{L}[w'(t)]$.

Rearranging and multiplying by $V\rho/\overline{w}$ yields

$$\left(\frac{V\rho}{\overline{w}} s + 1\right) T'(s) = \frac{1}{\overline{w}C} Q'(s) + T_i'(s) + \frac{\overline{T}_i - \overline{T}}{\overline{w}} W'(s) \qquad (4\text{-}61)$$

Recognizing that

$$\tau = \frac{V\rho}{\overline{w}} \qquad \text{and } K = \frac{1}{\overline{w}C}$$

and defining

$$K_w = \frac{\overline{T}_i - \overline{T}}{\overline{w}}$$

then

$$T'(s) = \frac{K}{\tau s + 1} Q'(s) + \frac{1}{\tau s + 1} T_i'(s) + \frac{K_w}{\tau s + 1} W'(s) \qquad (4\text{-}62)$$

Three input–output transfer functions can be identified from (4-62):

$$G_1(s) = \frac{T'(s)}{Q'(s)} = \frac{K}{\tau s + 1} \tag{4-63}$$

$$G_2(s) = \frac{T'(s)}{T_i'(s)} = \frac{1}{\tau s + 1} \tag{4-64}$$

$$G_3(s) = \frac{T'(s)}{W'(s)} = \frac{K_w}{\tau s + 1} \tag{4-65}$$

Note that all three transfer functions have the same first-order dynamics but different gains, and that the formulas for G_1 and G_2 are the same as those derived earlier for the linear model (Eqs. 4-16 and 4-17). While K is positive, the gain K_w is negative since an increase in the flow rate of a liquid that is cooler than the desired outlet temperature naturally will reduce the tank temperature.

This example shows that individual transfer functions for several inputs can be obtained by linearization. In the next example we present the basis for the linearized equations used earlier in Example 4.4.

EXAMPLE 4.6

Consider a single-tank liquid-level system where the outflow passes through a valve. Recalling Eq. 2-22, assume now that the valve discharge rate is related to the square root of liquid level:

$$q = C_v \sqrt{h} \tag{4-66}$$

where C_v depends on the fixed opening of the valve (see Chapter 9). Derive an approximate dynamic model for this process.

Solution

The material balance for the process (Eq. 2-20) after substituting the nonlinear relation for the valve is

$$A \frac{dh}{dt} = q_i - C_v \sqrt{h} \tag{4-67}$$

To obtain the system transfer function, we must linearize (4-67) about the steady-state conditions $(\bar{h}, \bar{q}_i)$. The deviation variables are

$$h' = h - \bar{h}$$

$$q_i' = q_i - \bar{q}_i$$

Applying Eq. (4-55) where $y = h$ and $x = q_i$, and $f(h, q_i)$ is the right side of (4-67), the linearized differential equation is

$$A \frac{dh'}{dt} = q_i' - \frac{C_v}{2\sqrt{h}} h' \tag{4-68}$$

If we define the valve resistance R using the relation

$$\frac{1}{R} = \frac{C_v}{2\sqrt{h}} \tag{4-69}$$

the resulting dynamic equation is analogous to the linear model presented earlier as Eqs. 2-23 and 4-44:

$$A \frac{dh'}{dt} = q_i' - \frac{1}{R} h' \tag{4-70}$$

The transfer function corresponding to (4-70) was derived earlier as (4-46).

___

In the next example we consider the situation that arises when the accumulation term is nonlinear.

EXAMPLE 4.7

A horizontal cylindrical tank shown in Fig. 4.5a is used to slow the propagation of liquid flow surges in a processing line. Develop a model for the height of liquid in the tank at any time using the inlet and outlet volumetric flow rates as model inputs. Linearize the model assuming that the process initially is at steady state.

Solution
Note that the primary complication in modeling this process is that the liquid surface area varies as the level varies. The accumulation term must represent this feature. Figure 4.5b illustrates an end view of the tank.

If density is constant, a mass balance yields

$$\frac{dm}{dt} = \rho q_i - \rho q \tag{4-71}$$

The mass accumulation term in (4-71) can also be written as[2]

$$\frac{dm}{dt} = \rho \frac{dV}{dt} = \rho w_t L \frac{dh}{dt} \tag{4-72}$$

where $w_t L$ represents the changing surface area of the liquid. Substituting (4-72) in (4-71) and simplifying gives

$$w_t L \frac{dh}{dt} = q_i - q \tag{4-73}$$

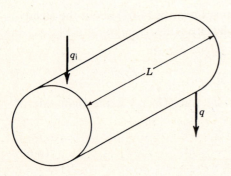

Figure 4.5.(a)   A horizontal cylindrical liquid surge tank.

___

[2]From symmetry in Fig. 4.5(b), $V(h) = 2 \int_0^h L(w_t/2)dh^*$.

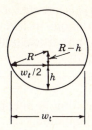

Figure 4.5.(b) The end view of a cylindrical surge tank.

The geometric construction in Fig. 4.5b indicates that $w_t/2$ is the length of one side of a right triangle whose hypotenuse is R. Thus, $w_t/2$ is related to the level h by

$$\frac{w_t}{2} = \sqrt{R^2 - (R - h)^2} \tag{4-74a}$$

After rearrangement, we obtain

$$w_t = 2\sqrt{(D - h)h} \tag{4-74b}$$

with $D = 2R$ the diameter of the tank. Substituting (4-74b) into (4-73) yields a nonlinear dynamic model for the tank with q_i and q as inputs:

$$\frac{dh}{dt} = \frac{1}{2L\sqrt{(D - h)h}} (q_i - q) \tag{4-75}$$

To linearize (4-75), let

$$f = \frac{q_i - q}{2L\sqrt{(D - h)h}}$$

Then

$$\left(\frac{\partial f}{\partial q_i}\right)_s = \frac{1}{2L\sqrt{(D - \bar{h})\bar{h}}}$$

$$\left(\frac{\partial f}{\partial q}\right)_s = \frac{-1}{2L\sqrt{(D - \bar{h})\bar{h}}}$$

$$\left(\frac{\partial f}{\partial h}\right)_s = (\bar{q}_i - \bar{q}) \left[\frac{\partial}{\partial h}\left(\frac{1}{2L\sqrt{(D - h)h}}\right)\right]_s = 0$$

Note that $\bar{q}_i - \bar{q} = 0$ from the steady-state relation. The derivative term in brackets is finite for all $0 < h < D$ and need not be evaluated. Consequently, the linearized model of the process is, after substitution of deviation variables,

$$\frac{dh'}{dt} = \frac{1}{2L\sqrt{(D - \bar{h})\bar{h}}} (q_i' - q') \tag{4-76}$$

Recall that the term $2L\sqrt{(D - h)h}$ in (4-75) represents the variable surface area of the tank. The linearized model (4-76) treats this quantity as a constant $(2L\sqrt{(D - \bar{h})\bar{h}}$ that depends on the nominal (steady-state) operating level. Consequently, operation of the horizontal cylindrical tank for small variations in level around the steady-state value would be much like that of any tank with equivalent but constant liquid surface, for example, a vertical cylindrical tank with diameter D' where surface area of liquid in the tank $= \pi(D')^2/4 = 2L\sqrt{(D - \bar{h})\bar{h}}$. Note that the coefficient $1/2L\sqrt{(D - \bar{h})\bar{h}}$ is infinite for $\bar{h} = 0$ or $\bar{h} = D$ and is a minimum at $\bar{h} = D/2$. Hence, for large variations in level, Eq. 4-76 would not be a good approximation, and the horizontal and vertical tanks would operate very differently.

Finally, we examine the application of linearization methods when the model involves more than one nonlinear equation.

EXAMPLE 4.8

As shown in Chapter 2, the continuous stirred-tank reactor, when used to carry out a first-order chemical reaction, has the following material and energy balance equations

$$V\frac{dc_A}{dt} = q(c_{Ai} - c_A) - Vkc_A \tag{2-30}$$

$$V\rho C\frac{dT}{dt} = wC(T_i - T) + (-\Delta H)Vkc_A + UA(T_s - T) \tag{2-31}$$

If the reaction rate coefficient k is given by the Arrhenius equation,

$$k = k_0e^{-E/RT} \tag{2-29}$$

this model is nonlinear. However, it is possible to find approximate transfer functions relating the inputs and outputs. For the case where the flow rate (q or w) and inlet conditions (c_{Ai} and T_i) are assumed to be constant, calculate the transfer function relating changes in the reactor concentration c_A to changes in the steam (or coolant) temperature T_s.

Solution
The input variables in this case are c_{Ai}, T_i, and T_s, while the output variables are c_A and T. First, the steady-state operating point must be determined. Note that such a determination will require iterative solution of two nonlinear algebraic equations; this may be done using a Newton-Raphson or similar algorithm [1]. Normally we would specify $\bar{T}_i$, $\bar{c}_{Ai}$, and $\bar{c}_A$ and then determine $\bar{T}$ and $\bar{T}_s$ that satisfy (2-30) and (2-31) at steady state. Once the steady state has been determined, we can proceed with the linearization of (2-30) and (2-31). In this case, it is necessary to perform linearization of each equation with respect to all input and output variables. Here, only the input T_s is considered to change; inputs c_{Ai} and T_i remain constant. Defining deviation variables c_A', T', and T_s', the following equations are obtained:

$$\frac{dc_A'}{dt} = a_{11}c_A' + a_{12}T' \tag{4-77}$$

$$\frac{dT'}{dt} = a_{21}c_A' + a_{22}T' + b_2T_s' \tag{4-78}$$

where

$$a_{11} = -\frac{q}{V} - k_0e^{-E/R\bar{T}}$$

$$a_{12} = -k_0 e^{-E/R\bar{T}} \bar{c}_A \left(\frac{E}{R\bar{T}^2} \right)$$

$$a_{21} = \frac{(-\Delta H)k_0 e^{-E/R\bar{T}}}{\rho C}$$

$$a_{22} = \frac{1}{V\rho c} \left[-(wC + UA) + (-\Delta H)\bar{V}c_A k_0 e^{-E/R\bar{T}} \left(\frac{E}{R\bar{T}^2} \right) \right]$$

$$b_2 = \frac{UA}{V\rho C}$$

Note that Eq. 2-30 does not contain the input variable T_s explicitly, so no input term appears in (4-77). We can rearrange (4-77) and (4-78) into a transfer function between the steam jacket temperature $T_s'(s)$ and the tank outlet concentration $C_A'(s)$ via Laplace transformation:

$$(s - a_{11})C_A'(s) = a_{12}T'(s) \tag{4-79}$$

$$(s - a_{22})T'(s) = a_{21}C_A'(s) + b_2 T_s'(s) \tag{4-80}$$

Substituting for $T'(s)$, Eq. (4-79) becomes

$$(s - a_{11})(s - a_{22})C_A'(s) = a_{12}a_{21}C_A'(s) + a_{12}b_2 T_s'(s) \tag{4-81}$$

yielding

$$\frac{C_A'(s)}{T_s'(s)} = \frac{a_{12}b_2}{s^2 - (a_{11} + a_{22})s + a_{11}a_{22} - a_{12}a_{21}} \tag{4-82}$$

which is a second-order transfer function without numerator dynamics. The coefficients can be evaluated for a particular operating state and the transfer function can be put into standard gain and time constant form.

SUMMARY

In this chapter we have defined an important concept, the transfer function. Ordinarily, the transfer function that relates changes in any process output to changes in any process input can be found from the original differential equation model using Laplace transformation methods. The transfer function contains key information about the steady-state and dynamic relations between input and output, namely the process gain and the process time constants, respectively. Zero initial conditions are assumed in deriving the transfer function. These are obtained when deviation variables are used and the process initially is at steady state.

The transfer function is defined only for linear equations. Hence, if the differential and algebraic equations constituting the model contain nonlinearities, an approximate, linearized model must be obtained before attempting to develop the transfer function(s) of the nonlinear model. All of these considerations have been summarized in Fig. 4.6, which indicates the procedure for finding the transfer function(s) that represent a particular process.

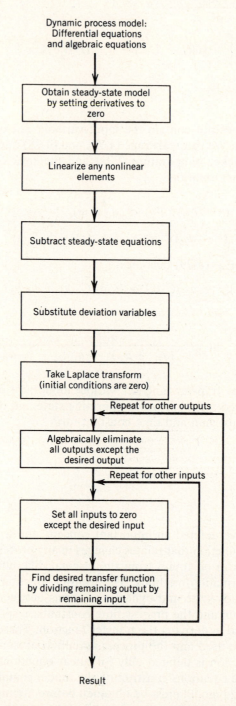

Figure 4.6. Procedure for developing transfer function models.

REFERENCE

1. Constantinides, A., *Applied Numerical Methods with Personal Computers*, McGraw-Hill, New York, 1987.

EXERCISES

4.1. The constant-volume stirred-tank heater model given by Eq. 2-2 can be easily modified to account for heat losses to the ambient. Assume that w, T_i, and T_a (the ambient temperature) are constant. The area available for heat losses is A and an appropriate overall heat transfer coefficient is U.

    (a) Obtain a transfer function that relates liquid outlet temperature T to q, the heating rate of the element.

    (b) Does the inclusion of ambient losses increase or decrease the speed of the system response to changes in q (i.e., make the time constant smaller or larger for $T_a < T$? For $T_a > T$)?

4.2. A perfectly stirred, *constant mass* liquid system is heated by an electrical immersion heater element as shown in the drawing. If the thermal capacitance of the element is not small or the effective heat transfer coefficient to the liquid is not large, the dynamics of the heater must be considered in developing a model.

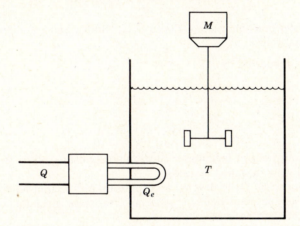

    (a) Write an unsteady-state model for this system. You may assume that:

        i. The only input is the electrical heating rate of the element Q, which initially is zero.

        ii. The rate of energy transfer to the liquid Q_e is given by the usual transport expression.

        iii. Losses to ambient are zero.

    (b) Find transfer functions relating Q_e to Q and T to Q. Put the relations in standard form.

    (c) Check the units on all gains and time constants in the transfer functions.

    (d) Determine how the transfer function for the liquid temperature is related to Eq. 4-29 for a similar system but with flow.

    Notation and Units

Q:	Electrical heating rate, kW
T_e:	Temperature of the heater, °F
T:	Temperature of the liquid, °F
m_e:	Mass of the heater, lb
m:	Mass of the liquid, lb

C_e: Specific heat of the heater, Btu/lb °F
C: Specific heat of liquid, Btu/lb °F
P: Conversion factor, kW → Btu/min
h_e: Heat transfer coefficient for the heater, Btu/°F min ft²
A: Heat transfer area of heater, ft²
Q_e: Heat transfer rate, Btu/min

4.3. I. M. Appelpolscher has asked you to investigate the dynamic operating characteristics of a stirred-tank heater with variable holdup of the type described by Eqs. 2-11 and 2-12. He is convinced that changes in the inlet flow rate w_i will affect the outlet temperature T according to a transfer function of the form

$$\frac{T'(s)}{W_i'(s)} = \frac{K\,(1 - \tau_a s)}{s(\tau s + 1)}$$

when the outlet flow rate w and other inputs T_i and Q all are maintained constant.

(a) Appelpolscher claims to have forgotten more about process dynamics than most people know. What specifically did he forget this time?
(b) While you are working on this assignment, find $V'(s)/W_i'(s)$.

4.4. A process follows the differential equation model:

$$\frac{d^3y}{dt^3} + 5\frac{d^2y}{dt^2} + 8\frac{dy}{dt} + 4y = 2\frac{dx}{dt} + 3x$$

Find the transfer function relating $Y(s)$ to $X(s)$. Both y and x are deviation variables with zero initial values.

4.5. For the process modeled by

$$2\frac{dy_1}{dt} = -2y_1 - 3y_2 + 2x_1$$

$$\frac{dy_2}{dt} = 4y_1 - 6y_2 + 2x_1 + 4x_2$$

(a) Find the four transfer functions relating the outputs (y_1, y_2) to the inputs (x_1, x_2). The x and y are deviation variables.
(b) Use Cramer's rule [1] to develop an expression for the transfer functions relating n output variables to any number m of input variables, given that the n differential equations are written in the above form.

4.6. The Calrod heating element shown in the drawing transfers heat largely by a radiation mechanism. If the rate of electrical energy input to the heater is Q and the rod temperature and ambient temperatures are, respectively, T and T_a, then an appropriate unsteady-state model for the system is

$$mC\frac{dT}{dt} = Q - k(T^4 - T_a^4)$$

Find the transfer functions relating T' to Q' and T' to T_a'. (Be sure they are both in standard form.)

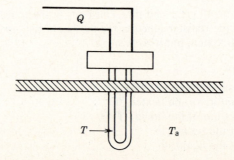

4.7. A single equilibrium stage in a distillation column is shown in the drawing. The model that describes this stage is

$$\frac{dH}{dt} = L_0 + V_2 - (L_1 + V_1)$$

$$\frac{d(Hx_1)}{dt} = L_0x_0 + V_2y_2 - (L_1x_1 + V_1y_1)$$

$$y_1 = a_0 + a_1x_1 + a_2x_1^2 + a_3x_1^3$$

(a) Assuming that the molar holdup H in the stage is constant and that equimolal overflow holds—for a mole of vapor that condenses, one mole of liquid is vaporized—simplify the model as much as possible.
(b) Linearize the resulting model and introduce deviation variables.
(c) For constant liquid and vapor flowrates, derive transfer functions relating outputs x_1 and y_1 to inputs x_0 and y_2 (four total). Put in standard form.

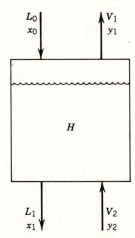

4.8. The jacketed vessel described in Exercise 2.3 contains a heat transfer coefficient that depends on the coolant flowrate in the jacket. Derive a transfer function that relates (approximately) the temperature in the tank to (a) the coolant flowrate q_J, and (b) the temperature of the tank feed T_F.

4.9. For the steam-heated stirred-tank system modeled by Eqs. 2-18 and 2-19, assume that the steam temperature T_s is constant.

(a) Find a transfer function relating tank temperature T to inlet liquid temperature T_i.
(b) What is the gain for this choice of input and output?
(c) Based on physical arguments only, should the gain for this process be unity? Why or why not?

4.10. The contents of the stirred-tank heater shown in the figure are heated at a constant rate of $Q(\text{Btu/h})$ using a gas-fired heater. The flow rate $w(\text{lb/h})$ and volume $V(\text{ft}^3)$ are constant, but the heat loss to the surroundings $Q_L(\text{Btu/h})$ varies with the wind velocity v according to the expressions

$$Q_L = UA(T - T_a)$$

$$U(t) = \overline{U} + bv(t)$$

where $\overline{U}$, A, b, and T_a are constant. Derive the transfer function between exit temperature T and wind velocity v. List any additional assumptions that you make.

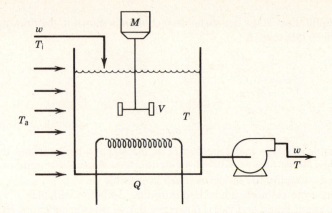

4.11. For the reactor system described in Exercise 2.10, develop transfer functions for c_A, c_B, and c_C in terms of the inlet concentration c_{Bi}.

4.12. For the stirred-tank reactor of Exercise 3.20, derive transfer functions for the outlet concentration c_A with respect to (a) c_{Ai}, and (b) w_i (assuming that w remains equal to w_i).

4.13. An irreversible reaction of A to B in an otherwise inert medium proceeds according to the relation

$$2A \xrightarrow{k} 2B$$

The reaction is exothermic with a heat of reaction $= -\Delta H$. The reaction rate (based on reaction of A) in a stirred-tank reactor of volume V is

$$r_A = kc_A^2 \; [=] \; \frac{\text{mass A reacted}}{\text{time} \cdot \text{volume}}$$

$$\text{where} \; c_a \; [=] \; \frac{\text{mass A}}{\text{volume}}$$

$$k = k_0 e^{-E/RT} \; [=] \; \frac{\text{volume}}{\text{mass} \cdot \text{time}}$$

The reaction system is shown schematically in the drawing. Assume that fluid properties are constant. A cooling coil with coolant at temperature T_c and heat transfer area A_c is used to remove energy from the reacting mixture.

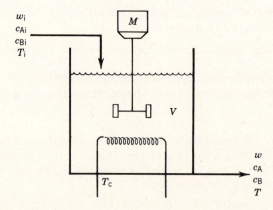

(a) Write the unsteady-state equations for the system variables, V, c_A, c_B, and T. Be sure to specify the inputs for your model.

(b) Develop transfer functions $C_A'(s)/T_i'(s)$ and $C_A'(s)/T_c'(s)$.

where $K = 1/wC$, $\tau = V\rho/w$. For a general first-order transfer function with output $Y(s)$ and input $X(s)$,

$$Y(s) = \frac{K}{\tau s + 1} X(s) \tag{5-3}$$

a general time-domain solution can be found once the input change is specified. This solution applies to the stirred-tank system for the same change in Q or T_i or to any other process with a first-order transfer function, for example, the liquid surge tanks of Eqs. 4-46 and 4-48.

We will exploit this ability to develop general process dynamic formulas as much as possible, concentrating on transfer functions that commonly arise in describing the dynamic behavior of processes. This chapter covers the simplest transfer functions: first-order systems, integrating units, and second-order systems. In Chapter 6 the responses of more complicated transfer functions will be discussed. To keep the results as general as possible we now consider several standard process inputs that have important features or uses.

5.1 STANDARD PROCESS INPUTS

We have previously discussed outputs and inputs for process models; we now introduce more precise working definitions. The word output generally refers to a controlled variable in a process, a process variable to be maintained at a desired value. For example, the output from the stirred-tank heater just discussed is the temperature T of the effluent stream. The word input refers to any variable that influences the process output when it changes in value, such as the temperature of the stream flowing into the stirred-tank heater. The characteristic feature of all inputs is that they influence the output variables that we wish to control.

In analyzing process dynamics and in designing control systems, it is important to know how the process outputs will respond to changes in the process inputs. For purposes of analysis we can choose any arbitrary input change; however, over the years, control engineers have tended to use input changes that are typical of those seen in industrial practice. We consider six important types of input changes, several of which we later use extensively in designing control systems.

1. Step Input. One characteristic of industrial processes is that they can be subjected to sudden and sustained input changes; for example, a reactor feedstock may be changed quickly from one supply to another, causing a corresponding change in important input variables such as feed concentration and feed temperature. Such a change can be approximated by the step change

$$x_S(t) = \begin{cases} 0 & t < 0 \\ M & t \geq 0 \end{cases} \tag{5-4}$$

where *zero time*, as noted earlier, is taken to be the time at which the sudden change of magnitude M occurs. Note that $x_S(t)$ is defined as a deviation variable. Also, we can always express a physical step change of any magnitude in terms of a unit step change, that is, Eq. 5-4 with $M = 1$. Suppose the heat input to a stirred-tank heating unit suddenly is changed from 8000 to 10,000 kcal/h, by changing the electrical heater input. Then

$$Q(t) = 8000 + 2000\, \mathbf{S}(t) \tag{5-5a}$$

$$Q'(t) = 2000\, \mathbf{S}(t) \tag{5-5b}$$

where $S(t)$ is the unit step function. In Chapter 3 we developed the Laplace transform of the unit step function (Eq. 3-4). The transform of a step of magnitude M, Eq. 5-4, is

$$X_S(s) = \frac{M}{s} \tag{5-6}$$

2. *Ramp Input.*   Industrial processes often are subjected to inputs that *drift*, that is, they change relatively slowly upward or downward for some period of time with a roughly constant slope. For example, ambient conditions (air temperature and relative humidity) can change slowly so that the temperature of plant cooling water from the cooling tower also changes slowly. It is also common practice to ramp set points from one value to another rather than making a step change. We can approximate such a change in an input variable by means of the ramp function:

$$x_R(t) = \begin{cases} 0 & t < 0 \\ at & t \geq 0 \end{cases} \tag{5-7}$$

where $x_R(t)$ is a deviation variable. The Laplace transform of a ramp input with a slope of 1 is given in Table 3.1 as $1/s^2$. Hence, transforming Eq. 5-7 yields

$$X_R(s) = \frac{a}{s^2} \tag{5-8}$$

3. *Rectangular Pulse.*   Processes sometimes are subjected to a sudden but *unsustained* type of disturbance. Suppose that a feed to a reactor is shut off for a certain period of time or a natural-gas-fired furnace experiences a brief interruption in fuel gas. We might approximate this type of input change as a rectangular pulse:

$$x_{RP}(t) = \begin{cases} 0 & t < 0 \\ h & 0 \leq t < t_w \\ 0 & t \geq t_w \end{cases} \tag{5-9}$$

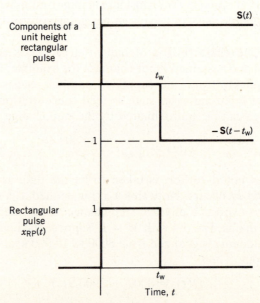

Figure 5.1.   How two step inputs can be combined to form a rectangular pulse.

where the pulse width t_w can range from very short (approximation to an impulse) to very long. An alternative way of expressing (5-9) utilizes the shifted unit step input $\mathbf{S}(t - t_w)$ which is defined to be equal to unity for $t \geq t_w$ and equal to zero for $t < t_w$. Equation 5-9 can be interpreted as the sum of two steps, one step of magnitude equal to 1 occurring at $t = 0$ combined with a second step of magnitude equal to -1 occurring at $t = t_w$ (see Fig. 5.1). Mathematically, this combination can be expressed as

$$x_{RP}(t) = h[\mathbf{S}(t) - \mathbf{S}(t - t_w)]$$

Since only $t \geq 0$ is of concern (recall that the Laplace transform is only defined for $t \geq 0$), this expression can be simplified to

$$x_{RP}(t) = h[1 - \mathbf{S}(t - t_w)] \qquad t \geq 0 \tag{5-10}$$

which can be Laplace transformed to yield

$$X_{RP}(s) = \frac{h}{s}(1 - e^{-t_w s}) \tag{5-11}$$

the same result given in (3-22).

The three important inputs discussed above—step, ramp, rectangular pulse—are depicted in Fig. 5.2. Note that many types of inputs can be represented as combinations of step and ramp inputs. For example, a unit height (isosceles) triangular pulse of width t_w can be constructed from three ramp inputs, as shown in Fig. 5.3. In this case, we write a single expression for the triangular pulse function

$$x_{TP}(t) = \frac{2}{t_w}[t\mathbf{S}(t) - 2(t - t_w/2)\mathbf{S}(t - t_w/2) + (t - t_w)\mathbf{S}(t - t_w)]$$

$$= \frac{2}{t_w}[t - 2(t - t_w/2)\mathbf{S}(t - t_w/2) + (t - t_w)\mathbf{S}(t - t_w)] \qquad t \geq 0$$

$$\tag{5-12}$$

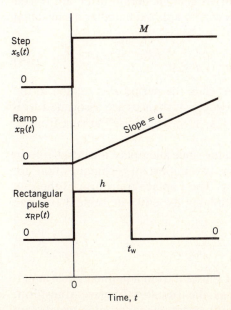

Figure 5.2. Three important examples of deterministic inputs.

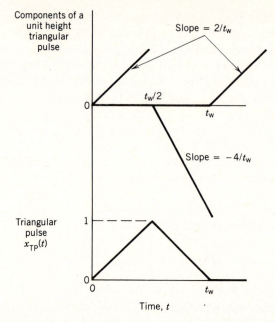

Figure 5.3.   How three ramp inputs can be combined to form a triangular pulse.

where the second relation is valid only for $t \geq 0$. Equation 5-12 can be Laplace transformed term-by-term to obtain

$$X_{TP}(s) = \frac{2}{t_w} \left(\frac{1 - 2e^{-t_w s/2} + e^{-t_w s}}{s^2} \right) \tag{5-13}$$

Note that Eq. 5-12 written without the unit step function multipliers is incorrect.

4. Sinusoidal Input.   Processes are also subjected to inputs that vary periodically. As an example, the drift in cooling water temperature discussed earlier can often be closely tied to diurnal (day-to-night-to-day) fluctuations in ambient conditions. Cyclic process changes within a twenty-four hour period often are traced to just such a variation in cooling water which can be approximated as a sinusoidal function:

$$x_{\sin}(t) = \begin{cases} 0 & t < 0 \\ A\sin \omega t & t \geq 0 \end{cases} \tag{5-14}$$

The amplitude of the sinusoidal function is A while the period P is related to the angular frequency by $P = 2\pi/\omega$. High-frequency disturbances are associated with mixing and pumping operations and with 60-Hz "electrical noise" arising from ac electrical equipment and instrumentation.

Sinusoidal inputs are particularly important since they play a central role in frequency response analysis and pulse testing, two important techniques that are developed in Chapters 14 and 15, respectively. The Laplace transform of the sine function in Eq. 5-14 can be obtained by multiplying the entry in Table 3.1 by the amplitude A to obtain

$$X_{\sin}(s) = \frac{A\omega}{s^2 + \omega^2} \tag{5-15}$$

5. Impulse Input.   One other type of input is useful from a mathematical standpoint—the unit impulse function—which has the simplest Laplace transform, $(= 1)$, (Eq. 3-24). However, impulse functions are not encountered in usual plant operations, nor are they easy to generate physically by the engineer. To obtain an impulse input it is necessary to inject a finite amount of energy or material into a process in an infinitesimal length of time. Sometimes this type of input can be approximated through the injection of a concentrated dye or other tracer into the process (see Example 3.7).

6. *Random Inputs.*   Many process inputs change with time in such a complex manner that it is not possible to describe them as deterministic functions of time. If an input x exhibits apparently random fluctuation, it is convenient to characterize it in statistical terms, that is to specify its mean value μ_x and standard deviation σ_x. The mathematical treatment of such random or stochastic processes is beyond the scope of this book. The interested reader may refer to the books by Maybeck [1], Box and Jenkins [2], and Sage and Melsa [3]. Even though many processes would appear to have largely random inputs, control systems designed assuming deterministic inputs usually perform satisfactorily. Since controller design is considerably easier and more straightforward for deterministic inputs, we restrict later discussion to them except for a discussion of statistical quality control in Chapter 18.

Having discussed transfer functions in Chapter 4 and important types of forcing functions (process inputs) here, we are now in a position to evaluate the dynamic behavior of processes on an organized basis. We begin with first-order processes, that is, processes that can be modeled as first-order transfer functions. Then integrating elements are considered and finally second-order processes. Despite their simplicity, these transfer functions are quite important because they represent building blocks for modeling more complicated processes. In addition, many important industrial processes can be adequately approximated by first- and second-order transfer functions. In Chapter 6 the dynamic characteristics of more complicated systems, for example, those that contain time delays, numerator terms, or that are of order higher than two, are considered.

5.2 RESPONSE OF FIRST-ORDER SYSTEMS

In Section 4.1 we developed a relation for the dynamic response of the simple stirred-tank heater (Eq. 4-12). To find how the outlet temperature changes when either of the inputs, $Q'(s)$ or $T_i'(s)$, is changed, we use the general first-order transfer function

$$\frac{Y(s)}{X(s)} = \frac{K}{\tau s + 1} \tag{5-16}$$

where K is the process gain and τ is the time constant. Now we investigate some particular forms of input $X(s)$, deriving expressions for $Y(s)$ and the response, $y(t)$.

Step Response

For a step input of magnitude M, $X(s) = M/s$, and

$$Y(s) = \frac{KM}{s(\tau s + 1)} \tag{5-17}$$

Using Table 3.1, the time domain response is

$$y(t) = KM(1 - e^{-t/\tau}) \tag{5-18}$$

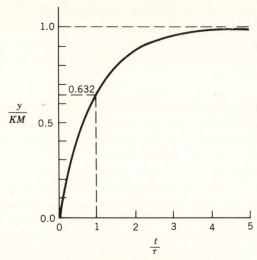

Figure 5.4.   Step response of a first-order process.

A plot of this relation (Fig. 5.4), shows that a first-order process does not respond instantaneously to a sudden change in its input. In fact, after a time interval equal to the process time constant (i.e., for $t = \tau$), we see that the process response is still only 63.2% complete. Theoretically the process output never reaches the new steady-state value; it does approximate the new value when t equals 3 to 5 process time constants, as shown in Table 5.1. Notice that Fig. 5.4 has been drawn in normalized form, with time divided by the process time constant and the output change divided by the product of the process gain and magnitude of the input change. Now we look at a more specific example.

Table 5.1   Response of a First-Order Process to a Step Input

t	$y(t)/KM = 1 - e^{-t/\tau}$
0	0
τ	0.6321
2τ	0.8647
3τ	0.9502
4τ	0.9817
5τ	0.9933

EXAMPLE 5.1

A stirred-tank heater described by Eq. 4-12 is used to preheat a reactant containing a suspended solid catalyst at a constant flow rate of 1000 kg/h. The volume in the tank is 2 m³ and the density and specific heat of the suspended mixture are, respectively, 900 kg/m³ and 1 cal/g °C. The process initially is operating with inlet and outlet temperatures of 100 and 130 °C. The following questions concerning process operations are posed:

1. What is the heater input at the initial steady state?
2. If the heater input is increased by +30%, how long will it take for the process to reach the new steady-state outlet temperature (99% of the response) and what will that temperature be?

3. If the inlet temperature is increased suddenly from 100 to 120 °C, how long will it take before the outlet temperature changes from 130 to 135 °C (assuming original steady-state conditions, i.e., no change in heater input)?

Solution

First calculate the process steady-state operating conditions and then the gain and time constant in Eq. 4-12. Assuming no heat losses, the energy input from the heater at the initial steady state is equal to the enthalpy increase between the inlet and outlet streams. Thus, the steady-state energy balance provides the answer to Question 1:

$$\bar{Q} = \bar{w}C(\bar{T} - \bar{T}_i)$$

$$= \left(10^6 \frac{g}{h}\right)\left(\frac{1 \text{ cal}}{g \text{ °C}}\right)(130 \text{ °C} - 100 \text{ °C})$$

$$= 3 \times 10^7 \text{ cal/h}$$

Using Eq. 4-12, the gain and time constants can be determined:

$$K = \frac{1}{wC} = \frac{1}{\left(10^6 \frac{g}{h}\right)\left(\frac{1 \text{ cal}}{g \text{ °C}}\right)}$$

$$= 10^{-6} \frac{°C}{cal/h}$$

$$\tau = \frac{V\rho}{w} = \frac{(2 \text{ m}^3)\left(9 \times 10^5 \frac{g}{m^3}\right)}{10^6 \frac{g}{h}}$$

$$= 1.8 \text{ h}$$

Question 2 now can be answered immediately. The time required to attain a new steady state (99% response) following a step change of any magnitude in heater input will be 5 process time constants, that is, 9 h. The change in temperature due to a change of +30% in Q (9×10^6 cal/h) can be found from the Final Value Theorem, Eq. 3-94:

$$T'(t \to \infty) = \lim_{s \to 0} s\left(\frac{10^{-6}}{1.8s + 1}\frac{9 \times 10^6}{s}\right) = 9 \text{ °C}$$

Note that we have calculated the outlet temperature *change* as a result of the input change; hence the actual outlet temperature at the final steady state will be 130 °C + 9 °C = 139 °C. However, the use of the Final Value Theorem is an unnecessary formality when a process transfer function is written in the standard form, that is, in terms of gain and time constants. As we have seen in Eq. 4-6, the input change need only be multiplied by the process gain to obtain the ultimate change in the process output, assuming that the final value does in fact exist and is finite.

Figure 5.4 can be used to obtain an approximate answer to Question 3. The input temperature change is 20 °C and the gain of the appropriate transfer function (between T' and T_i') is 1. Thus, we know the total outlet temperature change will be 20 °C. The time required for the output to change by 5 °C, or 25% of the ultimate steady-state change can be read from the figure as $t/\tau = 0.3$ or $t = 0.54$ h. Rearranging Eq. 5-18 furnishes a more accurate way to calculate this value:

$$\frac{y(t)}{KM} = 1 - e^{-t/\tau}$$

$$\frac{5\ °C}{(1)(20\ °C)} = 1 - e^{-t/\tau}$$

$$e^{-t/\tau} = 0.75$$

$$-\frac{t}{\tau} = \ln 0.75 = -0.288$$

$$t = 0.52\ h$$

Ramp Response

We now evaluate how a first-order system responds to the ramp input, $X(s) = a/s^2$ of Eq. 5-8. Performing a partial fraction expansion yields

$$Y(s) = \frac{Ka}{(\tau s + 1)s^2} = \frac{\alpha_1}{\tau s + 1} + \frac{\alpha_2}{s} + \frac{\alpha_3}{s^2} \tag{5-19}$$

The Heaviside expansion (Chapter 3) gives

$$Y(s) = \frac{Ka\tau^2}{\tau s + 1} - \frac{Ka\tau}{s} + \frac{Ka}{s^2} \tag{5-20}$$

Inverting yields

$$y(t) = Ka\tau(e^{-t/\tau} - 1) + Kat \tag{5-21}$$

The above expression has the interesting property that for large values of time $(t \gg \tau)$

$$y(t) \approx Ka(t - \tau) \tag{5-22}$$

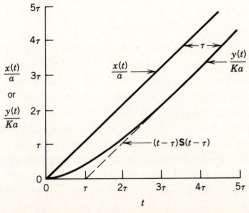

Figure 5.5. Ramp response of a first-order process (comparison of input and output).

Equation 5-22 implies that after an initial transient period, the ramp input yields a ramp output with slope equal to Ka, but shifted in time by the process time constant τ (see Fig 5.5). An unbounded ramp input will ultimately cause some process component to saturate, so the duration of the ramp input ordinarily is limited. A process input frequently will be "ramped" from one value to another in a fixed amount of time so as to avoid the sudden change associated with a step change. Ramp inputs of this type are particularly useful during the start-up of a continuous process or in operating a batch process.

Sinusoidal Response

As a final example of the response of first-order processes, consider a sinusoidal input $x_{sin}(t) = A \sin \omega t$, with transform given by Eq. (5-15):

$$Y(s) = \frac{KA\omega}{(\tau s + 1)(s^2 + \omega^2)} \tag{5-23}$$

$$= \frac{KA}{\omega^2 \tau^2 + 1} \left(\frac{\omega \tau^2}{\tau s + 1} - \frac{s\omega\tau}{s^2 + \omega^2} + \frac{\omega}{s^2 + \omega^2} \right) \tag{5-24}$$

Hence

$$y(t) = \frac{KA}{\omega^2 \tau^2 + 1} (\omega\tau e^{-t/\tau} - \omega\tau \cos \omega t + \sin \omega t) \tag{5-25}$$

or, by using trigonometric identities,

$$y(t) = \frac{KA\omega\tau}{\omega^2 \tau^2 + 1} e^{-t/\tau} + \frac{KA}{\sqrt{\omega^2 \tau^2 + 1}} \sin(\omega t + \phi) \tag{5-26}$$

where $\phi = -\tan^{-1}(\omega\tau)$ $\qquad\qquad$ (5-27)

Notice that in both (5-25) and (5-26) the exponential term goes to zero as $t \to \infty$ leaving a pure sinusoidal response. In Chapter 14 this property is exploited in developing the frequency response method of analysis.

Students often have difficulty imagining how a real process variable might change sinusoidally. How can the flow rate into a reactor be negative as well as positive? Remember that we have defined the input x and output y in these relations to be deviation variables. An actual input might be

$$x(t) = 0.4 \, \frac{m^3}{s} + \left(0.1 \, \frac{m^3}{s} \right) \sin \omega t \tag{5-28}$$

where the amplitude of the deviation input signal A is $0.1 \, m^3/s$. After a long period of time, the output response (5-26) also will be the sum of a constant (steady-state) value and a sinusoidal deviation, similar to that given in Eq. 5-28.

EXAMPLE 5.2

A single liquid surge tank similar to the one described by Eq. 4-43 is operating under conditions such that the transfer function of Eq. 4-46 approximates the actual process

$$\frac{H'(s)}{Q_i'(s)} = \frac{10}{50s + 1}$$

where h is the tank level (m), q is the flow rate (m³/s), the gain has units m/m³/s, or s/m² and the time constant has units of seconds. The system is operating at steady state with $\bar{q} = 0.4$ m³/s and $\bar{h} = 4$ m when a sinusoidal perturbation in inlet flow rate begins with amplitude = 0.1 m³/s and frequency 0.002 cycles/s. What are the maximum and minimum values of the tank level after the flow rate disturbance has occurred for 6 min or more? What are the largest level perturbations one would expect as a result of sinusoidal variations in flow rate with this amplitude?

Solution

Note that the actual input signal $q(t)$ is given by Eq. 5-28, but only the amplitude of the input deviation (0.1) is required. From Eq. 5-26 we note that the value of the exponential term 6 min after the start of sinusoidal forcing is $e^{-360/50} = e^{-7.2} < 10^{-3}$. Hence, the effect of the exponential transient term is less than 0.1% of the disturbance amplitude and can be safely neglected. Consequently, from Eq. 5-26 the amplitude of the output (level) perturbation is

$$\frac{KA}{\sqrt{\omega^2\tau^2 + 1}}$$

where A is the input amplitude and ω is the frequency (in radians) = (2π) (cyclic frequency) = $(6.28)(0.002)$ radians/s. The amplitude of the perturbation in the liquid level is

$$\frac{10 \ (\text{s/m}^2)(0.1 \ \text{m}^3/\text{s})}{\sqrt{[(6.28 \ \text{rad/cyc})(0.002 \ \text{cyc/s})(50 \ \text{s})]^2 + 1}}$$

or 0.85 m. Hence the actual tank level varies from a minimum of 3.15 m to a maximum of 4.85 m.

The largest deviations that can result from sinusoidal variations of amplitude 0.1 m³/s occur for $\omega \to 0$, that is, for very low frequencies. In this case, the deviations would be $\pm KA = \pm(10 \ \text{s/m}^2) \ (0.1 \ \text{m}^3/\text{s}) = \pm 1$ m. Hence, the minimum and maximum values of level would be 3 and 5 m, respectively.

5.3 RESPONSE OF INTEGRATING PROCESS UNITS

In Section 2.5 we discussed the model of a liquid-level system with a pump attached to the outflow line. Assuming that the outflow rate q can be set at any time by the speed of the pump or by the valve in the effluent line, Eq. 2-20 becomes

$$A \frac{dh(t)}{dt} = q_i(t) - q(t) \tag{5-29}$$

If at initial time $t = 0$, $q_i = q = \bar{q}$, then $dh/dt = 0$ and the system is at steady state with $h = \bar{h}$. After subtracting the steady-state version of (5-29),

$$A \frac{dh'(t)}{dt} = q_i'(t) - q'(t) \tag{5-30}$$

where the primed deviation variables are all zero at time $t = 0$. Taking Laplace transforms

$$sAH'(s) = Q_i'(s) - Q'(s) \tag{5-31}$$

and rearranging gives

$$H'(s) = \frac{1}{As} [Q_i'(s) - Q'(s)] \tag{5-32}$$

Both of the transfer functions, $H'(s)/Q_i'(s) = 1/As$ and $H'(s)/Q'(s) = -1/As$, represent integrating units, characterized by the term $1/s$. Any transfer function containing a $1/s$ term will exhibit this characteristic integrating behavior. By returning to Eq. 5-29 we can clearly see why: Rearranging (5-29) and integrating yields

$$\int_{\bar{h}}^{h} dh^* = \frac{1}{A} \int_0^t [q_i(t^*) - q(t^*)] \, dt^*$$

or

$$h(t) - \bar{h} = \frac{1}{A} \int_0^t [q_i(t^*) - q(t^*)] \, dt^* \qquad (5\text{-}33)$$

Note that, for integrating processes, the gain is undefined in the usual sense. For such a process operating at steady state, any positive step change in q_i (increase in q_i above q) will cause the tank level to increase linearly with time in proportion to the difference $q_i(t) - q(t)$, while a positive step change in q will cause the tank level to decrease linearly. Hence no new steady state will be attained.

EXAMPLE 5.3

A vented cylindrical tank is used for storage between a tank car unloading facility and a continuous reactor that uses the tank car contents as feedstock (Fig. 5.6). The reactor feed exits the storage tank at a constant flow rate of $0.02 \ \text{m}^3/\text{s}$. During some periods of operation, feedstock is simultaneously transferred from the tank car to the feed tank and from the tank to the reactor. The operators have to be particularly careful not to let the feed tank overflow or empty. The feed tank is 5 m high (distance to the vent) and has an internal cross-sectional area of $4 \ \text{m}^2$.

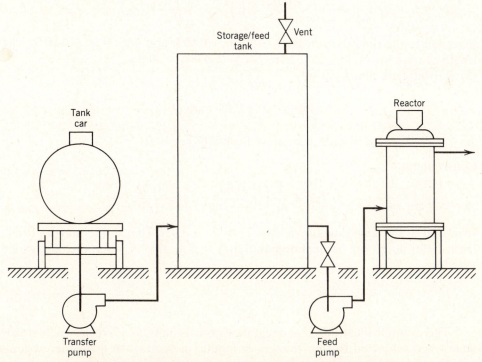

Figure 5.6. Unloading and storage facility for a continuous reactor.

(a) Suppose after a long period of operation, the tank level is 2 m at the time the tank car empties. How long can the reactor be operated before the feed tank is depleted?

(b) If another tank car is moved into place and connected, and flow is introduced into the feed tank just as the tank level reaches 1 m, how long can the transfer pump from the tank car be operated? Assume that it pumps at a constant rate of 0.1 m³/s when switched on.

Solution

(a) For such a system, there is no unique steady-state level corresponding to a particular value of input and output flow rate. We might choose the initial level, $h = 2$ m and the constant flow rate from the feed pump, $q = 0.02$ m³/s as the basis for defining deviation variables. Then

$$\bar{h} = 2 \text{ m}$$
$$\bar{q}_i = \bar{q} = 0.02 \text{ m}^3/\text{s}$$

and, from Eq. 5-32, the process model is

$$H'(s) = \frac{1}{4s} [Q_i'(s) - Q'(s)]$$

At the time the tank car empties

$$q_i = 0 \Longrightarrow q_i' = -0.02 \Longrightarrow Q_i'(s) = -\frac{0.02}{s}$$

$$q = 0.02 \Longrightarrow q' = 0 \Longrightarrow Q'(s) = 0$$

Hence

$$H'(s) = \frac{1}{4s} \left(-\frac{0.02}{s} - 0 \right) = -\frac{0.005}{s^2}$$

Inversion to the time domain gives $h'(t) = -0.005t$ and $h(t) = 2 - 0.005t$. The length of time for $h(t)$ to go to zero is $t = 2/0.005 = 400$ s.

(b) For the tank filling period

$$q_i = 0.1 \Longrightarrow q_i' = +0.08 \Longrightarrow Q_i'(s) = \frac{0.08}{s}$$

$$q = 0.02 \Longrightarrow q' = 0 \Longrightarrow Q'(s) = 0$$

Consequently,

$$H'(s) = \frac{1}{4s} \left(+\frac{0.08}{s} - 0 \right) = \frac{0.02}{s^2}$$

Inversion to the time domain yields $h(t) = 1 + 0.02t$. Hence, the length of time the transfer pump can be operated until $h(t) = 5$ m, that is, when the tank would overflow, is 200 s.

This example illustrates that integrating process units do not reach a new steady state when subjected to step changes in inputs, in contrast to first-order systems (cf. Eq. 5-18). Integrating systems represent an example of *non-self-regulating* systems. Closed pulse inputs, where the initial and final values of the input are equal, do lead to a new steady state. For example, the rectangular pulse with height h given in Eq. 5-9 has the Laplace transform given in Eq. 5-10. The response of an integrating system with transfer function

$$\frac{Y(s)}{X(s)} = \frac{K}{s} \tag{5-34}$$

to a rectangular pulse input is

$$Y(s) = \frac{Kh(1 - e^{-t_w s})}{s^2} = Kh\left(\frac{1}{s^2} - \frac{e^{-t_w s}}{s^2}\right) \tag{5-35}$$

From (5-35) note that there are two regions for the solution, depending on the value of t compared to the pulse width t_w. For $0 \le t < t_w$, the second term in the parentheses of (5-35) is 0, hence

$$y(t) = Kht \tag{5-36}$$

corresponding to a linear increase with respect to time. Inverting (5-35) for $t \ge t_w$ gives

$$y(t) = Kh[t - (t - t_w)] = Kht_w \tag{5-37}$$

which is a constant value. Combining the solutions yields

$$y(t) = \begin{cases} Kht & t < t_w \\ Kht_w & t \ge t_w \end{cases} \tag{5-38}$$

Equation 5-38 shows that the change in y at any time is proportional to the area under the input pulse curve, an intuitive result. Note that unit step functions were not needed in (5-36)–(5-38) because each solution was written for a particular time interval.

5.4 RESPONSE OF SECOND-ORDER SYSTEMS

As noted earlier, a second-order transfer function can arise physically whenever two first-order processes are connected in series. For example, two stirred-tank heaters, each with first-order transfer function relating inlet to outlet temperature, might be physically connected so the outflow stream of the first heater is used as the inflow stream of the second tank. Figure 5.7 illustrates the signal flow relation for such a process. Here

$$G(s) = \frac{Y(s)}{X(s)} = \frac{K_1 K_2}{(\tau_1 s + 1)(\tau_2 s + 1)} = \frac{K}{(\tau_1 s + 1)(\tau_2 s + 1)} \tag{5-39}$$

where $K = K_1 K_2$. Alternatively, a second-order process transfer function will arise upon transforming either a second-order differential equation process model such as the one given in Eq. 2-15 for the electrically heated stirred-tank unit, or two coupled first-order differential equations, such as for the CSTR (cf. Eq. 4-82). In this chapter we consider the case where the second-order transfer function has the

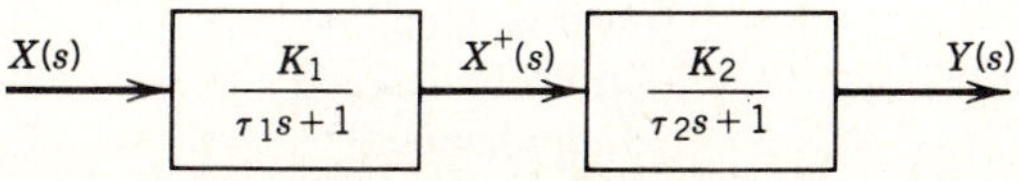

Figure 5.7. Two first-order systems in series yield an overall second-order system.

standard form

$$G(s) = \frac{K}{\tau^2 s^2 + 2\zeta\tau s + 1} \tag{5-40}$$

We defer discussion of the more general cases with a time-delay term in the numerator or with *input dynamics* present (cf. Eq. 4-31) until Chapter 6.

In Eq. 5-40, K and τ have the same importance as for a first-order transfer function. K is the process gain, τ determines the speed of response (or, equivalently, the response time) of the system. The new parameter ζ (zeta) is dimensionless. It provides a measure of the amount of *damping* in the system, that is, the degree of oscillation in a process response after a perturbation. Small values of ζ imply little damping, as, for example, in an automobile suspension system with bad or ineffective shock absorbers. Hitting a bump causes such a vehicle to bounce up and down dangerously. In some textbooks, Eq. 5-40 is written in terms of $\omega_n = 1/\tau$, the undamped natural frequency of the system. This name arises because it represents the frequency of oscillation of the system for the case where there is no damping ($\zeta = 0$).

There are three important subcases as shown in Table 5.2. The case where $\zeta < 0$ is omitted since it corresponds to an unstable second-order system that would have an **unbounded** response to any input. The overdamped and critically damped forms of the second-order transfer function most often appear when two first-order systems occur in series. The transfer functions given by Eqs. 5-39 and 5-40 differ only in the denominators. Equating them yields the relation between the two alternative forms for the *overdamped* second-order case. Note that when $\zeta \geq 1$, the denominator of (5-40) can be factored as:

$$\tau^2 s^2 + 2\zeta\tau s + 1 = (\tau_1 s + 1)(\tau_2 s + 1) \tag{5-41}$$

Expanding the right side of (5-41) and equating coefficients of the s terms, yields

$$\tau^2 = \tau_1 \tau_2$$

$$2\zeta\tau = \tau_1 + \tau_2$$

from which we obtain

$$\tau = \sqrt{\tau_1 \tau_2} \tag{5-42}$$

$$\zeta = \frac{\tau_1 + \tau_2}{2\sqrt{\tau_1 \tau_2}} \tag{5-43}$$

Alternatively, the left side of (5-41) can be factored:

$$\tau^2 s^2 + 2\zeta\tau s + 1 = \left(\frac{\tau s}{\zeta - \sqrt{\zeta^2 - 1}} + 1\right)\left(\frac{\tau s}{\zeta + \sqrt{\zeta^2 - 1}} + 1\right) \tag{5-44}$$

Table 5.2   The Three Forms of Second-Order Transfer Functions

Case	Range of Damping Coefficient	Characterization of Response	Roots of Characteristic Equation
a	$\zeta > 1$	Overdamped	Real and unequal
b	$\zeta = 1$	Critically damped	Real and equal
c	$0 < \zeta < 1$	Underdamped	Complex conjugates (of the form $a + jb$ and $a - jb$)

from which expressions for τ_1 and τ_2 are obtained

$$\tau_1 = \frac{\tau}{\zeta - \sqrt{\zeta^2 - 1}} \qquad (\zeta \geq 1) \qquad\qquad (5\text{-}45)$$

$$\tau_2 = \frac{\tau}{\zeta + \sqrt{\zeta^2 - 1}} \qquad (\zeta \geq 1) \qquad\qquad (5\text{-}46)$$

EXAMPLE 5.4

A system consists of two first-order processes operating in series ($\tau_1 = 4$, $\tau_2 = 1$). Find the equivalent overdamped second-order representation (τ, ζ) for this system.

Solution

From Eqs. 5-42 and 5-43,

$$\tau = \sqrt{(4)(1)} = 2$$

$$\zeta = \frac{4 + 1}{(2)(2)} = 1.25$$

A check of Eqs. 5-45 and 5-46 can be made with these results:

$$\tau_1 = \frac{2}{1.25 - \sqrt{(1.25)^2 - 1}} = \frac{2}{1.25 - 0.75} = 4$$

$$\tau_2 = \frac{2}{1.25 + \sqrt{(1.25)^2 - 1}} = \frac{2}{1.25 + 0.75} = 1$$

The underdamped form of (5-40) can arise from some mechanical systems, from flow or other processes such as a pneumatic (air) instrument line with too little line capacity, or from a mercury manometer [4], where inertial effects are important.

For process control problems the underdamped form is frequently encountered in investigating the properties of *controlled* processes. Control systems are often designed so that the controlled process responds in a manner similar to that of an underdamped second-order system. Before discussing the way control system designers view the response of such systems, we will develop the relation for the step response of all three classes of second-order systems.

Step Response

For the step input with transform given by Eq. 5-6

$$Y(s) = \frac{KM}{s(\tau^2 s^2 + 2\zeta\tau s + 1)} \qquad\qquad (5\text{-}47)$$

After some manipulation, and inverting to the time domain, three forms of response are obtained:

Case a ($\zeta > 1$)

If the denominator of Eq. 5-47 is factored using Eqs. 5-45 and 5-46, then the response can be written

$$y(t) = KM \left(1 - \frac{\tau_1 e^{-t/\tau_1} - \tau_2 e^{-t/\tau_2}}{\tau_1 - \tau_2} \right) \qquad\qquad (5\text{-}48)$$

If the denominator of Eq. 5-47 is left unfactored, then the response can be written in the equivalent form

$$y(t) = KM \left\{ 1 - e^{-\zeta t/\tau} \left[\cosh\left(\frac{\sqrt{\zeta^2 - 1}}{\tau} t\right) + \frac{\zeta}{\sqrt{\zeta^2 - 1}} \sinh\left(\frac{\sqrt{\zeta^2 - 1}}{\tau} t\right) \right] \right\}$$

$$(5\text{-}49)$$

Case b ($\zeta = 1$)

$$y(t) = KM \left[1 - \left(1 + \frac{t}{\tau}\right) e^{-t/\tau} \right] \tag{5-50}$$

Case c ($0 \leq \zeta < 1$)

$$y(t) = KM \left\{ 1 - e^{-\zeta t/\tau} \left[\cos\left(\frac{\sqrt{1 - \zeta^2}}{\tau} t\right) + \frac{\zeta}{\sqrt{1 - \zeta^2}} \sin\left(\frac{\sqrt{1 - \zeta^2}}{\tau} t\right) \right] \right\}$$

$$(5\text{-}51)$$

Plots of the step responses for different values of ζ are shown in Figs. 5.8 and 5.9. The abscissas are normalized with respect to τ. When τ is small, a rapid response is signified, implying a large value for the undamped natural frequency, $\omega_n = 1/\tau$.

Several general remarks can be made concerning the responses shown in Figs. 5.8 and 5.9:

1. Responses exhibiting oscillation and overshoot ($y/KM > 1$) are obtained only for values of ζ less than one.
2. Large values of ζ yield a sluggish (slow) response.
3. The fastest response without overshoot is obtained for the critically damped case ($\zeta = 1$).

Control system designers often attempt to make the set-point step response of the controlled variable approximate the step response of an underdamped sec-

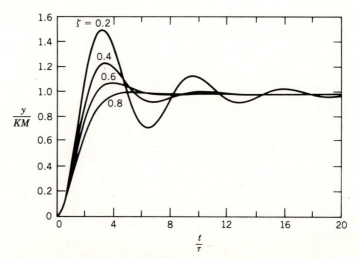

Figure 5.8. Step response of underdamped second-order processes.

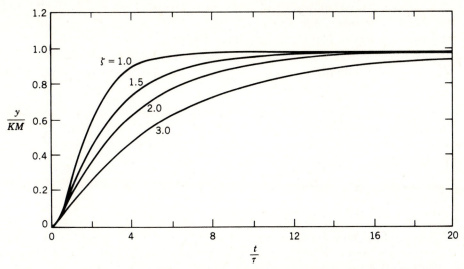

Figure 5.9. Step response of critically-damped and overdamped second-order processes.

ond-order system, that is, make it exhibit a prescribed amount of overshoot and oscillation as it settles at the new operating point. Values of ζ in the range 0.4 to 0.8 often are suitable for specifying a desired control system response, assuming that it can be approximated as an underdamped second-order system. In this range, the controlled variable y reaches the new operating point faster than with $\zeta = 1.0$ or 1.5, but the response is much less oscillatory (it settles faster) than with $\zeta = 0.2$.

Figure 5.10 illustrates the step response of a second-order underdamped process. We can define a number of terms that are commonly used to describe approximately the dynamics of underdamped processes, whether controlled or uncontrolled:

1. **Rise Time.** t_r is the time the process output takes to first reach the new steady-state value.
2. **Time to First Peak.** t_p is the time required for the output to reach its first maximum value.
3. **Settling Time.** t_s is defined as the time required for the process output to reach and remain inside a band whose width is equal to $\pm5\%$ of the total change in

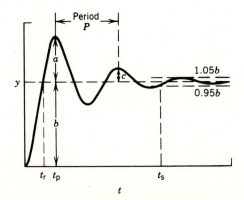

Figure 5.10. Performance characteristics for the step response of an underdamped process.

y. The term 95% response time sometimes is used to refer to this case. Also, values of $\pm 1\%$ sometimes are used.

4. Overshoot. OS $= a/b$ (% overshoot is $100a/b$).

5. Decay Ratio. DR $= c/a$ (where c is the height of the second peak).

6. Period of Oscillation. P is the time between two successive peaks or two successive valleys of the response.

Note that the above definitions generally apply to the step response of any underdamped process. For the particular case of an underdamped *second-order* process, we can develop analytical expressions for some of these characteristics. Using Eq. 5-51

$$\text{Overshoot: } OS = \exp\left(-\pi\zeta/\sqrt{1-\zeta^2}\right) \tag{5-52}$$

$$\text{Decay ratio: } DR = (OS)^2 = \exp(-2\pi\zeta/\sqrt{1-\zeta^2}) \tag{5-53}$$

Both of the above characteristics are functions of ζ only. For a second-order system, the decay ratio is constant for each successive pair of peaks. Figure 5.11 illustrates the dependence of overshoot and decay ratio on damping coefficient.

$$\text{Period: } P = \frac{2\pi\tau}{\sqrt{1-\zeta^2}} \tag{5-54}$$

Note that if it is possible to describe a particular process by an underdamped second-order transfer function, Figs. 5.8 and 5.11 can be used to obtain estimates of ζ and τ based on a plot of the step response.

EXAMPLE 5.5

A stirred-tank reactor has an internal cooling coil to remove heat liberated in the reaction. A proportional controller is used to regulate coolant flow rate so as to keep the reactor temperature reasonably constant. The controller has been designed so that the controlled reactor exhibits typical underdamped second-order temperature response characteristics when it is disturbed, either by feed flow rate or by coolant temperature changes.

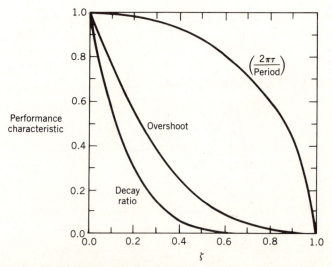

Figure 5.11. Relation between some performance characteristics of an underdamped second-order process and the process damping coefficient.

(a) An operator changes the feed flow rate to the reactor suddenly from 0.4 to 0.5 kg/s and notes that the temperature of the reactor contents, initially 100 °C, changes eventually to 102 °C. What is the gain of the process transfer function (under feedback control) that relates changes in reactor temperature to changes in feed flow rate? (Be sure to specify the units.)

(b) The operator notes that the resulting response is slightly oscillatory with maxima equal to 102.5 and 102.1 °C occurring at times 1000 and 3060 s after the change is initiated. What is the complete process transfer function?

(c) The operator failed to note the rise time. What was it?

Solution

Note that in this example we are dealing with a controlled process that causes the oscillatory response observed by the operator; however, this feature makes no difference in the way the analysis is carried out.

(a) The gain of the controlled process is obtained by dividing the steady-state change in output by the input change:

$$K = \frac{102 - 100}{0.5 - 0.4} = 20 \ \frac{°C}{kg/s}$$

(b) The oscillatory characteristics of the response can be used to find the dynamic elements in the transfer function relating temperature to feed flow rate. Assuming the step response is due to an underdamped second-order process, Figs. 5.8 and 5.11 can be used to obtain estimates of ζ and τ. Alternatively, analytical expressions can be used, which is the approach that we take here. Either Eq. 5-52 or 5-53 can be employed to find ζ independently of τ. Since the second peak value of temperature (102.1 °C) is so close to the final value (102 °C), the calculated value of peak height c may be subject to appreciable measurement error. Instead, we use the relation for overshoot (rather than decay ratio) to take advantage of the potentially greater precision of the first peak measurement. Rearranging (5-52) gives

$$\zeta = \sqrt{\frac{[\ln(OS)]^2}{\pi^2 + [\ln(OS)]^2}}$$

$$OS = \frac{102.5 - 102}{102 - 100} = \frac{0.5}{2} = 0.25 \quad (i.e., 25\%) \tag{5-55}$$

$$\zeta = 0.4037 \approx 0.4$$

Equation 5-54 can be rearranged to find τ:

$$\tau = \frac{\sqrt{1 - \zeta^2}}{2\pi} P$$

$$P = 3060 - 1000 = 2060 \ s \tag{5-56}$$

$$\tau = 300 \ s$$

(c) Thus the rise time t_r can be calculated from Eq. 5-51. When $t = t_r$, $y(t)$ is equal to its final steady-state value, KM. In other words, the bracketed quantity is identically zero at $t = t_r$:

$$\cos\left(\frac{\sqrt{1 - \zeta^2}}{\tau} t_r\right) + \frac{\zeta}{\sqrt{1 - \zeta^2}} \sin\left(\frac{\sqrt{1 - \zeta^2}}{\tau} t_r\right) = 0 \tag{5-57}$$

Solving for the rise time gives

$$t_r = \frac{\tau}{\sqrt{1 - \zeta^2}} (\pi - \cos^{-1} \zeta) \tag{5-58}$$

where the result of the inverse cosine computation must be in radians. Since $\tau = 300$ s and $\zeta = 0.40$

$$t_r = 649 \text{ s}$$

Note that an infinite number of values of t_r satisfy (5-57). The general solution for any value of t that satisfies $y(t) = KM$ is given by

$$t = \frac{\tau}{\sqrt{1 - \zeta^2}} (n\pi - \cos^{-1} \zeta) \qquad n = 1, 2, \ldots \tag{5-59}$$

The rise time corresponds to the first time that response $y(t) = KM$, the final value. It occurs for $n = 1$.

In summary, the process transfer function between feed flow rate and outlet temperature while under feedback control is

$$\frac{T'(s)}{W'(s)} = \frac{20}{(300)^2 s^2 + 2(0.4)(300)s + 1}$$

$$= \frac{20}{90{,}000 s^2 + 240 s + 1}$$

where the process gain has units of °C/kg/s.

Sinusoidal Response

When a linear second-order system is forced by a sinusoidal input $A \sin \omega t$, the output for large values of time (after exponential terms have disappeared) is also a sinusoidal signal given by

$$y(t) = \frac{KA}{\sqrt{[1 - (\omega\tau)^2]^2 + (2\zeta\omega\tau)^2}} \sin(\omega t + \phi) \tag{5-60}$$

where $\phi = -\tan^{-1} \left[\dfrac{2\zeta\omega\tau}{1 - (\omega\tau)^2} \right]$ \hfill (5-61)

The output amplitude $\hat{A}$ is obtained directly from Eq. 5-60:

$$\hat{A} = \frac{KA}{\sqrt{[1 - (\omega\tau)^2]^2 + (2\zeta\omega\tau)^2}} \tag{5-62}$$

The ratio of output to input amplitudes is the *amplitude ratio* AR. When normalized by the process gain, it is called the *normalized amplitude ratio* AR_N

$$AR_N = \frac{\hat{A}}{KA} = \frac{1}{\sqrt{[1 - (\omega\tau)^2]^2 + (2\zeta\omega\tau)^2}} \tag{5-63}$$

AR_N represents the effect of the process dynamic parameters (ζ, τ) on the sinusoidal response; that is, AR_N is independent of steady-state gain K and the amplitude of the forcing function A. The maximum value of AR_N can be found (if it exists) by

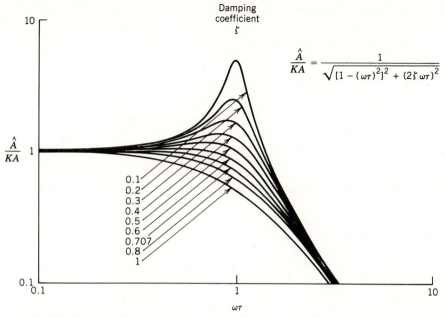

Figure 5.12. Sinusoidal response amplitude of a second-order system after exponential terms have become negligible.

differentiating (5-63) with respect to ω and setting the derivative to zero. Solving for ω_{max} gives

$$\omega_{max} = \frac{\sqrt{1 - 2\zeta^2}}{\tau} \quad \text{for } 0 < \zeta < 0.707 \tag{5-64}$$

For $\zeta \geq 0.707$, there is no maximum, as Fig. 5.12 illustrates. Substituting (5-64) into (5-63) yields an expression for the maximum value of AR_N:

$$AR_N\bigg|_{max} = \frac{\hat{A}}{KA}\bigg|_{max} = \frac{1}{2\zeta\sqrt{1 - \zeta^2}} \quad 0 < \zeta < 0.707 \tag{5-65}$$

We see from (5-65) that the maximum output amplitude for a second-order process that has no damping ($\zeta = 0$) is undefined. The *small damping* case is invariably avoided in the design of processes, as well as in designing control systems. Equation 5-65 indicates that a process with little damping can exhibit very large output oscillations if it is perturbed by periodic signals with frequency near ω_{max}. Since, in general, no one can guarantee what disturbances will occur to a particular process, damping coefficients for controlled processes should be established conservatively, generally around 0.5.

EXAMPLE 5.6

An engineer uses a pressure bulb sensor and pneumatic transmitter to measure the temperature in a CSTR. The output of the transmitter, a signal in the range of 3 to 15 psig, is sent over a relatively long pneumatic line to a recorder in the control room. The displayed output has been calibrated directly in temperature units, °C. The temperature sensor/transmitter combination operates approximately as a first-order dynamic system with time constant equal to 3 s. The pneumatic transmission line similarly acts like a first-order process with time constant of 10 s. While observing the recorder output, the engineer notes that the measured reactor temperature has been cycling approximately sinusoidally between 180 and 183 °C with a period of 30 s for at least several minutes. What can be concluded concerning the actual temperature in the reactor?

Solution

First, note that the sensor/transmitter and the transmission line act as two first-order processes in series (Eq. 5-39) with overall gain K equal to 1. Hence, the approximate transfer function is

$$\frac{T_{\text{meas}}(s)}{T_{\text{reactor}}(s)} = \frac{1}{(3s + 1)(10s + 1)} \tag{5-66}$$

From the reported results, we conclude that some disturbance has caused the reactor temperature to vary sinusoidally, which, in turn, has caused the recorded output to oscillate. The cycling has continued for a period of time that is much longer than the time constants of the process, that is, the instrumentation system. Hence, we can infer the conditions in the reactor from the measured results, using Eq. 5-63 for the sinusoidal response of a second-order system. From (5-66), $\tau_1 = 3$ s and $\tau_2 = 10$ s; τ and ζ of the process are calculated from Eqs. 5-42 and 5-43:

$$\tau = \sqrt{(3)(10)} = 5.48 \text{ s}$$

$$\zeta = \frac{13}{2(5.477)} = 1.19$$

The frequency of the perturbing sinusoidal signal (reactor temperature) is calculated from the observed period of 30 s:

$$\omega = \frac{2\pi}{P} = \frac{6.28}{30} = 0.2093 \text{ s}^{-1}$$

The amplitude of the output perturbation also is obtained from observed results as

$$\hat{A} = \frac{183 - 180}{2} = 1.5 \text{ °C}$$

Equation 5-62 now can be rearranged to calculate the amplitude of the actual reactor temperature

$$A = \frac{\hat{A}}{K} \sqrt{[1 - (\omega\tau)^2]^2 + (2\zeta\omega\tau)^2}$$

from which $A = (1.5)(2.75) = 4.12$ °C. Hence the actual reactor temperature is varying between $181.5 - 4.12 = 177.38$ °C and $181.5 + 4.12 = 185.62$ °C, nearly three times the variation indicated by the recorder.

Since the second-order process in this example is overdamped, that is, $\zeta = 1.19 > 1$, with $K = 1$, we expect that sinusoidal perturbations in the reactor temperature always will be attenuated (reduced in amplitude) regardless of the frequency of the perturbation. This is not true for underdamped systems with unity gain. Further discussion of sinusoidal forcing is contained in Chapter 14 on frequency response techniques.

SUMMARY

Transfer functions can be used conveniently to obtain process responses to any type of input forcing function. In this chapter we have focused on first- or second-order transfer functions and integrating processes. Because a relatively small number of inputs have industrial or analytical significance, we have studied in detail the responses of these basic process transfer functions to the important types of inputs.

If the control engineer knows from prior analysis that a process can be modeled as a first-order or second-order transfer function, then how the process will respond to any standard input change also can be found. When a fundamental model is not available, as occurs in many plant situations, it is usually possible to observe a process output response to some observable input. Using these data the engineer can formulate an approximate process transfer function and calculate values of the model parameters. The model permits predictions of how a process will react to other types of disturbances or input changes. Thus, it gives a framework for decisions that can be made quickly and without recourse to more complicated models, computer simulations, and so forth.

Unfortunately, not all processes can be modeled by such simple transfer functions. Hence, in Chapter 6 several additional elements are introduced and combined into more complicated transfer functions. However, the emphasis there remains on showing how complex process behavior can be explained with combinations of simple transfer function elements.

REFERENCES

1. Maybeck, P. S., *Stochastic Models, Estimation, and Control,* Academic, New York, 1979.
2. Box, G. E. P. and G. M. Jenkins, *Time Series Analysis: Forecasting and Control,* Holden Day, San Francisco, 1976.
3. Sage, A. P. and J. L. Melsa, *System Identification,* Academic, New York, 1971.
4. Coughanowr, D. R. and L. B. Koppel, *Process Systems Analysis and Control,* McGraw-Hill, New York, 1965.

EXERCISES

5.1. In addition to the standard inputs discussed in Section 5.1, there are other input functions that occasionally are useful for special purposes. One, the so-called doublet pulse, is shown in the drawing.

(a) Find the Laplace transform of this function by first expressing it as a composite of functions whose transforms you already know.

(b) What would be the response of a process having a first-order transfer function $K/(\tau s + 1)$ to this input? Of the integrating process K/s?

(c) From these results, can you determine what special property this input offers?

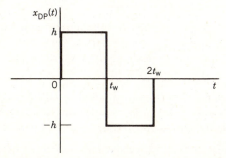

5.2. The caustic concentration of a process stream is to be measured using a conductivity cell. To determine the dynamic response characteristics, a step change of 3 lb/ft^3 in the caustic concentration passing through the cell $c(t)$ is made at time $t = 0$. The measured concentration $c_m(t)$ is shown in the figure. Determine the transfer function between c_m and c.

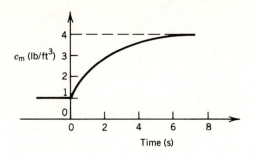

5.3. The transfer function

$$G(s) = \frac{Y(s)}{X(s)} = \frac{10}{s + 1}$$

represents the model of a process written in deviation variable form. Derive an expression for the response $y(t)$ to the input, $x(t) = 1 + t$, for an initial value, $y(0) = 2$, corresponding to a value of $x(0) = 1$.

5.4. The dynamic behavior of a stirred-tank reactor can be represented by the transfer function

$$\frac{C'(s)}{C'_F(s)} = \frac{0.3}{4s + 1}$$

where C' is the exit concentration, mol/L, C'_F is the feed concentration, mol/L. Obtain the response $c(t)$ for the c_F disturbance in the drawing.

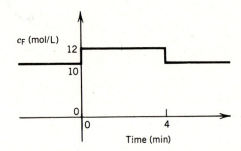

5.5. A thermocouple has the following characteristics when it is immersed in a stirred bath:

Mass of thermocouple = 1 g
Heat capacity of thermocouple = 0.25 cal/g °C
Heat transfer coefficient = 20 cal/cm² h °C (for thermocouple and bath)
Surface area of thermocouple = 3 cm²

(a) Derive a transfer function model for the thermocouple relating the change in its indicated output T to the change in the temperature of its surroundings T_s assuming uniform temperature (no gradients in the thermocouple bead), no conduction in the leads, constant physical properties, and that the millivolt-level output is converted directly to a °C reading by a very fast meter.
(b) If the thermocouple is initially out of the bath and at room temperature (23 °C), what is the maximum temperature that it will register if it is suddenly plunged into the bath (80 °C) and held there for 20 s.

5.6. A bare thermocouple, when suddenly moved from a temperature of 82 °F into a stream of hot air at a constant temperature of 105 °F, reads 93 °F after 1 s. If the same thermocouple is used to record the temperature of the same airstream when this temperature is falling at a constant rate of 2 °F per second, what will be the error in the thermocouple reading at the end of 10 s? You may assume that the thermocouple has first-order dynamics with gain equal to one.

5.7. Appelpolscher has just left a meeting with Stella J. Smarly, IGC's Vice-President for Process Operations and Development. Smarly is concerned about an upcoming extended plant test of a method intended to improve the yields of a large packed-bed reactor. The basic idea, which came from IGC's university consultant and was recently tested for feasibility in a brief run, involves operating the reactor cyclically so that nonlinearities in the system cause the time-average yield at the exit to exceed the steady-state value. Smarly is worried about the possibility of sintering the catalyst during an extended run, particularly in the region of the "hotspot" (axially about one-third of the way down the bed and at the centerline) where temperatures invariably peak. Appelpolscher, who plans to leave the next day on a two-week big game safari, doesn't want to cancel his vacation. On the other hand, Smarly has told him he faces early, unexpected retirement in Botswanaland if the measurement device (located near the hot spot) fails to alert operating people and the reactor catalyst sinters. Appelpolscher likes Botswanaland but doesn't want to retire there. He manages to pull together the following data and assumptions before heading for the airport and leaves them with you for analysis with the offer to use his swimming pool while he is gone. What do you report to Smarly?

Data:
Frequency of cyclic operation = 0.1 cycles/min
Amplitude of thermal wave (temperature) at the measurement point obtained experimentally in the recent brief run = 15 °C
Average operating temperature at measurement point = 350 °C
Time constant of temperature sensor and thermowell = 1.5 min
Temperature at the reactor wall = 200 °C
Temperature at which the catalyst sinters if operated for several hours = 700 °C
Temperature at which the catalyst sinters instantaneously = 715 °C

Assumptions:
The reactor operational cycle is approximately sinusoidal at the measurement point.
The thermowell is located near the reactor wall so as to measure a "radial average" temperature rather than the centerline temperature.
The approximate relation is

$$T_{meas} = \frac{T_{center} + 2T_{wall}}{3}$$

which also holds during transient operation.

5.8. A liquid storage system is shown below. The normal operating conditions are

$\bar{q}_1 = 10 \text{ ft}^3/\text{min}$, $\bar{q}_2 = 5 \text{ ft}^3/\text{min}$, $\bar{h} = 4 \text{ ft}$.

The tank is 6 ft in diameter and the density of each stream is 60 lb/ft^3. Suppose that a pulse change in q_1 occurs as shown in the drawing.

(a) What is the transfer function relating H' to Q_i'?
(b) Derive an expression for $h(t)$ for this input change.
(c) What is the new steady-state value of liquid level $\bar{h}_2$?
(d) Repeat (b) and (c) for the doublet pulse input of Exercise 5.1 where the changes in q_1 are from 10 to 15 to 5 to 10 ft^3/min and where $t_w = 12$ min.

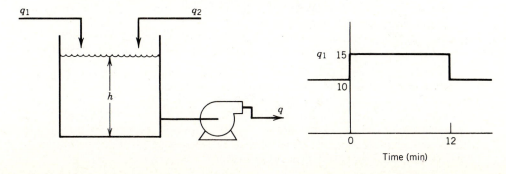

5.9. Two liquid storage systems are shown in the drawing. Each tank is four feet in diameter. For System I, the valve acts as a linear resistance with the flow–head relation $q = 8.33\,h$, where q is in gal/min and h is in feet. For System II, variations in liquid level h do not affect exit flow rate q. Suppose that each system is initially at steady state with $\bar{h} = 6$ ft and $\bar{q}_i = 50$ gal/min and that at time $t = 0$ the inlet flow rate suddenly changes from 50 to 70 gal/min. For each system, determine the following information:

(a) The transfer function $H'(s)/Q_i'(s)$ where the primes denote deviation variables.
(b) The transient response $h(t)$.
(c) The new steady-state levels.
(d) If each tank is 8 ft tall, which tank overflows first? When?

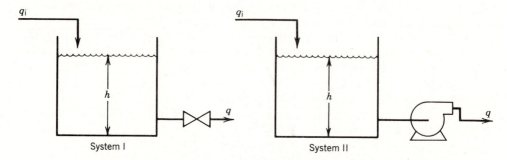

System I   System II

5.10. A force–momentum balance on a mercury manometer results in the following equation:

$$4\frac{d^2x}{dt^2} + 0.8\frac{dx}{dt} + x = p(t)$$

where x is the displacement of the mercury column from its equilibrium position and $p(t)$ is the time-varying pressure acting on the manometer.

(a) Find the transfer function relating x to p, assuming that the system initially is in equilibrium.
(b) What can you determine about the response time of the manometer? About its damping characteristics?
(c) Calculate the response of the manometer to a pressure change, $p(t) = 2e^{-4t}$.

5.11. An expression has been found for the transient response of a second-order process:

$$\frac{d^2y}{dt^2} = -W^2\frac{dy}{dt} - \frac{KV}{W}y + Qx^{3/2}$$

(a) What is the effect on the system characteristic time (τ) and damping coefficient of doubling V while keeping W constant?
(b) Of doubling W while keeping V constant?
(c) Find a transfer function relating y to x.

5.12. For the equation

$$\frac{d^2y}{dt^2} + K\frac{dy}{dt} + 4y = x$$

(a) Find the transfer function and put it in standard gain/time constant form.
(b) Discuss the qualitative form of the response of this system (independent of the input forcing) over the range $-10 \le K \le 10$.
   Specify regions where the response will converge and where it will not. Write the form of the response without evaluating any coefficients.

5.13. The dynamic behavior of a physical process can be represented by the transfer function:

$$\frac{Y(s)}{X(s)} = \frac{18}{s^2 + 3s + 9}$$

(a) After a step change of $x(t) = 3S(t)$, what is the new steady-state value of y?

(b) For physical reasons, it is required that $y \leq 10$. What is the largest step change in x that the process can tolerate without exceeding this limit?

5.14. A step change from 15 to 31 psi in actual pressure results in the measured response from a pressure-indicating element shown in the drawing.

(a) Assuming second-order dynamics, calculate all important parameters and write an approximate system transfer function in the form

$$\frac{R'(s)}{P'(s)} = \frac{K}{\tau^2 s^2 + 2\zeta\tau s + 1}$$

where R' is the instrument output deviation (mm),

P' is the actual pressure deviation (psi).

(b) Write an equivalent differential equation model in terms of actual (not deviation) variables.

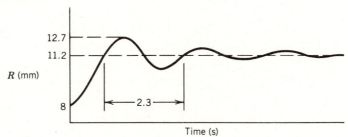

5.15. A step change in the steam pressure to a heat exchanger is made from 20 to 24 psig. The system outlet temperature response (see figure) appears to be approximately second order. Someone forgot, however, to note the recorder speed. A second test of the system using a sinusoidal variation of steam pressure to the exchanger with an amplitude of 2 psig and a frequency of 2 cycles/min yields an output temperature variation with amplitude of 4 °F. Determine an approximate transfer function for this system. Is the transfer function unique?

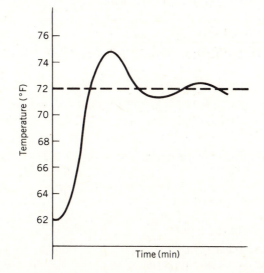

5.16. Starting with Eq. 5-51, derive expressions for the response characteristics of the underdamped second-order system. In particular, find expressions for:

(a) The time to first peak t_p.

(b) The fraction overshoot (Eq. 5-52).

(c) The decay ratio (Eq. 5-53).
(d) The settling time (t_s, defined in Fig. 5.10). Can a *single* expression be used for t_s over the full range of ζ, $0 < \zeta < 1$?

5.17. Derive Eq. 5-64, the value of frequency ω which yields a maximum amplitude for sinusoidal forcing of underdamped systems with $0 \le \zeta < 0.707$. (Verify this limit.) Show that this value of ω (i.e., ω_{max}) yields the normalized amplitude ratio in Eq. 5-65.

5.18. A two-tank flow surge system is constructed as shown in Fig. 4.3. A block diagram representing the approximate dynamics of the system is shown in the figure.

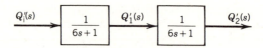

(a) What is the response $q_2(t)$ to a ramp input of slope 0.4 m³/min in $q_i(t)$ if the system is initially operating at a steady state corresponding to

$$\bar{q}_i = \bar{q}_1 = \bar{q}_2 = 1 \text{ m}^3/\text{min}$$

(b) If the head–outflow relations for the two tanks are

$$q_1 = \frac{4 \text{ m}^3/\text{min}}{\text{m}} h_1$$

$$q_2 = \frac{2 \text{ m}^3/\text{min}}{\text{m}} h_2$$

determine the two system level responses after 1 m³ of liquid is suddenly added to the first tank.

5.19. A surge tank system is to be installed as part of a pilot plant facility. The initial proposal calls for the configuration shown in Fig. 4.3. Each tank is 5 ft high and 3 ft in diameter. The design flow rate is $q_i = 100$ gal/min. It has been suggested that an improved design will occur if the two-tank system is replaced by a single tank that is 4 ft in diameter and has the same total volume (i.e., $V = V_1 + V_2$).

(a) Which surge system (original or modified) can handle larger step disturbances in q_i? (Justify your answer.)
(b) Which system provides the best damping of step disturbances in q_i? (Justify your answer).

In your analysis you may assume that:

i. The valves on the exit lines act as linear resistances.
ii. The valves are adjusted so that each tank is half full at the nominal design condition of $q_i = 100$ gal/min.

5.20. The caustic concentration of the mixing tank shown in the drawing is measured using a conductivity cell. The total volume of solution in the tank is constant at 7 ft³ and the density ($\rho = 70$ lb/ft³) can be considered to be independent of concentration. Let c_m denote the caustic concentration measured by the conductivity cell. The dynamic response of the conductivity cell to a step change (at $t = 0$) of 3 lb/ft³ in the actual concentration (passing through the cell) is also shown in the drawing.

(a) Determine the transfer function $C'_m(s)/C'_1(s)$ assuming the flow rates are equal and constant: ($w_1 = w_2 = 5$ lb/min):
(b) Find the response for a step change in c_1 from 14 to 17 lb/ft³.
(c) If the transfer function $C'_m(s)/C'(s)$ were approximated by 1 (unity) what would be the step response of the system for the same input change?
(d) By comparison of (b) and (c), what can you say about the dynamics of the conductivity cell? Plot both responses, if necessary.

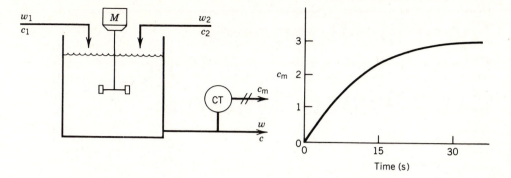

5.21. An exothermic reaction, A → 2B, takes place adiabatically in a stirred-tank system. This liquid phase reaction occurs at constant volume in a 100-gal reactor. The reaction can be considered to be first order and irreversible with the rate constant given by

$$k = 2.4 \times 10^{15} \, e^{-20,000/T} \quad (\text{min}^{-1})$$

where T is in °R. Using the information below, derive a transfer function relating the exit temperature T to the inlet concentration c_{Ai}. State any assumptions that you make. Simplify the transfer function by making a first-order approximation and show that the approximation is valid by comparing the step responses of both the original and the approximate models.

Available Information

i. Nominal steady-state conditions are:

$$\overline{T} = 150 \, °F, \qquad \overline{c}_{Ai} = 0.8 \text{ lb mole/ft}^3$$

$$q = 20 \text{ gal/min} = \text{flow rate in and out of the reactor}$$

ii. Physical property data for the mixture at the nominal steady state:

$$C_p = 0.8 \text{ Btu/lb °F}, \qquad \rho = 52 \text{ lb/ft}^3, \qquad -\Delta H = 500 \text{ kJ/lb mole}$$

— CHAPTER 6 —

Dynamic Response Characteristics of More Complicated Systems

In Chapter 5 we discussed the dynamics of relatively simple processes, those that can be modeled dynamically as first- or second-order transfer functions or as an integrator. Obviously, there are more complicated processes than the ones we have discussed. Additional complexity can appear in the transfer function model as a higher order (more than two) denominator and/or in the appearance of functions of s in the transfer function numerator. In this chapter we first show how the form of the numerator and denominator of a transfer function qualitatively affects the dynamic response of the system. Other topics covered include the time delay and its approximation as a rational transfer function, interacting lower-order process elements, staged processes, distributed processes, and multiple-input, multiple-output processes.

6.1 POLES AND ZEROS AND THEIR EFFECT ON SYSTEM RESPONSE

One feature of the simple process elements discussed in Chapter 5 is that their response characteristics are determined by the factors of the transfer function denominator, the characteristic polynomial. Consider a particular transfer function,

$$G(s) = \frac{K}{s(\tau_1 s + 1)(\tau_2^2 s^2 + 2\zeta\tau_2 s + 1)} \qquad (6\text{-}1)$$

where $0 \leq \zeta < 1$. Using the principle of partial fraction expansion followed by the inverse transformation operation, we know that the response of the system to *any* input will contain the following functions of time:

- A constant term resulting from the s factor
- An e^{-t/τ_1} term resulting from the $(\tau_1 s + 1)$ factor

$$\left. \begin{aligned} &\bullet\ e^{-\zeta t/\tau_2} \sin \frac{\sqrt{1 - \zeta^2}}{\tau_2} t \\ &\text{and} \\ &\bullet\ e^{-\zeta t/\tau_2} \cos \frac{\sqrt{1 - \zeta^2}}{\tau_2} t \end{aligned} \right\} \text{terms resulting from the } (\tau_2^2 s^2 + 2\zeta\tau_2 s + 1) \text{ factor}$$

Additional terms determined by the specific input forcing term will also appear in the response, but the intrinsic dynamic features of the process, the so-called response modes or natural modes, are determined by the process itself. Each of the above response modes is determined from the factors of the denominator polynomial (roots of the characteristic equation):

$$s_1 = 0$$

$$s_2 = -\frac{1}{\tau_1}$$

$$s_3 = -\frac{\zeta}{\tau_2} + j\frac{\sqrt{1 - \zeta^2}}{\tau_2}$$

$$s_4 = -\frac{\zeta}{\tau_2} - j\frac{\sqrt{1 - \zeta^2}}{\tau_2}$$

(6-2)

Roots s_3 and s_4 are found by applying the quadratic formula.

Control engineers refer to the values of s that are roots of the characteristic equation as *poles* of transfer function $G(s)$. Sometimes it is useful to plot the roots (poles) and to discuss process response characteristics in terms of root locations in the complex s plane. In Fig. 6.1 the ordinate expresses the imaginary part of each root; the abscissa expresses the real part. Figure 6.1 is based on Eq. 6-2 and indicates the presence of an integrating element (pole at the origin), one real pole (at $-1/\tau_1$), and a complex pair of poles, s_3 and s_4. Note that the real pole is closer to the imaginary axis than the complex pair, indicating a slower response mode (e^{-t/τ_1} decays slower than $e^{-\zeta t/\tau_2}$).

Historically, plots such as Fig. 6.1 have played an important role in the design of mechanical and electrical control systems, but they are rarely used in designing process control systems. However, it is helpful to develop some intuitive feeling for the influence of pole locations. A pole to the right of the imaginary axis (a so-called right-half plane pole), for example, at $s = +1/\tau$, indicates that one of the

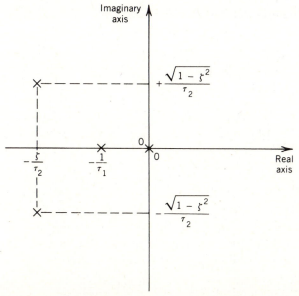

Figure 6.1.   Roots of the denominator of $G(s)$ (Eq.6-1) plotted in the complex s plane.

system response modes is $e^{t/\tau}$. This mode obviously grows without bound as t becomes large, a characteristic of unstable systems. As a second example, a complex pole will always appear as part of a conjugate pair, such as s_3 and s_4 in (6-2). The presence of complex conjugate poles indicates that the response will contain sine and cosine modes, that is, it will exhibit oscillatory modes.

All of the transfer functions discussed so far can be made to represent more complex process dynamics simply by including the effect of *input dynamics*. For example, some control systems contain a *lead–lag element*. The differential equation for this element is

$$\tau_1 \frac{dy}{dt} + y = K \left(\tau_a \frac{dx}{dt} + x \right) \tag{6-3}$$

In Eq. 6-3 the standard first-order dynamics have been modified by the presence of the derivative of the input x weighted by a time constant τ_a. The transfer function for this dynamic element is given by

$$G(s) = \frac{K(\tau_a s + 1)}{\tau_1 s + 1} \tag{6-4}$$

Transfer functions with numerator terms such as $\tau_a s + 1$ above are said to exhibit *numerator dynamics*. Suppose that the integral of x is included in the input terms:

$$\tau_1 \frac{dy}{dt} + y = K \left(x + \frac{1}{\tau_a} \int_0^t x(t^*) \, dt^* \right) \tag{6-5}$$

The transfer function for (6-5), assuming zero initial conditions, would be

$$G(s) = \frac{K(\tau_a s + 1)}{\tau_a s(\tau_1 s + 1)} \tag{6-6}$$

In this example, integration of the input introduces a pole at the origin (the $\tau_a s$ term in the denominator), an important point that will be discussed later.

The dynamics of a process are affected not only by the poles of $G(s)$ but also by the values of s that cause the numerator of $G(s)$ to become 0. These values are called the *zeros* of $G(s)$.

Before discussing zeros it is useful to show several equivalent ways in which transfer functions can be written. In Chapter 4, a particular transfer function form was discussed:

$$G(s) = \frac{\displaystyle\sum_{i=0}^{m} b_i s^i}{\displaystyle\sum_{i=0}^{n} a_i s^i} = \frac{b_m s^m + b_{m-1} s^{m-1} + \cdots + b_0}{a_n s^n + a_{n-1} s^{n-1} + \cdots + a_0} \tag{4-34}$$

which can also be written as

$$G(s) = \frac{b_m}{a_n} \frac{(s - z_1)(s - z_2) \cdots (s - z_m)}{(s - p_1)(s - p_2) \cdots (s - p_n)} \tag{6-7}$$

where the z_i and p_i are zeros and poles, respectively. We often employ transfer functions in *gain/time constant form*, that is, b_0 is factored out of the numerator of Eq. 4-34 and a_0 out of the denominator to show the steady-state gain explicitly ($K = b_0/a_0$). Then the resulting expressions are factored to give

$$G(s) = K \frac{(\tau_a s + 1)(\tau_b s + 1) \cdots}{(\tau_1 s + 1)(\tau_2 s + 1) \cdots} \tag{6-8}$$

for the case where all factors represent real roots. Obviously, the relations between poles and zeros and the time constants are given by

$$z_1 = -1/\tau_a, \quad z_2 = -1/\tau_b, \quad \cdots \tag{6-9}$$

$$p_1 = -1/\tau_1, \quad p_2 = -1/\tau_2, \quad \cdots \tag{6-10}$$

The presence or absence of system zeros has no effect on the number and location of the poles and their associated response modes in Eq. 6-7 unless there is an exact cancellation of a pole by a zero with the same numerical value. However, the zeros exert a profound effect on the coefficients of the response modes (i.e., how they are weighted) in the system response. Such coefficients are found by partial fraction expansion.

EXAMPLE 6.1

Calculate the response of the lead–lag element (Eq. 6-4) to a step change of magnitude M in its input.

Solution
For this case we have

$$Y(s) = \frac{KM(\tau_a s + 1)}{s(\tau_1 s + 1)} \tag{6-11}$$

which can be expanded into partial fractions

$$Y(s) = KM \left(\frac{1}{s} + \frac{\tau_a - \tau_1}{\tau_1 s + 1} \right) \tag{6-12}$$

yielding the solution

$$y(t) = KM \left[1 - \left(1 - \frac{\tau_a}{\tau_1} \right) e^{-t/\tau_1} \right] \tag{6-13}$$

Figure 6.2a shows the response of such a system for $\tau_1 = 4$ and different values of τ_a.

Case a: $\quad 0 < \tau_1 < \tau_a$
Case b: $\quad 0 < \tau_a < \tau_1$
Case c: $\quad \tau_a < 0 < \tau_1$

Figure 6.2b is a pole–zero plot showing the location of the single system zero, $s = -1/\tau_a$, for each of these three cases. If $\tau_a = \tau_1$, the transfer function simplifies to K as a result of cancellation of numerator and denominator elements, that is, a *pole–zero cancellation*.

From inspection of Eq. (6-13) and Fig. 6.2a, the presence of a zero in the first-order system causes a jump discontinuity in $y(t)$ at $t = 0$ when the step input is applied. Such an instantaneous step response is possible only when the numerator and denominator polynomials have the same order. Most processes have higher-order dynamics in the denominator causing them to exhibit some degree of *inertia*. This feature usually prevents them from responding instantaneously to any input,

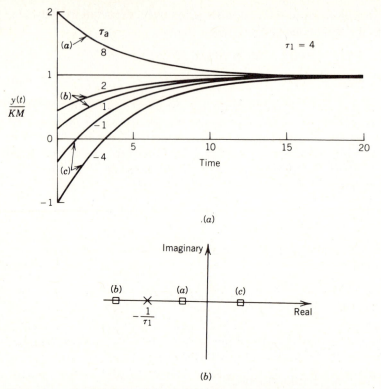

Figure 6.2. (a). Step response to a first-order system (Eq. 6-13) with a single zero [$y(t=0) = \tau_a/\tau_1$]. (b). Pole–zero plot for a first-order system showing alternative locations of the single zero. ×, pole location; □, location of single zero.

including an impulse input. These physical arguments can be summarized in terms of the numerator and denominator orders of Eq. 4-34, that is, $m \leq n$ for a system to be physically realizable.

EXAMPLE 6.2

For the case of a single zero in an overdamped second-order transfer function,

$$G(s) = \frac{K(\tau_a s + 1)}{(\tau_1 s + 1)(\tau_2 s + 1)} \tag{6-14}$$

calculate the response to a step input of magnitude M and plot the results qualitatively.

Solution

The response of this system to a step change in input is

$$y(t) = KM\left(1 + \frac{\tau_a - \tau_1}{\tau_1 - \tau_2} e^{-t/\tau_1} + \frac{\tau_a - \tau_2}{\tau_2 - \tau_1} e^{-t/\tau_2}\right) \tag{6-15}$$

Note that $y(t \rightarrow \infty) = KM$ as expected; hence, the effect of including the single zero does not change the final value nor does it change the number or location of the response modes. But the zero does affect how the response modes (exponential terms) are weighted in the solution, Eq. 6-15.

A certain amount of mathematical analysis (see Exercises 6.4, 6.5, and 6.6) will show that there are three types of responses involved here:

Case a: $\tau_a > \tau_1$

Case b: $0 < \tau_a \leq \tau_1$

Case c: $\tau_a < 0$

for the situation where $\tau_1 > \tau_2$ is arbitrarily chosen. Figure 6.3 shows step responses for eight values of τ_a. Case (a) shows that *overshoot* can occur if the zero $(-1/\tau_a)$ is large enough, that is, if τ_a is sufficiently large. Case (b) looks much like an ordinary first-order process response as a result of the single zero. Case (c) exhibits a so-called *inverse response* initially, an infrequently encountered yet important process response to step inputs.

The phenomena of overshoot or inverse response seen in the above example will not occur for a simple overdamped second-order transfer function, one containing two poles but no zero. These features arise from competing dynamic effects that operate on two different time scales (τ_1 and τ_2 in Example 6.2). For example, an inverse response can occur for a distillation column, when the steam pressure to the reboiler is suddenly changed. An increase in steam pressure ultimately will decrease the reboiler level (in the absence of level control) by boiling off more of the liquid. However, the initial effect usually is to increase the amount of frothing on the trays above the reboiler, causing a rapid spillover of liquid from these trays into the reboiler. This effect can result in an initial *increase* in reboiler liquid level, which is later overwhelmed by the increased vapor boil-up.

As a second physical example, tubular catalytic reactors with exothermic chemical reactions exhibit an inverse response in exit temperature when the feed temperature is increased. The initial effect is that the exit temperature decreases as a result of increased conversion in the entrance region of the bed, which momentarily depletes reactants at the exit end of the bed. Subsequently, higher reaction rates lead to a higher exit temperature, as would be expected. Conversely, if the feed

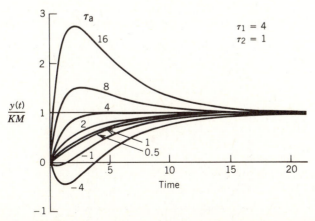

Figure 6.3. Step response of an overdamped second-order system (Eq. 6-14) with a single zero.

temperature is decreased, the inverse response initially results in a higher exit temperature.

Inverse response or overshoot can be expected whenever there are two physical effects that act on the process output variable in opposite ways and with different time scales. For the case of reboiler level mentioned above, the *fast* effect is to spill liquid off of the stages above the reboiler immediately as the vapor flow increases. The *slow* effect is to remove significant amounts of the liquid mixture from the reboiler through increased boiling. Hence, the process dynamic relation—reboiler level/reboiler steam pressure—might be represented approximately as an overdamped second-order transfer function with a right-half plane zero.

Figure 6.4 shows two simple first-order processes in a parallel configuration. In this case, the transfer function can be expressed as

$$\frac{Y(s)}{X(s)} = \frac{K_1}{\tau_1 s + 1} + \frac{K_2}{\tau_2 s + 1}$$

$$= \frac{K_1(\tau_2 s + 1) + K_2(\tau_1 s + 1)}{(\tau_1 s + 1)(\tau_2 s + 1)} \tag{6-16}$$

or, after rearranging the numerator into standard gain/time constant form, we have

$$\frac{Y(s)}{X(s)} = \frac{(K_1 + K_2)\left(\dfrac{K_1\tau_2 + K_2\tau_1}{K_1 + K_2}s + 1\right)}{(\tau_1 s + 1)(\tau_2 s + 1)} \tag{6-17}$$

Equation 6-17 can be put into the form of (6-14) if

$$K = K_1 + K_2 \tag{6-18}$$

and

$$\tau_a = \frac{K_1\tau_2 + K_2\tau_1}{K_1 + K_2} \tag{6-19}$$

$$= \frac{K_1\tau_2 + K_2\tau_1}{K} \tag{6-20}$$

The condition for an inverse response to exist is $\tau_a < 0$ or

$$\frac{K_1\tau_2 + K_2\tau_1}{K} < 0 \tag{6-21}$$

For either positive or negative K, (6-21) can be rearranged to the convenient form

$$-\frac{K_2}{K_1} > \frac{\tau_2}{\tau_1} \tag{6-22}$$

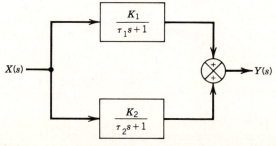

Figure 6.4. Two first-order process elements acting in parallel.

Note that Eq. 6-22 indicates that K_1 and K_2 have opposite signs since $\tau_1 > 0$ and $\tau_2 > 0$. It is left to the reader to show that $K > 0$ when $K_1 > 0$ and that $K < 0$ when $K_1 < 0$. In other words, the sign of the overall transfer function gain is the same as that of the slower process.

EXAMPLE 6.3

A process has the transfer function

$$G(s) = \frac{-0.5s + 1}{(4s + 1)(s + 1)}$$

What pair of first-order process elements acting in parallel is equivalent to this process transfer function?

Solution
Note that $K = 1$, $\tau_1 = 4$, $\tau_2 = 1$, $\tau_a = -0.5$ in this example. Gains K_1 and K_2 must satisfy Eqs. 6-18 and 6-20,

$$K_1 + K_2 = K = 1$$

$$K_1 + 4K_2 = -0.5$$

which, on solving, yield

$$K_1 = 1.5$$

$$K_2 = -0.5$$

By substituting these values into Eq. 6-16, the first-order elements with time constants of 1 and 4 act in opposition to each other but in parallel to produce an inverse response.

In the above example, the conditions on K_1 and K_2 essentially ensure that the fast mode (e^{-t}) will act (with a suitably negative gain) to make the output variable negative before the slow mode $(e^{-t/4})$ can respond significantly to a positive step input. A similar analysis for the overshoot case $(\tau_a > \tau_1)$ is left for the reader in Exercise 6.6.

EXAMPLE 6.4

Show that the step response of any process described by Eq. 6-14 will have a negative slope initially (at $t = 0$) if the product of gain and step change magnitude is positive $(KM > 0)$, τ_a is negative, and τ_1 and τ_2 are both positive.

Solution
For $X(s) = M/s$,

$$Y(s) = G(s)X(s) = \frac{KM(\tau_a s + 1)}{s(\tau_1 s + 1)(\tau_2 s + 1)} \tag{6-23}$$

Noting that differentiation in the time domain corresponds to multiplication by s in the Laplace domain, we let $z(t)$ denote dy/dt. Then

$$Z(s) = sY(s) = G(s)M = \frac{KM(\tau_a s + 1)}{(\tau_1 s + 1)(\tau_2 s + 1)} \tag{6-24}$$

Applying the Initial Value Theorem

$$
\begin{aligned}
z(0) = \left.\frac{dy}{dt}\right|_{t=0} &= \lim_{s\to\infty}\left[s\,\frac{KM(\tau_a s + 1)}{(\tau_1 s + 1)(\tau_2 s + 1)}\right] \\
&= \lim_{s\to\infty}\left[\frac{K(\tau_a + 1/s)}{(\tau_1 + 1/s)(\tau_2 + 1/s)}\right] = \frac{KM\tau_a}{\tau_1\tau_2}
\end{aligned}
\tag{6-25}
$$

which has the sign of τ_a if the other parameters (KM, τ_1, and τ_2) are positive. Note that if τ_a is zero, the initial slope is zero. Evaluation of Eq. 5-48 for $t = 0$ yields the same result.

6.2   TIME DELAYS

Whenever material or energy is physically moved in a process or plant there is a *time delay* associated with the movement. For example, if a fluid is transported through a pipe in plug flow, as shown in Fig. 6.5, then the transportation time between points 1 and 2 is given by

$$
\begin{aligned}
\theta &= \frac{\text{length of pipe}}{\text{fluid velocity}} \\
&= \frac{\text{volume of pipe}}{\text{volumetric flowrate}}
\end{aligned}
\tag{6-26}
$$

where length and volume both refer to the pipe segment between 1 and 2. The first relation in (6-26) indicates why a time delay sometimes is referred to as a *distance–velocity lag*. Other synonyms are *transportation lag* and *dead time*. If the plug flow assumption does not hold, for example, with laminar flow or for non-Newtonian liquids, approximation of the bulk transport dynamics using a time delay still may be useful, as discussed below.

If x is some fluid property at point 1, such as temperature or concentration, and y is the same property at point 2, then y and x are related by a simple time-delay θ

$$y(t) = \begin{cases} 0 & t < \theta \\ x(t - \theta) & t \geq \theta \end{cases} \tag{6-27}$$

Thus the output $y(t)$ is simply the same input function shifted back in time by the

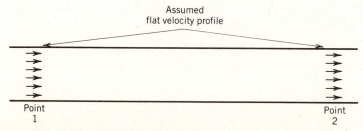

Figure 6.5.   Transportation of fluid in a pipe for turbulent flow.

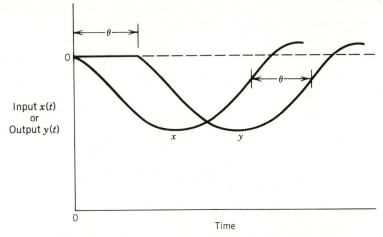

Figure 6.6. The effect of a pure time delay is a translation of the function in time.

amount of the delay. Figure 6.6 shows this simple *translation in time* for an arbitrary change in the input.

In Chapter 3 we showed (Eq. 3-104) that the Laplace transform of a function shifted in time by t_0 units is simply $e^{-t_0 s}$. Hence the transfer function of a time delay of θ units is given by

$$\frac{Y(s)}{X(s)} = G(s) = e^{-\theta s} \qquad (6\text{-}28)$$

Besides the physical movement of liquid and solid materials, there are other sources of time delays in process control problems. For example, the use of a chromatograph to measure concentration in liquid or gas stream samples taken from a process introduces a time delay, the analysis time. One distinct characteristic of the process control field, as compared to control of most mechanical and electrical systems, is the common occurrence of time delays.

Even when the plug flow assumption is not valid, transportation processes usually can be modeled *approximately* by a transfer function of the type given in Eq. 6-28. For liquid flow in a pipe, the plug flow assumption is most nearly satisfied when the axial velocity profile is flat, a condition that occurs for Newtonian fluids in turbulent flow. For non-Newtonian fluids and/or laminar flow, the fluid transport process still might be modeled by including a pure time delay based on the average fluid velocity. A more general approach is to model the transportation process as a first-order plus time-delay transfer function

$$G(s) = \frac{e^{-\theta_m s}}{\tau_m s + 1} \qquad (6\text{-}29)$$

where τ_m is a time constant associated with the degree of mixing in the pipe or channel. Both τ_m and θ_m may have to be determined from empirical relations or by experiment. Note that the process gain is unity for a material property such as composition.

EXAMPLE 6.5

For the pipe section illustrated in Fig. 6.5, find the transfer functions: **(a)** relating the mass flow rate of liquid at 2, w_2, to the mass flow rate of liquid at 1, w_1,

(b) relating the concentration of a chemical species at 2 to the concentration at 1. Assume that the liquid is incompressible.

Solution

(a) First we make an overall material balance on the pipe segment in question. Since there can be no accumulation (incompressible fluid),

$$\text{material in} = \text{material out} \Rightarrow w_1(t) = w_2(t) \tag{6-30}$$

Putting (6-30) in deviation form and taking Laplace transforms yields the transfer function,

$$\frac{W_2'(s)}{W_1'(s)} = 1 \tag{6-31}$$

(b) Observing a very small cell of material passing point 1 at time t, we note that it contains $mc_1(t)$ units of the chemical species of interest where m is the total mass of material in the cell. If, at time $t + \theta$, the cell passes point 2, it contains $mc_2(t + \theta)$ units of the species. If the material moves in plug flow, not mixing at all with adjacent material, then the amount of species in the cell is constant:

$$mc_2(t + \theta) = mc_1(t) \tag{6-32}$$

or

$$c_2(t + \theta) = c_1(t) \tag{6-33}$$

An equivalent way of writing (6-33) is

$$c_2(t) = c_1(t - \theta) \tag{6-34}$$

if the flow rate is constant. Putting (6-34) in deviation form and taking Laplace transforms yields

$$\frac{C_2'(s)}{C_1'(s)} = e^{-\theta s} \tag{6-35}$$

Equation 6-31 indicates that flow rate changes at 1 propagate instantaneously to any other point in the pipe when the fluid is incompressible. For compressible fluids such as gases, the simple expression of (6-31) will not be applicable, except as an approximation. A time delay is associated with fluid properties such as concentration or temperature, as indicated by Eq. 6-34. However, note that use of a fixed time delay implies constant flow rate.

One major advantage of the use of Laplace transform techniques for modeling processes is the relatively simple form of the transfer function describing time delays. The simple exponential form of Eq. 6-28 is somewhat deceptive, however, since it is a nonrational transfer function, one that analytically cannot be put in the form of a quotient (ratio) of two polynomials in s. Historically, there has been some incentive[1] to find rational approximations to $e^{-\theta s}$. The presence of a time

---

[1] When electronic analog computers were widely used to solve process simulation problems, the only practical way to simulate a pure time delay was to use circuits designed on the basis of rational polynomial approximations [1]. With digital computer simulations, however, the time delay can be obtained by simply shifting, in time, a computed number.

delay in a process means that we cannot factor the process transfer function in terms of simple poles and zeros only. As is demonstrated below (Example 6.6), analytical solutions for system responses where recycle of material or energy is involved requires rational approximations.

Polynomial Approximations to $e^{-\theta s}$

The simplest approach to approximate a time delay by rational functions is to choose the quotient of two polynomials specifically designed to match the terms of a truncated Taylor series expansion of $e^{-\theta s}$:

$$e^{-\theta s} = 1 - \theta s + \frac{\theta^2 s^2}{2!} - \frac{\theta^3 s^3}{3!} + \frac{\theta^4 s^4}{4!} - \frac{\theta^5 s^5}{5!} + \cdots \qquad (6\text{-}36)$$

Suppose only first-order numerator and denominator terms (the simplest pole–zero approximation) are used:

$$e^{-\theta s} \approx G_1(s) = \frac{1 - \dfrac{\theta}{2} s}{1 + \dfrac{\theta}{2} s} \qquad (6\text{-}37)$$

Performing the indicated long division in (6-37) gives

$$G_1(s) = 1 - \theta s + \frac{\theta^2 s^2}{2} - \frac{\theta^3 s^3}{4} + \cdots \qquad (6\text{-}38)$$

A comparison of Eqs. 6-36 and 6-38 indicates that $G_1(s)$ is correct through the first three terms. Because the error occurs only in terms involving the third power of s and higher, (6-37) should serve as a reasonable approximation of $e^{-\theta s}$ for small values of θs.[2] Note that in Eq. 6-37, a single, right-half plane zero has been used along with the pole at $-2/\theta$ to provide the approximation. Equation 6-37 is known as a 1/1 Padé approximation because it is first order in both numerator and denominator. There are higher-order Padé approximations, as well; for example, the 2/2 Padé approximation is

$$e^{-\theta s} \approx G_2(s) = \frac{1 - \dfrac{\theta s}{2} + \dfrac{\theta^2 s^2}{12}}{1 + \dfrac{\theta s}{2} + \dfrac{\theta^2 s^2}{12}} \qquad (6\text{-}39)$$

In this case, a *complex* pair of right-half plane zeros has been combined with a complex pair of poles to achieve a more accurate approximation (i.e., error is $O(s^5)$) than can be obtained with $G_1(s)$. Tables of Padé approximations up to third order in numerator and fourth order in denominator have been published by Truxal [2].

Figure 6.7*a* illustrates the response of the first- and second-order Padé approximations to a unit step input. The first-order approximation exhibits the same sort of discontinuous response discussed in Section 6.1 in connection with a first-order system with a right-half plane zero. (Why?) The second-order approximation is somewhat more accurate; the discontinuous response and the oscillatory behavior

---

[2]In Chapter 14 we will discuss what is meant by small values of θs.

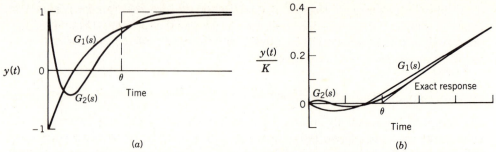

Figure 6.7.   (a). Step response of first- and second-order Padé approximations of a pure time delay ($G_1(s)$ and $G_2(s)$, respectively). (b). Step response of a first-order plus time-delay process ($\theta = 0.25\tau$) using a first- or second-order Padé approximation of $e^{-\theta s}$.

are features expected for a second-order system (both numerator and denominator) with a pair of complex poles. (Why?). Neither approximation can accurately represent the discontinuous change in the step input very well; however, if the response of a first-order system with time delay is considered,

$$G_p(s) = \frac{Ke^{-\theta s}}{\tau s + 1} \tag{6-40}$$

Fig. 6.7b shows that the approximations are satisfactory for a step response, especially if $\theta \ll \tau$, which is often the case.

EXAMPLE 6.6

The trickle-bed catalytic reactor shown in Fig. 6.8 utilizes product recycle to obtain satisfactory operating conditions for temperature and conversion. Use of a high recycle rate eliminates the need for mechanical agitation. Concentrations of the single reactant and the product are measured at a point in the recycle line where the product stream is removed. A first-order reaction is involved.

Under normal operating conditions, the following assumptions may be made:

i. The reactor operates isothermally so that the reaction rate k is constant.
ii. Inlet flow rate q and recycle flow rate αq are constant.

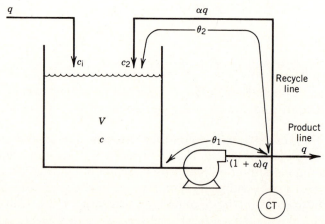

Figure 6.8.   Schematic diagram of a trickle-bed reactor with recycle line. CT, composition transmitter; θ_1, time delay associated with material flow from reactor outlet to transmitters; θ_2, time delay associated with material flow from transmitter to reactor inlet.

iii. No reaction occurs in the piping. The dynamics of the exit line may be simply approximated as constant time delays θ_1 and θ_2, as indicated in the figure.
iv. Because of the high recycle flow rate, mixing in the reactor is complete.

Do the following:
(a) Find the transfer function $C_1'(s)/C_i'(s)$.
(b) Using the following information, calculate $c_1'(t)$ for a step change in $c_i'(t) = 2000\ \mathrm{kg/m^3}$

Parameter Values

$$V = 5\ \mathrm{m^3} \qquad\qquad \alpha = 12$$
$$q = 0.05\ \mathrm{m^3/min} \qquad \theta_1 = 0.9\ \mathrm{min}$$
$$k = 0.04\ \mathrm{min^{-1}} \qquad \theta_2 = 1.1\ \mathrm{min}$$

Solution
(a) Equation 2-30 is applicable only to an isothermal stirred-tank reactor *without* recycle. Hence, we make a material balance around the reactor on component A to yield

$$V\frac{dc}{dt} = qc_i + \alpha qc_2 - (1 + \alpha)qc - Vkc \tag{6-41}$$

where the concentration of species A is denoted by c without subscript A for convenience. Equation 6-41 is linear with constant coefficients. Subtracting the steady-state equation and substituting deviation variables yields

$$V\frac{dc'}{dt} = qc_i' + \alpha qc_2' - (1 + \alpha)qc' - Vkc' \tag{6-42}$$

Additional relations are needed for $c_2'(t)$ and $c_1'(t)$. These can be obtained from assumption iii which states that the recycle lines act as simple time-delay elements:

$$c_1'(t) = c'(t - \theta_1) \tag{6-43}$$

$$c_2'(t) = c_1'(t - \theta_2) \tag{6-44}$$

Equations 6-42 through 6-44 provide the process model for the isothermal reactor with recycle. Taking the Laplace transform of each equation yields

$$sVC'(s) = qC_i'(s) + \alpha qC_2'(s) - (1 + \alpha)qC'(s) - VkC'(s) \tag{6-45}$$

$$C_1'(s) = e^{-\theta_1 s}C'(s) \tag{6-46}$$

$$C_2'(s) = e^{-\theta_2 s}C_1'(s)$$

$$\qquad = e^{-(\theta_1 + \theta_2)s}C'(s)$$

$$\qquad = e^{-\theta_3 s}C'(s) \tag{6-47}$$

Substitute (6-47) into (6-45) and solve for the output $C'(s)$:

$$C'(s) = \frac{q}{sV - \alpha qe^{-\theta_3 s} + (1 + \alpha)q + Vk}\, C_i'(s) \tag{6-48}$$

Equation 6-48 can be rearranged to the following form

$$C'(s) = \frac{K}{\tau s + 1 + \alpha K(1 - e^{-\theta_3 s})}\, C_i'(s) \tag{6-49}$$

where
$$K = q/(q + Vk) \tag{6-50}$$

$$\tau = V/(q + Vk) \tag{6-51}$$

Note that, in the limit as $\theta_3 \to 0$, $e^{-\theta_3 s} \to 1$ and

$$C'(s) = \frac{K}{\tau s + 1} C_i'(s) \tag{6-52}$$

hence K and τ can be interpreted as the process gain and time constant, respectively, of a recycle reactor with no time delay in the recycle line or, equivalently, as a stirred isothermal reactor with no recycle.

The desired transfer function $C_1'(s)/C_i'(s)$ is found by combining Eqs. 6-49 and 6-46 to obtain

$$\frac{C_1'(s)}{C_i'(s)} = \frac{K e^{-\theta_1 s}}{\tau s + 1 + \alpha K(1 - e^{-\theta_3 s})} \tag{6-53}$$

(b) To find $c_1'(t)$ when $c_i'(t) = 2000 \ kg/m^3$, we multiply (6-53) by $2000/s$

$$C_1'(s) = \frac{2000 K e^{-\theta_1 s}}{s[\tau s + 1 + \alpha K(1 - e^{-\theta_3 s})]} \tag{6-54}$$

and invert. From inspection of (6-54) it is clear that the numerator time delay causes no problems; however, there is no transform in Table 3.1 that contains a time delay term in the denominator. To obtain an analytical solution the denominator delay element must be eliminated by some rational approximation, for example the 1/1 Padé approximation

$$e^{-\theta_3 s} \approx \frac{1 - \dfrac{\theta_3}{2} s}{1 + \dfrac{\theta_3}{2} s} \tag{6-55}$$

Substituting (6-55) and rearranging yields

$$C_1'(s) \approx \frac{2000 K \left(\dfrac{\theta_3}{2} s + 1 \right) e^{-\theta_1 s}}{s \left[\tau \dfrac{\theta_3}{2} s^2 + \left(\tau + \dfrac{\theta_3}{2} + \alpha K \theta_3 \right) s + 1 \right]} \tag{6-56}$$

This expression can be written in the form

$$C_1'(s) = \frac{2000 K (\tau_a s + 1) e^{-\theta_1 s}}{s(\tau_1 s + 1)(\tau_2 s + 1)} \tag{6-57}$$

where $\tau_a = \theta_3/2$ and τ_1 and τ_2 are obtained by factoring the expression in brackets. For $\alpha K \theta_3 > 0$, τ_1 and τ_2 will be real and distinct.

The numerical parameters in (6-57) are:

$$K = \frac{q}{q + Vk} = \frac{0.05}{0.05 + (5)(0.04)} = 0.2$$

$$\tau = \frac{V}{q + Vk} = 20 \ min$$

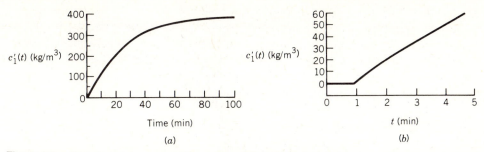

Figure 6.9. Recycle reactor composition measured at analyzer: (a) complete response; (b) detailed view of short-term response.

Substituting, we obtain

$$C_1'(s) = \frac{400(s + 1)e^{-0.9s}}{s[20s^2 + (20 + 1 + (24)(0.2)(1))s + 1]}$$

$$= \frac{400(s + 1)e^{-0.9s}}{s(25s + 1)(0.8s + 1)} \tag{6-58}$$

Inverting gives

$$c_1'(t) = 400(1 - 0.99174e^{-(t - 0.9)/25} - 0.00826e^{-(t - 0.9)/0.8})S(t - 0.9) \tag{6-59}$$

which is plotted in Fig. 6.9. A numerical solution of Eqs. 6-42 through 6-44 that uses no approximation for the total recycle delay is indistinguishable, within the resolution of these plots. Note that in obtaining (6-59), we did not have to approximate the numerator delay. It can be dealt with exactly and appears finally as a time shift of 0.9 min, that is, as $(t - 0.9)$ terms in (6-59).

6.3   APPROXIMATION OF HIGHER-ORDER SYSTEMS

As discussed earlier, a time delay can be used to approximate high-order model dynamics. Consider the step response of a hypothetical nth-order system with n equal time constants

$$G_n(s) = \frac{K}{\left(\dfrac{\tau}{n} s + 1\right)^n} \tag{6-60}$$

that is plotted in Fig. 6.10. This transfer function could represent a series of n stages, each described by a first-order transfer function, with a total residence time τ equally divided among the stages. As the number of stages becomes larger, the system response to a step of magnitude M,

$$y(t) = KM\left[1 - e^{-nt/\tau} \sum_{i=0}^{n-1} \frac{(nt/\tau)^i}{i!}\right] \tag{6-61}$$

is well approximated by a time delay of magnitude τ, as shown in Fig. 6.10. In fact, as $n \to \infty$, $G_n(s)$ approaches $e^{-\tau s}$ [3].

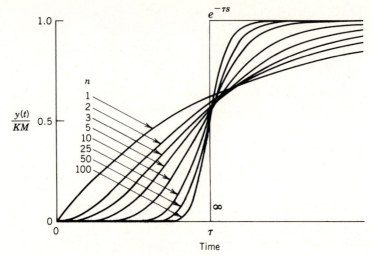

Figure 6.10. A pure time delay element can approximate a large number of first-order systems in series.

Many processes that do not contain an explicit time delay consist of a large number of process units connected in series, for example, the trays in a distillation or absorption column. A change in the distillation column reflux rate affects conditions in the reboiler only after some period of time because this flow change must cascade through a large number of trays before reaching the reboiler. Both staged processes and distributed parameter processes (to be discussed in Section 6.5) exhibit this type of apparent time delay.

In such cases a time-delay term can be used in a process model as an approximation for a number of small time constants. For example, if an nth-order process composed of first-order processes in series ($\tau_1, \tau_2, \ldots, \tau_n$) is dominated[3] by two of these processes (τ_1 and τ_2), then an approximate transfer function $\hat{G}(s)$ for the system is

$$\hat{G}(s) = \frac{Ke^{-\theta s}}{(\tau_1 s + 1)(\tau_2 s + 1)} \tag{6-62}$$

$$\text{where } \theta = \sum_{i=3}^{n} \tau_i \tag{6-63}$$

Apparent time delays result when actual transportation lags (delays) or measurement delays are present *or* when high-order systems are approximated by means of lower-order transfer functions. Friedly [3, p. 187] has described a method for fitting simple time-delay transfer function models to high-order models based on a Taylor series expansion.

Although the mathematical analysis of uncontrolled processes containing a time delay is quite simple, a time delay within a recycle loop or a feedback control loop leads to considerable mathematical difficulty. In general, the presence of a large time delay in a controlled system reduces the effectiveness of the controller

---

[3]As used here, "dominated" implies that the overall process transient response is determined primarily by a few large time constants, that is, response modes that are much slower than the others.

(see Chapter 18). Hence, in a well-designed control system time delays should be minimized as much as possible, for example, by selecting appropriate locations for sensors and actuators.

6.4 INTERACTING AND NONINTERACTING PROCESSES

Most of the systems considered so far have been simple ones whose elements could be isolated and treated individually. Unfortunately, for many common processes this cannot be done. Typically, processes with variables that interact with each other or that contain internal feedback of material or energy (recycle streams) will exhibit so-called interacting behavior. The result of interacting behavior usually is a more complicated process transfer function.

An example of a higher-order system that does *not* exhibit interaction was discussed in Example 4.4. There two storage tanks were connected in series in such a way that liquid level in the second tank did not influence the level in the first tank (Fig. 4.3). Transfer functions for changes in tank levels and flows were derived as follows:

$$\frac{H_1'(s)}{Q_i'(s)} = \frac{K_1}{\tau_1 s + 1} \tag{4-43}$$

$$\frac{Q_1'(s)}{H_1'(s)} = \frac{1}{K_1} \tag{4-44}$$

$$\frac{H_2'(s)}{Q_1'(s)} = \frac{K_2}{\tau_2 s + 1} \tag{4-45}$$

$$\frac{Q_2'(s)}{H_2'(s)} = \frac{1}{K_2} \tag{4-46}$$

where $K_1 = R_1$, $K_2 = R_2$, $\tau_1 = A_1 R_1$, $\tau_2 = A_2 R_2$. Each tank level has first-order dynamics with respect to its inlet flow rate. Tank 2 level is related to Q_i by a second-order transfer function that can be obtained by simple multiplication

$$\frac{H_2'(s)}{Q_i'(s)} = \frac{H_2'(s)}{Q_1'(s)} \frac{Q_1'(s)}{H_1'(s)} \frac{H_1'(s)}{Q_i'(s)}$$

$$= \frac{K_2}{(\tau_1 s + 1)(\tau_2 s + 1)} \tag{6-64}$$

A simple generalization of the dynamic expression in Eq. 6-64 is applicable to n tanks in series shown in Fig. 6.11:

$$\frac{H_n'(s)}{Q_i'(s)} = \frac{K_n}{\prod_{i=1}^{n} (\tau_i s + 1)} \tag{6-65}$$

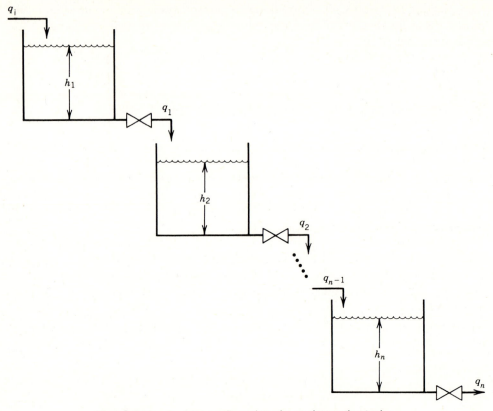

Figure 6.11.   A series configuration of n noninteracting tanks.

and

$$\frac{Q_n'(s)}{Q_i'(s)} = \frac{1}{\prod_{i=1}^{n} (\tau_i s + 1)} \tag{6-66}$$

For the special case where all τ_i are equal, the cascade of tanks represented by Eq. 6-66 reacts to a step change in input flow as shown in Fig. 6.10.

Next consider an example of an interacting system that superficially resembles the two-tank process discussed in Chapter 4. The system shown in Fig. 6.12 is called an *interacting system* because Tank 1 level depends on Tank 2 level (and vice versa) as a result of the interconnecting stream with flow rate q_1. Therefore, the equation for flow from Tank 1 to Tank 2 must be written to reflect that physical feature:

$$q_1 = \frac{1}{R_1} (h_1 - h_2) \tag{6-67}$$

For the Tank 1 level transfer function, a much more complicated expression than (4-43) results:

$$\frac{H_1'(s)}{Q_i'(s)} = \frac{(R_1 + R_2)\left(\dfrac{R_1 R_2 A_2}{R_1 + R_2} s + 1\right)}{R_1 R_2 A_1 A_2 s^2 + (R_2 A_2 + R_1 A_1 + R_2 A_1) s + 1} \tag{6-68}$$

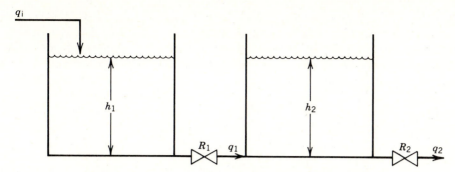

Figure 6.12.   Two tanks in series whose liquid levels interact.

which is of the form

$$\frac{H_1'(s)}{Q_i'(s)} = \frac{K_1'(\tau_a s + 1)}{\tau^2 s^2 + 2\zeta \tau s + 1} \tag{6-69}$$

In Exercise 6.15, the reader can show that $\zeta > 1$ by analyzing the denominator of (6-68); hence, the transfer function is overdamped, second order, and has a negative zero (at $-1/\tau_a$ where $\tau_a = R_1 R_2 A_2/(R_1 + R_2)$). The transfer function relating the two tank levels is

$$\frac{H_2'(s)}{H_1'(s)} = \frac{\dfrac{R_2}{R_1 + R_2}}{\dfrac{R_1 R_2 A_2}{R_1 + R_2} s + 1} \tag{6-70}$$

and is of the form $K_2'/(\tau_a s + 1)$. Consequently, the overall transfer function between H_2' and Q_i' is

$$\frac{H_2'(s)}{Q_i'(s)} = \frac{R_2}{\tau^2 s^2 + 2\zeta \tau s + 1} \tag{6-71}$$

The above analysis of the interacting two-tank system is more complicated than that for the noninteracting system of Example 4.4. The denominator polynomial can no longer be factored into two separate first-order factors, each associated with a single tank. Also, the numerator of the first tank transfer function in (6-69) contains a zero that will modify dynamic response along the lines suggested in Section 6.1. For processes containing multiple interacting units, such as staged systems (distillation, extraction, absorption columns, etc.) with countercurrent internal flows, or for distributed parameter systems (described by partial differential equations) that have been converted to ordinary differential equation form, the dynamic characteristics of any stage will depend on the dynamics of all other stages. These cases, which are treated briefly in the next two sections, give a glimpse of how model complexity and large numbers of process variables often hinder the derivation of useful results via theoretical analysis.

6.5 STAGED SYSTEMS

In Chapter 2, a model for a three-stage absorber is derived; it consists of three differential equations (Eqs. 2-38a through 2-38c) involving the three liquid-stage compositions. At that point we noted that the equations are coupled, since each output (x_1, x_2, x_3) appears in more than one equation. However, Eqs. 2-38a through 2-38c are linear and can be put into deviation variable form. The derivation of transfer functions for the gas concentration in each stage resulting from changes in the gas feed concentration to the column (y_f) is straightforward and is left to the reader as an exercise:

$$G_1(s) = \frac{Y_1'(s)}{Y_f'(s)} = \frac{s + s^2 + s^3}{1 + s + s^2 + s^3}$$

$$\times \left\{ \frac{\dfrac{\tau^2}{1 + s + s^2} s^2 + \dfrac{2(1 + s)\tau}{1 + s + s^2} s + 1}{\left(\dfrac{\tau}{1 + s} s + 1\right)\left[\dfrac{\tau^2}{1 + s^2} s^2 + \dfrac{2(1 + s)\tau}{1 + s^2} s + 1\right]} \right\} \tag{6-72a}$$

$$G_2(s) = \frac{Y_2'(s)}{Y_f'(s)} = \frac{s^2}{1 + s^2}\left[\frac{1}{\dfrac{\tau^2}{1 + s^2} s^2 + \dfrac{2(1 + s)\tau}{1 + s^2} s + 1} \right] \tag{6-72b}$$

$$G_3(s) = \frac{Y_3'(s)}{Y_f'(s)} = \frac{s^3}{1 + s + s^2 + s^3}$$

$$\times \left\{ \frac{1}{\left(\dfrac{\tau}{1 + s} s + 1\right)\left[\dfrac{\tau^2}{1 + s^2} s^2 + \dfrac{2(1 + s)\tau}{1 + s^2} s + 1\right]} \right\} \tag{6-72c}$$

In these transfer functions s is the stripping factor and τ is the stage residence time (Section 2.5). The interactions that result from the counterflowing liquid and gas streams give rise to rather complicated, high-order transfer functions (n stages would yield an nth-order model):

1. For $G_1(s)$, the three poles are all real, and two real zeros are present that influence the dynamic response (as explained below).
2. For $G_2(s)$, a factor in the denominator $[(\tau/(1 + s)) s + 1]$ exactly cancels a factor in the numerator, leaving a relatively simple second-order transfer function after this pole-zero cancellation.
3. For $G_3(s)$ the transfer function is third order with no zeros.

A numerical example of an absorber model has been discussed by Wong and Luus [4]. Using their parameters for the three-stage process ($H = 75.72$ lb, $L = 40.8$ lb/min, $G = 66.7$ lb/min, $a = 0.72$, $b = 0$ yields the following transfer functions:

$$\frac{Y_1'(s)}{Y_f'(s)} = \frac{0.807\,(1.69s + 1)(0.57s + 1)}{(2.89s + 1)(0.85s + 1)(0.5s + 1)} \tag{6-73a}$$

$$\frac{Y_2'(s)}{Y_f'(s)} = \frac{0.581}{(2.89s + 1)\,(0.5s + 1)} \tag{6-73b}$$

$$\frac{Y_3'(s)}{Y_f'(s)} = \frac{0.314}{(2.89s + 1)(0.85s + 1)(0.5s + 1)} \tag{6-73c}$$

Figure 6.13 shows the responses of this system to a 10% step increase in gas phase concentration of the absorbed component (for initial operating conditions:

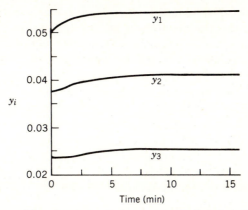

Figure 6.13.   Response of the absorber gas phase concentrations to a 10% step increase in y_f.

$x_f = 0.01$, $y_f = 0.06$). In plotting Fig. 6.14 the three responses have been put in deviation form and normalized (divided by $\Delta y_i = y_i(\infty) - y_i(0)$) so that the dynamic response characteristics can be seen more clearly. The transfer function for the first stage, Eq. 6.73a, yields a response that is nearly first order. The transfer function of (6.73a) is of "second-over-third order" type, and the numerator dynamics partially compensate (cancel) the third-order effects from the denominator. Exact cancellation of the numerator/denominator factors, as with the second-stage transfer function, would yield a simple first-order transfer function for $Y_1'(s)$.

6.6   TRANSFER FUNCTION MODELS FOR DISTRIBUTED-PARAMETER SYSTEMS

In Chapter 2 we analyzed a simple distributed system, a double-pipe heat exchanger. The process model selected was a partial differential equation to describe the spatial variations of the wall and fluid temperatures. Section 2.5 discusses how the original partial differential equations can be converted to a set of ordinary differential equations describing the wall and fluid temperatures at particular locations, the nodes. Once an ordinary differential equation model is available, it can be linearized

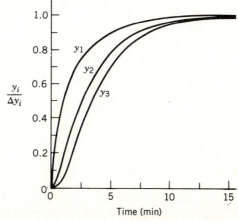

Figure 6.14.   Normalized step responses of the gas phase concentrations.

and, in principle, put into transfer function form. However, the algebraic manipulations required for this latter step may be quite difficult.

Return to the set of four equations, (2-48a)–(2-48d) that describe two fluid temperatures and two wall temperatures of the double-pipe heat exchanger. Because of the interactions that are inherent in these equations—in particular, the effects of energy and fluid flows between nodes—the expressions for the important transfer functions, $T_{L2}(s)/T_F(s)$ and $T_{L2}(s)/T_s(s)$, are quite involved. Coughanowr and Koppel [1] have given complete theoretical derivations of these transfer functions. Here we develop second-order transfer functions for the two-lump approximate models. But first we derive an important analytical expression for the denominator factors (poles) for this case. Let $\Delta z/v = \tau_{FL}$ denote the characteristic residence time for plug flow of liquid through a single lump. To simplify the algebra, let

$$\left.\begin{array}{c} \dfrac{1}{\tau_1} = \dfrac{1}{\tau_{FL}} + \dfrac{1}{\tau_{HL}} \\[3mm] \dfrac{1}{\tau_2} = \dfrac{1}{\tau_{sw}} + \dfrac{1}{\tau_{wL}} \\[3mm] K_1 = \dfrac{\tau_{HL}}{\tau_{HL} + \tau_{FL}} \\[3mm] K_2 = \dfrac{\tau_{wL}}{\tau_{sw} + \tau_{wL}} \end{array}\right\} \qquad (6\text{-}74)$$

All transfer functions for this two-lump process approximation have the same denominator

$$D(s) = \left(\frac{\tau_1\tau_2}{K_1 + K_2 - K_1K_2} s^2 + \frac{\tau_1 + \tau_2}{K_1 + K_2 - K_1K_2} s + 1\right)^2 \qquad (6\text{-}75)$$

We note that the form of $D(s)$ after factoring is $(\tau_1's + 1)^2(\tau_2's + 1)^2$ with τ_1' and τ_2' both positive since $K_1 + K_2 - K_1K_2 > 0$ for any choice of parameters. Hence responses of the liquid or wall temperatures at the node points *appear to be* fourth order (as from two overdamped systems in series). However, before drawing any conclusions, we must evaluate the effect of the numerator of each transfer function.

Consider the parameters reported by Cohen and Johnson [5] for a pilot-scale exchanger with steam condensing in the outer tube and liquid flowing in the inner tube:

$$\tau_{FL} = 3.69 \text{ s}$$
$$\tau_{wL} = 2.65 \text{ s}$$
$$\tau_{sw} = 1.05 \text{ s}$$
$$L/v = 3 \text{ s}$$

where L is the length of the concentric heat exchanger tubes. The second-order approximate transfer functions are [5]

$$\frac{T_{L2}'(s)}{T_F'(s)} = \frac{0.60 \, (0.76s + 1)^2}{(1.32s + 1)^2(0.66s + 1)^2} \qquad (6\text{-}76a)$$

$$\frac{T_{L2}'(s)}{T_s'(s)} = \frac{0.40 \, (0.89s + 1)(0.55s + 1)}{(1.32s + 1)^2(0.66s + 1)^2} \qquad (6\text{-}76b)$$

In this case, substitution of numerical values considerably simplifies calculation of the numerator and denominator factors of s. We note that there are two system zeros, which could be expected from the interactive nature of this approximate lumped model. If there had been a counterflowing hot fluid in the outer tube instead of condensing steam, or if the effects of an outer wall and/or of axial conduction had been included, the level of interaction would have been even higher. In addition, the algebra associated with derivation of the approximate transfer functions would have been much more complicated.

Figure 6.15 illustrates the system response for a 10°C step increase in steam temperature. In this case the apparent *second-order* response dynamics ("second-over-fourth-order") are obvious. The effect of changing the number of lumps in the approximate model is also shown. For this system, $N = 2, 4, 10$ corresponds to systems of 4th, 8th, 20th order, respectively. Note that the gain in this lumped model is also a function of the number of lumps chosen.

An empirical time-delay term is characteristic of distributed parameter models usually because of the time required for material and/or energy transport. For the double-tube heat exchanger, the time delay effect is much more apparent when Eqs. 2-48a through 2-48d are solved for a step change in fluid entrance temperature, particularly as N (the number of lumps in the approximate model) is increased (Fig. 6.16). In this case a value of N should be chosen to give a good approximation to the actual fluid mixing conditions in the inner tube. The limiting cases are $N = 1$ for perfectly mixed and $N = \infty$ for plug flow. Since finite difference methods often are inefficient for situations involving large changes in a dependent variable over a short distance, as occurs for a step change in the flowing fluid, more efficient techniques such as orthogonal collocation [6] may be necessary.

In some distributed-parameter systems, such as packed-bed reactors, the time associated with transport of energy through a catalyst bed can be two to three orders of magnitude longer than the time required for reactants and products to flow through the reactor. An overdamped response, sometimes with a right-half plane zero, often is observed experimentally for distributed reaction processes. Hence, a general empirical model of the form

$$G(s) = \frac{K(\tau_a s + 1) e^{-\theta s}}{(\tau_1 s + 1)(\tau_2 s + 1)} \tag{6-77}$$

may be appropriate for these systems.

Although lumped models can be derived from fundamental distributed process models, such models are difficult to obtain and invariably contain parameters that

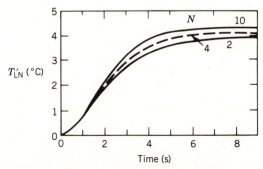

Figure 6.15. Response of the heat exchanger exit temperature to a 10° step increase in steam temperature (no. of nodes, $N = 2, 4, 10$).

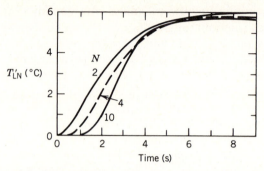

Figure 6.16. Response of the heat exchanger exit temperature to a 10° step increase in feed temperature (no. of nodes, N = 2, 4, 10).

must be found by experiment. Hence, they are most useful in trying to understand the qualitative behavior of the process or in research projects. Equation 6-77 with parameters obtained by experiment (see Chapter 7) can be used to good advantage in industrial applications.

6.7 MULTIPLE-INPUT, MULTIPLE-OUTPUT (MIMO) PROCESSES

Most industrial process control applications involve a number of input (manipulated) and output (controlled) variables. These applications often are referred to as MIMO systems to distinguish them from the simpler single-input/single-output (SISO) systems that have been emphasized so far. Modeling MIMO processes is no different conceptually than modeling SISO processes. A degree of freedom analysis (Chapter 2) can be employed to select the model structure, namely the manipulated, controlled, and disturbance variables. For example, consider the system illustrated in Fig. 6.17. Here the level h in the stirred tank and the temperature T are to be controlled by adjusting the flow rates of the hot and cold streams w_h and w_c, respectively. The temperatures of the inlet streams T_h and T_c represent potential disturbance or load variables. Note that the outlet flow rate w is maintained constant and the liquid properties are assumed to be constant in the following derivation.

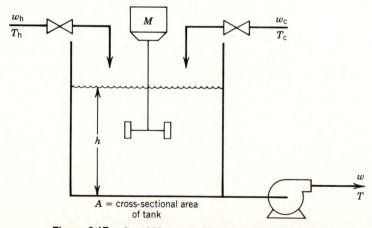

Figure 6.17. A multi-input, multi-output mixing process.

Writing the energy and material balances for this process and noting that the volume in the tank can change, gives

$$\rho C \frac{d[V(T - T_{\text{ref}})]}{dt} = w_h C(T_h - T_{\text{ref}}) + w_c C(T_c - T_{\text{ref}})$$

$$- wC(T - T_{\text{ref}}) \tag{6-78}$$

$$\rho \frac{dV}{dt} = w_h + w_c - w \tag{6-79}$$

The energy balance has been written using a thermodynamic reference temperature T_{ref} (see Section 2.3). Expanding the derivative of the product $V(T - T_{\text{ref}})$ gives,

$$\frac{d[V(T - T_{\text{ref}})]}{dt} = (T - T_{\text{ref}}) \frac{dV}{dt} + V \frac{dT}{dt} \tag{6-80}$$

Equation 6-80 can be substituted for the left side of Eq. 6-78. Following substitution of the material balance (6-79), a simpler set of equations results

$$\frac{dT}{dt} = \frac{1}{\rho A h} [w_h T_h + w_c T_c - (w_h + w_c)T] \tag{6-81}$$

$$\frac{dh}{dt} = \frac{1}{\rho A} (w_h + w_c - w) \tag{6-82}$$

After linearizing (6-81) and (6-82), putting them in deviation form, and taking Laplace transforms, we obtain a set of eight transfer functions that describe the effect of the input variables (w_h', w_c', T_h', T_c') on each output variable, T' and h':

$$\frac{T'(s)}{W_h'(s)} = \frac{(\overline{T}_h - \overline{T})/\overline{w}}{\tau s + 1} \tag{6-83}$$

$$\frac{T'(s)}{W_c'(s)} = \frac{(\overline{T}_c - \overline{T})/\overline{w}}{\tau s + 1} \tag{6-84}$$

$$\frac{T'(s)}{T_h'(s)} = \frac{\overline{w}_h/\overline{w}}{\tau s + 1} \tag{6-85}$$

$$\frac{T'(s)}{T_c'(s)} = \frac{\overline{w}_c/\overline{w}}{\tau s + 1} \tag{6-86}$$

$$\frac{H'(s)}{W_h'(s)} = \frac{1/A\rho}{s} \tag{6-87}$$

$$\frac{H'(s)}{W_c'(s)} = \frac{1/A\rho}{s} \tag{6-88}$$

$$\frac{H'(s)}{T_h'(s)} = 0 \tag{6-89}$$

$$\frac{H'(s)}{T_c'(s)} = 0 \tag{6-90}$$

where $\tau = \dfrac{\rho A \overline{h}}{\overline{w}}$ and the overbar denotes a nominal steady-state value.

Equations 6-83 through 6-86 indicate that all inputs affect the tank temperature through a first-order dynamic expression whose time constant is the nominal residence time of the tank τ. Equations 6-87 and 6-88 show that the inlet flow rates affect level through integrating transfer functions due to the pump on the exit line. Finally, it is clear from Eqs. 6-89 and 6-90 as well as physical intuition that inlet temperature changes have no effect on tank level.

A very compact way of expressing Eqs. 6-83 through 6-90 is by means of a transfer function matrix:

$$
\begin{bmatrix} T'(s) \\ H'(s) \end{bmatrix} = \begin{bmatrix} \dfrac{(\overline{T}_h - \overline{T})/\overline{w}}{\tau s + 1} & \dfrac{(\overline{T}_c - \overline{T})/\overline{w}}{\tau s + 1} & \dfrac{\overline{w}_h/\overline{w}}{\tau s + 1} & \dfrac{\overline{w}_c/\overline{w}}{\tau s + 1} \\ \dfrac{1/A\rho}{s} & \dfrac{1/A\rho}{s} & 0 & 0 \end{bmatrix} \begin{bmatrix} W'_h(s) \\ W'_c(s) \\ T'_h(s) \\ T'_c(s) \end{bmatrix}
$$

(6-91)

Equivalently, two transfer function matrices can be used to separate the variables selected for manipulation, w_h and w_c, from the disturbance variables, T_h and T_c:

$$
\begin{bmatrix} T'(s) \\ H'(s) \end{bmatrix} = \begin{bmatrix} \dfrac{(\overline{T}_h - \overline{T})/\overline{w}}{\tau s + 1} & \dfrac{(\overline{T}_c - \overline{T})/\overline{w}}{\tau s + 1} \\ \dfrac{1/A\rho}{s} & \dfrac{1/A\rho}{s} \end{bmatrix} \begin{bmatrix} W'_h(s) \\ W'_c(s) \end{bmatrix}
$$

$$
+ \begin{bmatrix} \dfrac{\overline{w}_h/\overline{w}}{\tau s + 1} & \dfrac{\overline{w}_c/\overline{w}}{\tau s + 1} \\ 0 & 0 \end{bmatrix} \begin{bmatrix} T'_h(s) \\ T'_c(s) \end{bmatrix}
$$

(6-92)

The block diagram in Figure 6.18 illustrates how the four input variables affect the two output variables.

Two points are worth mentioning in conclusion:

1. A transfer function matrix or, equivalently, the set of individual transfer functions, allows the control system designer to construct a more complex control system that deals with the interactions between inputs and outputs. For example, in the variable-level stirred-tank system the designer can devise control strategies that minimize or eliminate the effect of flow changes on temperature and level. This type of multivariable control system is considered in Chapter 19.
2. The development of physically based MIMO models can be quite difficult. Hence the caveat given at the end of Section 6.6 applies here as well: Empirical models rather than theoretical models often must be used for complicated systems.

SUMMARY

In this chapter we have considered the dynamics of processes that cannot be described by simple transfer functions. Models for these processes include numerator dynamics such as time delays or process zeros. Often an observed time delay can be a manifestation of higher order dynamics. Using such a term provides a way to simplify some high-order models. We have seen that staged and distributed-

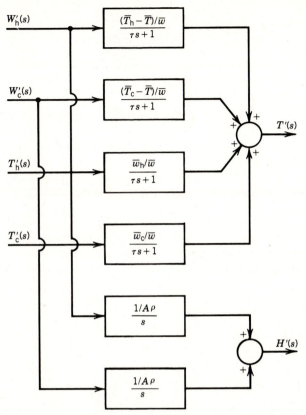

Figure 6.18. Block diagram of the MIMO mixing system with variable level.

parameter processes lead to high-order, complicated transfer functions when or-dinary, lumped models are used to describe them. Processes with multiple inputs and outputs also lead to complicated transfer function models except in the simplest cases. Control engineers generally do not like to work with such models, nor do they need them to design efficient control systems. The key idea is to capture the important steady-state and dynamic characteristics in the simplest possible model. Hence, we often use a two-part modeling approach, first analyzing the dynamics of a complicated process to determine the *form* of its model. Then the methods of Chapters 7 or 14 can be applied to fit empirical parameters to that form using experimental input–output data.

REFERENCES

1. Coughanowr, D.R., and L.B. Koppel, *Process Systems Analysis and Control,* McGraw-Hill, New York, 1965.
2. Truxal, J.G., *Automatic Feedback Control System Synthesis,* McGraw-Hill, New York, 1955.
3. Friedly, J.C., *Dynamic Behavior of Processes,* Prentice-Hall, Englewood Cliffs, NJ, 1972.
4. Wong, K.T., and R. Luus, Model reduction of high-order multistage systems by the method of orthogonal collocation, *Can. J. Chem. Eng.* **58,** 382 (1980).
5. Cohen, W.C., and E.F. Johnson, Dynamic characteristics of double pipe heat exchangers, *IEC* **48,** 1031 (1956).
6. Villadsen, J., and M.L. Michelsen, *Solution of Differential Equation Models by Polynomial Approximation,* Prentice-Hall, Englewood Cliffs, NJ, 1977.

EXERCISES

6.1. A process was modeled yielding the transfer function

$$G(s) = \frac{30}{24s^3 + 20s^2 + 10s + 2}$$

which appears not to be in standard form.

(a) Plot the poles of this transfer function in the complex plane. What is the steady-state gain?

(b) Will the process output be bounded for all time, assuming that the input is bounded? Justify your answer.

6.2. The locations of the poles of a transfer function $G(s)$ yield general information about the dynamics of that process *independent* of the form of the input(s). By extension, we know that all the modes of the output response arise from the poles of the output, $Y(s) = G(s)X(s)$, that is, the exact form of the output response (the solution, $y(t)$) is known once the input(s) are known. Sketch the location of the poles in the complex plane for the output variable y of the following differential equations. Without solving explicitly, state the type of response you expect as $t \to \infty$. Why do you expect this response?

(a) $\dfrac{dy}{dt} + 120y = 6e^{-120t}, \quad y(0) = 0$

(b) $\dfrac{d^3y}{dt^3} + 3\dfrac{d^2y}{dt^2} + 3\dfrac{dy}{dt} + y = 1, \quad y(0) = 0, \quad \left.\dfrac{dy}{dt}\right|_{t=0} = 0, \quad \left.\dfrac{d^2y}{dt^2}\right|_{t=0} = 0$

(c) $\dfrac{d^2y}{dt^2} + 4\dfrac{dy}{dt} + 3y = 16, \quad y(0) = 0, \quad \left.\dfrac{dy}{dt}\right|_{t=0} = 0$

6.3. For a lead-lag unit,

$$\frac{Y(s)}{X(s)} = \frac{K(\tau_a s + 1)}{\tau_1 s + 1}$$

show that for a step response of magnitude M:

(a) The value of y at $t = 0^+$ is given by $y(0^+) = \dfrac{\tau_a}{\tau_1} KM$.

(b) Overshoot occurs only for $\tau_a > \tau_1$, in which case dy/dt is always negative.

(c) Inverse response occurs only for $\tau_a < 0$.

6.4. For a second-order system that has a single zero,

$$\frac{Y(s)}{X(s)} = \frac{K(\tau_a s + 1)}{(\tau_1 s + 1)(\tau_2 s + 1)}$$

with $\tau_1 > \tau_2$, show that:

(a) $y(t)$ can exhibit an extremum (maximum or minimum value) in the step response only if

$$\frac{1 - \tau_a/\tau_2}{1 - \tau_a/\tau_1} > 1$$

(b) Overshoot occurs only for $\tau_a > \tau_1$.

(c) Inverse response occurs only for $\tau_a < 0$.

(d) If an extremum in y exists, the time at which it occurs can be found analytically. (What is it?)

(e) When $\tau_a = 0$, the values of $y(t)$ and dy/dt are 0 at $t = 0$.

(f) When $\tau_a \neq 0$, $y(0) = 0$ but $\left.\dfrac{dy}{dt}\right|_{t=0} \neq 0$.

6.5. A process has the transfer function of Eq. 6-14 with $K = 4$, $\tau_1 = 5$, $\tau_2 = 1$. If τ_a has the following values

$$\text{Case i:} \quad \tau_a = 10$$
$$\text{Case ii(a):} \; \tau_a = 2$$
$$\text{Case ii(b):} \; \tau_a = 0.5$$
$$\text{Case iii:} \quad \tau_a = -1$$

calculate the responses for a step input of magnitude 0.5 and plot in a single figure. What conclusions can you make concerning the effect of the zero location? Is the location of the pole corresponding to τ_2 important so long as $\tau_1 > \tau_2$?

6.6. The condition for a unity gain system consisting of two first-order processes connected in parallel (Fig. 6.4) to exhibit an inverse response is given by Eq. 6-21. What, if any, conditions would lead to overshoot in the step response of such a system? (*Note:* Maintain the convention that $\tau_1 > \tau_2$.)

6.7. A pressure measuring device has been analyzed and found to be described by a model with the structure shown in the figure. In other words, the device responds to changes in input as if it were a first-order process in parallel with a second-order process. Preliminary tests have shown that the gain of the first-order process is -3 and the time constant equals 20, as shown in the figure. An additional test is made on the overall system. The actual output P_m (not P'_m) resulting from a step change in P from 2 to 4 is plotted below.

(a) Compute $Q'(t)$.
(b) What are the values of K, τ, and ζ?
(c) What is the overall transfer function for the system $P'_m(s)/P'(s)$? Solve for the general case $Q'(s)/P'(s) = K'/(\tau's + 1)$.
(d) Find an expression for the overall process gain.

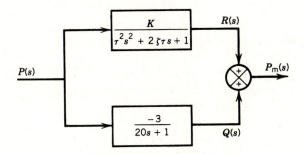

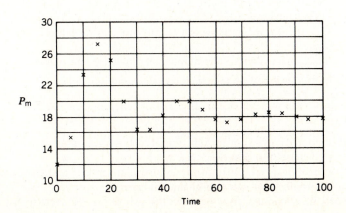

6.8. For plug flow in a perfectly insulated transfer line, the dynamic behavior of the transfer line can be represented by

$$\frac{T'_{out}(s)}{T'_{in}(s)} = e^{-Vs/q}$$

where T'_{out} and T'_{in} denote the exit and inlet temperatures of the transfer line and V/q is the theoretical time delay (volume/flow rate).

(a) Suppose that laminar flow rather than plug flow occurs in the transfer line. What is the theoretical time delay? (Justify your answer.)
(b) For a transfer line that is not insulated, heat losses can occur and thus the liquid temperature will change as the liquid passes through the line. Derive a dynamic model that shows how liquid temperature T varies with axial distance z and time t. In your derivation, assume that plug flow exists in the line and that differential heat losses can be expressed as

$$dQ_L = U[T(z, t) - T_a]dA$$

where U = overall heat transfer coefficient
        dA = surface area of a differential section of the transfer line
        T_a = ambient temperature

6.9. The following transfer function has been widely used as an approximate mathematical model for many types of processes:

$$\frac{Y(s)}{X(s)} = \frac{Ke^{-\theta s}}{(\tau_1 s + 1)(\tau_2 s + 1)}$$

where Y is the output variable, X is the input variable, and K, τ_1, τ_2, and θ are constants. Determine the output response $y(t)$ to a step change in input of magnitude = 3 when

$$K = 1.5 \qquad \tau_2 = 10 \text{ min}$$
$$\tau_1 = 20 \text{ min} \qquad \theta = 3 \text{ min}$$

6.10. A process consists of five perfectly stirred tanks in series. The volume in each tank is 50 L and the volumetric flow rate through the system is 5 L/min. At some particular time, the inlet concentration of a nonreacting species is changed from 0.70 to 0.85 (mass fraction) and held there.

(a) Write an expression for c_5 (the concentration leaving the fifth tank) as a function of time.
(b) Determine $c_1, c_2, \ldots, c_5$ at $t = 30$ min.

6.11. A process similar to that of Exercise 3.16 has equal volumes of liquid in each tank and time constant (V/q) equal to 3 min. The process initially is operating at steady-state conditions when a ramp change in c_i is initiated, $c_i = 2t$.

(a) Verify that after sufficient time has elapsed and assuming that no constraints are encountered, the output concentrations c_1, c_2, and c_3 will have the form of a similar ramp function.
(b) Determine by how much time each output concentration lags input c_i under these conditions.
(c) Generalize your results to any number of equal-volume tanks.

6.12. For the process described by the exact transfer function

$$G(s) = \frac{10}{(5s + 1)(2s + 1)(0.5s + 1)(0.1s + 1)}$$

(a) Find an approximate transfer function of second-order plus time-delay form (i.e., that of Exercise 6.9) that describes this process.
(b) Plot the response $y(t)$ of both the approximate model and the exact model on the same graph for a unit step change in input $x(t)$.
(c) What is the maximum error between the two responses? Where does it occur?

6.13. Find the transfer functions $P_1'(s)/P_d'(s)$ and $P_2'(s)/P_d'(s)$ for the compressor-surge tank system of Exercise 2.7 when it is operated isothermally. Put the results in standard (gain/time constant) form. For second-order results, determine whether the system is over- or underdamped.

6.14. In Example 4.8 we linearized the transient equations for a stirred-tank chemical reactor and derived a transfer function relating changes in the output concentration c_A to changes in the steam temperature T_s. For this same system:

(a) Derive transfer functions $T'(s)/T_i'(s$ and $T'(s)/C_{Ai}'(s)$.

(b) After putting Eq. 4-82 and the transfer functions from part (a) into standard (gain/time constant) form, what can you say about interactions in this system? Is the system overdamped or underdamped?

6.15. Show that the liquid-level system consisting of two interacting tanks (Fig. 6.12) exhibits overdamped dynamics; that is, show that the damping coefficient in Eq. 6-69 is larger than 1.

6.16. The storage tank shown in the drawing is equipped with a sight glass to provide a visual indication of liquid level. As part of a level control system, a level transmitter will be connected to either the tank or the sight glass. The sight glass is considered because the tank level measurement is quite noisy due to turbulence, splashing liquid, etc. To determine how the process dynamics are affected by the choice of transmitter location, do the following:

(a) Derive the transfer function $H_1'(s)/Q_i'(s)$, assuming that both valve and piping between the tank and the sight glass act as linear resistances with values R_1 and R_2, respectively.

(b) Derive the transfer function $H_2'(s)/Q_i'(s)$ assuming linear resistances.

(c) Provide a physical interpretation for the two limiting cases where $R_2 \to 0$ and $R_2 \to \infty$.

(d) Which level transmitter connection would be preferred on the basis of dynamic considerations? Justify your answer.

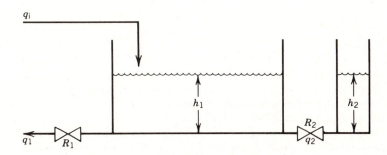

6.17. The liquid storage system shown in the drawing has the following flow–head relations:

$$q_1 = 3.3 \sqrt{h_1 + 2} \qquad q_1, q_2 \ [=] \ \text{cfm}$$
$$q_2 = 2(h_1 - h_2) \qquad h_1, h_2 \ [=] \ \text{ft}$$

The normal steady-state inlet flow rate is $\bar{q}_i = 10$ cfm. Each tank is 5 ft in diameter.

(a) Derive the transfer function $H_1'(s)/Q_i'(s)$.

(b) Suppose that the valve between the two tanks is completely shut. Describe *qualitatively* how the response $h_1(t)$ to a + 2 cfm step change in q_i would differ from the response when the valve is open. Would the final steady state be the same?

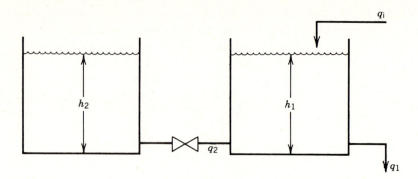

6.18. Appelpolscher and Arrhenius Quirk, head of IGC's Chemical Kinetics and Reactor Design Group, are engaged in a disputation concerning the model developed earlier for the inadequately agitated reactor (Exercise 2.8). Quirk, who really likes to beat Appelpolscher at his own game, is attempting to combine the two differential equations in the model into a single equation relating outlet composition (c_T) to inlet tracer composition (c_{Ti}). Quirk wants to analyze the resulting expression to be sure that it makes physical sense in the limiting cases. However, derivatives always gave Quirk trouble, and he is having a bad time with the calculus. Appelpolscher, in an unusual fit of magnanimity, has offered to help out. He personally is sold on the idea of transfer function analysis, but is having his own troubles with the algebra. After several hours of wrestling with their individual methods, both decide that supervising other people is more productive than working. They dump the problem in your lap with the following specific instructions:

(a) Find suitable dynamic relations for both c_T and c_{T1} in terms of input c_{Ti}, only. (Appelpolscher suggests that, since you report to him, they ought to be transfer functions.)

(b) For the situation $V_1 = \gamma V$ ($0 \le \gamma \le 1$) evaluate the limiting cases of your dynamic relations for

    i. $\gamma \to 0$
    ii. $\gamma \to 1$
    iii. $q_R \to 0$
    iv. $q_R \to \infty$

(c) For each case (i–iv), explain how the system dynamics are affected in the limit and discuss how this relates to the physical situation.

(d) Determine whether this system is overdamped or underdamped and whether it can exhibit overshoot or inverse response.

(e) Determine mathematical relations for $c_T(t)$ and $c_{T1}(t)$ if $c_{Ti}(t)$ is a rectangular pulse of magnitude h and width t_w. Under what circumstances could the pulse response of this system be approximated reasonably well by the impulse response?

6.19. Transfer functions describing transient operation of a three-stage absorber are given by Eqs. 6-72a through 6-72c. These expressions relate the exit gas concentration of each stage to concentration of the absorbed species in the feed gas. Show how these transfer functions can be obtained from the differential equation model given in Chapter 2 (Eqs. 2-38a–2-38c).

6.20. Distributed parameter systems such as tubular reactors and heat exchangers often can be modeled as a set of lumped parameter equations. In this case an alternative (approximate) physical interpretation of the process is used to obtain an o.d.e. model directly rather than by converting a p.d.e. model to o.d.e. form by means of a lumping method such as finite differences. As an example, consider a single concentric-tube heat exchanger with energy exchange between two liquid streams flowing in opposite directions, as shown in the drawing. We might model this process as if it were three

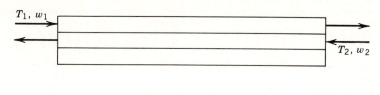

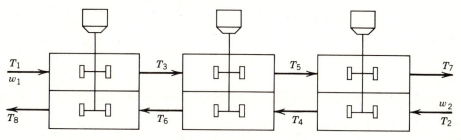

small, perfectly stirred tanks with heat exchange. If the mass flowrates w_1 and w_2 and the inlet temperatures T_1 and T_2 are known functions of time, derive transfer function expressions for the exit temperatures T_7 and T_8 in terms of the inlet temperature T_1. Assume that all liquid properties (ρ_1, ρ_2, C_{p1}, C_{p2}) are constant, that the area for heat exchange in each stage is A, that the overall heat transfer coefficient U is the same in each stage, and that the wall between the two liquids has negligible thermal capacitance.

6.21. For the case of the double-pipe heat exchanger with condensing steam in the annulus, we developed a finite difference equation model from which transfer functions (relating liquid exit temperature both to liquid feed temperature and to steam temperature) were obtained (Eqs. 6-76a, 6-76b). Use the techniques discussed in Exercise 6.20 to develop an alternative two-cell transfer function model for this system. (Recall that, in this case, the thermal mass of the separating wall could not be neglected.) How is your model similar to the finite-difference model derived in the text?

6.22. A two-input/two-output process involving simultaneous heating and liquid-level changes is illustrated in Fig. 2.1. Find the transfer function models and expressions for the gains and the time constant τ for this process. What is the output response for a unit step change in Q? For a unit step change in w? *Note:* Transfer function models for a somewhat similar process depicted in Fig. 6.18 are given in Eqs. 6-83 through 6-90. These can be compared with your results.

6.23. Exercise 2.4 dealt with a jacketed vessel used to heat a liquid by means of condensing steam. Liquid level in the tank could vary, thus changing the area available for heat transfer.

(a) Find transfer functions relating the two primary output variables h (level) and T (liquid temperature) to inputs q_F, q, and T_S. You should obtain six separate transfer functions.

(b) Briefly interpret the form of each transfer function using physical arguments, as much as possible.

(c) Discuss qualitatively the form of the response of each output to a step change in each input.

CHAPTER 7

Development of Empirical Dynamic Models from Step Response Data

As discussed in Chapter 2, a number of modeling approaches are used in process control applications. Theoretical models based on the chemistry and physics of the system represent one alternative. However, the development of rigorous theoretical models may not be practical for complex processes if the model requires a large number of differential equations with a significant number of unknown parameters (e.g., chemical and physical properties). An alternative approach is to develop an empirical model directly from experimental data. Empirical models are sometimes referred to as *black box* models. The process being modeled is likened to an opaque box where the inputs and outputs are known but the inner workings of the box are unknown. The development of empirical steady-state and dynamic models is the subject of this chapter.

Steady-state empirical models are used for instrument calibration and process optimization, and typically consist of simple polynomials relating an output to an input. Dynamic empirical models can be employed to understand process behavior during upset conditions or to analyze the performance of a control system with the process. An empirical dynamic model typically is a low-order transfer function (first- or second-order, perhaps including a time-delay element), with as many as five unspecified parameters to be determined from experimental data.

In this chapter several methods are presented for determining both steady-state and dynamic empirical models. We first develop some general model-fitting techniques that can be used to calculate parameters for any type of model. Then we discuss simple but very useful methods for obtaining first-order and second-order dynamic models from step response data. These methods yield models suitable for the design of effective control systems; however, the resulting models are usually accurate only in a narrow range of operating conditions close to the nominal steady state.

7.1 DEVELOPMENT OF MODELS BY LINEAR AND NONLINEAR REGRESSION

To develop an empirical steady-state or dynamic relation between two variables (for example, a process input x and output y), it is instructive to first plot the data (e.g., y vs. x for steady-state data and y and x vs. t for transient response data). From such a plot it is easy to visualize overall trends in the data and to select a reasonable form for the model. Once the model form is specified, the unknown

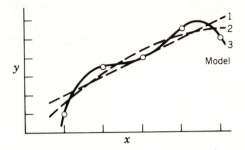

Figure 7.1. Three models for scattered data.

model parameters can be evaluated so that it accurately represents the data. This parameter estimation procedure is usually referred to as *fitting* the model, that is, minimizing some overall measure of the differences beween model predictions and data. However, the problem of fitting a model to a set of input–output data becomes complicated when the model relation is not simple or involves multiple inputs and outputs.

First consider steady-state models. Suppose a set of steady-state input–output data is obtained as shown in Fig. 7.1. Variable y represents a process output (e.g., a reactor yield), while x represents an input variable (e.g., an operating condition such as temperature). While a two-coefficient linear model (Model 1) approximately matches the results, the higher-order polynomial relations (Models 2 and 3) exhibit smaller errors between the data and the curve representing the empirical model. Models 2 and 3 provide better agreement with the data at the expense of greater complexity since more model parameters must be determined. Sometimes the model form may be known through some fundamental insight into how the process operates. In Fig. 7.1, if the true model is linear, the errors between model (the straight line of Model 1) and data could arise because of inconsistencies in establishing the operating conditions or errors in making the measurements. In empirical modeling, it is preferable to choose the simplest model structure that yields a reasonable fit of the data.

Linear Regression

Statistical analysis can be used to fit unknown model parameters and to evaluate the uncertainty associated with the fitted model as well as to compare several candidate models [1, 2]. One common approach is to use optimization theory to derive *least-squares* estimates for the model parameters. Referring to the linear model (Model 1 of Fig. 7.1), let y_i represent the data points ($i = 1$ to 5) while $\hat{y}_i$ is the model prediction for $x = x_i$. Then

$$\hat{y}_i = a_0 + a_1 x_i \tag{7-1}$$

where a_0 and a_1 are the model parameters to be calculated. The error or residual between model prediction $\hat{y}_i$ and the corresponding measurement y_i is given by

$$\epsilon_i = y_i - \hat{y}_i = y_i - a_0 - a_1 x_i \tag{7-2}$$

A commonly accepted method of determining the *best* fit is to calculate the values of a_0 and a_1 that minimize the sum of the squares of the errors $\mathcal{E}$, namely

$$\min_{a_0, a_1} \mathcal{E} = \sum_{i=1}^{5} \epsilon_i^2 = \sum_{i=1}^{5} (y_i - a_0 - a_1 x_i)^2 \tag{7-3}$$

Note that, in (7-3), the values of y_i and x_i are known, while a_0 and a_1 are to be found so as to minimize $\mathcal{E}$, the sum of squares of the errors. The estimates of a_0 and a_1 found for a specific data set are designated as $\hat{a}_0$ and $\hat{a}_1$ [1].

For a linear model and r data points, values of $\hat{a}_0$ and $\hat{a}_1$ that minimize (7-3) are found by first setting the derivatives of $\mathscr{E}$ with respect to a_0 and a_1 equal to zero. Since $\mathscr{E}$ is a quadratic function, this approach leads to two linear equations in two unknowns, $\hat{a}_1$ and $\hat{a}_0$. The analytical solution is [3]

$$\hat{a}_0 = \frac{S_{xx}S_y - S_{xy}S_x}{rS_{xx} - (S_x)^2}$$

$$\hat{a}_1 = \frac{rS_{xy} - S_xS_y}{rS_{xx} - (S_x)^2} \tag{7-4}$$

$$\text{where } S_x = \sum_{i=1}^{r} x_i \qquad S_{xx} = \sum_{i=1}^{r} x_i^2$$

$$S_y = \sum_{i=1}^{r} y_i \qquad S_{xy} = \sum_{i=1}^{r} x_i y_i$$

This least-squares estimation approach (also called linear regression) can be extended to more general models with

1. Any number of input variables.
2. Any functional form of the inputs, such as polynomials and exponentials.

A general nonlinear SISO model will have the form

$$\hat{y} = \sum_{j=0}^{n} a_j g_j(x) \tag{7-5}$$

where the a_j are the unknown parameters to be estimated and the g_j are the chosen functions of x (a total of $n + 1$ functions). Note that the unknown *parameters* appear *linearly* in the model. When the expression for $\mathscr{E}$ (the sum of squares of the errors) is differentiated with respect to the unknown a_j, a set of $n + 1$ linear algebraic equations in $n + 1$ unknowns results [3]:

$$a_0 S_{00} + a_1 S_{01} + \cdots + a_n S_{0n} = T_0$$

$$a_0 S_{10} + a_1 S_{11} + \cdots + a_n S_{1n} = T_1 \tag{7-6}$$

$$.$$
$$.$$
$$.$$

$$a_0 S_{n0} + a_1 S_{n1} + \cdots + a_n S_{nn} = T_n$$

$$\text{where } S_{jm} = \sum_{i=1}^{r} g_{ji} g_{mi} \qquad (m = 0, \ldots, n)$$

$$T_j = \sum_{i=1}^{r} g_{ji} y_i$$

and g_{ji} denotes the jth function of the specified model ($j = 0, \ldots, n$) evaluated at the ith data point ($i = 1, \ldots, r$).

When the number of data points is equal to the number of model parameters, that is, when $r = n + 1$, a single solution of (7-6) exists that matches all the data points exactly.[1] For $r > n + 1$ a *least-squares solution* results that minimizes the

---

[1]Note that for this unique solution to exist, the $(n + 1) \times (n + 1)$ matrix formed of the S_{jm} elements must be invertible.

sum of the squared deviations between each of the data points and the model predictions. The corresponding set of linear equations can be solved rapidly using a computer or, for simpler models, by using a hand calculator. Standard computer packages are available for model fitting that require the user to specify only the input and output variables in the model, the form of the model equation, and the values of the data points; the computer program performs all necessary calculations. Usually extensive statistical diagnostics on the goodness of fit and confidence intervals for the estimated parameters are given [1, 2]. These results can help determine the best model form, but cannot completely eliminate the need to use good engineering judgment.

Common applications of model fitting in the field of process control occur in calculating steady-state models suitable for supervisory control (Chapter 20), in fitting frequency response data (Chapter 14), and in determining difference equation models useful for digital control systems (Chapter 23). Next we consider the development of a steady-state performance model, such as might be used in the supervisory control of an electrical power generator.

EXAMPLE 7.1

An experiment has been run to determine the steady-state power delivered by a gas turbine-driven generator as a function of fuel rate. The following data were obtained:

Fuel Flow Rate x_i	Power Generated y_i (MW)
1.0	2.0
2.3	4.4
2.9	5.4
4.0	7.5
4.9	9.1

A straight line model should be satisfactory since a plot of the data reveals a generally linear pattern. Find the best linear model and compare it with the best quadratic model.

Solution

To solve for the linear model, Eq. 7-4 could be applied directly. However, to illustrate the use of Eq. 7-6, first define the terms in the linear model: $g_0(x) = 1$, $g_1(x) = x$. The following equations are then obtained:

$$a_0 S_{00} + a_1 S_{01} = T_0 \tag{7-7a}$$

$$a_0 S_{10} + a_1 S_{11} = T_1 \tag{7-7b}$$

$$\text{where } S_{00} = \sum_{i=1}^{5} (1)^2 \qquad T_0 = \sum_{i=1}^{5} (1)(y_i)$$

$$S_{01} = \sum_{i=1}^{5} (1)(x_i) \qquad T_1 = \sum_{i=1}^{5} (x_i)(y_i)$$

$$S_{10} = \sum_{i=1}^{5} (x_i)(1)$$

$$S_{11} = \sum_{i=1}^{5} (x_i)^2$$

Table 7.1   A Comparison of Results from Example 7.1

x_i	y_i	Linear Model $\hat{y}_{1i} = \hat{a}_0 + \hat{a}_1 x_i$	Quadratic Model $\hat{y}_{2i} = \hat{a}_0 + \hat{a}_1 x_i + \hat{a}_2 x_i^2$
1.0	2.0	1.998	2.003
2.3	4.4	4.368	4.364
2.9	5.4	5.461	5.457
4.0	7.5	7.466	7.465
4.9	9.1	9.107	9.111
		$(\mathcal{E}_1 = 0.0060)$	$(\mathcal{E}_2 = 0.0059)$

Solution of Eq. 7-7 yields the same results given in Eq. 7-4, with $\hat{a}_0 = 0.175$ and $\hat{a}_1 = 1.823$.

To determine how much the model accuracy might be improved with a quadratic model, Eq. 7-6 is again applied, this time with $g_0(x) = 1$, $g_1(x) = x$, $g_2(x) = x^2$. Equation 7-6 now represents three equations in three unknowns, yielding $\hat{a}_0 = 0.192$, $\hat{a}_1 = 1.809$, and $\hat{a}_2 = 0.002376$. The predicted values of y are compared with the measured values (actual data) in Table 7.1 for both linear and quadratic models. It is evident from this comparison that the linear model is adequate and little improvement results from the more complicated quadratic model.

Nonlinear Regression

If the model is nonlinear with respect to the model parameters, then nonlinear regression rather than linear regression needs to be used. For example, suppose that a reaction rate expression of the form $r_A = kc_A{}^n$ is to be fit to experimental data. Here r_A is the reaction rate of component A, c_A is the reactant concentration, and k and n are model parameters. This model is *linear* with respect to rate constant k but is *nonlinear* with respect to reaction order n. A general nonlinear model can be written as

$$\hat{y} = f(x_1, x_2, x_3, \ldots, a_0, a_1, a_2, \ldots) \tag{7-8}$$

where $\hat{y}$ is the empirical model output, the x_i are inputs, and a_j are the parameters to be estimated. In this case, the a_j are not linear multipliers of the input functions, as in Eq. 7-5. However, we still can define a sum of squares error criterion that is to be minimized by selecting the parameter set a_j :

$$\min_{a_j} \mathcal{E} = \sum_{i=1}^{r} (y_i - \hat{y}_i)^2 \tag{7-9}$$

where y_i and $\hat{y}_i$ denote the ith output measurement and model prediction corresponding to the ith data point, respectively. The values of the parameters a_j that minimize $\mathcal{E}$ are the *least-squares estimates*.

Consider the problem of estimating the time constants for first-order and overdamped second-order dynamic models based on a measured step response of the process. Equations for the step response of these types of processes were developed in Chapter 5:

Transfer Function   *Step Response*

$$\frac{K}{\tau s + 1} \tag{5-3} \qquad y(t) = KM(1 - e^{-t/\tau}) \tag{5-18}$$

$$\frac{K}{(\tau_1 s + 1)(\tau_2 s + 1)} \tag{5-39} \qquad y(t) = KM\left(1 - \frac{\tau_1 e^{-t/\tau_1} - \tau_2 e^{-t/\tau_2}}{\tau_1 - \tau_2}\right) \tag{5-48}$$

In the step response equations, t is the independent variable instead of the input x used earlier, and y is the dependent variable expressed in deviation form. While K appears linearly in both response equations, τ in (5-18) and τ_1 and τ_2 in (5-48) are contained within nonlinear functions, ruling out the use of linear regression to estimate them. Since K can be calculated directly from the steady-state data, we assume that it can be determined prior to estimation of τ (or τ_1 and τ_2).

Sometimes a transformation can be employed to modify a nonlinear model so that linear regression can be used. For example, if K is assumed to be known, the first-order model (5-18) can be rearranged to give

$$\ln\left(1 - \frac{y_i}{KM}\right) = -\frac{t_i}{\tau} \tag{7-10}$$

Since $\ln(1 - y_i/KM)$ can be evaluated at each time t_i, this model is linear in the parameter $1/\tau$.

The transformation in Eq. 7-10 leads to the *fraction incomplete response* method of determining first-order models discussed in the next section. However, for higher-order step response models, such as Eq. 5-48, the transformation approach is not feasible. In these cases, we must use an iterative optimization method to find the least-squares estimates of the time constants [3]. There are many optimization techniques that can be used for nonlinear regression; computer packages are available specifically for performing parameter estimation. On the other hand, there are a number of graphical correlations that can be used quickly to find approximate values of τ_1 and τ_2. The accuracy of models obtained in this way is usually sufficient for controller design. In the next two sections, we present several methods for estimating transfer function parameters based on graphical analysis.

7.2 GRAPHICAL FITTING OF FIRST-ORDER MODELS USING STEP TESTS

The output response of a process to a step change in input, when plotted, is sometimes referred to as the *process reaction curve*. If the process of interest can be approximated by a first- or second-order linear differential equation, the model parameters can be obtained by inspection of the reaction curve. For example, recall the dynamic model for a first-order process,

$$\tau \frac{dy}{dt} + y = Kx \tag{7-11}$$

where the system is initially at rest ($y(0) = 0$, corresponding to $x = 0$). If the input x is abruptly changed from zero to M at time $t = 0$, the step response in Eq. 5-18 results. The normalized graphical response is shown in Fig. 7.2. The response of $y(t)$ reaches 63.2% of its final value at $t = \tau$. The steady-state output deviation value is $y(\infty) = KM$. From Eq. 5-18 or 7-11, after rearranging and evaluating the limit at $t = 0$, the initial slope of the normalized step response is

$$\frac{d}{dt}\left(\frac{y}{KM}\right)\bigg|_{t=0} = \frac{1}{\tau} \tag{7-12}$$

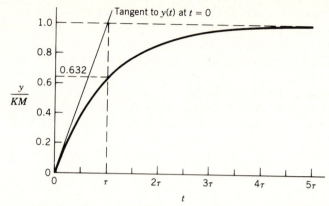

Figure 7.2. Step response of a first-order system and graphical constructions used to estimate the time constant, τ.

From the graphical interpretation in Fig. 7.2, the intercept of the tangent at $t = 0$ with $y/KM = 1$ occurs at $t = \tau$. Hence, for a simple first-order system, there are two quick ways to calculate the time constant τ from a step response plot:

1. Find the value of t at which the response is 63.2% complete.
2. Construct the tangent to the initial response curve and find its intercept with the final steady-state value.

Note that these procedures do not explicitly require the output variable to be expressed in deviation or normalized form.

EXAMPLE 7.2

Figure 7.3 gives the response of the temperature T in a continuous stirred-tank reactor to a step change in feed flow rate w from 120 to 125 kg/min. Find an approximate first-order model for the process for these operating conditions.

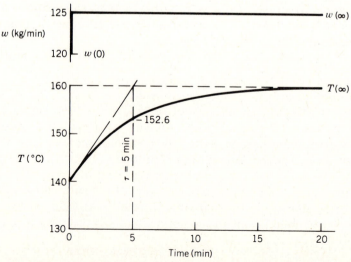

Figure 7.3. Temperature response of a stirred-tank reactor for a step change in feed flow rate.

Solution

First note that $\Delta w = M = 125 - 120 = 5$ kg/min. Since $\Delta T = T(\infty) - T(0) = 160 - 140 = 20$ °C, the process gain is

$$K = \frac{\Delta T}{\Delta w} = \frac{20 \text{ °C}}{5 \text{ kg/min}} = 4 \frac{\text{°C}}{\text{kg/min}}$$

The time constant obtained from the graphical construction shown in Fig. 7.3 is $\tau = 5$ min. Note that this result agrees with the time when 63.2% of the response is complete, that is,

$$T = 140 + 0.632(20) = 152.6 \text{ °C}$$

Consequently, the desired process model is

$$\frac{T'(s)}{W'(s)} = \frac{4}{5s + 1}$$

where the gain has units of °C/kg/min.

Very few experimental plots of the step response show purely first-order behavior because:

1. It is difficult to form a perfect step input. Usually process equipment, such as pumps and the control valves discussed in Chapter 9, cannot be changed instantaneously from one setting to another but must be ramped over a finite time. However, if the ramp time is small compared to the process time constant, a reasonably good approximation to a step input may be obtained.
2. The true process model *is neither first order nor linear*. Only the simplest processes will exhibit such ideal dynamics.
3. The output data are usually corrupted with noise, that is, they contain a random component in the measured output. Noise can arise from normal operation of the process, for example, inadequate mixing that produces eddies of higher and lower concentration (or temperature), or it can result from electronic instrumentation. If noise is completely random, a first-order response plot may still be drawn that fits the output data well in some time-averaged sense. However, nonrandom noise, such as linear drift arising in the measurement instrumentation, can cause problems in the analysis.
4. Another input (disturbance) may change during the step test without the operator's knowledge.

Hence, normally there will be some departure from the curve in Fig. 7.2.

To account for higher-order dynamics that are neglected in a simple first-order model, a time-delay term can be included. This additional dynamic element will improve the agreement between model and experimental responses. The fitting of a first-order plus time-delay model,

$$G(s) = \frac{Ke^{-\theta s}}{\tau s + 1} \tag{7-13}$$

to the actual step response requires the following steps, as shown in Fig. 7.4 [3]:

1. The process gain K for the model is found by calculating the ratio of the change in the steady-state value of y to the size of the step change M in x.

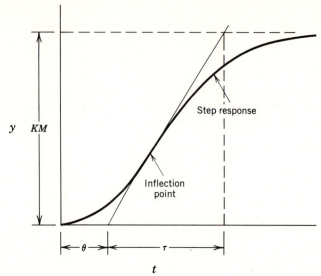

Figure 7.4. Graphical analysis of the process reaction curve to obtain parameters of a first-order plus time delay model.

2. A tangent is drawn at the point of inflection of the step response; the intersection of the tangent line and the time axis (where $y = 0$) is the time delay.
3. If the tangent is extended to intersect the steady-state response line (where $y = KM$), the point of intersection corresponds to time $t = \theta + \tau$. Therefore τ can be found by subtracting θ from the point of intersection.

The tangent method presented above for obtaining the time constant suffers from using only a single point to estimate the time constant. Use of multiple points may provide a better estimate. Returning to Eq. 7-10, note that introducing θ and rearranging gives the expression

$$\ln \left[\frac{y(\infty) - y_i}{y(\infty)} \right] = - \frac{t_i - \theta}{\tau} \tag{7-14}$$

since $y(\infty)$, the final steady-state value, equals KM. In (7-14), $y(\infty) - y_i$ can be interpreted as the *incomplete response;* dividing by $y(\infty)$ yields the *fraction incomplete response.* A semilog plot of $[y(\infty) - y_i]/y(\infty)$ vs. $(t_i - \theta)$ will then yield a straight line with slope $= -1/\tau$, from which an average value of the time constant is obtained. An equation equivalent to (7-14) for the variables of Example 7.2 (not in deviation form) would be

$$\ln \left[\frac{T(\infty) - T(t)}{T(\infty) - T(0)} \right] = -\frac{t - \theta}{\tau} \tag{7-15}$$

The major disadvantage of the time-delay estimation method in Fig. 7.4 is that it is difficult to find the point of inflection, due to measurement noise (errors) and small-scale recorder charts. The method of Sundaresan and Krishnaswamy [4] avoids use of the point of inflection construction entirely to estimate the dead time. They proposed that two times (t_1 and t_2) be estimated from a step response curve, corresponding to 35.3 and 85.3% response times, respectively. The time delay and time constant are then estimated from the following equations:

$$\theta = 1.3t_1 - 0.29t_2$$
$$\tau = 0.67(t_2 - t_1) \tag{7-16}$$

These values of τ and θ approximately minimize the difference between the measured response and the model, in a least-squares sense [4]. Note that, using actual step response data, the parameters K, τ, and θ calculated for a first-order model approximation can vary considerably depending on the operating conditions of the process, the size of the input step change, and the direction of the change. These variations usually can be attributed to process nonlinearities.

The methods considered above yield first-order models, only. If a first-order model is not sufficiently accurate, a second-order model (two time constants) can be estimated, as discussed in the next section. It is difficult to identify more than two time constants using graphical techniques.

7.3 FITTING SECOND-ORDER MODELS USING STEP TESTS

Normally a better approximation to the step response can be obtained by fitting a second-order model to the data. Figure 7.5 shows the range of shapes that can occur for the step response assuming an overdamped model with no numerator dynamics. The transfer function used in Fig. 7.5,

$$G(s) = \frac{K}{(\tau_1 s + 1)(\tau_2 s + 1)} \tag{5-39}$$

includes two limiting cases: $\tau_2/\tau_1 = 0$, where the system becomes first order, and $\tau_2/\tau_1 = 1$, the critically damped case. The larger of the two time constants, τ_1, is called the dominant time constant. The S-shaped response becomes more pronounced as the ratio of τ_2/τ_1 becomes closer to one.

Time constants for second-order systems which include time delays can be estimated using several graphical methods. Harriott [5] and Oldenbourg and Sartorius [6] provide estimation procedures for the parameters of the following model:

$$G(s) = \frac{Ke^{-\theta s}}{(\tau_1 s + 1)(\tau_2 s + 1)} \tag{7-17}$$

Thus, these methods apply only to overdamped systems. A method due to Smith [7] is more general since it yields a model of the form

$$G(s) = \frac{Ke^{-\theta s}}{\tau^2 s^2 + 2\zeta\tau s + 1} \tag{7-18}$$

which includes both overdamped and underdamped systems.

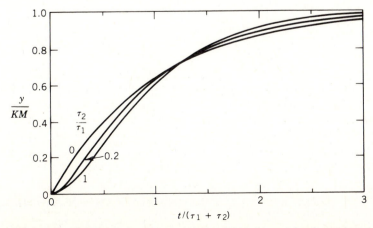

Figure 7.5. Step response for several overdamped second-order systems.

For fitting second-order models, some caution must be exercised in estimating θ. A second-order model with no time delay exhibits a point of inflection (see Fig. 7.5 when $\tau_1 \approx \tau_2$). If the point-of-inflection construction (Fig. 7.4) is applied to this case, however, a nonzero time delay is indicated. To avoid this conflict, visual determination of θ is recommended, but estimation of θ by trial and error may be required to obtain a good fit. Van Cauwenberghe and Willequet [8] have discussed more sophisticated methods for estimating θ. In the methods presented below, the time delay is subtracted from the actual time value prior to determination of τ_1 and τ_2; that is, an adjusted time, $t' = t - \theta$, is employed for the actual graphical analysis.

Here we consider only the methods of Harriott and Smith. The Oldenbourg-Sartorius method [6,8] requires estimating a point of inflection, which may be extremely difficult with noisy process data.

Harriott's Method

Harriott plotted the fractional response of a second-order system (without time delay) against $t/(\tau_1 + \tau_2)$ for various ratios of τ_2/τ_1. He found that all the curves intersected approximately at 73% of the final steady-state value where $t/(\tau_1 + \tau_2)$ = 1.3, as shown in Fig. 7.5. The actual range is $0.7275 < y < 0.7326$. Thus, by measuring the time required for the system to reach 73% of its final value t_{73}, the sum of the two time constants can be calculated: $\tau_1 + \tau_2 = t_{73}/1.3$.

Harriott then plotted the fractional response at $t/(\tau_1 + \tau_2) = 0.5$ against $\tau_1/(\tau_1 + \tau_2)$ since the curves in Fig. 7.5 show the greatest deviation at this point. The fractional response is shown in Fig. 7.6. The value of the fractional response when $t = 0.5(\tau_1 + \tau_2)$ can be determined from the experimental data, and the value of $\tau_1/(\tau_1 + \tau_2)$ can be read from Fig. 7.6. If the fractional response is less than 0.26 or greater than 0.39 at this point, the method is not applicable, which generally indicates that the process requires a model that is higher than second order or that is underdamped. Harriott's method generally becomes less accurate as τ_2/τ_1 approaches unity, and τ_1 and τ_2 are fairly sensitive [7] to the estimate of K (obtained from the steady-state response).

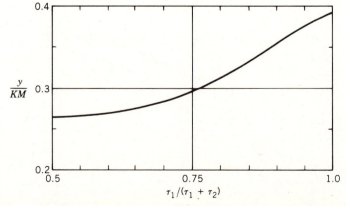

Figure 7.6.   Harriot's method: fractional response of overdamped second-order system at $t/(\tau_1 + \tau_2) = 0.5$

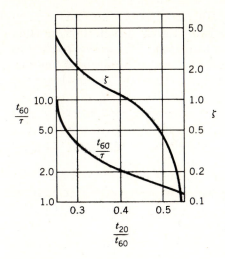

Figure 7.7. Smith's method: relationship of ζ and τ to t_{20} and t_{60}.

Smith's Method

Smith [7] has reported a method for the model in (7-18) based on two points of the fractional response of the system, at 20% and 60% of its final value. Smith's method requires the times (with apparent time delay removed) at which the normalized value of the response reaches 20% and 60%, respectively. Using Fig. 7.7, the ratio of t_{20}/t_{60} gives the value of ζ. An estimate of τ can be obtained from the plot of t_{60}/τ vs. t_{20}/t_{60}.

EXAMPLE 7.3

Use Harriott's and Smith's second-order methods as well as nonlinear regression to fit the normalized step response curve shown in Fig. 7.8. For all three methods assume $\theta = 0$, since the response curve becomes nonzero immediately after $t = 0$.

Solution by Harriott's Method

From Figure 7.8, $t_{73} = 6.4$ min, hence $\tau_1 + \tau_2 = t_{73}/1.3 = 6.4/1.3 = 4.92$ min. For $t/(\tau_1 + \tau_2) = 0.5$, $t = 2.46$ min and $y/KM = 0.295$. From Fig. 7.6, $\tau_1/(\tau_1 + \tau_2) = 0.74$, hence $\tau_1 = 3.64$ min and $\tau_2 = 1.28$ min.

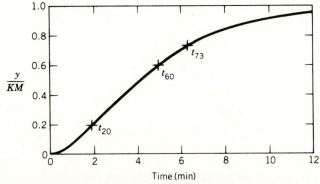

Figure 7.8. Normalized experimental step response.

Solution by Smith's Method

The two points of interest are the 20% response, $t_{20} = 1.85$ min, and the 60% response, $t_{60} = 5.0$ min. Hence, $t_{20}/t_{60} = 0.37$. From Fig. 7.7, $t_{60}/\tau = 2.8$; hence $\tau = 5.0/2.8 = 1.79$ min and $\zeta = 1.3$. Since the model is overdamped, we can calculate two time constants:

$$\tau_1 = \tau\zeta + \tau\sqrt{\zeta^2 - 1}$$

$$\tau_2 = \tau\zeta - \tau\sqrt{\zeta^2 - 1}$$

Solving gives $\tau_1 = 3.80$ min and $\tau_2 = 0.85$ min.

Solution by Nonlinear Regression

Using a computer program based on Powell's optimization method [3], the time constants in Eq. 5-48 were selected to minimize the sum of the squares of the errors between data and model predictions (see Eq. 7-9). The computed time constants are $\tau_1 = 3.00$ min and $\tau_2 = 1.92$ min.

In summary, the three methods give similar results for the dominant time constant τ_1 but there is considerable variation in τ_2, as shown below.

	τ_1 (min)	τ_2 (min)	Sum of Squares
Harriott	3.64	1.28	0.00267
Smith	3.80	0.85	0.01905
Nonlinear regression	3.00	1.92	0.00084

The responses are plotted in Fig. 7.9; all three models give an acceptable fit to the step response curve. Note that the sum of the two time constants is about the same in each case. For general use, all of the above methods are satisfactory although the nonlinear regression method is less subject to graphical approximation errors and it provides a better fit to the data. Also, the nonlinear regression method permits the step test to be terminated before the final steady state is reached;

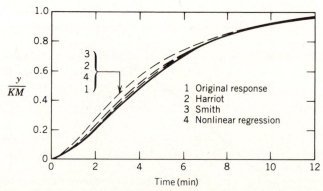

Figure 7.9. Comparison of step responses of three fitted second-order models.

however, sufficient response data must be obtained for the regression method to distinguish the steady-state effect of K from the dynamic effects of τ_1 and τ_2. For monitoring with process control computers, nonlinear regression is preferred.

SUMMARY

When theoretical models are impractical or impossible to obtain—for example, when the process is complicated or when there are many unknown parameters in the theoretical model—empirical process models provide a viable alternative. In this case, a model that is sufficiently accurate for control system design often can be obtained from step response data. The process output can be analyzed graphically or by computer (nonlinear regression) to obtain a first- or second-order model. Usually a time-delay term will have to be included to make the resulting model satisfactory for complicated processes, that is, processes containing many elements (stages) or processes that are distributed in nature, such as packed columns and heat exchangers. Sometimes a step input to the process is not permissible, due to safety considerations or the possibility of producing off-standard material while the process output deviates substantially from the desired value. In these situations, the sinusoidal or pulse testing methods of Chapter 14, use of advanced optimization codes for estimation of parameters in differential equations [9], or statistical identification techniques [10] must be employed.

REFERENCES

1. Box, G. E. P., and N. R. Draper, *Empirical Model-Building and Response Surfaces*, Wiley, New York, 1987.
2. Guttman, I., S. S. Wilks, and J. S. Hunter, *Introductory Engineering Statistics*, 2d Ed., Wiley, New York, 1971.
3. Edgar, T. F., and D. M. Himmelblau, *Optimization of Chemical Processes*, McGraw-Hill, New York, 1988.
4. Sundaresan, K. R., and P. R. Krishnaswamy, Estimation of time delay, time constant parameters in time, frequency, and Laplace domains, *Can. J. Chem. Eng.*, **56**, 257 (1977).
5. Harriott, P., *Process Control*, McGraw-Hill, New York, 1964.
6. Oldenbourg, R. C., and H. Sartorius, The dynamics of automatic control, *Trans. ASME*, 77 (1948).
7. Smith, C. L., *Digital Computer Process Control*, Intext, Scranton, PA, 1972.
8. van Cauwenberghe, A. R., and S. Willequet, Parameter evaluation for a second-order dead-time model from the system's step response, *Quart. Appl. Control, A* **15**(3), 114 (1974).
9. Bard, Y., *Nonlinear Parameter Estimation*, Academic, New York, 1974.
10. Eykhoff, P., *System Identification: Parameter and State Estimation*, Wiley, New York, 1974.

EXERCISES

7.1. An operator introduces a step change in the flow q_i to a particular process at 3:05 a.m., changing the flow from 500 to 520 gal/min. The first change in the process temperature T (initially at 120 °F) comes at 3:09 a.m. After that, the response in T is quite rapid, slowing down gradually until it appears to reach a steady-state value of 124.7 °F. The operator notes in the logbook that there is no change after 3:34 a.m. What approximate transfer function might be used to relate temperature to flow for this process in the absence of more accurate information? What should the operator do next time to obtain a better estimate?

7.2. A single-tank process has been operating for a long period of time with the inlet flow rate q_i equal to 29.4 ft^3/min. The operator increases the flow rate suddenly by 10%, resulting in a level change in the tank recorded in the following table.

Time (min)	h (ft)	Time (min)	h (ft)
0	5.50	1.4	6.37
0.2	5.75	1.6	6.40
0.4	5.93	1.8	6.43
0.6	6.07	2.0	6.45
0.8	6.18	3.0	6.50
1.0	6.26	4.0	6.51
1.2	6.32	5.0	6.52

Assuming that liquid-level dynamics follow a first-order model, calculate the process gain and the time constant using four methods:

(a) From the time required for the output to reach 63.2% of the total change.
(b) From the initial slope of the response curve.
(c) Using the time at which the fraction incomplete response is 0.368.
(d) From the slope of the fraction incomplete response curve.

7.3. A process consists of two stirred tanks with input q and outputs T_1 and T_2 (see drawing). To test the hypothesis that the dynamics in each tank are basically first order, a step change in q is made from 82 to 84, with output results given in the table.

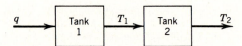

(a) By means of the fraction incomplete response method, find the transfer functions $T_1'(s)/Q'(s)$ and $T_2'(s)/T_1'(s)$. Assume they are of the form $K_i/(\tau_i s + 1)$
(b) Obtain an approximate differential equation relating T_1 to q and one relating T_2 to q.
(c) Solve both equations for the same step change in q and plot along with the experimental data to indicate how well the equations model the process.

Time	T_1	T_2	Time	T_1	T_2
0	10.00	20.00	11	17.80	25.77
1	12.27	20.65	12	17.85	25.84
2	13.89	21.79	13	17.89	25.88
3	15.06	22.83	14	17.92	25.92
4	15.89	23.68	15	17.95	25.94
5	16.49	24.32	16	17.96	25.96
6	16.91	24.79	17	17.97	25.97
7	17.22	25.13	18	17.98	25.98
8	17.44	25.38	19	17.99	25.98
9	17.60	25.55	20	17.99	25.99
10	17.71	25.68	50	18.00	26.00

7.4. For a process described by the transfer function

$$G(s) = \frac{2}{(6s + 1)(4s + 1)(2s + 1)}$$

Calculate the response to a step input change of magnitude $= 1.5$.

(a) Obtain an approximate first-order plus delay model using the fraction incomplete response method.
(b) Find an approximate second-order model using the method of Section 7.3.
(c) Calculate the responses of both approximate models using the same step input as for the third-order model. Plot all three responses on the same graph. What can you conclude concerning the approximations?

7.5. The step response for a second-order overdamped system is given by Eq. 5-48.

(a) Show that the slope of the step response is zero at $t = 0$.
(b) Find an expression for the value of t where the inflection point occurs.
(c) Solve the expression for t/τ_1, expressing the right side in terms of the ratio τ_2/τ_1. Repeat for t/τ_2.
(d) Do the forms of these expressions suggest a way to use them and Eq. 5-48 to find both time constants from one point on the response curve and the initial and final values of the output? If so, what would be the method? Can you see a potential difficulty in using this approach, that is, in having to find the inflection point?

7.6. For the unit step response shown in the drawing, estimate the following models using graphical methods:

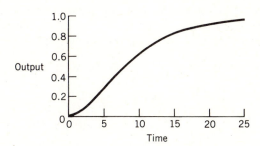

(a) First-order plus time delay.
(b) Second-order using
  i. Harriott's method
  ii. Smith's method

Plot all three predicted model responses on the same graph.

7.7. A heat exchanger used to heat a glycol solution with a hot oil is known to exhibit first-order plus time delay behavior. $(G_1(s) = T'(s)/Q'(s)$ where T' is the outlet temperature deviation and Q' is the hot oil flow rate deviation.) A thermocouple is placed downstream from the outlet of the heat exchanger at a distance of 3 m. The average velocity

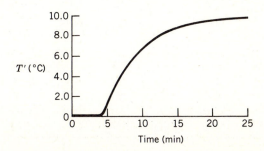

of the glycol in the outlet pipe is 0.05 m/s. The thermocouple also is known to exhibit first-order behavior; however, its time constant is expected to be considerably smaller than the heat exchanger time constant.

(a) Data from a unit step test in Q' on the complete system are shown in the drawing. Using a method of your choice, compute the empirical time constants of this process from the step response.

(b) From your empirical model, find transfer functions for the heat exchanger, for the pipe, and for the thermocouple. What assumptions do you have to make to obtain these individual empirical transfer functions from the overall transfer function?

PART THREE

FEEDBACK CONTROL

— CHAPTER 8 —————————————
Feedback Controllers

In previous chapters we considered the dynamic behavior of representative processes and developed some of the mathematical tools required to analyze process dynamics. We are now in a position to consider the important topic of feedback control.

We first introduce feedback control systems via the stirred-tank heater example. The control algorithms for standard pneumatic (air pressure operated) and electronic (electrical current or voltage operated) controllers are then developed, with emphasis on those algorithms that are widely used in industrial equipment. The proportional–integral–derivative (PID) controller and the on–off controller represent the most important commercial types. A detailed discussion of controller options and special features is also given. Finally, although the major development of digital control begins in Chapter 21, we introduce digital PID control algorithms in this chapter to emphasize the obvious parallels between digital and standard analog (pneumatic or electronic) controller implementations. The remaining elements in the feedback control loop—sensors, transmitters, and control valves—will be considered in Chapter 9.

8.1 STIRRED-TANK HEATER EXAMPLE

As an example of a feedback control system, consider the stirred-tank heater system shown in Fig. 8.1. The control objective is to keep the tank temperature T at the desired value T_R by adjusting the rate of heat input Q from the electrical heater. The thermocouple measurement (in millivolts) is sent to a temperature transmitter where the measurement is amplified before being transmitted to the controller. The transmitter is selected so that its output signal is compatible with the input range of the electronic controller, for example, 1–5 V dc. The controller output signal p is sent to a silicon-controlled rectifier (SCR), which converts it to a form compatible with the electric heater.

This example illustrates that the basic components in the feedback control loop are

- Process being controlled (stirred tank)
- Sensor and transmitter
- Controller

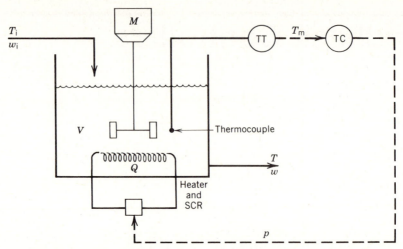

Figure 8.1. Schematic diagram for a stirred-tank heater control system.

- SCR and final control element (electrical heater)
- Transmission lines (electrical cables) between the various instruments.

The stirred-tank heater example employs electronic instrumentation. However, pneumatic instruments are also widely used in the process industries. For a pneumatic instrument, the input and output signals are air pressures in the range of 3 to 15 psig. Metal or plastic tubing (usually 1/4 or 3/8 in. o.d.) is used to interconnect the various pneumatic instruments, for example, to connect a transmitter to a controller. In Chapter 9, control instrumentation will be considered in greater detail. But first, we consider the heart of a feedback control system, the controller itself.

8.2 CONTROLLERS

Historical Perspective

We tend to regard automatic control devices as a modern development. However, ingenious feedback control systems for water-level control were used by the Greeks as early as 250 B.C. [1] with their mode of operation being very similar to that of the level regulator in the modern flush toilet. The fly-ball governor, which was first applied by James Watt to his new steam engine in 1788, played a key role in the development of steam power. Feedback control was essential for the development in the 1930s of high-gain, operational amplifiers that are widely used in electronic equipment.

During the 1930s *three-mode* controllers with proportional, integral, and derivative (PID) feedback control action became commercially available [2]. The first theoretical papers on process control were published during this same period [3,4]. Pneumatic PID controllers gained widespread industrial acceptance during the 1940s, and their electronic counterparts entered the market in the 1950s. The first computer control applications in the process industries were reported in the late 1950s and early 1960s [5]. During the past 20 years digital computer hardware has been used on a routine basis and has had a tremendous impact on process control [5,6]. Process control computers and the analysis of digital control systems will be considered in Part Six of this book. However, we introduce digital versions of PID control algorithms in Section 8.3.

PID Controllers

The three basic feedback control modes that are employed are proportional (P), integral (I), and derivative (D) control. Consider the flow control system shown in Fig. 8.2. where the process stream flow rate is measured and transmitted pneumatically to the flow controller. The controller compares the measured value to the set point and takes the appropriate corrective action by sending an output signal to the control valve. Pneumatic signals are denoted by the standard symbol, a line with parallel lines through it (—#—).

Figure 8.3 is a block diagram for the feedback controller. The set point is shown as a dashed line since it is normally specified by a dial setting or lever position on the controller. In addition to this *local* set point, some controllers have a *remote* set point option that permits them to receive a set-point signal from an external device such as another controller or a digital computer. The controller input and output signals are continuous signals that are either pneumatic or electrical.

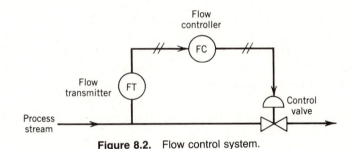

Figure 8.2. Flow control system.

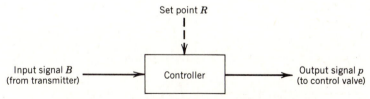

Figure 8.3. Schematic diagram of a feedback controller.

Proportional Control

In feedback control the objective is to reduce the error signal $e(t)$ to zero where

$$e(t) = R(t) - B(t) \tag{8-1}$$

and

$$R(t) = \text{set point}$$

$$B(t) = \text{measured value of the controlled variable}$$
$$\text{(or equivalent signal from the transmitter)}$$

Although Eq. 8-1 indicates that the set point can be time–varying, in most process control problems it is kept constant for long periods of time.

For proportional control, the controller output is proportional to the error signal,

$$p(t) = \bar{p} + K_c e(t) \tag{8-2}$$

where $p(t)$ = controller output
$\bar{p}$ = bias value
K_c = controller gain (usually dimensionless)

The key concepts behind proportional control are the following: (1) The controller gain can be adjusted to make the controller output changes as sensitive as desired to deviations between set point and controlled variable. (2) The sign of K_c can be chosen to make the controller output increase (or decrease) as the deviation increases. For the stirred-tank heater example, we want the heat input to the tank to decrease as T increases; hence, K_c is chosen to be a positive number.

For proportional controllers the bias $\bar{p}$ can be adjusted. Since the controller output equals $\bar{p}$ when the error is zero, $\bar{p}$ is adjusted so that the controller output (and consequently the manipulated variable) are at their nominal steady-state values. For example, if the final control element is a control valve, $\bar{p}$ is adjusted so that the flow rate through the control valve is equal to the nominal, steady-state value when $e = 0$. The controller gain K_c is adjustable and is usually tuned (i.e., adjusted) after the controller has been installed and brought into service. For general purpose controllers, K_c is dimensionless. This situation will occur if p and e in Eq. 8-2 have the same units. For example, the units could be associated with electronic or pneumatic instruments (milliamperes, volts, psi, etc.) or p and e could be expressed as numbers between 0 and 100%. The latter representation is especially convenient for graphical displays and computer control software. On the other hand, in analyzing control systems it is often convenient to express the error signal in engineering units such as °C or mol/L. For these situations, K_c will not be dimensionless. As an example, consider the stirred-tank heater system. Suppose that $e \, [=] \,$ °C and $p \, [=] \,$ volts, then Eq. 8.2 implies that $K_c \, [=] \,$ V/°C. When a controller gain is not dimensionless, it includes the steady-state gain for another component of the control loop such as a transmitter or control valve. This characteristic is discussed in Chapter 10.

Some controllers, especially older models, have a proportional band setting instead of a controller gain. The proportional band PB (in %) is defined as

$$\text{PB} = \frac{100\%}{K_c} \tag{8-3}$$

This definition applies only if K_c is dimensionless. Note that a small (narrow) proportional band corresponds to a large controller gain, while a large (wide) PB value implies a small value of K_c.

The ideal proportional controller in Eq. 8-2 and Fig. 8.4 does not include physical limits on the controller output. A more realistic representation is shown in Fig. 8.5. We say that the controller *saturates* when its output reaches a physical limit, either p_{max} or p_{min}.

To derive the transfer function for a proportional controller, define a deviation variable $p'(t)$ as

$$p'(t) = p(t) - \bar{p} \tag{8-4}$$

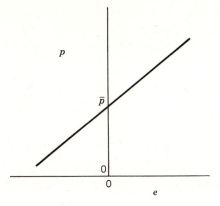

Figure 8.4. Proportional control: ideal behavior (slope of line = K_c).

Then Eq. 8-2 can be written as

$$p'(t) = K_c e(t) \tag{8-5}$$

It is unnecessary to define a deviation variable for the error signal, since e is already in deviation form and its nominal steady-state value is $\bar{e} = 0$. Taking Laplace transforms and rearranging (8-5) gives the transfer function

$$\frac{P'(s)}{E(s)} = K_c \tag{8-6}$$

An inherent disadvantage of proportional-only control is its inability to eliminate the steady-state errors that occur after a set-point change or a sustained load disturbance. In Chapter 10 we demonstrate that offset will occur with proportional-only control *regardless* of the value of K_c that is employed. In principal, offset can be eliminated by manually resetting the set point R or bias $\bar{p}$ after an offset occurs. However, this approach is inconvenient since it requires operator intervention and the new value of R (or $\bar{p}$) must usually be found by trial and error. In practice, it is more convenient to use a controller that contains integral control action; this mode provides automatic reset, as discussed below.

In some control applications where offsets can be tolerated, proportional-only control is attractive because of its simplicity. For example, in some level control problems, maintaining the liquid level close to the set point is not important as long as the storage tank does not overflow or run dry.

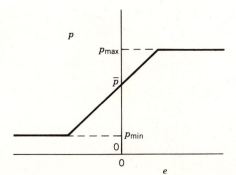

Figure 8.5. Proportional control: actual behavior.

Integral Control

Integral control action is also referred to as *reset* or *floating control*. Here the controller output depends on the integral of the error signal over time,

$$p(t) = \bar{p} + \frac{1}{\tau_I} \int_0^t e(t^*) \, dt^* \tag{8-7}$$

where τ_I is referred to as the integral time or reset time and has units of time. For commercial controllers, controller parameter τ_I is adjustable.

Integral control action is widely used because it provides an important practical advantage, the elimination of offset. To understand why offset is eliminated, consider Eq. 8-7. If the process is at steady state, then the error signal e and controller output p are constant. But Eq. 8-7 implies that p will change with time until $e(t^*) = 0$. Thus, when integral action is used, p will attain a value that causes the steady-state error to be zero, after a set-point change or sustained load disturbance has occurred. This desirable situation occurs unless the controller output or final control element saturates (reaches a limiting value) and remains there, unable to bring the controlled variable back to the desired set point. In this case, the saturation persists because the disturbance or set-point change is beyond the range of the manipulated variable.

While the elimination of offset is usually an important control objective, the integral controller in Eq. 8-7 is seldom used by itself since little control action occurs until the error signal has persisted for some time. In contrast, proportional control action takes immediate corrective action as soon as an error is detected. Consequently, integral control action is normally employed in conjunction with proportional control as the popular proportional–integral (PI) controller:

$$p(t) = \bar{p} + K_c \left[e(t) + \frac{1}{\tau_I} \int_0^t e(t^*) \, dt^* \right] \tag{8-8}$$

The corresponding transfer function for the PI controller in Eq. 8-8 is given by

$$\frac{P'(s)}{E(s)} = K_c \left(1 + \frac{1}{\tau_I s} \right) = K_c \left(\frac{\tau_I s + 1}{\tau_I s} \right) \tag{8-9}$$

The response of the PI controller to a unit step change in $e(t)$ is shown in Fig. 8.6. At time zero, the controller output changes instantaneously due to the pro-

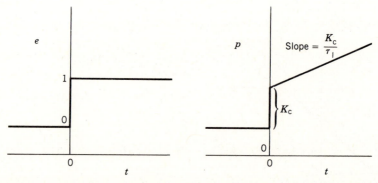

Figure 8.6. Response of proportional-integral controller to unit step change in $e(t)$.

portional action. Integral action causes the ramp increase in $p(t)$ for $t > 0$. When $t = \tau_I$, the integral term has contributed the same amount to the controller output as the proportional term. Thus, the integral action has "repeated" the proportional action once. Some commercial controllers are calibrated in terms of $1/\tau_I$ (repeats per minute) rather than τ_I (minutes). For example, if $\tau_I = 0.2$ min, this corresponds to $1/\tau_I$ having a value of 5 repeats/minute.

One disadvantage of using integral action is that it tends to produce oscillatory responses of the controlled process and thus reduces system stability. A limited amount of oscillation can usually be tolerated since it often is associated with a faster response. The undesirable effects of too much integral action can be avoided by proper tuning of the controller or by including derivative action which tends to counteract the destabilizing effects.

Reset Windup

An inherent disadvantage of integral control action is a phenomenon known as *reset windup*. Recall that the integral mode causes the controller output to change as long as $e(t^*) \neq 0$ in Eq. 8-8. When a sustained error occurs, the integral term becomes quite large and the controller output eventually saturates. Further buildup of the integral term while the controller is saturated is referred to as reset windup or *integral windup*. Figure 8.7 shows a typical response to a step change in set point when a PI controller is used. Note that the indicated areas under the curve provide either positive or negative contributions to the integral term depending on whether the controlled variable is below or above the set point R. The large overshoot occurs because the integral term continues to increase until the error signal changes sign at $t = t_1$. Only then does the integral term begin to decrease. After the integral term becomes sufficiently small, the controller output moves away from the saturation limit and has the value determined by Eq. 8-8.

Reset windup occurs when a PI or PID controller encounters a sustained error, for example, during the start-up of a batch process or after a large set-point change. It can also occur as a consequence of a large sustained load disturbance that is beyond the range of the manipulated variable. In this situation a physical limitation (control valve fully open or completely shut) prevents the controller from reducing the error signal to zero. Clearly, it is undesirable to have the integral term continue to build up after the controller output saturates since the controller is already doing

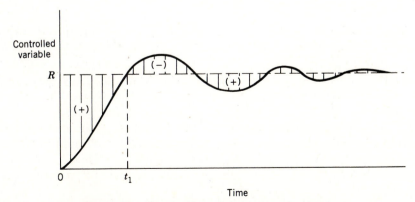

Figure 8.7. Reset windup during a set-point change.

all it can to reduce the error. Fortunately, commercial controllers are available which provide *antireset windup*. This feature reduces reset windup by temporarily halting the integral control action whenever the controller output saturates. The integral action resumes when the output is no longer saturated. The antireset windup feature is sometimes referred to as a *batch unit* because it is required when batch processes are started up automatically [7].

Derivative Control

Derivative control action is also referred to as rate action, pre-act, or anticipatory control. Its function is to anticipate the future behavior of the error signal by considering its rate of change. For example, suppose that a reactor temperature increases by 10 °C in a short period of time, say 3 min. This clearly is a more rapid increase in temperature than a 10 °C rise in 30 min, and it could indicate a potential *runaway* situation for an exothermic reaction. If the reactor were under manual control, an experienced plant operator would anticipate the consequences and quickly take appropriate corrective action to reduce the temperature. Such a response would not be obtainable from the types of automatic controllers discussed so far. Note that a proportional controller reacts to a deviation in temperature only, making no distinction as to the time period over which the deviation develops. Addition of the integral mode would also be ineffective. In fact, the integral mode builds up corrective action proportional to the length of time a disturbance persists, so a slower disturbance would generate the larger corrective action.

The anticipatory strategy used by the experienced operator can be incorporated in automatic controllers by making the controller output proportional to the rate of change of the controlled variable. Thus, for *ideal* derivative action,

$$p(t) = \bar{p} + \tau_D \frac{de}{dt} \tag{8-10}$$

where τ_D, the derivative time, has units of time. Note that the controller output is equal to the nominal value $\bar{p}$ as long as the error is constant (that is, as long as $de/dt = 0$). Thus, derivative action is never used alone; it is always used in conjunction with proportional or proportional–integral control. In combination with proportional control, for example, the PD controller has the ideal transfer function

$$\frac{P'(s)}{E(s)} = K_c(1 + \tau_D s) \tag{8-11}$$

By providing anticipatory control action, the derivative mode tends to stabilize the controlled process. It is often used to counteract the destabilizing tendency of the integral mode (see Chapter 10).

Derivative control also tends to improve the dynamic response of the controlled variable by decreasing the process settling time, the time it take the process to reach steady state. But if the process measurement is *noisy,* that is, if it contains high-frequency, random fluctuations, then the derivative of the measured (controlled) variable will change wildly and derivative action will amplify the noise unless the measurement is *filtered.* Consequently, derivative action is seldom used for flow control since flow control loops respond quickly and flow measurements tend to be noisy.

Derivative action can be combined with proportional and integral action to form the ideal three-mode or PID controller. The ideal PID controller equation is formed by summing the three modes:

$$p(t) = \bar{p} + K_c \left[e(t) + \frac{1}{\tau_I} \int_0^t e(t^*) \, dt^* + \tau_D \frac{de}{dt} \right] \qquad (8\text{-}12)$$

which has the transfer function,

$$\frac{P'(s)}{E(s)} = K_c \left(1 + \frac{1}{\tau_I s} + \tau_D s \right) \qquad (8\text{-}13)$$

However, an electronic or pneumatic device that provides ideal derivative action cannot be built (is *physically unrealizable*). Commercial controllers approximate the ideal behavior in Eq. 8-13 by using transfer functions of the form

$$\frac{P'(s)}{E(s)} = K_c \left(\frac{\tau_I s + 1}{\tau_I s} \right) \left(\frac{\tau_D s + 1}{\alpha \tau_D s + 1} \right) \qquad (8\text{-}14)$$

where α is a small number, typically between 0.05 and 0.2 [7,8]. The effect of making this approximation is dealt with in Exercise 14.13.

One disadvantage of the ideal controller in Eq. 8-12 is that a sudden change in set point (and hence e) will cause the derivative term to become very large and thus provide a *derivative kick* to the final control element. This sudden jolt is undesirable and can be avoided by basing the derivative action on the measurement B, rather than on the error signal, e. Thus, replacing de/dt by $-dB/dt$ in Eq. 8-12 gives

$$p(t) = \bar{p} + K_c \left[e(t) + \frac{1}{\tau_I} \int_0^t e(t^*) \, dt^* - \tau_D \frac{dB}{dt} \right] \qquad (8\text{-}15)$$

This method of eliminating derivative kick has become a standard feature of most commercial controllers. A few authors [9,10] have advocated eliminating the set point from the proportional term, as well as the derivative term, to eliminate the *proportional kick* that occurs after a step change in set point. However, this approach has not been widely adopted.

It should be emphasized that the dial settings for K_c, τ_I, and τ_D on standard pneumatic and electronic controllers are only nominal values. Ideally, one would like to adjust these three parameters independently, but in practice there are interactions among the control modes on standard PID controllers. For these reasons the effective values may differ from the nominal values by as much as 30%. By contrast, in digital control systems the controller settings can be specified as accurately as desired, with no interaction among the modes.

Reverse or Direct Action

The controller gain can be made either negative or positive. When $K_c > 0$, the controller output $p(t)$ increases as the input signal $B(t)$ decreases. This is a *reverse-acting* controller. When $K_c < 0$, the controller is said to be *direct-acting* since the controller output increases as the input increases. Note that these definitions are based on the physical input signal $B(t)$ rather than the error signal $e(t)$. Direct- and reverse-acting proportional controllers are compared in Fig. 8.8.

To illustrate why both reverse- and direct-acting controllers are needed, again consider the flow control loop in Fig. 8.2. Suppose that the flow transmitter is direct acting in the sense that its output signal increases as the flow rate increases.

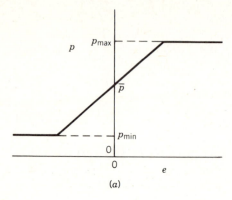

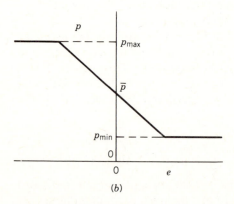

Figure 8.8.   Reverse and direct-acting proportional controllers. (a) reverse acting $K_c > 0$ (b) direct acting ($K_c < 0$).

(Most transmitters are designed to be direct acting.) Also assume that the flow control valve is designed so that the flow through the valve increases as the signal to the valve $p(t)$ increases. In this case the valve is designated as *air-to-open* (see Chapter 9). The question is: Should the flow controller have direct or reverse action? Clearly, when the measured flow is higher than the set point we want to reduce the flow by closing the control valve a bit. For an air-to-open valve, this implies that the controller output signal should be decreased. Thus, the controller should be reverse-acting.

But what if the control valve is *air-to-close* rather than air-to-open? Now when the process flow rate is too high, the controller output should increase to further close the valve. Here a direct-acting controller is required.

It is extremely important that the controller action be set correctly since an incorrect choice usually results in loss of control. For the flow control example, having the wrong controller action would force the control valve to stay fully open or fully closed (why?). Thus, the controller action must be carefully checked after a controller is installed or when a troublesome control loop is being analyzed.

Automatic/Manual Control Modes

Equations 8-12 and 8-13 describe how the ideal controllers perform during the *automatic mode* of operation. However, in certain situations the plant operator may wish to override the automatic mode and adjust the controller output manually.

This *manual mode* of operation is very useful during a plant start-up, shutdown, or emergency situation. Conventional controllers have a manual/automatic switch that is used to transfer the controller from the automatic mode to the manual mode, and vice versa. However, the controller output can change abruptly during the manual/automatic transfer and "bump" the process. Newer controllers allow *bumpless transfers,* which do not upset the process.

A controller may be left *on manual* for long periods of time (or indefinitely) if the operator is not satisfied with the control system performance. Consequently, if a significant percentage of the controllers in a control room are on manual, it is an indication that the automatic control systems are not performing well, or that the plant operators do not have much confidence in them. The topic of trouble-shooting poorly performing control loops is covered in Chapter 13.

On–Off Controllers

On–off controllers are simple, inexpensive feedback controllers that are commonly used as thermostats in heating systems and domestic refrigerators. They are also used in noncritical industrial applications such as some level control loops and heating systems. However, on–off controllers are less widely used than the various versions of PID controllers since they are not as versatile or effective.

For ideal on–off control, the controller output has only two possible values:

$$p(t) = \begin{cases} p_{max} & \text{if } e \geq 0 \\ p_{min} & \text{if } e < 0 \end{cases} \tag{8-16}$$

where p_{max} and p_{min} denote the on and off values, respectively. For example, for a pneumatic controller, we could specify $p_{max} = 15$ psi and $p_{min} = 3$ psi. On-off controllers can be modified to include a dead band for the error signal to reduce sensitivity to measurement noise [7]. Equation 8-16 also indicates why on–off control is referred to as *two-position* or *bang-bang* control. Note that on–off control can be considered to be a special case of proportional control with a very high controller gain (see Fig. 8.5).

The disadvantages of on–off control are that it results in continuous cycling of the controlled variable and produces excessive wear on the final control element. The latter disadvantage is significant if a control valve is used but less of a factor for solenoid valves or solenoid switches.

Typical Responses of Feedback Control Systems

The responses shown in Fig. 8.9 illustrate the typical behavior of a controlled process after a step change in a load variable occurs. The controlled variable C is shown as a deviation from the initial steady-state value. If no feedback control is used, the process slowly reaches a new steady state. Proportional control speeds

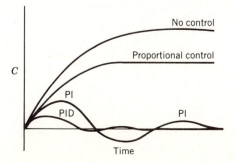

Figure 8.9. Typical process responses with feedback control.

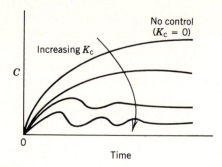

Figure 8.10. Proportional control: effect of controller gain.

up the process response and reduces the offset. The addition of integral control action eliminates offset but tends to make the response more oscillatory. Adding derivative action reduces both the degree of oscillation and the response time. It should be emphasized that the use of P, PI, and PID controllers does not always result in oscillatory process responses; this depends on the choice of the controller settings (K_c, τ_I, and τ_D) and the particular process dynamics. However, the responses in Fig. 8.9 are typical of what occurs in practice.

The qualitative effects of changing the individual controller settings are shown in Figs. 8.10–8.12. In general, increasing the controller gain tends to make the process response less sluggish; however, if too large a value of K_c is used, the response may exhibit an undesirable degree of oscillation or even become unstable. Thus, an intermediate value of K_c usually results in the best control. These guidelines are also applicable to PI and PID control, as well as to the proportional controller shown in Fig. 8.10.

Increasing the reset time τ_I usually makes PI and PID control more conservative (sluggish) as shown in Fig. 8.11. Theoretically, offset will be eliminated for all values of τ_I between 0 and ∞. But, for extremely large values of τ_I, the controlled variables will return to the set point very slowly after a load upset or set-point change occurs.

It is more difficult to generalize about the effect of the derivative time τ_D. For small values, increasing τ_D tends to improve the response by reducing the maximum deviation, response time, and degree of oscillation, as shown in Fig. 8.12. However, if τ_D is too large, measurement noise tends to be amplified and the response may become oscillatory. Thus, an intermediate value of τ_D is desirable. A more detailed discussion of how PID controller settings should be determined is presented in Chapters 12, 13, and 16.

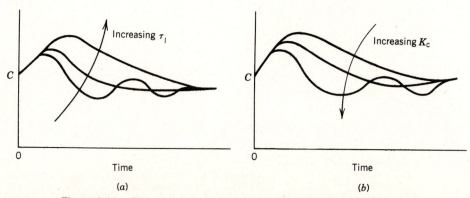

Figure 8.11. PI control: (a) effect of reset time (b) effect of controller gain.

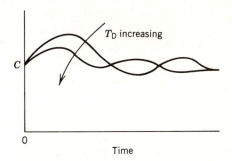

Figure 8.12.   PID control: effect of derivative time.

8.3 DIGITAL VERSIONS OF PID CONTROLLERS

So far in this chapter we have assumed that the input and output signals of the controller are continuous functions of time. This will be the situation for a conventional pneumatic or electronic controller. During the past two decades, there has been widespread application of digital control systems due to their flexibility, computational power, and cost effectiveness. In this section we briefly introduce digital control techniques by considering digital versions of PID control. A more complete discussion of digital computer control will be presented in Part 6.

When a feedback control strategy is implemented digitally, the controller input and output must be digital (or sampled) signals rather than continuous (or analog) signals. Thus, the continuous signal from the transmitter is sampled and converted periodically to a digital signal by an analog-to-digital converter (ADC). A digital control algorithm is then used to calculate the controller output, a digital signal. Before the controller output is sent to a final control element (such as a control valve), this digital signal is converted to a corresponding continuous signal by a digital-to-analog converter (DAC). Alternatively, the digital signal can be converted to a sequence of pulses representing the *change* in controller output. The pulses are then sent directly to a final control element that utilizes pulse inputs to change its position. Control valves driven by pulsed stepping motors are often used with digital controllers.

A straightforward way of deriving a digital version of the ideal PID control law in Eq. 8-12 is to replace the integral and derivative terms by their discrete equivalents. Thus, approximating the integral by a summation and the derivative by a first-order backward difference gives

$$p_n = \bar{p} + K_c \left[e_n + \frac{\Delta t}{\tau_I} \sum_{k=1}^{n} e_k + \frac{\tau_D}{\Delta t} (e_n - e_{n-1}) \right] \tag{8-17}$$

where Δt = the sampling period (the time between successive samples of the controlled variable)
  p_n = controller output at the nth sampling instant, $n = 1, 2, \ldots$
  e_n = error at the nth sampling instant.

The other symbols in Eq. 8-17 have the same meaning as in Eq. 8-12. Equation 8-17 is referred to as the *position form* of the PID control algorithm since the actual controller output is calculated.

An alternative approach is to use a *velocity form* of the algorithm in which the *change* in controller output is calculated. It can be derived by writing the position form of the algorithm (8-17) for the $(n - 1)$ sampling instant:

$$p_{n-1} = \bar{p} + K_c \left[e_{n-1} + \frac{\Delta t}{\tau_I} \sum_{k=1}^{n-1} e_k + \frac{\tau_D}{\Delta t} (e_{n-1} - e_{n-2}) \right] \tag{8-18}$$

Note that the summation still begins at $k = 1$ since it is assumed that the system is at the desired steady state for $k \leq 0$, that is, $e_k = 0$ for $k \leq 0$. Subtracting (8-18) from (8-17) gives the velocity form of the ideal PID algorithm:

$$\Delta p_n = p_n - p_{n-1} = K_c \left[(e_n - e_{n-1}) + \frac{\Delta t}{\tau_I} e_n + \frac{\tau_D}{\Delta t} (e_n - 2e_{n-1} + e_{n-2}) \right]$$

(8-19)

Note that in the velocity form, the incremental change in controller output is calculated. The velocity form has three advantages over the position form:

1. It inherently contains some provision for antireset windup since the summation of errors is not explicitly calculated.
2. The output Δp_n is in a form to be utilized directly by final control elements that require an input specifying change in position, such as the valve driven by a pulsed stepping motor mentioned above.
3. For velocity control algorithms, putting the controller in automatic mode, that is, switching it from manual operation, does not require any initialization of the output ($\bar{p}$ in Eq. 8-17). Presumably, the valve (or other final control element) has been placed in the appropriate position during the start-up procedure.

Certain types of advanced control strategies, such as cascade control and feedforward control, often require that the actual controller output p_n be explicitly calculated. These strategies are discussed in Chapters 17 and 18. However, the actual controller output can easily be calculated by rearranging Eq. 8-19 to solve for p_n.

$$p_n = p_{n-1} + K_c \left[(e_n - e_{n-1}) + \frac{\Delta t}{\tau_I} e_n + \frac{\tau_D}{\Delta t} (e_n - 2e_{n-1} + e_{n-2}) \right] \quad (8\text{-}20)$$

A minor disadvantage of the velocity form, even if rearranged as in (8-20), is that the integral mode *must* be included. Note that the set point cancels out in both the proportional and derivative error expressions, except momentarily after a change in set point is made. Hence, application of the velocity form without including the integral mode would yield a controlled process that is likely to drift away from the set point.

Digital versions of PID control are widely used in industry. For a more detailed discussion of digital control algorithms, see Chapter 26.

SUMMARY

In this chapter we have discussed the most commonly employed types of feedback controllers. Although there are potentially many types of feedback controllers, the process industries have chosen variations of the PID (or three-mode) controller and the on–off controller as standards. The remaining important elements within the control loop—sensors, transmitters, and final control elements—are discussed in detail in the next chapter. Once the steady-state and dynamic characteristics of these elements are understood, we can investigate the dynamic characteristics of the controlled process.

REFERENCES

1. Mayr, O., *The Origins of Feedback Control*, MIT Press, Cambridge, MA, 1970.
2. Ziegler, J. G., Those magnificent men and their controlling machines, *J. Dynamic Systems, Measurement and Control, Trans. ASME* **97**, 279 (1975).

3. Grebe, J. J., R. H. Boundy, and R. W. Cermak, The control of chemical processes, *Trans. AIChE* **29,** 211 (1933).

4. Ivanoff, A., Theoretical foundations of the automatic regulation of temperature, *J. Inst. Fuel* **7,** 117 (1934).

5. Williams, T. J., Two decades of change—A review of the 20-year history of computer control, *Control Eng.* **24** (9), 71 (1977).

6. Mellichamp, D. A. (Ed.), *Real-Time Computing with Applications to Data Acquisition and Control,* Van Nostrand Reinhold, New York, 1983.

7. Shinskey, F. G., *Process Control Systems,* 3d ed., McGraw-Hill, New York, 1988.

8. Luyben, W. L., *Process Modeling, Simulation and Control for Chemical Engineers,* McGraw-Hill, New York, 1973.

9. Bernard, J. W., and J. F. Cashen, Direct digital control, *Instruments and Control Systems* **38** (9), 151 (1965).

10. Phelan, R. M., *Automatic Control System,* Cornell Univ. Press, Ithaca, NY, 1977.

EXERCISES

8.1. Analog proportional–derivative (PD) controllers are constructed with an actual transfer function of the form

$$G_a(s) = K_c \left(\frac{\tau_D s + 1}{\alpha \tau_D s + 1} \right)$$

where $\alpha \approx 0.05$ to 0.2. The ideal PD transfer function is obtained when $\alpha = 0$.

$$G_i(s) = K_c(\tau_D s + 1)$$

(a) Analyze the accuracy of this approximation for step and ramp responses. Treat α as a parameter and let $\alpha \to 0$.

(b) Why might it be difficult to construct an analog device with exactly this ideal transfer function?

(c) Is there any advantage in *not* being able to obtain the ideal transfer function?

8.2. The ideal three-mode controller has the transfer function given by Eq. 8-13. Many commercial analog controllers can be described by the so-called interacting transfer function, the form given by Eq. 8-14.

(a) For the simplest case, $\alpha \to 0$, find the relations between the settings for the ideal controller $(K_c^\dagger, \tau_I^\dagger, \tau_D^\dagger)$ and the settings for the interacting controller (K_c, τ_I, τ_D).

(b) Does the effect of interaction make each controller setting (K_c, τ_I, or τ_D) larger or smaller than would be expected from use of an ideal controller?

(c) What are the magnitudes of these interaction effects for $K_c = 4$, $\tau_I = 10$ min, $\tau_D = 2$ min?

(d) What can you say about the effect of nonzero α on these relations? (Discuss only first-order effects.)

8.3. A liquid-level control system can be configured in either of two ways: with a control valve manipulating flow of liquid into the holding tank (as in the left figure), or with

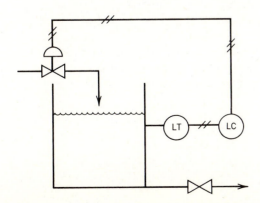
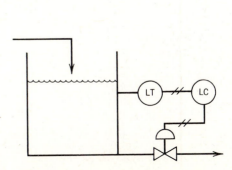

a control valve manipulating the flow of liquid from the tank (right figure). Assuming that the liquid-level transmitter always is direct acting,

(a) For each configuration what control action should a proportional pneumatic controller have if the control valve is air-to-open.

(b) Repeat part (a) for an air-to-close control valve.

8.4. (a) Find an expression for the amount of derivative kick that will be applied to the process when using the position form of the PID digital algorithm (Eq. 8-17) if a step set-point change of magnitude Δr is made between the $n - 1$ and n sampling instants.

(b) Repeat for the proportional kick, that is, the sudden change caused by the proportional mode.

(c) Plot the sequence of controller outputs at the $n - 1$, n, . . . sampling times for the case of a step set-point change of Δr magnitude made just after the $n - 1$ sampling time if the controller receives a constant measurement $\overline{B}$ and the initial set point is $\overline{r} = \overline{B}$. Assume that the controller output initially is $\overline{p}$.

(d) How can Eq. 8-17 be modified to eliminate derivative kick?

8.5. Repeat Exercise 8.4 for the case of the velocity form of the PID algorithm (Eqs. 8-19 and 8-20). In this case calculate and plot both Δp_n and p_n.

8.6. (a) For the case of the digital *velocity* P and PD algorithms, show how the set point enters into calculation of Δp_n on the assumption that it is not changing, that is, $r_{n-2} = r_{n-1} = r_n = \overline{r}$.

(b) What do the results indicate about use of the velocity form of P and PD digital control algorithms?

(c) Are similar problems encountered if the integral mode is present, that is, with PI and PID forms of the velocity algorithm? Explain.

CHAPTER 9

Control System Instrumentation

Having discussed analog (continuous signal) and digital PID controllers in Chapter 8, we turn our attention to the remaining elements in the control loop. As an illustrative example, consider the stirred-tank heating system of Fig. 8.1 which shows the use of a thermocouple to convert the temperature of the liquid into a millivolt-level signal representing the temperature. This measurement is then amplified to a voltage level and transmitted continuously to the controller. The controller output (volts) is transmitted to the process where it is converted to electrical energy (watts) in an SCR unit; the electrical energy in turn is dissipated into heat energy via the final control element, in this case an electrical resistance heater. This figure illustrates three important functions that must be carried out in each control loop: (1) measurement of one or more process output variables, (2) manipulation of a process input variable, and (3) signal transmission.

For an analog controller and analog instrumentation of the type shown in Fig. 8.1, the controller/process interconnection can be considered to be an *interface*, a term that is more commonly used when computer control is employed (Chapter 21). The interconnection is required with a single controller (Fig. 9.1*a*) or with a number of controllers (Fig. 9.1*b*). In each case the interface consists of all measurement, manipulation, and transmission instruments.

The interface elements in Fig. 9.1 all contain a common feature: Each involves the conversion of information, for example, temperature to a voltage-level signal. Devices that convert physical or chemical information in one form into an alternative physical form are called *transducers;* hence, this chapter largely consists of a discussion of transducers. We first present some general information about transducers before discussing specific methods of measuring and transmitting process variables. Final control elements, those devices that are used to manipulate the process, are discussed with an emphasis on pneumatic control valves. The chapter ends with a general discussion of instrumentation accuracy.

This chapter is intended to introduce some of the key ideas of instrumentation practice and to point out how the choice of measurement and manipulation hardware affects the choice of controllers, and vice versa. Many of the assumptions that are commonly used to simplify the design of control systems—linear behavior of instruments and manipulators, negligible instrumentation and signal transmission dynamics—depend on the proper design and specification of control loop instru-

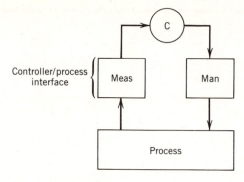

Meas = Measurement
Man = Manipulation

(*a*) Single loop control system

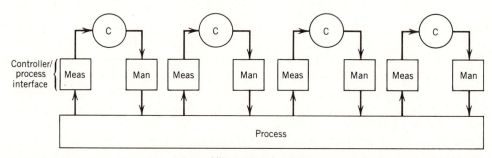

(*b*) multiloop control system

Figure 9.1.   The controller/process interface.

mentation. A number of general references and handbooks can be utilized for specification of instrumentation [1–8].

Note that Chapter 21 specifically covers digital computer control and digital instrumentation systems. Although a significant amount of instrumentation used with digital control systems now is based on digital technology, for example, pulse output flow meters and pulse input (stepping motor driven) valves, traditional analog instrumentation is still commonly used. Consequently, we consider it in detail in this chapter.

9.1 TRANSDUCERS AND TRANSMITTERS

Figure 9.2 illustrates the general configuration of a measurement transducer; it typically consists of a sensing element combined with a driving element (transmitter). Transducers for process control measurements convert the magnitude of a process variable—flow rate, pressure, temperature, level, concentration, and so on—into a signal that can be sent directly to the controller. The sensing element is required to convert the measured quantity, that is, the process variable, into some quantity more appropriate for mechanical or electrical processing within the transducer. Some process variables, such as liquid flow rate, pressure, and level, are measured relatively easily. Others, such as chemical composition, and flow rate of solid materials, are inherently difficult to sense (measure) in a quantitative way. The lack of appropriate sensing elements for many process variables is a key limitation to the application of process control.

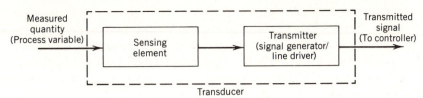

Figure 9.2. A typical process transducer.

The transmitter usually is required to convert the sensor output to a form compatible with the controller input and to drive the transmission lines connecting the two. Since analog automatic controllers are usually located in a control room that is remote from the process, the term transmitter is an appropriate designation for the combined functions of signal generation and line driving. Line driving refers to the ability of the transmitter to furnish sufficient air (for pneumatic systems) or current (for electrical systems) to overcome the inherent resistance and capacitance of the tubing or electrical lines connecting the transmitter to the controller. Process engineers often use the terms transmitter and transducer interchangeably; there is, however, the distinction noted above.

Standard Instrumentation Signal Levels

Until the 1950s, instrumentation in the process industries utilized pneumatic (air pressure) signals to transmit measurement and control information almost exclusively. These devices make use of mechanical force–balance elements to generate signals in the range of 3 to 15 psig., which has become an industry standard. Since about 1960, electronic instrumentation has come into widespread use. At one time or another signal ranges of 1 to 5 mA,[1] 4 to 20 mA, 10 to 50 mA, 0 to 5 VDC,[2] ±10 VDC, and several others have been used. Most industrial instrumentation now has standardized on a 4 to 20 mA range, although controllers and transmitters often are available with multiple outputs, such as 4 to 20 mA and 1 to 5 VDC.

The selection of appropriate process control instrumentation is closely linked to the signal transmission medium itself. For example, the choice of a sensor will be influenced significantly by whether pneumatic or electrical signals are used. In selecting a temperature transducer for the system of Fig. 8.1, the use of a thermocouple is attractive because it can be transduced easily to yield a current-level or voltage-level output. If the desired output is a 3 to 15 psig pneumatic signal, a current-to-pressure (I/P) or voltage-to-pressure (E/P) transducer can be added to the thermocouple/amplifier unit to obtain the appropriate pressure range. However, this hybrid approach would probably be more expensive than using a sensing technique specifically designed for pneumatic instrumentation, such as a pressure bulb.

Pneumatic instrumentation has evolved over a period of more than fifty years, and a wide variety of measurement techniques is available due to the clever design of mechanical elements. Pneumatic instruments are relatively inexpensive compared to electronic instruments, and they are intrinsically safe (even in hazardous or explosive environments). Although pneumatic devices continue to be used in

---

[1]mA, milliamperes (a measure of electrical current).
[2]VDC, volts, direct current.

many industrial applications, both analog and digital electronic instruments offer more features (functions), a greater degree of flexibility, and the potential for much wider areas of application. If one considers some of these intangible benefits in an economic analysis, electronic instruments usually more than justify any cost premium over pneumatic instruments. For use with electronic and digital controllers, electronic instruments almost invariably are preferred.

Sensors

Here we briefly discuss commonly used sensors for the most important process variables and indicate some of the ways that these quantities are sensed (measured) in standard instruments. Additional information is available in the references and handbooks cited earlier [1–8].

Temperature. Sensing techniques that are widely used for temperature measurement include the emf generated by the hot junction of a thermocouple, temperature-dependent electronic components such as resistors or thermistors, and pyrometry (radiation) for very hot objects. For pneumatic instruments, the expansion of a fixed amount of fluid within a closed pressure bulb provides a mechanical deflection that can be used to generate a 3 to 15 psig signal.

Pressure. With pneumatic instrumentation, pressure sensing is quite straightforward. A bellows, bourdon tube, or diaphragm isolates process liquid or gas from the instrument, at the same time furnishing a deflection to a force–balance element which generates a proportional signal in the 3 to 15 psig range. With electronic instrumentation, a strain gauge often is used to convert the deflection into a millivolt-level emf that can be amplified to an appropriate voltage or current range.

Differential Pressure. A pressure difference can be measured similarly by placing the two process pressures on either side of a diaphragm. Electronic measurements typically use a strain gauge to convert the diaphragm deflection in differential pressure instruments in the same way as in pressure measurement instruments. For many processing units, the liquid or gas streams cannot be brought into direct contact with the sensing element (diaphragm) because of high temperature or corrosion considerations. In these cases an inert fluid, usually an inert gas, is used to isolate the sensing element. Figure 9.3 indicates how this typically is done. In this case the valves are adjusted to maintain a small purge stream through the process connections. The differential pressure transmitter is mounted above the assembly.

Liquid or Gas Flow. Flow rate usually is measured indirectly, using the pressure drop across an orifice or venturi as the input signal to conventional differential pressure instrumentation. In this case, the volumetric flow rate is proportional to the square root of the pressure drop. The orifice plate is normally sized to provide a pressure drop in the range of 20 to 200 in. of water. Some means of pneumatically or electronically extracting the square root is usually available as an option on analog differential pressure transmitters; for digital systems, this function is usually performed in the computer rather than in the transmitter. Volumetric flow rates can also be measured using turbine flowmeters. The pulse output signal can be modulated to give an electronic signal, or it can be *totalized* in a counter and sent periodically to a digital controller. Deflection of a vane inserted in the pipe or channel also can be used as a flow sensor, as can a vortex shedding meter. For many process control problems, very accurate flow measurements are not required. In Example 9.3 (discussed later), the objective is to linearize the flow characteristic of a nonlinear control valve. In this case the measurement signal only needs to be reproducible; that is, the same output signal should result for a given flow rate. In situations where accuracy is required, temperature compensation (and pressure compensation, for gas flow) is utilized.

Liquid Level.   The position of a free float or the buoyancy effects on a fixed float can be detected and converted to level if the liquid density is known. The difference in pressure between the vapor above the liquid and the bottom of the liquid can be similarly used. Pressure taps (tubes connected from the transmitter to the appropriate process locations) can be kept from plugging by maintaining very low flows of inert gas through the taps to the process. The attenuation of high-energy radiation (e.g., from nuclear sources) by the liquid also can be used when solid material or gas streams cannot be put in contact with process liquids.

pH.   The acidity (basicity) of process fluids is determined by measurement of hydrogen ion concentration for which specially designed electrodes are utilized.

Viscosity.   Liquid and gas flow measurement techniques that are based on a pressure drop across a venturi or a vane deflection can be adapted to measure viscosity simply by ensuring that the flow rate past the sensor is constant. For example, a metering pump might force a constant flow stream of molten polymer through a capillary tube. Viscosity is then related to the measured pressure drop, assuming that other physical properties are constant.

Chemical Composition.   Many composition measurements are both difficult and expensive to obtain. Physical means of measuring concentrations in process liquid and gas streams are almost always preferable to chemical techniques, for example,

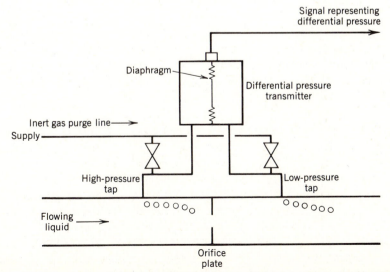

Figure 9.3.   Use of inert gas to isolate process liquid from a differential pressure sensor (transmitter).

relating the mole or mass fraction of a key liquid component to pH or conductivity, or the concentration of one component in a vapor stream to its IR or UV absorption. Often an indirect measure is used to *infer* composition; for example, the liquid temperature on a plate near the top of a distillation column might be used to indicate composition of the key overhead variable. Chromatographic methods for measurement of chemical composition have been adopted more widely in recent years as microcomputer-controlled chromatographs have become available. Such equipment can be designed and programmed to generate accurate measurements of all components in a stream. However, since results are obtained only intermittently and sometimes with significant time delays, advanced control techniques may be required to utilize the measurements effectively (see Chapter 18).

Transmitters

As noted above, a transmitter usually converts the sensor output to a signal level appropriate for input to a controller, such as 4 to 20 mA. Transmitters are generally designed to be direct-acting, that is, the output signal increases as the measured variable increases. In addition, most commercial transmitters have adjustable input ranges. For example, a temperature transmitter might be adjusted so that the input range of a platinum resistance element (the sensor) is 50 to 150 °C. In this case, the following correspondence is obtained:

Input	Output
50 °C	4 mA
150 °C	20 mA

This instrument (transducer) has a lower limit or *zero* of 50 °C and a range or *span* of 100 °C. Note that the transmitter is designed for a specific type of sensor; hence, the zero and span of the overall sensor/transmitter are adjustable. Figure 9.4

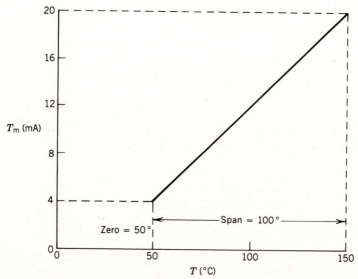

Figure 9.4. A linear instrument calibration showing its zero and span.

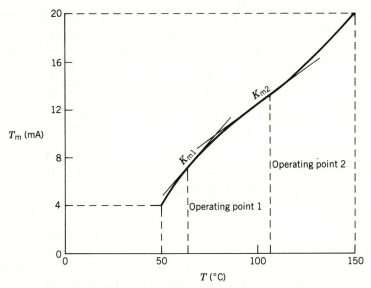

Figure 9.5. Gain of a nonlinear transducer as a function of operating point.

illustrates the concepts of zero and span. In this example, the relation between temperature and the transmitted (measured) signal is linear.

For the temperature transmitter discussed above, the relation between transducer output and input is

$$T_m(\text{mA}) = \left(\frac{20\ \text{mA} - 4\ \text{mA}}{150\ ^\circ\text{C} - 50\ ^\circ\text{C}}\right)(T - 50\ ^\circ\text{C}) + 4\ \text{mA}$$

$$= \left(0.16\ \frac{\text{mA}}{^\circ\text{C}}\right) T(^\circ\text{C}) - 4\ \text{mA}$$

The gain of the measurement element K_m is 0.16 mA/°C. For any linear instrument

$$K_m = \frac{\text{range of instrument output}}{\text{range of instrument input}} \qquad (9\text{-}1)$$

For a nonlinear instrument, the gain at any particular operating point is the tangent to the characteristic input–output relation at the operating point. Figure 9.5 illustrates a typical case. Note that the gain of such an instrument will change whenever the operating point changes; hence, it is preferable to utilize linear instruments.

Most transmitters respond rapidly. If the sensor response is also fast, the measurement dynamics can be neglected in comparison to the dynamics of the process itself, although in Section 9.4 we consider a situation where the measurement lag is appreciable. Ignoring measurement dynamics can lead to large dynamic errors.

9.2 FINAL CONTROL ELEMENTS

Every process control loop contains a final control element (actuator), the device that enables a process variable to be manipulated. For most chemical and petroleum processes, the final control elements adjust the flow rates of materials—solid, liquid, and gas feeds and products—and, indirectly, the rates of energy transfer to and from the process. Figure 8.1 illustrates the use of an electrical resistance heater as

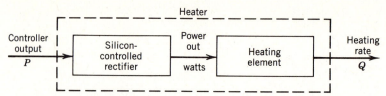

Figure 9.6.   Use of a transducer (SCR) to match controller output to the final control element yielding a linear overall relationship between Q and P.

the final control element. In this case the controller output, a voltage signal, cannot be applied directly to the terminals of the heater because the controller is not designed to supply the electrical energy requirements of the heater. Hence, a transducer must be placed between the controller and the heater, as shown in Fig. 9.6. A suitable choice might be a silicon-controlled rectifier (SCR) for relatively small heater capacities (e.g., several kilowatts). An SCR is designed to provide a nearly linear relation between the voltage input to its solid-state circuits and its power output [9]. As in the case of the resistance element/transmitter combination, a single gain would usually be employed for the overall SCR/heater unit (see Fig. 9.7).

Control Valves

There are many different ways to manipulate the flows of material and energy into and out of a process; for example, the speed of a pump drive, screw conveyer, or blower may be varied. However, a simple and widely used method of accomplishing this result with fluids is to use a control valve, also called an automatic control valve. Such valves typically utilize some type of mechanical driver to move the valve plug into and out of its seat, thus opening or closing the area for fluid flow. The mechanical driver can be either (1) a dc motor or a stepping motor that screws the valve stem in and out in much the same way as a hand valve would be operated, or (2) a pneumatically operated diaphgram device that moves the stem or the baffle against the opposing force of a fixed spring. As an alternative, the drive can rotate a baffle rather than turn a screw. Motor drivers are used for very large valves and with some electronic controllers.

The stepping motor is particularly useful with digital controllers utilizing a velocity algorithm since it rotates a small fraction of a turn (2 or 3°) for each pulse sent to its drive circuitry. Hence Δp_n in Eq. 8-19 is easily converted to the number of pulses required to open or close the valve the needed amount. Pulses are sent out by the computer over two electrical input lines, one to open the valve and one to close it.

Despite the growing use of motor-driven valves, most control applications utilize pneumatically driven control valves of the type shown schematically in Fig. 9.8. As the pneumatic controller output signal increases, increased pressure on the diaphragm compresses the spring, thus pulling the stem out and opening the valve further. As discussed briefly in Section 8.3, such a valve is termed *air-to-open* (A–O). By reversing *either* the plug/seat or the spring/air inlet orientation, the valve becomes *air-to-close* (A–C). For example, if the spring is located below the

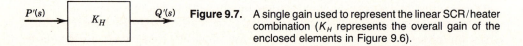

$P'(s)$ → K_H → $Q'(s)$ **Figure 9.7.**   A single gain used to represent the linear SCR/heater combination (K_H represents the overall gain of the enclosed elements in Figure 9.6).

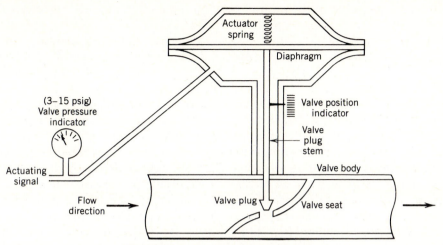

Figure 9.8. A pneumatic control valve.

diaphragm and the air inlet is placed above the diaphragm, an air-to-close valve results. Normally, the choice of A–O or A–C valve is based on safety considerations. We choose the way the valve should operate (full flow or no flow) in case of a transmitter failure. Hence A–C and A–O valves often are referred to as *fail-open* and *fail-closed,* respectively.

EXAMPLE 9.1

Pneumatic control valves are to be specified for the applications listed below. State whether an A–O or A–C valve should be specified for the following manipulated variables and give reason(s).

(a) Steam pressure in a reactor heating coil.
(b) Flow rate of reactants into a polymerization reactor.
(c) Flow of effluent from a wastewater treatment holding tank into a river.
(d) Flow of cooling water to a distillation condenser.

Solution

(a) A–O (fail closed) to make sure that a transmitter failure will not cause the reactor to overheat, which is usually more serious than having it operate at too low a temperature.
(b) The choice would depend on the application: A–O (fail closed) to prevent the reactor from being flooded with excessive reactants, A–C (fail open) if the reactor flow rate normally is close to the maximum flow rate of the valve and opening it fully would cause relatively little change in operating conditions.
(c) A–O (fail closed) to prevent excessive and perhaps untreated waste from entering the stream.
(d) A–C (fail open) to ensure that overhead vapor is completely condensed before it reaches the receiver.

Pneumatic control valves can be equipped with a valve positioner, a type of mechanical feedback device (mechanical control loop) that senses the actual stem position, compares it to the desired position, and adjusts the air pressure to the valve accordingly. Valve positioners are used to increase the relatively small mechanical force that can be exerted by a 3 to 15 psig pressure signal operating directly on the valve diaphragm. Valve positioners also eliminate valve hysteresis, flow rate loading (the effect of back pressure on the valve opening), and other undesirable characteristics.

Control valves are specified by first considering both properties of the process fluid and the desired flow chracteristics in order to choose the valve body material and type. Then the desired characteristics for the *topworks* (actuator) are considered. The choice of construction material depends on the corrosive properties of the process fluid at operating conditions. Commercial valves made of brass, carbon steel, and stainless steel can be ordered off-the-shelf, at least in smaller sizes. For large valves and more exotic materials of construction, special orders usually are required.

A design equation used for sizing control valves relates valve lift ℓ to the actual flow rate q by means of the *valve coefficient* C_v, the proportionality factor which depends predominantly on valve size or capacity:

$$q = C_v f(\ell) \sqrt{\frac{\Delta P_v}{g_s}} \qquad (9\text{-}2)$$

where q is the flow rate, $f(\ell)$ is the flow characteristic, ΔP_v is the pressure across the valve, and g_s is the specific gravity of the fluid. This relation is valid for non-flashing liquids. Reference [6] gives detailed information for other situations.

Specification of the valve size is dependent on the so-called *valve characteristic* f. Three control valve characteristics are mainly used. For a fixed pressure drop across the valve, the flow characteristic $f(0 \leq f \leq 1)$ is related to the lift ℓ ($0 \leq \ell \leq 1$), that is, the extent of valve opening, by one of the following relations:

Linear:	$f = \ell$
Quick opening:	$f = \sqrt{\ell}$
Equal percentage:	$f = R^{\ell-1}$

where R is a valve design parameter that is usually in the range 20 to 50. Figure 9.9 illustrates these three flow/lift characteristics graphically. Two minor points that are related to terminology should be noted: (1) The quick-opening valve above is referred to as a *square root* valve (valves with quicker-opening characteristics are available). (2) The equal percentage valve is given that name because the slope of the f versus ℓ curve, $df/d\ell$, is a constant fraction of f, leading to an equal percentage change in flow for a particular change in ℓ anywhere in the range.

Sizing Control Valves.

Unfortunately, sizing of control valves depends on the fluid processing units, such as pumps, heat exchangers, filters, that are placed in series with the valve. Considering only *control* objectives, the valve would be sized to take most of the pressure drop in the line. This choice would give the valve maximum influence over process changes that disturb the flow rate, such as upstream (supply) pressure changes. It also would yield the smallest (least expensive) valve. However, the most *economical operating conditions* require the valve to introduce as little pressure drop as possible, thus minimizing pumping costs. A common design compromise is to size the valve to take approximately one-quarter to one-third of the total pressure drop at the design flow rate.

To illustrate the trade-offs, consider the following example adapted from Luyben [10]. A control valve with linear characteristics is placed in series with a heat

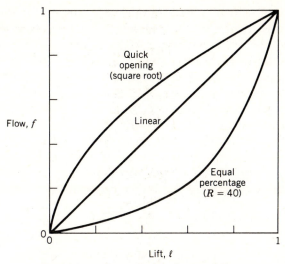

Figure 9.9.   Control valve characteristics.

exchanger, both supplied by a pump with a constant discharge pressure at 40 psi (although its discharge flow rate varies). If the heat exchanger has already been sized to give a 30-psi pressure drop (ΔP_{he}) for a 200-gal/min flow of liquid (specific gravity equal to 1), then the valve would take a 10-psi drop (ΔP_v). Figure 9.10 shows the equipment configuration. Luyben suggests that the linear control valve be sized so that it is half open ($f = \ell = 0.5$) at these conditions. Hence

$$C_v = \frac{200}{0.5\sqrt{10}} = 127 \tag{9-4}$$

which, using manufacturers' data books, would require a 4-in. control valve.

   Luyben also discusses the problem of control valve rangeability, defined for instruments generally as the ratio of maximum to minimum signal level. For control valves, rangeability translates to the need to operate the valve within the range $0.05 \le f \le 0.95$ or a rangeability of $0.95/0.05 = 19$. For the case where the flow is reduced to 25% of design, that is, 50 gal/min, the heat exchanger pressure drop will be reduced approximately to 1.9 psi [$30 \times (0.25)^2$], leaving the valve to supply the remaining 38.1 psi. The valve operating value of f, obtained by rearranging (9-2), is $50/127\sqrt{38.1}$ or 0.06; hence the valve is barely open. If the valve exhibits any hysteresis or other undesirable behavior due to internal *stiction* (combination of mechanical sticking and friction), it likely will *cycle* (close completely, then open too wide) as the controller output signal attempts to maintain this reduced flow

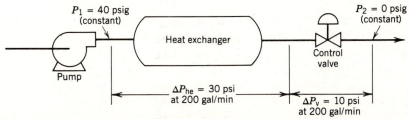

Figure 9.10.   A control valve placed in series with a pump and a heat exchanger. Pump discharge pressure is constant.

setting. A valve positioner will reduce the cycling somewhat. If a centrifugal pump is used, the pump discharge pressure will actually increase at the lower flow rate, leading to a lower value of f and even worse problems.

The choice of valve characteristic and its effect on valve sizing deserve some discussion at this point. A valve with linear behavior would appear to be the most desirable; however, the designer's objective is to obtain an *installed* flow characteristic that is as linear as possible, that is, to have the flow through the valve and all connected process units vary linearly with ℓ. Because ΔP_v usually varies with flow rate, a nonlinear valve often will yield a more linear flow relation *after installation* than will a linear valve characteristic. In particular, the equal percentage valve is designed to compensate, at least approximately, for changes in ΔP_v with flow rate. In the heat exchanger case, the valve coefficient C_v should be selected to match the choice of $\Delta P_{he}/\Delta P_v$ at design operating conditions. The objective is to obtain a nearly linear relation for q with respect to ℓ over the normal operating range of the valve.

Control valves must be sized quite carefully. This can be particularly difficult because many of the published recommendations are ambiguous or conflicting, or do not match control system objectives. For example, Luyben [10] recommends that the valve be half open at nominal operating conditions, while Moore [11] recommends that the required C_v not exceed 90% of the valve's rated C_v. The latter approach may result in poor control of the process if the controller output often exceeds the "required" (design) conditions as a result of disturbances.[3] On the other hand, Moore's recommendation does reduce the valve size considerably compared to the use of Luyben's criterion. Moore [11] suggests that, for some newer valve designs, the valve should take 33% of the pressure drop at nominal operating conditions; figures as low as 5 to 10% have also been recommended [12]. The lower figures will yield larger valves, therefore higher equipment costs but lower pumping costs (energy costs) due to lower pressure loss.

The objective of installed valve linearity (constant valve gain) complicates the picture. Will the valve generally operate in a narrow region around design conditions? If not, the design must provide a much larger region of linearity to ensure that the valve gain will be approximately constant. Some general guidelines are given below:

1. If the pump characteristic (discharge pressure vs. flow rate) is fairly flat and system frictional losses are quite small over the entire operating region, choose a linear valve. However, this situation occurs infrequently because it results from an overdesigned process (pump and piping capacity too large).
2. To select an equal percentage valve:

   a. Plot the pump characteristic curve and ΔP_s, the system pressure drop curve without the valve, as shown in Figure 9.11. The difference between these two curves is ΔP_v. The pump should be sized to obtain the desired value of $\Delta P_v/\Delta P_s$, for example, 25 to 33%, at the design flow rate q_d.[4]
   b. Calculate the valve's rated C_v, the value that yields at least 100% of q_d with the available pressure drop at that higher flow rate.

---

[3]At least one process construction company uses the more conservative (relative to disturbance control) criteria that the required valve C_v be sized at 70% the valve's rated C_v and that the C_v required to accommodate the maximum expected flow rate (not the design flow rate) equal 90% of the valve's rated C_v.
[4]Note that q_d is used to denote a particular value of $\bar{q}$. It is the steady-state flow rate for which the system is designed.

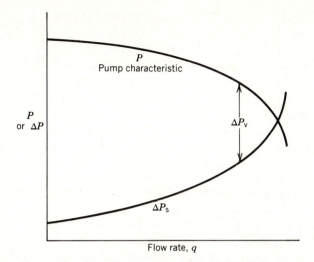

Figure 9.11. Calculation of the valve pressure drop (ΔP_v) from the pump characteristic curve and the system pressure drop without the valve (ΔP_s).

c. Compute q as a function of ℓ using Eq. 9-2, the rated C_v, and ΔP_v from (a). A plot of the valve characteristic (q versus ℓ) should be reasonably linear in the operating region of interest (at least around the design flow rate). If it is not suitably linear, adjust the rated C_v and repeat.

EXAMPLE 9.2

In Luyben's example, the pump furnishes a constant head of 40 psi over the entire flow rate range of interest. The heat exchanger pressure drop is 30 psig at 200 gal/min (q_d) and can be assumed to be proportional to q^2. Select the rated C_v of the valve and plot the installed characteristic for the following cases:

(a) A linear valve that is half open at the design flow rate.
(b) An equal percentage valve ($R = 50$ in Eq. 9-3) that is sized to be completely open at 110% of the design flow rate.
(c) As in **(b)** except with a C_v that is 20% higher than calculated.
(d) As in **(b)** except with a C_v that is 20% lower than calculated.

Solution

First we write an expression for the pressure drop across the heat exchanger

$$\frac{\Delta P_{he}}{30} = \left(\frac{q}{200}\right)^2 \tag{9-5}$$

$$\Delta P_s = \Delta P_{he} = 30\left(\frac{q}{200}\right)^2 \tag{9-6}$$

Since the pump head is constant at 40 psi, the pressure drop available for the valve is

$$\Delta P_v = 40 - \Delta P_{he} = 40 - 30\left(\frac{q}{200}\right)^2 \tag{9-7}$$

Figure 9.12 illustrates these relations. Note that in all four design cases $\Delta P_v/\Delta P_s = 10/30 = 33\%$ at q_d.

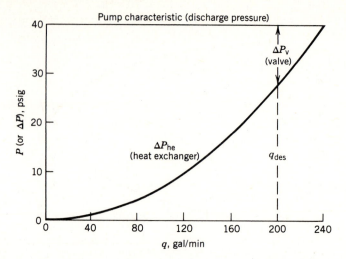

Figure 9.12.   Pump characteristic and system pressure drop for Example 9.2.

(a) First calculate the rated C_v.

$$C_v = \frac{200}{0.5\sqrt{10}} = 126.5. \qquad (9\text{-}8)$$

We will use $C_v = 125$. For a linear characteristic valve, use the relation between ℓ and q from Eq. 9-2:

$$\ell = \frac{q}{C_v\sqrt{\Delta P_v}} \qquad (9\text{-}9)$$

Using Eq. 9-9 and values of ΔP_v from Eq. 9-7, the installed valve characteristic curve can be plotted (Fig. 9.13).

(b) Again calculate the rated C_v (valve fully open) at 110% of q_d

$$C_v = \frac{220}{\sqrt{3.7}} = 114.4$$

Use a value of $C_v = 115$. For the equal percentage valve, rearrange Eq. 9.2 as follows:

$$\frac{q}{C_v\sqrt{\Delta P_v}} = R^{\ell-1} \qquad (9\text{-}10)$$

or

$$\ell = 1 + \log\left(\frac{q}{C_v\sqrt{\Delta P_v}}\right)\bigg/\log R \qquad (9\text{-}11)$$

Substituting $C_v = 115$, $R = 50$, and values of q and ΔP_v yields the installed characteristic curve in Fig. 9.13.

(c) $C_v = 1.2(115) = 138$

(d) $C_v = 0.8(115) = 92$

Again, the installed characteristics are given in Fig. 9.13. Note that the maximum flow rate that could be achieved in this system (negligible pressure drop across the valve) would be with a pressure drop of 40 psi across the heat exchanger:

$$\left(\frac{q_{max}}{200}\right)^2 = \frac{40}{30} \qquad (9\text{-}12)$$

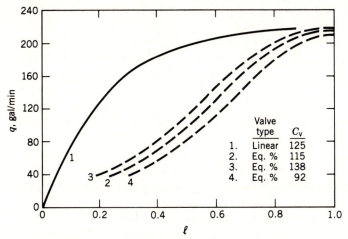

Figure 9.13. Installed valve characteristics for Example 9.2.

The effect of including each of the valves can be seen from the plots of the installed valve characteristics. From these results we conclude that an equal percentage valve with $C_v \approx 115$ would give a reasonably linear installed characteristic over a large range of flows and has sufficient capacity to accommodate flows as high as 110% of the design flow rate.

Sometimes valve nonlinearities can be partially compensated by choosing a sensor with nonlinear characteristics. A differential pressure transmitter used in conjunction with an orifice plate to measure flow rate yields a signal that is proportional to the square of the flow rate unless square root compensation is included. Such a flow transmitter might be paired with a valve having a square root characteristic (quick-opening valve) in a flow control loop. In this case the valve characteristic will tend to compensate for the nonlinearity in the flow transmitter; however, there would be no inherent compensation for pressure drop (ΔP_v) changes in such an arrangement, as discussed earlier.

The sizing and selection of control valves is complicated by many side issues not noted above. Lipták [13] discusses a number of such points, including the discrepancies between idealized valve characteristics and those of actual valves as presently manufactured. The reader requiring more specific information should refer to Refs. [11–13] or to the general references at the beginning of this chapter.

9.3 TRANSMISSION LINES

Pneumatic pressure signals between controllers and instruments are transmitted by means of tubing, usually PVC-coated copper tubing of $\frac{1}{4}$- or $\frac{3}{8}$-in. diameter although polyethylene tubing also can be used in short, noncritical applications. The propagation of a signal changing in time through such a medium is limited by dynamic accuracy considerations (Section 9.4) to one or two hundred meters at most. Hence, pneumatic controllers will usually be located relatively close to their associated sensors and actuators, even when placed in an isolated (remote) control room.

Electronic controllers using two-wire current loop (e.g., 4 to 20 mA) signal transmission can be located relatively far from their instruments with little concern for the impedance of the intervening transmission lines or for the time of transmission, which, for all practical purposes, is instantaneous. Multipair shielded cable is usually used for this purpose. A further advantage of such two-wire systems is that the power supply can be located *in the loop;* thus, separate wiring is not required. Voltage-level control and instrumentation signals (e.g., 1 to 5 VDC) are better restricted to laboratory environments, where short distances are normally encountered. Otherwise transmission line resistances have to be taken into account in calibrating the instruments. Whenever millivolt-level signals such as thermocouple outputs (emf) are transmitted over any significant distance (more than several meters), particular care must be taken. Such a situation might arise when all of the thermocouple transmitters for a particular process are placed in a single rack, which introduces a variable distance for each sensor. Very careful practice must be followed in wiring and terminating these signals to prevent biasing, attenuation, or inducing noise (e.g. 60-Hz noise) in the transmission lines. Usually shielded coaxial cable is required to minimize these effects [9].

Signals from digital instruments and controllers are usually transmitted in digital format as a sequence of on–off pulses. The transmissions can be made over a number of parallel wires as we noted in Chapter 8, for example, when a digital controller operates a stepping motor to open or close a valve by sending a sequence of pulses over one of two signal lines to the motor. Newer instrumentation systems utilize digital transmission over a single *data highway,* usually a coaxial cable that is linked in serial or daisy-chain (serial with a complete loop) fashion to all instruments and controllers. A second coaxial cable is usually installed for backup. A microcomputer built into each instrument or controller is responsible for communicating periodically over the highway, either directing information to, or requesting information from, some other device. Digital transmission techniques, discussed more fully in Chapter 21, have revolutionized the way plants and control rooms are wired. Replacing the many wires, cables, and tubes typically found in older control systems with several coaxial or fiber optic cables has greatly simplified the installation and maintenance problems. This approach has reduced the associated costs of newer digital instrumentation systems.

9.4 ACCURACY IN INSTRUMENTATION

The subject of accuracy or inaccuracy of control instrumentation is important; unfortunately, it also is confusing. One reason is that there are so many terms relating to the accuracy of instruments. Also, the definition and interpretation of these measures are not obvious to most students and beginning engineers. Another reason is that accuracy requirements are inherently tied up with control system objectives. For example, cooling water flow errors on the order of 10% (of the measured flow rate) might be acceptable in a control loop regulating the temperature of a liquid leaving a condenser, so long as the measurements are simply biased from the true value by this constant amount. Errors in the feed flow to a process on the order of 1 or 2% might be completely unacceptable if throughput/inventory calculations must be made with the measured data or if the measurement errors are random.

Process applications often involve the direct control of secondary or environmental variables (process operating conditions such as flow rates, temperatures, and pressures). Their effect on primary process variables (compositions, finished product properties such as the mechanical strength of a polymeric material, etc.)

can be measured only indirectly or after long delays in an analytical laboratory. The relation between primary and secondary variables may be vaguely known. In these situations it is not so important that the measurements of the secondary variables be made accurately as that they be made consistently. Manual adjustment of the controller set points based on information from the laboratory or from product testing will eventually bring the secondary variables to the proper or best values regardless of biases in the instruments. Hence, temperature transmitters that are 5 or 10 degrees in error, or flow transmitters that are in error by 10% of the measurement are occasionally used with relatively little detrimental effect on the process.

However, the number of control applications where such errors can be tolerated is small and is decreasing. In particular, as computers replace people in the supervision of processes, instrument accuracy has grown more important.

Terms Used to Describe Instrumentation Accuracy

Lipták [14] notes that accuracy designations for control instruments are often misused. We say that a transmitter has $\pm X\%$ accuracy when we should say that it is $\pm X\%$ inaccurate. Confusion also exists among the terms *precision, resolution, accuracy,* and *repeatability.*

To make these definitions clear, *error* should be defined. Error is the difference between a perfect measurement and the measurement that actually is made. Since an instrument is designed to operate over a particular input range, error is often expressed as a percentage of full scale (% FS), or, less commonly, as referred to the input (RTI), that is, as a fraction of the input value. Consider a hypothetical experiment with some device measuring the flow rate of a liquid. Under conditions that should yield a constant flow rate, we make a number of measurements and record them, assuming that the instrument can be read to the nearest 0.01 flow unit. Figure 9.14 might result. For this instrument, the *precision* is limited to (no better than) ± 0.01 flow units since we cannot read the instrument any more precisely. Precision is related to *resolution,* which is defined as the smallest change in

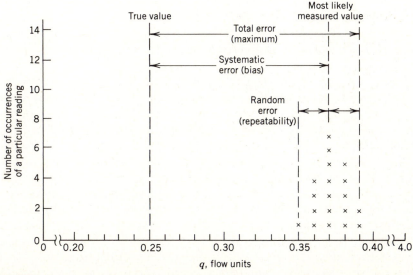

Figure 9.14. Analysis of types of error for a flow instrument whose range is 0 to 4 flow units.

the input that will result in a significant change in transducer output. In this case the resolution can be no better than 0.01 flow units, although we can verify this value only by changing the input by this amount and observing if the measured value changes by an equivalent amount.

Referring to the figure, systematic error or bias gives an average (probable) measured value that is $0.37 - 0.25 = 0.12$ flow units too high at these conditions. The maximum error can be as large as 0.14 units; hence, the *accuracy* (inaccuracy) is no worse than 0.14 at these conditions. Similarly, the *repeatability* of the measured value is ±0.02 flow units at the constant conditions of this experiment. Note that the precision (resolution) of a transducer can be quite good while its accuracy is quite poor. The converse is not true.

By performing this experiment at a number of conditions over the full range of the instrument we can evaluate its resolution, accuracy, and repeatability and express them appropriately; for example, accuracy can be expressed as a percentage of full scale. If the instrument provides a continuous measurement, that is, if it furnishes an estimate of flow rate that is continuous in time, we can obtain much the same information from a recording rather than from discrete data points. However, in this case, the interpretation of precision and repeatability is not as straightforward.

The transducer output signal for an actual flow measuring instrument might appear as in Fig. 9.15*a,* where the ideal, linear relation between measured and actual flow also is shown. If the ideal relation is taken to be true, the discrepancy, plotted in Fig. 9.15*b* as a percentage of full scale (% FS), represents the transducer error. From this plot we observe that the *limit of error* is 5% of full scale. Figure 9.15*c* represents the transducer error as a percentage of the actual input (RTI); note that the constant error limits on a % FS plot become hyperbolic on an RTI plot. Hence, transducer accuracy potentially can be quite poor in the lower range if calculated on an RTI basis [14].

As a final point on terminology, the *total error* of a series of instruments should be measured or estimated by summing the relative (% FS) errors of each of the instrument components over the range of interest or by performing a standard error analysis using estimates of the component errors. For a transducer consisting of orifice plate, flow (ΔP) transmitter, and square root extraction, each element might have an error of 0.5% FS. Hence, their total error could be as high as 1.5% FS leading to errors as large as 15% RTI at flow levels around 10% of full scale (i.e., 1.5%/0.10 = 15%). Obviously, this is a worst-case calculation, but errors of this magnitude are often seen with instrumentation operated in the low end of its design range.

There are a number of types (or sources) of instrument error, among which nonlinearity, hysteresis (backlash), drift, and dynamic (lag or time delay) deserve additional comment. Nonlinearity was a significant source of error prior to the availability of digital instrumentation. Now, as discussed below, digital instruments can self-compensate to yield an accurate output signal. Instruments connected to computer data logging and/or control systems can also be compensated within the digital computer. The other three sources of error are not so easily dealt with. Hysteresis, characterized by an output result that depends on the direction of change of the instrument input, usually results from nonideal magnetic or electrical components. Backlash, the mechanical equivalent of hysteresis, usually results from friction effects or gears with play. Drift is characterized by a slowly changing instrument output when the input is constant; it often results from faulty or temperature-sensitive electrical components, particularly operational amplifiers.

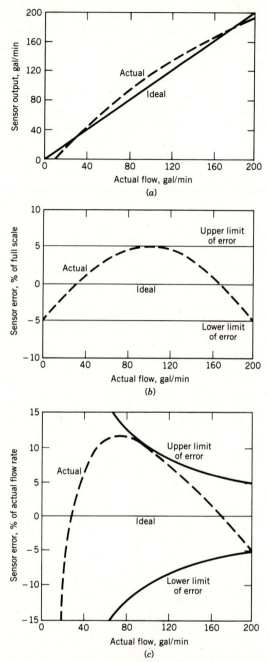

Figure 9.15. Analysis of instrument error showing the increased error at low readings (from Lipták[14]).

Hysteresis, drift, and dynamic errors (discussed below) can be avoided altogether by proper design and maintenance of instrument systems.

Calibration of Instruments

Any measurement instrument from which a high degree of accuracy is expected should be calibrated both initially (before commissioning) and periodically (as it remains in service). The same is true of any nonlinear transducer used in a computer control system. In the latter case, calibration of the instrument includes fitting an equation to the calibration data. This expression can be used to increase the accuracy of the control loop and, in some cases, to modify the controller gain on-line as operating conditions change. In the simplest method, the computer calculates the transducer gain from time to time and adjusts the controller gain so that the product of the two remains approximately constant. This approach can help to ensure consistent operation of the controlled process (as will be discussed in Section 18.4) if it is operated over a wide operating range.

In recent years, the use of so-called *smart sensors* has become more widespread. These devices incorporate a microcomputer as part of the sensor/transmitter, giving a transducer that can calibrate itself periodically, furnish a linearized output signal, and check for malfunctions. Such instruments can greatly reduce the need for in-service calibration and checkout.

Dynamic Errors Resulting from Measurement and Transmission Lags

The measuring elements, transmission lines, and final control element introduce dynamic lags into the control loop. For example, Fig. 9.16 shows a thermocouple placed in a metal thermowell with mass m and specific heat C. The dynamic lag introduced by the thermowell/thermocouple combination can be easily estimated if several simplifying assumptions are made. In particular, assume that the well and thermocouple are always at the same temperature T_m, which can be different from the surrounding fluid temperature T. Further assume that heat is transferred only between the fluid and the well (there are no ambient losses from the thermowell due to conduction along its length to the environment). An energy balance on the thermowell gives

$$mC \frac{dT_m}{dt} = UA(T - T_m) \tag{9-13}$$

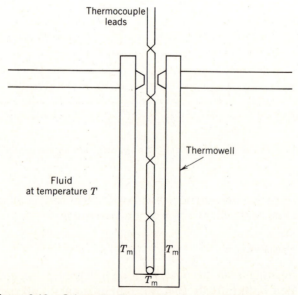

Figure 9.16. Schematic diagram of a thermowell/thermocouple.

Here U is the heat transfer coefficient and A is the area for heat transfer. Rearranging gives,

$$\frac{mC}{UA}\frac{dT_m}{dt} + T_m = T \tag{9-14}$$

which, after converting to deviation variables and transforming, becomes

$$\frac{T'_m(s)}{T'(s)} = \frac{1}{\tau s + 1} \tag{9-15}$$

with $\tau = mC/UA$.

Based on the transfer function in (9-15), the dynamic measurement lag of the sensor will be minimized if the thermal capacitance of the well (mC) is made as small as possible and if UA is made large. The combined effect will be to make τ small. Thus, we should make the thermowell as thin as possible, consistent with maintaining isolation between the thermocouple and the process fluid. At the same time, since U will be strongly dependent on the fluid velocity, the thermowell should be placed in a region of maximum fluid velocity, near the centerline of a pipe or in the vicinity of a mixing impeller. The model indicates that materials such as a plastic, which have a lower specific heat C than a metal, will yield a somewhat faster dynamic response. However, such a material typically has low heat conductivity which may invalidate the assumption that the entire thermowell is at the same temperature. In this case, a more rigorous model incorporating the effect of heat conduction in the thermowell must be used to study the effect of heat capacitance/conduction trade-offs.

Final control elements (such as pneumatic valves) and signal transducers introduce dynamic lags as well. Often these are small and can be modeled as first-order transfer functions with the time constants estimated or measured through simple experiment. With electronic transmission of measurement and control signals, any dynamic lags caused by the transmission processes can be safely neglected in most process control applications. However, with pneumatic signals this often cannot be done. A long pneumatic signal line can introduce a considerable dynamic lag [15]. This lag can be estimated using the total volume of the line and the maximum bleed rates of the driver and receiver instruments; alternatively, a simple experiment can be made to evaluate the time constant.

Any measurement transducer output will contain some dynamic error; an estimate of the error can be obtained if transducer time constant τ and the maximum expected rate of change of the variable to be measured are known. For a ramp input $x(t) = at$ and a first-order dynamic model (see Eq. 9-15), it follows that

$$Y(s) = \frac{1}{\tau s + 1} X(s) = \frac{1}{\tau s + 1}\frac{a}{s^2} \tag{9-16}$$

The time-domain solution for the ramp response of a first-order system was obtained in Eqs. 5-19 through 5-21. The maximum deviation between input and output is $a\tau$ (obtained when $t \gg \tau$), which is shown graphically in Fig. 5.5. Hence as a general result, we can say that the maximum dynamic error that can occur for any instrument with *first-order* dynamics is

$$\epsilon_{max} = |y(t) - x(t)|_{max} = a\tau$$

Clearly, reducing the time constant to zero can make the dynamic error negligibly small.

For cases where the overall measurement dynamics are not first-order, such as a first-order pneumatic transmitter in series with a first-order pneumatic transmission line, an estimate of the worst-case dynamic error can be made in similar fashion.

In general, measurement and transmission time constants should be less than one-tenth the largest process time constant, preferably much less, to keep dynamic measurement errors low. The dynamics of measurement, transmission, and final control elements also significantly limit the speed of response of the controlled process. Hence, it is important that the dynamics of these components be made as fast as is practical or economic.

SUMMARY

In this chapter we have discussed the instrumentation required in all process control applications—the sensor/transmitters that provide information about key process output variables (in a form that can be transmitted to the controllers) and the final control elements that manipulate key process input variables based on signals from the controllers—with an emphasis on analog instrumentation. The technology in this field is changing rapidly, with major new developments being the trend for more digital and microcomputer-based instrumentation and with digital techniques (data highways) being used to transmit information in digital form. Both of these topics are covered in more detail in Chapter 21. The trend toward higher data transmission rates will be enhanced by the increasing use of fiber optics to replace traditional electronic cabling.

Another major trend is the increasing integration of sensing elements into silicon chip microcircuitry [16]. Using this approach, it now is possible to measure pressure, temperature, ion and gas concentration, radiation level, and other important process variables with sensors that directly incorporate all circuitry needed to self-compensate for environmental changes and to yield a linear output that is suitably amplified for transmission to standard electronic controllers. These new sensors offer the advantage of small size, greatly reduced prices, and virtually no mechanical parts to wear out.

REFERENCES

1. Johnson, C. D., *Process Control Instrumentation Technology,* Wiley, New York, 1977.
2. Lipták, B. G., *Instrument Engineers Handbook,* Vol. 1, *Process Measurement,* Chilton, Philadelphia, 1969.
3. Ibid., Vol. 2, *Process Control,* Chilton, Philadelphia, 1970.
4. Ibid., Supplement 1, Chilton, Philadelphia, 1972.
5. Harvey, G. F. (Ed.), *ISA Transducer Compendium,* 2d ed., Part 1, IFI/Plenum, New York, 1969; Part 2, ISA Press, Pittsburgh, 1970; Part 3, ISA Press, Pittsburgh, 1972.
6. Hutchinson, J. W., *ISA Handbook of Control Valves,* 2nd ed., ISA Press, Pittsburgh, 1976.
7. Spink, L. K., *Principles and Practice of Flow Meter Engineering,* 9th ed., Plimpton, Norwood, MA, 1967.
8. Harrison, T. J., *Handbook of Industrial Control Computers,* Wiley-Interscience, New York, 1972.
9. Wright, J. D., Measurements, transmission, and signal processing, in D. A. Mellichamp, (Ed.), *Real-Time Computing with Applications to Data Acquisition and Control,* Chap. 4, Van Nostrand Reinhold, New York, 1983.
10. Luyben, W. L., *Process Modeling, Simulation and Control for Chemical Engineers,* McGraw-Hill, New York, 1973.
11. Moore, R. W., Allocating pressure drops to control valves, *Instrum. Tech.* **24,** 102, (Oct. 1977).
12. Wolter, D. G., Control valve selection—A practical guide, *Instrum. Tech.* **24,** 55, (Oct. 1977).
13. Lipták, B. G., Control valves in optimized systems, *Chem. Eng.* **90,** 104, (Sept. 1983).

14. Lipták, B. G., Flow metering accuracy, *Instrum. Tech.* **18**, 35, (July 1971).
15. Buckley, P. S., Dynamic design of pneumatic control loops, *Instrum. Tech.* **22**, 33, 39, (Apr./June 1975).
16. Allen, R., Sensors in silicon, *High Tech.* 43, (Sept. 1984).

EXERCISES

9.1. Several linear transmitters have been installed and calibrated as follows:

Flow rate: 400 gal/min → 15 psig
0 gal/min → 3 psig } pneumatic transmitter

Pressure: 30 in. Hg → 20 mA
10 in. Hg → 4 mA } current transmitter

Level: 20 m → 5 VDC
0.5 m → 1 VDC } voltage transmitter

(a) Develop an expression for the output of each transmitter as a function of its input. Be sure to include appropriate units.
(b) What is the gain of each transmitter? Is it a constant or variable quantity?

9.2. An orifice plate and differential pressure transmitter are used to measure the flow rate of a liquid stream. The pressure drop across the orifice is related to flow rate by the relation

$$\Delta P_0 = \left(\frac{q}{\bar{q}}\right)^2$$

where ΔP_0 = pressure drop

q = flow rate

$\bar{q}$ = a characteristic flow rate related to geometry of the

orifice, fluid properties, etc.

A differential pressure transmitter that converts pressure drop across the orifice to a useful signal follows either of the following relations:

i. $P_T = P_0 + K\Delta P_0$ without square root extraction
ii. $P_T = P_0 + K\sqrt{\Delta P_0}$ with square root extraction

(a) Write an expression for the transmitter output as a function of flow rate for each of the two cases.
(b) If a pneumatic transmitter has been adjusted to yield a 9 psig output at a flow rate of 100 gal/min and 3 psig at 0 gal/min, what expressions would be obtained?
(c) What would be the maximum flow rate that could be measured in each case (i and ii) using a standard pneumatic instrument range?
(d) Sketch the transmitter output vs. liquid flow rate over the full range of the two instruments.

9.3. (Adapted from Luyben [10]).

(a) Calculate the gain of an orifice plate and differential-pressure transmitter without square root extraction for flow rates of 10, 50, 75, and 90% of full scale.
(b) Calculate the gains of linear, equal-percentage, and square root valves at the same operating points, assuming constant pressure drop over the valve.
(c) For each type of valve, calculate the overall gain of the valve and sensor–transmitter system at each operating point. Which combination would yield the most linear characteristics over the range 10% to 90% of full-scale flow?
(d) Evaluate the validity of the assumptions made in this analysis; that is, to what extent would the conclusions from part (c) be expected to apply to the real situation of an orifice plate placed in series with a control valve?

Notes

i. The gain of an orifice plate and differential pressure transmitter is

$$K_t = \frac{dP_t}{dq}$$

where P_t = transmitter output
q = flow rate

ii. The gain of a control valve (constant pressure drop across the valve) is

$$K_v = \frac{dq}{dP}$$

where q = flow rate
P = pressure applied to the valve.

iii. You may assume that the valve lift ℓ is proportional to P.

9.4. Chilled ethylene glycol (sp gr = 1.11) is pumped through the shell side of a condenser and a control valve at a nominal flow rate of 200 gal/min. The total pressure drop over the entire system is constant. The pressure drop over the condenser is proportional to the square of the flow rate and is 30 psi at the nominal flow rate. Make plots of flow rate versus valve stem position ℓ for linear and equal percentage control valves, assuming that the valves are set so that $f(\ell) = 0.5$ at the nominal flow rate. Prepare these plots for the situations where the pressure drop over the control value at the design flow is

(a) 5 psi
(b) 30 psi
(c) 90 psi

What can you conclude concerning the results from these three sets of design conditions? In particular, for each case comment on linearity of the installed valve, ability to handle flow rates greater than nominal, and pumping costs.

9.5. A pneumatic control valve is used to adjust the flow rate of a petroleum fraction (sp gr = 0.9) that is used as fuel in a cracking furnace. A centrifugal pump is used to supply the fuel and an orifice meter/differential pressure transmitter is used to monitor flow rate. The nominal fuel rate to the furnace is 320 gal/min. Select an equal percentage valve that will be satisfactory to operate this system. You may use the following data (all pressures in psi; all flow rates in gal/min):

(a) Pump characteristic (discharge pressure):

$$P = (1 - 2.44 \times 10^{-6}q^2)P_{de}$$

where P_{de} is the pump discharge pressure when dead ended (no flow).
(b) Pressure drop across orifice:

$$\Delta P_0 = 1.953 \times 10^{-4}q^2$$

(c) Pressure drop across the furnace burners:

$$\Delta P_b = 40$$

(d) R for the valve: 50
(e) Operating region of interest:

$$250 \leq q \leq 350$$

Your design attempt should minimize pumping costs by keeping the pump capacity (related to P_{de}) as low as possible. In no case should $\Delta P_v/\Delta P_s$ be greater than 0.33 at the nominal flow rate. Show, by means of a plot of the installed valve characteristic (q vs. ℓ), just how linear the final design is.

9.6. An engineer sets the pressure in a supply tank using a very accurate manometer as a guide and then reads the output of a 20-psig pressure gauge attached to the tank as 10.2 psig. Sometime later she repeats the procedure and obtains values of 10.4 and 10.3 psig. What can she say about the gauge's

Precision?
Accuracy?
Resolution?
Repeatability?

Express these answers on a percentage of full scale basis.

9.7. A process temperature sensor/transmitter exhibits critically damped second-order dynamics with time constants of 1 s. If the quantity being measured changes at a constant rate of 0.1 °C/s, what is the maximum error that this instrument combination will exhibit? Show your result on a response plot.

CHAPTER 10

Dynamic Behavior of Closed-Loop Control Systems

In this chapter we consider the dynamic behavior of processes that are operated using feedback control. The combination of the process, the feedback controller, and the instruments is referred to as a *feedback control loop* or a *closed-loop system*. We begin by demonstrating that block diagrams and transfer functions provide a useful description of closed-loop systems. We then use block diagrams to analyze the dynamic behavior of several simple closed-loop systems.

10.1 BLOCK DIAGRAM REPRESENTATION

In Chapters 1 and 8 we have seen that block diagrams provide a convenient representation of the flow of information around a feedback control loop. The previous discussion of block diagrams was qualitative rather than quantitative since the blocks were labeled but did not indicate the relations between process variables. However, quantitative information can also be included by showing the transfer function for each block.

To illustrate the development of a block diagram, we consider a previous example, temperature control of a stirred-tank heater. The schematic diagram in Fig. 10.1 shows a stirred-tank heater system with steam as the heating medium. The control objective is to regulate tank temperature T by adjusting the steam pressure P_s in the heating coil. The primary load variable is assumed to be inlet temperature T_i. The tank temperature is measured by a thermocouple whose output is conditioned by a temperature transmitter before being sent to an electronic controller. Since a pneumatic control valve is used, the controller output (an electrical signal in the range of 4 to 20 mA) must be converted to an equivalent pneumatic signal by a current-to-pressure transducer. The transducer output signal is then used to adjust the steam control valve.

Next, we derive a transfer function for each component in the feedback control loop.

Process

In Section 2.4 the following dynamic model of a steam-heated, stirred tank was developed:

$$mC \frac{dT}{dt} = wC(T_i - T) + h_p A_p(T_w - T) \tag{10-1}$$

224

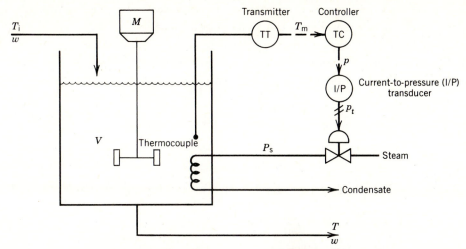

Figure 10.1 Temperature control system for a stirred-tank heater with steam heating. (— — — electrical signal; —//— pneumatic signal)

$$m_w C_w \frac{dT_w}{dt} = h_s A_s (T_s - T_w) - h_p A_p (T_w - T) \qquad (10\text{-}2)$$

Suppose that the dynamics of the heating coil are fast compared to the dynamics of the tank contents. Then as an approximation assume that the dynamics of Eq. 10-2 are negligible, which implies that the wall temperature T_w closely tracks changes in the steam temperature T_s. Setting the left side of Eq. 10-2 equal to zero,

$$0 = h_s A_s (T_s - T_w) - h_p A_p (T_w - T) \qquad (10\text{-}3)$$

Solving for T_w,

$$T_w = \frac{h_s A_s T_s + h_p A_p T}{h_s A_s + h_p A_p} \qquad (10\text{-}4)$$

and substituting into Eq. 10-1 gives

$$mC \frac{dT}{dt} = wC(T_i - T) + U_A(T_s - T) \qquad (10\text{-}5)$$

where the overall heat transfer coefficient $\times$ effective area, U_A, is defined by

$$U_A = \frac{h_p A_p h_s A_s}{h_p A_p + h_s A_s} \qquad (10\text{-}6)$$

Assume that the steam temperature T_s and condensation pressure P_s for saturated steam are related by the following expression:

$$T_s = a + bP_s \qquad (10\text{-}7)$$

Parameters a and b in Eq. 10-7 can be calculated by fitting P_s and T_s data from the steam tables for the region of interest. Substituting (10-7) into (10-5) gives

$$mC \frac{dT}{dt} = wC(T_i - T) + U_A(a + bP_s - T) \qquad (10\text{-}8)$$

The dynamic model in Eq. 10-8 relates the output variable T to two inputs, T_i and P_s, and a number of constant parameters. Next we derive the transfer functions between T and T_i and between T and P_s. Since this derivation is analogous to the one in Section 4.1 for the electrically heated, stirred tank, several inter-

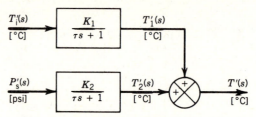

Figure 10.2. Block diagram of the process.

mediate steps will be omitted. Again, we assume that flow rate w is constant and that the system is initially at the nominal steady state. Taking the Laplace transform of Eq. 10-8 and introducing deviation variables gives

$$mCsT'(s) = wC[T_i'(s) - T'(s)] + U_A[bP_s'(s) - T'(s)] \qquad (10\text{-}9)$$

where the primes denote deviation variables. Rearranging gives

$$T'(s) = \frac{K_1}{\tau s + 1} T_i'(s) + \frac{K_2}{\tau s + 1} P_s'(s) \qquad (10\text{-}10)$$

where

$$\tau = \frac{mC}{wC + U_A} \qquad (10\text{-}11)$$

$$K_1 = \frac{wC}{wC + U_A} \qquad (10\text{-}12)$$

$$K_2 = \frac{U_A b}{wC + U_A} \qquad (10\text{-}13)$$

The desired transfer functions are then obtained directly from Eq. 10-10:

$$\frac{T'(s)}{T_i'(s)} = \frac{K_1}{\tau s + 1} \qquad \text{for } P_s'(s) = 0 \qquad (10\text{-}14)$$

$$\frac{T'(s)}{P_s'(s)} = \frac{K_2}{\tau s + 1} \qquad \text{for } T_i'(s) = 0 \qquad (10\text{-}15)$$

Figure 10.2 provides a block diagram representation of the information in Eq. 10-10 and indicates the units for each variable. A new deviation variable, $T_1'(s)$, denotes the change in exit temperature due to a change in inlet temperature. Similarly, $T_2'(s)$ is a deviation variable that denotes the change in T' due to a change in steam pressure. The effects of these changes are additive, because $T'(s) = T_1'(s) + T_2'(s)$. This is a direct consequence of the Superposition Principle for linear systems, which was discussed in Chapter 3. Recall that the transfer function representation is valid only for linear systems or for nonlinear systems that have been linearized.

Thermocouple and Transmitter

We assume that the dynamic behavior of the thermocouple and transmitter can be approximated by a first-order transfer function:

$$\frac{T_m'(s)}{T'(s)} = \frac{K_m}{\tau_m s + 1} \qquad (10\text{-}16)$$

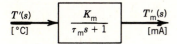

Figure 10.3. Block diagram for the thermocouple and temperature transmitter.

This element has negligible dynamics when $\tau \gg \tau_m$. For a change in one of the inputs, the measured temperature T_m' rapidly follows the true temperature T', even while T' is slowly changing with time constant τ. Hence, the dynamic error associated with the measurement can be neglected (cf. Section 9.4). A useful approximation is to set $\tau_m = 0$ in Eq. 10-16. The steady-state gain K_m depends on the input and output ranges of the thermocouple–transmitter combination, as indicated in Eq. 9-1. The block diagram is shown in Fig. 10.3.

Controller

Suppose that a proportional plus integral controller is used. Then from Chapter 8, the ideal controller transfer function is

$$\frac{P'(s)}{E(s)} = K_c \left(1 + \frac{1}{\tau_I s} \right) \tag{10-17}$$

where $P'(s)$ and $E(s)$ are the Laplace transforms of the controller output $p'(t)$ and the error signal $e(t)$. Note that p' and e are electrical signals which have units of mA while K_c is dimensionless. The error signal is expressed as

$$e(t) = \widetilde{T}_R'(t) - T_m'(t) \tag{10-18a}$$

or after taking Laplace transforms,

$$E(s) = \widetilde{T}_R'(s) - T_m'(s) \tag{10-18b}$$

The symbol $\widetilde{T}_R'(t)$ denotes the set point expressed as an electrical current signal. This signal is used internally by the controller. $\widetilde{T}_R'$ is related to the actual temperature set point T_R' by the thermocouple–transmitter gain K_m:

$$\widetilde{T}_R'(t) = K_m T_R'(t) \tag{10-19a}$$

Thus

$$\widetilde{T}_R'(s) = K_m T_R'(s) \tag{10-19b}$$

The block diagram corresponding to Eqs. 10-17 through 10-19 is shown in Fig. 10.4. The symbol that represents the subtraction operation is called a *comparator*.

In general, if a reported controller gain is not dimensionless, it includes at least one other gain of another device (such as an actuator) in addition to the dimensionless controller gain. For example, the apparent controller gain in Eq. 1-3 actually includes the gains for both the thermocouple–transmitter combination and the electrical heater–SCR combination.

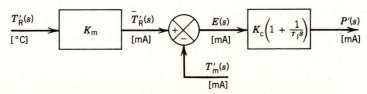

Figure 10.4. Block diagram for the controller.

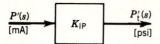

Figure 10.5. Block diagram for the I/P transducer.

Current-to-Pressure (I/P) Transducer

Since transducers are usually designed to have linear characteristics and negligible (fast) dynamics, assume that the transducer transfer function merely consists of a steady-state gain K_{IP}:

$$\frac{P'_t(s)}{P'(s)} = K_{IP} \tag{10-20}$$

In Eq. 10-20 P'_t denotes the output signal from the I/P transducer in deviation form. The corresponding block diagram is shown in Fig. 10.5.

Control Valve

As discussed in Section 9.2, control valves are usually designed so that the flow through the valve is a nonlinear function of the signal to the valve actuator. However, a first-order transfer function usually provides an adequate model for operation of an installed valve in the vicinity of a nominal steady state. Thus, we assume that the steam control valve can be modeled as

$$\frac{P'_s(s)}{P'_t(s)} = \frac{K_v}{\tau_v s + 1} \tag{10-21}$$

The block diagram for a pneumatic control valve is shown in Fig 10.6.

Now that transfer functions and block diagrams have been developed for the individual components of the feedback control system, we can combine this information to obtain the complete block diagram shown in Fig. 10.7. Note that Fig. 10.7 is merely a composite of the block diagrams for the individual components in Figs. 10.2 to 10.6.

10.2 CLOSED-LOOP TRANSFER FUNCTIONS

The block diagrams considered so far have been specifically developed for the stirred-tank heating system. The more general block diagram in Fig. 10.8 contains the standard notation:

C = controlled variable
M = manipulated variable
L = load variable
P = controller output
E = error signal
B = measured value of C
R = set point
$\tilde{R}$ = internal set point (used by the controller)
X_1 = change in C due to L

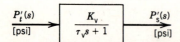

Figure 10.6. Block diagram for the control valve.

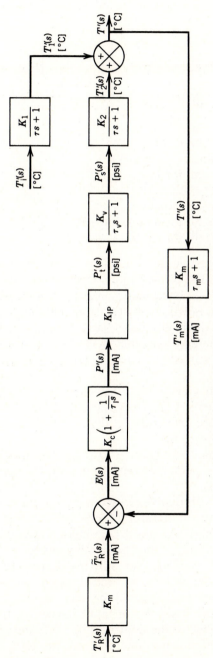

Figure 10.7. Block diagram for the entire control system.

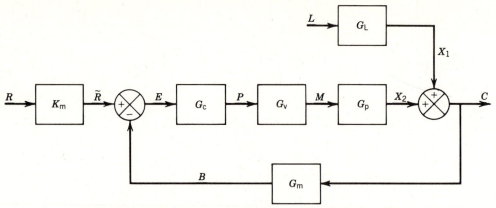

Figure 10.8. Standard block diagram of a feedback control system.

X_2 = change in C due to M
G_c = controller transfer function
G_v = transfer function for final control element
G_p = process transfer function
G_L = load transfer function
G_m = transfer function for measuring element and transmitter
K_m = steady-state gain for G_m

In Fig. 10.8, each variable is the Laplace transform of a deviation variable. To simplify the notation, the primes and s dependence have been omitted; thus, C is used rather than $C'(s)$. Because the final control element is often a control valve, its transfer function is denoted by G_v. Note that the process transfer function G_p indicates the effect of the manipulated variable on the controlled variable. The load transfer function G_L represents the effect of the load variable upon the controlled variable. For the stirred-tank heater, G_L and G_p are given in Eqs. 10-14 and 10-15, respectively.

The standard block diagram in Fig. 10.8 can be used to represent a wide variety of practical control problems. Other blocks can be added to the standard diagram to represent additional elements in the feedback control loop such as the current-to-pressure transducer in Fig. 10.7.

In Fig. 10.8, the signal path from E to C through blocks G_c, G_v, and G_p is referred to as the *forward path*. The path from C to the comparator through G_m is called the *feedback path*.

Figure 10.9 shows an alternative representation of the standard block diagram that is also used in the control literature. Since the load transfer functions appear in different locations in Figs. 10.8 and 10.9, different symbols are used. For these two block diagrams to be equivalent, the relation between C and L must be preserved. Thus, G_L and G_L^* must be related by the expression, $G_L = G_L^* G_p$.

To evaluate the performance of the control system, we need to know how the controlled process responds to changes in the load variable L and the set point R. Note that R and L are the independent input signals for the controlled process since they are not affected by the control loop. In contrast, M and L are the independent inputs for the uncontrolled process. The term closed-loop system is used to denote the controlled process. In the next section, we derive expressions for the *closed-loop transfer functions*, $C(s)/R(s)$ and $C(s)/L(s)$. But first, we review some block diagram algebra.

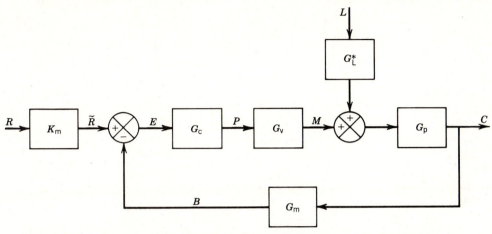

Figure 10.9. Alternative form of the standard block diagram of a feedback control system.

Block Diagram Reduction

In deriving closed-loop transfer functions it is often convenient to combine several blocks into a single block. For example, consider the three blocks in series in Fig. 10.10. The block diagram indicates the following relations:

$$X_1 = G_1 U$$
$$X_2 = G_2 X_1 \tag{10-22}$$
$$X_3 = G_3 X_2$$

By successive substitution,

$$X_3 = G_3 G_2 G_1 U \tag{10-23}$$

or

$$X_3 = GU \tag{10-24}$$

where $G \overset{\Delta}{=} G_3 G_2 G_1$. Equation 10-24 indicates that the block diagram in Fig. 10.10 can be reduced to the equivalent block diagram in Fig. 10.11.

Set-Point Changes

Next we derive the closed-loop transfer function for set-point changes. The closed-loop system behavior for set-point changes is also referred to as the servomechanism (*servo*) *problem* in the control literature because early applications were concerned with positioning devices called servomechanisms. We assume for this case that no load change occurs and thus $L = 0$. From Fig. 10.8 it follows that

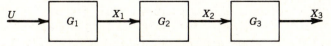

Figure 10.10. Three blocks in series.

Figure 10.11. Equivalent block diagram.

$$C = X_1 + X_2 \tag{10-25}$$

$$X_1 = G_L L = 0 \quad \text{(since } L = 0\text{)} \tag{10-26}$$

$$X_2 = G_p M \tag{10-27}$$

Combining gives

$$C = G_p M \tag{10-28}$$

Figure 10.8 also indicates the following input/output relations for the individual blocks:

$$M = G_v P \tag{10-29}$$

$$P = G_c E \tag{10-30}$$

$$E = \tilde{R} - B \tag{10-31}$$

$$\tilde{R} = K_m R \tag{10-32}$$

$$B = G_m C \tag{10-33}$$

Combining the above equations gives

$$C = G_p G_v P = G_p G_v G_c E \tag{10-34}$$

$$C = G_p G_v G_c (\tilde{R} - B) \tag{10-35}$$

$$C = G_p G_v G_c (K_m R - G_m C) \tag{10-36}$$

Rearranging gives the desired closed-loop transfer function,[1]

$$\frac{C}{R} = \frac{K_m G_c G_v G_p}{1 + G_c G_v G_p G_m} \tag{10-37}$$

Load Changes

Now consider the case of load changes, which is also referred to as the *regulator problem* since the process is to be *regulated* at a constant set point. From Fig. 10.8 for $R = 0$,

$$C = X_1 + X_2 = G_L L + G_p M \tag{10-38}$$

Substituting (10-29) through (10-33) gives

$$C = G_L L + G_p G_v G_c (K_m R - G_m C) \tag{10-39}$$

Since $R = 0$ we can rearrange (10-39) to give the closed-loop transfer function for load changes:

$$\frac{C}{L} = \frac{G_L}{1 + G_c G_v G_p G_m} \tag{10-40}$$

A comparison of Eqs. 10-37 and 10-40 indicates that both closed-loop transfer functions have the same denominator, $1 + G_c G_v G_p G_m$. The denominator is often

---

[1]In both the numerator and denominator of Eq. 10-37, the transfer functions are listed in the order in which they are encountered in the feedback control loop. This convention makes it easy to determine which transfer functions are present or missing in subsequent problems.

written as $1 + G_{OL}$ where G_{OL} is the *open-loop transfer function*, $G_{OL} \triangleq G_c G_v G_p G_m$. The term open-loop transfer function (or *open-loop system*) is used because G_{OL} relates B to $\tilde{R}$ if the feedback loop is opened just before the comparator.

At different points in the above derivations, we assumed that $L = 0$ or $R = 0$, that is, that one of the two inputs was constant. But suppose that $L \neq 0$ *and $R \neq 0$* as would be the case if a disturbance occurs during a set-point change. To analyze this situation, we rearrange Eq. 10-39 and substitute the definition of G_{OL} to obtain

$$C = \frac{G_L}{1 + G_{OL}} L + \frac{K_m G_c G_v G_p}{1 + G_{OL}} R \qquad (10\text{-}41)$$

Thus, the response to simultaneous load variable and set-point changes is merely the sum of the individual responses, as can be seen by comparing Eqs. 10-37, 10-40, and 10-41. This result is a consequence of the Superposition Principle for linear systems.

General Expression for Feedback Control Systems

Closed-loop transfer functions for more complicated block diagrams can be written in the general form [1]:

$$\frac{Y}{X} = \frac{\pi_f}{1 + \pi_e} \qquad (10\text{-}42)$$

where Y and X are any two variables in the block diagram and

$\pi_f \equiv$ product of the transfer functions in the path from X to Y

$\pi_e \equiv$ product of all the transfer functions in the entire feedback loop

Thus, for the previous servo problem we have $X = R$, $Y = C$, $\pi_f = K_m G_c G_v G_p$, and $\pi_e = G_{OL}$. For the regulator problem, $X = L$, $Y = C$, $\pi_f = G_L$, and $\pi_e = G_{OL}$.

It is important to note that Eq. 10-42 is only applicable to portions of the block diagram that include a *feedback loop* with a negative sign in the comparator.

EXAMPLE 10.1

Find the closed-loop transfer function C/R for the complex control system in Fig. 10.12. Notice that this block diagram has two feedback loops and two load variables. This configuration occurs when the cascade control scheme of Chapter 18 is employed.

Solution

Using the general rule in (10-42), we first reduce the inner loop to a single block, as shown in Fig. 10.13. We also set $L_1 = 0$ and $L_2 = 0$ because we are interested in the servo problem. Since Fig. 10.13 contains a single feedback loop, we can again use (10-42) to obtain Fig. 10.14a. The final block diagram is shown in Fig. 10.14b with $C/R = K_{m1} G_5$. Substitution for G_4 and G_5 gives the desired closed-loop transfer function:

$$\frac{C}{R} = \frac{K_{m1} G_{c1} G_{c2} G_1 G_2 G_3}{1 + G_{c2} G_1 G_{m2} + G_{c1} G_2 G_3 G_{m1} G_{c2} G_1}$$

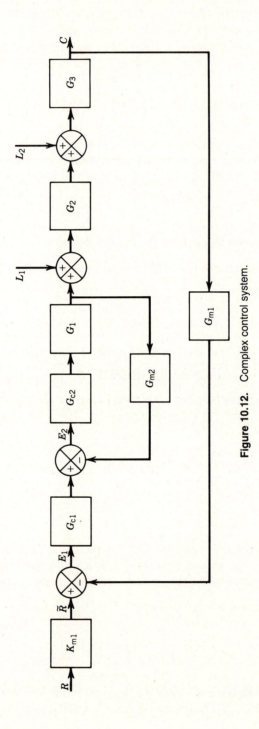

Figure 10.12.   Complex control system.

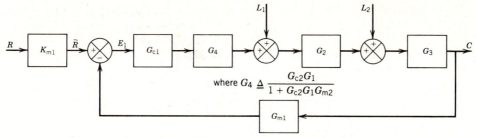

Figure 10.13. Block diagram for reduced system.

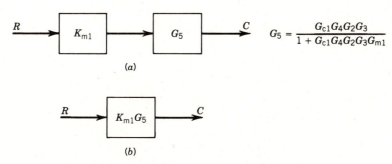

(a)

(b)

Figure 10.14. Final block diagrams for Example 10.1.

10.3 CLOSED-LOOP RESPONSES OF SIMPLE CONTROL SYSTEMS

In this section we consider the dynamic behavior of several elementary control problems for load variable and set-point changes. The transient responses can be determined in a straightforward manner if the closed-loop transfer functions are available.

Consider the liquid-level control system shown in Fig. 10.15. The liquid level is measured and the level transmitter (LT) output is sent to a feedback controller (LC) which controls liquid level by adjusting volumetric flow rate q_2. A second inlet flow rate q_1 is the load variable. We assume that:

1. The liquid density ρ and the cross-sectional area of the tank A are constant.
2. The flow–head relation is linear, $q_3 = h/R$.
3. The transmitter and control valves have negligible dynamics.
4. Pneumatic instruments are used.

To derive the process and load transfer functions, consider the unsteady-state mass balance for the tank contents:

$$\rho A \frac{dh}{dt} = \rho q_1 + \rho q_2 - \rho q_3 \tag{10-43}$$

Substituting the flow–head relation, $q_3 = h/R$, and introducing deviation variables gives

$$A \frac{dh'}{dt} = q_1' + q_2' - \frac{h'}{R} \tag{10-44}$$

By employing the methodology developed in Section 10.2, we obtain the transfer

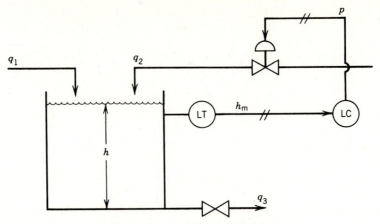

Figure 10.15. Liquid-level control system.

functions

$$\frac{H'(s)}{Q_2'(s)} = G_p(s) = \frac{K_p}{\tau s + 1} \tag{10-45}$$

$$\frac{H'(s)}{Q_1'(s)} = G_L(s) = \frac{K_p}{\tau s + 1} \tag{10-46}$$

where $K_p = R$ and $\tau = RA$. Note that $G_p(s)$ and $G_L(s)$ are identical since q_1 and q_2 are both inlet flow rates and thus have the same effect on h. These transfer functions were also derived in Example 4.4.

Since we have assumed that the level transmitter and control valve have negligible dynamics, the corresponding transfer functions can be written as $G_m(s) = K_m$ and $G_v(s) = K_v$. The block diagram for the level control system with pneumatic components is shown in Fig. 10.16. The symbol H_R' denotes the desired value of liquid level (in feet) while $\tilde{H}_R'$ denotes the corresponding value (in psi) that is used internally by the pneumatic controller. Note that these two set points are related by the level transmitter gain K_m as was discussed in Section 10.1.

The block diagram in Fig. 10.16 is in the *alternative form* of Fig. 10.9 with $G_L^*(s) = 1$.

Proportional Control and Set-Point Changes

If a proportional controller is used, then $G_c(s) = K_c$. From Fig. 10.16 and the material in the previous section, it follows that the closed-loop transfer function for set-point changes is given by

$$\frac{H'(s)}{H_R'(s)} = \frac{K_c K_v K_p K_m/(\tau s + 1)}{1 + K_c K_v K_p K_m/(\tau s + 1)} \tag{10-47}$$

This relation can be rearranged in the standard form for a first-order transfer function,

$$\frac{H'(s)}{H_R'(s)} = \frac{K_1}{\tau_1 s + 1} \tag{10-48}$$

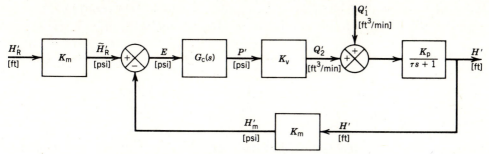

Figure 10.16. Block diagram for level control system.

where

$$K_1 = \frac{K_{OL}}{1 + K_{OL}} \tag{10-49}$$

$$\tau_1 = \frac{\tau}{1 + K_{OL}} \tag{10-50}$$

and K_{OL} is the *open-loop gain*,

$$K_{OL} = K_c K_v K_p K_m \tag{10-51}$$

Equations 10-48 to 10-51 indicate that the closed-loop process is a first-order system with a time constant τ_1 that is smaller than the process time constant τ. We assume here that $K_{OL} > 0$; otherwise the control system would not function properly, as will be apparent from the stability analysis in Chapter 11. Since $\tau_1 < \tau$, one consequence of feedback control is that it enables the controlled process to respond more quickly than the uncontrolled process.

From Eq. 10-48 it follows that the closed-loop response to a unit step change of magnitude M in set point is given by

$$h'(t) = K_1 M (1 - e^{-t/\tau_1}) \tag{10-52}$$

This response is shown in Fig. 10.17. Note that a steady-state error or *offset* exists since the new steady-state value is $K_1 M$ rather than the desired value of M. The offset is defined as

$$\text{offset} \overset{\Delta}{=} h'_R(\infty) - h'(\infty) \tag{10-53}$$

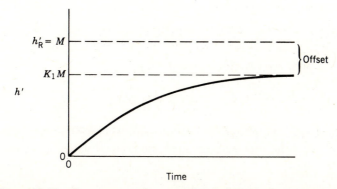

Figure 10.17. Step response for proportional control (set-point change).

For a step change of magnitude M in setpoint, $h'_R(\infty) = M$. From (10-52), it is clear that $h'(\infty) = K_1 M$. Substituting these values and (10-49) into (10-53) gives

$$\text{offset} = M - K_1 M = \frac{M}{1 + K_{OL}} \qquad (10\text{-}54)$$

EXAMPLE 10.2

Consider the level control system shown in Fig. 10.15. The tank is 1 m in diameter while the valve on the exit line acts as a linear resistance with $R = 6.37$ min/m². The pneumatic level transmitter has a span of 0.5 m and an output range of 3 to 15 psi. The flow characteristic f of the equal percentage control valve is related to the fraction of lift ℓ by the relation $f = (30)^{\ell - 1}$. The air-to-open control valve receives a 3 to 15 psi signal from the pneumatic, proportional-only controller. When the control valve is fully open ($\ell = 1$), the flow rate through the valve is 0.2 m³/min. At the nominal operating condition the control valve is half open ($\ell = 0.5$). Using the dynamic model in the block diagram of Fig. 10.16, calculate the closed-loop responses to a unit step change in the set point for three values of the controller gain: $K_c = 1$, 2, and 5.

Solution

From the given information, we can calculate the cross-sectional area of the tank A, the process gain K_p, and the time constant:

$$A = \pi \left(\tfrac{1}{2} \text{ m}\right)^2 = 0.785 \text{ m}^2$$

$$K_p = R = 6.37 \text{ min/m}^2 \qquad (10\text{-}55)$$

$$\tau = RA = 5 \text{ min}$$

The transmitter gain K_m can be calculated from Eq. 9-1:

$$K_m = \frac{\text{output range}}{\text{input range (span)}} = \frac{15 - 3 \text{ psi}}{0.5 \text{ m}} = 24 \text{ psi/m} \qquad (10\text{-}56)$$

Next, we calculate the gain for the control valve K_v. The valve relation between flow rate q and fraction of lift ℓ can be expressed as (cf. Eqs. 9-2 and 9-3)

$$q = 0.2 \, (30)^{\ell - 1} \qquad (10\text{-}57)$$

Thus

$$\frac{dq}{d\ell} = 0.2 \ln 30 \, (30)^{\ell - 1} \qquad (10\text{-}58)$$

At the nominal operating condition, $\ell = 0.5$ and

$$\frac{dq}{d\ell} = 0.124 \text{ m}^3/\text{min} \qquad (10\text{-}59)$$

The control valve gain K_v provides a steady-state relation between the controller output p (in psi) and the flow rate through the control valve, q. Thus,

$$K_v = \frac{dq}{dp} \qquad (10\text{-}60)$$

Using the chain rule for differentiation, we can write

$$K_v = \frac{dq}{dp} = \left(\frac{dq}{d\ell}\right)\left(\frac{d\ell}{dp}\right) \tag{10-61}$$

Assuming that the valve actuator is designed so that the fraction of lift ℓ varies linearly with the controller output p, it follows that

$$\frac{d\ell}{dp} = \frac{\Delta\ell}{\Delta p} = \frac{1 - 0}{15 - 3 \text{ psi}} = 0.0833 \text{ psi}^{-1} \tag{10-62}$$

Then, from Eqs. 10-59, 10-61, and 10-62

$$K_v = 0.0103 \text{ m}^3/\text{min psi} \tag{10-63}$$

An alternative method to estimate K_v is to use the tangent to the valve characteristic curve (see Chapter 9), estimated either graphically or numerically. Now that all of the gains and the time constant in Fig. 10.16 have been calculated, we can calculate the closed-loop gain K_1 and time constant τ_1 from Eqs. 10-49 and 10-50. Substituting these numerical values into Eq. 10-52 for the three values of K_c gives

K_c	τ_1 (min)	K_1
1	1.94	0.612
2	1.20	0.759
5	0.56	0.887

The closed-loop responses are shown in Fig. 10.18. Increasing K_c reduces both the offset and the time required to reach the new steady state.

Equation 10-54 suggests that offset can be reduced by increasing K_c. However, for most control problems, making K_c too large can result in oscillatory or unstable responses due to the effect of additional lags and time delays that have been neglected in the present analysis. For example, we have neglected the dynamics

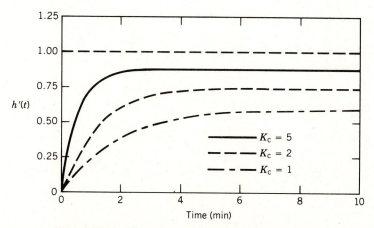

Figure 10.18. Set-point responses for Example 10.2.

associated with the control valve, level transmitter, and pneumatic transmission lines. A more rigorous analysis including the dynamics of these components would reveal the possibility of oscillations or instability. Stability problems associated with feedback control systems are analyzed in Chapter 11.

For many liquid-level control problems, a small offset can be tolerated since the vessel serves as a surge capacity (or intermediate storage volume) between processing units. If offset is not acceptable, then integral control action should be used.

Proportional Control and Load Changes

From Fig. 10.16 and Eq. 10-40 the closed-loop transfer function for load changes with proportional control is

$$\frac{H'(s)}{Q_i'(s)} = \frac{K_p/(\tau s + 1)}{1 + K_{OL}/(\tau s + 1)} \tag{10-64}$$

Rearranging gives

$$\frac{H'(s)}{Q_i'(s)} = \frac{K_2}{\tau_1 s + 1} \tag{10-65}$$

where τ_1 is defined in (10-50) and K_2 is given by

$$K_2 = \frac{K_p}{1 + K_{OL}} \tag{10-66}$$

A comparison of (10-65) and (10-48) indicates that both closed-loop transfer functions are first order and have the same time constant. However, the steady-state gains, K_1 and K_2, are different.

From Eq. 10-65 it follows that the closed-loop response to a step change in load of magnitude M is given by

$$h'(t) = K_2 M(1 - e^{-t/\tau_1}) \tag{10-67}$$

The offset can be determined from Eq. 10-67. Now $h_R' = 0$ since we are considering load changes and $h'(\infty) = K_2 M$ for a step change of magnitude M. Thus

$$\text{offset} = 0 - h'(\infty) = -K_2 M = -\frac{K_p M}{1 + K_{OL}} \tag{10-68}$$

As was the case for set-point changes, increasing K_c reduces the amount of offset for load changes.

EXAMPLE 10.3

For the liquid-level control system and numerical parameter values of Example 10.2, calculate the closed-loop response to a step change in the load variable of 0.05 m³/min. Calculate the offsets and plot the results for $K_c = 1, 2,$ and 5.

Solution

The closed-loop responses in Fig. 10.19 indicate that increasing K_c reduces the offset and speeds up the closed-loop response. The offsets are:

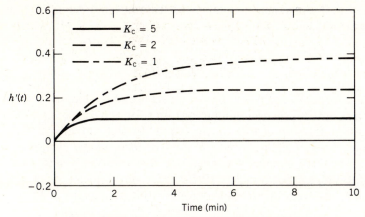

Figure 10.19.   Load responses for Example 10.3.

K_c	Offset
1	-0.124
2	-0.077
5	-0.036

The negative values of offset indicate that the controlled variable is greater than the set point.

PI Control and Load Changes

For ideal PI control, $G_c(s) = K_c(1 + 1/\tau_I s)$. The closed-loop transfer function for load changes can then be derived from Fig. 10.16:

$$\frac{H'(s)}{Q_1'(s)} = \frac{K_p/(\tau s + 1)}{1 + K_{OL}(1 + 1/\tau_I s)/(\tau s + 1)} \tag{10-69}$$

Clearing terms in the denominator gives

$$\frac{H'(s)}{Q_1'(s)} = \frac{K_p \tau_I s}{\tau_I s(\tau s + 1) + K_{OL}(\tau_I s + 1)} \tag{10-70}$$

Further rearrangement allows the denominator to be placed in the standard form for a second-order transfer function:

$$\frac{H'(s)}{Q_1'(s)} = \frac{K_3 s}{\tau_3^2 s^2 + 2\zeta_3 \tau_3 s + 1} \tag{10-71}$$

where $K_3 = \tau_I / K_c K_v K_m$ (10-72)

$$\zeta_3 = \frac{1}{2}\left(\frac{1 + K_{OL}}{\sqrt{K_{OL}}}\right)\sqrt{\tau_I/\tau} \tag{10-73}$$

$$\tau_3 = \sqrt{\tau \tau_I / K_{OL}} \tag{10-74}$$

For a unit step change in load, $Q_1'(s) = 1/s$, and (10-70) becomes

$$H'(s) = \frac{K_3}{\tau_3^2 s^2 + 2\zeta_3 \tau_3 s + 1} \tag{10-75}$$

For $0 < \zeta_3 < 1$, the response is a damped oscillation that can be described by

$$h'(t) = \frac{K_3}{\tau_3\sqrt{1 - \zeta_3^2}} \, e^{-\zeta_3 t/\tau_3} \sin[\sqrt{1 - \zeta_3^2} \, t/\tau_3] \qquad (10\text{-}76)$$

It is clear from (10-76) that $h'(\infty) = 0$ because of the exponential term. Thus, the addition of integral action eliminates offset for a step change in load. It also eliminates offset for step changes in set point. In fact, integral action eliminates offset not only for step changes but for *any type of sustained change* in load or set point. By a sustained change, we mean one that eventually settles out at a new steady-state value, as shown in Fig. 10.20. However, it does not necessarily follow that integral action will eliminate offset for an arbitrary change in set point or load. (See Exercise 10.7.)

Equation (10-76) and Fig. 10.21 indicate that increasing K_c or decreasing τ_I tends to speed up the response. Also, the response becomes more oscillatory as either K_c or τ_I decreases. But in general, closed-loop responses become more oscillatory as K_c is *increased* (see Example 11.1). The anomalous result for Eq. (10-76) is due to having neglected the small dynamic lags associated with the control valve and transmitter. If these lags are included, the transfer function in (10-71) is no longer second order and then increasing K_c makes the response more oscillatory.

PI Control of an Integrating Process

Consider the liquid-level control system shown in Fig. 10.22. This system differs from the previous example in two ways: (1) the exit line contains a pump, and (2) the manipulated variable is the exit flow rate rather than an inlet flow rate. In Section 5.3 we saw that a tank with a pump in the exit stream acts as an integrator with respect to flow-rate changes since

$$\frac{H'(s)}{Q_3'(s)} = G_p(s) = -\frac{1}{As} \qquad (10\text{-}77)$$

$$\frac{H'(s)}{Q_1'(s)} = G_L(s) = \frac{1}{As} \qquad (10\text{-}78)$$

If the level transmitter and control valve in Fig. 10.22 have negligible dynamics, then $G_m(s) = K_m$ and $G_v(s) = K_v$. For PI control, $G_c(s) = K_c(1 + 1/\tau_I s)$.

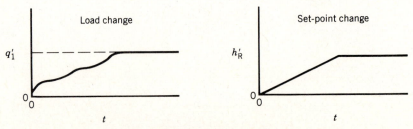

Figure 10.20. Sustained changes in load and set point.

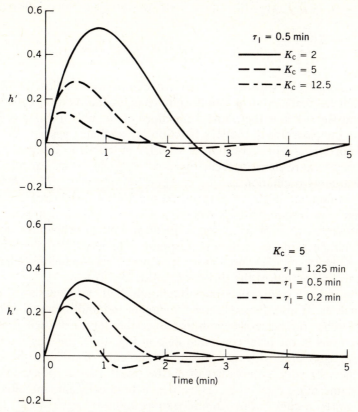

Figure 10.21. Effect of controller settings on load responses (τ = 5 min, K_p = 6.37, K_m = 24, K_v = 1/60).

Substituting these expressions into the closed-loop transfer function for load changes,

$$\frac{H'(s)}{Q_1'(s)} = \frac{G_L}{1 + G_cG_vG_pG_m} \tag{10-79}$$

and rearranging gives

$$\frac{H'(s)}{Q_1'(s)} = \frac{K_4 s}{\tau_4^2 s^2 + 2\zeta_4\tau_4 s + 1} \tag{10-80}$$

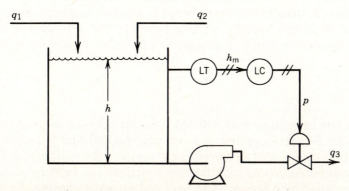

Figure 10.22. Liquid-level control system with pump in exit line.

where $K_4 = -\tau_I / K_c K_v K_m$ (10-81)

$\tau_4 = \sqrt{\tau_I / K_{OL}}$ (10-82)

$\zeta_4 = 0.5 \sqrt{K_{OL} \tau_I}$ (10-83)

and $K_{OL} = K_c K_v K_p K_m$ with $K_p = -1/A$. A comparison of Eqs. 10-78 and 10-80 indicates that feedback control significantly changes the relation between Q_1 and H. Note that Eq. 10-78 is the transfer function for the uncontrolled process while Eq. 10-80 is the closed-loop transfer function for load changes.

From our analysis of second-order transfer functions in Chapter 5, we know that the closed-loop response is oscillatory for $0 < \zeta_4 < 1$. Thus, Eq. 10-83 indicates that the degree of oscillation can be reduced by increasing either K_c or τ_I. The effect of τ_I is familiar since we have noted previously that increasing τ_I tends to make closed-loop responses less oscillatory. However, the effect of K_c is just the *opposite* of what we normally observe. In most control problems, increasing K_c tends to produce a more oscillatory response. However, (10-83) indicates that increasing K_c results in a *less* oscillatory response. This anomalous behavior is due to the integrating nature of the process (cf. Eq. 10-77).

This liquid-level system illustrates the insight that can be obtained from block diagram analysis. It also demonstrates the danger in blindly using a rule of thumb such as "decrease the controller gain to reduce the degree of oscillation."

The analysis of the level control system in Fig. 10.22 has neglected the small dynamic lags associated with the transmitter and control valve. If these lags were included, then for very large values of K_c the closed-loop response would indeed tend to become more oscillatory. Thus, if τ_I is held constant, the effect of K_c on the higher-order system can be summarized as follows:

Value of K_c	Closed-Loop Response
Small	Oscillatory
Moderate or large	Overdamped (nonoscillatory)
Very large	Oscillatory or unstable

Since the liquid-level system in Fig. 10.22 acts as an integrator, the question arises as to whether the controller must also contain integral action to eliminate offset. This question is considered further in Exercise 10.6.

In the previous examples, the denominator of the closed-loop transfer function was either a first- or second-order polynomial in s. Consequently, the transient responses to specified inputs were easily determined. In many control problems, the order of the denominator polynomial is three or higher and the roots of the polynomial have to be determined numerically. Furthermore, for higher-order $(n > 2)$ systems, feedback control can result in unstable responses if inappropriate values of the controller settings are employed. In the next chapter we analyze system stability and show how to determine whether a closed-loop system will be stable or not.

SUMMARY

This chapter has been concerned with the dynamic behavior of processes that are operated under feedback control. A block diagram provides a convenient representation for analyzing control system performance. By using block diagram algebra, expressions for closed-loop transfer functions can be derived and used to calculate the closed-loop responses to load and set-point changes. Several liquid-level control problems have been considered to illustrate key features of proportional and proportional–integral control. Proportional control results in offset for sustained load or set-point changes; however, these offsets can be eliminated by including integral control action.

REFERENCE

1. Coughanowr, D.R., and L.B. Koppel, *Process Systems Analysis and Control,* McGraw-Hill, New York, 1965, p. 134.

EXERCISES

10.1. A temperature control system for a distillation column is shown schematically in the drawing. The temperature T of a plate near the top of the column is controlled by adjusting the reflux flow rate R. Draw a block diagram for this feedback control system. You may assume that both feed flow rate F and feed composition x_F are load variables and that all of the instrumentation, including the controller, is pneumatic.

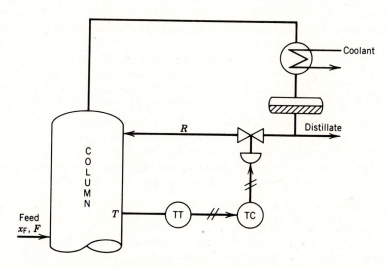

10.2. Consider the liquid-level, PI control system shown in Fig. 10.15 with the following parameter values: $A = 3$ ft^2, $R = 1.0$ min/ft^2, $K_v = 0.2$ cfm/psi, $K_m = 1.7$ psi/ft, $K_c = 4$, and $\tau_I = 3$ min. Suppose that the system is initially at the nominal steady state with a liquid level of 2 ft. If the set point is suddenly changed from 2 to 3 ft, how long will it take the system to reach (a) 2.5 ft and (b) 3 ft?

10.3. Consider proportional-only control of the stirred-tank heater control system in Fig. 10.7. The temperature transmitter has a span of 50 °F and a zero of 55 °F. The nominal design conditions are $\overline{T} = 80$ °F and $\overline{T}_i = 65$ °F. The controller has a gain of 5 while the gains for the control valve and current-to-pressure transducer are $K_v = 1.2$ (dimensionless) and $K_{IP} = 0.75$ psi/mA, respectively. The time constant for the tank is $\tau = 5$ min. After the set point is changed from 80 to 85 °F, the tank temperature eventually reaches a new steady-state value of 84.14 °F, which was measured with a highly accurate thermometer.

(a) What is the offset?
(b) What is the process gain K_2 (cf. Fig. 10.7)?
(c) What is the pressure signal p_t to the control valve at the final steady state?

10.4. It is desired to control the exit concentration c_3 of the liquid blending system shown in the drawing. Using the information given below, do the following:

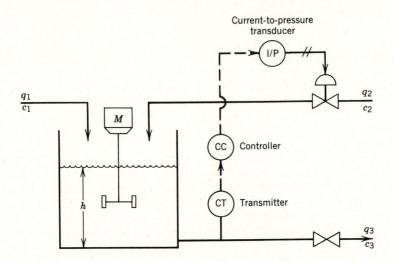

(a) Draw a block diagram for the composition control scheme. (Use the symbols in the figure as much as possible.)
(b) Derive an expression for each transfer function and substitute numerical values.
(c) Suppose that the PI controller has been tuned for the nominal set of operating conditions below. Indicate whether the controller should be retuned for *each* of the following situations. (Briefly justify your answers.)

  i. The nominal value of c_2 changes to $\bar{c}_2 = 8.5$ lb solute/ft^3.
 ii. The span of the composition transmitter is adjusted so that the transmitter output varies from 4 to 20 mA as c_3 varies from 3 to 14 lb solute/ft^3.
iii. The zero of the composition transmitter is adjusted so that the transmitter output varies from 4 to 20 mA as c_3 varies from 4 to 10 lb solute/ft^3.

Available Information

1. The tank is perfectly mixed.
2. The volumetric flow rate and solute concentration of stream 2, q_2 and c_2, vary with time while those of stream 1 are constant.
3. The flow–head relation for the valve on the exit line is given by

$$q_3 = C_v \sqrt{h}$$

4. The densities of all three streams are identical and do not vary with time.
5. A 2-min time delay is associated with the composition measurement. The transmission output signal varies linearly from 4 to 20 mA as c_3 varies from 3 to 9 lb solute/ft^3.
6. The pneumatically actuated control valve has negligible dynamics. Its steady-state behavior is summarized below where p_t is the air pressure signal to the control valve from the I/P transducer:

p_t(psi)	q_2(gal/min)
6	20
9	15
12	10

7. An electronic, direct-acting, PI controller is used.
8. The current-to-pressure transducer has negligible dynamics and a gain of 0.3 psi/mA.

9. The nominal operating conditions are:

$\rho = 75$ lb/ft^3 $\bar{c}_3 = 5$ lb of solute/ft^3
$\bar{q}_1 = 10$ gal/min $\bar{c}_2 = 7$ lb of solute/ft^3
$\bar{q}_2 = 15$ gal/min $C_v = 12.5$ gal/min/(ft)$^{1/2}$
                   $D = $ tank diameter $= 4$ ft.

10.5. A block diagram of a feedforward–feedback control system used in Chapter 17 is shown in the drawing where G_f is the feedforward controller transfer function.

(a) Derive an expression for the closed-loop transfer function for load changes, $C(s)/L(s)$.

(b) Assume that perfect control is desired for load changes, that is, $C(s) = 0$ when $L(s) \neq 0$. Derive an expression for the ideal feedforward controller transfer function G_f that will theoretically provide perfect control.

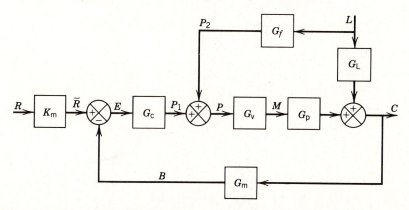

10.6. For a liquid-level control system similar to that in Fig. 10.22, Appelpolscher has argued that integral control action is not required since the process acts as an integrator (cf. Eq. 10-77). To evaluate his assertion, determine whether proportional-only control will eliminate offset for step changes in (a) set point and (b) load variable.

10.7. Consider the liquid-level control system of Fig. 10.16. Derive an expression for the offset that will occur for a ramp change in set point, $h_R'(t) = at$, if PI control is used. What type of control action would eliminate offset for ramp set-point changes?

10.8. A block diagram for internal model control, a control technique that is considered in Chapter 12, is shown in the drawing. Transfer function $\tilde{G}_p$ denotes the process model, while G_p denotes the actual process transfer function. It has been assumed

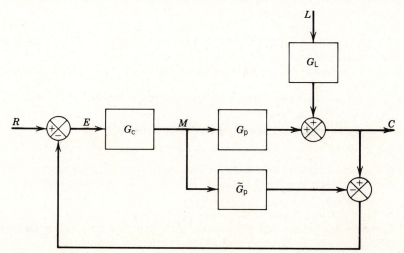

that $G_v = G_m = 1$ for simplicity. Derive closed-loop transfer functions for both the servo and regulator problems.

10.9. An electrically heated, stirred-tank system is shown in the drawing. Using the given information, do the following:

(a) Draw a block diagram for the case where T_3 is the controlled variable and voltage signal V_2 is the manipulated variable. Derive an expression for each transfer function.

(b) Repeat part (a) using V_1 as the manipulated variable.

(c) Which of these two control configurations would provide the better control? Justify your answer.

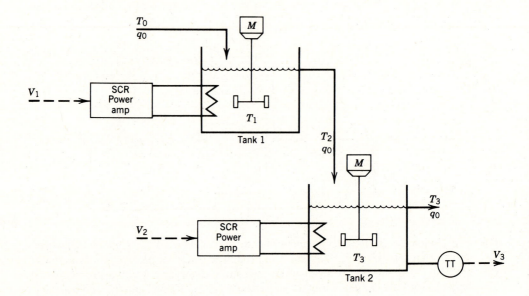

Available Information

1. The volume of liquid in each tank is kept constant using an overflow line.
2. Temperature T_0 is constant.
3. A 0.75-gal/min decrease in q_0 ultimately makes T_1 increase by 3 °F. Two-thirds of this total temperature change occurs in 12 min. This change in q_0 ultimately results in a 5 °F increase in T_3.
4. A change in V_1 from 10 to 12 volts ultimately causes T_1 to change from 70 to 78 °F. A similar test for V_2 causes T_3 to change from 85 to 90 °F. The apparent time constant for these tests is 10 min.
5. A step change in T_2 produces a transient response in T_3 that is essentially complete in 50 min.
6. The thermocouple output is amplified to give $V_3 = 0.15T_3 + 5$, where $V_3 [=]$ volts and $T_3 [=]$ °F.
7. The pipe connecting the two tanks has a mean residence time of 30 s.

10.10. For the feedback control system shown in the drawing:

(a) Calculate expressions for ζ and τ for the closed-loop transfer function $C(s)/R(s)$.

(b) If $\tau_p = 1$ min and $\tau_v = 12$ s, find K_c so that $\zeta = 0.7$ for the two cases:
  i. $\tau_D = 0$
  ii. $\tau_D = 3$ s

(c) Compare the offsets and the periods that occur for both cases. Discuss the advantages and disadvantages of adding the derivative mode.

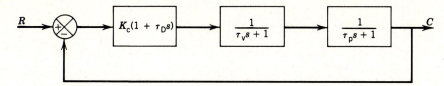

10.11. Consider the standard block diagram in Fig. 10.8 with the following transfer functions: $G_c = K_c$, $G_v = G_m = 1$, $G_p = 1/[(5s + 1)(s + 1)]$, $G_L = 1/(s + 2)$.

(a) If a rectangular pulse disturbance enters the system, $L(s) = (1 - e^{-3s})/s$ and the set point is kept constant, what is the offset?

(b) Simultaneous unit step changes in set point and load variable occur. Calculate the offset.

10.12. For the block diagram shown in the drawing:

(a) Derive an expression for the closed-loop transfer function, $C(s)/L(s)$.

(b) For the transfer functions given below, what is the smallest value of K_c that can be used and have an offset of no more than 0.4 after a step change of -2 in the load variable. The following information is available:

$$G_1(s) = \frac{0.3}{s + 1} \qquad G_3(s) = \frac{0.9e^{-s}}{12s + 1}$$

$$G_2(s) = \frac{1}{10s + 1} \qquad K_m = 0.6$$

State any additional assumptions that you make.

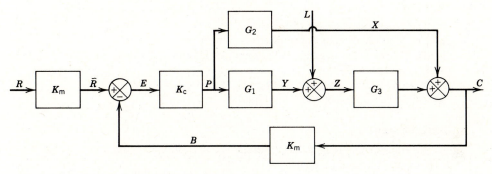

10.13. For the control system in Exercise 10.11:

(a) Draw a mathematically equivalent block diagram that has the disturbance entering the control loop between the controller and the final control element.

(b) What values of the controller gain will result in an offset of less than 0.2 after a unit step change in set point?

10.14. For the control system shown in the drawing, design the *simplest* form of the three-mode controller that results in zero steady-state error when the following set-point change is applied:

$$R(t) = \begin{cases} 0 & t < 0 \\ 1 & 0 \le t < 1 \\ 0 & t \ge 1 \end{cases}$$

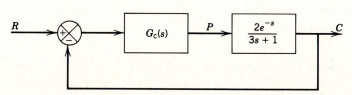

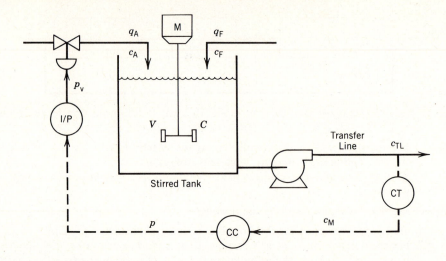

10.15. A mixing process consists of a single stirred tank instrumented as shown in the drawing. The concentration of a single species A in the feed stream varies. The controller attempts to compensate for this by varying the flow rate of pure A through the control valve.

(a) Draw a block diagram for the controlled process.
(b) Derive a transfer function for each block in your block diagram.

Process

  i. The volume is constant (5 m^3).
  ii. The feed flow rate is constant ($\bar{q}_F = 7 \text{ m}^3/\text{min}$).
 iii. The flow rate of the A stream varies but is small compared to $\bar{q}_F$ ($\bar{q}_A = 0.5 \text{ m}^3/\text{min}$).
 iv. $\bar{c}_F = 50 \text{ kg/m}^3$ and $\bar{c}_A = 800 \text{ kg/m}^3$.
  v. All densities are constant and equal.

Transfer Line

  i. The transfer line is 20 m long and has 0.5 m inside diameter.
  ii. Pump volume can be neglected.

Composition Transmitter Data:

$c(\text{kg/m}^3)$	$c_m(\text{mA})$
0	4
200	20

Controller

   i. Ideal PID controller (cf. Eq. 8-13)
  ii. Derivative on process output only
 iii. Direct or reverse acting, as required
 iv. Current (mA) input and output signals

I/P Transducer Data:

$p(\text{mA})$	$p_v(\text{psig})$
4	3
20	15

Control Valve

An equal percentage valve is used which has the following relation:

$$q_A = 0.17 + 0.03 \, (20)^{\frac{p_v - 3}{12}}$$

For a step change in input pressure, the valve requires approximately 1 min to move to its new position.

CHAPTER 11

Stability of Closed-Loop Control Systems

An important consequence of feedback control is that it can cause oscillatory responses. If the oscillation has a small amplitude and damps out quickly, then the control system performance is generally considered to be satisfactory. However, under certain circumstances the oscillations may be undamped or even have an amplitude that increases with time until a physical limit is reached, such as a control valve being fully open or completely shut. In these situations, the closed-loop system is said to be *unstable*.

In this chapter we analyze the stability characteristics of closed-loop systems and present several useful criteria for determining whether a system will be stable. Additional stability criteria based on frequency response analysis are discussed in Chapter 16. But first we consider an illustrative example of a closed-loop system that can become unstable.

EXAMPLE 11.1

Consider the feedback control system shown in Fig. 11.1 with the following transfer functions:

$$G_c = K_c \qquad\qquad G_v = \frac{1}{2s + 1} \qquad\qquad (11\text{-}1)$$

$$G_p = G_L = \frac{1}{5s + 1} \qquad G_m = \frac{1}{s + 1} \qquad (11\text{-}2)$$

Show that the closed-loop system produces unstable responses if controller gain K_c is too large.

Solution

To determine the effect of K_c on the closed-loop response $c(t)$, we consider a unit step change in set point, $R(s) = 1/s$. In Section 10.2 we derived the closed-loop transfer function for set-point changes:

$$\frac{C}{R} = \frac{K_m G_c G_v G_p}{1 + G_c G_v G_p G_m} \qquad (11\text{-}3)$$

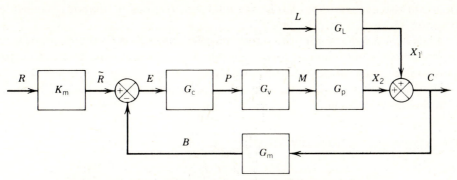

Figure 11.1.   Standard block diagram of a feedback control system.

Substituting (11-1) and (11-2) into (11-3) and rearranging gives

$$C(s) = \frac{1}{s} \frac{K_c(s + 1)}{10s^3 + 17s^2 + 8s + 1 + K_c} \qquad (11\text{-}4)$$

After K_c is specified, $c(t)$ can be determined from the inverse Laplace transform of Eq. 11-4. But first the roots of the cubic polynomial in s must be determined before performing the partial fraction expansion. These roots can be calculated using standard root-finding techniques [1]. Figure 11.2 demonstrates that as K_c increases, the response becomes more oscillatory and is unstable for $K_c = 15$. More details on the actual stability limit of this control system are given in Example 11.7.

The unstable response for Example 11.1 is an oscillation where the amplitude grows in each successive cycle. In contrast, for an actual physical system the amplitudes would increase until a physical limit is reached or an equipment failure occurs. Since the final control element usually has saturation limits (see Chapter 9), the unstable response would manifest itself as a sustained oscillation with a constant amplitude instead of a continually increasing amplitude. Sustained oscil-

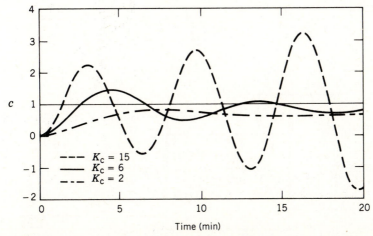

Figure 11.2.   Effect of controller gains on closed-loop response to a unit step change in set point (example 11.1).

lations can also occur without having the final control element *saturate,* as indicated in Section 11.3.

Clearly, a feedback control system must be stable as a prerequisite for satisfactory control. Consequently, it is of considerable practical importance to be able to determine under what conditions a control system becomes unstable. For example, what values of the PID controller parameters K_c, τ_I, and τ_D keep the controlled process stable?

11.1 GENERAL STABILITY CRITERION

Most industrial processes are stable without feedback control. Thus, they are said to be *open-loop stable* or *self-regulating.* An open-loop stable process will return to the original steady state after a transient disturbance (one that is not sustained) occurs. By contrast there are a few processes, such as exothermic chemical reactors, that can be *open-loop unstable.* These processes are extremely difficult to operate without feedback control.

Before presenting various stability criteria, we introduce the following definition for unconstrained linear systems. We use the term "unconstrained" to refer to the ideal situation where there are no physical limits on the output variable.

> ***Definition of Stability.***   *An unconstrained linear system is said to be stable if the output response is bounded for* all *bounded inputs. Otherwise it is said to be unstable.*

By a bounded input, we mean an input variable that stays within upper and lower limits for all values of time. For example, consider a variable $x(t)$ that varies with time. If $x(t)$ is a step or sinusoidal function, then it is bounded. However, the functions $x(t) = t$ and $x(t) = e^{3t}$ are not bounded.

EXAMPLE 11.2

A liquid storage system is shown in Fig. 11.3. Show that this process is not self-regulating by considering its response to a step change in inlet flow rate.

Solution

The transfer function relating liquid level h to inlet flow rate q_i was derived in Section 5.3:

$$\frac{H'(s)}{Q_i'(s)} = \frac{1}{As} \tag{11-5}$$

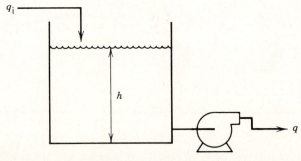

Figure 11.3.   A liquid storage system which is not self-regulating.

where A is the cross-sectional area of the tank. For a step change of magnitude M_0, $Q_i' = M_0/s$, and thus

$$H'(s) = \frac{M_0}{As^2} \tag{11-6}$$

Taking the inverse Laplace transform gives the transient response,

$$h'(t) = \frac{M_0}{A} t \tag{11-7}$$

Since this response is unbounded, we conclude that the liquid storage system is open-loop unstable (or *non-self-regulating*) since a bounded input has produced an unbounded response. However, if the pump in Fig. 11.3 were replaced by a valve, then the storage system would be self-regulating (cf. Example 4.4).

Characteristic Equation

As a starting point for the stability analysis, consider the block diagram in Fig. 11.1. Using block diagram algebra that was developed in Chapter 10, we obtain

$$C = \frac{K_m G_c G_v G_p}{1 + G_{OL}} R + \frac{G_L}{1 + G_{OL}} L \tag{11-8}$$

where G_{OL} is the open-loop transfer function, $G_{OL} = G_c G_v G_p G_m$.

For the moment consider set-point changes only, in which case Eq. 11-8 reduces to the closed-loop transfer function,

$$\frac{C}{R} = \frac{K_m G_c G_v G_p}{1 + G_{OL}} \tag{11-9}$$

If G_{OL} is a ratio of polynomials in s (i.e., a *rational function*), then the closed-loop transfer function in Eq. 11-9 is also a rational function. After a rearrangement it can be factored into poles (p_i) and zeroes (z_i) as

$$\frac{C}{R} = K' \frac{(s - z_1)(s - z_2) \cdots (s - z_m)}{(s - p_1)(s - p_2) \cdots (s - p_n)} \tag{11-10}$$

where K' is a multiplicative constant selected to give the correct steady-state gain. To have a physically realizable system, the number of poles must be greater than or equal to the number of zeroes, that is, $n \geq m$ [2, 3]. Note that a *pole–zero cancellation* occurs if a zero and a pole have the same numerical value.

Comparing Eqs. 11-9 and 11-10 indicates that the poles are also the roots of the following equation which is referred to as the *characteristic equation* of the closed-loop system:

$$1 + G_{OL} = 0 \tag{11-11}$$

The characteristic equation plays a decisive role in determining system stability, as discussed later.

For a unit change in set point, $R(s) = 1/s$, and Eq. 11-10 becomes

$$C = \frac{K'}{s} \frac{(s - z_1)(s - z_2) \cdots (s - z_m)}{(s - p_1)(s - p_2) \cdots (s - p_n)} \tag{11-12}$$

If there are no repeated poles (i.e., if they are all *distinct poles*), then the partial fraction expansion of Eq. 11-12 has the form considered in Section 6.1,

$$C(s) = \frac{A_0}{s} + \frac{A_1}{s - p_1} + \frac{A_2}{s - p_2} + \cdots + \frac{A_n}{s - p_n} \qquad (11\text{-}13)$$

where the $\{A_i\}$ can be determined using the methods of Chapter 3. Taking the inverse Laplace transforms of Eq. 11-13 gives

$$c(t) = A_0 + A_1 e^{p_1 t} + A_2 e^{p_2 t} + \cdots + A_n e^{p_n t} \qquad (11\text{-}14)$$

Suppose that one of the poles is a positive real number, that is, $p_k > 0$. Then it is clear from Eq. 11-14 that $c(t)$ is unbounded and thus the closed-loop system in Fig. 11.1 is unstable. If p_k is a complex number, $p_k = a_k + jb_k$, with a positive real part ($a_k > 0$), then the system is also unstable. By constrast, if *all* of the poles are negative (or have negative real parts), then the system is stable. These considerations can be summarized in the following stability criterion:

> ***General Stability Criterion.*** *The feedback control system in Fig. 11.1 is stable if and only if* all *roots of the characteristic equation are negative or have negative real parts. Otherwise, the system is unstable.*

Figure 11.4 provides a graphical interpretation of this stability criterion. Note that all of the roots of the characteristic equation must lie to the left of the imaginary axis in the complex plane for a stable system to exist. The qualitative effects of these roots on the transient response of the closed-loop system are shown in Fig. 11.5. The left portion of each part of Fig. 11.5 shows representative root locations in the complex plane. The corresponding figure on the right shows the contributions these poles make to the closed-loop response to a step change in set point. Similar responses would occur for a step change in load. A system that has all negative real roots will have a stable, nonoscillatory response, as shown in Fig. 11.5a. On the other hand, if one of the real roots is positive, then the response is unbounded, as shown in Fig. 11.5b. A pair of complex conjugate roots results in oscillatory responses as shown in Figs. 11.5c and d. If the complex roots have negative real

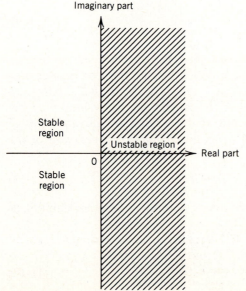

Figure 11.4. Stability regions in the complex plane for roots of the characteristic equation.

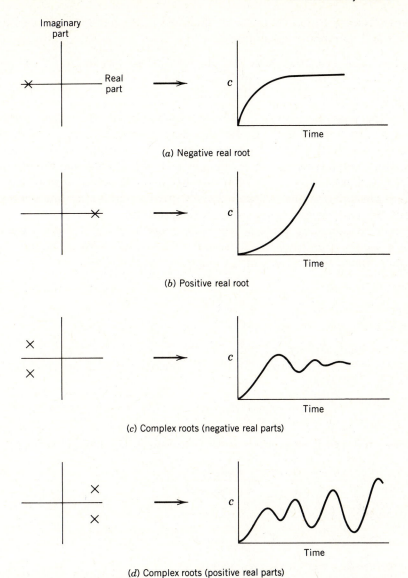

(a) Negative real root

(b) Positive real root

(c) Complex roots (negative real parts)

(d) Complex roots (positive real parts)

Figure 11.5. Contributions of characteristic equation roots to closed-loop response.

parts, the system is stable; otherwise it is unstable. Recall that complex roots always occur as complex conjugate pairs.

The root locations also provide an indication of how rapid the transient response will be. A real root at $s = -a$ corresponds to a closed-loop time constant of $\tau = 1/a$, as is evident from Eqs. 11-13 and 11-14. Thus, real roots close to the imaginary axis result in slow responses. Similarly, complex roots near the imaginary axis correspond to slow response modes. The farther the complex roots are away from the real axis, the more oscillatory the transient response will be (See Example 11.11). However, the process zeros influence the response as well, as discussed in Chapter 6.

Note that the same characteristic equation occurs for both load and set-point changes since the term, $1 + G_{OL}$, appears in the denominator of both terms in Eq. 11-8. Thus, if the closed-loop system is stable for load disturbances, it will also be stable for set-point changes.

The analysis in Eqs. 11-8 to 11-14 which led to the general stability criterion was based on a number of assumptions:

1. Set-point changes (rather than load changes) were considered.
2. The closed-loop transfer function was a ratio of polynomials.
3. The poles in Eq. 11-10 were all distinct.

However, the general stability criterion is valid even if these assumptions are removed. In fact, this stability criterion is valid for *any* linear control system.[1] By contrast, for nonlinear systems rigorous stability analyses tend to be considerably more complex and involve special techniques such as Liapunov and Popov stability criteria [2,4]. Fortunately, a stability analysis of a linearized system using the techniques presented in this chapter normally provides useful information for nonlinear systems operating near the point of linearization.

From a mathematical point of view, the general stability criterion presented above is a *necessary and sufficient condition*. Thus, linear system stability is completely determined by the roots of the characteristic equation.

EXAMPLE 11.3

Consider the feedback control system in Fig. 11.1 with $G_v = K_v$, $G_m = 1$, and $G_p = K_p/(\tau_p s + 1)$. Determine the stability characteristics if a proportional controller is used, $G_c = K_c$.

Solution
Substituting the transfer functions into the characteristic equation in (11-11) gives

$$1 + \frac{K_c K_v K_p}{\tau_p s + 1} = 0$$

which reduces to

$$\tau_p s + 1 + K_c K_v K_p = 0 \tag{11-15}$$

This characteristic equation has a single root,

$$s = -\frac{K_c K_v K_p + 1}{\tau_p} \tag{11-16}$$

The closed-loop system will be stable if this root is negative. Since time constants are always positive, $\tau_p > 0$ and the feedback control system will be stable if $K_c K_v K_p > -1$. This means that as long as the controller has the correct control action (i.e., reverse or direct acting, as per Section 8.2), then the system will be stable. For example, if $K_p > 0$ and $K_v > 0$, then the controller must be made reverse acting so that $K_c > 0$. By contrast, if $K_p < 0$, then a direct-acting controller ($K_c < 0$) is required.

---

[1]By a linear control system we mean that each component in the feedback loop obeys the Principle of Superposition and consequently can be represented by a transfer function.

EXAMPLE 11.4

Consider the feedback control system in Example 11.1, but now assume that $G_m = 1$. Determine the range of K_c values that result in a stable closed-loop system.

Solution

Substituting these transfer functions into Eq. 11-11 gives

$$1 + \frac{K_c}{(2s + 1)(5s + 1)} = 0 \tag{11-17}$$

which can be rearranged as

$$10s^2 + 7s + K_c + 1 = 0 \tag{11-18}$$

Applying the quadratic formula yields the roots,

$$s = \frac{-7 \pm \sqrt{49 - 40(K_c + 1)}}{20} \tag{11-19}$$

To have a stable system, both roots of this characteristic equation must have negative real parts. Equation 11-19 indicates that the roots will be negative if $40(K_c + 1) > 0$ since this means that the square root will have a value less than 7. If $40(K_c + 1) > 0$, then $K_c + 1 > 0$ and $K_c > -1$. Thus, we conclude that the closed-loop system will be stable if $K_c > -1$.

The stability analyses for Examples 11.3 and 11.4 have indicated that these closed-loop systems will be stable for all positive values of K_c, no matter how large. However, this result is *not* typical since it occurs only for the special case where the open-loop system is stable and the open-loop transfer function G_{OL} is first or second order with no time delay. In more typical problems, K_c must be below an upper limit to have a stable closed-loop system.[2] (See examples in the next section.)

EXAMPLE 11.5

Consider a first-order process that has the transfer function $G_p = 0.2/(s - 1)$ and thus is open-loop unstable. If $G_v = G_m = 1$, determine whether a proportional controller can stabilize the system.

Solution

The characteristic equation for this system is

$$s + 0.2K_c - 1 = 0 \tag{11-20}$$

which has the single root, $s = 1 - 0.2K_c$. Thus, the stability requirement is that $K_c > 5$. This example illustrates the important fact that feedback control can be used to stabilize a system that is not stable without control.

---

[2]If a direct-acting controller is used (i.e., $K_c < 0$), then stability considerations place an upper limit on $-K_c$ rather than on K_c.

In Examples 11.3 to 11.5 the characteristic equations were either first or second order and thus we could find the roots analytically. For higher-order polynomials, this is not possible and numerical root-finding techniques [1] must be employed. Fortunately, an attractive alternative, the Routh Stability Criterion, is available to evaluate stability without requiring calculation of the roots of the characteristic equation.

11.2 ROUTH STABILITY CRITERION

In 1905 Routh [5] published an analytical technique for determining whether any roots of a polynomial have positive real parts. According to the General Stability Criterion, a closed-loop system will be stable only if all of the roots of the characteristic equation have negative real parts. Thus, by applying Routh's technique to analyze the coefficients of the characteristic equation, we can determine whether the closed-loop system is stable. This approach is referred to as the Routh Stability Criterion. It can be applied only to systems whose characteristic equations are polynomials in s. Thus, the Routh Stability Criterion is not directly applicable to systems containing time delays, since an $e^{-\theta s}$ term appears in the characteristic equation where θ is the time delay. However, if $e^{-\theta s}$ is replaced by a Padé approximation (see Section 6.2) then an *approximate* stability analysis can be performed. An ***exact*** stability analysis of systems containing time delays can be performed by direct root-finding or by using a frequency response analysis and the Bode Stability Criterion presented in Chapter 16.

The Routh Stability Criterion is based on a characteristic equation that has the form

$$a_n s^n + a_{n-1} s^{n-1} + \cdots + a_1 s + a_0 = 0 \qquad (11\text{-}21)$$

We arbitrarily assume that $a_n > 0$. If $a_n < 0$, we simply multiply Eq. 11-21 by -1 to generate a new equation that satisfies this condition. A *necessary* (but not sufficient) condition for stability is that all of the coefficients $(a_0, a_1, \cdots, a_n)$ in the characteristic equation must be positive. If any coefficient is negative or zero, then at least one root of the characteristic equation lies to the right of, or on, the imaginary axis, and the system is unstable. If all of the coefficients are positive, we next construct the following Routh array:

Row				
1	a_n	a_{n-2}	a_{n-4}	$\cdots$
2	a_{n-1}	a_{n-3}	a_{n-5}	$\cdots$
3	b_1	b_2	b_3	$\cdots$
4	c_1	c_2	$\cdots$	
$\vdots$	$\vdots$			
$n+1$	z_1			

The Routh array has $n + 1$ rows where n is the order of the characteristic equation, Eq. 11-21. The Routh array has a roughly triangular structure with only a single element in the last row. The first two rows are merely the coefficients in the characteristic equation, arranged according to odd and even powers of s. The elements in the remaining rows are calculated from the formulas:

$$b_1 = \frac{a_{n-1} a_{n-2} - a_n a_{n-3}}{a_{n-1}} \qquad (11\text{-}22)$$

$$b_2 = \frac{a_{n-1}a_{n-4} - a_n a_{n-5}}{a_{n-1}} \qquad (11\text{-}23)$$

$$\vdots$$

$$c_1 = \frac{b_1 a_{n-3} - a_{n-1}b_2}{b_1} \qquad (11\text{-}24)$$

$$c_2 = \frac{b_1 a_{n-5} - a_{n-1}b_3}{b_1} \qquad (11\text{-}25)$$

$$\vdots$$

Note that the expressions in the numerators of Eqs. 11-22 to 11-25 are similar to the calculation of a determinant for a 2×2 matrix except that the order of subtraction is reversed. Having constructed the Routh array, we can now state the Routh Stability Criterion:

>   *Routh Stability Criterion.*   *A necessary and sufficient condition for all roots of the characteristic equation in Eq. 11-21 to have negative real parts is that all of the elements in the left column of the Routh array are positive.*

Next we present three examples that show how the Routh Stability Criterion can be applied.

EXAMPLE 11.6

Determine the stability of a system that has the characteristic equation

$$s^4 + 5s^3 + 3s^2 + 1 = 0 \qquad (11\text{-}26)$$

Solution
Since the s term is missing, its coefficient is zero. Thus, the system is unstable. Recall that a necessary condition for stability is that all of the coefficients in the characteristic equation must be positive.

EXAMPLE 11.7

Find the values of controller gain K_c that make the feedback control system of Example 11.1 stable.

Solution
From Eq. 11-4, the characteristic equation is

$$10s^3 + 17s^2 + 8s + 1 + K_c = 0 \qquad (11\text{-}27)$$

All coefficients are positive provided that $1 + K_c > 0$ or $K_c > -1$. The Routh array is

$$
\begin{array}{cc}
10 & 8 \\
17 & 1 + K_c \\
b_1 & b_2 \\
c_1 &
\end{array}
$$

$$\text{where } b_1 = \frac{17(8) - 10(1 + K_c)}{17} = 7.41 - 0.588 K_c$$

$$b_2 = 0$$

$$c_1 = \frac{b_1(1 + K_c) - 17(0)}{b_1} = 1 + K_c$$

To have a stable system, each element in the left column of the Routh array must be positive. Element b_1 will be positive if $K_c < 7.41/0.588 = 12.6$. Similarly, c_1 will be positive if $K_c > -1$. Thus, we conclude that the system will be stable if

$$-1 < K_c < 12.6 \qquad (11\text{-}28)$$

This example illustrates that stability limits for controller parameters can be derived analytically using the Routh array; that is, it is not necessary to compute the roots of the characteristic equation nor specify a numerical value for K_c before performing the stability analysis.

EXAMPLE 11.8

Consider a feedback control system with $G_c = K_c$, $G_v = 2$, $G_m = 0.25$, and $G_p = 4e^{-2s}/(5s + 1)$. The characteristic equation is

$$1 + 5s + 2K_c e^{-2s} = 0 \qquad (11\text{-}29)$$

Since this characteristic equation is not a polynomial in s, the Routh criterion is not directly applicable. However, if a polynomial approximation to e^{-2s} is introduced, such as a Padé approximation or power series expansion (see Chapter 6), then the Routh criterion can be used to determine approximate stability limits. For simplicity, use the 1/1 Padé approximation,

$$e^{-2s} \simeq \frac{1 - s}{1 + s} \qquad (11\text{-}30)$$

and determine the stability limits for the controller gain.

Solution
Substituting Eq. 11-30 into 11-29 gives

$$1 + 5s + 2K_c \left(\frac{1 - s}{1 + s} \right) = 0 \qquad (11\text{-}31)$$

Multiplying both sides by $1 + s$ and rearranging gives

$$5s^2 + (6 - 2K_c) s + (1 + 2K_c) = 0 \qquad (11\text{-}32)$$

For stability, each coefficient in this characteristic equation must be positive. This situation occurs if $K_c < 3$ and $K_c > -0.5$. The Routh array is

$$
\begin{array}{ll}
5 & 1 + 2K_c \\
6 - 2K_c & 0 \\
1 + 2K_c &
\end{array}
$$

In this example, the Routh array provides no additional information but merely confirms that the system with the Padé approximation is stable if $-0.5 < K_c < 3$.

An exact time-delay analysis, without the Padé approximation and based on the Bode Stability Criterion in Example 16.3, indicates that the actual upper limit on K_c is 2.29, which is 24% lower than the approximate value of 3.0 from the Routh Stability Criterion. If the 2/2 Padé approximation in Eq. 6-39 had been used with the Routh Stability Criterion, a maximum value of $K_c = K_{cm} = 2.32$ would be obtained where K_{cm} denotes the upper limit. (This derivation is left as an exercise for the reader.)

11.3 DIRECT SUBSTITUTION METHOD

The imaginary axis divides the complex plane into stable and unstable regions for the roots of the characteristic equation, as indicated in Fig. 11.4. On the imaginary axis, the real part of s is zero and thus we can write $s = j\omega$. Substituting $s = j\omega$ into the characteristic equation allows us to find a stability limit such as the maximum value of K_c [6]. As the gain K_c is increased, the roots of the characteristic equation cross the imaginary axis when $K_c = K_{cm}$.

EXAMPLE 11.9

Use the direct-substitution method to determine K_{cm} for the system with the characteristic equation given by Eq. 11-27.

Solution

Substitute $s = j\omega$ and $K_c = K_{cm}$ into Eq. 11-27:

$$-10j\omega^3 - 17\omega^2 + 8j\omega + 1 + K_{cm} = 0$$

or   (11-33)

$$(1 + K_{cm} - 17\omega^2) + j(8\omega - 10\omega^3) = 0$$

Equation 11-33 is satisfied if both real and imaginary parts of (11-33) are identically zero:

$$1 + K_{cm} - 17\omega^2 = 0 \tag{11-34a}$$

$$8\omega - 10\omega^3 = \omega(8 - 10\omega^2) = 0 \tag{11-34b}$$

Therefore,

$$\omega^2 = 0.8 \Rightarrow \omega = \pm 0.894 \tag{11-35}$$

and from (11-34a),

$$K_{cm} = 12.6$$

Thus, we conclude that $K_c < 12.6$ for stability. Equation 11-35 indicates that at the stability limit (where $K_c = K_{cm} = 12.6$), a sustained oscillation occurs that has a frequency of $\omega = 0.894$ radians per min if the time constants have units of minutes. (Recall that a pair of complex roots on the imaginary axis, $s = \pm j\omega$, results in an undamped oscillation of frequency ω.) The corresponding period P is $P = 2\pi/0.894 = 7.03$ min.

The direct substitution method is related to the Routh Stability Criterion in Section 11.2. If the characteristic equation has a pair of roots on the imaginary axis, equidistant from the origin, and all other roots are in the left hand plane, the single element in the next-to-last row of the Routh Array will be zero. Then the location of the two imaginary roots can be obtained from the solution of the equation

$$Cs^2 + D = 0$$

where C and D are the two elements in the $(n - 1)$ row of the Routh Array, as read from left to right [7].

The direct substitution method is also related to the frequency response approach of Chapters 14 through 16 because both techniques are based on the substitution, $s = j\omega$.

11.4 ROOT LOCUS DIAGRAMS

In the previous section we have seen that the roots of the chacteristic equation play a crucial role in determining system stability and the nature of the closed-loop responses. In the design and analysis of control systems, it is instructive to know how the roots of the characteristic equation change when a particular system parameter such as a controller gain changes. A root locus diagram provides a convenient graphical display of this information, as indicated in the following example.

EXAMPLE 11.10

Consider a feedback control system that has the open-loop transfer function [7],

$$G_{OL}(s) = \frac{2K_c}{(s + 1)(s + 2)(s + 3)} \tag{11-36}$$

Plot the root locus diagram for $0 \le K_c \le 40$.

Solution

The characteristic equation is $1 + G_{OL} = 0$ or

$$(s + 1)(s + 2)(s + 3) + 2K_c = 0 \tag{11-37}$$

The root locus diagram in Fig. 11.6 shows how the three roots of this characteristic equation vary with K_c. When $K_c = 0$, the roots are merely the poles of the open-loop transfer function, -1, -2, and -3. These are designated by an x symbol in Fig. 11.6. As K_c increases, the root at -3 decreases monotonically. The other two roots converge and then form a complex conjugate pair when $K_c = 0.2$. When $K_c = 30$, the complex roots cross the imaginary axis and enter the unstable region. This illustrates why the substitution of $s = j\omega$ (Section 11.3) determines the unstable controller gain. Thus, the root locus diagram indicates that the closed-loop system is unstable for $K_c > 30$. It also indicates that the closed-loop response will be nonoscillatory for $K_c < 0.2$.

The root locus diagram can be used to provide a quick estimate of the transient response of the closed-loop system. The roots closest to the imaginary axis correspond to the slowest response modes. If the two closest roots are a complex conjugate pair (as in Fig. 11.7), then the closed-loop system can be approximated by an underdamped second-order system. This approximation will now be developed.

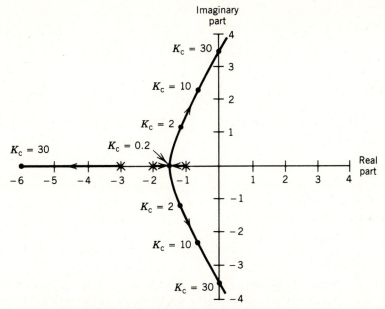

Figure 11.6. Root locus diagram for third-order system. X denotes an open-loop pole. Dots denote locations of the closed-loop poles for different values of K_c. Arrows indicate change of pole locations as K_c increases.

Consider the standard second-order transfer function of Chapter 6,

$$G(s) = \frac{K}{\tau^2 s^2 + 2\zeta\tau s + 1} \tag{11-38}$$

which has the following roots when $0 \leq \zeta < 1$:

$$s = -\frac{\zeta}{\tau} \pm j\frac{\sqrt{1 - \zeta^2}}{\tau} \tag{11-39}$$

These roots are shown graphically in Fig. 11.7. Note that the length d in Fig. 11.7

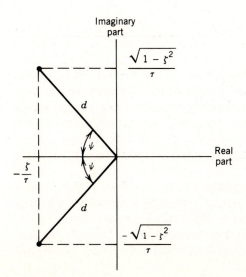

Figure 11.7. Root locations for underdamped second-order system.

is given by

$$d = \sqrt{\left(-\frac{\zeta}{\tau}\right)^2 + \left(\frac{1-\zeta}{\tau}\right)^2} = \sqrt{\frac{1}{\tau^2}} = \frac{1}{\tau} \tag{11-40}$$

Consequently,

$$\cos \psi = \frac{\zeta/\tau}{1/\tau} = \zeta \tag{11-41}$$

and

$$\psi = \cos^{-1}(\zeta) \tag{11-42}$$

This information provides the basis for a second-order approximation to a higher-order system, as shown in Example 11.11.

EXAMPLE 11.11

Consider the root locus diagram in Fig. 11.6 for the third-order system of Example 11.10. For $K_c = 10$, determine values of ζ and τ that can be used to characterize the transient response approximately.

Solution

For $K_c = 10$, there is one real root and two complex roots. By measuring the angle ψ and the distance d to the complex root, we obtain

$$\psi = \cos^{-1} \zeta = 75°$$

$$d = 2.3$$

Then it follows from Eqs. 11-36 and 11-37 that

$$\zeta = 0.25 \quad \text{and} \quad \tau = 0.43$$

Thus, the third-order system can be approximated by an underdamped second-order system with the ζ and τ values given above. This information (and the material in Chapter 6) provides a useful characterization of the transient response.

The utility of root locus diagrams has been illustrated by the third-order system of Examples 11.10 and 11.11. The major disadvantage of root locus analysis is that time delays cannot be handled conveniently and they require iterative solution of the nonlinear and nonrational characteristic equation. Nor is it easy to display simultaneous changes in more than one parameter (e.g., controller parameters K_c and τ_I). For this reason the root locus technique has not found much use as a design tool in process control.

Root locus diagrams can be quickly generated by using a calculator or a computer and root-finding techniques. Computer listings of standard root-finding techniques are available in Refs. 1, 6, and 8. Root locus diagrams can also be constructed using graphical techniques [7,9]. For example, the graphical construction of the diagram in Fig. 11.6 has been described in detail by Coughanowr and Koppel [7].

SUMMARY

In this chapter we have considered several stability criteria for linear systems that can be described by transfer function models. The various steps involved in performing a stability analysis are shown in Fig. 11.8.

If the process model is nonlinear, then advanced stability theory can be used [2,4] or an approximate stability analysis can be performed based on a linearized, transfer function model. If the transfer function model includes time delays, then an exact stability analysis can be performed using root-finding or, preferably, the frequency response methods of Chapter 16. A less desirable alternative is to approximate the $e^{-\theta s}$ terms and apply the Routh Stability Criterion.

Having considered the stability of closed-loop systems, we are ready for our next topic, the design of feedback control systems. This important subject is considered in Chapters 12 and 16. We will see that a number of prominent design procedures and controller tuning techniques are based on stability criteria.

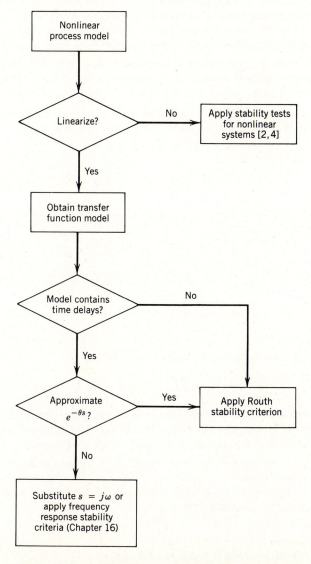

Figure 11.8. Flowchart for performing a stability analysis.

REFERENCES

1. Constantinides, A., *Applied Numerical Methods with Personal Computers*, McGraw-Hill, New York, 1987.
2. Koppel, L. B., *Introduction to Control Theory*, Prentice-Hall, Englewood Cliffs, NJ, 1968.
3. Kwakernaak, H., and R. Sivan, *Linear Optimal Control Systems*, Wiley, New York, 1972, p. 39.
4. Vidyasagar, M., *Nonlinear Systems Analysis*, Prentice-Hall, Englewood Cliffs, NJ, 1978.
5. Routh, E. J., *Dynamics of a System of Rigid Bodies, Part II*, MacMillan, London, 1905.
6. Luyben, W. L., *Process Modeling, Simulation and Control for Chemical Engineers*, McGraw-Hill, New York, 1973, Chapter 11.
7. Coughanowr, D. R., and L. B. Koppel, *Process Systems Analysis and Control*, McGraw-Hill, New York, 1965, Chapters 14 and 17.
8. Melsa, J. L., *Computer Programs for Computational Assistance in the Study of Linear Control Theory*, McGraw-Hill, New York, 1970.
9. Kuo, B. C., *Automatic Control Systems*, 4th ed., Prentice-Hall, Englewood Cliffs, NJ, 1982.

EXERCISES

11.1. A PI controller is to be used in a temperature control loop. For nominal conditions, it has been determined that the closed-loop system is stable when $\tau_I = 10$ min and $-10 < K_c < 0$. Would you expect these stability limits to change for any of the following instrumentation changes? Justify your answers using qualitative arguments.

(a) The span on the temperature transmitter is reduced from 40 to 20 °C.

(b) The zero on the temperature transmitter is increased from 110 to 130 °C.

(c) The control valve "trim" is changed from linear to equal percentage.

11.2. The block diagram of a feedback control system is shown in the drawing. Determine the values of K_c that result in a stable closed-loop system.

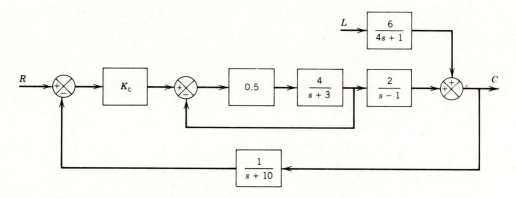

11.3. An open-loop unstable process has the transfer function

$$G = G_v G_p G_m = \frac{K}{(\tau_1 s + 1)(\tau_2 s - 1)}$$

Can this process be made closed-loop stable by using a proportional feedback controller, $G_c(s) = K_c$? Can closed-loop stability be achieved using an ideal proportional-derivative controller, $G_c(s) = K_c(1 + \tau_D s)$? Justify your answers.

11.4. For the liquid-level control system in Figure 10.22, determine the numerical values of K_c and τ_I that result in a stable closed-loop system. The level transmitter has negligible dynamics while the control valve has a time constant of 10 s. The following numerical values are available:

$A = 3 \text{ ft}^2$

$\bar{q}_3 = 10 \text{ gal/min}$

$K_v = -1.3 \text{ gal/min/mA}$ (for control valve)

$K_m = 4 \text{ mA/ft}$ (for level transmitter)

11.5. Consider the block diagram in Fig. 11.1 with a proportional controller, $G_c(s) = K_c$, and the following transfer functions:

$$G_v(s) = 0.8 \qquad G_p(s) = \frac{5e^{-2s}}{10s + 1}$$

$$G_L(s) = \frac{4e^{-s}}{15s + 1} \qquad G_m(s) = \frac{1.2}{2s + 1}$$

Which of the following changes would have the more significant effect on the stability limits for the closed-loop system. Justify your answer.

(a) The process time delay is reduced from 2 to 1.
(b) The measurement time constant is reduced from 2 to 1.

11.6. It is desired to control the exit temperature T_2 of the heat exchanger shown in the drawing by adjusting the steam flow rate w_s. Unmeasured disturbances occur in inlet temperature T_1. The dynamic behavior of the heat exchanger can be approximated by the transfer functions:

$$\frac{T_2'(s)}{W_s'(s)} = \frac{2.5e^{-s}}{10s + 1} \; [=] \; \frac{°F}{lb/s}$$

$$\frac{T_2'(s)}{T_1'(s)} = \frac{0.9e^{-2s}}{5s + 1} \; [=] \; \text{dimensionless}$$

where the time constants and time delays have units of seconds. The control valve has the following steady-state characteristics:

$$w_s = 0.6\sqrt{p - 4}$$

where p is the controller output expressed in milliamps. At the nominal operating condition, $p = 12$ mA. After a sudden change in the controller output, w_s reaches a new steady-state value in 20 s (assumed to take five time constants). The temperature transmitter has negligible dynamics and is designed so that its output signal varies linearly from 4 to 20 mA as T_2 varies from 120 to 160 °F.

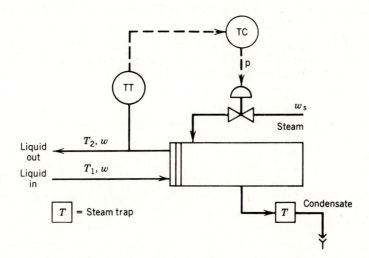

(a) If a proportional feedback controller is used, what is K_{cm}? What is the frequency of the resulting oscillation when $K_c = K_{cm}$? (*Hint:* Use the direct substitution method and Euler's identity.)
(b) Estimate K_{cm} using the Routh criterion and a 1/1 Padé approximation for the time-delay term. Does this analysis provide a satisfactory approximation?

11.7. Consider the block diagram in Fig. 11.1 with the following transfer functions:

$$G_c = K_c \qquad\qquad G_v = 2$$

$$G_p = \frac{0.4(s - a)}{(2s + 1)(10s + 1)} \qquad G_L = \frac{3}{5s + 1}$$

$$G_m = \frac{1}{s + 1}$$

Does the presence of a "right half plane zero" (i.e., $a > 0$) in the process transfer function affect the stability of the closed-loop system? (*Hint:* Consider the two situations where $a = 0$ and a > 0.)

11.8. A second-order process plus measuring element is controlled by a proportional controller as shown in the drawing.

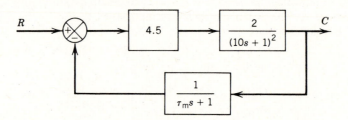

(a) For what range(s) of τ_m will the resulting system be stable?

(b) What practical arguments might be used to restrict the range(s) of acceptable τ_m even further?

(c) If a proportional-derivative controller with $\tau_D = \tau_m$ is used, how would your answers to (a) and (b) be affected?

11.9. An open-loop unstable process is described by the transfer function

$$G(s) = \frac{B(s)}{P(s)} = \frac{e^{-2s}}{3s - 1}$$

Can a proportional feedback controller stabilize such a process? If so, what values of K_c result in a stable closed-loop system?

11.10. A process, including control valve and transmitter, consists of four first-order systems with time constants of 1, 2, 2, and 5 min. The steady-state gain is 3. Calculate the range of controller gains that result in a stable closed-loop system if a PD controller is used with $\tau_D = 2$ min.

11.11. A block diagram of a feedforward-feedback control system is shown. For what values of controller gains, K_f and K_c, is the closed-loop system stable?

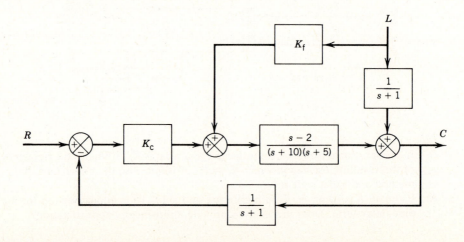

11.12. A feedback control system has the open-loop transfer function, $G_{OL}(s) = 0.5K_c e^{-3s}/(10s + 1)$. Determine the values of K_c for which the closed-loop system is stable using two approaches:

(a) An approximate analysis using the Routh Stability Criterion and a 1/1 Padé approximation for e^{-3s}.

(b) An exact analysis based on substitution of $s = j\omega$. (*Hint:* Recall Euler's identity.)

11.13. To determine information about the response time of a closed-loop system, it would be desirable to determine whether any roots of the characteristic equation are located in certain regions of the complex plane. In particular, it would be useful to know if any roots lie in the region where $s > -\sigma$ ($\sigma > 0$). Discuss how the Routh approach can be modified so as to provide this information.

11.14. A feedback control system has the open-loop transfer function

$$G_{OL}(s) = \frac{K_c(1 + \tau_D s)}{s^2(\tau s + 1)}$$

Determine the stability limits on controller parameters K_c and τ_D for the following cases:

(a) Proportional-only control ($\tau_D = 0$)
(b) Proportional-derivative control ($\tau_D \neq 0$).

CHAPTER 12

Controller Design Based on Transient Response Criteria

In Chapter 11 we considered several examples that illustrated how controller settings affect closed-loop stability. For most control problems, the closed-loop system will be stable for a range of controller settings. Thus, it is possible to select numerical values that result in desired closed-loop system performance. Although this problem is frequently referred to as control system design, it would be more accurate to say that we are choosing controller settings for conventional PID controllers.

In this chapter we consider the selection of PID controller settings based on transient response criteria. Controller settings based on frequency response criteria are considered in Chapter 16. Advanced feedback control systems that are not merely PID controllers are discussed in Chapters 18 and 19.

12.1 PERFORMANCE CRITERIA FOR CLOSED-LOOP SYSTEMS

The function of a feedback control system is to ensure that the closed-loop system has desirable dynamic and steady-state response characteristics. Ideally, we would like the closed-loop system to satisfy the following performance criteria:

1. The closed-loop system must be stable.
2. The effects of disturbances are minimized.
3. Rapid, smooth responses to set-point changes are obtained.
4. Offset is eliminated.
5. Excessive control action is avoided.
6. The control system is robust, that is, it is insensitive to changes in process conditions and to errors in the assumed process model.

In typical control problems, it is not possible to achieve all of these goals since they involve inherent conflicts and trade-offs. For example, PID controller settings that minimize the effects of disturbances tend to produce large overshoots for set-point changes. On the other hand, if the controller is adjusted to provide a rapid, smooth response to a set-point change, it usually results in sluggish control for disturbances. Thus, a trade-off is required in selecting controller settings that are satisfactory for both load and set-point changes. This type of trade-off will be illustrated in Section 12.4.

A second type of trade-off is required between robustness and performance. A control system can normally be made robust by choosing conservative values for the controller settings (e.g., small K_c, large τ_I), but this choice results in sluggish responses to load and set-point changes, in other words, *high-performance control* is not obtained.

A number of alternative approaches for specifying controller settings are available:

1. Direct Synthesis method
2. Internal Model Control
3. Tuning relations
4. Frequency response techniques
5. Computer simulation using physically-based models
6. Field tuning after installation.

The first five methods are based on process models and thus can be used to determine suitable controller settings before the control system is installed. However, field tuning of the controller after installation is often required since the process models are rarely exact. Consequently, the objective in the first five methods is to provide approximate values for the PID controller settings which can then be used as the starting point for field tuning. Since field tuning can be both tedious and time-consuming, it is very useful to have good initial estimates for the controller settings.

Methods 1–3 above are based on simple transfer function models and will be considered in Sections 12.2 and 12.3, respectively. Design techniques based on frequency response characteristics can be applied to linear models of any order and are the subject of Chapter 16. The fifth method, computer simulation, can provide considerable insight into dynamic behavior and control system performance. However, a significant amount of engineering effort is usually required to develop a physically-based model, and controller settings are difficult to obtain for such a complicated model. In contrast, Methods 1–3 require much less information, a simple transfer function model. The last approach, field tuning, is considered in Chapter 13.

12.2 DIRECT SYNTHESIS METHOD

In principle, a feedback controller can be designed by using a process model and specifying the desired closed-loop response [1, 2]. This *Direct Synthesis* approach is valuable because it provides insight about the relation between the process and the resulting controller. A disadvantage of this approach is that the resulting controllers may not have a PID structure. However, as is shown below, PI or PID controllers are obtained for particular types of process models.

As a starting point for the analysis, consider the standard block diagram of a feedback control system in Fig. 12.1. The following closed-loop transfer function for set-point changes was derived in Section 10.2:

$$\frac{C}{R} = \frac{K_m G_c G_v G_p}{1 + G_c G_v G_p G_m} \tag{12-1}$$

For simplicity, let $G \overset{\Delta}{=} G_v G_p G_m$ and assume that $G_m = K_m$. Then Eq. 12-1 reduces to

$$\frac{C}{R} = \frac{G_c G}{1 + G_c G} \tag{12-2}$$

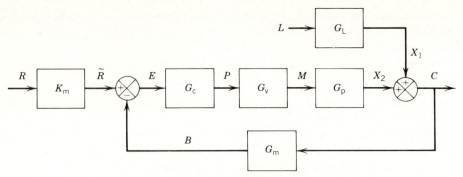

Figure 12.1.   Block diagram for a standard feedback control system.

Rearranging gives an expression for the feedback controller:

$$G_c = \frac{1}{G}\left(\frac{C/R}{1 - C/R}\right) \tag{12-3a}$$

Equation 12-3a does not provide a practical design equation for two reasons: (1) The process transfer function G is not known and (2) the closed-loop transfer function C/R is unknown since G_c has not yet been designed. However, a practical design procedure can be obtained from (12-3a) if the unknown G is replaced by an assumed process model $\tilde{G}$ ($= \tilde{G}_v\tilde{G}_p\tilde{G}_m$) and if C/R is replaced by a desired closed-loop transfer function $(C/R)_d$:

$$G_c = \frac{1}{\tilde{G}}\left[\frac{(C/R)_d}{1 - (C/R)_d}\right] \tag{12-3b}$$

The specification of $(C/R)_d$ is an important design decision that will be addressed later in this section. For the ideal case where a perfect model is available ($\tilde{G} = G$), the design equation becomes

$$G_c = \frac{1}{G}\left(\frac{(C/R)_d}{1 - (C/R)_d}\right) \tag{12-3c}$$

Thus, an expression for G_c can be obtained if a process model is available and a desired closed-loop transfer function $(C/R)_d$ is specified.

Note that the controller contains the inverse of the process model due to the $1/G$ term. This situation is a common feature of many model-based controller design techniques [3, 4]. Since the controller in (12-3b) contains the inverse of the process model, pole–zero cancellation is used to achieve the desired closed-loop transfer function $(C/R)_d$. Thus, Eqs. 12-2 and 12-3c indicate that for the product G_cG, controller poles cancel process zeros while the controller zeros cancel the process poles. Since these pole–zero cancellations are seldom exact due to modeling errors, the Direct Synthesis approach should be used with caution for processes that have unstable poles or zeros (see Exercise 12.8). In particular, this approach should be used with care to design controllers for open-loop unstable processes unless the process is first stabilized via an additional feedback control loop. (See Exercise 12.8.)

Perfect Control

Ideally, we would like the controlled variable to track set-point changes instantaneously without any error. This situation, referred to as *perfect control,* occurs when $(C/R)_d = 1$, or $C = R$. Unfortunately, perfect control is impossible since the required controller in Eq. 12-3 would require an infinite gain. But perfect control can be approximated by the following approach.

Suppose that

$$G_c = \frac{K_c}{G} \tag{12-4}$$

where K_c is the controller gain. Substituting this expression into Eq. 12-2 gives

$$\frac{C}{R} = \frac{K_c}{1 + K_c} \tag{12-5}$$

Thus, perfect control is approached in the limit as $K_c \to \infty$ since $C/R \to 1$.

Note that the controller in Eq. 12-4 attempts to cancel the poles and zeros of the process transfer function since the controller contains an inverse of the process model. This controller is not physically realizable if G contains a time delay or has more poles than zeros. Another difficulty arises when G contains a right-half plane zero since the controller in Eq. 12-4 then contains a pole in the right-half plane and thus is inherently unstable.

EXAMPLE 12.1

Consider a second-order process model

$$G(s) = \frac{K}{(\tau_1 s + 1)(\tau_2 s + 1)} \tag{12-6}$$

Design a controller using Eq. 12-4 and discuss its feasibility.

Solution
The controller obtained from Eq. 12-4 is

$$G_c(s) = \frac{K_c}{K} [\tau_1 \tau_2 s^2 + (\tau_1 + \tau_2)s + 1] \tag{12-7}$$

The s^2 term in Eq. 12-7 implies that the control action is based on the second derivative of the error signal. In the presence of noisy process measurements, this controller will overreact to small changes in the error signal. However, we conclude that perfect control cannot be practically achieved for second-order processes because the perfect controller is not physically realizable (See Chapter 6, pp. 133–134).

Controllers Designed to Give Finite Settling Times

The controllers considered in the previous section are unrealistic since the design objective was to have the process reach a new set point instantaneously. A more practical approach is to specify the closed-loop transfer function C/R so that realistic settling times are achieved. For example, suppose that the desired closed-loop transfer function is $(C/R)_d = 1/(\tau_c s + 1)$ where time constant τ_c is a design parameter. This choice means that the desired closed-loop response to a step change in set point would respond as a first-order process with time constant τ_c. It would not have any offset since the steady-state gain is one. Thus, in this approach

$$\left(\frac{C}{R}\right)_d = \frac{G_c G}{1 + G_c G} = \frac{1}{\tau_c s + 1} \tag{12-8}$$

Solving for G_c gives

$$G_c = \frac{1}{G}\frac{1}{\tau_c s} \tag{12-9}$$

Note that the $1/\tau_c s$ term provides integral control action, which arises from the zero offset requirement in (12-8). This design procedure is illustrated in Example 12.2.

EXAMPLE 12.2

Use Eq. 12-9 to design feedback controllers for the following processes:

$$\textbf{(a)}\quad G(s) = \frac{K}{\tau s + 1} \tag{12-10}$$

$$\textbf{(b)}\quad G(s) = \frac{K}{(\tau_1 s + 1)(\tau_2 s + 1)} \tag{12-11}$$

Note that $K = K_v K_p K_m$ since $G = G_v G_p G_m$.

Solution
(a) Substituting Eq. 12-10 into 12-9 gives

$$G_c(s) = \frac{\tau s + 1}{K \tau_c s} = \frac{\tau}{\tau_c K}\left(1 + \frac{1}{\tau s}\right) \tag{12-12}$$

which is equivalent to the standard form for a PI controller,

$$G_c(s) = K_c\left(\frac{\tau_I s + 1}{\tau_I s}\right) \tag{12-13}$$

with $K_c = \tau/\tau_c K$ and $\tau_I = \tau$. These expressions indicate that the controller gain K_c is related inversely to the process gain K and directly with the ratio of the open-loop and closed-loop time constants, τ/τ_c. Thus, if a short response time (small τ_c) is desired, a large value of K_c must be employed. Large controller gains are typically associated with high-performance controllers which provide small settling times.
(b) For the second-order process, substituting Eq. 12-11 into 12-9 and rearranging gives

$$G_c(s) = \frac{(\tau_1 + \tau_2)}{K \tau_c}\left[1 + \frac{1}{(\tau_1 + \tau_2)s} + \frac{\tau_1 \tau_2}{\tau_1 + \tau_2}s\right] \tag{12-14}$$

which is in the standard form for an ideal PID controller:

$$G_c(s) = K_c\left(1 + \frac{1}{\tau_I s} + \tau_D s\right) \tag{12-15}$$

with

$$K_c = \frac{1}{K} \frac{\tau_1 + \tau_2}{\tau_c} \qquad \tau_I = \tau_1 + \tau_2 \qquad \tau_D = \frac{\tau_1 \tau_2}{\tau_1 + \tau_2} \tag{12-16}$$

Note that the expressions for K_c and τ_I are equivalent to those obtained for the PI controller in part (**a**) if $\tau_2 = 0$.

Processes with Time Delays

If the process transfer function contains a time delay θ, then a reasonable choice for the desired closed-loop transfer function is

$$\left(\frac{C}{R}\right)_d = \frac{e^{-\theta_c s}}{\tau_c s + 1} \tag{12-17}$$

where θ_c and τ_c are design parameters. Parameter θ_c must be selected so that $\theta_c \geq \theta$ because the controlled variable cannot respond to a set-point change in less than θ time units due to the process time delay θ.

Combining Eqs. 12-17 and 12-3 and setting $\theta_c = \theta$ gives a design equation for G_c:

$$G_c = \frac{1}{G} \frac{e^{-\theta s}}{\tau_c s + 1 - e^{-\theta s}} \tag{12-18}$$

This controller does not have the standard PID form but is physically realizable. The $1 - e^{-\theta s}$ term in the denominator provides *time-delay compensation,* as discussed in Chapter 18.

The controller transfer function in Eq. 12-18 is difficult to implement with analog components due to the $e^{-\theta s}$ terms. However, digital versions such as Dahlin's algorithm (Chapter 26) are easy to apply and have been widely used in the pulp and paper industry.

Next, it will be shown that the design equation in Eq. 12-18 can be used to derive PID controller settings. This derivation, due to Smith et al. [2], is based on approximating the time-delay term in the denominator by a first-order Taylor series expansion:

$$e^{-\theta s} \approx 1 - \theta s \tag{12-19}$$

Substituting this approximation into the denominator of Eq. 12-18 and rearranging gives

$$G_c(s) = \frac{1}{G(s)} \frac{e^{-\theta s}}{(\tau_c + \theta)s} \tag{12-20}$$

Note that in this approach it is not necessary to approximate the time-delay term in the numerator because it is canceled by the identical term in $G(s)$. Next we consider two special cases.

Case a.   First-Order plus Time-Delay Model
Consider the process model in Eq. 12-21,

$$G(s) = \frac{Ke^{-\theta s}}{\tau s + 1} \tag{12-21}$$

Substituting Eq. 12-21 into 12-20 and rearranging gives the PI controller in Eq. 12-13 with the following controller settings:

$$K_c = \frac{1}{K}\frac{\tau}{\theta + \tau_c} \qquad \tau_I = \tau \qquad (12\text{-}22)$$

A comparison of Eqs. 12-16 and 12-22 indicates that K_c should be reduced when the process contains a time delay. Also, time delay θ imposes an upper limit on K_c even for the case where $\tau_c \to 0$. By contrast, Eq. 12-16 indicates that there is no limit on K_c for $\theta = 0$ and $\tau_c \to 0$.

Case b. Second-Order plus Time-Delay Model

For the second-order plus time-delay model in Eq. 12-23,

$$G(s) = \frac{Ke^{-\theta s}}{(\tau_1 s + 1)(\tau_2 s + 1)} \qquad (12\text{-}23)$$

substitution into Eq. 12-20 gives the ideal PID controller in Eq. 12-15 with

$$K_c = \frac{1}{K}\frac{\tau_1 + \tau_2}{\tau_c + \theta} \qquad \tau_I = \tau_1 + \tau_2 \qquad \tau_D = \frac{\tau_1 \tau_2}{\tau_1 + \tau_2} \qquad (12\text{-}24)$$

The PI and PID controllers in Eqs. 12-22 and 12-24 provide approximate time-delay compensation. A comparison of Eqs. 12-24 and 12-16 indicates that the effect of time delay θ is to reduce K_c without changing τ_I or τ_D. Again, the time delay imposes an upper limit on K_c even when $\tau_c \to 0$.

Controller settings obtained from the Direct Synthesis method can be made more conservative by increasing the desired closed-loop time constant τ_c. A conservative choice of τ_c is prudent when θ/τ is significant, since the controller design equations, Eqs. 12-22 and 12-24, were derived based on set-point responses and first-order approximations for the time-delay term.

12.3 INTERNAL MODEL CONTROL

In recent years, Morari and coworkers [3, 4] have developed an important new control system design strategy, Internal Model Control (IMC). As in the Direct Synthesis approach, the IMC design method is based on an assumed process model and relates the controller settings to the model parameters in a straightforward manner. The IMC approach has two important additional advantages: (1) It explicitly takes into account model uncertainty, and (2) it allows the designer to trade-off control system performance against control system robustness to process changes and modeling errors. The IMC approach is based on the simplified block diagram shown in Fig. 12.2b. Transfer function G denotes the actual process plus related control instrumentation as defined in Eq. 12-2. A process model $\tilde{G}$ and the controller output P are used to calculate a model response $\tilde{C}$. The model response is subtracted from the actual response C and the difference, $C - \tilde{C}$, is used as the input signal to the controller, which has transfer function G_c^*. In general, $\tilde{C} \neq C$ due to modeling errors (i.e., $\tilde{G} \neq G$) and unknown disturbances ($L \neq 0$) that are not accounted for in the model.

The block diagrams for conventional feedback control and IMC are shown in Fig. 12.2. The two block diagrams are identical if controllers G_c and G_c^* satisfy the relation

$$G_c = \frac{G_c^*}{1 - G_c^* \tilde{G}} \qquad (12\text{-}25)$$

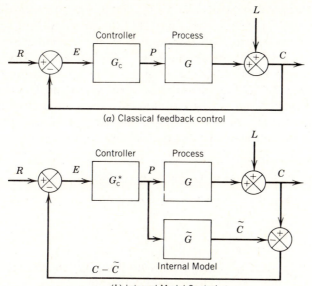

(a) Classical feedback control

(b) Internal Model Control

Figure 12.2.   Feedback control strategies.

Thus, an IMC controller G_c^* can be represented as a standard feedback controller G_c using (12-25).

The following closed-loop relation for IMC can be derived from Fig. 12.2*b*:

$$C = \frac{G_c^* G}{1 + G_c^*(G - \tilde{G})} R + \frac{1 - G_c^* \tilde{G}}{1 + G_c^*(G - \tilde{G})} L \qquad (12\text{-}26)$$

For the special case of a perfect model ($\tilde{G} = G$), Eq. 12-26 reduces to

$$C = G_c^* G R + (1 - G_c^* G) L \qquad (12\text{-}27)$$

The IMC controller is designed in two steps:

Step 1.   The process model is factored as

$$\tilde{G} = \tilde{G}_+ \tilde{G}_- \qquad (12\text{-}28)$$

where $\tilde{G}_+$ contains any time delays and right-half plane zeros. $\tilde{G}_+$ is specified so that its steady-state gain is one.

Step 2.   The controller is specified as

$$G_c^* = \frac{1}{\tilde{G}_-} f \qquad (12\text{-}29)$$

where f is a *low-pass filter* with a steady-state gain of one.

The IMC filter f typically has the form [4]

$$f = \frac{1}{(\tau_c s + 1)^r} \qquad (12\text{-}30)$$

where τ_c is the desired closed-loop time constant. Parameter r is a positive integer that is selected so that G_c^* is either a proper transfer function (i.e., its denominator is at least the same order as the numerator) or if ideal derivative action is allowed, r can be chosen so that the order of the numerator exceeds the order of the denominator by one.

Note that the IMC controller in Eq. 12-29 includes the inverse of $\tilde{G}_-$ rather than the inverse of the entire process model $\tilde{G}$. In contrast, if $\tilde{G}$ had been used, the controller would contain a prediction term $e^{+\theta s}$ (if $\tilde{G}_+$ contained a time delay θ) or an unstable pole (if $\tilde{G}_+$ contained a right-half plane zero). Thus, by employing the factorization given in (12-28) and using a filter of the form of (12-30), the resulting controller G_c^* is guaranteed to be physically realizable and stable. Since the IMC controller in Eq. 12-29 is based on pole–zero cancellation, the IMC approach should not be used for processes that are open-loop unstable.

For the ideal situation where the process model is perfect ($\tilde{G} = G$), substituting Eq. 12-29 into 12-27 gives the relation

$$C = \tilde{G}_+ f R + (1 - f\tilde{G}_+)L \qquad (12\text{-}31)$$

The closed-loop transfer function for set-point changes ($L = 0$) is

$$\frac{C}{R} = \tilde{G}_+ f \qquad (12\text{-}32)$$

In certain situations the IMC and Direct Synthesis (DS) approaches produce equivalent controllers and identical closed-loop performance. For example, if the filter f in Eq. 12-32 is specified so that C/R in (12-32) is equal to the desired transfer function $(C/R)_d$ in (12-3c), then the IMC and DS controllers are equivalent. Identical closed-loop performance results even when modeling errors are present. Recall that Eq. 12-25 shows how to convert G_c^* to G_c.

In general, the IMC approach does not necessarily result in a PID controller for either G_c or G_c^* in Fig. 12.2. However, Rivera et al. [4] have shown that the IMC approach can be used to derive PID settings for controller G_c for a wide variety of process models. Some of their results are shown in Table 12.1. For each case in Table 12.1 they assumed that the desired closed-loop transfer function is given by Eqs. 12-30 and 12-32 with $r = 1$. Additional results are available for other process models and desired closed-loop transfer functions [4].

Table 12.1   IMC-Based PID Controller Settings for $G_c(s)$ [4][a]

Case	Model	$K_c K$	τ_I	τ_D
A	$\dfrac{K}{\tau s + 1}$	$\dfrac{\tau}{\tau_c}$	τ	—
B	$\dfrac{K}{(\tau_1 s + 1)(\tau_2 s + 1)}$	$\dfrac{\tau_1 + \tau_2}{\tau_c}$	$\tau_1 + \tau_2$	$\dfrac{\tau_1 \tau_2}{\tau_1 + \tau_2}$
C	$\dfrac{K}{\tau^2 s^2 + 2\zeta\tau s + 1}$	$\dfrac{2\zeta\tau}{\tau_c}$	$2\zeta\tau$	$\dfrac{\tau}{2\zeta}$
D	$\dfrac{K(-\beta s + 1)}{\tau^2 s^2 + 2\zeta\tau s + 1}, \beta > 0$	$\dfrac{2\zeta\tau}{\tau_c + \beta}$	$2\zeta\tau$	$\dfrac{\tau}{2\zeta}$
E	$\dfrac{K}{s}$	$\dfrac{1}{\tau_c}$	—	—
F	$\dfrac{K}{s(\tau s + 1)}$	$\dfrac{1}{\tau_c}$	—	τ

[a]Based on Eq. 12-30 with $r = 1$.

In Table 12.1 the controller gain K_c is always inversely proportional to the process gain K, as would be expected from stability considerations (cf. Chapter 11). The IMC controllers for Cases A and B are identical with the controllers in Eqs. 12-13 and 12-15, which were designed using the Direct Synthesis Approach. The controllers are identical because both design methods used the same desired closed-loop transfer function, $1/(\tau_c s + 1)$.

Table 12.1 indicates that the IMC approach can be used to derive PID controller settings for a wide variety of process models including those that contain right-half plane zeros or integrating elements. Left-half plane zeros can also be accommodated since they merely add corresponding poles to the controller [4]. Although the process models in Table 12.1 do not explicitly contain time delays, such models can be accommodated by introducing Padé approximations or power series expansions for the time-delay terms [4]. This procedure is illustrated by the following example.

EXAMPLE 12.3

Use the IMC approach to derive PID tuning relations for a first-order plus time-delay model.

Solution
Following Rivera et al. [4], substitute a 1/1 Padé approximation (Eq. 6-37) for the time delay in (12-21):

$$\tilde{G}(s) = \frac{K\left(1 - \dfrac{\theta}{2}s\right)}{\left(1 + \dfrac{\theta}{2}s\right)(\tau s + 1)} \tag{12-33}$$

Factor this model as $\tilde{G} = \tilde{G}_+\tilde{G}_-$ where

$$\tilde{G}_+ = 1 - \frac{\theta}{2}s \tag{12-34}$$

and

$$\tilde{G}_- = \frac{K}{\left(1 + \dfrac{\theta}{2}s\right)(\tau s + 1)} \tag{12-35}$$

For the IMC filter, choose $f = 1/(\tau_c s + 1)$. Substituting into Eq. 12-29 gives

$$G_c^* = \frac{\left(1 + \dfrac{\theta}{2}s\right)(\tau s + 1)}{K(\tau_c s + 1)} \tag{12-36}$$

The equivalent controller, G_c in Fig. 12.2a, can be obtained from Eq. 12-25:

$$G_c = \frac{\left(1 + \dfrac{\theta}{2}s\right)(\tau s + 1)}{K\left(\tau_c + \dfrac{\theta}{2}\right)s} \tag{12-37}$$

This equation can be rearranged into the ideal PID controller of Eq. 12-15 with

$$
K_c = \frac{1}{K} \frac{2\left(\dfrac{\tau}{\theta}\right) + 1}{2\left(\dfrac{\tau_c}{\theta}\right) + 1} \qquad \tau_I = \frac{\theta}{2} + \tau \qquad \tau_D = \frac{\tau}{2\left(\dfrac{\tau}{\theta}\right) + 1} \qquad (12\text{-}38)
$$

Note that each controller setting depends on model parameters θ and τ. In contrast, when the Direct Synthesis Approach and the Taylor series approximation, $e^{-\theta s} \cong 1 - \theta s$, were used to derive the controller settings in Eqs. 12-22 and 12-24, the time delay appeared only in the expressions for K_c. Equation 12-38 indicates that an upper limit exists for K_c even for the case where $\tau_c \to 0$.

Rivera et al. [4] recommend that τ_c be selected for the process model in Eq. 12-21 so that both $\tau_c/\theta > 0.8$ and $\tau_c > \tau/10$.

12.4 DESIGN RELATIONS FOR PID CONTROLLERS

The previous sections dealt with design methods that are applicable to a wide range of process models. In this section we consider some well-known controller design relations that are based on a specific model, namely the first-order plus time-delay model in Eq. 12-21. Recall that the three model parameters in a first-order plus time-delay model (K, θ, and τ) can easily be determined from experimental step response data, as indicated in Section 7.2.

In 1953 Cohen and Coon [5] reported design relations that were developed empirically to provide closed-loop responses with a decay ratio of 1/4. The decay ratio is the ratio of two successive peaks of a damped oscillation, as shown in Fig. 5.10. For a second-order system, a 1/4 decay ratio corresponds to a damping coefficient of $\zeta \cong 0.2$ and a 50% overshoot for set-point changes, as shown in Fig. 5.11 and Eq. 5-53. The Cohen and Coon design relations are shown in Table 12.2.

Table 12.2 Cohen and Coon Controller Design Relations

Controller	Settings	Cohen–Coon
P	K_c	$\dfrac{1}{K}\dfrac{\tau}{\theta}\,[1 + \theta/3\tau]$
PI	K_c	$\dfrac{1}{K}\dfrac{\tau}{\theta}\,[0.9 + \theta/12\tau]$
	τ_I	$\dfrac{\theta[30 + 3(\theta/\tau)]}{9 + 20(\theta/\tau)}$
PID	K_c	$\dfrac{1}{K}\dfrac{\tau}{\theta}\left[\dfrac{16\tau + 3\theta}{12\tau}\right]$
	τ_I	$\dfrac{\theta[32 + 6(\theta/\tau)]}{13 + 8(\theta/\tau)}$
	τ_D	$\dfrac{4\theta}{11 + 2(\theta/\tau)}$

Design Relations Based on Integral Error Criteria

The design relations in the previous section were developed to provide a closed-loop response with a 1/4 decay ratio. This performance criterion has several disadvantages:

1. Responses with 1/4 decay ratios are often judged to be too oscillatory by plant operating personnel.
2. The criterion considers only two points of the closed-loop response $c(t)$, namely the first two peaks.

An alternative approach is to develop controller design relations based on a performance index that considers the entire closed-loop response. Three popular performance indices are:

Integral of the absolute value of the error (IAE)

$$\text{IAE} = \int_0^\infty |e(t)| \, dt \qquad (12\text{-}39)$$

where the error signal $e(t)$ is the difference between the set point and the measurement.

Integral of the squared error (ISE)

$$\text{ISE} = \int_0^\infty [e(t)]^2 \, dt \qquad (12\text{-}40)$$

Integral of the time-weighted absolute error (ITAE)

$$\text{ITAE} = \int_0^\infty t|e(t)| \, dt \qquad (12\text{-}41)$$

Design relations for PID controllers have been developed that minimize these integral error criteria for simple process models and a particular type of load or set-point change. A graphical interpretation of the IAE performance index is shown in Fig. 12.3. The ISE criterion tends to place a greater penalty on large errors than

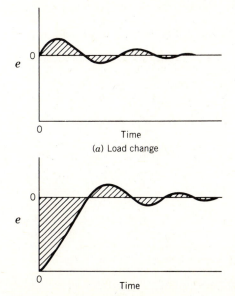

e

0

Time

(a) Load change

e

0

Time

(b) Set-point change

Figure 12.3. Graphical interpretation of IAE. The shaded area is the IAE value.

the IAE or ITAE criteria. The ITAE criterion penalizes errors that persist for long periods of time. In general, ITAE is the preferred integral error criterion since it results in the most conservative controller settings [6].

Design relations that minimize the ITAE performance index are shown in Table 12.3. These relations are based on the first-order plus time-delay model of Eq. 12-21 and the ideal PID controller in Eq. 12-15. Note that the optimal controller settings are different depending on whether step responses to load or set point are considered. For load changes, the load and process transfer functions in Fig. 12.1 are assumed identical, that is, $G_L = G_p$ [6, 7].

Design relations of similar form have been developed for the ISE and IAE performance indices [6–8] and for random load disturbances [9]. Fertik [10] has reported design relations based on second-order plus time-delay models and the interacting form of the PID controller in Eq. 8-14.

EXAMPLE 12.4

Use the integral error approach to obtain alternative controller settings for a process with transfer function:

$$G(s) = \frac{10e^{-s}}{2s + 1} \tag{12-42}$$

In particular, compare PI controllers designed for load changes.

Solution
From Table 12.3 it follows that

$$KK_c = (0.859) \left(\frac{1}{2}\right)^{-0.977} = 1.69$$

or

$$K_c = \left(\frac{1}{10}\right)(1.69) = 0.169$$

Table 12.3   Controller Design Relations Based on the ITAE Performance Index and a First-Order plus Time-Delay Model [6–8][a]

Type of Input	Type of Controller	Mode	A	B
Load	PI	P	0.859	−0.977
		I	0.674	−0.680
Load	PID	P	1.357	−0.947
		I	0.842	−0.738
		D	0.381	0.995
Set point	PI	P	0.586	−0.916
		I	1.03[b]	−0.165[b]
Set point	PID	P	0.965	−0.85
		I	0.796[b]	−0.1465[b]
		D	0.308	0.929

[a]Design relation: $Y = A(\theta/\tau)^B$ where $Y = KK_c$ for the proportional mode, τ/τ_I for the integral mode, and τ_D/τ for the derivative mode.
[b]For set-point changes, the design relation for the integral mode is $\tau/\tau_I = A + B(\theta/\tau)$. [8]

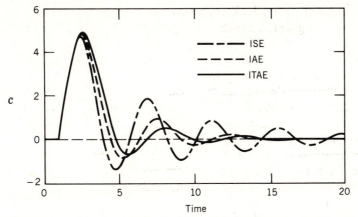

Figure 12.4. Comparison of load responses for PI controllers designed using integral error criteria. (Example 12.4)

and

$$\frac{\tau}{\tau_I} = 0.674 \left(\frac{1}{2}\right)^{-0.680} = 1.08$$

or

$$\tau_I = \frac{2}{1.08} = 1.85$$

The IAE AND ISE controller settings can be calculated from the formulas given in Refs. 6 and 7, The resulting PI controller settings are summarized below:

Method	K_c	τ_I
IAE	0.195	2.02
ISE	0.245	2.44
ITAE	0.169	1.85

The load responses are compared in Fig. 12.4. ISE tuning results in an oscillatory response with the longest settling time. ITAE tuning provides the best damping and the smallest settling time. The IAE response is slightly worse than the ITAE response based on degree of oscillation and settling time.

EXAMPLE 12.5

For the process model,

$$G(s) = \frac{4e^{-3.5s}}{7s + 1} \qquad (12\text{-}43)$$

compare PI and PID controller settings based on the ITAE tuning relations for both load and set-point changes.

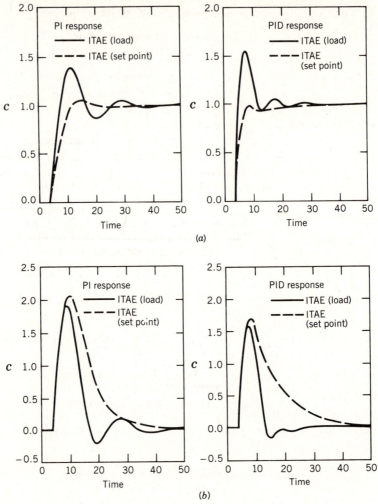

Figure 12.5. Comparison of controllers designed using ITAE criteria for (a) set-point and (b) load changes. (Example 12.5)

Solution

The ITAE controller settings are shown below:

Controller/Design Method	K_c	τ_I	τ_D
PI/load	0.423	6.48	—
PI/set point	0.276	7.39	—
PID/load	0.654	4.98	1.34
PID/set point	0.435	9.69	1.13

Figure 12.5 compares the ITAE controllers. Design for load changes results in large overshoots for set-point changes, while set-point design produces sluggish responses to load disturbances. If set-point changes and load disturbances are both likely to occur, then a compromise in the controller settings should be employed.

This example has demonstrated that, in general, integral error criteria for set-point changes results in more conservative controller settings than for load changes.

12.5 COMPARISON OF CONTROLLER DESIGN RELATIONS

Although the design relations of the previous sections are based on different performance criteria, several general conclusions can be drawn:

1. The controller gain K_c should be inversely proportional to the product of the other gains in the feedback loop (i.e., $K_c \propto 1/K$ where $K = K_v K_p K_m$).
2. K_c should decrease as the ratio of the time delay to the dominant time constant, θ/τ, increases. In general, the quality of control decreases as θ/τ increases because longer settling times and larger maximum deviations occur.
3. The reset time τ_I and derivative time τ_D should increase as θ/τ increases. For these and other design relations considered in Chapters 13 and 16, the ratio τ_D/τ_I typically is between 0.1 to 0.3 [11]. As a rule of thumb, set $\tau_D/\tau_I = 0.25$.
4. When integral control action is added to a proportional-only controller, K_c should be reduced. The further addition of derivative action allows K_c to be increased to a value greater than that for proportional-only control.
5. The Cohen–Coon design relations tend to yield oscillatory closed-loop responses since the design objective is a 1/4 decay ratio. If less oscillatory responses are desired, K_c should be reduced and τ_I increased.
6. Of the three integral error criteria, ITAE provides the most conservative settings while ISE gives the least conservative.

Similar trends will be evident for the design methods based on frequency response criteria, which will be considered in Chapter 16.

Although the tuning relations and design equations in the previous sections were developed for ideal (noninteracting) PID controllers, many industrial analog controllers are constructed so that the derivative and integral modes interact. Shinskey [11] and McMillan [12] have reported equations that express the noninteracting controller settings in terms of the equivalent interacting controller settings (see Exercise 8.2).

Controller settings obtained using various design criteria are compared in the following examples.

EXAMPLE 12.6

Consider a process that can be modeled as

$$G(s) = \frac{2e^{-s}}{s + 1} \tag{12-44}$$

Compare PI controller settings and closed-loop responses for a load change obtained using IMC ($\tau_c = 0.0, 0.8$), Cohen–Coon, and ITAE (load) methods. Then repeat the simulations using the same PI controllers for a change in process gain to 3 (model error) to check the robustness of the controllers.

Solution

The controller settings in Table 12.4 indicate the range of values that result from the various design methods. Closed-loop responses are shown in Fig. 12.6. The IMC ($\tau_c = 0.8$) and ITAE settings provide the least oscillatory responses, while the Cohen and Coon and IMC ($\tau_c = 0$) settings result in the most oscillatory responses. For this process model the PI controllers obtained from Direct Synthesis and IMC methods are identical, and $\tau_c = 0.8$ is the recommended setting [4].

Table 12.4 Controller Settings for Example 12.6

Method	K_c	τ_I
IMC ($\tau_c = 0$)[a]	0.50	1.00
IMC ($\tau_c = 0.8$)[a]	0.28	1.00
Cohen–Coon	0.49	1.14
ITAE (load)	0.43	1.48

[a]It is assumed that the IMC controller is designed using the approximation in Eq. 12-19. It has the same equation as the Direct Synthesis controller (12-22) for the model given by (12-44).

Figure 12.7 shows the results when the process gain changes, clearly indicating the lack of robustness of the Cohen–Coon settings. Both ITAE and IMC give satisfactory performance for the case of model error. Similar results (not shown) are obtained for set-point changes and for a 50% increase in the time delay as the model error.

EXAMPLE 12.7

A process with the transfer function,

$$G(s) = \frac{2e^{-s}}{(10s + 1)(5s + 1)} \tag{12-45}$$

can be approximated by

$$\tilde{G}(s) \cong \frac{2e^{-4.7s}}{12s + 1} \tag{12-46}$$

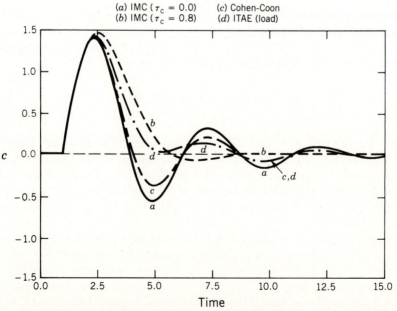

(a) IMC ($\tau_c = 0.0$) (c) Cohen-Coon
(b) IMC ($\tau_c = 0.8$) (d) ITAE (load)

Figure 12.6. Comparison of load responses for Example 12.6 (no model error).

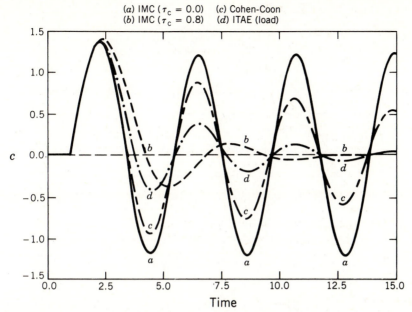

(a) IMC ($\tau_c = 0.0$)   *(c)* Cohen-Coon
(b) IMC ($\tau_c = 0.8$)   *(d)* ITAE (load)

Figure 12.7. Comparison of load responses for Example 12.6 (+50% model error in process gain).

This approximation was obtained using the process reaction curve approach of Section 7.2. Determine PID controller settings for the ITAE (set point) and ITAE (load) correlations using this approximate model and evaluate their performance using the original model in Eq. 12-45. Assume that the load transfer function is also given by Eq. 12-45.

Solution

The PID controller settings are shown in Table 12.5. The ITAE (load) settings are less conservative than the ITAE (set point) settings. The corresponding closed-loop responses are shown in Fig. 12.8.

This example illustrates that design methods based on first-order plus time-delay models can provide satisfactory control even though the process models are only approximate. However, such approaches can produce anomalous results when the estimated time delay becomes small, so they should be used with care.

SUMMARY

A variety of controller design methods is available that are based on transient response criteria. The Cohen–Coon and integral error approaches provide design relations for PID controllers based on first-order plus time-delay transfer function models. Other approaches such as Direct Synthesis and Internal Model Control are more general since they can be used for arbitrary transfer function models and

Table 12.5   Controller Settings for Example 12.7

Method	K_c	τ_I	τ_D
ITAE (load)	1.65	7.12	1.80
ITAE (set point)	1.07	16.25	1.55

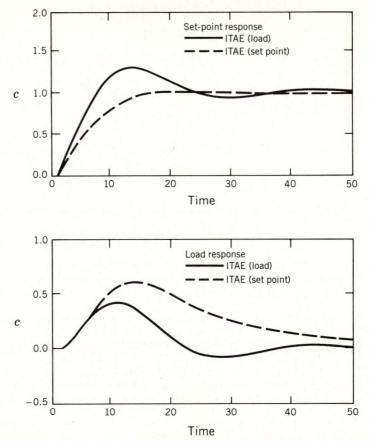

Figure 12.8. Comparison of PID controllers with ITAE settings for Example 12.7.

do not necessarily result in a conventional PID structure. The last two methods are especially useful since the controller settings are easily calculated from the model parameters and only a single tuning parameter, τ_c, need be specified.

The PID controller design relations considered in this chapter contain a number of common characteristics. In general, the controller gain K_c should be inversely proportional to K where $K = K_v K_p K_m$. The reset time τ_I and derivative time τ_D should increase as θ/τ increases for a first-order plus time-delay model. The ratio τ_D/τ_I ranges between 0.1 and 0.3 and typically is 0.25.

REFERENCES

1. Ragazzini, J. R., and G. F. Franklin, *Sampled-Data Control Systems,* McGraw-Hill, New York, 1958.
2. Smith, C. L., A. B. Corripio, and J. Martin, Jr., Controller Tuning from Simple Process Models, *Instrum. Technol.* **22** (12), 39 (1975).
3. Morari, M. and E. Zafiriou, *Robust Process Control,* Prentice-Hall, Englewood Cliffs, NJ, 1989.
4. Rivera, D. E., M. Morari, and S. Skogestad, Internal Model Control, 4. PID Controller Design, *Ind. Eng. Process Design Dev.* **25,** 252 (1986).
5. Cohen, G. H., and G. A. Coon, Theoretical Considerations of Retarded Control, *Trans. ASME* **75,** 827 (1953).
6. Lopez, A. M., P. W. Murrill, and C. L. Smith, Controller Tuning Relationships Based on Integral Performance Criteria, *Instrum. Technol.* **14** (11), 57 (1967).

7. Murrill, P. W., *Automatic Control of Processes,* International Textbook, Scranton, PA, 1967, Chapter 17.
8. Rovira, A. A., P. W. Murrill, and C. L. Smith, Tuning Controllers for Setpoint Changes, *Instrum. Control Systems* **42** (12), 67 (1969).
9. Sood, M., and H. T. Huddleston, Tuning PID Controllers for Random Disturbances, *Instrum. Technol.* **20** (3), 47 (1973).
10. Fertik, H. A., Tuning Controllers for Noisy Processes, *ISA Trans.* **14,** 4 (1975).
11. Shinskey, F. G., *Process Control Systems,* 2d ed., McGraw-Hill, New York, 1979, Chapter 4.
12. McMillan, G. K., *Tuning and Control Loop Performance,* Instrum. Soc. America, Research Triangle Park, NC, 1983.

EXERCISES

12.1. A process has the transfer function,

$$G(s) = \frac{K}{(10s + 1)(5s + 1)}$$

where K has a nominal value of one. PID controller settings are to be calculated using the Direct Synthesis approach with $\tau_c = 5$ min. Suppose that these controller constants are employed and that K changes unexpectedly from 1 to $1 + \alpha$.

(a) For what values of α will the closed-loop system be stable?
(b) Suppose that the PID controller constants are calculated using the nominal value of $K = 1$ but it is desired that the resulting closed-loop system be stable for $|\alpha| \leq 0.2$. What is the smallest value of τ_c that can be used?
(c) What conclusions can be made concerning the effect that the choice of τ_c has on the robustness of the closed-loop system to changes in steady-state gain K?

12.2. The liquid level in a reboiler of a steam-heated distillation column is to be controlled by adjusting the control valve on the steam line, as shown in the drawing. The process transfer function has been empirically determined to be

$$\frac{H(s)}{P_s(s)} = \frac{1.6(0.5s - 1)}{s(3s + 1)}$$

where H denotes the liquid level (in inches) and P_s is the steam pressure (in psi). The level transmitter and control valves have negligible dynamics and steady-state gains of $K_m = 0.5$ psi/in. and $K_v = 2.5$ (dimensionless), respectively.

(a) It is proposed to design a PI level controller using the Direct Synthesis approach and Eq. 12-9. Is this a satisfactory design? Briefly justify your answer.
(b) Suggest a modification of the Direct Synthesis approach used in part (a) which will result in an improved control system.

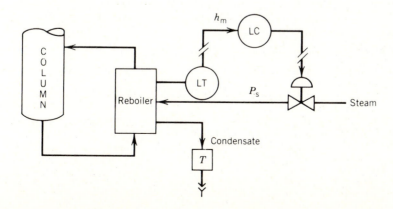

12.3. A process has the transfer function, $G(s) = 2e^{-0.2s}/(s + 1)$. Compare the PI controller settings that result from the following approaches:

(a) Direct Synthesis method ($\tau_c = 0.2$)
(b) Direct Synthesis method ($\tau_c = 1.0$)
(c) Cohen–Coon tuning relations
(d) ITAE performance index (load)
(e) ITAE performance index (set point)

Which approach gives the most conservative controller settings? Which gives the least conservative? (i.e., the most overdamped or underdamped responses?)

12.4. Consider the tuning relations for the last three design methods of Exercise 12.3. If a process model is in the form of Eq. 12-21, what general conclusions can be drawn concerning the relative conservatism of the resulting PI controller settings?

12.5. For the distillation column of Exercise 12.2, design a feedback control system using the IMC approach.

12.6 The IMC controller design in Example 12.3 was based on a 1/1 Padé approximation for the time delay. Suppose that the approximation $e^{-\theta s} \simeq 1 - \theta s$ was used instead. Compare the resulting IMC controller settings with those obtained in Example 12.3 and with the settings in Eq. 12-22 which were obtained from the Direct Synthesis approach.

12.7. A process stream is heated using a shell and tube heat exchanger. The exit temperature is controlled by adjusting the steam control valve shown in the drawing. During

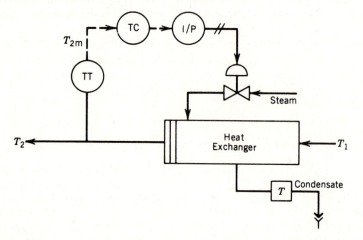

an open-loop experimental test, the steam pressure P_s was suddenly changed from 18 to 20 psig and the temperature data shown below were obtained. At the nominal conditions the control valve and current-to-pressure transducers have gains of $K_v = 0.9$ psi/psi and $K_{IP} = 0.75$ psi/mA, respectively. Determine appropriate PID controller settings using the following approaches:

(a) Direct Synthesis (select a reasonable value of τ_c)
(b) Cohen and Coon
(c) ITAE (load)

Experimental Data

$t(min)$	$T_{2m}(mA)$
0	12.0
1	12.0
2	12.5
3	13.1
4	14.0
5	14.8
6	15.4
7	16.1
8	16.4
9	16.8
10	16.9
11	17.0
12	16.9

12.8. To investigate the applicability of the Direct Synthesis approach to open-loop unstable problems, do the following:

(a) Derive an expression for the closed-loop transfer function when the controller is designed using a process model $\tilde{G}$ while the actual process has a transfer function G.

(b) Theoretically, will the resulting controller yield a stable closed-loop system if the process model is perfect (i.e., $\tilde{G} = G$)? Justify your answer.

(c) Suppose that $\tilde{G} = 1/(s - 5)$ and that $G = 1/(s - 3)$ and that the Direct Synthesis controller is designed using Eq. 12-9. For what values of τ_c is the resulting closed-loop system stable?

(d) Rework part (c) for $\tilde{G} = 1/(s + 5)$ and $G = 1/(s - 3)$. Explain the results in parts (b) and (c).

12.9. Repeat Exercise 12.8 for the IMC approach.

12.10. Suggest a modification of the Direct Synthesis approach that will allow it to be applied to open-loop unstable processes. (*Hint*: First stabilize the process using a proportional-only feedback controller.) Draw a block diagram for your proposed control scheme.

CHAPTER 13

Controller Tuning and Troubleshooting Control Loops

After a control system is installed the controller settings must usually be adjusted until the control system performance is considered to be satisfactory. This activity is referred to as *controller tuning* or *field tuning* of the controller. Because controller tuning is usually done by trial and error, it can be quite tedious and time-consuming. Consequently, it is desirable to have good preliminary estimates of satisfactory controller settings. A good first guess may be available from experience with similar control loops. Alternatively, if a process model or frequency response data are available, the design methods in Chapters 12 and 16 can be employed to calculate controller settings. However, field tuning may still be required to fine tune the controller, especially if the available process information is incomplete or not very accurate.

This chapter has the following organization. In Section 13.1, guidelines for controller design and tuning are presented for common types of control loops. Common controller tuning techniques are considered in Sections 13.2 through 13.4, and strategies for troubleshooting control loops that are not performing satisfactorily are considered in Section 13.5.

In recent years there has been considerable interest in *adaptive controllers*, that is, controllers that have the capability of automatically finding suitable controller settings in the field. Adaptive controllers are considered in Chapter 18.

13.1 GUIDELINES FOR COMMON CONTROL LOOPS

General guidelines for selection of controller type (P, PI, etc.) and choice of settings are available for commonly encountered process variables: flow rate, liquid level, gas pressure, temperature, and composition. The guidelines discussed below are useful for situations where a process model is not available. However, they should be used with caution because exceptions do occur. Similar guidelines are available for selecting the initial controller settings for the startup of a new plant [1].

Flow Control

Flow and liquid pressure control loops are characterized by fast responses (on the order of seconds), with essentially no time delays. The process dynamics are due to compressibility (in a gas stream) or inertial effects (in a liquid). The sensor and signal transmission line may introduce significant dynamic lags if pneumatic instruments are used. Disturbances in flow-control systems tend to be frequent but generally not of large magnitude. Most of the disturbances are high-frequency noise (periodic or random) due to stream turbulence, valve changes, and pump vibration. PI flow controllers are generally used with intermediate values of the controller gain K_c. The presence of recurring high-frequency noise rules out the use of derivative action.

Liquid Level

A typical non-self-regulating liquid-level process has been discussed in Chapters 2 and 10. Because of its integrating nature, a relatively high-gain controller can be used with little concern about instability of the control system. In fact, as shown in Section 10.3, an increase in controller gain often brings an increase in system stability, while low controller gains can increase the degree of oscillation. Integral control action is normally used but is not necessary if small offsets in the liquid level ($\pm 5\%$) can be tolerated. Derivative action is not normally employed in level control, since the level measurements often contain noise due to the splashing and turbulence of the liquid entering the tank.

In many level control problems, the liquid storage tank is used as a surge tank to damp out fluctuations in its inlet streams. If the exit flow rate from the tank is used as the manipulated variable, then conservative controller settings should be applied to avoid large, rapid fluctuations in the exit flow rate. This strategy is referred to as *averaging control* [2]. If level control also involves heat transfer, such as in a vaporizer or evaporator, the process model and controller design become much more complicated. In such situations special control methods can be advantageous [3].

Gas Pressure

Gas pressure is relatively easy to control, except when the gas is in equilibrium with a liquid. A gas pressure process is self-regulating: the vessel (or pipeline) admits more feed when the pressure is too low and reduces the intake when the pressure becomes too high. PI controllers are normally used with only a small amount of integral control action (i.e., τ_I large). Usually the vessel volume is not large, leading to relatively small residence times and time constants. Derivative action is normally not needed because the process response times are usually quite small compared to other process operations.

Temperature

General guidelines for temperature control loops are difficult to state because of the wide variety of processes and equipment involving heat transfer (and their different time scales). For example, the temperature control problems are quite different for heat exchangers, distillation columns, chemical reactors, and evaporators. Due to the presence of time delays and/or multiple thermal capacitances, there will usually be a stability limit on the controller gain. PID controllers are commonly employed to provide more rapid responses than can be obtained with PI controllers.

Composition

Composition loops generally have characteristics similar to temperature loops, but with several differences:

1. Measurement noise is a more significant problem in composition loops.
2. Time delays due to analyzers may be a significant factor.

These two factors can limit the effectiveness of derivative action. Due to their importance and the difficulty of control, composition and temperature loops often are prime candidates for the advanced control strategies discussed in Chapters 17 and 18.

13.2 TRIAL AND ERROR TUNING

Controller field tuning is often performed using trial and error procedures suggested by controller manufacturers. A typical approach for PID controllers can be summarized as follows [4]:

Step 1. Eliminate integral and derivative action by setting τ_D at its minimum value and τ_I at its maximum value.
Step 2. Set K_c at a low value (e.g., 0.5) and put the controller on automatic.
Step 3. Increase the controller gain K_c by small increments until continuous cycling occurs after a small set-point or load change. The term "continuous cycling" refers to a sustained oscillation with constant amplitude.
Step 4. Reduce K_c by a factor of two.
Step 5. Decrease τ_I in small increments until continuous cycling occurs again. Set τ_I equal to three times this value.
Step 6. Increase τ_D until continuous cycling occurs. Set τ_D equal to one-third of this value.

The value of K_c that results in continuous cycling in Step 3 is referred to as the *ultimate gain* and will be denoted by K_{cu}. In performing the experimental test, it is important that the controller output does not saturate. If saturation does occur, then a sustained oscillation can result even though $K_c > K_{cu}$. Typical results for trial and error tuning of a self-regulating process are shown in Fig. 13.1. If $K_c < K_{cu}$, the closed-loop response $c(t)$ is usually overdamped or slightly oscillatory. Increasing K_c to a value of K_{cu} results in continuous cycling as shown in Fig. 13.1*b*. For $K_c > K_{cu}$, the closed-loop system is unstable and will theoretically have an unbounded response if controller saturation is ignored. However, controller saturation usually prevents the response from going unstable and produces a sustained oscillation instead (cf. Fig. 13.1*d*). The sustained oscillation in Fig. 13.1*d* can lead to an estimate of K_{cu} that is too large. For example, suppose that the response in Fig. 13.1*d* occurs when a controller gain has a numerical value denoted by K_{c1}, and the person performing the experimental test is not aware that the controller has saturated. It would be reasonable for him or her to conclude that $K_{cu} = K_{c1}$ when, in fact, $K_{cu} < K_{c1}$. This overestimate of K_{cu} could result in poor control since the controller gain calculated in Step 4 will be too large.

Because the concept of an ultimate gain plays such a key role in control system design and analysis, we present a more formal definition:

> **Definition.** *The* ultimate gain K_{cu} *is the largest value of the controller gain* K_c *that results in closed-loop stability when a proportional-only controller is used.*[1]

---
[1]If a direct-acting controller is used, then $K_c < 0$ and the definition refers to $-K_c$ rather than K_c.

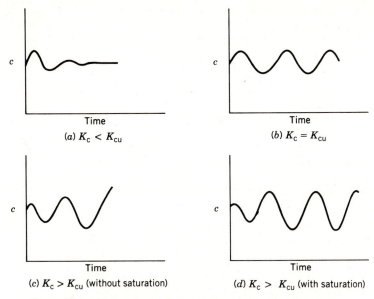

Figure 13.1. Experimental determination of the ultimate gain K_{cu}.

If a process model is available, then K_{cu} can be calculated theoretically using the stability criteria in Chapters 11 and 16. The trial and error tuning procedure described above has a number of disadvantages:

1. It is quite time-consuming if a large number of trials is required to optimize K_c, τ_I, and τ_D or if the process dynamics are quite slow. Unit control loop testing may be expensive because of lost productivity or poor product quality.
2. Continuous cycling may be objectionable since the process is pushed to the stability limit. Consequently, if external disturbances or a change in the process occurs during controller tuning, unstable operation or a hazardous situation could result (e.g., a "runaway" chemical reactor).
3. This tuning procedure is not applicable to processes that are open-loop unstable because such processes typically are unstable at both high and low values of K_c, but are stable for an intermediate range of values.
4. Some simple processes do not have an ultimate gain (e.g., processes that can be accurately modeled by first-order or second-order transfer functions without time delays).

13.3 CONTINUOUS CYCLING METHOD

Trial and error tuning methods based on a sustained oscillation can be considered to be variations of the famous *continuous cycling* method that was published by Ziegler and Nichols [5] in 1942. This classic approach is probably the best known method for tuning PID controllers. The continuous cycling approach has also been referred to as *loop tuning* or the *ultimate gain method*. The first step is to experimentally determine K_{cu} as described in the previous section. The period of the resulting sustained oscillation is referred to as the *ultimate period P_u*. The PID controller settings are then calculated from K_{cu} and P_u using the Ziegler–Nichols (Z–N) tuning relations in Table 13.1. The Z–N tuning relations were empirically developed to provide a 1/4 decay ratio. These tuning relations have been widely

Table 13.1 Ziegler–Nichols Controller Settings Based on the Continuous Cycling Method

Controller	K_c	τ_I	τ_D
P	$0.5K_{cu}$	—	—
PI	$0.45K_{cu}$	$P_u/1.2$	—
PID	$0.6K_{cu}$	$P_u/2$	$P_u/8$

used in industry [6] and serve as a convenient base case for comparing alternative control schemes. However, controller tuning examples presented later in this section and in Chapter 16 indicate that Z–N tuning can be inferior to settings obtained by other methods and should be used with caution.

Note that the Z–N setting for proportional control provides a significant safety margin since the controller gain is one-half of the stability limit K_{cu}. When integral action is added, K_c is reduced to $0.45K_{cu}$ for PI control. However, the addition of derivative action allows the gain to be increased to $0.6K_{cu}$ for PID control.

For some control loops, the degree of oscillation, associated with the 1/4 decay ratio and the corresponding large overshoot for set-point changes are undesirable. Thus, more conservative settings are often preferable, such as the modified Z–N settings in Table 13.2 [7].

Although widely applied, the Ziegler–Nichols continuous cycling method has some of the same disadvantages as the trial and error method listed in Section 13.2. However, the continuous cycling method is less time-consuming than the trial and error method because it requires only one trial and error search. Again, we wish to emphasize that the controller settings in Tables 13.1 and 13.2 should be regarded as first estimates. Subsequent fine tuning via trial and error is often required, especially if the "original settings" in Table 13.1 are selected. Alternatively, the continuous cycling autotuning method discussed at the end of this section may be used.

EXAMPLE 13.1

Compare the performance of PID controllers that are tuned using the original and modified Z–N settings in Table 13.2 for the process model of Example 12.5:

$$G(s) = \frac{4e^{-3.5s}}{7s + 1}$$

Consider both set-point and load responses and compare them with the ITAE responses of Fig. 12.5.

Solution

The ultimate gain and ultimate period are determined by trial and error to be $K_{cu} = 0.95$ and $P_u = 12$. The Z–N controller settings calculated from Table 13.2

Table 13.2 Original and Modified Ziegler–Nichols Settings for PID Controllers [7]

	K_c	τ_I	τ_D
Original (1/4 decay ratio)	$0.6K_{cu}$	$P_u/2$	$P_u/8$
Some overshoot	$0.33K_{cu}$	$P_u/2$	$P_u/3$
No overshoot	$0.2K_{cu}$	$P_u/2$	$P_u/3$

are as follows

Method	K_c	τ_I	τ_D
Original Z–N	0.57	6.0	1.5
Some overshoot	0.31	6.0	4.0
No overshoot	0.19	6.0	4.0

The closed-loop responses for unit step changes in set point are shown in Fig. 13.2. Smaller overshoots result for the modified Z–N settings but even the "no overshoot" settings do not completely eliminate overshoot. It is somewhat surprising that the "some overshoot" settings produce a more oscillatory response than the original Z–N settings despite the lower K_c value. This anomaly is due to the larger τ_D value for the "some overshoot" case.

A comparison of the load and set-point responses in Fig. 13.2 indicates that improved load responses can be obtained providing that some overshoot for set-point changes can be tolerated. This same trade-off occurred for the ITAE controller settings of Fig. 12.5. A comparison of Figs. 12.5 and 13.2 indicates that the original Z–N settings and the ITAE (load) settings are judged to be superior to the other settings for this example based on consideration of both set-point and load responses.

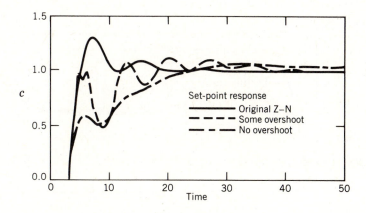

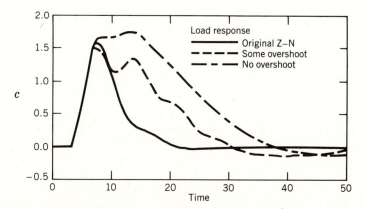

Figure 13.2. Controller comparison for Example 13.1.

EXAMPLE 13.2

Repeat the comparison of Example 13.1 for the process model of Example 12.7:

$$G(s) = \frac{2e^{-s}}{(10s + 1)(5s + 1)}$$

Solution

The controller settings shown below are based on the numerical values of $K_{cu} = 7.88$ and $P_u = 11.6$ which were obtained by trial and error:

Method	K_c	τ_I	τ_D
Original Z–N	4.73	5.8	1.45
Some overshoot	2.60	5.8	3.87
No overshoot	1.58	5.8	3.87

The closed-loop responses in Fig. 13.3 indicate that the modified Z–N settings result in more conservative responses, as would be expected. Somewhat surprisingly, the "no overshoot" settings result in more overshoot than the "some overshoot" settings for the change in set point. A comparison of Figs. 12.8 and

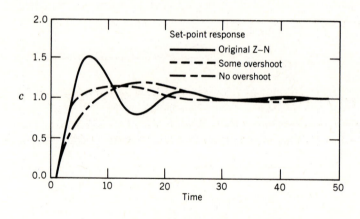

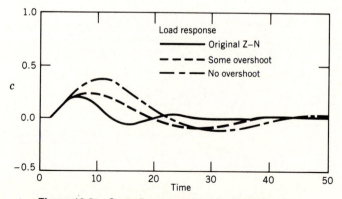

Figure 13.3. Controller comparison for Example 13.2.

13.3 indicates that the original Z–N settings are the least conservative, while the ITAE (set-point) settings are the most conservative.

Examples 13.1 and 13.2 demonstrate that the original Z–N settings tend to produce oscillatory responses. The modified Z–N settings [7] tend to be more conservative but do not necessarily eliminate overshoot for set-point changes. These two examples indicate that anomalies can occur due to the simple correlations that are employed.

Autotuning

Åström and Hägglund [8] describe an automatic tuning (autotuning) method that is an alternative to Ziegler–Nichols continuous cycling discussed above. Their method has the following features:

1. The system is forced by a relay controller which causes the system to oscillate with a small amplitude. The amplitude of the oscillation can be constrained by adjusting the amplitude of the input variations.
2. Usually a single closed-loop experiment is sufficient to find the dynamic model, and the experiment does not require a priori information about the process dynamic model.

The autotuner uses a relay with a dead zone to generate the process oscillation, as shown in Fig. 13.4. The period P_u is found simply by measuring the period of the process oscillation. The ultimate gain is given by

$$K_{cu} = \frac{4d}{\pi a} \tag{13-1}$$

where d is the relay amplitude set by the operator, and a is the measured amplitude of the process oscillation. The controller settings are then found using Table 13.1. This strategy is the basis for the SattControl adaptive autotuner, discussed in Section 18.4. The autotuner has a single user-specified parameter, relay amplitude d.

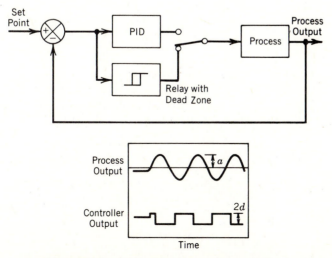

Figure 13.4.   Autotuning using a relay controller.

13.4   PROCESS REACTION CURVE METHOD

In their famous paper [5], Ziegler and Nichols proposed a second on-line tuning technique, the process reaction curve method. This method is based on a single experimental test that is made with the controller in the manual mode. A small step change in the controller output is introduced and the measured process response $B(t)$ is recorded. This step response is also referred to as the *process reaction curve* as noted in Section 8.2. It is characterized by two parameters: S, the slope of the tangent through the inflection point, and θ, the time at which the tangent intersects the time axis. The graphical determination of S and θ was described in Section 8.2.

Two different types of process reaction curves are shown in Fig. 13.5 for step changes occurring at $t = 0$. The response for Case a is unbounded, which indicates that this process is not self-regulating. In contrast, the hypothetical process considered in Case b is self-regulating since the process reaction curve reaches a new steady state. Note that the slope–intercept characterization can be used for both types of process reaction curves.

The Ziegler–Nichols tuning relations for the process reaction curve method are shown in Table 13.3. S^* denotes the normalized slope, $S^* = S/\Delta p$ where Δp is the magnitude of the step change that was introduced in controller output p. These tuning relations were developed empirically to give closed-loop responses with 1/4 decay ratios. The tuning relations in Table 13.3 can be used for both self-regulating and non-self-regulating processes.

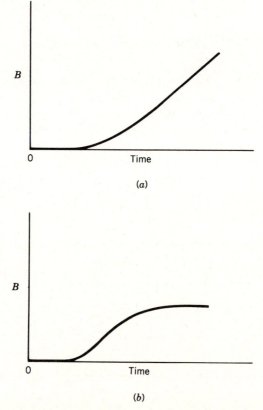

(a)

(b)

Figure 13.5.   Typical process reaction curves: (a) non-self-regulating process, (b) self-regulating process.

Table 13.3   Ziegler–Nichols Tuning Relations (Process Reaction Curve Method)

Controller Type	K_c	τ_I	τ_D
P	$\dfrac{1}{\theta S^*}$	—	—
PI	$\dfrac{0.9}{\theta S^*}$	3.33θ	—
PID	$\dfrac{1.2}{\theta S^*}$	2θ	0.5θ

If the process reaction curve has the typical sigmoidal shape shown in Case b of Fig. 13.5, the following model usually provides a satisfactory fit:

$$\frac{B'(s)}{P'(s)} = G_v G_p G_m = \frac{Ke^{-\theta s}}{\tau s + 1} \tag{13-2}$$

where B' is the measured value of the controlled variable and P' is the controller output, both expressed as deviation variables. Note that this model includes the transfer functions for the final control element and sensor–transmitter combination, as well as the process transfer function. Model parameters K, τ, and θ can be determined from the process reaction curve using the techniques presented in Section 7.2.

The PID controller settings can be calculated from this model using any of the design methods in Chapter 12. In fact, the Cohen and Coon tuning relations in Table 12.2 were originally developed as a modification of the process reaction curve approach for cases where the process could be adequately modeled by (13-2) (cf. Table 13.3).

The process reaction curve (PRC) method offers several significant advantages:

1. Only a single experimental test is necessary.
2. It does not require trial and error.
3. The controller settings are easily calculated.

However, the PRC method also has several disadvantages:

1. The experimental test is performed under open-loop conditions. Thus, if a significant load change occurs during the test, no corrective action is taken *and* the test results may be significantly distorted.
2. It may be difficult to determine the slope at the inflection point accurately, especially if the measurement is noisy and a small recorder chart is used.
3. The method tends to be sensitive to controller calibration errors. By contrast, the Z–N method is less sensitive to calibration errors in K_c since the controller gain is adjusted during the experimental test.
4. The recommended settings in Tables 12.2 and 13.3 tend to result in oscillatory responses since they were developed to provide a 1/4 decay ratio.
5. The method is not recommended for processes that have oscillatory open-loop responses since the process model in Eq. 13-2 will be quite inaccurate.

Closed-loop versions of the process reaction curve method have been proposed as a partial remedy for the first disadvantage [9, 10]. In this approach, a process reaction curve is generated by making a step change in set point during proportional-only control. The model parameters in Eq. 13-2 are then calculated in a novel manner from the closed-loop response. A minor disadvantage of these closed-loop process reaction methods is that the model parameter calculations are more complicated than for the standard open-loop method.

The following example illustrates the application of the process reaction curve method.

EXAMPLE 13.3

Consider the feedback control system for the stirred-tank heater shown in Fig. 10.1. With the controller in manual, the controller output is suddenly changed from 10 to 12 mA. The process reaction curve is shown in Fig. 13.6. The electronic temperature transmitter has a span of 50 °C. Its output signal is sent to a recorder as well as a controller. The recorded signal varies linearly from 0 to 100% as the transmitter output changes from 4 to 20 mA. The recorder chart speed is 0.75 in. per minute. Determine an approximate process model and calculate controller settings using the Cohen–Coon relations in Table 12.2.

Solution

A block diagram for the closed-loop system is shown in Fig. 13.7. This block diagram is similar to the one in Fig. 10.8 but differs in two important respects:

1. The feedback loop has been broken between the controller and the current-to-pressure (I/P) transducer. A step change of $M = 12 - 10 = 2$ mA is introduced in the controller output p.

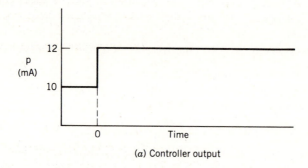

(a) Controller output

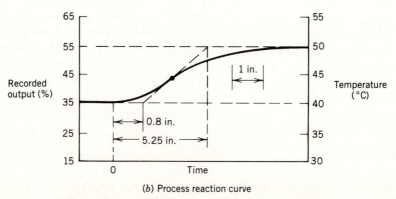

(b) Process reaction curve

Figure 13.6. Process reaction curve for Example 13.3.

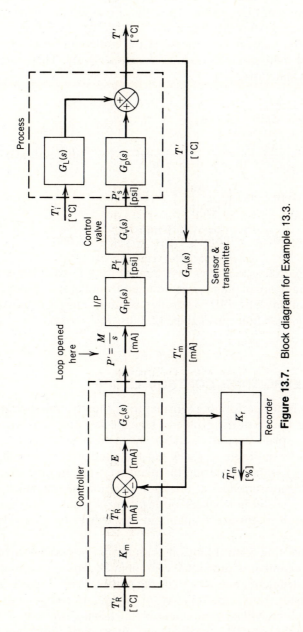

Figure 13.7. Block diagram for Example 13.3.

2. The transfer functions for the individual blocks are not known with the exception of gains K_m and K_r. From the given information, $K_m = (20 - 4 \text{ mA})/50\,°C = 0.32 \text{ mA}/°C$. The recorder gain K_r can be calculated as $K_r = (100 - 0\%)/(20 - 4 \text{ mA}) = 6.25\%/\text{mA}$.

To develop a first-order plus time-delay model, consider the process reaction curve in Fig. 13.6. First inspect the recorder chart which plots the measured output (in %) as a function of time. In Fig. 13.6, the corresponding temperature is shown along the right scale. After drawing a tangent through the inflection point, we can determine that the tangent line intersects the horizontal lines for the initial and final temperature values at 0.8 and 5.25 in., respectively. Thus, the model parameters can be calculated as

$$K = \frac{\Delta T_m}{\Delta p} = \frac{\Delta \tilde{T}_m}{\Delta p} \frac{1}{K_r}$$

$$K = \frac{55 - 35\%}{12 - 10 \text{ mA}} \frac{1}{6.25\%/\text{mA}} = 1.6 \quad \text{(dimensionless)}$$

$$\theta = (0.8 \text{ in.}) \left(\frac{1 \text{ min}}{0.75 \text{ in.}} \right) = 1.07 \text{ min}$$

$$\tau = (5.25 - 0.8 \text{ in.}) \left(\frac{1 \text{ min}}{0.75 \text{ in.}} \right) = 5.93 \text{ min}$$

Note that the recorder speed, 0.75 in./min, is used in calculating θ and τ. Also the apparent time delay of 0.8 in. is subtracted from the intercept value of 5.25 in. for the τ calculation. The PID controller settings can then be calculated from the Cohen–Coon relations in Table 12.2 as $K_c = 4.8$, $\tau_I = 2.4$ min, and $\tau_D = 0.38$ min.

13.5 TROUBLESHOOTING CONTROL LOOPS

If a control loop is not performing satisfactorily, then troubleshooting is necessary to determine the source of the problem. Unexpected problems with control loops often occur during plant start-up or after a change in plant capacity has been made. Ideally, it would be desirable to check out the control loop over the full range of process operating conditions during the plant start-up. In practice, this is seldom possible. This section provides a brief introduction to the basic principles and strategies that are useful in troubleshooting control loops. More detailed analyses that provide useful insight are available in the articles by Cho [11] and Buckley [12].

In troubleshooting control loops it is important to remember that the control loop consists of a number of individual components: sensor/transmitter, controller, control valve, instrument lines, as well as the process itself. Clearly, serious control problems can result from a malfunction of any single component. On the other hand, even if each individual component is functioning properly by itself, there is no guarantee that the overall system will perform properly. For example, the control loop may produce undesirable oscillations in the controlled variable due to a poor choice of controller settings.

As Buckley [12] has noted, there is an unfortunate tendency among operating and maintenance personnel to use controller retuning as a cure-all for control loop problems. Buckley states that in his experience the most common factors that cause a control loop that once operated satisfactorily to become either unstable or excessively sluggish are

a. Changing process conditions, usually changes in throughput rate.
b. Sticking control valve stem.
c. Plugged line in a pressure or differential pressure transmitter.
d. Fouled heat exchangers, especially reboilers for distillation columns.
e. Cavitating pumps (usually due to a suction pressure that is too low).

Note that only items a and d provide valid reasons for retuning the controller.

An important part of any troubleshooting strategy is to obtain enough background information to clearly define the control problem. A number of questions need to be answered [11]:

1. What is the process to be controlled?
2. What is the controlled variable?
3. What are the control objectives?
4. Is there any recorded information or visible results of the control loop problem available?
5. Is the controller in the manual or automatic mode? Is it reverse or direct acting?
6. If the process is cycling, what is the cycling frequency?
7. What are the controller settings for K_c, τ_I, and τ_D?
8. Is the process open-loop stable?
9. What additional documentation is available? (Control loop summary sheets, instrumentation diagrams, etc.)

After obtaining this information, the next step is to check out the individual components in the control loop. In particular, one should determine that the measurement device and control valve are in proper working condition. Typically, transmitters and control valves located in the field require more maintenance than controllers located in the central control room.

If the instruments in the control loop are working properly, the next step is to ensure that the process is functioning properly. Finally, controller retuning may be necessary if the control loop exhibits undesirable oscillations or an excessively sluggish response.

SUMMARY

In this chapter, we have considered some of the practical problems associated with feedback control systems, namely, field tuning controllers and troubleshooting control loops. Of the available field tuning methods, the process reaction curve is the easiest to apply since it requires only a single step test. If the continuous cycling method is employed, the modified Z–N settings in Table 13.2 are recommended since they usually provide more conservative control than the original Z–N settings. The advantage of the step test is that other model-based design methods in Chapter 12 (e.g., IMC) can be applied. Note that all design methods furnish guidelines for setting only; field adjustment still may be necessary.

When troubleshooting a malfunctioning control loop, retuning the controller should be considered only after proper operation of all components in the control loop has been verified.

Computer-based "expert systems" have the potential of providing valuable assistance during control loop troubleshooting. Expert systems are discussed in Section 18.2.

REFERENCES

1. Anderson, G. D., Initial Controller Settings to Use at Plant Startup, *Chem. Eng.*, **90** (7), 113 (1983).
2. Buckley, P., *Techniques of Process Control,* John Wiley, New York, 1964, Chapter 18.
3. Shinskey, F. G., *Process Control Systems,* 3d ed., McGraw-Hill, New York, 1988, Chapter 3.
4. Jury, F. D., Fundamentals of Three-Mode Controllers, Fisher Controls Company Technical Monograph, No. 28 (1973).
5. Ziegler, J. G., and N. B. Nichols, Optimum Settings for Automatic Controllers, *Trans. ASME* **64,** 759 (1942).
6. Luyben, W. L., *Process Modeling, Simulation and Control,* McGraw-Hill, New York, 1973, Chapter 11.
7. Perry, R. H., and C. H. Chilton (Ed.), *Chemical Engineers' Handbook,* 5th ed., McGraw-Hill, New York, 1973, pp. 22–25.
8. Åström, K. J., and T. Hägglund, *Automatic Tuning of PID Controllers,* Instrum. Society of America, Research Triangle Park, NC, 1988.
9. Yuwana, M., and D. E. Seborg, A New Method for On-Line Controller Tuning, *AIChE J.* **28,** 434 (1982).
10. Jutan, A., and E. S. Rodriguez II, Extension of a New Method for Online Controller Tuning, *Can. J. Chem. Eng.* **62,** 802 (1984).
11. Cho, C., Troubleshooting Process Control Loops, *Instrum. Technol.* **23** (3), 31 (1976).
12. Buckley, P., A Modern Perspective on Controller Tuning, Proc., Texas A&M Instrum. Sympos., pp. 80–88 (January 1973).

EXERCISES

13.1. A PID controller is used to control the temperature of a jacketed batch reactor by adjusting the flow rate of coolant to the jacket. The temperature controller has been tuned to provide satisfactory control at the nominal operating conditions. Would you anticipate that the temperature controller may have to be retuned for any of the following instrumentation changes? Justify your answers.

(a) The span of the temperature transmitter is reduced from 30 to 15 °C.
(b) The zero of the temperature transmitter is increased from 50 to 60 °C.
(c) The control valve "trim" is changed from linear to equal percentage.
(d) The temperature of the coolant leaving the jacket is used as the controlled variable instead of the temperature in the reactor.

13.2. Suppose that a process can be adequately modeled by the first-order plus time-delay model in Eq. 13-2. Use the Z–N tuning relations in Table 13.3 to derive an equivalent set of tuning relations expressed in terms of model parameters K, θ, and τ. Compare these relations with the Cohen and Coon tuning relations in Table 12.2. Which would you expect to be more conservative?

13.3. Consider the experimental step response data for the heat exchanger of Exercise 12.7. Determine PID controller settings using the process reaction curve method and two tuning relations:

(a) Ziegler–Nichols settings in Table 13.3.
(b) Cohen–Coon settings in Table 12.2.

13.4. IGC's operations area personnel are experiencing problems with a particular feedback control loop on an interstage cooler. Appelpolscher has asked you to assess the situation and report back what remedies, if any, are available. The control loop is exhibiting an undesirable sustained oscillation which the operations people are sure is caused by the feedback loop itself (e.g., poor controller tuning). They want assistance in retuning the loop. Appelpolscher thinks that the oscillations are caused by external disturbances (e.g., cyclic load disturbances such as cycling of the cooling water temperature); he wants the operations people to deal with the problem themselves. Suggest a simple procedure that will allow you to determine quickly what is causing the oscillations. How will you explain your logic to Appelpolscher? (Recall that he won't be happy if he is wrong!)

13.5. A problem has arisen in the level control loop for the flash separation unit shown in the drawing. The level control loop had functioned in a satisfactory manner for a long period of time. However, the liquid level is gradually increasing with time even though the PI level controller output has saturated. Furthermore, the liquid flow rate is well above the nominal value, while the feed flow rate is at the nominal value, according to the recorded measurements from the two flow transmitters. The accuracy of the level transmitter measurement has been confirmed by comparison with sight glass readings for the separator. The two flow measurements are obtained via orifice plates and differential pressure transmitters, as described in Chapter 9.

Suggest possible causes for this problem and describe how you would troubleshoot this situation.

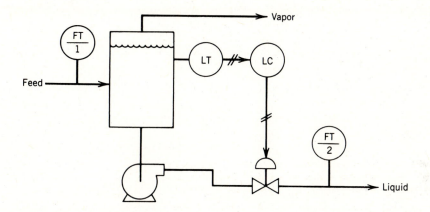

13.6 Consider the transfer function model in Eq. (13-2) with $K = 2$, $\tau = 5$ and $\theta = 1$. Compare the PID controller settings obtained from the original and modified Z-N tuning relations in Table 13.2. Plot the closed-loop responses for unit step changes in the set point.

PART FOUR

FREQUENCY RESPONSE METHODS

— CHAPTER 14 —————

Frequency Response Analysis

In previous chapters, Laplace transform techniques have been used to express mathematical models in terms of transfer functions in order to calculate transient responses. This chapter focuses on an alternative way to interpret the transfer function model, referred to as the complex transfer function or the *frequency response*. We start with the response properties of a first-order process when forced by a sinusoidal input and show how the output response characteristics depend on the frequency of the input signal. This is the origin of the term, frequency response. Next we introduce a simplified procedure to calculate the frequency-related input–output characteristics from the transfer function of any linear process. This procedure yields a powerful tool both for analyzing dynamic systems and for designing controllers. Graphical methods for displaying frequency response results are then developed. These graphical techniques form the basis for several empirical process modeling methods discussed in Chapter 15 and the frequency domain controller design techniques of Chapter 16.

14.1 SINUSOIDAL FORCING OF A FIRST-ORDER PROCESS

The responses for first- and second-order processes forced by a sinusoidal input were presented in Chapter 5. Recall that these responses consisted of sine, cosine, and exponential terms. Specifically, for a first-order transfer function with gain K and time constant τ, the response to a general sinusoidal input, $x(t) = A \sin \omega t$, is

$$y(t) = \frac{KA}{\omega^2 \tau^2 + 1} (\omega \tau e^{-t/\tau} - \omega \tau \cos \omega t + \sin \omega t) \qquad (5\text{-}25)$$

Note that in (5-25) both input and output variables are in deviation form.

If the sine wave is continued for a long time, the exponential term becomes negligible. The remaining sine and cosine terms can be combined via a trigonometric identity to yield

$$y_\ell(t) = \frac{KA}{\sqrt{\omega^2 \tau^2 + 1}} \sin(\omega t + \phi) \qquad (14\text{-}1)$$

where $\phi = -\tan^{-1}(\omega \tau)$. The *long-time response* $y_\ell(t)$ is called the frequency response of the first-order system and has two distinctive features (see Fig. 14.1).

1. The output sine wave has the same frequency but its phase is shifted relative to the input sine wave by the angle ϕ (referred to as the phase shift or the phase angle); the amount of phase shift depends on the forcing frequency.
2. The output sine wave has an amplitude $\hat{A}$ that also is a function of the forcing frequency:

$$\hat{A} = \frac{KA}{\sqrt{\omega^2\tau^2 + 1}} \tag{14-2}$$

Dividing both sides of (14-2) by the input signal amplitude A yields the *amplitude ratio* AR:

$$AR = \frac{\hat{A}}{A} = \frac{K}{\sqrt{\omega^2\tau^2 + 1}} \tag{14-3a}$$

which can, in turn, be divided by the process gain to yield the *normalized amplitude ratio* AR_N

$$AR_N = \frac{1}{\sqrt{\omega^2\tau^2 + 1}} \tag{14-3b}$$

Since the process steady-state gain K is constant, the normalized amplitude ratio often is used to analyze the frequency-dependent characteristics of the amplitude ratio.

Next we examine the physical significance of the above equations, with specific reference to the stirred-tank heater example we have discussed earlier. In Chapter 4 the transfer function model for the electrically heated system was found to be

$$T'(s) = \frac{K}{\tau s + 1} Q'(s) + \frac{1}{\tau s + 1} T'_i(s) \tag{4-12}$$

Suppose the heating rate is varied sinusoidally about some constant value while the inlet temperature is fixed at the nominal value; that is, $T'_i(s) = 0$. Since $Q'(t)$ is sinusoidal, the output temperature deviation $T'(t)$ will eventually become sinusoidal according to Eq. 5-25. However, there will be a phase shift in the response $T'(t)$ relative to the input as shown in Fig. 14.1, due to the thermal *inertia* of the tank. If the heating rate oscillates very slowly relative to the residence time τ ($\omega \ll 1/\tau$), the phase shift is very small, approaching $0°$, while the normalized amplitude ratio ($\hat{A}/KA$) is very nearly unity. For the case of a low-frequency input the output is seen to be in phase with the input, tracking the sinusoidal input as if the process model were $G(s) = K$.

On the other hand, suppose the heating rate is varied rapidly by increasing the input signal sinusoidal frequency. For $\omega \gg 1/\tau$, Eq. 14-1 indicates that the phase shift approaches a value of $-\pi/2$ radians ($-90°$). The presence of the

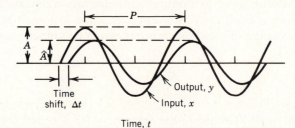

Figure 14.1. Attenuation and time shift between input and output sine waves ($K = 1$). The phase angle ϕ of the output signal is given by $\phi = -\Delta t/P \times 180°$, where Δt is the time (period) shift and P is the period of oscillation.

negative sign indicates that the output follows or lags the input by 90°; in other words the *phase lag* is 90°. The amplitude ratio approaches zero as the frequency becomes large, indicating that the input signal is almost completely attenuated; in other words, the sinusoidal deviation in the output signal is very small.

These results indicate that positive and negative deviations in the heating rate are essentially canceled by the thermal capacitance of the stirred-tank heater if the frequency of input deviations is high enough. In this case, high frequency implies $\omega \gg 1/\tau$. Most processes behave qualitatively like the stirred-tank heater. For high-frequency input changes, the process output deviations are so completely attenuated that the corresponding periodic variation in the output is difficult (perhaps impossible) to detect or measure.

Input–output phase shift and attenuation (or amplification) are observed for any stable transfer function, regardless of its complexity. In all cases the phase shift and amplitude ratio are related to the frequency ω of the input sinusoidal signal. In developments up to this point, expressions for the amplitude ratio and phase shift have been found using the process transfer function. However, the frequency response of a process also can be obtained experimentally. By performing a series of tests in which a sinusoidal input is applied to the process, the resulting amplitude ratio and phase shift can be measured for a number of different forcing signal frequencies. In this case, the frequency response is expressed as a table of measured amplitude ratios and phase shifts for selected values of frequency. Such an experimental approach can be employed to determine an empirical model of an operating process, as is discussed in Chapter 15. In this chapter, we develop a powerful analytical method to calculate the process frequency response for any process transfer function, as shown below.

14.2 SINUSOIDAL FORCING OF AN *n*TH-ORDER PROCESS

This section develops a general approach for deriving the frequency response of any stable transfer function. We show that a rather simple procedure can be employed to find the sinusoidal response.

For a general transfer function $G(s)$, multiplication by the Laplace transform of a sine wave input with amplitude A and frequency ω (see Table 3.1) gives

$$Y(s) = G(s) \frac{A\omega}{s^2 + \omega^2} \tag{14-4}$$

Suppose that the denominator of $G(s)$ can be written as a product of n distinct factors $(s + b_1)(s + b_2) \cdots (s + b_n)$, where b_i can be real or imaginary and $\operatorname{Re} b_i > 0$. Then $Y(s)$ can be expressed as

$$Y(s) = \frac{N(s)}{(s + b_1)(s + b_2) \cdots (s + b_n)} \frac{A\omega}{s^2 + \omega^2} \tag{14-5}$$

where $N(s)$ is the numerator polynomial. A partial fraction expansion gives,

$$Y(s) = \frac{\alpha_1}{s + b_1} + \frac{\alpha_2}{s + b_2} + \cdots + \frac{\alpha_n}{s + b_n} + \frac{Cs + D}{s^2 + \omega^2} \tag{14-6}$$

If we take the inverse Laplace transform of both sides of the above equation, the first n terms become a sum of exponential terms (possibly involving some damped sine and cosine expressions). Since $\operatorname{Re} b_i > 0$, all of the negative exponentials approach zero as t becomes large. On the other hand, the term $(Cs + D)/(s^2 + \omega^2)$ transforms to $C \cos \omega t + (D/\omega) \sin \omega t$, which is unaffected by the length of time that the sinusoidal forcing has been applied. In the sequel we focus on the *long time* response, ignoring the exponential terms as in Eq. 14-1.

The Heaviside expansion (Chapter 3) can be employed to find the constants C and D; we do not bother to calculate the α_i since those terms can be neglected for large times. Multiplying $Y(s)$ by $s^2 + \omega^2$ and setting $s = j\omega$,

$$G(s)A\omega|_{s=j\omega} = (Cs + D)|_{s=j\omega} \tag{14-7}$$

Then

$$G(j\omega) = \frac{1}{A\omega}(Cj\omega + D)$$

$$= \frac{D}{\omega A} + j\frac{C}{A} \tag{14-8}$$

If $G(j\omega)$ is expressed as a complex number, $G(j\omega) = R + jI$, we can equate the real and imaginary parts given above with $R + jI$:

$$\frac{C}{A} = I \tag{14-9a}$$

$$\frac{D}{\omega A} = R \tag{14-9b}$$

Hence, to determine the coefficients C and D, we set $s = j\omega$ in $G(s)$ and then by algebraic manipulation convert $G(j\omega)$ into a complex number $R + jI$. Since the response for large values of time is $C \cos \omega t + (D/\omega) \sin \omega t$, the coefficients of $\cos \omega t$ and $\sin \omega t$ are IA and RA, respectively.

$$y_\ell(t) = A(I \cos \omega t + R \sin \omega t) \tag{14-10}$$

Using the trigonometric identity

$$\sin(\omega t + \phi) = \sin \phi \cos \omega t + \cos \phi \sin \omega t$$

we can express the response in the alternative form

$$y_\ell(t) = \hat{A} \sin(\omega t + \phi) \tag{14-11}$$

where $\hat{A}$ and ϕ are related to $I(\omega)$ and $R(\omega)$ by the following relations:

$$\hat{A} = A\sqrt{R^2 + I^2} \tag{14-12a}$$

$$\phi = \tan^{-1}(I/R) \tag{14-12b}$$

Next we show that the preceding equations lead to a simple but elegant relation for the frequency response. The polar form of the complex function, $G(j\omega)$, is

$$G(j\omega) = |G|\,e^{j\psi} = |G|(\cos \psi + j \sin \psi) \tag{14-13}$$

In the complex plane $|G|$ is the magnitude of $G(j\omega)$ (also called the modulus) and ψ is its angle (or the argument). Equating the polar form to Eq. 14-12 shows that the amplitude ratio is given by

$$\text{AR} = \frac{\hat{A}}{A} = |G| = \sqrt{R^2 + I^2} \tag{14-14}$$

and the phase difference between the output and input sine waves is given by

$$\phi = \angle G = \tan^{-1}(I/R) \tag{14-15}$$

Because $R(\omega)$ and $I(\omega)$ (and hence AR and ϕ) can be found without calculating the transient response, using these characteristics is a shortcut method to determine the frequency response of any stable transfer function $G(s)$.

Equations (14-14) and (14-15) provide a general and powerful technique for calculating the frequency response characteristics of any stable $G(s)$, including those with time delay terms. Note that an unstable $G(s)$ does not have a "frequency response" because a sinusoidal input produces an unstable output response.

Shortcut Method for Finding the Frequency Response

The shortcut method consists of the following steps:

Step 1.   Set $s = j\omega$ in $G(s)$ to obtain $G(j\omega)$.
Step 2.   Rationalize $G(j\omega)$: Express $G(j\omega)$ as $R + jI$, where R and I are functions of ω and possibly model parameters, using complex conjugate multiplication. (Find the complex conjugate of the denominator of $G(j\omega)$ and multiply both numerator and denominator of $G(j\omega)$ by this quantity.)
Step 3.   The output sine wave has amplitude $\hat{A} = A\sqrt{R^2 + I^2}$ and phase angle $\phi = \tan^{-1}(I/R)$. The amplitude ratio is AR $= \sqrt{R^2 + I^2}$ and it is independent of the value of A.

EXAMPLE 14.1

Find the frequency response of a first-order system, with

$$G(s) = \frac{1}{\tau s + 1} \tag{14-16}$$

Solution

First substitute $s = j\omega$ in the transfer function

$$G(j\omega) = \frac{1}{\tau j\omega + 1} = \frac{1}{j\omega\tau + 1} \tag{14-17}$$

Then multiply both numerator and denominator by the complex conjugate of the denominator, that is, $-j\omega\tau + 1$:

$$G(j\omega) = \frac{-j\omega\tau + 1}{(j\omega\tau + 1)(-j\omega\tau + 1)} = \frac{-j\omega\tau + 1}{\omega^2\tau^2 + 1}$$

$$= \frac{1}{\omega^2\tau^2 + 1} + j\frac{(-\omega\tau)}{\omega^2\tau^2 + 1} = R + jI \tag{14-18}$$

From (14-18)

$$R = \frac{1}{\omega^2\tau^2 + 1} \tag{14-19a}$$

and

$$I = \frac{-\omega\tau}{\omega^2\tau^2 + 1} \tag{14-19b}$$

both of which are functions of ω. From (14-12)

$$\text{AR} = |G(j\omega)| = \sqrt{\left(\frac{1}{\omega^2\tau^2 + 1}\right)^2 + \left(\frac{-\omega\tau}{\omega^2\tau^2 + 1}\right)^2}$$

$$= \sqrt{\frac{(1 + \omega^2\tau^2)}{(\omega^2\tau^2 + 1)^2}} = \frac{1}{\sqrt{\omega^2\tau^2 + 1}} \tag{14-20a}$$

and

$$\phi = \angle G(j\omega) = \tan^{-1}(-\omega\tau) = -\tan^{-1}(\omega\tau) \tag{14-20b}$$

If the process gain had been K instead of 1,

$$\text{AR} = \frac{K}{\sqrt{\omega^2\tau^2 + 1}} \tag{14-21}$$

and the phase angle would be unchanged (Eq. 14-20b). Both the amplitude ratio and phase angle are identical to those values calculated in Section 14.1 using the time-domain expression.

From this example we conclude that direct analysis of the complex transfer function $G(j\omega)$ is computationally easier than solving for the actual process response. The computational advantages are even greater when dealing with more complicated processes as shown below. Suppose a general transfer function

$$G(s) = \frac{G_a(s)G_b(s)G_c(s) \cdots}{G_1(s)G_2(s)G_3(s) \cdots} \tag{14-22}$$

is converted to the complex transfer function $G(j\omega)$ by the substitution $s = j\omega$:

$$G(j\omega) = \frac{G_a(j\omega)G_b(j\omega)G_c(j\omega) \cdots}{G_1(j\omega)G_2(j\omega)G_3(j\omega) \cdots} \tag{14-23}$$

As a consequence of Eq. 14-13 we can express the magnitude and angle of $G(j\omega)$ as follows:

$$|G(j\omega)| = \frac{|G_a(j\omega)|\,|G_b(j\omega)|\,|G_c(j\omega)| \cdots}{|G_1(j\omega)|\,|G_2(j\omega)|\,|G_3(j\omega)| \cdots} \tag{14-24a}$$

$$\angle G(j\omega) = \angle G_a(j\omega) + \angle G_b(j\omega) + \angle G_c(j\omega) + \cdots$$
$$- [\angle G_1(j\omega) + \angle G_2(j\omega) + \angle G_3(j\omega) + \cdots] \tag{14-24b}$$

Equations 14-24a and 14-24b greatly simplify the computation of $|G(j\omega)|$ and $\angle G(j\omega)$ and consequently of AR and ϕ in the frequency response calculations. Such expressions eliminate much of the complex arithmetic associated with the rationalization of complicated functions. Hence, use of the factored transfer function (Eq. 14-22) is strongly preferred over one that is not factored when calculating the frequency response.

EXAMPLE 14.2

Calculate the amplitude ratio and phase angle for the overdamped second-order transfer function

$$G(s) = \frac{K}{(\tau_1 s + 1)(\tau_2 s + 1)}$$

Solution

Let the components of $G(s)$ be

$$G_a = K$$

$$G_1 = \tau_1 s + 1$$

$$G_2 = \tau_2 s + 1$$

Substituting $s = j\omega$

$$G_a(j\omega) = K$$

$$G_1(j\omega) = j\omega\tau_1 + 1$$

$$G_2(j\omega) = j\omega\tau_2 + 1$$

Calculating the magnitudes and angles of each of the components of the complex transfer function

$$|G_a| = K \qquad\qquad \angle G_a = 0$$

$$|G_1| = \sqrt{\omega^2\tau_1^2 + 1} \qquad \angle G_1 = \tan^{-1}(\omega\tau_1)$$

$$|G_2| = \sqrt{\omega^2\tau_2^2 + 1} \qquad \angle G_2 = \tan^{-1}(\omega\tau_2)$$

Combining these expressions via Eqs. 14-24a and 14-24b yields

$$AR = |G(j\omega)| = \frac{|G_a|}{|G_1||G_2|}$$

$$= \frac{K}{\sqrt{\omega^2\tau_1^2 + 1}\,\sqrt{\omega^2\tau_2^2 + 1}} \qquad (14\text{-}25a)$$

$$\phi = \angle G(j\omega) = \angle G_a - (\angle G_1 + \angle G_2)$$

$$= -\tan^{-1}(\omega\tau_1) - \tan^{-1}(\omega\tau_2) \qquad (14\text{-}25b)$$

14.3 BODE DIAGRAMS

If $G(s)$ is known from an engineering analysis of the system, $G(j\omega)$ can be determined directly by substituting $s = j\omega$. A special graph, called the Bode diagram or Bode plot, is used to display $G(j\omega)$ where AR and ϕ are each plotted as a function of ω. Ordinarily, ω is expressed in units of radians/time to simplify inverse tangent calculations (cf. Eq. 14-15) where the arguments must be dimensionless, that is, in radians. Occasionally a dimensional frequency f ($\omega = 2\pi f$) with units of cycles/time is used. ϕ is normally expressed in degrees rather than radians. For reasons that will become apparent in the development below, the Bode diagram consists of (1) a log–log plot of AR versus ω, and (2) a semilog plot of ϕ versus ω. Such plots are particularly useful for rapid analysis of the response characteristics and stability of closed-loop systems. Prior to the development of computer graphics, frequency response plots were generated by hand, and such plots were of limited utility. Now that graphics terminals are widely available, graphical techniques provide a powerful design strategy. See the Appendix for a listing of software for frequency response calculations.

First-Order System

For a first-order system $K/(\tau s + 1)$, Fig. 14.2 shows a log–log plot of the normalized amplitude ratio versus $\omega\tau$. Note that with these coordinates the figure applies for all values of K and τ. Also shown is a semilog plot of ϕ versus $\omega\tau$. In Figure 14.2 the abscissa $\omega\tau$ has units of radians. If τ and K are known, specific plots of AR_N (or AR) and ϕ can be made against ω. Because of the variety of conventions, graph labels should be carefully used and observed.

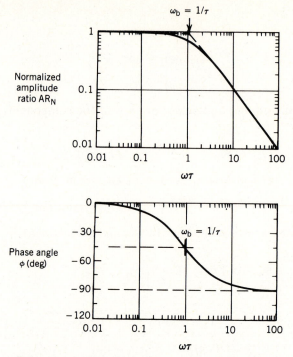

Figure 14.2. Bode diagram for a first-order process.

Next we examine some properties of the Bode plot of the first-order system. At low frequencies, as $\omega \to 0$ ($\omega \ll 1/\tau$),

$$AR_N \approx 1 \qquad (AR \approx K) \tag{14-26a}$$

$$\phi \approx 0 \tag{14-26b}$$

Hence, the amplitude ratio approaches the process gain and the phase shift becomes quite small. At high frequencies, as $\omega \to \infty$ ($\omega \gg 1/\tau$),

$$AR_N \approx 1/\omega\tau \tag{14-27a}$$

$$\phi \approx -90° \tag{14-27b}$$

Here the amplitude ratio drops to an infinitesimal level and the *phase lag* (defined as a positive value) approaches a maximum of 90°. In Fig. 14.2 both the low-frequency and high-frequency asymptotes are shown:

$$\text{Low frequencies: } AR_N = 1 \tag{14-28}$$

$$\text{High frequencies: } AR_N = 1/\omega\tau \tag{14-29}$$

Note that the asymptotes intersect at $\omega = \omega_b = 1/\tau$, known as the *break frequency* or *corner* frequency where the value of AR_N from (14-21) is

$$AR_N(\omega = \omega_b) = \frac{1}{\sqrt{1+1}} = 0.707 \tag{14-30}$$

An important feature of the log–log AR plot for the first-order system is that the slope of the high-frequency asymptote is -1 since

$$\log AR_N = \log 1 - \log \omega\tau = -\log \omega\tau \tag{14-31}$$

The phase angle always lies between 0 and $-90°$. The phase angle at ω_b is

$$\phi(\omega = \omega_b) = \tan^{-1}(-1) = -45° \qquad (14\text{-}32)$$

Table 14.1 gives the values of ϕ as a function of $\omega\tau$. Asymptotic representations can be used to sketch approximate frequency response plots by hand. Using the above guidelines, a Bode plot for a first-order process can be drawn quite rapidly. In this case we assume that K and τ are known explicitly.

To summarize the procedure for constructing a first-order system AR plot:

Step 1. Locate the break frequency, $\omega_b = 1/\tau$. AR at this point is $0.707K$.

Step 2. Draw a horizontal asymptote (AR $= K$) for the low-frequency region, $\omega \leq \omega_b$.

Step 3. Draw the high-frequency asymptote with slope $= -1$, intersecting with AR $= K$ at $\omega = \omega_b$.

Step 4. Sketch the actual frequency response, using the shape shown in Fig. 14.2.

For the ϕ plot:

Step 5. At $\omega = \omega_b$, $\phi = -45°$. Locate this point and draw horizontal asymptotes at $\phi = 0$ and $-90°$. Sketch the phase angle curve using Fig. 14.2 as a pattern or employing the data points from Table 14.1.

Some books define AR differently, namely in terms of decibels. The amplitude ratio in decibels AR_d is defined as

$$AR_d = 20 \log AR \qquad (14\text{-}33)$$

The use of decibels merely requires rescaling the ordinate of the amplitude ratio portion of a Bode plot. The decibel unit is an artifact of electrical communication and acoustic theory and is seldom used in the process control field.

Second-Order System

A general transfer function, describing any underdamped, critically damped, or overdamped second-order system, is

$$G(s) = \frac{K}{\tau^2 s^2 + 2\zeta\tau s + 1} \qquad (14\text{-}34)$$

Substituting $s = j\omega$ and rearranging into real and imaginary parts (see Example 14.1) yields

Table 14.1 Phase Angle as a Function of $\omega\tau$, First-Order System

$\omega\tau$ (radians)	ϕ (°)
0.01	-0.6
0.1	-5.7
0.5	-26.6
1.0	-45.0
2.0	-63.4
10.0	-84.3
100.0	-89.4

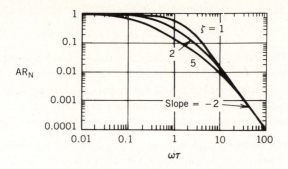

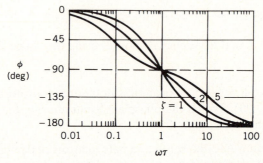

Figure 14.3. Bode diagrams for critically damped and overdamped second-order processes.

$$AR = \frac{K}{\sqrt{(1 - \omega^2\tau^2)^2 + (2\omega\tau\zeta)^2}} \qquad (14\text{-}35a)$$

$$\phi = \tan^{-1}\left[\frac{-2\omega\tau\zeta}{1 - \omega^2\tau^2}\right] \qquad (14\text{-}35b)$$

Note that, in evaluating the phase angle, multiple results are obtained since Eq. 14-35b has infinitely many solutions, each differing by the addition of $n180°$, where n is an integer. The appropriate solution of (14-35b) for the second-order system yields $-180° < \phi < 0$.

Figures 14.3 and 14.4 show the Bode plots for overdamped ($\zeta > 1$) and underdamped ($0 < \zeta < 1$) processes as a function of $\omega\tau$. The low-frequency characteristics of the second-order system are identical with those of the first-order system, for example, Eq. 14-26. At high frequencies

$$AR_N \approx 1/(\omega\tau)^2 \qquad (14\text{-}36a)$$

$$\phi \approx -180° \qquad (14\text{-}36b)$$

The high-frequency amplitude ratio asymptote has a slope of -2 in the log–log plot since

$$\log AR_N = \log 1 - 2 \log \omega\tau$$

$$= -2 \log \omega\tau \qquad (14\text{-}37)$$

For overdamped systems, the normalized amplitude ratio is attenuated ($\hat{A}/KA < 1$) for all ω. For underdamped systems, the amplitude ratio plot exhibits a maximum (for values of $0 < \zeta < \sqrt{2}/2$) at the resonant frequency

$$\omega_r = \frac{\sqrt{1 - 2\zeta^2}}{\tau} \qquad (14\text{-}38)$$

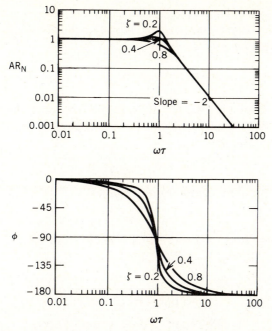

Figure 14.4. Bode diagrams for underdamped second-order processes.

$$(AR_N)_{max} = \frac{1}{2\zeta\sqrt{1 - \zeta^2}} \qquad (14\text{-}39)$$

These expressions are left for the student to develop (Exercise 14.15). ω_r is called the resonant frequency because at that frequency the sinusoidal output response has the maximum amplitude for a given sinusoidal input. Figure 14.5 illustrates how ω_r and $(AR_N)_{max}$ depend on ζ. The principles of resonance are used in designing organ pipes to create sounds at particular frequencies. However, excessive resonance usually is an undesired phenomenon, for example, in automobiles where a particular vibration is noticeable only at a certain speed. For industrial processes operated without feedback control, resonance is seldom encountered, although some measurement instruments are designed to exhibit a limited amount of resonant

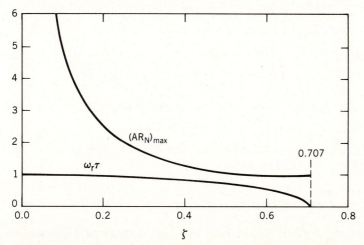

Figure 14.5. Dependence of the normalized amplitude ratio and resonant frequency on the damping coefficient, ζ.

behavior. On the other hand, feedback controllers often are tuned to give the controlled process a slight amount of resonant behavior so as to speed up the controlled system response (e.g., see Fig. 12.6).

EXAMPLE 14.3

Use the results of Example 14.2 to calculate the frequency response of the over-damped second-order system (time constants in minutes),

$$G(s) = \frac{2}{(10s + 1)(2.5s + 1)}$$

Prepare the Bode plot and compare the results with Eqs. 14-35a,b and Fig. 14.3.

Solution

First obtain the amplitude ratio and phase angle from Eqs. 14-25a and 14-25b, when $K = 2$, $\tau_1 = 10$, and $\tau_2 = 2.5$:

$$AR = \frac{2}{\sqrt{(10\omega)^2 + 1}\sqrt{(2.5\omega)^2 + 1}} \tag{14-40a}$$

$$\phi = -\tan^{-1}(10\omega) - \tan^{-1}(2.5\omega) \tag{14-40b}$$

Figure 14.6 is a Bode plot of the above results including both the individual components of the AR and ϕ curves as well as the composite functions. Note that the use of log–log and semilog plots, respectively, for the amplitude ratio and phase angle, allows the component contributions to be summed in each case. For rapid manual plotting of results, this is an important advantage because the asymptotic approximations of AR and ϕ can be obtained by graphical addition.

The second-order transfer function can also be put in the general form of Eq. 14-34:

$$G(s) = \frac{2}{25s^2 + 12.5s + 1}$$

from which $\tau = 5$ and $\zeta = 1.25$.

Hence, in this alternative form,

$$G(j\omega) = \frac{2}{1 - 25\omega^2 + j12.5\omega}$$

and

$$AR = \frac{2}{\sqrt{(1 - 25\omega^2)^2 + (12.5\omega)^2}} \tag{14-41a}$$

$$\phi = \tan^{-1}\left(\frac{-12.5\omega}{1 - 25\omega^2}\right) \tag{14-41b}$$

The equivalence of equation sets (14-40) and (14-41) is left for the reader (Exercise 14.16).

In Example 14.3 we mentioned that the asymptotic representations of the component amplitude ratio curves could be added graphically in the log–log rep-

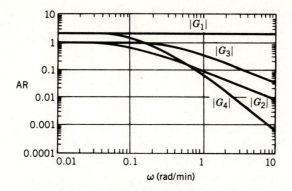

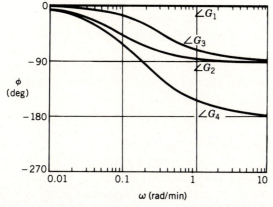

Figure 14.6. Bode diagram for a second-order process and its individual components.

$$G_1 = 2, \quad G_2 = \frac{1}{10s + 1}, \quad G_3 = \frac{1}{2.5s + 1}, \quad G_4 = \frac{2}{(10s + 1)(2.5s + 1)}$$

resentation, thus making the sketching of composite curves quite simple. For this process model

$$\log AR = \log 2 - \tfrac{1}{2} \log[(10\omega)^2 + 1] - \tfrac{1}{2} \log[(2.5\omega)^2 + 1] \quad (14\text{-}42)$$

from which appropriate asymptotic representations can be obtained as follows:

At low frequencies ($\omega \ll 1/10$),

$$\log AR = \log 2$$

or

$$AR = 2 \quad (14\text{-}43a)$$

At middle frequencies ($1/10 \ll \omega \ll 1/2.5$),

$$\log AR = \log 2 - \log(10\omega)$$

or

$$AR = \frac{2}{10\omega} \quad (14\text{-}43b)$$

Since $1/10$ and $1/2.5$ are of the same magnitude, it is impossible to meet the ($\ll$) conditions. Hence, Eq. 14-43b does not yield an approximate result but rather represents the asymptotic behavior only.

At high frequencies ($\omega \gg 1/2.5$),

$$\log AR = \log 2 - \log(10\omega) - \log(2.5\omega)$$

or

$$AR = \frac{1}{25\omega^2} \qquad (14\text{-}43c)$$

From the results of Eqs. 14-43 we develop the following procedure for drawing the asymptotic representations of the Bode plot AR curve. Ordinarily a computer would be used to obtain and plot frequency response representations, but these rules can be used to sketch the composite curve for "back-of-the-envelope" design calculations.

To construct the AR plot of an overdamped second-order system:

Step 1. Draw the low-frequency asymptote at K.
Step 2. At $\omega_{b1} = 1/\tau_1$ (τ_1 is the larger time constant) the slope of the asymptote changes to -1.
Step 3. At $\omega_{b2} = 1/\tau_2$ the slope of the asymptote changes to -2.
Step 4. Calculate several points on the actual curve and sketch it using the asymptotic representations as guidelines.

For the AR plot of an underdamped second-order system:

Step 1. Draw the low-frequency asymptote at K.
Step 2. At $\omega_{b3} = 1/\tau$ the slope of the asymptote changes to -2.
Step 3. Calculate several points on the actual curve and sketch it using the asymptotic representations as guidelines. A maximum value in AR will be observed only if $0 < \zeta < \sqrt{2}/2$.

The general representation of Eq. 14-35a yields only a low-frequency asymptotic representation ($AR = K$) and high-frequency representation ($AR = K/\omega^2\tau^2$). In Example 14.3 the high-frequency asymptote breaks downward with a slope of -2 at $\omega_{b3} = 1/\tau = 1/\sqrt{(10)(2.5)} = 1/5$ rad/min yielding an equivalent representation to that of Eq. 14-43c.

EXAMPLE 14.4

Find the frequency response and draw the Bode plot for the third-order transfer function where the time constants are in minutes

$$G(s) = \frac{K}{(10s + 1)(5s + 1)(s + 1)}$$

Solution

Note that the transfer function denominator, which is already in factored form, should not be multiplied out. This guideline applies to both analytical and computer-based evaluations of AR and ϕ. Application of Eqs. 14-24 yields

$$AR = \frac{K}{\sqrt{(10\omega)^2 + 1}\,\sqrt{(5\omega)^2 + 1}\,\sqrt{\omega^2 + 1}} \qquad (14\text{-}44a)$$

$$\phi = -\tan^{-1}(10\omega) - \tan^{-1}(5\omega) - \tan^{-1}\omega \qquad (14\text{-}44b)$$

from which the Bode plot shown in Fig. 14.7 has been drawn. For sketching the normalized amplitude ratio, the composite asymptote is drawn in with a value of

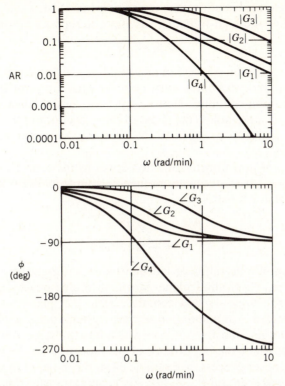

Figure 14.7. Bode diagram for a third-order process and its individual components.

$$\left(G_1 = \frac{1}{10s + 1}, \quad G_2 = \frac{1}{5s + 1}, \quad G_3 = \frac{1}{s + 1}, \quad G_4 = G_1 G_2 G_3 \right)$$

1 at low frequencies. The slope of the composite asymptote changes to -1 at $\omega_{b1} = 1/10$, to -2 at $\omega_{b2} = 1/5$, and to -3 at $\omega_{b3} = 1$. Note that as $\omega \to \infty$, the maximum phase lag approaches $270°$ ($\phi = -270°$), or one $90°$ increment for each first-order element.

Process Zeros

Terms of the form $\tau s + 1$ in the *denominator* of a transfer function are called process *lags* because they cause the process output to lag the input (phase angle is negative), as we have seen. Similarly, process zeros of the form $\tau s + 1$ ($\tau > 0$) in the *numerator* (see Section 6.1) cause the sinusoidal output of the process to lead the input; hence, such left-half plane zeros are called process *leads*. Next we consider the amplitude ratio and phase angle for such a term.

Substituting $s = j\omega$ gives

$$G(j\omega) = j\omega\tau + 1 \tag{14-45}$$

from which

$$AR = |G(j\omega)| = \sqrt{\omega^2\tau^2 + 1} \tag{14-46a}$$

$$\phi = \angle G(j\omega) = +\tan^{-1}(\omega\tau) \tag{14-46b}$$

Therefore, the presence of this term in a process transfer function introduces a positive phase angle contribution that varies between 0 and $+90°$. The amplitude ratio is seen to have the high-frequency asymptotic representation

$$AR \approx \omega\tau \tag{14-47}$$

that is, an upward slope of $+1$ beginning at $\omega_b = 1/\tau$. This result implies that the output signal amplitude becomes very large at high frequencies (AR $\rightarrow \infty$ as $\omega \rightarrow \infty$), a physical impossibility. Hence, a process zero is always found in combination with one or more poles (the order of the numerator of the process transfer function must be less than or equal to the order of the denominator as noted in Section 6.1.)

If the numerator of a transfer function contains the term $1 - \tau s$, that is, if a right-half plane zero associated with an inverse step response is present, the frequency response characteristics are

$$AR = \sqrt{\omega^2\tau^2 + 1} \tag{14-48a}$$

$$\phi = -\tan^{-1}(\omega\tau) \tag{14-48b}$$

Hence, the amplitude ratios of left-half plane and right-half plane zeros are identical. However, a right-half plane zero contributes phase *lag* to the overall frequency response. Processes that contain a right half plane zero or time delay are sometimes referred to as *nonminimum phase* systems because they exhibit more phase lag than another transfer function which has the same AR plot [1, p. 34]. Exercise 14.11 illustrates the importance of zero location on the phase angle.

For the third-order process of Example 14.4, the slope of the amplitude ratio curve for large ω was seen to be -3. For any general transfer function (Eq. 4-34), the high-frequency asymptotic slope of the AR curve is given by

high-frequency slope = (numerator order) $-$ (denominator order)

or

$$s_{\text{hf}} = m - n \tag{14-49}$$

The phase angle for large ω approaches

$$\phi_{\text{hf}} = (m_{\text{rhp}} + n - m_{\text{lhp}})(-90°) \tag{14-50}$$

where m_{rhp} and m_{lhp} represent the number of right-half plane and left-half plane zeros, respectively. Note that Eq. 14-50 holds only for rational transfer functions; it does not hold for the case where a time delay is present in the transfer function, which is discussed below.

Time Delay

The time delay $e^{-\theta s}$ is the remaining important dynamic element to be analyzed. Its frequency response characteristics can be obtained by substituting $s = j\omega$

$$G(j\omega) = e^{-j\omega\theta} \tag{14-51}$$

which can be written in rational form by substitution of the Euler identity

$$G(j\omega) = \cos \omega\theta - j \sin \omega\theta \tag{14-52}$$

From (14-52)

$$AR = |G(j\omega)| = \sqrt{\cos^2 \omega\theta + \sin^2 \omega\theta} = 1 \tag{14-53}$$

$$\phi = \angle G(j\omega) = \tan^{-1}\left(-\frac{\sin \omega\theta}{\cos \omega\theta}\right)$$

or

$$\phi = -\omega\theta \tag{14-54}$$

Note that, with ω expressed in radians, the phase angle in degrees is $-180\omega\theta/\pi$. Figure 14.8 illustrates the Bode plot for a time delay. The phase angle is unbounded, that is, it is not restricted to be smaller than some multiple of $-90°$. The unbounded phase lag is an important feature of a time delay. This characteristic is detrimental to closed-loop system stability as is seen in later chapters dealing with controller design.

EXAMPLE 14.5

In Chapter 6 we discussed polynomial approximations for the irrational time-delay transfer function, $e^{-\theta s}$. For the first- and second-order Padé approximations of $e^{-\theta s}$ given in Eqs. 6-37 and 6-39, find the frequency responses and plot them. Compare the accuracy of these approximations.

Solution

For Eq. 6-37 the frequency response is

$$AR_1 = |G_1(j\omega)| = \frac{\sqrt{\left(-\dfrac{\omega\theta}{2}\right)^2 + 1}}{\sqrt{\left(\dfrac{\omega\theta}{2}\right)^2 + 1}} = 1 \tag{14-55a}$$

$$\phi_1 = \angle G_1(j\omega) = \tan^{-1}\left(-\frac{\omega\theta}{2}\right) - \tan^{-1}\left(\frac{\omega\theta}{2}\right) = -2\tan^{-1}\left(\frac{\omega\theta}{2}\right) \tag{14-55b}$$

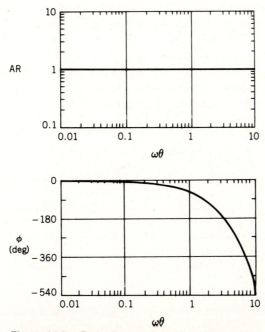

Figure 14.8. Bode diagram for a time delay, $e^{-\theta s}$.

For Eq. 6-39

$$AR_2 = |G_2(j\omega)| = \frac{\sqrt{\left(1 - \frac{\omega^2\theta^2}{12}\right)^2 + \left(-\frac{\omega\theta}{2}\right)^2}}{\sqrt{\left(1 - \frac{\omega^2\theta^2}{12}\right)^2 + \left(\frac{\omega\theta}{2}\right)^2}} = 1 \tag{14-56a}$$

$$\phi_2 = \angle G_2(j\omega) = \tan^{-1}\left(\frac{-\frac{\omega\theta}{2}}{1 - \frac{\omega^2\theta^2}{12}}\right) - \tan^{-1}\left(\frac{\frac{\omega\theta}{2}}{1 - \frac{\omega^2\theta^2}{12}}\right)$$

$$= -2\tan^{-1}\left(\frac{\frac{\omega\theta}{2}}{1 - \frac{\omega^2\theta^2}{12}}\right) \tag{14-56b}$$

Observe that the Padé approximations are exact for the amplitude ratio portion of the frequency response. The phase angle representations (Eqs. 14-54, 14-55b, and 14-56b) are plotted in Fig. 14.9. The 1/1 Padé approximation gives accurate results for $\omega\theta \leq 1$, whereas the 2/2 approximation is satisfactory for $\omega\theta \leq 2$.

In the discussion of Padé approximations in Chapter 6, we raised the following question: What is meant by small values of θs? For the 1/1 approximation, this condition is met when $0 < \omega < 1/\theta$. Most physical processes act as *low-pass filters*, passing low-frequency signals and attenuating high-frequency signals. For example, the attenuation of a first-order process with gain equal to one is $1/\sqrt{(\omega\tau)^2 + 1}$, where τ is the process time constant. For $\omega > 10/\tau$, all input signals are attenuated to less than 10% $(1/\sqrt{(10)^2 + 1})$ of the corresponding input. Consequently, if $\theta < 0.1\tau$, any errors in the approximation will be within engineering accuracy (<10%). With the 2/2 Padé approximation, the equivalent condition is $\theta < 0.2\tau$. Figure 6.7b can now be interpreted in terms of physical parameters: Since $\theta = 0.25\tau$, the 2/2 approximation is quite good, as expected, and the 1/1 approximation is reasonable. In both cases shown in Fig. 6.7b, the transient step response is accurate to better than 5% of the total output change

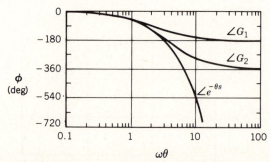

Figure 14.9.   Phase angle plots for $e^{-\theta s}$ and for 1/1 and 2/2 Padé approximations ($G_1 = 1/1$, $G_2 = 2/2$).

$(y(t \rightarrow \infty)/K = 1)$, reflecting the fact that only the phase angle is approximated. The amplitude ratio characteristic is exact for both Padé approximations.

Experimental frequency response data can be obtained from direct sinusoidal forcing of the process or from pulse tests, as is discussed in Chapter 15. If such data are available, the observations made above in connection with Eqs. 14-49, 14-50, and 14-54 can help identify the form of the model [2]. Note that the steady-state gain of the transfer function can be obtained directly from the low-frequency asymptote to the amplitude ratio curve, as long as the denominator contains no multiplicative terms of the form s^i (where i is a positive integer). These terms arise from the presence of pure integration elements; frequency response characteristics of integrating elements are covered in the exercises.

EXAMPLE 14.6

As a final example of the use of Bode plots to represent frequency response characteristics, plot the individual and overall amplitude and phase angle curves for the transfer function

$$G(s) = \frac{5(0.5s + 1)e^{-0.5s}}{(20s + 1)(4s + 1)}$$

where the time constants and time delay have units of minutes.

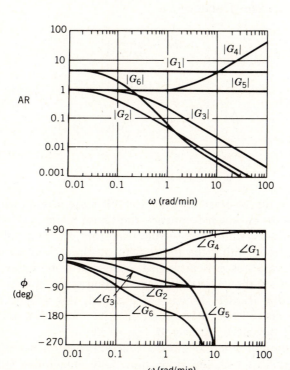

Figure 14.10. Bode plot of the transfer function in Example 14.6:

$$G(s) = \frac{5(0.5s + 1)e^{-0.5s}}{(20s + 1)(4s + 1)}$$

$$G_1 = 5, \quad G_2 = \frac{1}{20s + 1}, \quad G_3 = \frac{1}{4s + 1}, \quad G_4 = 0.5s + 1, \quad G_5 = e^{-0.5s}, \quad G_6 = G_1 G_2 G_3 G_4 G_5$$

Solution

Figure 14.10 shows the results of these calculations. The steady-state gain ($K = 5$) is the value of AR when $\omega \to 0$. All other components of the transfer function have unity gain at low frequencies. Also, the amplitude ratio exhibits an intermediate region where the slope is -2, but for higher frequencies the ultimate slope is -1 due to the effect of the left-half plane zero. The phase angle at high frequencies is dominated by the time delay.

14.4 NYQUIST DIAGRAMS

The Nyquist diagram is an alternative representation of frequency response information. Essentially it is a polar plot of $G(j\omega)$ in which frequency ω appears as an implicit parameter. Since the Bode plot consists of $|G(j\omega)|$ and $\angle G(j\omega)$ versus ω, results for different values of ω can be taken directly from a Bode plot and used to construct the Nyquist diagram. The advantage of a Bode plot is that frequency is plotted explicitly on the abscissa. In addition, the log–log and semilog cordinate systems facilitate block multiplication such as is used in generating the closed-loop transfer function (Chapters 10 and 11). The Nyquist diagram, on the other hand, is more compact and is sufficient for many important analytical techniques, for example, determining system stability. Most of the recent interest in Nyquist diagrams has been in connection with designing multiloop controllers [3]. For single-loop controllers, Bode plots are used more often.

Consider the transfer function

$$G(s) = \frac{1}{2s + 1} \tag{14-57}$$

with

$$AR = |G(j\omega)| = \frac{1}{\sqrt{(2\omega)^2 + 1}} \tag{14-58a}$$

and

$$\theta = \angle G(j\omega) = -\tan^{-1}(2\omega) \tag{14-58b}$$

Figure 14.11 gives the Nyquist diagram for this system. Note that $|G(j\omega)|$ is a maximum at $\omega = 0$, at which frequency the phase angle is 0°. As frequency is increased, $|G(j\omega)| \to 0$ while $\phi \to -90°$. Therefore, the polar plot (Nyquist diagram) remains entirely within the lower right quadrant. Arrows on the diagram show how $G(j\omega)$ varies as ω is increased. Values of ω corresponding to specific points on the curve are shown, in this case indicating a nonuniform frequency scale. If the gain K of the transfer function is changed, this will be reflected in the magnitude of the curve since K is a multiplicative factor that scales the distance from the origin to the curve. At $\omega = 0$, $|G| = K$.

The combination of different transfer function components is not as easy with the Nyquist diagram as with the Bode plot. Only qualitative guidelines can be stated (see Luyben [4] for more details):

1. For no time delay, and denominator order four or less, the Nyquist diagram terminates at the origin without encircling the origin. The point of termination will depend on numerator and denominator orders, but the total angle (phase shift) will be greater in absolute value than 360° ($\phi < -360°$).

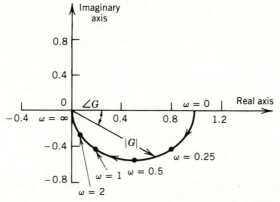

Figure 14.11. The Nyquist diagram for $G(s) = 1/(2s + 1)$ plotting $Re(G(j\omega))$ and $Im(G(j\omega))$.

2. If the transfer function contains a time delay in addition to poles and zeros, there will be an infinite number of encirclements of the origin. An encirclement occurs if the Nyquist plot completely encloses the origin ($\phi \le -360°$). This is because of the unbounded phase shift for the time delay discussed earlier in connection with Bode plots.

EXAMPLE 14.7

As a more complicated example of a Nyquist diagram, reconsider the transfer function in Example 14.6, this time displaying the results as a polar plot.

Solution

Figure 14.12 gives the resulting Nyquist diagram that is obtained directly from the magnitude and phase information in Fig. 14.10. In this case the presence of a time-delay term results in an infinite number of encirclements of the origin. However, because of the rapid attenuation of $G(j\omega)$ with increasing frequency, little of this detail can be observed.

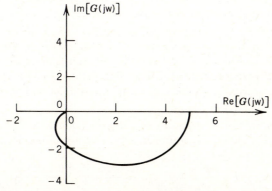

Figure 14.12. The Nyquist diagram for the transfer function in Example 14.7:

$$G(s) = \frac{5(0.5s + 1)e^{-0.5s}}{(20s + 1)(4s + 1)}$$

One approach used to overcome the attenuation problem is to plot the so-called *inverse Nyquist curve*, the Nyquist diagram for $1/G(j\omega)$ [3]. In this case a logarithmic representation of the magnitude will permit as much high-frequency information to be displayed as is required.

SUMMARY

At the beginning of this chapter, we used transient response methods to find the response of a process to a sinusoidal input. The long-time response, after experimental terms have died out, is also sinusoidal and is referred to as the frequency response. The frequency response can be obtained directly by converting the process transfer function $G(s)$ to $G(j\omega)$. Analysis of the complex transfer function permits the control engineer to infer how the process will respond to inputs other than sinusoids. A key point is that the complex transfer function $G(j\omega)$ or, equivalently, its graphical representation furnishes a very compact way to represent the dynamics of a process. This is true regardless of whether $G(j\omega)$ is obtained by the methods of this chapter or by the experimental techniques developed in Chapter 15. A number of powerful controller design methods are based on frequency response characteristics and are discussed in Chapter 16.

REFERENCES

1. Franklin, G. F., and J. D. Powell, *Digital Control of Dynamic Systems,* Addison-Wesley, Reading, MA, 1980.
2. Hougen, J. O. *Measurements and Control Applications,* 2d ed., Instrument Society of America, Research Triangle Park, NC, 1979.
3. Munro, N. (Ed.), *Modern Approaches to Control System Design,* Peter Peregrinus, Stevenage, England, 1979.
4. Luyben, W. L., *Process Modeling, Simulation and Control for Chemical Engineers,* McGraw-Hill, New York, 1973.

EXERCISES

14.1. Find the response of the first-order system

$$G(s) = \frac{2}{0.2s + 1}$$

to a sinusoidal input $x(t) = \sin 2t$ and plot it, showing the initial transient behavior. Find the long-time behavior and show that the amplitude and phase angle are equal to values given by $|G(j\omega)|$ and $\angle G(j\omega)$.

14.2. A high-speed recorder monitors the temperature of an airstream as sensed by a thermocouple. It shows an essentially sinusoidal variation after about 15 s. The maximum recorded temperature is 127 °F and the minimum is 119 °F at 1.8 cycles per minute. It is estimated that the thermocouple under these conditions has a time constant of 4.5 seconds. Estimate the actual maximum and minimum air temperatures.

14.3. A perfectly stirred tank is used to heat a flowing liquid. The dynamics of the system have been determined to be approximately as shown in the drawing.

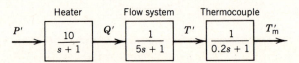

P is the power applied to the heater,
Q is the heating rate of the system,
T is the actual temperature in the tank,
T_m is the measured temperature.

A test has been run with P' varied sinusoidally as

$$P' = 0.5 \sin 0.2t.$$

Under these conditions the measured temperature is

$$T'_m = 3.464 \sin(0.2t + \phi).$$

Find a value for the maximum error bound between T' and T'_m if the sinusoidal input has been applied for a long time.

14.4. Find expressions for $|G(j\omega)|$ and $\angle G(j\omega)$ for transfer functions representing pure integration and differentiation elements:

(a) $G(s) = 1/s \cdot$
(b) $G(s) = s$

14.5. For each of the following transfer functions, sketch the asymptotic Bode diagram for the amplitude ratio. For each case, find the actual amplitude ratio and phase angle at $\omega = 0.1$, 1, and 10. *Note:* It is not necessary to use log–log graph paper; simply rule off decades on rectangular paper.

(a) $\dfrac{10}{(10s + 1)(s + 1)}$

(b) $\dfrac{10}{(10s + 1)(s + 1)^2}$

(c) $\dfrac{10(s + 1)}{(10s + 1)(0.1s + 1)}$

(d) $\dfrac{10(-s + 1)}{(10s + 1)(0.1s + 1)}$

(e) $\dfrac{10}{s(10s + 1)}$

(f) $\dfrac{10(s + 1)}{s(10s + 1)(0.1s + 1)}$

14.6. A second-order process transfer function is given by

$$G(s) = \frac{K(\tau_a s + 1)}{\tau^2 s^2 + 2\zeta\tau s + 1}$$

(a) Find $|G(j\omega)|$ and $\angle G(j\omega)$ when $\zeta = 0.2$.
(b) Plot $|G(j\omega)|$ versus $\omega\tau$ and $\angle G(j\omega)$ versus $\omega\tau$ for the range $0.01 \le \omega\tau \le 100$ and values of $\tau_a/\tau = (0, 0.1, 1, 10)$.

14.7. Plot the Bode diagram ($0.1 \le \omega \le 100$) for the third-order transfer function,

$$G(s) = \frac{3}{(8s + 1)(2s + 1)(s + 1)}$$

by making graphs for each component and adding them together.

14.8. For the transfer function

$$G(s) = \frac{5(s + 1)e^{-2s}}{(3s + 1)(2s + 1)}$$

evaluate the amplitude ratio and phase angle of each component of $G(s)$ and plot them along with the composite functions.

14.9. Two thermocouples, one of them a known standard unit, are placed in an airstream whose temperature is varying sinusoidally. The temperature responses of the two thermocouples are recorded at a number of frequencies, with the phase angle between the two measured as shown below. The standard unit is known to follow first-order dynamics and to have a time constant of 0.15 min when operating in the airstream. From the data, show that the unknown unit also is first order and find its time constant under these same conditions.

Frequency (cycles/min)	Phase Difference (degrees)
0.05	4.5
0.1	8.7
0.2	16.0
0.4	24.5
0.8	26.5
1.0	25.0
2.0	16.7
4.0	9.2

14.10. Exercise 5.19 considered whether a two-tank liquid surge system provided better damping of step disturbances than a single-tank system with the same total volume. Reconsider this situation, this time with respect to sinusoidal disturbances; that is, determine which system better damps sinusoidal inputs of frequency ω. Does your answer depend on the value of ω?

14.11. For the process described in Exercise 6.5, plot the composite amplitude ratio and phase angle curves on a single Bode plot for each of the four cases of numerator dynamics. What can you conclude concerning the importance of the zero location on the amplitude and phase characteristics of this second-order system?

14.12. Develop the frequency response and plot AR and ϕ for the following forms of the three-mode controller:

(a) P
(b) PI (Eq. 8-9)
(c) PD

14.13. Develop expressions for the amplitude ratio of each of the two forms of the PID controller:

(a) The ideal controller transfer function of Eq. 8-13.
(b) The commercial controller transfer function of Eq. 8-14.

Put the results on a single plot along with asymptotic representations of each AR curve. You may assume that $\tau_I = 4\tau_D$ and $\alpha = 0.1$.
For what region(s) of ω are the differences significant? By how much?

14.14. A process has the transfer function,

$$\frac{Y(s)}{X(s)} = \frac{K}{\tau s + e^{-\theta s} + 1}$$

with $K = 10$, $\tau = 2$ min, $\theta = 1$ min.

(a) Derive expressions for the amplitude ratio and phase angle. Plot the AR function over the frequency range of interest. Comment on the difficulty of plotting the phase angle.
(b) Using a 1/1 Padé approximation for $e^{-\theta s}$, repeat part (a).

(c) Derive an approximate, analytical expression for the response $y(t)$ to an impulse in x (i.e., $x(t) = \delta(t)$).

(d) Can you find an exact response for $y(t)$? Why or why not?

14.15. Show that for an underdamped second-order system, the amplitude ratio will exhibit a maximum and a resonant frequency only if $0 < \zeta < \sqrt{2}/2$. Derive the expression for the maximum value of AR_N given in Eq. 14-39.

14.16. Show that equation sets (14-40) and (14-41) are mathematically equivalent. How must Eq. 14-41b be evaluated over the range $0 < \omega < \infty$ to yield correct values for ϕ? For critically damped or overdamped systems ($\zeta \geq 1$), what would be the advantage of using Eq. 14-40b rather than 14-41b?

14.17. Appelpolscher has just left IGC's Monday-morning status meeting with Smarly, Quirk, and the operations people. Over the weekend, a major element in the hydrogen production process, a centrifugal compressor supplying carbon monoxide to a reactor, failed. The operations people were able to switch in a standby unit, an old reciprocating compressor that can barely meet minimum requirements for throughput and discharge pressure.

Smarly has questioned the possibility that discharge pressure oscillations from the temporary unit may damage the reactor product during the interim operating period by causing flow rate changes in the feed to the reactor. Since Appelpolscher has suggested the idea of putting one or two surge tanks between the compressor and reactor to damp out any oscillations, his group was chosen to come up with a quick design. Some assumptions can be made:

i. The proposed piping arrangement would be similar to that shown in Exercise 2.7. If two tanks are used, they should be sized identically.

ii. The valves before the surge tanks exhibit approximate linear pressure/flow relations.

iii. The ideal gas law holds approximately.

iv. Pressure perturbations are caused by reciprocating action of the pistons in the compressor. These perturbations are approximately sinusoidal.

v. Because of the present low-pressure limitations, no more than 10% of the compressor's nominal discharge pressure P_d can be dissipated in the surge system excluding the valve at the entrance to the reactor. If two surge tanks are included in the design, the allowable drop across each tank is $0.1P_d/2$.

vi. The pressure controller for the reactor is able to maintain its pressure essentially constant.

Data You May Use

i. The nominal discharge pressure P_d is 200 psig (gauge). The nominal throughput of carbon monoxide is 6000 pounds per hour.

ii. The compressor contains four cylinders (pistons) driven by a common shaft that rotates at 600 rpm. The cylinders are spaced equally around the shaft to balance shaft loading.

iii. Maximum (estimated) pressure fluctuation is 2 psig; that is, the amplitude of the fluctuation caused by reciprocation of the pistons is 2 psig.

iv. Operating people want no more than a 0.02-psig variation in pressure before the valve at the entrance to the reactor.

v. The nominal discharge temperature of the compressor is 300 °F. The surge tanks probably will operate isothermally.

Appelpolscher is always worried about losing his year-end bonus; this time he thinks it will disappear if any of the reactor product is out of spec and has to be burned.

He asks you to make a two-step analysis of the situation:

(a) Would a single tank or two equal-sized surge tanks be better to damp the pressure fluctuations? (If two tanks are used, they would each have to be about one-fourth the volume of a single tank to keep total system costs roughly equivalent.)

(b) How large would the tank or tanks have to be to reduce pressure fluctuations to an acceptable level?

Your answers should deal with how the proposed design can incorporate a 20-ft-long, 3-in.-i.d. pipe that now connects the compressor to the reactor.

CHAPTER 15

Development of Empirical Models from Frequency Response Data

Chapter 14 dealt predominantly with the calculation of frequency response information directly from a transfer function model. In this chapter we deal with the reverse problem: Given a set of frequency response data, how can one determine an empirical (approximate) transfer function to represent the process?

Experimental frequency response data can be obtained from an operating process by several different techniques:

1. Direct sinusoidal forcing.
2. Pulse tests.
3. Random inputs.

As discussed in Chapter 14, the first method involves forcing a process with an input signal consisting of a constant value plus an additive sine wave. A series of tests is made, using sinusoidal inputs with different frequencies, and the values of the amplitude ratio and phase shift are determined directly from the input and output signals. The test results can then be displayed in a Bode (or Nyquist) plot as a function of the frequency ω. The second method, pulse testing, is a more efficient approach that potentially can yield the same information from a single experimental test but requires computational analysis. The third method involves advanced mathematical analysis and thus is not treated here.

We also discuss the graphical fitting of a transfer function model to frequency response data. No assumption need be made as to the source of the data; whether they come from sinusoidal forcing or from pulse tests is unimportant. The goal is to determine how key features of the Bode plot can be analyzed to select the form of the appropriate low-order transfer function. This type of model should represent the process with accuracy sufficient for design of control systems in cases where a transfer function model of the process is required (cf. Chapter 16). Subsequently, we discuss operational procedures for conducting the required experimental tests.

15.1 GENERAL RULES FOR GRAPHICALLY FITTING TRANSFER FUNCTION MODELS

In Section 14.3 we briefly discussed the general characteristics of frequency response results. From these general results, some rules of thumb can be devised to identify the type of low-order process model that is likely to fit a specific set of frequency response data:

1. If the low-frequency AR data indicate zero slope, the transfer function contains no $1/s$ terms. A low-frequency slope of approximately -1 indicates the presence of a $1/s$ term; a slope of -2, the presence of a $1/s^2$ term, and so on. Process models rarely contain more than a single integration element ($1/s$ term), if any.
2. The slope of the high-frequency AR data can be used to obtain an estimate of $m - n$, the difference between numerator and denominator orders, if no integration elements are present. For example, a slope of high-frequency AR data of -2 indicates that the transfer function is second order, or first over third order, and so forth. Experimentally obtained frequency response data are seldom accurate enough to exhibit a slope more negative than -2; hence, empirical models obtained from such data generally are no higher than second or third order.
3. A low-frequency phase angle of $-90°$ confirms the presence of a single integration element. A high-frequency phase angle approaching some multiple of $-90°$ ($-90°$, $-180°$, etc.) indicates that no time delay is present. In such a case it is possible to determine for a second-order model whether a transfer function zero is present and, if so, whether it falls in the left-half plane ($\tau s + 1$ term) or right-half plane ($-\tau s + 1$ term). Often, a time delay will be present, indicated by a phase shift that becomes negative without bound.

In the event that a time delay is present, it may not be possible to determine the form of the transfer function any more precisely than first order plus delay or overdamped second order plus delay. However, these two standard forms are satisfactory for most applications. Below we develop specific graphical fitting approaches for first-order and second-order model types without an integrating element. A general computer-based fitting method, described in Section 15.2, can be used for more complicated transfer functions.

Fitting a First-Order Plus Time-Delay Transfer Function

Determine $\hat{K}$, $\hat{\tau}$, and $\hat{\theta}$ (if present) as follows (the caret denotes estimated values as used in Chapter 7):

Step 1.   Draw a low-frequency asymptote for the AR data; set $\hat{K}$ equal to the value of the asymptote.

Step 2.   Draw a high-frequency AR asymptote with slope equal to -1. The frequency ω_b, where the two asymptotes intersect, determines the time constant, $\hat{\tau} = 1/\omega_b$. A useful consistency check is $AR(\omega_b) = 0.707\hat{K}$.

Step 3.   Since the time delay affects only the phase angle and not the AR results, it can be estimated by first calculating a residual phase shift ϕ_{res} using $\hat{\tau}$ calculated in Step 2:

$$\phi_{res}(\omega_i) = \phi(\omega_i) - \hat{\phi}_1(\omega_i)$$

$$= \phi(\omega_i) - [-\tan^{-1}(\omega_i\hat{\tau})]$$

$$= \phi(\omega_i) + \tan^{-1}(\omega_i\hat{\tau}) \qquad i = 1, \cdots, r \qquad (15\text{-}1)$$

where $\phi(\omega_i)$ is the experimentally determined value of the phase shift at ω_i, $\hat{\phi}_1(\omega_i)$ represents the amount of phase shift in degrees contributed by the $1/(\hat{\tau}s + 1)$ term, and $\omega_i (i = 1, \cdots, r)$ represent the r different frequencies for which phase angle data are available. According to Eq. 14-54, the empirical time delay θ can be obtained graphically by plotting ϕ_{res} against ω; $\hat{\theta}$ is then the slope of the straight line that best fits the plotted data, estimated visually or computed using least squares. Once $\hat{\tau}$ and $\hat{\theta}$ are obtained, the modified residual

$$\phi'_{res}(\omega_i) = \phi(\omega_i) + \tan^{-1}(\omega_i\hat{\tau}) + \omega_i\hat{\theta}(180/\pi) \qquad (15\text{-}2)$$

can be used as a measure of the suitability of the model since the second and third terms on the right side of (15-2) represent the phase shift of the complete model at each frequency.

EXAMPLE 15.1

Frequency response data obtained from a process are given in Table 15.1. Fit a first-order plus time-delay transfer function to these data.

Solution

1. The AR data appear to be approaching a value of 3.0 as $\omega \to 0$. Hence, $\hat{K} = 3$.
2. Since a first-order transfer function is to be used, a quick way to estimate the time constant is to find the frequency that yields an amplitude ratio 0.707 times the low-frequency value. An AR $= 0.707 \times 3 = 2.121$ is found at $\omega_b = 0.1$ rad/min. Hence, $\hat{\tau} = 1/\omega_b = 1/0.1 = 10$ min. In this case the required value (AR $= 2.121$) corresponds closely to a value in the table; more generally, graphical or numerical interpolation would have to be used to find ω_b.
3. The time delay can be found by evaluating Eq. 15-1 for each value of ω given in Table 15.1. Table 15.2 summarizes the calculation of ϕ_{res}. An estimate of the time delay can be obtained from the values of ϕ_{res}; for example, for $\omega = 0.1$ rad/min, $\hat{\theta} = 11.5° \times 1/(0.1$ rad/min$) \times (\pi$ rad$/180°) = 2.003$ min. An average value of 2.02 min is obtained from the table. Note that θ could also have been estimated graphically; that is, by plotting ϕ_{res} against ω and finding the slope of the line that best fits the data.

The estimated transfer function is

$$\hat{G}(s) = \frac{3e^{-2.02s}}{10s + 1}$$

Table 15.1   Experimental Frequency Response Data (Example 15.1)

ω (rad/min)	AR	ϕ (deg)
0.01	2.98	−6.9
0.02	2.94	−13.6
0.05	2.68	−32.3
0.1	2.12	−56.5
0.2	1.34	−86.4
0.5	0.59	−136.0
1.0	0.30	−198.9

Table 15.2   Calculation of the Time Delay for Example 15.1

ω (rad/min)	ϕ (deg)	$tan^{-1}(\omega\tau)$ (deg)	ϕ_{res} (deg)	θ (min)
0.01	−6.9	5.7	−1.2	2.094
0.02	−13.6	11.3	−2.3	2.007
0.05	−32.3	26.6	−5.8	2.024
0.1	−56.5	45.0	−11.5	2.003
0.2	−86.4	63.4	−23.0	2.007
0.5	−136.0	78.7	−57.3	2.000
1.0	−198.9	84.3	−114.6	2.000

Fitting a Second-Order Plus Time-Delay Transfer Function

Step 1. Draw a low-frequency asymptote for the AR data. Its magnitude is $\hat{K}$.

Step 2. Draw a high-frequency AR asymptote with slope equal to -2. The intersection frequency of the low-frequency and high-frequency asymptotes, ω_b, determines the product of the time constant(s). If the model is overdamped, $\hat{\tau}_1\hat{\tau}_2 = 1/\omega_b^2$; if underdamped, $\hat{\tau}^2 = 1/\omega_b^2$ and $\hat{\tau} = 1/\omega_b$.

Step 3. If a midfrequency asymptote with slope $= -1$ can be drawn, the intercepts with the low- and high-frequency asymptotes determine approximate values for the two time constants; $\hat{\tau}_1 = 1/\omega_{b1}$, $\hat{\tau}_2 = 1/\omega_{b2}$. Figure 15.1 shows these constructions, indicating that some trial-and-error fitting of the midfrequency asymptote may be necessary before the values of $\hat{\tau}_1$ and $\hat{\tau}_2$ that best fit the AR data are found.[1]

Step 4. If an underdamped representation is more suitable (no asymptotic representation of the midfrequency AR data is appropriate), the data can be compared with Fig. 14.4 and 14.5 to estimate a value of ζ. Some trial-and-error adjust of $\hat{\zeta}$ may be required if a very accurate fit of the AR data is required.

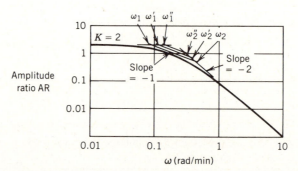

Figure 15.1. Fitting of AR asymptotes for a second-order model ω_1, ω_1', and ω_1'' (and the corresponding intersections for ω_2, ω_2', and ω_2'') result from three trial constructions. ω_1 and ω_2 are the best fit of the data.

---

[1]As an alternative to Steps 2 and 3, the dominant time constant τ_1 of an overdamped second-order system can first be found from the intersection of two asymptotes with slopes $= 0$ (low frequency) and -1 (midfrequency). The effect of the first-order lag, $1/(\tau_1 s + 1)$, can be removed from the experimental AR and ϕ plots, either numerically or by graphical addition. This corresponds to multiplication of the process transfer function by $1/(\tau_1 s + 1)$. The residual plots can then be treated as if they represented a first-order plus time-delay process.

Step 5. The time delay $\hat{\theta}$ may be found as for a first-order model using

$$\phi_{res}(\omega_i) = \phi(\omega_i) - \hat{\phi}_2(\omega_i)$$

$$= \phi(\omega_i) - [-\tan^{-1}(\omega_i\hat{\tau}_1) - \tan^{-1}(\omega_i\hat{\tau}_2)]$$

$$= \phi(\omega_i) + \tan^{-1}(\omega_i\hat{\tau}_1) + \tan^{-1}(\omega_i\hat{\tau}_2) \qquad (15\text{-}3)$$

where $\hat{\phi}_2(\omega_i)$ is the phase angle contribution of the empirical second-order model without time delay (Eqs. 5-39 or 5-40). The estimated time constants $\hat{\tau}_1$ and $\hat{\tau}_2$ are obtained from previous steps. For an underdamped model, an equation analogous to (15-3) would be used.

Graphical fitting of transfer function parameters has a number of important limitations: (1) the tediousness of performing calculations by hand, (2) the introduction of errors in fitting time constants for second-order models using this type of trial-and-error approach, and (3) the need to consider more general transfer function models, such as those with numerator dynamics. These disadvantages have led to the development of numerical methods for estimating model parameters directly from the frequency response data.

15.2 NUMERICAL TECHNIQUES FOR ESTIMATING PARAMETERS IN TRANSFER FUNCTION MODELS

Levy [1] developed a weighted least-squares method for fitting a model of the form

$$G(s) = \frac{N(s)}{D(s)} = \frac{\displaystyle\sum_{i=0}^{n-1} A_i s^i}{1 + \displaystyle\sum_{i=1}^{n} B_i s^i} \qquad (15\text{-}4)$$

Suppose $G(s)$ is second order; the frequency response is given by

$$G(j\omega) = \frac{A_0 + jA_1\omega}{1 + jB_1\omega - B_2\omega^2} \qquad (15\text{-}5)$$

If the data for $G(j\omega)$ are expressed as $R(\omega) + jI(\omega)$, the model error $\epsilon(j\omega)$ is given by

$$\epsilon(j\omega) = R(\omega) + jI(\omega) - \frac{A_0 + jA_1\omega}{1 + jB_1\omega - B_2\omega^2} \qquad (15\text{-}6)$$

Note that linear least squares fitting techniques (Ch. 7) cannot be used to find the A_i and B_i parameters because the parameters appear nonlinearly in this expression. Alternatively, define a weighted error, $\epsilon'(j\omega)$, which is obtained by multiplying both sides of (15-6) by $(1 + jB_1\omega - B_2\omega^2)$,

$$\epsilon'(j\omega) = (R + jI)(1 + jB_1\omega - B_2\omega^2) - (A_0 + jA_1\omega) \qquad (15\text{-}7)$$

$$= [(R - A_0) - B_1 I\omega - B_2 R\omega^2]$$

$$+ j[I + (B_1 R - A_1)\omega - B_2 I\omega^2] \qquad (15\text{-}8)$$

Now, minimization of $|\epsilon'(j\omega)|^2$ is a linear regression problem, since all unknown parameters in (15-8) appear linearly as coefficients on powers of ω. Levy showed that the parameter values that minimize the sum of weighted squared errors can be found by solving a set of linear algebraic equations for the A_i and B_i. Note, however, that minimizing $|\epsilon'|^2$ does not necessarily minimize $|\epsilon|^2$.

Levy's procedure works for the general transfer function in (15-4); however, it gives overly heavy weighting to high-frequency data, with a corresponding better fit, at the expense of fitting low-frequency data [2]. Sanathanan and Koerner [2] proposed an iterative least-squares technique (repeated solution of linear algebraic equations) to remove the unfavorable weighting of high-frequency data. Tests of this method with several known transfer functions demonstrated that the iterative technique was superior to Levy's method.

Note that the presence of a time-delay term in the process model prevents the development of an expression similar to Eq. 15-8 in which all of the unknown parameters appear linearly. Hence, Levy's approach cannot be used directly to fit such models unless a rational (Padé) approximation to the time-delay term is substituted. An alternative trial-and-error approach, where values of θ are selected iteratively and Levy's procedure is used to find the remaining process model parameters, is feasible but requires more computation. In Example 15.2 we show one way to estimate θ if time domain data are available.

The above least squares (linear regression) methods for fitting frequency response data are attractive because they can be used to estimate a relatively large number of parameters without large increases in computer time. On the other hand, the fitting of transfer function coefficients to time-domain data [3] based on a step input is a nonlinear regression problem (see Section 7.1) that potentially requires more computation time. However, in both cases a computer must be employed; the incremental difference in effort between the two techniques probably is not large for low-order models.

When frequency domain data indicate that a time delay is present, it is preferable to use nonlinear regression to estimate the transfer function parameters. This approach is computationally feasible for second-order plus time-delay models. The objective function to be minimized would be the squared magnitude of the complex error, similar to Eq. 15-6:

$$\mathcal{E} = \sum_{i=1} |\epsilon(j\omega_i)|^2 \qquad (15\text{-}9)$$

where

$$\epsilon(j\omega_i) = R(\omega_i) + jI(\omega_i) - \frac{(A_0 + jA_1\omega_i)e^{j\omega_i\theta}}{1 + jB_1\omega_i - B_2\omega_i^2}$$

Here the five model parameters in (15-9) would be estimated directly by an unconstrained optimization code. As in the linear regression case, one needs to be careful to achieve a good fit of low-frequency data.

At this point, the only experimental technique we have discussed for obtaining the frequency response data is direct sinusoidal forcing of the process input. Such an approach is not necessary and is an inefficient way to obtain frequency response data. A separate experiment must be run for each frequency at which data are desired. This method seldom can be employed in an industrial setting because of the extended time the process is disturbed. A much more efficient technique, pulse testing, is discussed next.

15.3   THE USE OF PULSE TESTS TO OBTAIN FREQUENCY RESPONSE DATA

The pulse testing method has proved satisfactory in many plant applications [4,5] because it is relatively easy to implement; in particular, it avoids the necessity of performing many different tests to obtain process input–output data at different frequencies. In conducting a pulse test, the input variable is changed from its steady-state value in a pulselike manner (see Fig. 15.2); the shape of this pulse does not necessarily match any prescribed function. Generally, the process output will respond in a similar, pulselike fashion, although over a longer duration. Since an input pulse contains a range of frequencies, in effect the process is forced simultaneously by each of these frequency components. Hence, although additional computational effort is required to generate the frequency response from input–output data, in theory a single experiment can yield the entire frequency representation of the process. In practice, several pulse tests might have to be performed to obtain a more complete data base.

A pulse test is performed by first permitting the process to come to steady state. A closed pulse, one that begins and ends at the steady-state operating value, is then introduced in an input variable of the process. Both the input and the output response data are recorded (see Fig. 15.2). The data necessary to calculate the frequency response are simply the time histories of the input and output pulses. The time at which the input variable first deviates from its steady-state value is specified as zero time. Normally the final time is chosen to be when both the input and output pulses have returned to and remain at their previous steady-state values. However, with systems containing pure integrator characteristics, the output variable will never return to its original steady-state value. In these cases the final time is taken to be when the input has returned to its initial steady-state value and the output has attained a constant value.

The process transfer function is related to the pulse input and output by the definition of the Laplace transforms for $Y(s)$ and $X(s)$:

$$G(s) = \frac{Y(s)}{X(s)} = \frac{\displaystyle\int_0^\infty y(t)e^{-st}\,dt}{\displaystyle\int_0^\infty x(t)e^{-st}\,dt} \tag{15-10}$$

Because $y(t)$ and $x(t)$ are deviation variables, the numerator and denominator integrals of Eq. 15-10 need only be evaluated over the output pulse duration T_y and the input pulse duration T_x, respectively, since y and x are zero everywhere

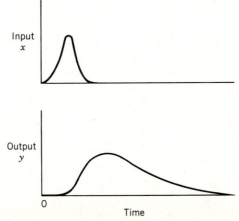

Figure 15.2.   Typical input and output pulses.

else. Then, substituting $j\omega$ for s in Eq. 15-10 gives

$$G(j\omega) = \frac{\int_0^{T_y} y(t)e^{-j\omega t} \, dt}{\int_0^{T_x} x(t)e^{-j\omega t} \, dt} \tag{15-11}$$

Note that the numerator and denominator of (15-11) are simplifications of one-sided Fourier transforms of $y(t)$ and $x(t)$ [3,6].

Numerical techniques can now be used to evaluate the integrals of Eq. 15-11 for each frequency value ω desired [5,6]; that is, once a value of ω is specified, the numerical integration can be performed. These evaluations yield complex values of $G(j\omega)$ at each desired frequency within the frequency range of interest, which can then be used to calculate the frequency response (amplitude ratio and phase angle as a function of frequency).

Hougen [5] presented a numerical method for approximating the integrals of Eq. 15-11 directly by subdividing the time axis into equal increments Δt, forming trapezoids using each increment as a base, and then summing the areas of the trapezoids. Clements and Schnelle [7] have expanded the two integrals of Eq. 15-11 into four integrals that are more conducive to numerical integration techniques. The approach they suggested is to substitute the Euler identity

$$e^{-j\omega t} = \cos \omega t - j \sin \omega t \tag{15-12}$$

into Eq. 15-11, yielding

$$G(j\omega) = \frac{A(\omega) - jB(\omega)}{C(\omega) - jD(\omega)} \tag{15-13}$$

where $A(\omega) = \int_0^{T_y} y(t)\cos \omega t \, dt$ \hfill (15-14a)

$$B(\omega) = \int_0^{T_y} y(t)\sin \omega t \, dt \tag{15-14b}$$

$$C(\omega) = \int_0^{T_x} x(t)\cos \omega t \, dt \tag{15-14c}$$

$$D(\omega) = \int_0^{T_x} x(t)\sin \omega t \, dt \tag{15-14d}$$

Finally, rationalization of (15-13) gives

$$R(\omega) = \frac{AC + BD}{C^2 + D^2} \tag{15-15}$$

$$I(\omega) = \frac{AD - BC}{C^2 + D^2} \tag{15-16}$$

where $R(\omega)$ and $I(\omega)$ denote the real and imaginary parts of $G(j\omega)$, respectively. The amplitude ratio and phase angle of $G(j\omega)$ may be obtained from

$$\text{AR} = |G(j\omega)| = \sqrt{R^2(\omega) + I^2(\omega)} \tag{15-17}$$

$$\phi = \angle G(j\omega) = \tan^{-1}(I/R) \tag{15-18}$$

Thus, the task of calculating frequency response data from pulse test data becomes one of numerically evaluating the integrals of Eqs. 15-14a through 15-14d for each value of ω of interest and using these results in Eqs. 15-15 through 15-18. Listings of computer programs that compute the frequency response for a pulse test are available in the literature [6,8].

An example of typical input and output pulses is shown in Fig. 15.3. Numerical techniques for approximating the integrals of Eqs. 15-14a through 15-14d require data for both $x(t)$ and $y(t)$ throughout their respective pulse durations. In gathering $y(t)$ data, it is convenient to use two different sampling periods [5]. A small sampling period Δt_1 is used for the first part of the output pulse where $y(t)$ changes rapidly with time, and a larger sampling period Δt_2 is used for that portion where $y(t)$ changes slowly with time, the tail of the output pulse. This practice allows better resolution of the data in the first part of the output pulse without requiring an excessive number of data points to represent the tail.

Another procedure that should be used in analyzing pulse test data is to estimate any apparent time delay in the output pulse before data processing. This step is accomplished by shifting the output pulse time scale by the amount of any apparent time delay θ'; that is, $t' = t - \theta'$. This strategy can be useful in identifying the model order since the resulting phase angle curve has most of the time delay removed.

If the recorded output pulse does not close but appears to be approaching zero, graphical extrapolation to zero can be employed. If the process contains a pure integrator (as in the liquid storage process of Section 5.3), the integrals in (15-14) will not be bounded. One technique found to be successful in this case is to differentiate the output data numerically with respect to time and then compute the frequency response for the differentiated data [5]. If $G_1(s)$ represents the original data set and $G_2(s)$ the differentiated data, then

$$G_1(s) = \frac{G_2(s)}{s} \tag{15-19}$$

An alternative approach has been suggested by Luyben [6].

The size and shape of the input pulse also affect the quality of resulting frequency response data [5,6]. While simple sinusoidal forcing in the input variable

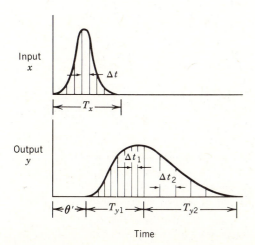

Figure 15.3. Typical pulse plots showing how numerical results are obtained for use in digital processing.

excites the system with a single particular frequency and amplitude, pulse forcing excites the system with a range of frequencies and a distribution of amplitudes. This input distribution is called the normalized frequency or harmonic content spectrum, when normalized by the value of the input transform at zero frequency. The normalized frequency content for an input pulse $x(t)$ at a specific frequency ω is expressed mathematically as

$$S_N(\omega) = \frac{S(\omega)}{S(0)} \tag{15-20}$$

where $S(\omega) = \left| \int_0^{T_x} x(t)e^{-j\omega t}\, dt \right|$ $\qquad$ (15-21)

Note that $S(0)$ corresponds to the area under the input pulse. Different input pulse shapes exhibit different normalized frequency content spectra. The best pulse shapes to use as input forcings are those that have large normalized frequency contents over a wide range of frequencies [5]. Hougen analyzed a number of pulse shapes and found that time-weighted displaced cosine pulses have more favorable frequency content spectra than rectangular pulses. Figure 15.4 shows $S(\omega)$ for three different input shapes.

The larger the normalized frequency content is at high frequencies, the more accurately the true high-frequency response of the process can be extracted from a noisy pulse test output signal. In addition, the computed frequency response at high frequencies often is altered by the inaccuracy of the numerical integration procedure for Eq. 15-11. To obtain meaningful results, frequency response data should be calculated only up to the frequency where the normalized frequency content of the input pulse becomes less than 0.3 [5]. Figure 15.4 illustrates how much of the harmonic content of the rectangular pulse is "wasted" at higher frequencies where $S_N(\omega) < 0.3$; the displaced cosine, on the other hand, exhibits a significant portion of its harmonic content in the low- and mid-frequency bands.

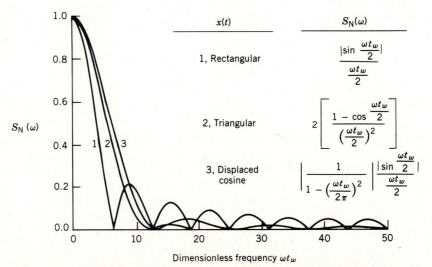

Figure 15.4. Normalized frequency content for three different input pulse shapes.

rectangular pulse: height = 1, width = t_w, $0 \le t \le t_w$ (Fig. 5.1)
triangular pulse: height = 1, width = t_w, $0 \le t \le t_w$ (Fig. 5.3)
displaced cosine pulse: $1 - \cos(2\pi t/t_w)$, $0 \le t \le t_w$

Consequently, the rectangular pulse yields useful results only out to a dimensionless frequency of $\omega t_w \approx 5$; the displaced cosine pulse is useful out to $\omega t_w \approx 8$, a considerable improvement.

Another potential difficulty experienced with pulse testing is the fact that the computed process gain can vary for different tests of the same physical system, depending upon the shape of the pulse. This unfortunate result can occur because the pulse testing method is designed to obtain information over a wide frequency range; consequently, its effectiveness suffers somewhat in obtaining the steady-state (low-frequency) gain. If desired, the data from several tests can be averaged to yield a value for K; alternatively, a step test can be performed to find the process gain more directly, as per Chapter 7.

A good input pulse is one that is smooth, large in magnitude, and of short duration. The pulse should not be so large as to cause any component within the system to reach a constraint nor so short in duration that there is difficulty in reading the output response data accurately. Luyben [6] has suggested that, as a general rule, the width of the input pulse should be less than half the magnitude of the smallest time constant to be estimated.

The reader should recognize that design of a pulse test involves some inherent performance conflicts. Large pulses of short duration excite the process satisfactorily but may cause violation of operating variable constraints or make the process behave more in a nonlinear (rather than linear) fashion. If the pulse is reduced in magnitude and made longer in duration to avoid these problems, the high-frequency characteristics of the process may not be excited sufficiently. An alternative approach that avoids this conflict is to use multi-frequency binary sequences as suggested by Harris and Mellichamp [9].

EXAMPLE 15.2

A laboratory-scale liquid heating process consisting of two constant-volume stirred tanks connected in series (see Fig. 15.5) was subjected to a pulse test. Figures 15.6a and b show the complete input and output pulses (the input is the voltage to the SCR, V, which is proportional to the heating rate in the first tank, Q, and the output is the second tank temperature, T_2) as taken from the process recorder chart (scale is 0 to 100%). Table 15.3 shows details of a portion of the original input–output data. The input, a time-weighted displaced cosine pulse, was logged every 7.5 s. The output was logged every 15 s for the first 240 s, then every 30 s for the remainder of the slowly varying tail of the signal. Determine the time constants and any time delay in the transfer function relating input V to output T_2.

Solution

An inspection of Fig. 15.6a and Table 15.3 indicates that, after data logging commenced, approximately 15 s elapsed before the input began to change. Hence, *both* input and output time scales should be shifted by 15 s to reflect this feature. Inspection of Fig. 15.6b and Table 15.3 indicates that the output does not change until after about 60 s have elapsed, indicating an *apparent* time delay in the process transfer function of $\theta' = 60 - 15 = 45$ s. We verify this value in later calculations. Table 15.4 shows how the time scales of input and output data are shifted prior to computer analysis (calculation of the frequency response). The 15 s subtracted from both input and output recorded times depends arbitrarily on the time at which data logging was initiated. Once removed, it can be neglected in further analysis.

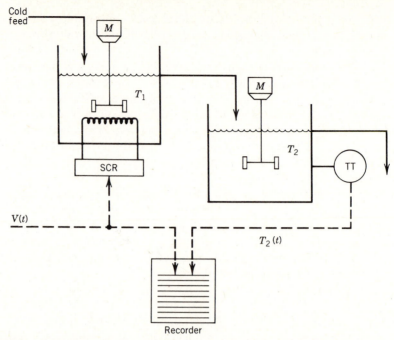

Figure 15.5. Experimental system consisting of two stirred-tank heaters in series, Example 15.2.

The input–output data of Table 15.4 were then processed using the pulse analysis computational methods discussed above. The resulting frequency response data are shown in Fig. 15.7. The amplitude ratio data have been normalized by the steady-state gain obtained in a separate step response test.

Analysis of the AR data yields values of $\omega_{b1} = 0.0055$ rad/s and $\omega_{b2} = 0.0125$ rad/s (as shown in Fig. 15.7) from which $\hat{\tau}_1 \approx 180$ s and $\hat{\tau}_2 \approx 80$ s. In fitting the phase angle data, we use Eq. 15-3 in the following form:

$$\phi_{\text{res}}(\omega_i) = \phi(\omega_i) + [\tan^{-1}(180\omega_i) + \tan^{-1}(80\omega_i)] \qquad (15\text{-}22)$$

Table 15.3   A Portion of the Recorded Data for Example 15.2

Time (s)	Input V (%)	Time (s)	Output T_2 (%)
0.0	19.5	0	38.2
7.5	19.5	15	38.2
15.0	19.5	.	.
22.5	19.7	60	38.2
30.0	20.0	75	38.3
37.5	21.3	90	38.3
45.0	23.9	105	38.4
52.5	28.2	120	38.5
60.0	37.1	135	39.0
.	.	150	39.8
.	.	.	.
105	93.9 (max value)	.	.
.	.	270	46.8 (max value)
.	.	.	.
.	.	.	.
150.0	20.2	840	38.5
157.5	19.5	870	38.4
165.0	19.5	900	38.2

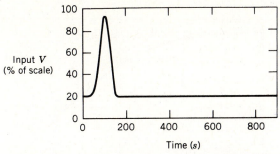

Figure 15.6a. Experimental input pulse, Example 15.2.

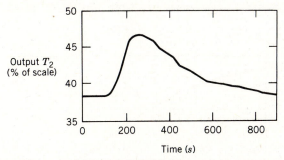

Figure 15.6b. Experimentally obtained output pulse, Example 15.2.

Figure 15.8 shows a plot of ϕ_{res} versus ω evaluated over the midfrequency range. Although some trends are apparent in this plot, a straight line serves as a good approximation. Recall that if ϕ_{res} is due solely to a time delay θ', then $\phi_{res} = -\theta'\omega$. Thus, the time delay of the empirical model can be obtained from the slope of this line and is found to be approximately 46 s, close to the apparent

Table 15.4   Shifting of Example 15.2 Input–Output Data Prior to Frequency Response Analysis

Time − 15 s	Input q	Time − 15 s	Output T_2
0.0	19.5	0	38.2
7.5	19.7	15	38.2
15.0	20.0	30	38.2
22.5	21.3	45	38.2
30.0	23.9	60	38.3
37.5	28.2	⋮	⋮
45.0	37.1	255	46.8
.	.	.	.
.	.	.	.
.	.	.	.
90.0	93.9	825	38.5
⋮	⋮	855	38.4
135.0	20.2	885	38.2
142.5	19.5		
150.0	19.5		

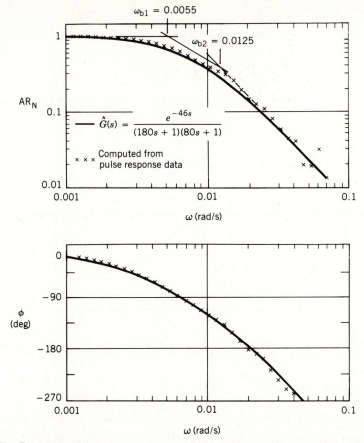

Figure 15.7. Experimental frequency response of the heating system and the empirical model, Example 15.2.

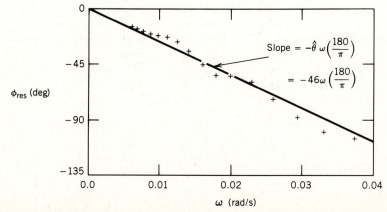

Figure 15.8. Determination of time delay in Example 15.2 from a plot of the residual phase angle.

value of 45 s (estimated from Table 15.3). Hence, the empirical model of the process, normalized by the steady-state gain, is

$$\hat{G}(s) = \frac{e^{-46s}}{(180s + 1)(80s + 1)} \tag{15-23}$$

Figure 15.7 shows that the approach given here does yield useful results, in particular, that the empirical model of Eq. 15-23 shown by solid lines gives a very good fit to the frequency response data. However, the scatter in the ϕ_{res} plot of Fig. 15.8 indicates that a second- or third-order model incorporating a single zero probably would provide even better agreement with the frequency response data. Such a model could be found by the numerical techniques discussed earlier in this chapter. Based on physical arguments, there is no justification for a zero in the model of this two-tank system. The most likely source of the trends in Fig. 15.8 would be slight errors in estimating τ_1 and τ_2.

SUMMARY

Most methods for fitting frequency response data are attractive because they avoid the use of iterative optimization procedures. Frequency-domain approaches also allow the user to employ any type of input forcing including arbitrary pulse inputs. The method chosen for empirical modeling, time domain or frequency domain, will depend on the experience of the user, the desired model complexity, the flexibility available in performing dynamic tests in the plant, and the ultimate use of the model (such as for controller design).

Recently, there has been an interest in identifying a process while it is operating under feedback control rather than by open-loop forcing using step, pulse, or sinusoidal functions. Yuwana and Seborg [10] have proposed an approach based on a small change in set point under proportional control. Another approach involves using random inputs such as binary sequences [9]. However, most of these topics are beyond the scope of this book, requiring sophisticated statistical analysis [11,12].

REFERENCES

1. Levy, E. C., Complex Curve Fitting, *IRE Trans. Auto. Conf.* **AC-4,** 37 (1959).
2. Sanathanan, C. K., and J. Koerner, Complex Function Synthesis as a Ratio of Two Complex Polynomials, *IEEE Trans. Auto. Contr.* **AC-8,** 37 (1963).
3. Seinfeld, J. H., and L. Lapidus, *Process Modeling, Estimation, and Identification,* Wiley, New York, 1974.
4. Hougen, J. O., *Measurements and Control Applications,* Instrument Society of America, Research Triangle Park, NC, 1979.
5. Hougen, J. O., Experiences and Experiments with Process Dynamics, *CEP Monogr. Ser.* **60,** No. 4 (1964).
6. Luyben, W. L., *Process Modeling, Simulation, and Control for Chemical Processes,* McGraw-Hill, New York, 1973.
7. Clements, W. C., and K. B. Schnelle, Pulse Testing for Dynamic Analysis, *IEC Proc. Des. Devel.* **2,** 94 (1963).
8. Dynamic Response Testing of Process Control Instrumentation, ISA-526, Standard, Instrument Society of America, Pittsburgh, PA October 1968.
9. Harris, S. L., and D. A. Mellichamp, "On-line Identification of Process Dynamics and Use of Multifrequency Binary Sequences", *IEC Proc. Des. Dev.,* 19, 166 (1980).
10. Yuwana M., and D. E. Seborg, "A New Method for On-Line Controller Tuning", *AICHE J.,* 28, 434 (1982).
11. Eykhoff, P., *System Identification: Parameter and State Estimation,* Wiley, London, 1974.
12. Box, G. E. P., and G. M. Jenkins, *Time Series Analysis: Forecasting and Control,* Holden-Day, San Francisco, 1970.

EXERCISES

15.1. Exercise 3.16 discussed the use of tracer tests with a process consisting of three perfectly mixed tanks in series. There we considered the tracer species to be injected suddenly; that is, an impulse input was assumed. For the case where the inlet concentration cannot be changed so rapidly but rather the input change occurs over a finite period of time:

(a) What should be the maximum width of the input pulse? Discuss in terms of characteristics of the process that can be assumed to be known, at least approximately, such as volumes and flow rate.

(b) How could you calculate the amount of tracer in the input pulse? (Assume that the input pulse closes.) In the pulse representing effluent of tracer from the third tank? What do you know about these two values?

15.2. Find the normalized frequency content $S_N(\omega)$ for the pulses illustrated in Fig. 15.4:

(a) A rectangular pulse with height $= 1$ and width $= t_w$.

(b) A triangular pulse with height $= 1$ and width $= t_w$.

(c) A displaced cosine pulse with amplitude $= 1$ and width $= t_w$, that is, $x(t) = 1 - \cos(2\pi t/t_w), \, 0 < t < t_w$.

15.3. Show that the expressions obtained in Exercise 15.2 for the normalized frequency content of rectangular, triangular, and displaced cosine pulses reduce to the proper values for $\omega = 0$; that is, show that

$$\lim_{\omega \to 0} S_N(\omega) = 1$$

for each case.

15.4. For the process described by the transfer function

$$G(s) = \frac{10}{(5s + 1)(2s + 1)(0.5s + 1)(0.1s + 1)}$$

(a) Find two second-order plus time delay models that approximate $G(s)$ and are of the form

$$\hat{G}(s) = \frac{Ke^{-\theta s}}{(\tau_1 s + 1)(\tau_2 s + 1)}$$

One of the approximate models can be found by using the method discussed in Section 6.3; the other, by graphical fitting of the frequency responses of G and $\hat{G}$.

(b) Compare all three models (one "exact" and two approximate) in the frequency domain and also by plotting their impulse responses.

15.5. Input pulses that have equal positive and negative components on either side of the steady-state value are not useful for estimating the process steady-state gain. The source of the problem is related to the value of $|X(j\omega)|$ at $\omega = 0$.

(a) Why is this value of the pulse frequency content important?

(b) Find $|X(j\omega)|$ for either a rectangular doublet pulse (shown in the drawing) or for a full-wave sine [i.e., $\sin(2\pi t/t_w)$ for $0 \leq t \leq t_w$] pulse. Show what happens when $\omega = 0$.

(c) Do you see any practical advantage to the use of these pulses, otherwise?

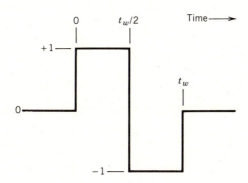

15.6. In Example 3.7 we found analytical expressions for the response of two stirred tanks in series to a rectangular pulse input.

(a) Using available expressions for the input (Eq. 3-117) and the output concentration from the second tank (Eqs. 3-138 and 3-139), obtain values of $x(t)$ and $y(t)$ corresponding to the pulse response plots in Fig. 15.2.

(b) Use a computer program, if available, to numerically evaluate the integrals in Eq. 15-11. Plot the resulting frequency response information as a Bode plot.

(c) Compare these results to an asymptotic representation of the AR results obtained from the differential equation model of the system. (Convert Eqs. 3-114 and 3-115 to a transfer function for the outlet concentration and plot its asymptotic amplitude ratio.)

15.7. A process that contains an integration element, that is, one whose transfer function contains a $1/s$ term, will exhibit special characteristics in pulse testing. In particular, a closed input pulse $x(t)$ will yield an output pulse $y(t)$ that does not close. By closed we mean that the pulse returns to its original value.

(a) For a rectangular pulse of unity magnitude and width t_w and for a second-order process containing an integration element

$$G(s) = \frac{K_p}{s(\tau s + 1)}$$

indicate the *form* of the response for $t_w < \tau$.

(b) Calculate exactly the final deviation in y.

(c) Repeat part (b) for a transfer function of the form

$$\frac{K_p e^{-\theta s} (\tau_a s + 1) \cdots (\tau_z s + 1)}{s(\tau_1 s + 1) \cdots (\tau_N s + 1)}$$

(d) One technique used to deal with output pulses that do not close is to numerically differentiate the output pulse response $y(t)$ and then use the resulting data for analysis. What is the theoretical basis for this approach? What potential problems do you foresee in using it?

(e) An implicit assumption in parts (a)–(d) is that a single integration element is present in the process. What would be the effect if more than one integration element were present?

15.8. A pulse test was run on an operating pilot plant in the semiworks area. Although the results obtained by computer analysis of the experimental data (see table) look somewhat noisy, the area supervisor insists that they be analyzed and that another test not be run.

(a) Find an empirical second-order plus time-delay model that fits the computer output below.

(b) Show how well the resulting model fits the amplitude ratio and phase angle results at $\omega = 0.05$ rad/min and $\omega = 0.5$ rad/min.

Frequency (radians/min)	AR	ϕ (deg)
0.0010	4.5	0
0.0025	4.6	0
0.0050	4.2	−5
0.0075	4.4	−5
0.0100	4.5	−10
0.0200	4.6	−12
0.0300	4.3	−23
0.0400	4.3	−27
0.0500	4.0	−32
0.0600	4.1	−43
0.0700	3.8	−50
0.0800	3.7	−52
0.0900	3.6	−60
0.1000	3.2	−68
0.2000	1.9	−108
0.3000	1.1	−139
0.4000	0.83	−162
0.5000	0.51	−162
0.6000	0.44	−184
0.7000	0.30	−192
0.8000	0.19	−200
0.9000	0.16	−206
1.0000	0.13	−220

15.9. An Ideal Gas Company temperature transmitter needed testing to determine its steady-state and dynamic characteristics. It supposedly has a transfer function of the form $Y(s)/X(s) = G(s) = K/(\tau s + 1)$. A control engineer assigned by Appelpolscher ran a pulse test on the instrument using a triangular pulse with height $= 1.5$ °C and width $= 4$ s. A computer-generated transform of the transmitter output in mA, $y(t)$, is shown in the drawing. The plot of $|Y(j\omega)|$ versus ω looked peculiar to the engineer, who quit working on the project until he had more time to study the results. Now Appelpolscher wants an immediate answer and has asked you to help out.

(a) What are good estimates of K and τ?

(b) What, if anything, was wrong with the input pulse chosen by the original engineer?

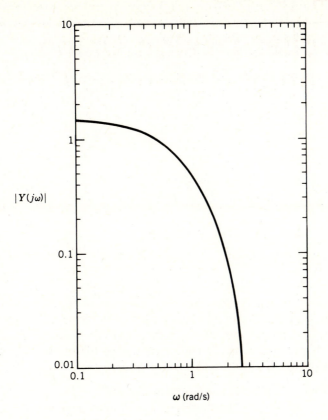

CHAPTER 16

Controller Design Using Frequency Response Criteria

In Chapter 12 we presented controller design equations for processes described by a first-order plus time-delay model. While such methods based on the time-domain response are simple to use, they have significant limitations, as discussed in Section 12.2. In this chapter we consider controller design based on frequency response analysis. This approach has several useful features:

1. It is applicable to dynamic models of any order.
2. The designer can specify the desirable closed-loop response characteristics.
3. Information on stability margins and sensitivity characteristics is provided.

The chief disadvantage of the frequency response approach is that it is generally iterative in nature; hence, it can be time-consuming. However, the use of an interactive computer graphics facility can greatly simplify and speed up the synthesis of feedback controllers. Interactive computer programs with graphical displays allow the designer to examine results quickly and make modifications so that the final control system satisfies the design objectives and constraints.

The first prerequisite of a controller is that it must be stable. In fact, conservative controller settings should be employed as protection against instability. In this chapter we consider a stability analysis for feedback control systems using frequency response. Since there will be a range of permissible controller parameters that satisfy the requirement of closed-loop stability, controller design specifically involves selecting from among the stable values those controller settings that yield the most desirable response. Recall that our objective is to design a closed-loop system that exhibits desirable (time-domain) characteristics. By establishing the relation between time domain and frequency response characteristics, we can identify some figures of merit for the frequency domain that correspond to desirable transient characteristics of a closed-loop system.

16.1 FREQUENCY RESPONSE CHARACTERISTICS OF FEEDBACK CONTROLLERS

The use of frequency response to design control systems first requires analysis of the input–output characteristics of feedback controllers with various mode combinations: P, PI, PD, and PID. As discussed in Chapter 14, the response of a system with transfer function $G(s)$ to a sinusoidal input of frequency ω is a sine wave after initial transients have diminished. This output sine wave has the same frequency ω but exhibits an amplitude ratio AR and a phase angle ϕ. The frequency response $G(j\omega)$ can be obtained mathematically by substitution of $s = j\omega$ into $G(s)$. Next we develop expressions for each controller type discussed in Chapter 8, assuming that the controller is reverse acting ($K_c > 0$).

Proportional Control.   Consider a proportional controller with positive gain $G_c(s) = K_c$. In this case $|G_c(j\omega)| = K_c$, which is independent of ω. Therefore, the amplitude ratio AR is directly proportional to K_c, and $\phi = 0°$ for all values of K_c.

Proportional–Integral Control.   A proportional–integral (PI) controller has the transfer function,

$$G_c(s) = K_c \left(1 + \frac{1}{\tau_I s} \right) = K_c \left(\frac{\tau_I s + 1}{\tau_I s} \right) \tag{16-1}$$

Substituting $s = j\omega$ gives

$$G_c(j\omega) = K_c \left(1 + \frac{1}{\tau_I j\omega} \right) = K_c \left(1 - \frac{j}{\omega\tau_I} \right) \tag{16-2}$$

Thus, the amplitude ratio and phase angle are

$$\text{AR} = |G_c(j\omega)| = K_c \sqrt{\frac{1}{\omega^2 \tau_I^2} + 1} \tag{16-3}$$

$$\phi = \angle G_c(j\omega) = \tan^{-1}(-1/\omega\tau_I) \tag{16-4}$$

Figure 16.1 shows the Bode plot for a PI controller with $K_c = 2$ and $\tau_I = 10$ min. At low frequencies the integral action dominates and the slope of the AR curve is -1. As $\omega \to 0$, $\text{AR} \to \infty$, and $\phi \to -90°$. At high frequencies $\text{AR} = K_c$ and $\phi = 0°$; neither is a function of ω in this region (cf. the proportional controller). Figure 16.1 also shows the break frequency, which is the intersection of low- and high-frequency asymptotes on the AR plot ($\omega_b = 1/\tau_I = 0.1$).

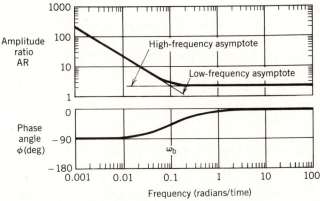

Figure 16.1.   Bode plot for PI controller, $G_c(s) = 2 \left(1 + \dfrac{1}{10s} \right)$.

Proportional–Derivative Control. For an ideal PD controller,

$$G_c(j\omega) = K_c(1 + \tau_D j\omega) \tag{16-5}$$

the frequency response characteristics

$$AR = K_c\sqrt{\omega^2\tau_D^2 + 1} \tag{16-6}$$

$$\phi = \tan^{-1}(\omega\tau_D) \tag{16-7}$$

are shown in Fig. 16.2 for $K_c = 2$ and $\tau_D = 4$ min. Both AR and ϕ are dominated by the derivative action term as $\omega \rightarrow \infty$. At high frequencies, AR is unbounded and has a slope of $+1$ while the phase angle approaches $+90°$. In the low-frequency region, the slope is zero (AR $= K_c$) and the phase angle approaches zero. The break frequency for the PD controller is $\omega_b = 1/\tau_D$.

Ideal PID Controller. The ideal PID controller exhibits features of both the PI and PD controllers:

$$G_c(s) = K_c\left(1 + \frac{1}{\tau_I s} + \tau_D s\right) = K_c\left(\frac{1 + \tau_I s + \tau_I\tau_D s^2}{\tau_I s}\right) \tag{16-8}$$

Substituting $s = j\omega$ and rearranging gives

$$AR = K_c\sqrt{\left(\omega\tau_D - \frac{1}{\omega\tau_I}\right)^2 + 1} \tag{16-9}$$

$$\phi = \tan^{-1}\left(\omega\tau_D - \frac{1}{\omega\tau_I}\right) \tag{16-10}$$

Figure 16.3 shows a Bode plot for a PID controller with $K_c = 2$, $\tau_I = 10$ min, and $\tau_D = 4$ min. Its asymptotic behavior at low frequencies conforms with that of a PI controller and is in agreement with Eqs. 16-3 and 16-4, while the high-frequency region is described by Eqs. 16-6 and 16-7, as was the case for the PD controller. The phase angle for the PID controller varies from $-90°$ ($\omega = 0$) to $+90°$ ($\omega \rightarrow \infty$). Note that Eq. 16-9 and Fig. 16.3 indicate AR is equal to K_c at $\omega = 1/\sqrt{\tau_I\tau_D}$ and larger than K_c at other frequencies.

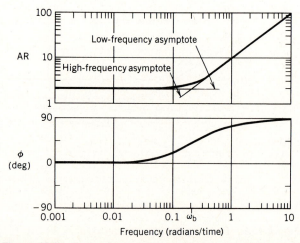

Figure 16.2. Bode plot of a PD controller, $G_c(s) = 2(1 + 4s)$.

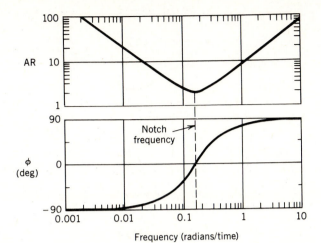

Figure 16.3. Bode plot of PID controller, $G_c(s) = 2 \left(1 + \dfrac{1}{10s} + 4s \right)$.

By adjusting the values of τ_I and τ_D, we can prescribe the shape and location of the "notch" for the controller frequency response. Decreasing τ_I and increasing τ_D narrows the notch, while the opposite changes broaden the notch. The center of the notch is located at $\omega = 1/\sqrt{\tau_I \tau_D}$ where $\phi = 0$. Varying K_c merely moves the amplitude ratio curve up or down, without affecting the width of the notch. Generally, the integral time τ_I should be larger than τ_D, typically $\tau_I \approx 4\tau_D$ (see Chapter 12).

Pneumatic PID Controller.   The transfer function for a pneumatic PID controller differs slightly from the ideal PID controller and is

$$G_c(s) = K_c \left[1 + \frac{1}{\tau_I s} + \frac{\tau_D s + 1}{\dfrac{\tau_D}{\gamma} s + 1} \right] \tag{16-11}$$

See Hougen [1] and Coughanowr and Koppel [2] for a derivation of (16-11) based on internal mechanical components (bellows, baffle, and nozzle) of the controllers. The values for γ typically range from 5 to 10. Equation 16-11 modifies the behavior of the ideal PID controller at high frequency, as shown in Fig. 16.4 ($\gamma = 10$). In (16-11) the asymptote of AR is bounded:

$$\text{AR}|_{\omega \to \infty} = \lim_{\omega \to \infty} |G_c(j\omega)| = \gamma K_c = 20$$

The implication of this constraint is that derivative mode amplitude ratio is limited at high frequencies rather than being unbounded. This feature is actually an advantage, since the ideal derivative action in (16-8) amplifies high-frequency input noise due to the large value of AR. Note that when $\tau_D = 0$, the pneumatic PI controller is the same as the PI controller of Eq. (16-1).

Electronic PID Controller.   The transfer function for the electronic three-mode controller is often stated in a multiplicative form [1]:

$$G_c(s) = K_c \left(\frac{\tau_I s + 1}{\tau_I s} \right) \left(\frac{\tau_D s + 1}{\beta \tau_D s + 1} \right) \tag{16-12}$$

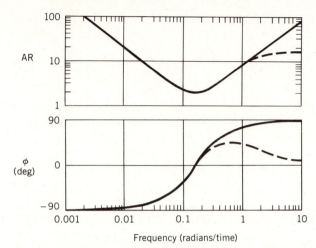

Figure 16.4.   Bode plots of pneumatic (——) and ideal (---) PID controllers.

$$\text{Pneumatic: } G_c(s) = 2\left(1 + \frac{1}{10s} + \frac{4s + 1}{0.4s + 1}\right)$$

$$\text{Ideal: } G_c(s) = 2\left(1 + \frac{1}{10s} + 4s\right)$$

where $0 < \beta \ll 1$. Equation (16-12) is essentially the product of PI and PD controllers. It can be arranged to yield

$$G_c(s) = \frac{1}{\beta\tau_D s + 1}\, K_c^\dagger \left(1 + \frac{1}{\tau_I^\dagger s} + \tau_D^\dagger s\right) \tag{16-13}$$

where $(K_c^\dagger, \tau_I^\dagger, \tau_D^\dagger)$ are functions of $(K_c, \tau_I, \tau_D, \beta)$. This derivation is left for the reader to perform in Exercise 16.1. Equation 16-13 implies that the electronic controller contains a first-order prefilter for noise with transfer function, $1/(\beta\tau_D s + 1)$. This effect is desirable since it makes the derivative mode less sensitive to noisy measurements. The amplitude ratio of (16-13) is bounded at high frequencies, as in Eq. 16-11. If a digital computer is used to implement the ideal three-mode controller, the designer can incorporate similar filtering action. In this case the filter can be tuned independently from the controller, which is an advantage for computer control (see Chapter 22).

If a direct-acting controller ($K_c < 0$) is employed, a change in the frequency response plots must be made. The AR plots in Figs. 16.1 to 16.4 are unchanged since $|K_c|$ is used in calculating the magnitude. However, all phase angles are shifted by $-180°$ due to the minus sign in K_c; for example, a proportional controller with $K_c < 0$ has a constant phase angle $\phi = -180°$. As a practical matter, it is possible to use the absolute value of K_c to calculate ϕ, since $K_c < 0$ only when $K_v K_p K_m < 0$. This convention gives $\phi = 0°$ for any proportional controller.

16.2   BODE AND NYQUIST STABILITY CRITERIA

The use of frequency response methods in the study and design of communications and control systems was originated by Bode during the 1920s [3]. The Bode stability criterion was a key result ensuing from that work and can be stated as follows:

Bode Stability Criterion: A closed-loop system is unstable if the frequency response of the open-loop transfer function $G_{OL} = G_c G_v G_p G_m$ has an amplitude ratio greater than one at the critical frequency. Otherwise the closed-loop system *is stable. The critical frequency ω_c is defined to be the frequency at which the open-loop phase angle is $-180°$.*

The Bode stability criterion allows the stability of *closed-loop* systems to be calculated from the *open-loop* transfer function G_{OL}. Because the criterion can be applied directly to systems that contain time delays, this method is preferred to stability criteria based on the characteristic equation (see Chapter 11). The Bode stability criterion is applicable only to open-loop stable systems with phase angle curves that exhibit a single critical frequency. This situation occurs in most process control problems. One exception is the so-called *conditional stability* case, covered later in this section, where multiple values of the critical frequency can occur.

When a closed-loop system is at the stability limit, that is, when the open-loop amplitude ratio is 1 at the critical frequency where the open-loop phase angle is $-180°$, the feedback control system produces a sustained oscillation in the controlled variable. This is the basis for the continuous cycling method of controller tuning discussed in Chapter 13. To understand how such sustained oscillations can occur, consider the analogy of pushing a child on a swing. The child will continue to swing in the same arc if a person pushes the child at the right time (in phase) and with the right amount of force. If the timing or the amount of force is incorrect, then the cyclic movement of the swing changes, as the child will quickly exclaim, and the arc will be reduced. A similar phenomenon occurs when a person bounces a ball.

Suppose that the feedback control system in Figure 16.5 is subjected to a sinusoidally varying set point $R(t) = A \sin \omega_c t$ for a period of time $0 < t < t_f$ and the comparator is disconnected during this time period. Assume that no load change occurs for $t > 0$ ($L(t) = 0$). The signal $R(t)$ oscillates at the critical frequency ω_c; after an initial transient period, this causes B to oscillate at the same frequency. At $t = t_f$, the set-point signal R is set to zero and the comparator connected again. If the feedback control system is *marginally stable*, the controlled variable C will exhibit a sustained sinusoidal oscillation (neither growing nor attenuating) with a frequency $\omega = \omega_c$. To understand why this special type of oscillation occurs only when $\omega = \omega_c$, note that the sinusoidal signal e passes through transfer functions G_c, G_v, G_p, and G_m before returning to the comparator. To maintain the oscillation, signal B must have the same amplitude as signal e and it also must have a $-180°$ phase shift, since the minus sign in the comparator also provides $-180°$ phase shift. After signal B passes through the comparator, it is identical to e and the oscillation continues indefinitely. At these conditions the open-loop transfer function has an amplitude ratio equal to unity at the critical frequency ω_c (which is also called the resonant frequency [4] or the phase crossover frequency [2]). The period of oscil-

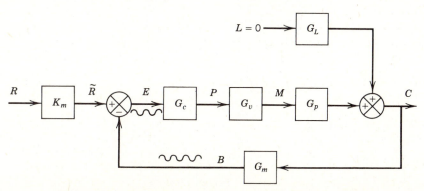

Figure 16.5. Sustained oscillation in a feedback control system.

lation will be $P_u = 2\pi/\omega_c$ (see Chapter 13). On the other hand, if the open-loop transfer function has an amplitude ratio greater than one at ω_c, then the amplitude of B is larger than the amplitude of e and the oscillation will grow with time. This behavior implies that the closed-loop system is unstable.

EXAMPLE 16.1

A process has the transfer function (time constant in minutes),

$$G_p(s) = \frac{2}{(0.5s + 1)^3}$$

The transfer functions for G_v and G_m are constants:

$$G_v = 0.1 \qquad G_m = 10$$

For a proportional controller, evaluate the stability of the closed-loop control system using the Bode stability criterion and three values of K_c: 1, 4, and 20.

Solution
First form G_{OL}:

$$G_{OL} = G_c G_v G_p G_m = (K_c)(0.1)\frac{2}{(0.5s + 1)^3}(10) = \frac{2K_c}{(0.5s + 1)^3}$$

Figure 16.6 shows a Bode plot of G_{OL} for three values of K_c. Note that for all three cases, the phase angle plot is the same, since the phase lag of a proportional controller is zero if $K_c > 0$. From the phase angle plot, we observe that $\omega_c = 3.46$ rad/min. This is the frequency a sustained oscillation will exhibit at the stability limit, as discussed above. Next examine the amplitude ratio of G_{OL}, AR_{OL}, for each value of K_c. Based on Fig. 16.6, we make the following observations:

K_c	AR_{OL} (*for* $\omega = \omega_c$)	*Classification*
1	0.25	Stable (same Bode plot as $G_p(s)$)
4	1	Marginally stable
20	5	Unstable

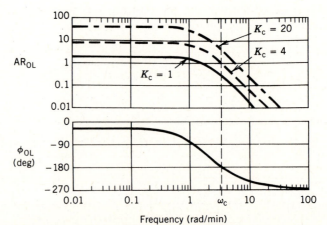

Figure 16.6. Bode plots for $G_{OL} = 2K_c/(0.5s + 1)^3$.

EXAMPLE 16.2

Find the critical frequency for the following process and PID controller assuming $G_v = G_m = 1$:

$$G_p(s) = \frac{e^{-0.3s}}{(9s + 1)(11s + 1)} \qquad G_c(s) = 20\left[1 + \frac{1}{2.5s} + s\right]$$

Solution

Figure 16.7 shows the open-loop magnitude and phase angle plots for G_{OL}. Note that the phase angle crosses $-180°$ at three points! Since there is more than one critical frequency, the Bode stability criterion (as stated above) cannot be applied; however, a related approach (the Nyquist stability criterion) can be used in this case, as discussed below.

Nyquist Stability Criterion

The Nyquist stability criterion is similar to the Bode criterion in that it can be used to determine closed-loop stability from open-loop frequency response characteristics. The Nyquist plot is a polar plot of the frequency response characteristics (see Chapter 14); consequently it conveys the same information as the Bode plot. Unlike the Bode criterion, the Nyquist criterion is applicable to open-loop unstable systems and to systems with more than one critical frequency. Thus it provides a more general approach.

> **Nyquist Stability Criterion:** *If* N *is the number of times that the Nyquist plot encircles the point* $(-1, 0)$ *in the complex plane in the clockwise direction, and* P *is the number of open-loop poles of* $G_{OL}(s)$ *that lie in the right-half plane, then* Z = N + P *is the number of unstable roots of the closed-loop characteristic equation* (*those roots lying in the right-half plane*).

Several comments regarding the Nyquist criterion can be made:

1. The reason that the $(-1, 0)$ point is so important follows from the characteristic equation, $1 + G_{OL}(s) = 0$, which can also be written as $G_{OL}(s) = -1$. This condition corresponds to a complex transfer function with an amplitude ratio of $+1$ and a phase angle of $-180°$.

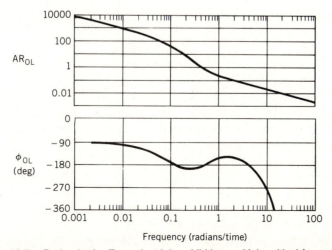

Figure 16.7. Bode plot for Example 16.2, exhibiting multiple critical frequencies.

2. Typically the open-loop system is stable and thus has no poles in the right-half plane (i.e., $P = 0$). For this situation, $Z = N$ and the closed-loop system is unstable if the Nyquist plot encircles the $(-1, 0)$ point one or more times.
3. A negative value of N indicates that the encirclements of the $(-1, 0)$ point occur in the opposite direction (i.e. counterclockwise).
4. The Nyquist diagram and associated theory [5] can be used to determine if an open-loop unstable process can be stabilized by feedback control. While a polar plot of $G_{OL}(j\omega)$ is meaningful in this case, the earlier physical interpretation of sinusoidal forcing is not, since the long-time response will be unbounded rather than sinusoidal.
5. The Nyquist stability criterion can be applied when multiple critical frequencies occur, which is the so-called conditional stability case. Figure 16.7 demonstrates that the phase angle for G_{OL}, ϕ_{OL}, may cross $-180°$ at more than one point. As shown by Luyben [5, pp. 402–403], the system can be closed-loop stable for two different ranges of controller gain, which is quite different from the normal result. In other words, increasing the absolute value of K_c can actually improve the stability of the closed-loop system for some ranges of K_c.

Details on use of the Nyquist plot for controller design are available elsewhere [5, 6].

One variation of the Nyquist diagram is the inverse Nyquist diagram, which is a polar plot of $G_{OL}^{-1}(j\omega)$. The critical point is also $(-1, 0)$ for this plot, but for stable open-loop systems the inverse Nyquist curve must be inside the critical point for the closed-loop system to be stable.

Now consider an example that illustrates the application of the Bode and Nyquist stability criteria.

EXAMPLE 16.3

Evaluate the stability of the closed-loop system in Figure 16.5 when

$$G_p(s) = \frac{4e^{-s}}{5s + 1}$$

(time constants and delay have units of minutes)

$$G_v = 2 \qquad G_m = 0.25 \qquad G_c = K_c$$

Find the critical frequency ω_c using a Bode plot. What controller gain makes $AR_{OL} = 1$ at $\omega = \omega_c$? Multiply this gain by 1.5 and draw the Nyquist plot for the resulting open-loop system.

Solution
Figure 16.8 shows the Bode plot for $K_c = 1$. At $\phi_{OL} = -180°$, the critical frequency is 1.69 rad/min and AR_{OL} is 0.235. If K_c is increased to 4.25 ($= 1/0.235$), $AR_{OL} = 1$ at $\omega = \omega_c$. This value is called the critical gain or ultimate controller gain K_{cu}. Next multiply K_{cu} by 1.5 to give $K_c = 6.38$. The resulting Nyquist plot for G_{OL} is shown in Fig. 16.9 (low frequency data for $\omega < 0.4$ have been omitted from this figure). Note that the $(-1, 0)$ point is encircled once. By applying the Nyquist stability criterion, $N = 1$, $P = 0$, and $Z = 1$; hence the increased controller gain causes the system to be unstable. Only gains lower than the critical controller gain of 4.25 will yield a stable closed-loop system.

In this chapter we emphasize use of the Bode plot for stability analysis of controlled systems. The Bode diagram provides more information than the Nyquist

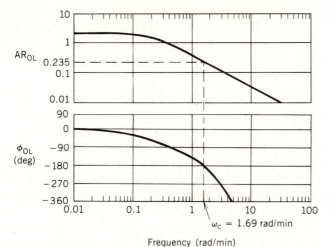

Figure 16.8. Bode plot for Example 16.3, $K_c = 1$.

plot because the frequency ω is shown explicitly. In addition, it facilitates analysis over a wide range of frequencies due to the log scale on the abscissa. Another advantage of the Bode diagram is that it allows the open-loop frequency response characteristic to be graphically constructed from the characteristics for the individual transfer functions, G_c, G_v, G_p and G_m. This feature is described in the next section.

16.3 EFFECT OF CONTROLLERS ON OPEN-LOOP FREQUENCY RESPONSE

The previous examples have illustrated that the open-loop frequency response can be adjusted by choosing G_c appropriately. Once G_v, G_p, and G_m are specified, the only remaining degree of freedom is incorporated in G_c.

In general the total phase angle and amplitude ratio for $G_{OL} = G_c G_v G_p G_m$ at any frequency are given by the rules for block multiplication presented in Chapter 14:

$$\text{AR}_{OL} = |G_c|\,|G_v|\,|G_p|\,|G_m| \tag{16-14}$$

$$\phi_{OL} = \phi_c + \phi_v + \phi_p + \phi_m \tag{16-15}$$

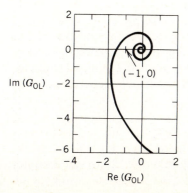

Figure 16.9. Nyquist plot for Example 16.3: $K_c = 1.5K_{cu} = 6.38$.

At the critical frequency $\phi_{OL} = -180°$ and $AR_{OL} = 1$; thus when $\omega = \omega_c$,

$$1 = |G_c| \, |G_v| \, |G_p| \, |G_m| \tag{16-16}$$

and

$$-180 = \phi_c + \phi_v + \phi_p + \phi_m \tag{16-17}$$

Equations 16-16 and 16-17 can be applied to find the stability limit K_{cu} for proportional-only control and a marginally stable system. Because ω_c can be found graphically (see Examples 16.1 and 16.3), controller gain K_c can be adjusted using Eq. 16-15 so that $AR_{OL} < 1$. We demonstrate this point in Example 16.4.

Integral action is normally included in a controller because it eliminates offset in the closed-loop response. However, it also adds phase lag to the open-loop system (see Fig. 16.1), making the system less stable. Derivative action adds phase lead to the open-loop system, improving stability and allowing higher gains to be used to improve the closed-loop response. Derivative action has a beneficial effect for systems with steep open-loop AR curves and flat phase angle curves near the $-180°$ point; for example, a second-order system with a small time delay exhibits such characteristics. For a process with an open-loop frequency response that exhibits a significant decrease in phase angle, derivative action will provide little improvement in the closed-loop system performance [1]. The latter situation usually occurs when the time delay is larger than the dominant time constants of the process or when the process is third or higher order.

Harriott [7, pp. 100–102] has shown that the product of the ultimate controller gain defined in Chapter 13 and the critical frequency can be used as a measure of control system performance. As a general rule, it is desirable to maximize this *gain–bandwidth* product $K_{cu}\omega_c$, since at least approximately, the integral absolute error is inversely proportional to both ω_c and AR_{OL} at the critical frequency. Therefore, AR_{OL} should be as large as possible at $\omega = \omega_c$ (up to a maximum value of unity), corresponding to a high value of K_c. A large value of ω_c is also desirable since it indicates a small closed-loop response time. These guidelines can be used as a shortcut procedure to determine how changes in process parameters or instrumentation affect the quality of control [7, 8].

EXAMPLE 16.4

Consider PI control of an overdamped second-order process with no time delay (time constants in minutes),

$$G_p(s) = \frac{5}{(s + 1)(0.5s + 1)} \tag{16-18}$$

$$G_m = G_v = 1$$

Find the ultimate controller gain for proportional control. Using the Bode plot, show that values of $K_c = 0.4$ and $\tau_I = 0.2$ min yield an unstable closed-loop system. What is the maximum value of K_c that can be used with $\tau_I = 0.2$ min and still have a stable system? Show that $\tau_I = 1$ min gives a stable controller regardless of the value of K_c.

Solution

The Bode plot for $G_p(s)$ is shown as curve A in Fig. 16.10. Note that a critical frequency does not exist since $\phi > -180°$. A proportional-only controller does not

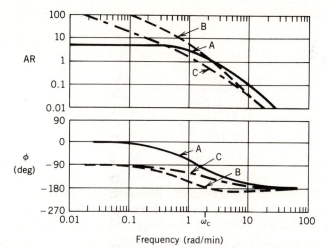

Figure 16.10. Bode plots for Example 16.3: Curve A: $G_p(s)$

Curve B: $G_{OL}(s)$; $G_c(s) = 0.4 \left(1 + \dfrac{1}{0.2f} \right)$

Curve C: $G_{OL}(s)$; $G_c(s) = 0.4 \left(1 + \dfrac{1}{s} \right)$

add any phase lag to $G_{OL}(s) = K_c G_p(s)$. Hence K_c can become extremely large and the closed-loop system will always be stable; K_{cu} in this case is infinite. On the other hand, the inclusion of integral action in the controller can cause the closed-loop system to become unstable. Curve B in Fig. 16.10 for $G_{OL}(s)$ indicates an unstable closed-loop system when $G_c(s) = 0.4 (1 + 1/0.2s)$; note that $AR_{OL} > 1$ when $\phi_{OL} = -180°$.

To find the maximum allowable value of K_c that provides stability when $\tau_I = 0.2$ min, first recognize that ω_c depends on τ_I but not on K_c because K_c affects only AR_{OL}. Using Fig. 16.10 (curve B), the critical frequency is $\omega_c = 2.2$ rad/min, which has a corresponding value of $AR_{OL} = 1.38$. The maximum allowable value of K_c ($AR_{OL} < 1$) can be found by multiplying K_c by a factor of $1/1.38 = 0.72$. Using the current value of 0.4 for K_c, the new value of K_c would be $0.4(0.72) = 0.29$.

If τ_I is increased to 1 min, curve C results, which is a stable closed-loop system for all values of K_c because there is no critical frequency. Now the PI controller can maintain stability without sacrificing performance.

In general, the maximum phase lag of $G_c G_v G_p G_m$ will be greater than 180° ($\phi_{OL} < -180°$); hence, the controller parameters must be selected carefully to ensure stability. As discussed earlier, ω_c is the critical frequency at which disturbances will be amplified. The controller gain must be low enough that the open-loop amplitude ratio is below unity at this critical frequency. But at other frequencies, $|G_c|$ can be increased to improve control system performance. Therefore, a controller with an AR that varies with frequency, such as the PID controller, can be used beneficially in most loops. The notch shape of the AR curve for PID control discussed earlier is well-suited to this task, with the notch occurring near the critical frequency [9]. Several techniques for selecting PID controller settings based on the frequency response are presented in subsequent sections.

16.4 GAIN AND PHASE MARGINS

If a poorly tuned control system operates with K_c near the stability limit K_{cu}, the closed-loop system could approach unstable operation. As measures of relative stability, the terms *gain margin* (GM) and *phase margin* (PM) often are used. Figure 16.11 illustrates the concepts of gain and phase margin. Let AR_c be the value of the open-loop amplitude ratio at the critical (or phase crossover) frequency ω_c. The gain margin GM is defined as

$$GM = \frac{1}{AR_c} \qquad (16\text{-}19)$$

From the Bode stability criterion, AR_c must be less than one to have a stable closed-loop system. Thus GM > 1 is a stability requirement.

Define ω_g as the frequency at which the open-loop gain is unity (the gain-crossover frequency). Let ϕ_g denote the phase angle at ω_g. Phase margin PM is defined as

$$PM = 180 + \phi_g \qquad (16\text{-}20)$$

Controller manufacturers recommend that a well-tuned controller have a gain margin between 1.7 and 2.0, while the phase margin should be between 30 and 45°. Recognize that these ranges are approximate and it is often not possible to choose controller settings that fix both GM and PM at arbitrary values. The GM and PM concepts are not meaningful when the open-loop system has multiple values for ω_c or ω_g [6].

The recommended values of phase and gain margins provide a compromise between performance and safety. Large values of GM and PM cause sluggish closed-loop response, while smaller values result in a less sluggish, more oscillatory response. The choice of GM and PM should also depend upon the level of confidence in the process model and how much the process parameters can change. For example, if the dominant time constant or time delay for a process depends on the flow rate (throughput), the phase margin will change if the flow rate changes.

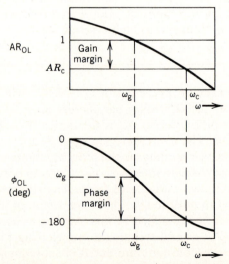

Figure 16.11. Gain and phase margins on Bode plot.

EXAMPLE 16.5

Calculate the gain and phase margins for the first-order plus time-delay process considered in Example 16.3. Use the PID controllers given in Table 13.2 with:

1. Ziegler–Nichols settings
2. Modified Ziegler–Nichols settings (some overshoot)

Solution

From Example 16.3 the ultimate controller gain is $K_{cu} = 4.25$ and the ultimate period is $P_u = 2\pi/1.69 = 3.72$ min. Therefore, the PID controllers have the following settings:

Controller Settings	K_c	τ_I (min)	τ_D (min)
Z–N	2.55	1.86	0.46
Modified Z–N	1.40	1.86	1.24

Figure 16.12 gives the frequency response of $G_{OL}(s)$ for the two controllers. The gain and phase margins by inspection of the two plots are

Controller	GM	PM
Z–N	1.8	40°
Modified Z–N	1.5	62°

The modified Z–N controller provides a larger phase margin but a smaller gain margin than the Z–N controller. We anticipate that the modified Z–N settings will provide a more conservative controller due to its lower value of K_c and larger value of PM. However, in view of its smaller GM value, a simulation of the closed-loop system will be required to verify this assertion.

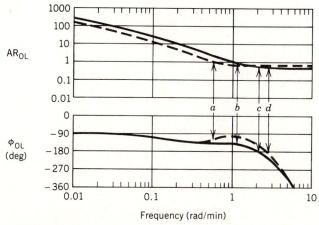

Figure 16.12. Bode plot of G_{OL}, Example 16.5. Vertical arrows mark phase and gain crossover frequencies.
- - - - - Modified Ziegler–Nichols settings
———— Ziegler–Nichols settings

16.5 CLOSED-LOOP FREQUENCY RESPONSE

In previous sections we have shown how the Bode plot of the open-loop transfer function $G_{OL}(s)$ is used to evaluate stability characteristics of the closed-loop system. However, it is difficult to infer the quality of the closed-loop transient response from the Bode plot. In this section we show how the closed-loop transfer function for set-point changes can be used in control systems analysis to evaluate the performance of a given controller.

First define M as the amplitude ratio of the closed-loop transfer function and ψ as its phase angle:

$$M = \left| \frac{C(j\omega)}{R(j\omega)} \right| \qquad (16\text{-}21)$$

$$\psi = \angle \frac{C(j\omega)}{R(j\omega)} \qquad (16\text{-}22)$$

The desired closed-loop frequency response amplitude ratio for set-point changes is shown in Fig. 16.13. Also shown is the corresponding time domain response, which has desirable rise time and overshoot characteristics. For set-point changes the closed-loop amplitude ratio M should have the following characteristics:

1. M should be unity as $\omega \to 0$, indicating no offset.
2. M should be maintained at unity up to as high a frequency as possible (implies $|C/R| = 1$). This ensures a rapid approach to the new steady state during a set-point change.
3. A resonant peak in M should be present similar to that which is exhibited by the frequency response of a second-order underdamped process (see Chapter 14). The *peak amplitude ratio M_p* should be no greater than 1.25, corresponding to a damping coefficient of $\zeta = 0.5$ for a second-order system. The controller parameters should be selected so that the peak frequency ω_p is as large as possible. A large value of ω_p implies a faster response for a set-point change.
4. The *bandwidth ω_{bw}* is the frequency at which $M = \sqrt{2}/2 = 0.707$. The bandwidth provides a measure of transient response characteristics. A large value of ω_{bw} indicates a relatively fast response with a short rise time.

For load changes, we can consider the closed-loop amplitude ratio $|C/L|$. In this case, the desired closed-loop frequency response minimizes $|C/L|$ over as wide

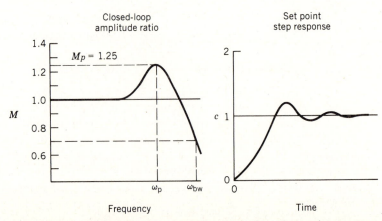

Figure 16.13. Desired closed-loop amplitude ratio M and set-point response.

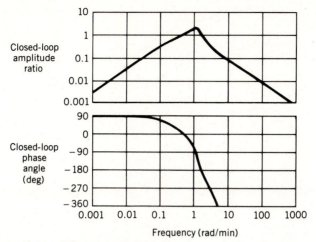

Figure 16.14. Closed-loop frequency response for load changes:

$$G_p(s) = G_L(s) = \frac{4e^{-s}}{1 + 5s} \qquad G_c(s) = 1.125 \left(1 + \frac{1}{3.33s}\right)$$

$$G_m = G_v = 1$$

a frequency range as possible. If the integral mode is employed in the controller, it will have zero offset ($|C/L| = 0$) as $\omega \to 0$. For control without integral action, $|C/L| > 0$ as $\omega \to 0$ and offset occurs. Figure 16.14 shows a typical plot of closed-loop frequency response for a load change with a PI controller.

When $G_m(s) = K_m$, the closed-loop frequency response for set-point changes can be calculated from the open-loop frequency response. Using Eq. 11-3, C/R is related directly to G_{OL} as follows:

$$\frac{C(s)}{R(s)} = \frac{G_c(s)G_v(s)G_p(s)K_m}{1 + G_c(s)G_v(s)G_p(s)K_m} = \frac{G_{OL}(s)}{1 + G_{OL}(s)} \qquad (16\text{-}23)$$

As shown in Chapter 14, we can calculate $C(j\omega)/R(j\omega)$ by separating $G_{OL}(j\omega)$ into real and imaginary parts and then rationalizing the resulting complex fraction given by (16-23). Note that this procedure also is applicable if $G_{OL}(j\omega)$ is available as a Bode plot.

We can also derive simple analytical relations between open-loop and closed-loop amplitude ratios (AR_{OL} and M) as well as phase angles (ϕ_{OL} and ψ), when $G_m(s) = K_m$. By algebraic manipulation (see Exercise 16.6) the closed-loop frequency response can be calculated from the corresponding open-loop quantities:

$$M = \frac{1}{\sqrt{[1 + (\cos\phi_{OL}/AR_{OL})]^2 + (\sin\phi_{OL}/AR_{OL})^2}} \qquad (16\text{-}24)$$

$$\psi = \tan^{-1}\left(\frac{\sin\phi_{OL}/AR_{OL}}{1 + \cos\phi_{OL}/AR_{OL}}\right) \qquad (16\text{-}25)$$

The Nichols chart in Fig. 16.15 provides a graphical display of the relations in Eqs. 16-24 and 16-25. In a typical application, the open-loop frequency response, AR_{OL} and ϕ_{OL}, are plotted on the Nichols chart as a series of points. Then the closed-loop quantities, M and ψ, are obtained by interpolation. For example, if at a certain frequency $AR_{OL} = 1$ and $\phi_{OL} = -100°$, then by interpolation of the

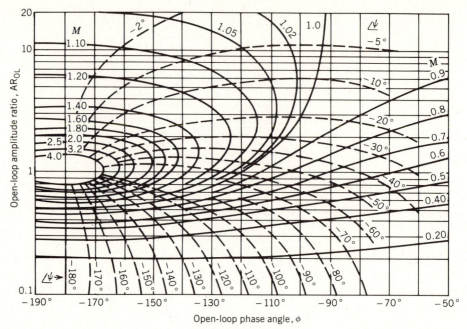

Figure 16.15.   A Nichols chart. [The closed-loop amplitude ratio M (———) and the phase angle ψ (– – –) are shown as families of curves.]

contours in Fig. 16.15, $M = 0.76$ and $\psi = -50°$ at the same frequency. As discussed in the next section, a computer graphics terminal facilitates these calculations.

The Nichols chart has been constructed for closed-loop systems with *unity feedback,* the structure of which is shown Fig. 16.16. Note that Fig. 16.16 corresponds exactly to Eq. 16-23 if $G_m(s) = K_m$ and $G_{OL}(s)$ is denoted as $G(s)$. If a control system does not have unity feedback, then a simple block diagram manipulation can be employed so that the closed-loop portion does have unity feedback [2]. This procedure is illustrated in Fig. 16.17, and it is applicable when the dynamics of $G_m = G_2$ cannot be ignored. The frequency response of the unity feedback system, $C'/R = G_1G_2/(1 + G_1G_2)$, can be obtained from the Nichols chart. This information can then be combined with the frequency response for $1/G_2$ to obtain the desired information for C/R:

$$|C/R| = |K_m|\,|C'/\tilde{R}|\,|1/G_2| \tag{16-26}$$

$$\angle(C/R) = \angle(C'/\tilde{R}) + \angle(1/G_2) \tag{16-27}$$

Note that both the original and modified block diagrams in Fig. 16.17 have the same characteristic equation, $1 + G_1G_2 = 0$. If there are significant measurement dynamics (i.e., if $G_m(s)$ is not well approximated by K_m), then block diagram manipulation should be done prior to design of the controller. This step will prevent excessive oscillation in the controlled variable which can occur if the measurement dynamics are not taken into account. In this case, set $G_2 = G_m$ and $G_1 = G_cG_vG_p$.

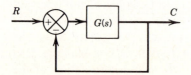

Figure 16.16.   A closed-loop system with unity feedback.

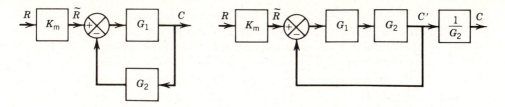

(a) Original block diagram   (b) Equivalent (unity feedback) block diagram

Figure 16.17. Block diagram rearrangement to generate a unity feedback system.

The Nichols chart can be used to estimate quickly the degree of oscillation and the amount of overshoot that will occur for set-point changes. Figure 16.18 shows a portion of the Nichols chart with the contours for $M = 1.0$ and $M = 1.25$. Recall that $M = 1.25$ is the desired overshoot for a feedback controller ($\zeta = 0.5$). Curve A (the dashed line) is the locus of points for AR_{OL} and ϕ_{OL} when a PID controller is used with a process model that is second-order plus time delay. Note that $M_p = 1.25$ because $1 < M \leq 1.25$ in Fig. 16.18 and $M < 1.25$ for large frequencies. Furthermore, $M \to 1$ as $\omega \to 0$ (not shown). Thus the closed-loop amplitude ratio has the desired shape in Fig. 16.13. Consequently, the PID controller should be satisfactory.

Robustness. In order for a control system to function properly, it should not be unduly sensitive to small changes in the process or to inaccuracies in the process model if a model is used to design the control system. A control system that satisfies this requirement is said to be *robust* or to exhibit a satisfactory degree of *robustness*. Let G = true process model and $\tilde{G}$ = approximate process model (that is used in controller design). The relative model error e_m is defined as

$$e_m = \frac{G - \tilde{G}}{\tilde{G}} \tag{16-28}$$

For the controlled process to be stable when the process model is inaccurate, the following condition must hold [10, 11]:

$$\max_{\omega} |e_m| < \frac{1}{M_p} \tag{16-29}$$

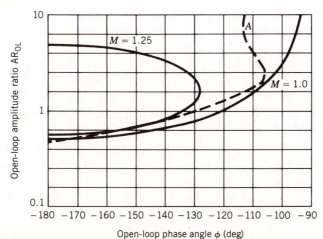

Open-loop phase angle ϕ (deg)

Figure 16.18. A Nichols chart, $K_c = 9.44$, $\tau_I = 9$ min, $\tau_D = 4$ min, $G_p = e^{-s}/(10s + 1)(5s + 1)$, $G_v = G_m = 1$.

Note that (16-29) provides a measure of how much modeling error can be tolerated when the controller is designed using model $\tilde{G}$ and a specified value of M_p. The bounds on the model error are narrower when M_p, the peak amplitude ratio, is large. Once M_p is determined for a given controller, (16-29) can be evaluated for errors in specific model parameters to determine the robustness of the controller. In addition, Rivera et al. [10] have shown that designing a controller to have a given value of M_p establishes lower bounds on gain margin (GM) and phase margin (PM) that can be realized for that controller:

$$\text{GM} \geq 1 + \frac{1}{M_p} \tag{16-30}$$

$$\text{PM} \geq 2 \sin^{-1}(1/2M_p) \tag{16-31}$$

When $M_p = 1.25$, then GM ≥ 1.8 and PM $\geq 47°$, which satisfy the guidelines given in Section 16.4. Equations 16-29 through 16-31 indicate that higher performance (corresponding to larger values of M_p) is obtained at the expense of robustness.

16.6 COMPUTER-AIDED DESIGN OF FEEDBACK CONTROLLERS

A major impediment to use of frequency response methods in controller design has been the trial and error graphical calculations, which can become tedious if done manually. Computer-aided design facilities, especially those employing interactive computer graphics terminals, can simplify and facilitate greatly the design of feedback controllers. The use of interactive computer programs allows the designer to visually examine results rapidly and make the necessary modifications so that the final control system complies with the design objectives. A compilation of software for computer-aided feedback controller design is given in the Appendix.

Hougen [1, p. 318] developed interactive frequency response synthesis procedures to design PID controllers which are described by the transfer function in Eq. 16-12. He has also presented correlations for electronic controller settings for a wide range of process models. These settings are based on achieving the desirable closed-loop response shown in Fig. 16.13. For the ideal three-mode controller, Eq. 16-8, Edgar et al. [12] developed a synthesis procedure based on Hougen's principles, while Jury [9] has presented an iterative method based on stability margins of the closed-loop system.

As shown by Edgar et al. [12], a set of six steps based on plots of the open-loop frequency response (OLFR), closed-loop frequency response (CLFR), OLFR superimposed on CLFR (Nichols chart), and time-domain set-point response can be used to evaluate the performance of a given controller. Because each of these plots changes when the controller settings are modified, the procedure is made most efficient by displaying the various plots on a graphics terminal.

Edgar et al. recommended the following steps that can be used iteratively to synthesize the PID controller:

Step 1. Initially set $K_c = 1$ and τ_I slightly less than the dominant time constant of the process. Then check the OLFR to see if $G_{OL}(s)$ has adequate gain and phase margins. Modify K_c accordingly.

Step 2. Next plot the OLFR on the Nichols chart. Adjust K_c so that the CLFR is tangent to $M = 1.25$, which is the desired peak amplitude ratio for $|C/R|$. An increase in K_c generally increases M_p, while a decrease in K_c usually reduces M_p.

Step 3. Check the CLFR diagram (*C/R* vs. ω) and note ω_p. If derivative action is to be used, a suggested first guess is $\tau_D = 2/\omega_p$ [13]. Inclusion of the derivative mode adds phase lead near the critical frequency.

Step 4. Iterate between Steps 2 and 3 until K_c and τ_D converge to settings that yield a satisfactory CLFR (cf. Fig. 16.13).

Step 5. The time-domain response to a set-point change can be checked at this point to ensure that it is satisfactory. If it is, proceed to Step 6. Otherwise set $\tau_I = 4\tau_D$ and return to Step 2.

Step 6. Choose a new value for τ_I by increasing or decreasing τ_I by 10%. Then return again to Step 2 to find appropriate values of K_c and τ_D. If no significant improvement in closed-loop transient response is detected from the previous best set of controller parameters, then stop.

The fact that the above procedure is iterative makes it somewhat unattractive to perform even with the availability of interactive computer graphics. However, relatively few iterations of Steps 2–6 are normally required (say three or four) before acceptable values of the controller parameters are obtained. Because the quality of the closed-loop response is somewhat subjective and depends on visual evaluation by the designer, viewing intermediate results during controller design is helpful. The interactive procedure can be performed in several minutes by an experienced user.

On the other hand, the complete design sequence can be performed automatically by computer if desired. What is required is a suitable objective function for closed-loop performance. Harris and Mellichamp [14] have proposed an objective function involving the peak amplitude ratio M_p, phase margin PM, and peak frequency ω_p. The goal is to maximize ω_p while maintaining M_p and PM close to desired values. Harris and Mellichamp used an optimization routine to find the best values of K_c, τ_I, and τ_D and achieved essentially the same results as obtained with human interaction (Edgar et al. [12]). In addition, the automated procedure is much faster than one based on human interaction and less prone to error.

EXAMPLE 16.6

Consider a second-order plus time delay process (time constants and time delay are in minutes)

$$G_p(s) = \frac{e^{-s}}{(10s + 1)(5s + 1)} \tag{16-32}$$

and assume $G_m = G_v = 1$ for simplicity. Design PI and ideal PID controllers using a frequency response analysis and compare their performance with controllers obtained by the Ziegler–Nichols method (process reaction curve version) and the modified Ziegler–Nichols method (continuous cycling), as presented previously in Tables 13.3 and 13.2.

Solution

The OLFR, CLFR, and Nichols plots for this process and details of the design procedure are given in Ref. 11. A process reaction curve analysis gives $K = 1$, $\tau = 12.9$ min, and $\theta = 2.8$ min. The PI and PID controller settings used in the three controllers tested are as follows (τ_I and τ_D have units of minutes).

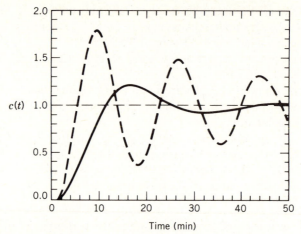

Figure 16.19. Set point change of PI controllers tuned by frequency response (———) and by Ziegler–Nichols correlation (- - -), Example 16.6.

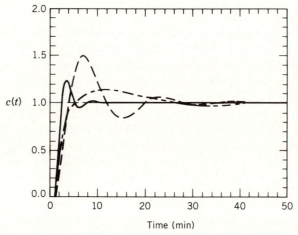

Figure 16.20. Set point change of PID controllers tuned by frequency response (———), Ziegler–Nichols settings (- - -), and modified Ziegler–Nichols settings (— - —), Example 16.6.

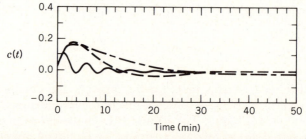

Figure 16.21. Load change for PID controllers tuned by frequency response (———), Ziegler–Nichols settings (- - - -), and modified Ziegler–Nichols settings (— - —), Example 16.6.

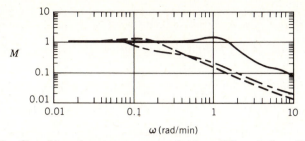

Figure 16.22.   Closed-loop frequency response for three PID controllers, Example 16.6.
————— Frequency response.
- - - - Ziegler–Nichols.
— · — Modified Ziegler–Nichols.

	PI		PID		
Method	K_c	τ_I	K_c	τ_I	τ_D
Frequency response (FR)	2.3	13.0	10.2	15.0	3.6
Modified Ziegler–Nichols	—	—	5.1	5.9	3.9
Ziegler–Nichols	1.8	15.7	5.5	5.6	1.4

Frequency response design provides a much higher controller gain. Figures 16.19 and 16.20 compare set-point changes for the two PI and three PID controllers, respectively. Figure 16.21 presents the closed-loop transient response of the controlled variable for load changes with the three PID controllers. The same pattern of controller peformance seen in Fig. 16.20 emerges here also. It is clear that the controllers designed using the synthesis procedure described in this section are superior to those obtained via the Z–N settings. Note how sensitive the closed-loop performance is to the value of τ_D. Figure 16.22 shows the CLFR for the three PID controllers. Note that the FR controller has the largest values of ω_{bw} and M_p. Hence it results in the fastest closed-loop response but is the least robust of the three controllers (cf. Eq. 16-29).

SUMMARY

Frequency response analysis of a feedback control system provides a general approach to design PID controllers. Using stability margins based on the open-loop frequency response, acceptable controller settings can be obtained which represent a reasonable compromise between performance and robustness to plant changes and modeling errors. An integrated design approach involving open-loop and closed-loop frequency response (Bode plot and Nichols chart) as well as the transient closed-loop response can yield controllers with superior performance for a wide variety of process models. Because such a technique is iterative, its use is facilitated greatly by computer-aided design.

REFERENCES

1. Hougen, J. O., *Measurements and Control Applications,* Instrum. Society of America, Pittsburgh, PA, 1979.
2. Coughanowr, D. R., and L. B. Koppel, *Process Systems Analysis and Control,* McGraw-Hill, New York, 1965.
3. MacFarlane, A. G. J., The Development of Frequency Response Methods in Automatic Control, *IEEE Trans. Auto. Contr.* **AC-24,** 250 (1979).

4. Shinskey, F. G., *Process Control Systems,* 3rd ed. McGraw-Hill, New York, 1988.
5. Luyben, W. L., *Process Modeling, Simulation and Control for Chemical Engineers,* McGraw-Hill, New York, 1973.
6. Kuo, B. C., *Automatic Control Systems,* 4th ed., Prentice Hall, Englewood Cliffs, NJ (1982), Chapter 9.
7. Harriott, P., *Process Control,* McGraw-Hill, New York, 1964.
8. Jeffreson, C. P., Controllability of Process Systems. Application of Harriott's Index of Controllability, *Ind. Eng. Chem. Proc. Des. Dev.* **15**(3), 171 (1976).
9. Jury, F. D., Fundamentals of Three-Mode Controllers, Fisher Controls Company Technical Monograph No. 28 (1973).
10. Rivera, D. E., M. Morari, and S. Skogestad, Internal Model Control: 4. PID Controller Design, *Ind. Eng. Chem. Proc. Des. Dev.* **25**, 252 (1986).
11. Morari, M., and E. Zafiriou, *Robust Process Control,* Prentice Hall, Englewood Cliffs, NJ, 1988.
12. Edgar, T. F., R. C. Heeb, and J. O. Hougen, Computer-aided Process Control System Design Using Interactive Graphics, *Comp. Chem. Eng.* **5**(4), 225 (1982).
13. Tyner, M., and F. P. May, *Process Engineering Control,* Ronald Press, New York, 1968.
14. Harris, S. L., and D. A. Mellichamp, Controller Tuning Using Optimization to Meet Multiple Closed-Loop Criteria, *AIChE J.* **31**, 484 (1985).

EXERCISES

16.1. For the electronic PID controller in Eqs. 16-12 and 16-13, do the following:

(a) Derive expressions for $K_c^\dagger$, $\tau_I^\dagger$, and $\tau_D^\dagger$ as functions of K_c, τ_I, τ_D, and β.
(b) Derive expressions for AR and ϕ.
(c) Sketch the Bode diagram for the following numerical values: $K_c = 2$, $\tau_I = 10$, $\tau_D = 4$, $\beta = 0.1$.

16.2. In Chapter 12, we considered a method for stability analysis based on substituting $s = j\omega$ into the characteristic equation. Demonstrate that this approach is (or is not) identical to the Bode Stability Criterion. *Hint:* Consider a special example such as $G_{OL}(s) = Ke^{-\theta s}/(\tau s + 1)$.

16.3. A process that can be modeled as a pure time delay is controlled using a proportional feedback controller. The control valve and measurement device have negligible dynamics and steady-state gains of $K_v = 0.5$ and $K_m = 1$, respectively. After a small set-point change is made, a sustained oscillation occurs which has as period of 10 min.

(a) What controller gain is being used?
(b) How large is the process time delay?

16.4. An open-loop transfer function is given by

$$G(s) = \frac{4(3s + 1)e^{-2s}}{s(10s + 1)}$$

Derive an expression for the amplitude ratio and the phase angle.

16.5. A Bode diagram for a process, valve, and sensor is shown in the drawing.

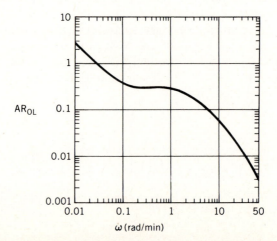

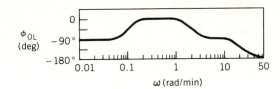

ϕ_{OL} (deg)

ω (rad/min)

(a) Determine an approximate transfer function for this system.

(b) Suppose that a proportional controller is used and that a value of K_c is selected so as to provide a gain margin of 1.7. What is the phase margin?

16.6. Derive the expressions for the closed-loop frequency response characteristics in Eqs. 16-24 and 16-25.

16.7. Consider the storage tank with sightglass in Exercise 6.16 of Chapter 6. The parameter values are $R_1 = 0.5$ min/ft², $R_2 = 2$ min/ft², $A_1 = 10$ ft², $A_2 = 0.8$ ft², $K_v = 2.5$ cfm/mA, $K_m = 1.5$ mA/ft, and $\tau_m = 0.5$ min.

(a) If R_2 were decreased to 0.5 min/ft², what effect would this have on the stability and performance of the control system? (Consider Harriott's gain–bandwidth criterion.)

(b) If PI controller settings are calculated using the Ziegler–Nichols rules, what are the gain and phase margins?

16.8. A process (including valve and transmitter) has the approximate transfer function, $G(s) = 2e^{-0.2s}/(s + 1)$ with time constant and time delay in minutes. Determine PI controller settings and the corresponding gain margins by two methods:

(a) Direct synthesis ($\tau_c = 0.3$ min).

(b) Phase margin $= 40°$ (assume $\tau_I = 0.5$ min).

16.9. For the feedback control system shown in the drawing, do the following:

(a) Plot a Bode diagram for the open-loop system, B/R.

(b) Calculate the value of K_c that will provide a phase margin of 30°.

(c) What is the gain margin when $K_c = 10$?

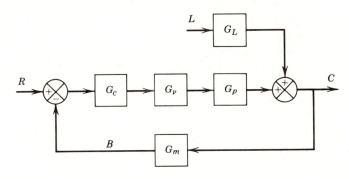

where $G_c = K_c \dfrac{2s + 1}{0.1s + 1}$ $G_v = \dfrac{2}{0.5s + 1}$

$G_p = \dfrac{0.4}{s(5s + 1)}$ $G_L = \dfrac{3}{5s + 1}$

$G_m = 1$

16.10. Frequency response data for a process are tabulated below. These results were obtained by introducing a sinusoidal change in the controller output (under manual control) and recording the measured response of the controlled variable. This procedure was repeated for various frequencies.

(a) If the PI controller is adjusted so that $\tau_I = 0.4$ min, what value of K_c will result in a phase margin of 45°?

(b) If the controller settings in part (a) are used, what is the gain margin?
(c) If the Ziegler–Nichols settings for a PI controller are used, estimate M_p, the peak value of the amplitude ratio for the closed-loop transfer function, $C(s)/R(s)$.

ω (rad/min)	AR	ϕ (deg)
0.01	2.40	−3
0.10	1.25	−12
0.20	0.90	−22
0.5	0.50	−41
1.0	0.29	−60
2.0	0.15	−82
5.0	0.05	−122
10.0	0.017	−173
15.0	0.008	−230

16.11. For the process in Exercise 16.8, the measurement is to be filtered using a noise filter with transfer function $G_F(s) = 1/(0.1s + 1)$. Would you expect this change to result in better or worse control system performance? Justify your argument by using Harriott's gain–bandwidth criterion.

16.12. The block diagram of a conventional feedback control system contains the following transfer functions:

$$G_c = K_c\left(1 + \frac{1}{5s}\right) \qquad G_v = 1$$

$$G_m = \frac{1}{s + 1}$$

$$G_p = G_L = \frac{5e^{-2s}}{10s + 1}$$

(a) Plot the Bode diagram for the open-loop system (G_{OL}).
(b) For what values of K_c is the system stable?
(c) If $K_c = 0.2$, what is the phase margin?
(d) What value of K_c will result in a gain margin of 1.7?

16.13. For the process in Exercise 16.10, calculate the Ziegler–Nichols settings for a PID controller. What are the resulting gain and phase margins?

16.14. The heat exchanger shown in the figure was found to have the following transfer functions (H. S. Wilson and L. M. Zoss, *ISA J.* **9**, 59 (1962)):

Control valve: $\dfrac{X'}{P'_c} = \dfrac{0.047 \text{ in./psi}}{0.083s + 1}$

$\dfrac{W'_s}{X'} = 112 \dfrac{\text{lb}}{\text{min-in.}}$

Process: $\dfrac{T'}{W'_s} = \dfrac{2°\text{F/lb-min}}{(0.432s + 1)(0.017s + 1)}$

Temperature transmitter: $\dfrac{P'_T}{T'} = \dfrac{0.12 \text{ psi/°F}}{0.024s + 1}$

x is the valve lift, measured in inches. Other symbols are defined in the figure.

(a) Find the Ziegler–Nichols settings for a PI controller.
(b) Calculate the corresponding gain and phase margins.

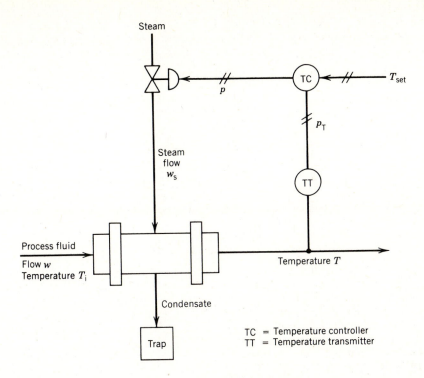

16.15. Consider a standard feedback control system with the following transfer functions:

$$G_m = e^{-0.5s} \qquad G_v = \frac{-10}{s+1}$$

$$G_p = \frac{1.5}{10s+1} \qquad G_L = \frac{2}{6s+1}$$

(a) Plot the Bode diagram for the transfer function $G = G_v G_p G_m$.
(b) Design a PI controller for this process and sketch the asymptotic Bode diagram for the open-loop transfer function $G_{OL} = G_c G$.
(c) Analyze the stability of the resulting feedback control system.
(d) Suppose that under *open-loop* conditions, a sinusoidal set-point change $R(t) = 1.5 \sin 0.5t$ is introduced. What is the amplitude of the measured output signal $B(t)$ that is also sinusoidal in nature?
(e) Repeat the same analysis for closed-loop conditions.
(f) Compare and discuss your results of parts (d) and (e).

PART FIVE

ADVANCED CONTROL TECHNIQUES

— CHAPTER 17 —
Feedforward and Ratio Control

In Chapter 8 it was emphasized that feedback control is an important technique that is widely used in the process industries. Its main advantages are:

1. Corrective action occurs as soon as the controlled variable deviates from the set point, regardless of the source and type of disturbance.
2. It requires minimal knowledge about the process to be controlled; in particular, a mathematical model of the process is *not* required, although it is useful for control system design.
3. The ubiquitous PID controller is both versatile and robust. If process conditions change, re-tuning the controller usually produces satisfactory control.

Feedback control also has certain inherent disadvantages:

1. No corrective action is taken until after a deviation in the controlled variable occurs. Thus, perfect control, where the controlled variable does not deviate from the set point during load or set-point changes, is theoretically impossible.
2. It does not provide predictive control action to compensate for the effects of known or measurable disturbances.
3. It may not be satisfactory for processes with large time constants and/or long time delays. If large and frequent disturbances occur, the process may operate continually in a transient state and never attain the desired steady state.
4. In some applications the controlled variable cannot be measured on-line and, consequently, feedback control is not feasible.

For situations in which feedback control by itself is not satisfactory, significant improvements in control can be achieved by adding feedforward control. But to use feedforward control, the disturbances must be measured (or estimated) on-line. In this chapter, we consider the design and analysis of feedforward control systems. We begin with an overview of feedforward control and ratio control, a special type of feedforward control, is introduced next. Then we discuss design and tuning techniques for feedforward controllers based on steady-state and dynamic models. Finally, alternative configurations for combined feedforward–feedback control systems are examined.

17.1 INTRODUCTION TO FEEDFORWARD CONTROL

With feedforward control the basic idea is to measure important load variables and take corrective action before they upset the process. In contrast, a feedback controller does not take corrective action until after the disturbance has upset the process and generated an error signal. Simplified block diagrams for feedforward and feedback control are shown in Fig. 17.1.

Feedforward control has several disadvantages:

1. The load disturbances must be measured on-line. In many applications this is not feasible.
2. To make effective use of feedforward control, at least a crude process model should be available. In particular, we need to know how the controlled variable responds to changes in both the load and manipulated variables. The quality of feedforward control depends on the accuracy of the process model.
3. Ideal feedforward controllers that are theoretically capable of achieving perfect control may not be physically realizable. Fortunately, practical approximations of these ideal controllers often provide very effective control.

Feedforward control was not widely used in the process industries until the 1960s [1,2]. However, the basic concept is much older and was applied as early as 1925 in the three-element level control system for boiler drums. We will use this control application to illustrate the use of feedforward control.

The boiler drum and a conventional feedback control system are shown in Fig. 17.2. The level of the boiling liquid is measured and used to adjust the feedwater flow rate. This control system tends to be quite sensitive to rapid changes in the load variable, steam flow rate, due to the small liquid capacity of the boiler drum. Rapid load changes can occur due to steam demands made by downstream processing units. Another difficulty is that large controller gains cannot be used since level measurements exhibit rapid fluctuations for boiling liquids. Thus, a high controller gain would tend to amplify the measurement noise and produce unacceptable variations in the feedwater flow rate.

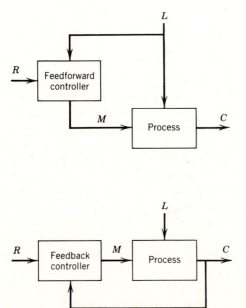

Figure 17.1. Block diagrams for feedforward and feedback control.

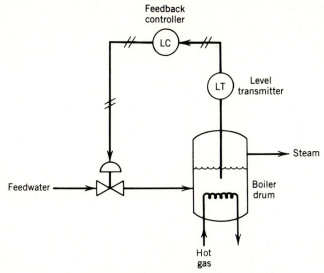

Figure 17.2. The feedback control of the liquid level in a boiler drum.

The feedforward control scheme in Fig. 17.3 can provide better control of the liquid level. Here the steam flow rate is measured and the feedforward controller adjusts the feedwater flow rate so as to balance the steam demand. Note that the controlled variable, liquid level, is not measured. As an alternative, steam pressure could be measured instead of steam flow rate.

In practical applications, feedforward control is normally used in combination with feedback control. Feedforward control is used to reduce the effects of measurable disturbances while *feedback trim* compensates for inaccuracies in the process model, measurement errors, and unmeasured disturbances. The feedforward and feedback controllers can be combined in several different ways, as discussed in Section 17.6. One popular configuration is shown in Fig. 17.4, where the outputs of the feedforward and feedback controllers are added together and the combined signal is sent to the control valve.

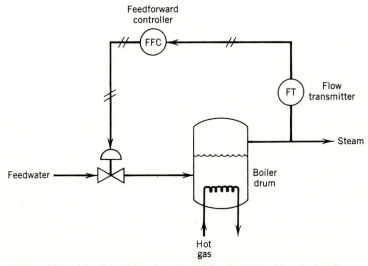

Figure 17.3. The feedforward control of the liquid level in a boiler drum.

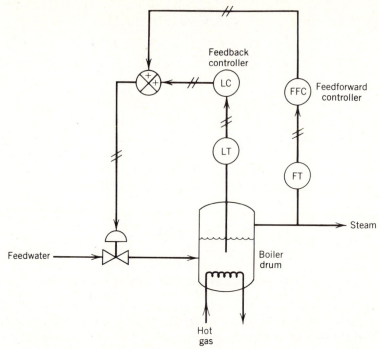

Figure 17.4. The feedforward–feedback control of the boiler drum level.

17.2 RATIO CONTROL

Ratio control is a special type of feedforward control that has been widely used in the process industries. In ratio control, the objective is to maintain the ratio of two variables at a specified value. Thus, the variable R_a, the actual ratio of the two process variables,

$$R_a = \frac{M}{L} \qquad (17\text{-}1)$$

is controlled rather than controlling the individual variables, M and L. The process variables are usually flow rates. The calculation of R_a in Eq. 17-1 is performed in terms of the original physical variables, rather than deviation variables.

Typical applications of ratio control include (1) blending operations, (2) maintaining a stoichiometric ratio of reactants to a reactor, (3) keeping a specified reflux ratio for a distillation column, and (4) holding the fuel–air ratio to a furnace at the optimum value.

Ratio control can be implemented in two basic schemes. In Method I, which is shown in Fig. 17.5, the flow rates for both the load stream and the manipulated stream are measured and the calculated ratio $R_m = M_m/L_m$ is computed using a *divider* element. Special computing elements such as dividers and multipliers are available as off-the-shelf items for both pneumatic and electronic control systems. The output of the divider is sent to a ratio controller (RC) which compares the calculated ratio R_m to the desired ratio R_d and adjusts the manipulated flow M accordingly. The ratio controller would normally be a PI controller with a set point equal to the desired ratio.

The main advantage of Method I is that the actual ratio R_m is calculated. A key disadvantage is that a divider element must be included in the loop and this

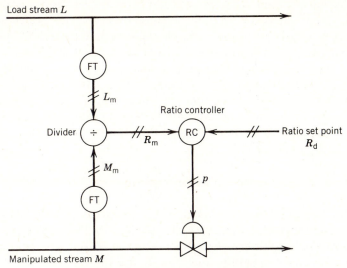

Figure 17.5. Ratio control, Method I.

element makes the process gain vary in a nonlinear fashion. From Eq. 17-1, the process gain is

$$K_p = \left(\frac{\partial R_a}{\partial M}\right)_L = \frac{1}{L} \qquad (17\text{-}2)$$

which is inversely related to load flow rate L. Because of this significant disadvantage, the preferred scheme for implementing ratio control is Method II, which is shown in Fig. 17.6.

In Method II the flow rate of the load stream is measured and the measurement is transmitted to the ratio station (RS), which multiplies this signal by an adjustable

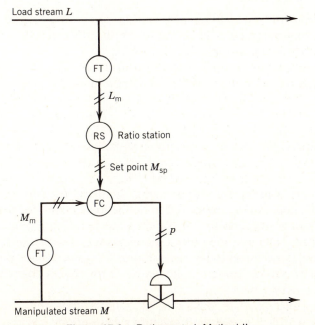

Figure 17.6. Ratio control, Method II.

gain. The value of this gain K_R is the desired ratio. The output signal from the ratio station is then used as the set point for the flow controller, which adjusts the flow rate of the manipulated stream. The chief advantage of Method II is that the open-loop gain remains constant since a divider is not used.

Note that load variable L is measured in both ratio control schemes (cf. Figs. 17.5 and 17.6). Thus, ratio control is, in essence, a very simple form of feedforward control. Regardless of how ratio control is implemented, the process variables must be scaled correctly. For example, in Method II the gain setting for the ratio station must take into account the spans of the two flow transmitters. Thus, the gain for the ratio station K_R should be set at

$$K_R = R_d \frac{K_m}{K_t} \tag{17-3}$$

where R_d is the desired ratio, and K_m and K_t are the spans of the flow transmitters for the manipulated and load streams, respectively. If orifice plates are used with differential-pressure transmitters, then the transmitter output is proportional to the square of the flow. Consequently, the gain of the ratio station should then be proportional to R_d^2 rather than R_d, unless square root extractors are used to convert each transmitter output to a signal that is proportional to flow (see Exercise 17.2).

EXAMPLE 17.1

A ratio control scheme is to be used to maintain a stoichiometric ratio of H_2 and N_2 as the feed to an ammonia synthesis reactor. Individual flow controllers will be used for both the H_2 and N_2 streams. Using the information given below, do the following:

(a) Draw a schematic diagram for the ratio control scheme.
(b) Specify the appropriate gain for the ratio station, K_R.

Available Information:

i. The electronic flow transmitters have built-in square root extractors. The spans of the flow transmitters are 30 L/min for H_2 and 15 L/min for N_2.
ii. The control valves have pneumatic actuators.
iii. Each required current-to-pressure (I/P) transducer has a gain of 0.75 psi/mA.
iv. The ratio station is an electronic instrument with 4–20 mA input and output signals.

Solution

The stoichiometric equation for the ammonia synthesis reaction is

$$3H_2 + N_2 \rightleftarrows 2NH_3$$

(a) The schematic diagram for the ammonia synthesis reacton is shown in Fig. 17.7. The H_2 flow rate is considered to be the load stream for purpose of ratio control, although this choice is arbitrary since both the H_2 and N_2 flow rates are controlled. Note that the ratio station is merely a device with an adjustable gain K_R. The input signal to the ratio station is L_m, the measured H_2 flow rate. The output signal M_{sp} serves as the set point for the N_2 flow control loop. It is calculated as $M_{sp} = K_R L_m$.
(b) From the stoichiometric equation, it follows that the desired ratio is $R_d = M/L = 1/3$. Because the transmitter gains are inversely proportional to

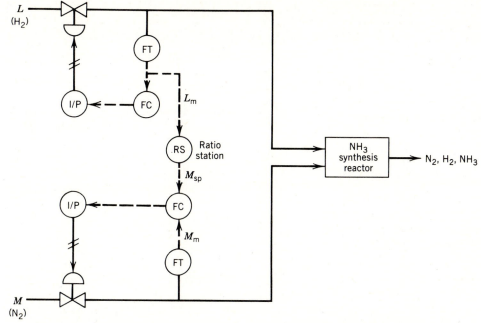

Figure 17.7. Ratio control scheme for an ammonia synthesis reactor of Example 17.1.

the spans, (cf. Section 9.1), Eq. 17-3 can be written in terms of spans S_L and S_m:

$$K_R = R_d \frac{S_L}{S_m}$$

$$K_R = \left(\frac{1}{3}\right)\left(\frac{30 \text{ L/min}}{15 \text{ L/min}}\right) = \frac{2}{3}$$

17.3 FEEDFORWARD CONTROLLER DESIGN BASED ON STEADY-STATE MODELS

A useful interpretation of feedforward control is that it continually attempts to balance the material or energy that must be delivered to the process against the demands of the load [1]. For example, the level control system in Fig. 17.3 adjusts the feedwater flow so that it balances the steam demand. Thus, it is natural to base the feedforward control calculations on material and energy balances. For simplicity, we will first consider controller designs based on steady-state balances using physical variables rather than deviation variables. Design methods based on dynamic models are considered in the next section.

To illustrate the design procedure, consider the distillation column shown in Fig. 17.8 which is used to separate a binary mixture. Feedforward control has gained widespread acceptance for distillation column control [2,3] due to the slow responses that typically occur with feedback control. In Fig. 17.8, the symbols B, D, and F denote molar flow rates while x, y, and z are the mole fractions of the more volatile component. The objective is to control y, despite measurable disturbances in feed flow rate F and feed composition z, by adjusting distillate flow rate D. It is assumed that measurements of x and y are not available.

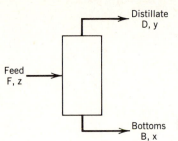

Figure 17.8. A simple schematic diagram of a distillation column.

The steady-state mass balances for the distillation column can be written as

$$F = D + B \tag{17-4}$$

$$Fz = Dy + Bx \tag{17-5}$$

Combining gives

$$D = \frac{F(z - x)}{y - x} \tag{17-6}$$

Since x and y are not measured, we replace these variables by their set points to yield the feedforward control law:

$$D = \frac{F(z - x_{sp})}{y_{sp} - x_{sp}} \tag{17-7}$$

Thus, the feedforward controller calculates the desired value of the manipulated variable D from measurements of the load variables, F and z, and knowledge of the composition set points x_{sp} and y_{sp}. Note that the control law is nonlinear due to the product of F and z.

Stirred-Tank Heater System

In order to illustrate the design method, consider the stirred-tank heater and the feedforward control system shown in Fig. 17.9. We wish to design a feedforward control scheme to maintain exit temperature T at its set point T_{sp}, despite disturbances in T_i and w. It is assumed that T_i and w are measured but T is not. If T were measured, then *feedback* control would be possible. The manipulated variable is the steam pressure P_s. It is also assumed that the volume of liquid in the tank is kept constant via an overflow line (not shown). Due to the overflow line, the inlet and exit flow rates to the tank are identical, as indicated in Fig. 17.9. Note that the feedforward controller has three input signals: measurements of w and T_i from the two transmitters plus the set point for the exit temperature T_{sp}. The set point could be entered manually or could come from another controller, as is discussed in Section 17.6.

The starting point for the controller design is the energy balance for the stirred tank that was derived in Chapter 10. The steady-state version of Eq. 10-8 is

$$0 = \overline{w}C(\overline{T}_i - \overline{T}) + U_A(a + b\overline{P}_s - \overline{T}) \tag{17-8}$$

where the bar over the variable denotes a steady-state value. Solving for $\overline{P}_s$ gives

$$\overline{P}_s = \frac{U_A(\overline{T} - a) + \overline{w}C(\overline{T} - \overline{T}_i)}{U_A b} \tag{17-9}$$

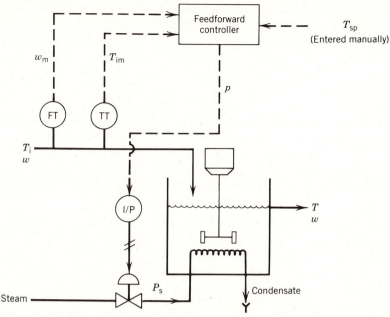

Figure 17.9. The feedforward control of the exit temperature in a continuous stirred-tank heater.

Equation 17-9 provides the basis for the feedforward control law,

$$P_s = K_1 T_{sp} + K_2 w T_{sp} - K_2 w T_i - K_3 \qquad (17\text{-}10)$$

where constants K_1, K_2 and K_3 are defined as

$$K_1 = \frac{1}{b} \qquad (17\text{-}11)$$

$$K_2 = \frac{C}{U_A b} \qquad (17\text{-}12)$$

$$K_3 = \frac{a}{b} \qquad (17\text{-}13)$$

Equation 17-10 was derived from (17-9) by making three modifications:

1. The steady-state values, $\overline{w}$, $\overline{T}$, and $\overline{P}_s$ were replaced by the current values, w, T, and P_s.
2. T was replaced by T_{sp}. (Recall that T is not measured.)
3. The control law was written in a more compact form by defining K_1, K_2, and K_3.

Note that the control law in Eq. 17-10 is nonlinear because it includes the product of w with T_i. This control law could be implemented either digitally or by using standard electronic (or pneumatic) analog components such as summers and multipliers.

The feedforward control law in Eq. 17-10 is not in the final form required for actual implementation since it ignores two important instrumentation considerations: (1) The actual values of T_i and w are not available; consequently, the measured values, T_{im} and w_m, must be used. (2) The controller output signal is p rather than steam pressure P_s. Thus, the feedforward control law should be expressed in terms of p rather than P_s.

Consequently, a more realistic feedforward control law should incorporate the following steady-state instrument relations (cf. Section 9.1):

Transmitters

$$w_m = K_{t1}(w - w_0) + 4 \tag{17-14}$$

$$T_{im} = K_{t2}(T_i - T_{i0}) + 4 \tag{17-15}$$

where K_{t1} and K_{t2} are the steady-state transmitter gains for each measurement and w_0 and T_{i0} denote the zeros of the transmitters, that is, the values of w and T_i that correspond to the minimum transmitter output signal of 4 ma.

Control valve and current-to-pressure transducer

$$P_s = K_v K_{IP}(p - 4) + P_{s0} \tag{17-16}$$

K_v and K_{IP} are the steady-state gains for the control valve and I/P transducer, respectively, while P_{s0} is the steam flow rate that corresponds to the minimum output signal of 3 psi from the I/P transducer. Note that all of the symbols in Eqs. 17-8 through 17-16 denote actual physical variables rather than deviation variables.

Rearranging Eqs. 17-14 and 17-15 gives

$$w = \frac{w_m - 4}{K_{t1}} + w_0 \tag{17-17}$$

$$T_i = \frac{T_{im} - 4}{K_{t2}} + T_{i0} \tag{17-18}$$

Substituting (17-16) through (17-18) into (17-10) and rearranging gives

$$p = K_4 T_{sp} + K_5 w_m T_{sp} - K_6 w_m T_{im} - K_7 w_m - K_8 T_{im} - K_9 \tag{17-19}$$

where

$$K_4 = \frac{1}{bK_v K_{IP}} + \frac{C}{bU_A K_{t1} K_v K_{IP}}(K_{t1} w_0 - 4) \tag{17-20}$$

$$K_5 = \frac{C}{bU_A K_{t1} K_v K_{IP}} \tag{17-21}$$

$$K_6 = \frac{K_5}{K_{t2}} \tag{17-22}$$

$$K_7 = K_6(K_{t2} T_{i0} - 4) \tag{17-23a}$$

$$K_8 = K_6(K_{t1} w_0 - 4) \tag{17-23b}$$

$$K_9 = \frac{K_7 K_8}{K_6} + \frac{a}{bK_v K_{IP}} + \frac{P_{s0}}{K_v K_{IP}} - 4 \tag{17-24}$$

An alternative feedforward control scheme for the stirred-tank heating system is shown in Fig. 17.10. Here the feedforward controller output signal serves as a set point to a feedback controller for steam pressure P_s. The advantage of this configuration is that it is less sensitive to valve sticking and upstream fluctuations in steam pressure. Since the feedforward controller calculates the P_s set point rather than the pneumatic signal to the control valve p, it is not necessary to incorporate Eq. 17-16 into the feedforward control law.

The stirred-tank heater and distillation column examples illustrate that feed-

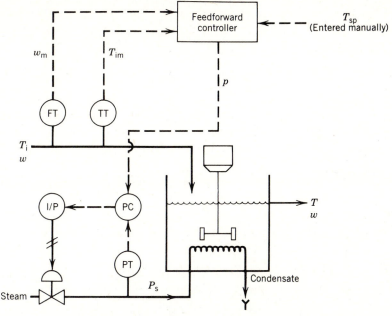

Figure 17.10. Feedforward control with a feedback control loop for steam pressure.

forward controllers can be designed using steady-state mass and energy balances. The advantages of this approach are that the required calculations are quite simple and a detailed process model is not required. However, a disadvantage is that process dynamics are neglected and consequently the control system may not perform well during transient conditions. The feedforward controllers can be improved by adding dynamic compensation, usually in the form of a lead–lag unit. This topic is discussed in Section 17.4. An alternative approach is to base the controller design on a dynamic model of the process, as discussed in the next section.

17.4 FEEDFORWARD CONTROLLER DESIGN BASED ON DYNAMIC MODELS

To illustrate how dynamic models can be used to design a feedforward control system, consider the block diagram shown in Fig. 17.11, which contains a single disturbance. This diagram is similar to Fig. 10.8 for feedback control but an additional signal path through G_t and G_f has been added. The load transmitter with transfer function G_t sends a measurement of the load variable to the feedforward controller G_f. The outputs of the feedforward and feedback controllers are then added together and the sum is sent to the control valve. In contrast to steady-state feedforward control, the block diagram in Fig. 17.11 is based on deviation variables.

The closed-loop transfer function for load changes in Eq. 17-25 can be easily derived using block diagram algebra:

$$\frac{C(s)}{L(s)} = \frac{G_L + G_t G_f G_v G_p}{1 + G_c G_v G_p G_m} \tag{17-25}$$

Ideally, we would like the control system to produce "perfect" control where the controlled variable remains exactly at the set point despite arbitrary changes in the

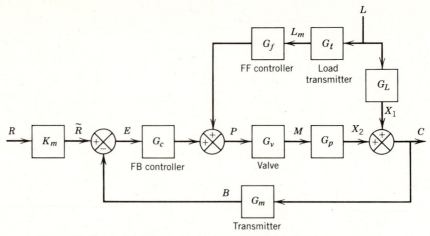

Figure 17.11.   A block diagram of a feedforward–feedback control system.

load variable L. Thus, if the set point is constant ($R(s) = 0$), we want $C(s) = 0$ even though $L(s) \neq 0$. Equation 17-25 indicates that this condition is satisfied if

$$G_L + G_t G_f G_v G_p = 0 \qquad (17\text{-}26)$$

Solving for G_f gives the ideal feedforward controller,

$$G_f = -\frac{G_L}{G_t G_v G_p} \qquad (17\text{-}27)$$

Figure 17.11 and Eq. 17-26 provide a useful interpretation of the ideal feedforward controller. Figure 17.11 indicates that a load disturbance has two effects: It tends to upset the process via the load transfer function, G_L; however, a corrective action is generated via the path through $G_t G_f G_v G_p$. Ideally, the corrective action compensates exactly for the upset so that signals X_1 and X_2 cancel each other and $C(s) = 0$.

Next, we consider several specific examples. In each example it is assumed for simplicity that the load transmitter and the control valve have negligible dynamics, that is, $G_t(s) = K_t$ and $G_v(s) = K_v$, where K_t and K_v denote steady-state gains. In the next three examples we derive feedforward controllers for various types of process models.

EXAMPLE 17.2

Suppose that

$$G_L = \frac{K_L}{\tau_L s + 1}, \qquad G_p = \frac{K_p}{\tau_p s + 1} \qquad (17\text{-}28)$$

Then from (17-27), the ideal feedforward controller is

$$G_f = -\left(\frac{K_L}{K_t K_v K_p}\right)\left(\frac{\tau_p s + 1}{\tau_L s + 1}\right) \qquad (17\text{-}29)$$

This controller is a lead–lag unit with a gain given by $K_f = -K_L/K_t K_v K_p$. The dynamic response characteristics of lead–lag units were considered in Example 6.1 of Chapter 6.

EXAMPLE 17.3

Now consider

$$G_L = \frac{K_L}{\tau_L s + 1}, \qquad G_p = \frac{K_p e^{-\theta s}}{\tau_p s + 1} \qquad (17\text{-}30)$$

From (17-17),

$$G_f = -\left(\frac{K_L}{K_t K_v K_p}\right)\left(\frac{\tau_p s + 1}{\tau_L s + 1}\right) e^{+\theta s} \qquad (17\text{-}31)$$

Since the term $e^{+\theta s}$ implies a negative time delay or a predictive element, the ideal feedforward controller in (17-31) is *physically unrealizable*. However, we can approximate the ideal controller by omitting the exponential term and perhaps adjusting the numerical values of the time constants for the lead and lag terms.

EXAMPLE 17.4

Finally, if

$$G_L = \frac{K_L}{\tau_L s + 1}, \qquad G_p = \frac{K_p}{(\tau_{p1} s + 1)(\tau_{p2} s + 1)} \qquad (17\text{-}32)$$

then the ideal feedforward controller,

$$G_f = -\left(\frac{K_L}{K_t K_v K_p}\right)\frac{(\tau_{p1} s + 1)(\tau_{p2} s + 1)}{(\tau_L s + 1)} \qquad (17\text{-}33)$$

is physically unrealizable since the numerator is a higher order polynomial in s than the denominator. Again, we could approximate this controller by a physically realizable one such as a lead–lag unit, where the lead term contains the sum of two time constants, $\tau_{p1} + \tau_{p2}$.

Stability Considerations

To analyze the stability of the closed-loop system in Fig. 17.11, we consider the closed-loop transfer function in Eq. 17-25. Setting the denominator equal to zero gives the characteristic equation,

$$1 + G_c G_v G_p G_m = 0 \qquad (17\text{-}34)$$

In Chapter 11 it was shown that the roots of the characteristic equation completely determine the stability of the closed-loop system. Since G_f does not appear in the characteristic equation, the feedforward controller has no effect on the stability of the feedback control system. This is a desirable situation which allows the feedback and feedforward controllers to be tuned individually.

Lead–Lag Units

The three examples in the previous section have demonstrated that lead–lag units can provide reasonable approximations to ideal feedforward controllers. Lead–lag units are easily implemented either digitally (see Exercise 17.11) or by using off-the-shelf analog components. Thus, if the feedforward controller consists of a lead–

lag unit with gain K_f we can write

$$G_f(s) = \frac{M(s)}{L(s)} = \frac{K_f(\tau_1 s + 1)}{\tau_2 s + 1} \qquad (17\text{-}35)$$

where K_f, τ_1, and τ_2 are adjustable parameters. For the controller to be stable, $\tau_2 > 0$. In the next section, we consider tuning procedures for this type of feedforward controller.

EXAMPLE 17.5

Consider the stirred-tank heating system of Sections 17.3 and 10.1. A feedforward–feedback control system is employed to compensate for load changes in inlet temperature T_i and mass flow rate w. Using the available information shown below, design the following control systems and compare the closed-loop responses for a $+2.3$ °F step change in T_i:

(a) Feedforward control based on a steady-state energy model.
(b) Feedforward control based on the dynamic model in Fig. 10.7 except that $G_m(s) = K_m/(\tau_m s + 1)$ and w can vary with time.
(c) Same as part (b) except that the feedforward controller is replaced by a PID feedback control with Z–N tuning based on "some overshoot" (See Table 13.2).
(d) A combination of feedforward and feedback controllers based on the controllers in parts (b) and (c).

Available Information

$m = 7.6$ lb	$P_{s0} = 14.7$ psi
$\overline{w} = 0.11$ lb/s	$a = 200.5$ °F
$C = 1.0$ Btu/lb °F	$b = 1.66$ °F/psi
$\overline{T}_i = 70$ °F	$h_p = 200$ Btu/ft²h °F
$T_{sp} = 95$ °F	$h_s = 2700$ Btu/ft²h °F
$U_A = 0.0167$ Btu/s °F	$K_m = 0.16$ mA/°F
$\tau_v = 5s$	$A_p = 0.330$ ft²
$\tau_m = 20s$	$A_s = 0.249$ ft²

$\tau_t = 20s$ (for the temperature transmitter and the flow transmitter with its noise filter)

Feed flow range: 0 to 0.2 lb/s
Inlet temperature range: 50 to 150 °F
Instrument signals: 4 to 20 mA, 3 to 15 psig
Steam supply pressure: 40 psig

Solution

(a) Using the given information, we can calculate the following steady-state gains:

$K_{IP} = 0.7$ psi/mA	$K_v = 3.33$
$K_{t1} = 80$ mA/lb/s	$K_{t2} = 0.16$ mA/°F

Substitution into Eqs. 17-19 to 17-24 gives the following feedforward control law:

$$p = 0.482 T_{sp} + 0.181 w_m T_{sp} - 1.13 w_m T_{im} \qquad (17\text{-}36)$$
$$- 4.62 w_m + 4.52 T_{im} - 35.1$$

(b) A block diagram for the feedforward–feedback control system is shown in Fig. 17.12. Note that there are two individual feedforward controllers: G_{f1} to com-

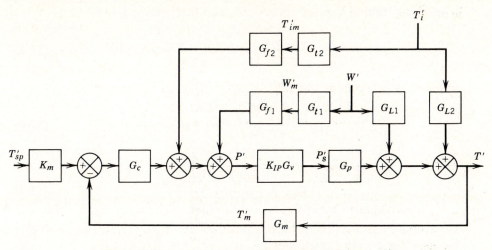

Figure 17.12. A block diagram for Example 17.5.

pensate for feed flow disturbances and G_{f2} to compensate for inlet temperature disturbances. Since the system is based on a linearized model developed below, the effects of the two disturbances in Fig. 17.12 are additive, which is consistent with the Principle of Superposition. The following expressions for the ideal feedforward control laws can be derived in analogy with Eq. 17-27:

$$G_{f1} = -\frac{G_{L1}}{K_{IP}G_{t1}G_vG_p} \tag{17-37}$$

$$G_{f2} = -\frac{G_{L2}}{K_{IP}G_{t2}G_vG_p} \tag{17-38}$$

The following transfer functions can be determined from the given information and Fig. 10.7:

$$G_{t1} = \frac{K_{t1}}{\tau_t s + 1} = \frac{80}{20s + 1} \tag{17-39}$$

$$G_{t2} = \frac{K_{t2}}{\tau_t s + 1} = \frac{0.16}{20s + 1} \tag{17-40}$$

$$G_v = \frac{K_v}{\tau_v s + 1} = \frac{3.57}{5s + 1} \tag{17-41}$$

Expressions for transfer functions G_p and G_{L2} were derived in Eqs. 10-14 and 10-15. Substituting numerical values gives

$$G_p = \frac{0.128}{60s + 1} \tag{17-42}$$

and

$$G_{L2} = \frac{0.868}{60s + 1} \tag{17-43}$$

To derive an expression for G_{L1}, we consider the energy balance of Eq. 10-8 and assume that $T_i = \overline{T}_i$:

$$mC\frac{dT}{dt} = wC(\overline{T}_i - T) + U_A(a + bP_s - T) \tag{17-44}$$

Linearizing, taking Laplace transforms, and rearranging gives

$$T'(s) = \frac{K_p}{\tau_p s + 1} P'(s) + \frac{K_{L1}}{\tau_p s + 1} W'(s) \qquad (17\text{-}45)$$

where $\tau_p = 60$ seconds and

$$K_{L1} = \frac{C(\overline{T}_i - \overline{T})}{\overline{w}C + U_A} = -197.4 \frac{°F}{lb/s} \qquad (17\text{-}46)$$

Thus

$$G_{L1} = \frac{T'(s)}{W'(s)} = \frac{-197.4}{60s + 1} \qquad (17\text{-}47)$$

Substituting the individual transfer functions into Eqs. 17-37 and 17-38 gives the ideal feedforward controllers:

$$G_{f1} = 4.52\,(20s + 1)(5s + 1) \qquad (17\text{-}48)$$

$$G_{f2} = -9.94\,(20s + 1)(5s + 1) \qquad (17\text{-}49)$$

Note that neither G_{f1} nor G_{f2} is physically realizable. However, we can approximate both controller transfer functions by a lead–lag unit that has a small time constant in the denominator:

$$G_{f1} = \frac{4.52\,(26s + 1)}{s + 1} \qquad (17\text{-}50)$$

$$G_{f2} = -\frac{9.94\,(26s + 1)}{s + 1} \qquad (17\text{-}51)$$

The numerator time constant, $26s$, was obtained by adding the assumed lag time constant of one second to the sum of the two time constants in Eqs. 17-48 and 17-49, that is, $20 + 5 + 1 = 26$. The small denominator time constant is consistent with practices for approximating ideal derivative control action (see Eq. 8-14).

(c) To find the ultimate gain K_{cu} and ultimate period P_u, we let $G_c = K_c$ and consider the characteristic equation for the system in Fig. 17.12:

$$1 + K_c K_{IP} G_v G_p G_m = 0 \qquad (17\text{-}52)$$

Substituting transfer functions and rearranging gives

$$6000s^3 + 1600s^2 + 85s + 1 + 0.0837K_c = 0 \qquad (17\text{-}53)$$

Substituting $s = j\omega$ and solving the resulting imaginary and real parts as per Section 11.3 gives

$$P_u = \frac{2\pi}{\omega_c} = 52.8 \text{ s} \qquad (17\text{-}54)$$

$$K_{cu} = 248 \qquad (17\text{-}55)$$

From Table 13.2, the Z–N controller settings for "some overshoot" are $K_c = 81.8$, $\tau_I = 26.4$ s, and $\tau_D = 17.6$ s.

(d) The combined feedforward–feedback controller is assumed to have the structure shown in Fig. 17.12 with controller transfer functions G_{f1}, G_{f2}, and G_c previously specified.

Various closed-loop responses to a $+2.3$ °F step change in inlet temperature T_i are shown in Figs. 17.13 and 17.14. In Fig. 17.13, the feedforward controller based on a steady-state model (case a) is quite sluggish in comparison with the standard PID controller (case c). It is likely that better performance could be obtained by adding dynamic compensation (e.g., lead–lag unit).

A comparison of Fig. 17.14 with Fig. 17.13 indicates that feedforward controllers designed using a dynamic model of the process (cases b and d) can result in significantly improved load responses even though the ideal feedforward controllers are approximated by lead–lag units. In particular, the combined feedforward–feedback controller (case d) is very effective. A comparison of the controller output signals in the bottom portions of Figs. 17.13 and 17.14 shows that the improved performance does not require excessive control action since the required controller output responses in Fig. 17.14 are comparable in magnitude to those in Fig. 17.13.

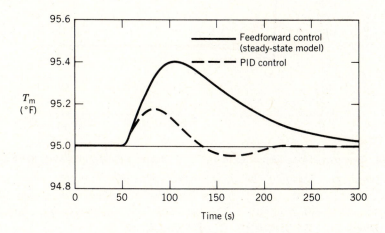

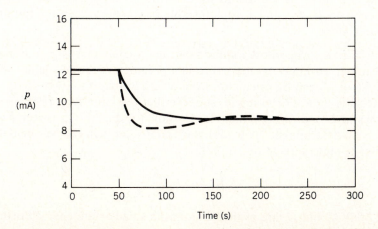

Figure 17.13. The closed-loop responses to a $+2.3$ °F change in feed temperature (cases a and c).

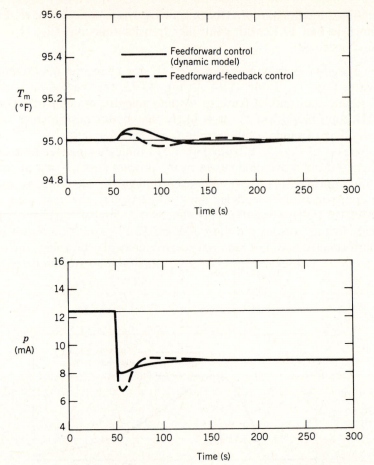

Figure 17.14. The closed-loop responses to a +2.3 °F change in feed temperature (cases b and d).

17.5 TUNING FEEDFORWARD CONTROLLERS

Feedforward controllers, like feedback controllers, usually require field tuning after installation in a plant. If the feedforward controller consists of the lead–lag unit in Eq. 17-35 with K_f, τ_1, and τ_2 as adjustable parameters, then the tuning can be done in three steps.

Step 1. *Adjust* K_f. The effort required to tune a controller is greatly reduced if good initial estimates of the controller parameters are available. An initial estimate of K_f can be calculated from a steady-state model of the process or steady-state data. For example, suppose that the open-loop responses to step changes in L and M are available, as shown in Fig. 17.15. After K_p and K_L have been determined, the feedforward controller gain can be calculated from the steady-state version of Eq. 17-27:

$$K_f = -\frac{K_L}{K_t K_v K_p} \qquad (17\text{-}56)$$

In (17-56) K_t and K_v are available from the steady-state characteristics of the transmitter and control valve.

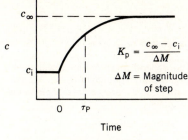

(a) Step change in M

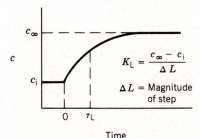

(b) Step change in L

Figure 17.15. The open-loop responses to step changes in M and L.

To tune the controller gain, K_f is set at the initial estimate and a small step change (3–5%) in the load variable L is introduced. If an offset results, then K_f is adjusted until the offset is eliminated. While K_f is being tuned, τ_1 and τ_2 should be set equal to their minimum values, ideally zero.

Step 2. *Determine initial values for τ_1 and τ_2.* Theoretical values for τ_1 and τ_2 can be calculated if a dynamic model of the process is available, as in Example 17.2. Alternatively, initial estimates can be determined from open-loop response data. For example, if the step responses have the shape shown in Fig. 17.15, a reasonable process model is

$$G_p(s) = \frac{K_p}{\tau_p s + 1} , \qquad G_L(s) = \frac{K_L}{\tau_L s + 1} \qquad (17\text{-}57)$$

where τ_p and τ_L can be calculated as shown in Fig. 17.15. A comparison of Eqs. 17-29 and 17-35 leads to the following expressions for τ_1 and τ_2:

$$\tau_1 = \tau_p \qquad (17\text{-}58)$$

$$\tau_2 = \tau_L \qquad (17\text{-}59)$$

These values can then be used as initial estimates for the fine tuning of τ_1 and τ_2 in Step 3.

If neither a process model nor experimental data are available, the relations $\tau_1/\tau_2 = 2$ or $\tau_1/\tau_2 = 0.5$ may be used, depending on whether the controlled variable responds faster to the load variable or to the manipulated variable. In view of Eq. 17-58, τ_1 should be set equal to the estimated dominant process time constant. To deal with complicated process interactions, for example, where time delay compensation is required, the more elaborate technique of Shinskey [1, 278–281] should be used.

Step 3. *Fine tune τ_1 and τ_2.* The final step is to use a trial-and-error procedure to fine tune τ_1 and τ_2 by making small step changes in L. The desired step response

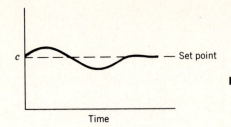

Figure 17.16. The desired response for a well-tuned feedforward controller. (Note approximately equal areas above and below the set point.)

consists of small deviations in the controlled variable with equal areas above and below the set point [1], as shown in Fig. 17.16. For simple process models, it can be proved theoretically that equal areas above and below the set point imply that the difference, $\tau_1 - \tau_2$, is correct (see Exercise 17.12). In subsequent tuning to reduce the size of the areas, τ_1 and τ_2 should be adjusted so that $\tau_1 - \tau_2$ remains constant.

As a hypothetical illustration of this trial-and-error tuning procedure, consider the set of responses shown in Fig. 17.17 for positive step changes in load variable L. It is assumed that $K_p > 0$, $K_L < 0$, and controller gain K_f has already been adjusted so that offset is eliminated. For the initial values of τ_1 and τ_2 in Fig. 17.17a, the controlled variable is below the set point, which implies that τ_1 should be increased to speed up the corrective action. (Recall that $K_p > 0$, $K_L < 0$, and that positive step changes in L are introduced.) Increasing τ_1 to a value of 2 gives the response in Fig. 17.17b, which has equal areas above and below the set point. Thus, in subsequent tuning to reduce the size of each area, $\tau_1 - \tau_2$ should be kept constant. Increasing both τ_1 and τ_2 by 0.5 reduces the size of each area, as shown in Fig. 17.17c. Since this

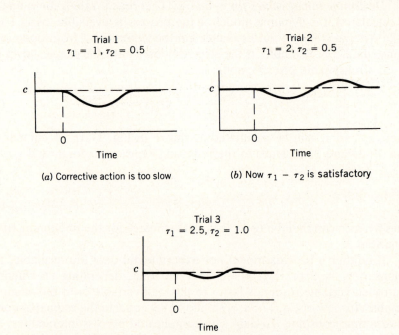

Figure 17.17. An example of feedforward controller tuning.

response is considered to be satisfactory, no further controller tuning is required.

17.6 CONFIGURATIONS FOR FEEDFORWARD–FEEDBACK CONTROL

As mentioned in Section 17.1, *feedback trim* is normally used in conjunction with feedforward control to compensate for modeling errors and unmeasured disturbances. Feedforward and feedback controllers can be combined in several different ways. One possible configuration for feedforward–feedback control is to add the outputs of the feedforward and feedback controllers together and send the resulting signal to the final control element. This configuration was introduced in Figs. 17.4 and 17.11. Its chief advantage is that the feedforward controller theoretically does not affect the stability of the feedback control loop. Recall that the feedforward controller transfer function $G_f(s)$ does not appear in the characteristic equation of Eq. 17-34.

A widely used configuration for feedforward–feedback control is to have the feedback controller output serve as the set point for the feedforward controller. It is especially convenient when the feedforward control law is designed using steady-state material and energy balances. For example, a feedforward–feedback control system for the stirred-tank heater system is shown in Fig. 17.18. Note that this control system is similar to the feedforward scheme in Fig. 17.9 except that the feedforward controller set point is now denoted as T_{sp}^*. It is generated as the output signal from the feedback controller. The actual set point T_{sp} is used as the set point for the feedback controller. In this configuration, the feedforward controller can affect the stability of the feedback control system since it is now an element in the feedback loop. If dynamic compensation is included, it should be introduced outside of the feedback loop. Otherwise it will interfere with the operation of the feedback loop, especially when the controller is placed in the manual mode [1].

A third way of incorporating feedback trim into a feedforward control system is to have the feedback controller output signal adjust the feedforward controller

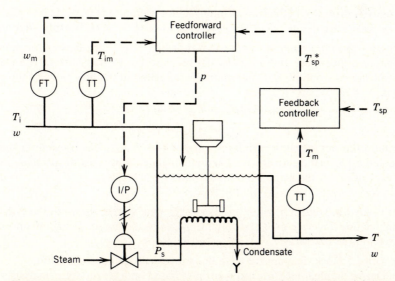

Figure 17.18. A feedforward–feedback control system for a stirred-tank heater.

gain. This configuration is a natural one for applications where the feedforward controller is merely a gain, such as for the ratio control systems of Section 17.2.

SUMMARY

Feedforward control is a powerful strategy for control problems in which important load variables can be measured on-line. By measuring disturbances and taking corrective action *before* the controlled variable is upset, feedforward control can provide dramatic improvements for regulatory control. The chief disadvantage of feedforward control is that the load variable(s) must be measured (or estimated) on-line, which is not always possible. Ratio control is a special type of feedforward control which is useful for applications such as blending operations where the ratio of two process variables is to be controlled.

Feedforward controllers tend to be custom designed for specific applications although lead–lag units are often used as a generic feedforward controller. The design of a feedforward controller requires knowledge of how the controlled variable responds to changes in the manipulated variable and the load variable(s). This knowledge is usually represented as a process model. Steady-state models can be used for controller design; however, it may be necessary to add a lead–lag unit to provide dynamic compensation empirically. Feedforward controllers can also be designed using dynamic models.

Feedforward control is normally implemented in conjunction with feedback control. Tuning procedures for combined feedforward–feedback control schemes have been described in Section 17.6. For these control configurations, the feedforward controller is usually tuned before the feedback controller.

REFERENCES

1. Shinskey, F. G., *Process Control Systems*, 3d ed., McGraw-Hill, New York, 1988, Chapter 7.
2. Nisenfeld, A. E., and R. K. Miyasaki, Applications of Feedforward Control to Distillation Control, *Automatica* **9**, 319 (1973).
3. Shinskey, F. G., *Distillation Control*, 2d ed., McGraw-Hill, New York, 1984, Chapter 6.

EXERCISES

17.1. In ratio control, would the control loop gain for Method I (Fig. 17.5) be less variable if the ratio were defined as $R_a = L/M$ instead of $R_a = M/L$? Justify your answer.

17.2. Consider the ratio control scheme shown in Fig. 17.6. Each flow rate is measured using an orifice plate and a differential pressure (D/P) transmitter. The electrical output signals from the D/P transmitters are related to the flow rates by the expressions

$$L_m = L_{m0} + K_1 L^2$$

$$M_m = M_{m0} + K_2 M^2$$

Each transmitter output signal has a range of 4 to 20 mA. The transmitter spans are denoted by S_L and S_m for the load and manipulated flow rates, respectively. Derive an expression for the gain of the ratio station K_R in terms of S_L, S_m, and the desired ratio R_d.

17.3. It is desired to control liquid h_2 in the storage tank system shown in the drawing by manipulating flow rate q_3. Load variable q_1 can be measured. Using the information below, do the following:

(a) Draw a block diagram for a feedforward–feedback control system.
(b) Derive an ideal feedforward controller based on a steady-state analysis.
(c) Suppose that the flow–head relation for the valve is $q_2 = C\sqrt{h_1 - h_2}$. Does the ideal feedforward controller of part (b) change?

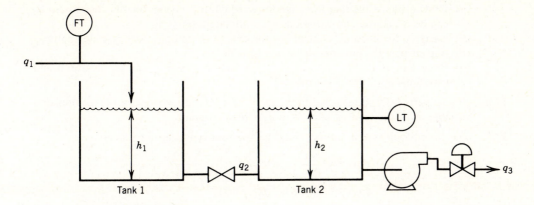

Available Information

   i. The two tanks have uniform cross-sectional areas, A_1 and A_2, respectively.

   ii. The valve on the exit line of Tank 1 acts as a linear resistance with a flow–head relation, $q_2 = (h_1 - h_2)/R$.

   iii. The transmitters and control valve are pneumatic instruments that have negligible dynamics.

   iv. The pump operates so that flow rate q_3 is independent of h_2 when the control valve stem position is maintained constant.

17.4. The following feedforward control law is proposed for the stirred tank heater system and the feedforward–feedback control configuration shown in Fig. 17.18:

$$p(t) = K\overline{w}[T_{sp}^*(t) - T_{im}(t)]$$

where K is an adjustable controller setting and $\overline{w}$ is the nominal value of the flow rate. Do the following:

(a) Draw a block diagram for the controlled process.

(b) Indicate whether the stability of the feedback control loop is influenced by the feedforward controller. Briefly justify your answer.

17.5. For the liquid storage system shown in the drawing, the control objective is to regulate liquid level h_2 despite load disturbances in q_1 and q_4. Flow rate q_2 can be manipulated. The two hand valves have the following flow–head relations:

$$q_3 = C_1\sqrt{h_1} \qquad q_5 = C_2\sqrt{h_2}$$

Do the following assuming that the flow transmitters and the control valve have negligible dynamics:

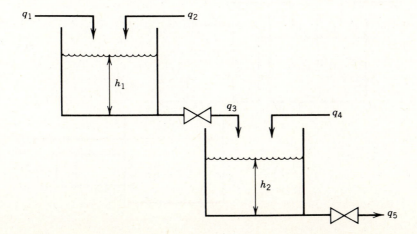

(a) Draw a block diagram for a feedforward control system for the case where q_1 can be measured and variations in q_4 are neglected.

(b) Design a feedforward control law for case (a) based on a steady-state analysis.

(c) Repeat part (b) but consider dynamic behavior.

(d) Repeat parts (a) through (c) for the situation where q_4 can be measured and variations in q_1 are neglected.

17.6. Repeat parts (a) through (c) of Exercise 17.5 for the case where q_4 is the manipulated variable, q_1 is the load variable, and q_2 is held constant.

17.7. The closed-loop system in Fig. 17.11 has the following transfer functions:

$$G_p(s) = \frac{1}{s + 1} \qquad G_L(s) = \frac{2}{(s + 1)(5s + 1)}$$

$$G_v(s) = G_m(s) = G_t(s) = 1$$

(a) Design a feedforward controller based on a steady-state analysis.

(b) Design a feedforward controller based on a dynamic analysis.

(c) Design a feedback controller based on the IMC approach of Chapter 12 and $\tau_c = 2$.

(d) Calculate and plot the closed-loop response to a unit step change in the load variable using feedforward control only and the controllers of parts (a) and (b).

(e) Repeat part (d) for the feedforward–feedback control scheme of Fig. 17.11 and the controllers of parts (a) and (b) as well as (b) and (c).

17.8. The distillation column in Fig. 17.8 has the following transfer function model:

$$\frac{Y(s)}{D(s)} = \frac{K_p e^{-20s}}{95s + 1} \qquad \frac{Y(s)}{F(s)} = \frac{K_L e^{-30s}}{60s + 1}$$

with $G_v = G_m = G_t = 1$. Gains K_p and K_L vary as the process operating conditions change due to the nonlinear nature of the column. Use this information and a steady-state material balance to develop a feedforward control law with dynamic compensation.

17.9. A feedforward control system is to be designed for the two-tank heating system shown in the drawing. The design objective is to regulate temperature T_4 despite

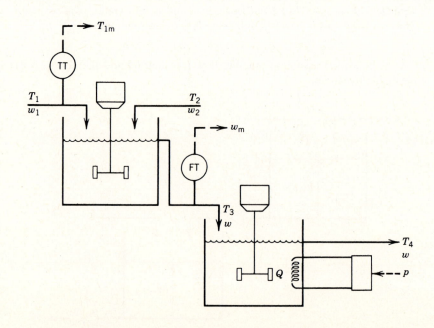

variations in load variables T_1 and w. The voltage signal to the heater, p, is the manipulated variable. Only T_1 and w are measured. Also it can be assumed that the heater and transmitter dynamics are negligible and that the heat duty is linearly related to voltage signal, p.

(a) Design a feedforward controller based on a steady-state analysis. This control law should relate p to T_{1m} and w_m.

(b) Is dynamic compensation desirable? Justify your answer.

17.10. Consider the liquid storage system of Exercise 17.5 but suppose that the hand valve for q_5 is replaced by a pump and a control valve (cf. Fig. 10.22). Repeat parts (a) through (c) of Exercise 17.5 for the situation where q_5 is the manipulated variable and q_2 is constant.

17.11. In practical applications feedforward control is usually implemented via digital control algorithms. Derive a discrete-time (digital) approximation to the lead–lag unit in Eq. 17-35 using a first-order backward difference approximation for derivatives with respect to time (cf. Section 8.3).

17.12. Shinskey [1, p. 279] has stated that in tuning a feedforward controller consisting of the lead–lag unit in Eq. 17-35, the integral of the error signal $\int_0^\infty e(t)\,dt$ is zero when the difference between the lead and lag parameters, $\tau_1 - \tau_2$, is correct. Verify this assertion for the process model of Example 17.2 and a unit step change in the load variable. Assume that the set point is $R = 0$. Assume that $G_t = G_v = 1$.

— CHAPTER 18 —

Advanced Control Strategies

In this chapter we introduce several advanced control strategies that provide improved process control beyond what can be obtained with conventional PID controllers. As processing plants become increasingly complex in order to increase efficiency or reduce costs, there are greater incentives for using advanced control. Although new methods are continually evolving and being field-tested [1,2], this chapter will emphasize six techniques that have been proven commercially:

> Cascade control
> Time-delay compensation
> Selective and override control
> Adaptive control
> Statistical quality control
> Expert systems

Each of these techniques has gained increased industrial acceptance with the advent of computer control. In particular, cascade control is routinely used. Feedforward and multivariable control are also classified as advanced control strategies, but they are discussed in separate chapters (Chapters 17 and 19) because of their importance.

18.1 CASCADE CONTROL

A disadvantage of conventional feedback control is that corrective action for disturbances does not begin until after the controlled variable deviates from the set point. As discussed in Chapter 17, feedforward control offers large improvements over feedback control for processes that have large time constants or time delays. However, feedforward control requires that the disturbances be measured explicitly and a model must be available for calculating the controller output. An alternative approach which improves the dynamic response to load changes is to use a secondary measurement point and a secondary feedback controller. The secondary measurement point is located so that it recognizes the upset condition sooner than the controlled variable, but the disturbance is not necessarily measured. This approach utilizes multiple feedback loops and is called *cascade control*. It is particularly useful when the disturbances are associated with the manipulated variable or when the final control element exhibits nonlinear behavior.

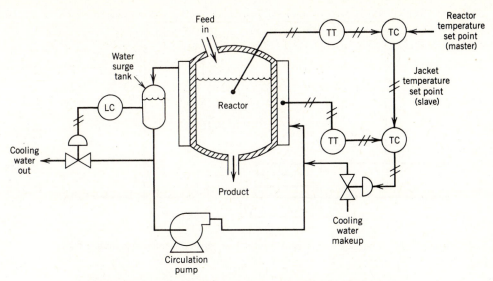

Figure 18.1. Cascade control of an exothermic chemical reactor.

Figure 18.1 shows a stirred chemical reactor where cooling water is passed through the reactor jacket to regulate the reactor temperature. The reactor temperature is affected by changes in disturbance variables such as reactant feed temperature or composition. The simplest control strategy would handle such disturbances satisfactorily by adjusting a control valve on the cooling water inlet stream. However, an increase in the inlet cooling water temperature may cause unsatisfactory performance. The resulting increase in the reactor temperature, due to a reduction in heat removal rate, may occur slowly. If there are appreciable dynamic lags in the jacket as well as in the reactor, the corrective action taken by the controller will be delayed. To circumvent this disadvantage, a feedback controller for the jacket temperature with its set point determined by the reactor temperature controller could be added to provide cascade control, as shown in Fig. 18.1. This approach measures the jacket temperature, compares it to a set point, and uses the resulting error signal as the input to a controller for the cooling water makeup, thus maintaining the heat removal rate from the reactor at a constant level. The controller set point and both measurements are used to adjust a single manipulated variable, the cooling water makeup rate. The principal advantage of the cascade control strategy is that a second measured variable is located close to a potential disturbance to improve the closed-loop response.

Cascade control is widely used in the process industries and has two distinguishing features:

1. The output signal of the *master* controller serves as the set point for the *slave* controller.
2. The two feedback control loops are nested, with the *secondary control loop* (for the slave controller) located inside the *primary control loop* (for the master controller).

In the reactor example the primary measurement is the reactor temperature, which is used by the master controller. The secondary measurement is the jacket temperature, which is transmitted to the slave controller.

As a second example of cascade control, consider the natural draft furnace temperature control problem reported by Johnson [3] and shown in Fig. 18.2. The

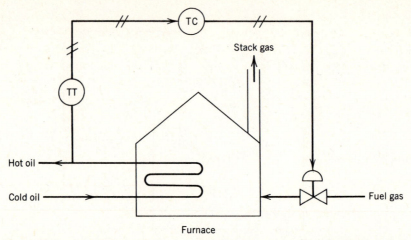

Figure 18.2.   A furnace temperature control that uses conventional feedback control.

conventional feedback control system in Fig. 18.2 may do a satisfactory job of regulating the hot oil temperature despite disturbances in oil flow rate or cold oil temperature. But if a disturbance occurs in the fuel gas supply pressure, the fuel gas flow changes, which upsets the furnace operation, thus changing the hot oil temperature. Only then will the temperature controller (TC) begin to take corrective action by adjusting the fuel gas flow via the control valve. Thus, we anticipate that conventional feedback control would result in very sluggish responses to changes in fuel gas supply pressure. This disturbance is clearly associated with the manipulated variable.

The cascade control scheme of Fig. 18.3 provides improved performance because the control valve will be adjusted as soon as the change in supply pressure is detected. The performance improvements for disturbances in oil flow rate or inlet temperature may not be as large, in which case feedforward control may be desirable for those disturbances. For the cascade control scheme, the master (or primary) controller is the temperature controller that adjusts the set point of the slave (or secondary) controller in the pressure control loop. If a disturbance in

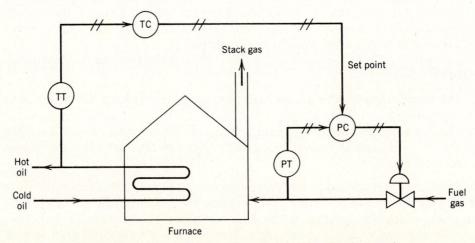

Figure 18.3.   A furnace temperature control using cascade control.

supply pressure occurs, the pressure controller will act very quickly to hold the fuel gas pressure at its set point. Since the pressure control loop responds rapidly, the supply pressure disturbance will have little effect on furnace operation and exit oil temperature. Alternatively, flow control, rather than pressure control, could be employed in the slave loop to achieve essentially the same result.

The block diagram for a general cascade control system is shown in Fig. 18.4. Subscript 1 refers to the primary control loop while subscript 2 refers to the secondary control loop. Thus, for the furnace temperature control example:

C_1 = hot oil temperature

C_2 = fuel gas pressure

L_1 = cold oil temperature (or cold oil flow rate)

L_2 = supply pressure of fuel gas

B_1 = measured value of hot oil temperature

B_2 = measured value of fuel gas pressure

R_1 = set point for C_1

R_2 = set point for C_2

All of the above variables represent deviations from the nominal steady state. Since load disturbances can occur in both the primary and secondary control loops, two load variables (L_1 and L_2) and two load transfer functions (G_{L1} and G_{L2}) are shown in Fig. 18.4.

It is clear from Figs. 18.3 and 18.4 that cascade control will effectively eliminate the effects of pressure disturbances entering the secondary loop (i.e., L_2 in Fig. 18.4). But what about the effects of disturbances such as L_1 which enter the primary loop? Cascade control can provide an improvement over conventional feedback control when both controllers are well tuned. The cascade arrangement will reduce the response time of the elements in the secondary loop, which will in turn affect the primary loop, but the improvement may be slight. As mentioned previously, feedforward control can be employed to reduce errors in C_1, but L_1 must be measured directly and a model relating L_1, C_1, and C_2 should be available.

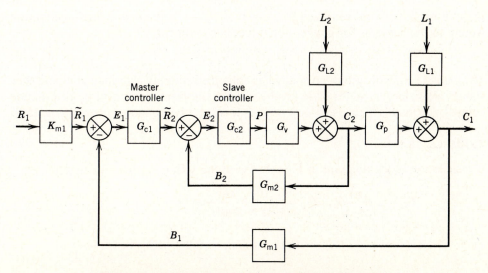

Figure 18.4. Block diagram of the cascade control system.

Design Considerations for Cascade Control

Cascade control can improve the response to a set-point change by using an intermediate measurement point and two feedback controllers. However, its performance for load changes is usually the principal concern [4, p. 212]. In Fig. 18.4 L_2 disturbances are compensated by feedback in the inner loop; the corresponding closed-loop transfer function (assuming $R_1 = L_1 = 0$) is obtained by block diagram algebra:

$$C_1 = G_p C_2 \tag{18-1}$$

$$C_2 = G_{L2} L_2 + G_v G_{c2} E_2 \tag{18-2}$$

$$E_2 = \tilde{R}_2 - B_2 = G_{c1} E_1 - G_{m2} C_2 \tag{18-3}$$

$$E_1 = -G_{m1} C_1 \tag{18-4}$$

Eliminating all variables except C_1 and L_2 gives

$$\frac{C_1}{L_2} = \frac{G_p G_{L2}}{1 + G_{c2} G_v G_{m2} + G_{c1} G_{c2} G_v G_p G_{m1}} \tag{18-5}$$

By similar analysis the servo transfer functions for the outer and inner loops are:

$$\frac{C_1}{R_1} = \frac{G_{c1} G_{c2} G_v G_p K_{m1}}{1 + G_{c2} G_v G_{m2} + G_{c1} G_{c2} G_v G_p G_{m1}} \tag{18-6}$$

$$\frac{C_2}{\tilde{R}_2} = \frac{G_{c2} G_v}{1 + G_{c2} G_v G_{m2}} \tag{18-7}$$

For disturbances in L_1, the load transfer function is

$$\frac{C_1}{L_1} = \frac{G_{L1} (1 + G_{c2} G_v G_{m2})}{1 + G_{c2} G_v G_{m2} + G_{c1} G_{c2} G_v G_p G_{m1}} \tag{18-8}$$

Several observations can be made about the above equations. First, the cascade control system has the characteristic equation

$$1 + G_{c2} G_v G_{m2} + G_{c1} G_{c2} G_v G_p G_{m1} = 0 \tag{18-9}$$

If the inner loop were removed ($G_{c2} = 1$, $G_{m2} = 0$), the characteristic equation would be the same as for conventional feedback control,

$$1 + G_{c1} G_v G_p G_{m1} = 0 \tag{18-10}$$

The cascade control system normally has enhanced stability characteristics and thus should allow larger values of K_{c1} to be used in the primary control loop. Cascade control also makes the closed-loop process less sensitive to model errors [5, p. 156].

EXAMPLE 18.1

Consider the block diagram in Fig. 18.4 with the following transfer functions:

$$G_v = \frac{5}{s + 1} \qquad G_p = \frac{4}{(2s + 1)(4s + 1)}$$

$$G_{L2} = 1 \qquad G_{m1} = 0.05 \qquad G_{m2} = 0.2 \qquad G_{L1} = \frac{1}{3s + 1}$$

where the time constants have units of minutes and the gains have consistent units. Determine the stability limits for a conventional proportional controller as well as for a cascade control system consisting of two proportional controllers. Assume $K_{c2} = 4$ for the secondary controller. Calculate the resulting offset for a unit step change in the secondary load variable L_2.

Solution

For the cascade arrangement, first analyze the inner loop. Substituting in Eq. 18-7 gives

$$\frac{C_2}{\tilde{R}_2} = \frac{4\left(\dfrac{5}{s + 1}\right)}{1 + 4\left(\dfrac{5}{s + 1}\right)(0.2)} = \frac{20}{s + 5} = \frac{4}{0.2s + 1} \qquad (18\text{-}11)$$

From Eq. 18-11 the closed-loop time constant for the inner loop is 0.2 min. In contrast, the conventional feedback control system has a time constant of 1 min since in this case $C_2(s)/\tilde{R}_2(s) = G_v = 5/(s + 1)$. Thus, cascade control significantly speeds up the response of C_2. Using a proportional controller in the primary loop ($G_{c1} = K_{c1}$), the characteristic equation becomes

$$1 + 4\left(\frac{5}{s + 1}\right)(0.2) + (K_{c1})(4)\left(\frac{5}{s + 1}\right)\left(\frac{4}{(2s + 1)(4s + 1)}\right)(0.05) = 0$$

$$(18\text{-}12)$$

which reduces to

$$8s^3 + 46s^2 + 31s + 5 + 4K_{c1} = 0 \qquad (18\text{-}13)$$

By use of the Routh array (Chapter 11), the ultimate controller gain for marginal stability is $K_{c1,u} = 43.3$.

For the conventional feedback system with proportional-only control, the characteristic equation in (18-9) reduces to

$$8s^3 + 14s^2 + 7s + 1 + K_{c1} = 0 \qquad (18\text{-}14)$$

The Routh array gives $K_{c1,u} = 11.25$. Therefore, the cascade configuration has increased the ultimate gain by nearly a factor of four. Increasing K_{c2} will result in even larger values for $K_{c1,u}$. For this example there is no theoretical upper limit for K_{c2} except that large values may cause the valve to saturate for small set-point or load changes.

The offset for C_1 for a unit step change in L_2 can be obtained by setting $s = 0$ in the right side of (18-5); equivalently the Final Value Theorem of Chapter 3 can be applied for a unit step change in L_2 ($R_1 = 0$):

$$e_1(t \to \infty) = r_1 - c_1(t \to \infty) = \frac{-2}{5 + 8K_{c1}} \qquad (18\text{-}15)$$

By comparison, the offset for conventional control ($G_{m2} = 0$, $G_{c2} = 1$) is

$$e_1(t \to \infty) = \frac{-2}{1 + 2K_{c1}} \qquad (18\text{-}16)$$

By comparing (18-15) and (18-16), it is clear that for the same gain K_{c1}, the offset is much smaller (in absolute value) for cascade control.

For a cascade control system to function properly, the response of the secondary control loop should be faster than the primary loop [6]. The secondary controller is normally a P or PI controller, depending upon the amount of offset that would occur with proportional-only control. Note that small offsets in the secondary loop can be tolerated since they will be compensated for by the primary loop. Derivative action is rarely used in the secondary loop. The primary controller is usually a PI or PID controller [7,8].

For processes with higher-order dynamics and/or delay time, the frequency response methods described in Chapter 16 can be employed to design the controllers. First the inner loop frequency response for a set-point change is calculated from (18-7) and a suitable value of K_{c2} is determined. The offset is checked to determine if PI control is required. After K_{c2} is specified, the outer loop frequency response can be calculated, as in conventional feedback controller design. The open-loop transfer function used in this part of the calculation is

$$G_{OL} = \frac{G_{c1}K_{c2}G_v}{1 + K_{c2}G_vG_{m2}} G_p G_{m1} \tag{18-17}$$

For the design of G_{c1} we should consider the closed-loop transfer function for set-point changes (C_1/R_1) as well as the ones for load changes C_1/L_2 and C_1/L_1 [9, p. 158]. The design objective in these last two cases is to reject disturbances; we want C_1/L_2 to be as small as possible for all frequencies. A resonant peak should appear in the amplitude ratio plot, and the controller should be adjusted to minimize this peak and force it to occur at as high a frequency as possible. Generally, cascade control is superior to conventional control in this regard, and also provides superior time domain responses [6,9]. Figures 18.5 and 18.6 show the closed-loop time-domain and frequency-domain responses for Example 18.1 and the disturbance variable L_2. The cascade configuration has a PI controller in the primary loop and a proportional controller in the secondary loop. The PI controllers in each case were tuned using frequency response analysis (see Section 16.6). Figures 18.5 and 18.6 demonstrate that the cascade control system is superior to a conventional PI controller for a secondary load disturbance. Figure 18.7 shows a similar comparison for a step change in the primary loop load variable L_1.

When a cascade control system is tuned after installation, the secondary controller should be tuned first (for set-point changes) with the primary controller in

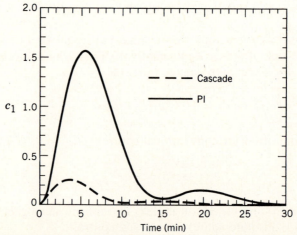

Figure 18.5. A comparison of L_2 responses ($L_2 = 1/s$) for cascade control ($K_{c2} = 4$, $K_{c1} = 3.5$, $\tau_{I1} = 5.3$ min) and conventional PI control ($K_c = 1.9$, $\tau_I = 6$ min).

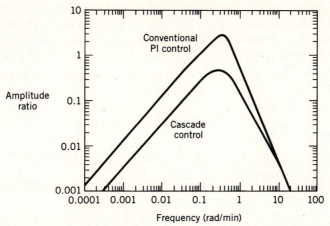

Figure 18.6. A comparison of the closed-loop frequency response (amplitude ratio of C_1/L_2) for secondary load change.

the manual mode. Then the primary controller is transferred to automatic and it is tuned. The on-line tuning techniques presented in Chapter 13 can be used for each control loop. If the secondary controller is retuned for some reason, usually the primary controller must also be retuned. When a cascade control system is transferred from the manual to the automatic mode, the secondary controller should be transferred before the primary controller. The order is reversed when transferring both loops from automatic to the manual mode.

18.2 TIME-DELAY COMPENSATION AND INFERENTIAL CONTROL

In this section we discuss two advanced control techniques that deal with problematic areas in process control; namely, the occurrence of significant time delays and the lack of on-line measurements for feedback control. The first problem is handled by *time-delay compensation* and the second problem can be solved by a technique called *inferential control*.

Time-Delay Compensation

Time delays commonly occur in the process industries because of the presence of distance velocity lags, recycle loops, and the *dead time* associated with composition

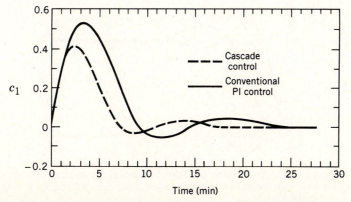

Figure 18.7. A comparison of primary load responses ($L_1 = 1/s$) for cascade and conventional PI control.

analysis. As discussed in Chapter 16, the presence of time delays in a process limits the performance of a conventional feedback control system. From a frequency response perspective, a time delay adds phase lag to the feedback loop, which adversely affects closed-loop stability. Consequently, the controller gain must be reduced below the value that could be used if no time delay were present, and the response of the closed-loop system will be sluggish compared to that of the control loop with no time delay.

EXAMPLE 18.2

Compare the set-point responses for a second-order process with a time delay and without the delay. The transfer function is

$$G_p(s) = \frac{e^{-\theta s}}{(3s + 1)(5s + 1)} \tag{18-18}$$

Assume $G_m = G_v = 1$ and time constants in minutes. For the time-delay case, let $\theta = 2$ min. Tune PI controllers by approximating (18-18) with a first-order plus time-delay model and then using the integral absolute error (IAE) criterion of Chapter 12.

Solution

The closed-loop responses are shown in Fig. 18.8. For $\theta = 2$ min, the controller gain must be reduced to meet stability requirements and the response is more sluggish. The deterioration of the closed-loop response for the time delay case is clear, with a 50% increase in response time (30 vs. 20 minutes), much longer than expected.

To improve the performance of time-delay systems, special control strategies have been developed to provide time-delay compensation. The Smith predictor technique is the best known strategy [10] and is discussed here. The Smith predictor is often referred to as a *model-based* controller, as is Internal Model Control (see Chapter 12). This is because the control strategy utilizes the model parameters directly. A related method, the analytical predictor, has been developed for digital control applications and is discussed in Chapter 26. Various investigators (e.g., [11,12]) have found that the performance of the Smith predictor for set-point

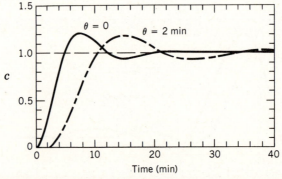

Figure 18.8.   A comparison of closed-loop set point changes of the second-order process ($\theta = 0$ and $\theta = 2$) with tuned PI controllers ($K_c = 3.02$, $\tau_I = 6.5$ min for $\theta = 0$; $K_c = 1.23$, $\tau_I = 7.0$ min for $\theta = 2$).

changes can be as much as 30% better than a conventional controller based on an integral squared error criterion.

A block diagram of the Smith predictor is shown in Fig. 18.9 with $G_v = G_m = 1$ for simplicity. Here the model of the process transfer function $\tilde{G}(s)$ is divided into two parts: the part without a time delay, $\tilde{G}^*(s)$, and the time-delay term, $e^{-\tilde{\theta}s}$. Thus, the total transfer function model is $\tilde{G}(s) = \tilde{G}^*(s)e^{-\tilde{\theta}s}$. The model of the process without the time delay, $\tilde{G}^*(s)$, is used to predict the effect of the control action on the process output. The controller uses this predicted response $\tilde{C}_1$ to calculate its output. The predicted process output is also delayed by the amount of the time delay $\tilde{\theta}$ for comparison with the actual process output C. This corrects for modeling errors and for load disturbances entering the process. This delayed model output is denoted by $\tilde{C}_2$ in Fig. 18.9. From the block diagram,

$$E' = E - \tilde{C}_1 = R - \tilde{C}_1 - (C - \tilde{C}_2) \tag{18-19}$$

If the process model is perfect and the disturbance is zero, then $\tilde{C}_2 = C$ and

$$E' = R - \tilde{C}_1 \tag{18-20}$$

In this case the controller acts on the error signal that would occur if no time delay were present.

Figure 18.10 shows an alternative (equivalent) configuration for the Smith predictor that includes an inner feedback loop, similar to that in cascade control. Assuming there is no model error ($\tilde{G} = G$), the inner loop has the effective transfer function

$$G_c' = \frac{P}{E} = \frac{G_c}{1 + G_c G^*(1 - e^{-\theta s})} \tag{18-21}$$

where G^* is defined analogous to $\tilde{G}^*$ that is, $G = G^* e^{-\theta s}$. After some rearrangement, the closed-loop servo transfer function is obtained:

$$\frac{C}{R} = \frac{G_c G^* e^{-\theta s}}{1 + G_c G^*} \tag{18-22}$$

By contrast, for conventional feedback control

$$\frac{C}{R} = \frac{G_c G^* e^{-\theta s}}{1 + G_c G^* e^{-\theta s}} \tag{18-23}$$

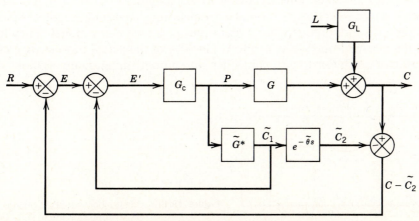

Figure 18.9. Block diagram of the Smith predictor.

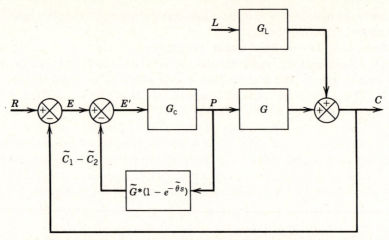

Figure 18.10. An alternative block diagram of a Smith predictor.

Comparison of Eqs. 18-22 and 18-23 indicates that the Smith predictor has the theoretical advantage of eliminating the time delay from the characteristic equation. Unfortunately, this advantage is lost if the process model is very inaccurate. However, the Smith predictor can still provide improvement over conventional feedback control [10–13] if the model errors are not too large (i.e., if the model parameters are within about ±30% of the actual values).

Figure 18.11 shows the closed-loop response for the Smith predictor and a step change in set point. The controller settings are the same as developed in Example 18.2 for $\theta = 0$. A comparison of Fig. 18.8 (dashed line) and Fig. 18.11 shows the improvement in performance that can be obtained with the Smith predictor. Note that Figs. 18.11 and 18.8 ($\theta = 0$) are identical, except for the initial time delay in Fig. 18.11. The time delay is due to the numerator term in (18-22). The process model $G^*(s)$ in this case is second order and thus readily yields a stable closed-loop system because the ultimate controller gain K_{cu} is infinity.

One drawback of the Smith predictor approach is that it is *model-based;* namely a dynamic model of the process is required. If the process dynamics change significantly, the predictive model will be inaccurate and the controller performance will deteriorate, perhaps to the point of instability. For such processes, the controller should be tuned conservatively to accommodate possible model errors. Schleck and Hanesian [12] performed a detailed study analyzing the effect of model errors on the Smith predictor for a first-order plus time-delay model. They found that if the assumed time delay is not within 30% of the actual process time delay, the predictor is inferior to a PI controller with no time-delay compensation. If the

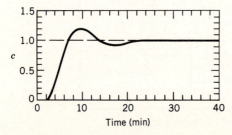

Figure 18.11. Closed-loop set-point change with Smith predictor for $\theta = 2$ ($K_c = 3.02$, $\tau_I = 6.50$).

time delay varies significantly, it may be necessary to use some sort of adaptive controller to achieve satisfactory performance (see Section 18.4).

The Smith predictor configuration generally is beneficial for load changes. However, the simulation study of Meyer et al. [13] indicated that a conventional PI controller can provide better regulatory control than the Smith predictor for certain conditions. This somewhat anomalous behavior can be attributed to the closed-loop transfer function, which is

$$\frac{C}{L} = \frac{G_L[1 + G_c G^*(1 - e^{-\theta s})]}{1 + G_c G^*} \tag{18-24}$$

where G_L is the load transfer function. The denominators of C/L and C/R in (18-22) are the same, but the numerator terms are quite different in form. Figure 18.12 shows load responses for Example 18.2 ($\theta = 2$) for PI controllers with and without the Smith predictor. A related method for time-delay compensation, the analytical predictor (see Ch. 26), is more effective for load changes than the Smith predictor.

The Smith predictor is seldom implemented as a continuous (analog) controller due to the difficulty of approximating time delays with analog components. However, this problem is avoided if a digital version of the Smith predictor is employed [13]. See Chapter 26 for further discussion of digital control algorithms employing time-delay compensation.

Inferential Control

The previous discussion of time-delay compensation assumed that measurements of the controlled variable were available. In some control problems, the process variable that is to be controlled cannot be conveniently measured on-line. For example, product composition measurement may require that a sample be sent to the plant analytical laboratory. In this situation, measurements of the controlled variable are not available frequently enough nor quickly enough to be used for feedback control.

One solution to this problem is to use inferential control, whereby process measurements that can be obtained rapidly are used to infer the value of the

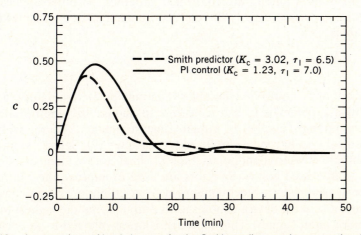

Figure 18.12.   A comparison of load changes for the Smith predictor and a conventional PI controller:

$$G_L(s) = \frac{1}{(3s + 1)(5s + 1)} \qquad G_p(s) = \frac{e^{-2s}}{(3s + 1)(5s + 1)}$$

controlled variable. For example, if the overhead product stream in a distillation column cannot be analyzed on-line, sometimes measurement of the top tray temperature can be used to infer the actual composition. For a binary mixture, the Gibbs phase rule indicates that there is a unique relation between composition and temperature if pressure is constant. Therefore, a thermodynamic equation could be employed to relate the temperature of the top tray to overhead composition.

On the other hand, for the separation of multicomponent mixtures, approximate methods to estimate compositions must be used. Based on process models and plant data, simple algebraic correlations can be developed to relate the mole fraction of the heavy key component to several different tray temperatures (usually near the top of the column). The overhead composition then can be inferred from the available temperature measurements and used in the control algorithm [14]. The parameters in the correlation may be updated, if necessary, as actual composition measurements become available. For example, if samples are sent to the plant's analytical laboratory once per hour, the correlation parameters can be adjusted so that the predicted values agree with the measured values.

The concept of inferential control can be employed for other process operations, such as chemical reactors, where composition is normally the controlled variable. Selected temperature measurements can be used to estimate the outlet composition if it cannot be measured on-line. However, when inferential control does not perform satisfactorily, incentive exists to introduce other on-line measurements for feedback control. Consequently, there is considerable interest in the development of new instrumentation, such as process analyzers, which can be used on-line and which exhibit very short response times.

18.3 SELECTIVE CONTROL/OVERRIDE SYSTEMS

Most process control problems have an equal number of controlled variables and manipulated variables. If there are fewer manipulated variables than controlled variables, then it is not possible to eliminate offset in all the controlled variables for arbitrary load or set-point changes. This assertion is evident from a steady-state model and a degrees-of-freedom analysis. For control problems with more controlled variables than manipulated variables, a strategy is needed for sharing the manipulated variables among the controlled variables. In other applications the opposite situation occurs, namely there are more manipulated variables than controlled variables.

Selectors

When there are more controlled variables than manipulated variables, a common solution to this problem is to use a *selector* to choose the appropriate process variable from among a number of available measurements. Alternatively, a multivariable control system can be employed (cf. Chapter 19). Selectors can be based on either multiple measurement points, multiple final control elements, or multiple controllers, as discussed below. Selectors are used to improve the control system performance as well as to protect equipment from unsafe operating conditions.

One type of selector device chooses as its output signal the highest (or lowest) of two or more input signals. This approach is often referred to as *auctioneering* [4, p. 232]. On instrumentation diagrams the symbol HS denotes high selector and LS a low selector. For example, a high selector can be used to determine the *hot spot* temperature in a fixed-bed chemical reactor as shown in Fig. 18.13. In this

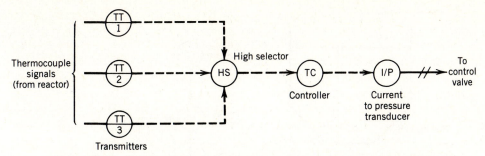

Figure 18.13.   Control of a reactor hot spot temperature by using a high selector.

case the output from the high selector is the input to the temperature controller. In an exothermic catalytic reaction, the process may "run away" due to disturbances or changes in the reactor. Immediate action should be taken to prevent a dangerous rise in temperature. Because a hot spot may potentially develop at one of several possible locations in the reactor, multiple (redundant) measurement points should be employed. This approach minimizes the time required to identify when a temperature has risen too high at some point in the bed. With a median selector, the selector output is the median of three or more input signals. These devices are useful for situations in which redundant sensors are used to measure a single process variable. By selecting the median input, maximum reliability is obtained since a single sensor failure will not result in loss of control.

The use of high or low limits for process variables is another type of selective control, called an *override*. The feature of anti-reset windup in feedback controllers, discussed in Chapter 8, is a type of override. Another example is a distillation column which has lower and upper limits on the heat input to the column reboiler. The minimum level ensures that liquid will remain on the trays [4,15], while the upper limit is determined by the onset of flooding. Overrides are also used in forced draft combustion control systems to prevent an imbalance between air flow and fuel flow, which could result in unsafe operating conditions [16].

Other types of selective systems employ multiple final control elements or multiple controllers. Stephanopoulos [17, p. 407] has described applications where several manipulated variables are used to control a single process variable (also called *split-range control*). Typical examples include the adjustment of both inflow and outflow from a chemical reactor in order to control reactor pressure or the use of both acid and base to control pH in wastewater treatment. For the multiple controller case, the controllers are generally employed in parallel rather than in series, as was the situation for the cascade configuration of Fig. 18.4. In this approach the selector chooses from several controller outputs which final control element should be adjusted.

Consider the selective control system shown in Fig. 18.14, which is used to regulate level and exit flow rate in a pumping system for a sand–water slurry. During normal operation, the level controller (LC) adjusts the slurry exit flow by changing the pump speed. A variable speed pump is used rather than a control valve due to the abrasive nature of the slurry. The slurry velocity in the exit line must be kept above a minimum value at all times to prevent the line from sanding up. Consequently, the selective control system is designed so that as the flow rate approaches the lower limit, the flow controller takes over from the level controller and speeds up the pump. The strategy is implemented in Fig. 18.14 using a high selector and a reverse-acting flow controller with a high gain. The set point and

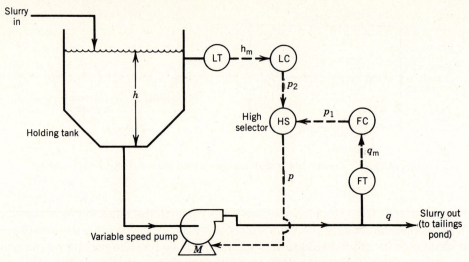

Figure 18.14.   A selective control system to handle a sand/water slurry.

gain of the flow controller are chosen so that the controller output is at the maximum value when the measured flow is near the constraint.

The block diagram for the selector control loop used in the slurry example is shown in Fig. 18.15. The selector compares signals P_1 and P_2, both of which have the same units (e.g., voltage or current). There are two parallel feedback loops. Note that G_v is the transfer function for the final control element, the variable speed drive pump. A stability analysis of Fig. 18.15 would be rather complicated because the high selector introduces a nonlinear element into the control system. Typically the second loop (pump flow) will be faster than the first loop (level) and uses proportional plus integral control (although reset windup protection will be required). Proportional control could be employed on the slower loop (liquid level) because tight level control is not required.

One alternative arrangement to Fig. 18.14 would be to employ a single controller, using the level and flow transmitter signals as inputs to a high selector, with its output signal sent to the controller. The controller output would then adjust

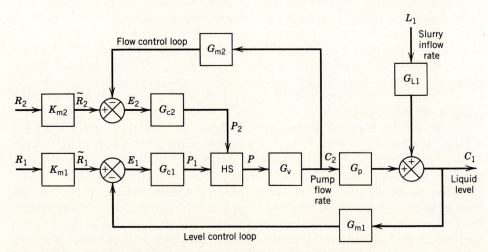

Figure 18.15.   Block diagram for the selective control loop with two measurements and two controllers (variables in Figure 18.14 are identified).

the pump speed. This scheme has a lower capital cost since only one controller is needed. However, it suffers from an important operational disadvantage, namely, it may not be possible to tune the single controller to meet the needs of both the level and flow control loops. In general, these control loops and their transmitters will have very different dynamic characteristics. A second alternative would be to replace the flow transmitter and controller with a constant (override) signal to the high selector whose value corresponds to the minimum allowable flowrate. However, this scheme would be susceptible to changing pump characteristics.

18.4 ADAPTIVE CONTROL SYSTEMS

Process control problems inevitably require on-line tuning of the controller settings to achieve a satisfactory degree of control. If the process operating conditions or the environment changes significantly, the controller may have to be retuned. If these changes occur frequently, then adaptive control techniques should be considered. An *adaptive control system* is one in which the controller parameters are adjusted automatically to compensate for changing process conditions. A variety of adaptive control techniques have been proposed for situations where the process changes are largely unknown or for the easier class of problems where the changes are known or can be anticipated. In this section we are principally concerned with automatic adjustment of feedback controller settings.

Catalytic reaction systems are a notable example of an important process where changes occur that are not directly measurable. As Lee and Weekman [18] have noted,

> *Catalytic processes are notorious for changes in catalyst behavior, and adaptive loops are found quite commonly in such applications. The ad hoc development of an adaptive loop has probably saved more advanced control projects from failure than any other single technique.*

Other causes of changing process conditions that may require controller retuning or adaptive control are:

1. Heat exchanger fouling
2. Unusual operational status, such as failures, start-up, and shutdown, or batch operations
3. Large, frequent disturbances (feed composition, fuel quality, etc.)
4. Ambient variations (rain storms, daily cycles, etc.)
5. Changes in product specifications (grade changes) or flow rates
6. Inherent nonlinear behavior (e.g., the dependence of chemical reaction rates on temperature)

It is convenient to distinguish between two general categories of adaptive control applications. The first category consists of situations where the process changes can be anticipated or measured directly. If the process is reasonably well understood, then it may be feasible to adjust the controller settings in a systematic fashion (called *programmed adaptation*) as process conditions change or as disturbances enter the system. The second category consists of situations where the process changes cannot be measured or predicted. In this more difficult situation the adaptive control strategy must be implemented in a feedback manner since there is little opportunity for a feedforward type of strategy such as programmed adaptation. Many such controllers are referred to as *self-tuning controllers* [19–21]; they are generally implemented via digital computer control.

Programmed Adaptation

If a process is operated over a range of conditions, improved control can be achieved by using a different set of controller settings for each operating condition. Alternatively, a relation can be developed between the controller settings and the process variables that characterize the process conditions. These strategies are examples of programmed adaptation [4,22,23]. Programmed adaptation is limited to applications where the process dynamics depend on known, measurable variables and the necessary controller adjustments are not too complicated. Usually the adaptation is simple enough in structure that it can be implemented with some analog and all digital controllers [4, p. 239]. The most popular type of programmed adaptation is gain scheduling, where the controller gain is adjusted so that the open-loop gain $K_{OL} = K_c K_v K_p K_m$ remains constant.

As an example of a control problem where programmed adaptation has been proposed [22], consider a once-through boiler. Here feedwater passes through a series of heated tube sections before emerging as superheated steam, the temperature of which must be accurately controlled. The feedwater flow rate has a significant effect on both the steady-state and dynamic behavior of the boiler. For example, Fig. 18.16 shows typical open-loop responses to a step change in flow rate for two different feedwater flow rates, 50 and 100% of the maximum flow. Suppose an empirical first-order plus delay model is chosen to approximate the process. The steady-state gain, time delay, and dominant time constant are all twice as large at 50% flow as the corresponding values at 100% flow. The proposed solution [22] to this control problem is to make the PID controller settings vary with w, the fraction of full-scale flow ($0 \le w \le 1$), in the following manner:

$$K_c = w\overline{K}_c$$

$$\tau_I = \overline{\tau}_I/w \qquad\qquad (18\text{-}25)$$

$$\tau_D = \overline{\tau}_D/w$$

where $\overline{K}_c$, $\overline{\tau}_I$, and $\overline{\tau}_D$ are the controller settings for 100% flow. Note that this recommendation for programmed adaptation is qualitatively consistent with the Cohen and Coon rules for tuning controllers given in Chapter 12 and assumes that the effect of flow changes is proportionally related to flow rate over the full range of operation.

In this example, step responses were available to categorize the process behavior for two different conditions. In other problems dynamic response data are not available but we have some knowledge of process nonlinearities. For pH control problems involving a strong acid and/or a strong base, the pH curve can

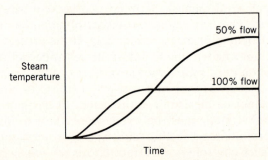

Figure 18.16. Open-loop step responses for a once-through boiler.

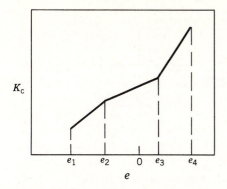

Figure 18.17. An adaptive controller where K_c varies with error signal e.

be very nonlinear, with gain variations over several orders of magnitude. Consequently, special nonlinear controllers, both adaptive and nonadaptive, have been developed for pH control problems [4, pp. 203, 242]. In this case the process gain changes dramatically with the operating conditions, necessitating the use of gain scheduling ($K_c K_p$ = constant) to maintain consistent stability margins [23].

For some types of adaptive control problems, the changes in steady-state and dynamic response characteristics can be related to the value of the controlled variable. For example, in a temperature control loop where the process gain varies with temperature, the controller gain could be made a function of the controlled variable, temperature. Feedback controllers are now commercially available which allow the user to vary K_c as a piecewise linear function of the error signal e, as shown in Fig. 18.17. If the process gain K_p varies in a known manner, we should vary K_c so that the product $K_c K_p$ is constant. This strategy would tend to keep the open-loop gain constant and thus maintain a specified margin of stability (assuming the process dynamics do not change also).

Self-tuning Control

If process changes can be neither measured nor anticipated, programmed adaptation cannot be used. An alternative approach is to update the parameters in a process model as new data are acquired (on-line estimation) and then base the control calculations on the updated model. For example, the controller settings could be expressed as a function of the parameters in the process model and the estimates of these parameters updated on-line as process input/output data are received. This type of controller is referred to as *self-tuning* or *self-adaptive*. Self-tuning controllers generally are implemented [20,24,25] as shown in Fig. 18.18.

In Fig. 18.18, three sets of computations are employed: estimation of the process model parameters, calculation of the controller settings, and implementation of the settings in a feedback loop. Most real-time parameter estimation techniques require that an external forcing signal occasionally be introduced to allow accurate estimation of process model parameters [20,25,26]. Such an input signal can be deliberately introduced through the set point or added to the controller output.

The first type of self-tuning adaptive controller, called the *self-tuning regulator*, was proposed in 1973 by Åström and Wittenmark and has since been implemented in several industrial applications [19–21]. Subsequent modifications, the *self-tuning controller* [27] and the *generalized predictive controller* [28], have also been used with industrial processes. These controllers are based on a difference equation

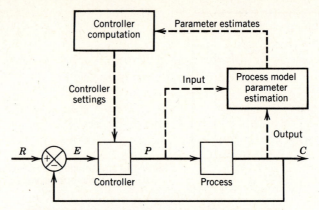

Figure 18.18. A block diagram for self-tuning control.

(digital) model of the process, and the self-tuning regulator and self-tuning controller utilize a minimum variance criterion to reduce the error in the controlled variable. The self-tuning regulator particularly is oriented towards applications where the process disturbances are stochastic in nature (i.e., random) rather than deterministic. Detailed discussion of these self-tuning methods is beyond the scope of this book.

A related type of adaptive control strategy, model reference adaptive control, attempts to achieve a closed-loop response that is as close as possible to a desired (or reference) response. The discrepancy or difference between actual response and the desired model reference ($\Delta C = C - C_m$) is used in the control law calculation (Fig. 18.19). The boldface blocks in Figs. 18.18 and 18.19 indicate functions carried out by a digital computer. While the controller can be implemented with analog or digital equipment, normally the latter is used.

Commercial Adaptive Control Systems

Several adaptive controllers have been field-tested and commercialized in the United States and abroad. The ASEA adaptive controller, called Novatune, was an-

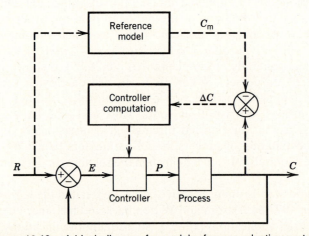

Figure 18.19. A block diagram for model-reference adaptive control.

nounced in 1983 and is based on the minimum variance–stochastic control algorithms mentioned above [19,27]. Both feedforward and feedback control capabilities reside in the hardware, which can be configured for as many as 16 control loops. Novatune has been tested successfully in reactor and paper machine control applications in Europe and in pH control of wastewater in the United States [29].

Another commercial adaptive controller (Leeds and Northrup) is based on making step changes in the set point [30] and has been used successfully in process heating applications. This self-tuning controller is designed to provide exponential responses to set-point changes, with no overshoot (see Eq. 12-8 for a similar type of controller algorithm). During the tuning procedure, a specified change in the set point (as large as feasible) is introduced to obtain the process gain and time constants. Controller settings are determined directly from a derived second-order process model and the desired closed-loop response time (time to reach 90% response) specified by the user. Additional testing may be required if the response exhibits some overshoot, which indicates that the model parameters are incorrect.

The Foxboro Company has developed a self-tuning PID controller that is based on an "expert system" approach for adjustment of the controller setting [31]. The on-line tuning of K_c, τ_I, and τ_D is based on the closed-loop transient response to a step change in set point. By evaluating the salient characteristics of the response (e.g., the decay ratio, overshoot, and closed-loop period), the controller settings can be updated without actually finding a new process model. The details of the technique, however, are proprietary.

The Satt controller, which is marketed in the United States by Fisher Controls, has an autotuning function that is based on placing the process in a controlled oscillation at very low amplitude, comparable with that of the noise level of the process. As shown in Fig. 13.4, this is done via a relay-type step function with a dead zone [21,32]. Using process data, the autotuner identifies the ultimate gain and period and automatically calculates K_c, τ_I, and τ_D using the Ziegler–Nichols rules. Gain scheduling can also be implemented with this controller, using up to three sets of PID controller parameters.

The subject of adaptive control is one of great current interest. Many new algorithms are presently under development [20,21], but these need to be field-tested before industrial acceptance can be expected. It is clear, however, that digital techniques are required for implementation of self-tuning controllers due to their complexity.

18.5 STATISTICAL QUALITY CONTROL

Statistical quality control (SQC), also called *statistical process control* (SPC), involves the application of statistical concepts to determine whether a process is operating satisfactorily. The basic concepts of statistical quality control are over fifty years old, but only recently with the growing worldwide focus on increased productivity have applications of SQC become widespread. If a process is operating satisfactorily (or *in control*), then the variation of product quality falls within acceptable bounds, usually the minimum and maximum values of a specific composition or property (product specification).

Figure 18.20 illustrates the variation of the controlled variable *c*, represented in the form of a *histogram*, that might be expected to occur under typical steady-state operating conditions. The mean $\bar{c}$ and root mean square (RMS) deviation σ are identified in Fig. 18.20 and can be calculated from a series of *J* observations

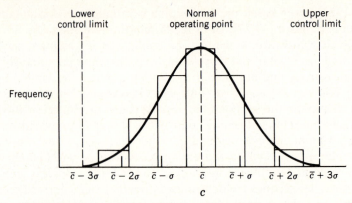

Figure 18.20. A histogram represents frequency of occurrence. ($\bar{c}$ = mean and σ = RMS deviation.) Also shown is a normal probability distribution fit to the data.

$c_1, c_2, \ldots, c_J$ as follows:

$$\bar{c} = \frac{1}{J} \sum_{i=1}^{J} c_i \tag{18-26}$$

$$\sigma = \left[\frac{1}{J} \sum_{i=1}^{J} (c_i - \bar{c})^2 \right]^{1/2} \tag{18-27}$$

The RMS deviation, also called the standard deviation, is a measure of the spread of observations around the mean. A large value of σ indicates that wide variations in c occur. The probability that the controlled variable lies between two arbitrary values, c_1 and c_2, is given by the area under the histogram between c_1 and c_2. If the histogram follows a normal probability distribution (the curve in Fig. 18.20), then 99.7% of all observations should lie within $\pm 3\sigma$ of the mean. These upper and lower control limits are used to determine whether the process is operating as expected. Note that the set point of the controlled variable should be selected near the mean $\bar{c}$ so that violations of the product specification are very unlikely (i.e., they have a low probability). In other words, both the upper and lower control limits should lie inside the operating constraints.

Figure 18.21 shows a process control chart for pH data taken over a time period of 50 days. Assume that $\bar{c}$ and σ have been calculated based on earlier observations. If all of the new data lie within the $\pm 3\sigma$ limits, then we conclude that nothing unusual has happened during the recorded time period. For this situation, the process environment is relatively unchanged and the product quality lies within specification. On the other hand, if repeated violations of the $\pm 3\sigma$ limits occur, then the process environment has changed and the process is *out of control*. This situation occurs in Fig. 18.21 at $t = 25$ days. Recall that if the normal prob-

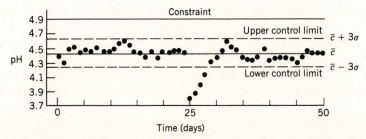

Figure 18.21. Process control chart for the average daily pH readings [33].

ability distribution is valid, only 0.3% of the observations should violate the $\pm 3\sigma$ window. There are important economic consequences of a process being "out of control," for example, wasted product and customer dissatisfaction. Hence, SQC provides a systematic way to continuously monitor process performance and to improve product quality.

Statistical quality control is a diagnostic tool, that is, an indicator of quality problems, but it does not identify the source of the problem nor the corrective action to be taken. For example, suppose the data in Fig. 18.21 were obtained from the monitoring of pH in a yarn-soaking kettle used in textile manufacturing. Because pH has a crucial influence on color and durability of the yarn, it is important to maintain pH within a range that gives the best results for both characteristics. The pH is considered to be in control between values of 4.25 and 4.64. At the 25th day, the data show that pH is out of control; this might imply that a property change in the raw material has occurred and must be corrected with the supplier [33]. However, a real-time correction would be preferable. In Fig. 18.21, the pH was adjusted by slowly adding more acid to the vats until it came back into control (Day 29).

In continuous processes where automatic feedback control has been implemented, the feedback mechanism theoretically ensures that product quality is at or near the set point, regardless of process disturbances. This, of course, requires that an appropriate manipulated variable has been identified for adjusting the product quality. However, even under feedback control, there may be daily variations of product quality because of disturbances or equipment or instrument malfunctions. These occurrences can be analyzed using the concepts of statistical quality control.

MacGregor [34] has shown that the underlying theory of statistical quality control based on time-series models proves that feedback control is appropriate under certain assumptions. If (1) the cost of corrections in the manipulated variable is small compared to the cost of quality deviations, (2) process dynamics are important, and (3) corrections are made frequently by feedback control, SQC theory shows that a PI controller is optimal. On the other hand, when changes in the manipulated variable are infrequent because of their large cost (e.g., a unit shutdown), and the process is at steady-state, making hypothesis testing appropriate, the use of process control charts such as Fig. 18.21 is optimal. Therefore process control charts should not be thought of as a general substitute for PID regulatory control and vice versa.

For a continuous plant some type of moving average can be employed to determine quality control [33]. An average of the quality value for the past several days may be more meaningful than a single data point because of the blending that occurs in large product storage tanks. If the moving average exceeds the $\pm 3\sigma$ limits on a process control chart, then we conclude that some significant change in the process *not correctable by existing feedback control* has occurred; for example,

- Persistent disturbances from the weather
- An undetected grade change in raw materials
- A malfunctioning instrument or control system

In any case, further engineering analysis of the problem is indicated.

The subject of statistical quality control is quite broad and we have presented only the simplest concepts above. More details on process control charts, probability distributions, sampling, grouping of observations, and other related items have been provided by Grant and Leavenworth [33] and Montgomery [35].

18.6 EXPERT SYSTEMS

In a manufacturing facility, the goal of process control is to maintain product quality under safe operating conditions. When operating conditions vary outside of acceptable limits, due to external causes, equipment malfunctions, or human error, then product quality deteriorates, energy consumption becomes suboptimal, and unsafe conditions can occur. Such process excursions may require plant shutdown or lead to catastrophic events such as explosions, fires, or discharge of toxic chemicals. In most existing control systems, abnormal measurements trigger alarms that alert the process operator. The operator then must take remedial action to either return the plant to normal levels of operation or shut the facility down.

The success of a manual strategy for handling abnormal conditions relies heavily on the operator being able to respond correctly to process alarms. However, the operator's response depends on many factors: the number of alarms and the frequency of occurrence of abnormal conditions, how information is presented to the operator, the complexity of the plant, and the operator's intelligence, training, experience, and reaction to stress. Because of the many factors involved in determining the appropriate response to an alarm situation, computational aids for the operator are crucial to the success of operating complex manufacturing plants. Such computer-based assistance can be developed as software systems. These so-called *expert systems* are based on emulating the actions of a human expert who is acknowledged to perform the required tasks at a high level of proficiency. The use of expert systems, also called *knowledge engineering,* is a branch of artificial intelligence (AI). AI is popularly defined as the science of enabling computer systems to learn, reason, and make judgments. Most expert systems utilize a set of procedures that simplify the application of inductive and deductive reasoning to the data base ("knowledge") of the system.

Figure 18.22 shows the architecture of an expert system. The expert system is usually written as part of a *shell,* which is a general software package designed to facilitate implementation. The shell contains the following components:

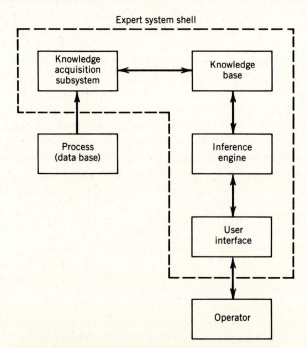

Figure 18.22.   General structure of an expert system.

1. **Knowledge Base Consisting of Data and Rules.** Data can be entered at system start-up or, via the knowledge acquisition system, in real time. Rules are structured with "if–then" statements.
2. **Inference Engine.** This software provides a means of scanning the available rules to draw conclusions or select the appropriate action to be taken.
3. **User Interface.** This component displays information, asks questions of the user, and so on.

A simple example illustrating the need for expert system fault detection and diagnosis is a situation where the actual flow in a process is significantly higher than the set point for a long period of time. Possible faults that could cause this event include (1) the sensor is malfunctioning, (2) the controller has failed in a saturated mode, or (3) the valve has failed open. The identification of the actual fault or failure in the system could be obtained by checking other measurements, performing material balances, or having the operator take certain actions and then observing the results. Interaction with the operator is employed to reach the correct decision. Expert systems for real-time process control have been described by Moore and Kramer [36] and Dhurjati et al. [37].

SUMMARY

In this chapter, we have presented a number of advanced control strategies that offer the potential of enhanced performance over what can be achieved with conventional PID controllers. These techniques are especially attractive for difficult control problems, which are characterized by unmeasured process variables and disturbances, long time delays, process constraints, changing operating conditions, and process nonlinearities and uncertainties. All of these characteristics can be treated by one or more of the advanced feedback control techniques discussed here: cascade control, time-delay compensation, inferential control, selective control, and adaptive control. Note that feedforward control (Chapter 17) and multivariable control (Chapter 19) augment the specialized methods treated here and also can be classified as advanced control strategies. All of these techniques can be readily implemented by a digital control computer.

Statistical quality control and expert systems control are other types of advanced control discussed in this chapter. These techniques do not usually employ feedback loops but rather are analysis tools to assist the operator or engineer in decision-making. The process control computer analyzes the data and presents it in a form that helps the plant operator make decisions. Increased utilization of both techniques in process plants is expected in the near future.

REFERENCES

1. Edgar, T. F., and D. E. Seborg (Eds.), *Chemical Process Control 2,* Engineering Foundation, New York, 1981.
2. Morari, M., and T. J. McAvoy (Eds.), *Chemical Process Control—CPC III,* Elsevier, NY, 1986.
3. Johnson, E. F., *Automatic Process Control,* McGraw-Hill, New York, 1967, p. 244.
4. Shinskey, F. G., *Process Control Systems,* 3d ed., McGraw-Hill, New York, 1988.
5. Perlmutter, D. D., *Introduction to Chemical Process Control,* Wiley, New York, 1965, p. 156.
6. Deshpande, P. B., and R. H. Ash, *Elements of Computer Process Control,* Instrum. Society of America, Research Triangle Park, NC, 1981, Ch. 16.
7. Clay, R. M., and C. D. Fournier, Cascade vs. Single Loop Control of Processes with Dead Time, *Proc. Joint Auto. Contr. Conf.,* 363 (1973).
8. Ward, T. J., Cascade and Ratio Control, AIChEM I Modular Instruction Series, AIChE, A4.4 (1983).
9. Harriott, P., *Process Control,* McGraw-Hill, New York, 1964.
10. Smith, O. J. M., Closer Control of Loops with Dead Time, *Chem. Eng. Prog.* **53,** 217 (1957).

11. Donoghue, J. F., Review of Control Design Approaches for Transport Delay Processes, *ISA Trans.* **16**(2), 27 (1977).
12. Schleck, J. R., and D. Hanesian, An Evaluation of the Smith Linear Predictor Technique for Controlling Deadtime Dominated Processes, *ISA Trans.* **17**(4), 39 (1978).
13. Meyer, C., D. E. Seborg, and R. K. Wood, A Comparison of the Smith Predictor and Conventional Feedback Control, *Chem. Eng. Sci.* **31**, 775 (1976).
14. Joseph, B., and C. Brosilow, Inferential Control of Processes, I. Steady State Analysis and Design, *AIChE J.* **24**, 485 (1978).
15. Buckley, P. S., W. L. Luyben, and J. P. Shunta, *Design of Distillation Column Control Systems,* Instrument Society of America, Research Triangle Park, NC, 1985.
16. Singer, J. G., (Ed.), *Combustion-Fossil Power Systems,* Combustion Engineering, Windsor, CT, 1981, p. 14–27.
17. Stephanopoulos, G., *Chemical Process Control,* Prentice-Hall, Englewood Cliffs, NJ, 1983.
18. Lee, W., and V. W. Weekman, Jr., Advanced Control Practice in the Chemical Process Industry: A View From Industry, *AIChE J.* **22**, 27 (1976).
19. Åström, K. J., and B. Wittenmark, *Adaptive Control Systems,* Addison-Wesley, Reading, MA, 1988.
20. Seborg, D. E., T. F. Edgar, and S. L. Shah, Adaptive Control Strategies for Process Control: A Survey, *AIChE J.* **32**, 881 (1986).
21. Åström, K. J., Adaptive Feedback Control, *Proc. IEEE* **75**(2), 185 (1987).
22. Liptak, B. G., *Instrument Engineers Handbook,* Vol. 2, Chilton, Philadelphia, 1970, p. 813.
23. Mellichamp, D. A., D. R. Coughanowr, and L. B. Koppel, Identification and Adaptation in Control Loops with Time-Varying Gain, *AIChE J.* **12**(1), 83 (1966).
24. Landau, Y. D., *Adaptive Control,* Dekker, New York, 1978.
25. Isermann, R., *Digital Control Systems,* Springer-Verlag, Berlin, 1981.
26. Eykhoff, P., *System Identification,* Wiley, London, 1974.
27. Clarke, D. W., and P. J. Gawthrop, Implementation and Application of Microprocessor-Based Self-Tuners, *Automatica* **17**, 233 (1981).
28. Clarke, D. W., C. Mohtadi, and P. J. Gawthrop, Generalized Predictive Control, Parts I and II, *Automatica* **23**, 137 (1987).
29. Piovoso, M. J., and J. M. Williams, Self-Tuning of pH, in *Advances in Instrum.,* Vol. 39, Instrum. Society of America, Research Triangle Park, NC, 1984, p. 705.
30. Hoopes, H. S., W. M. Hawk, and R. C. Lewis, A Self-tuning Controller, *ISA Trans.* **22**(3), 49 (1983).
31. Bristol, E. H., The Design of Industrially Useful Adaptive Controllers, *ISA Trans.* **22**(3), 17 (1983).
32. Åström, K. J., and T. Hagglund, *Automatic Tuning of PID Controllers,* Instrum. Society of America, Research Triangle Park, NC, 1987.
33. Grant, E. L., and R. S. Leavenworth, *Statistical Quality Control,* McGraw-Hill, New York, 1980.
34. MacGregor, J. F., On-Line Statistical Process Control, *Chem. Eng. Prog.,* **84**(10), 21 (1988).
35. Montgomery, D. C., *Introduction to Statistical Quality Control,* Wiley, New York, 1985.
36. Moore, R. L., and M. A. Kramer, Expert Systems in On-Line Process Control, in *Chemical Process Control—CPC III,* M. Morari and T. J. McAvoy, Eds., 839, Elsevier, New York, 1986.
37. Dhurjati, P. S., D. E. Lamb, and D. L. Chester, Experience in the Development of an Expert System for Fault Diagnosis in a Commercial Scale Chemical Process, in *Foundations of Computer-Aided Process Operations,* G. V. Reklaitis and H. D. Spriggs, Eds., 589, Elsevier, New York, 1988.

EXERCISES

18.1. Measurement devices and their dynamics influence the design of feedback controllers. Which of the two systems below would have its closed-loop performance enhanced significantly by application of cascade control (see Fig. 18.4 for notation)? Why?

System A	*System B*
$G_v = 5$	$G_v = 5$
$G_p = \dfrac{2}{10s + 1}$	$G_p = \dfrac{2}{10s + 1}$
$G_{m2} = \dfrac{0.5}{0.5s + 1}$	$G_{m2} = \dfrac{2}{5s + 1}$
$G_{m1} = \dfrac{1}{5s + 1}$	$G_{m1} = \dfrac{0.2}{0.5s + 1}$

All time constants are in minutes.

18.2. In Example 18.1, the ultimate gain for the primary controller was found to be 43.3 when $K_{c2} = 5$.

(a) Derive the closed-loop transfer functions for C_1/L_1 and C_1/L_2 as a function of K_{c1} and K_{c2}.

(b) Examine the effect of K_{c2} on the critical gain of K_{c1} by varying K_{c2} from 1 to 20. For what values of K_{c2} do the benefits of cascade control seem to be less important? Is there a stability limit on K_{c2}?

(c) Integral action was not included in either primary or secondary loops. First set $K_{c2} = 5$, $\tau_{I1} = \infty$, and $\tau_{I2} = 5$ min. Find the ultimate controller gain using the Routh array. Then repeat the stability calculation for $\tau_{I1} = 5$ min and $\tau_{I2} = \infty$ and compare the two results. Is offset for C_1 eliminated in both cases for step changes in L_1 or L_2?

18.3. Consider the cascade control system in the drawing.

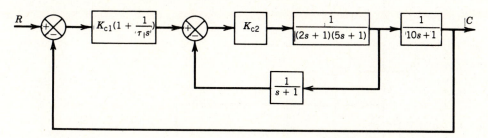

(a) Specify K_{c2} so that the gain margin ≥ 1.7 and phase margin $\geq 30°$ for the slave loop.

(b) Then specify K_{c1} and τ_I for the master loop using the Ziegler–Nichols tuning relation.

18.4. A jacketed well-mixed vessel is used to cool a process stream as shown in the drawing. The vessel temperature T is controlled using a cascade control scheme with the slave controller regulating the temperature of the coolant in the jacket T_c. It can be assumed that the thermal capacitance of the tank wall is negligible and that both the tank contents and the jacket contents are well mixed. Draw a block diagram for the closed-loop system, using the symbols in the schematic diagram. (Define any other symbols that you introduce.) The important load variables are feed temperature T_F and inlet coolant temperature T_{ci}. Derive an expression for the closed-loop transfer function between T and T_F in terms of the individual transfer functions. (To develop explicit transfer functions, write energy balances for the tank contents and jacket using a heat transfer coefficient U.)

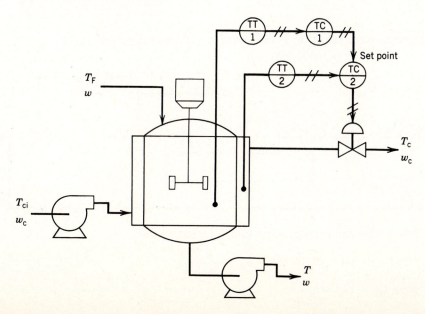

18.5. Consider the stirred-tank heating system shown in the drawing. It is desired to control temperature T_2 by adjusting the heating rate Q_1 (Btu/h) via voltage signal V_1 to the SCR. It has been suggested that measurements of T_1 and T_0, as well as T_2, could provide improved control of T_2.

(a) Briefly describe how such a control system might operate and sketch a schematic diagram. State any assumptions that you make.

(b) Indicate how you would classify your control scheme, for example, feedback, cascade, feedforward. Briefly justify your answer.

(c) Draw a block diagram for the control system.

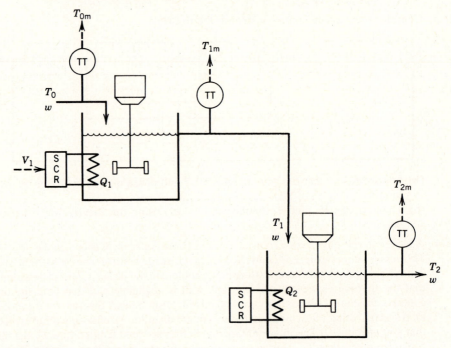

18.6 The block diagram of a feedback control system is shown. Derive the closed-loop transfer functions for

(a) X_1/P_1

(b) C/R

(c) C/L

Determine the values of K_c that result in a stable closed-loop system.

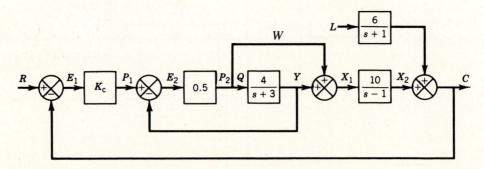

18.7. Find a time-delay compensator (Smith predictor) for

$$G_p = \frac{e^{-\theta s}}{5s + 1}$$

when $G_v = G_m = 1$.

18.8. Figure 18.8 shows the responses for tuned PI controllers applied to a second-order model with and without time delay. Compare values of the critical gain using a proportional controller for $\theta = 0$ and $\theta = 2$ min.

18.9. Applepolscher has designed a Smith predictor with proportional control for a stirred reactor temperature control loop. Based on simulation results for a first-order plus time delay model, he tuned the controller so that it will not oscillate. However, when the controller was implemented, severe oscillations occurred. He has verified through numerous step tests that the process model is linear. What explanations can be offered for this anomalous behavior?

18.10. The closed-loop transfer function for the Smith predictor in Eq. 18-22 was derived assuming no model error.

(a) Derive a formula for C/R when $G_p \neq \tilde{G}_p$. What is the characteristic equation?
(b) Let $G_p = 2e^{-2s}/(5s + 1)$. A proportional controller with $K_c = 15$ and a Smith predictor is used to control this process. Simulate set-point changes for $\pm 20\%$ errors in process gain (K_p), time constant (τ), and time delay (six different cases). Discuss the relative importance of each type of error.
(c) What controller gain would be satisfactory for $\pm 50\%$ changes in all three model parameters?
(d) For $K_c = 15$, how large a change in either K_p, τ, or θ can be tolerated before the loop goes unstable?

18.11. Equation 18-24 can be simplified using the approximation, $e^{-\theta s} \approx 1 - \theta s$. Derive an expression for C/L for a first-order plus time-delay process and a proportional controller. Compare this result with C/L derived for the conventional feedback control block diagram, that is, no time-delay compensation, and the same Taylor series approximation. Let $G_L = 1/(\tau_L s + 1)$.

18.12. In Chapter 12, we introduced the Direct Synthesis design method where the closed-loop servo response is specified and the controller transfer functions calculated using algebra.

(a) Show that if $(C/R)_d = P(s)e^{-\gamma s}$ the resulting controller is a Smith predictor when $\gamma = \theta$, the process time delay. Assume no model error.
(b) For an IMC controller (see Chapter 12), show that setting $G_+ = e^{-\theta s}$ leads to a Smith predictor controller structure when $G = \tilde{G}$ for a first-order plus time-delay process.

18.13. You are operating a CSTR to produce a specialty chemical. The reaction is exothermic and exhibits first-order kinetics. Laboratory analyses for the product quality are time consuming, requiring several hours to complete. No on-line composition measurement has been found satisfactory. It has been suggested that composition can be inferred from the exit temperature of the CSTR. Using the linearized CSTR model in Example 4.8, determine whether this inferential control approach would be feasible. Assume that measurements of feed flow rate, feed temperature, and coolant temperature are available.

18.14. The pressure of a reactor vessel can be adjusted by changing either the inlet or outlet gaseous flow rate. The outlet flow is kept fixed as long as the tank pressure remains between 100 and 120 psi, and pressure changes are treated by manipulating the inlet flow control valve. However, if the pressure goes outside these limits, the exit gas flow is then changed. Finally, if the pressure exceeds 200 psi, a vent valve on the vessel is opened and transfers the gas to a storage vessel. Design a selector control scheme that meets the performance objectives. Draw a process instrumentation diagram for the resulting control system.

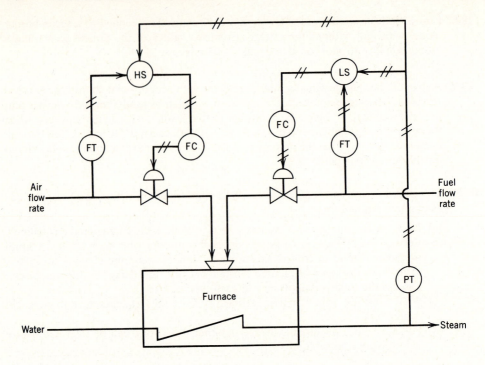

18.15. Selectors are normally used in combustion control systems to prevent unsafe situations from occurring. The diagram above shows the typical configuration for high and low selectors as they are applied to air and fuel flow rates. The energy demand signal comes from the steam pressure controller. Discuss how the selectors operate in this control scheme.

18.16. Buckley et al. [15, p. 207] discuss using a selector to control condensate temperature in a reflux drum, where the manipulated variable is the cooling water flow rate. If the condensate temperature becomes too low, the temperature controller reduces the cooling water flow rate, causing the cooling water exit temperature to rise. However, if the water temperature exceeds 50 to 60 °C, excessive fouling and corrosion can result. Draw a process instrumentation diagram that uses a selector to keep the exit temperature below 50 °C.

18.17. For many chemical processes, the steady-state gain changes when a process operating condition such as throughput changes. Consider a process where the steady-state gain K_p varies with the manipulated variable M according to the relation,

$$K_p = a + \frac{b}{M}$$

where $M > 0$ and a and b are constants that have been determined by fitting steady-state data. Suggest a modification for the standard PID controller to account for this variation in the process gain. Justify your answer. (In the above equation, M is *not* a deviation variable.)

18.18. The product quality from a catalytic tubular reactor is controlled using the flow rate of the entering stream, utilizing composition measurements from a process gas chromatograph. The catalyst decays over time and once its overall activity drops below 50%, it must be recharged. Deactivation usually takes two to three months to occur. One measure of catalyst activity is the average of three temperature measurements located near the peak temperature. Discuss how you would employ an adaptive control scheme to maintain product quality at acceptable levels. What transfer functions would need to be determined and why?

18.19. You wish to control a second-order process using a PID controller. The desired closed-loop servo transfer function is

$$\frac{C}{R} = \frac{1}{\tau_c s + 1} = \frac{G_p G_c}{1 + G_p G_c}$$

and the process model is

$$G_p = \frac{K_p}{(\tau_1 s + 1)(\tau_2 s + 1)}$$

(a) Derive a controller formula that shows how to adjust K_c, τ_I, and τ_D based on variations in K_p, τ_1, and τ_2 and the closed-loop time constant τ_c.

(b) Suppose $\tau_1 = 3$, $\tau_2 = 5$, and $K_p = 1$. Calculate values of K_c, τ_I and τ_D to achieve a $\tau_c = 1.5$. Show how the response deteriorates for changes in the model parameters when the controller remains unchanged, as follows:

   (1) $K_p = 2$
   (2) $K_p = 0.5$
   (3) $\tau_2 = 10$
   (4) $\tau_2 = 1$

18.20. The Ideal Gas Company has a process that requires an adaptive PI controller but the company capital budget has been frozen. Appelpolscher has been given the job to develop a homegrown, cheap adaptive controller. It has been suggested that the closed-loop response after a disturbance can be studied to determine how to adjust K_c and τ_I incrementally up or down, using measures such as settling time, peak error, and decay ratio. Appelpolscher has proposed the following algorithm: If decay ratio > 0.25, reduce K_c. If decay ratio < 0.25, increase K_c. He is not sure how to adjust τ_I. Critique his rule for K_c and propose a rule for changing τ_I.

18.21. Using the data in Fig. 18.21, calculate the mean, standard deviation, and lower and upper control limits. Compare the limits with those shown in the figure. Compute a histogram (frequency distribution) for the data using intervals of 0.1 in pH around the mean.

18.22. A portion of an ammonia plant consists of a gas purification unit, an NH_3 synthesis unit, and an air oxidation unit are shown in the drawing. On Friday you are assigned the job of setting up a statistical quality control on the N_2 stream concentration from the gas purifier. The last 60 analyses are listed in the table. One analysis is made each four hours (twice a shift). On Monday morning you are to report if a process control chart should be employed to analyze statistically the N_2 concentration. Is the N_2 stream in or out of control on Friday morning? (*Source:* D. M. Himmelblau.)

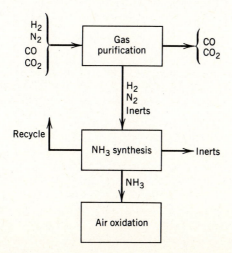

Sample Number	Percent N_2	Sample Number	Percent N_2
1	24.5	31	28.3
2	24.2	32	27.3
3	28.3	33	25.8
4	29.8	34	26.0
5	26.4	35	27.5
6	29.0	36	25.2
7	27.0	37	25.8
8	27.0	38	25.5
9	22.4	39	22.8
10	25.3	40	21.7
11	30.9	41	24.7
12	28.6	42	25.6
13	28.0	43	26.5
14	28.2	44	24.6
15	26.4	45	22.0
16	23.4	46	22.7
17	25.1	47	22.0
18	25.0	48	21.0
19	23.3	49	20.7
20	23.0	50	19.6
21	23.2	51	20.6
22	24.9	52	20.0
23	25.2	53	21.2
24	24.4	54	21.4
25	24.1	55	29.6
26	24.0	56	29.4
27	26.6	57	29.0
28	22.1	58	29.0
29	23.2	59	28.5
30	23.1	60 Friday 8 a.m.	28.7

CHAPTER 19

Control of Multiple-Input, Multiple-Output Processes

Thus far we have emphasized control problems that have only one controlled variable and one manipulated variable. Such problems are referred to as single-input, single-output (SISO) control problems. But in practical control problems there typically are a number of variables that must be controlled and a number of variables that can be manipulated. These problems are referred to as multiple-input, multiple-output (MIMO) control problems. It can be argued that for virtually any important process there are at least two variables that must be controlled, product quality and throughput [1].

Several examples of processes with two controlled variables and two manipulated variables are shown in Fig. 19.1. These examples illustrate a characteristic feature of MIMO control problems, namely, the presence of *process interactions*; that is, each manipulated variable can affect both controlled variables. Consider the in-line blending system shown in Fig. 19.1a. Two streams containing species A and B, respectively, are to be blended to produce a product stream with mass flow rate w and composition x, the mass fraction of A. Adjusting either manipulated flow rate, w_A or w_B, affects both w and x.

Similarly, for the distillation column in Fig. 19.1b, adjusting either reflux flow rate R or steam flow S will affect both distillate composition x_D and bottoms composition x_B. For the gas–liquid separator in Fig. 19.1c, adjusting gas flow rate G will have a direct effect on pressure P and a slower, indirect effect on liquid level h because changing the pressure in the vessel will tend to change the liquid flow rate L and thus affect h. In contrast, adjusting the other manipulated variable L directly affects h but has only a relatively small and indirect effect on P.

When significant process interactions are present, the selection of the most effective control configuration may not be obvious. For example, in the blending problem, suppose that a conventional feedback control strategy, consisting of two PI controllers, is to be used. This control system will be referred to as a *multiloop control system* because it employs two single-loop feedback controllers. Several questions come to mind. Should the composition controller adjust w_A and the flow controller adjust w_B or vice versa? How can we determine which of these two multiloop control configurations will be more effective? Will control loop interactions generated by the process interactions be a problem?

In this chapter, we consider systematic techniques for characterizing process interactions and for selecting an appropriate multiloop control configuration. If the

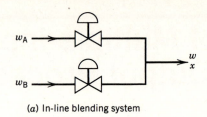

(a) In-line blending system

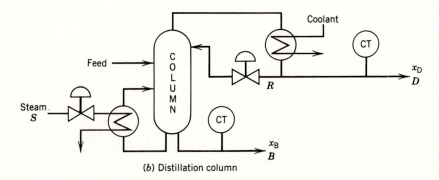

(b) Distillation column

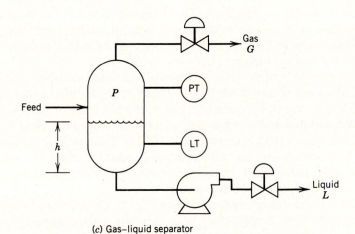

(c) Gas–liquid separator

Figure 19.1.   Physical examples of multivariable control problems.

process interactions are significant, even the best multiloop control system may not provide satisfactory control. In these situations there are incentives for considering *multivariable control strategies* such as decoupling control which will be discussed in Section 19.3. But first we examine the phenomena of process interactions and control loop interactions.

19.1   PROCESS INTERACTIONS AND CONTROL LOOP INTERACTIONS

A schematic representation of SISO and MIMO control problems is shown in Fig. 19.2. For convenience, it is assumed that the number of manipulated variables is equal to the number of controlled variables. Multiple-input, multiple-output control problems are inherently more complex than SISO control problems because process interactions occur between controlled and manipulated variables. In general, a change in a manipulated variable, say M_1, will affect all of the controlled variables

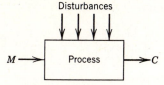

(a) Single-input, single-output process with multiple disturbances

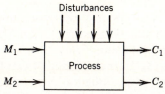

(b) Multiple-input, multiple-output process (2 × 2)

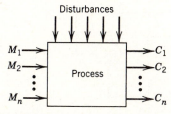

(c) Multiple-input, multiple-output process (n × n)

Figure 19.2. Types of control problems.

$C_1, C_2, \ldots, C_n$. Because of the process interactions, the selection of the best pairing of controlled and manipulated variables for a multiloop control scheme can be a difficult task. In particular, for a control problem with n controlled variables and n manipulated variables, there are $n!$ possible multiloop control configurations. Thus, if $n = 5$ there are 120 possible configurations.

Block Diagram Analysis

Consider the 2×2 control problem shown in Fig. 19.2b. Since there are two controlled variables and two manipulated variables, four process transfer functions are necessary to completely characterize the process dynamics:

$$\frac{C_1(s)}{M_1(s)} = G_{p11}(s) \qquad \frac{C_1(s)}{M_2(s)} = G_{p12}(s)$$

$$\frac{C_2(s)}{M_1(s)} = G_{p21}(s) \qquad \frac{C_2(s)}{M_2(s)} = G_{p22}(s) \tag{19-1}$$

The transfer functions in Eq. 19-1 can be used to determine the effect of a change in either M_1 or M_2. From the Principle of Superposition (Section 3.1), it follows that *simultaneous* changes in M_1 and M_2 have an additive effect on each controlled variable:

$$C_1(s) = G_{p11}(s)M_1(s) + G_{p12}(s)M_2(s) \tag{19-2}$$

$$C_2(s) = G_{p21}(s)M_1(s) + G_{p22}(s)M_2(s) \tag{19-3}$$

These input–output relations can also be expressed in vector–matrix notation as

$$\mathbf{C}(s) = \mathbf{G_p}(s)\mathbf{M}(s) \tag{19-4}$$

where $\mathbf{C}(s)$ and $\mathbf{M}(s)$ are vectors with two elements,

$$\mathbf{C}(s) = \begin{bmatrix} C_1(s) \\ C_2(s) \end{bmatrix} \qquad \mathbf{M}(s) = \begin{bmatrix} M_1(s) \\ M_2(s) \end{bmatrix} \tag{19-5}$$

and $\mathbf{G_p}(s)$ is the process transfer function matrix,

$$\mathbf{G_p}(s) = \begin{bmatrix} G_{p11}(s) & G_{p12}(s) \\ G_{p21}(s) & G_{p22}(s) \end{bmatrix} \tag{19-6}$$

The matrix notation in Eq. 19-4 provides a particularly compact representation for problems larger than 2×2. Recall that a transfer function matrix for an MIMO system, a stirred-tank heating system, was derived in Section 6.7.

Suppose that a conventional multiloop control scheme consisting of two feedback controllers is to be used. The two possible control configurations are shown in Fig. 19.3. In scheme (a), C_1 is controlled by adjusting M_1, while M_2 is used to

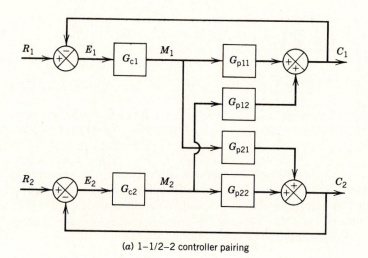

(a) 1–1/2–2 controller pairing

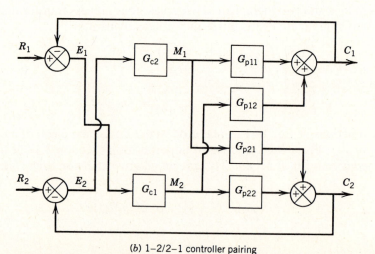

(b) 1–2/2–1 controller pairing

Figure 19.3. A block diagram for conventional multiloop control schemes.

control C_2. Consequently, this configuration will be referred to as the 1-1/2-2 control scheme. The alternative strategy is to pair C_1 with M_2 and C_2 with M_1, the 1-2/2-1 control scheme shown in Fig. 19.3b. Note that these block diagrams have been simplified by setting the transfer functions for the final control elements and the transmitters to unity. Also, the load variables have been omitted.

Figure 19.3 indicates that the process interactions can induce undesirable interactions between the control loops. For example, suppose that the 1-1/2-2 control scheme is used and a disturbance moves C_1 away from its set point R_1. Then the following events occur:

1. The controller for loop 1 (G_{c1}) adjusts M_1 so as to force C_1 back to the set point. However, M_1 also affects C_2 via transfer function G_{p21}.
2. Since C_2 has changed, the loop 2 controller (G_{c2}) adjusts M_2 so as to bring C_2 back to its set point, R_2. However, changing M_2 also affects C_1 via transfer function G_{p12}.

These controller actions proceed simultaneously until a new steady state is reached. Note that the initial change in M_1 has two effects on C_1: a *direct* effect (1) and an *indirect* effect via the control loop interactions (2).

It has been instructive to view this dynamic behavior as a sequence of events. However, in practice the process variables would change continuously and simultaneously.

As Shinskey [1] has noted, the control loop interactions in a 2×2 control problem are due to the presence of a third feedback loop which contains the two controllers and two of the four process transfer functions. Thus, for the 1-1/2-2 configuration, this *hidden feedback loop* contains G_{c1}, G_{c2}, G_{p12}, and G_{p21}, as shown in Fig. 19.4. A similar hidden feedback loop is also present in the 1-2/2-1 control scheme of Fig. 19.3b. The third feedback loop causes two potential problems:

1. It may destabilize the closed-loop system.
2. It tends to make controller tuning more difficult.

Next we show that the transfer function between a controlled variable and a manipulated variable depends on whether the other feedback control loops are open or closed. Consider the control system in Fig. 19.3a. If the controller for the second loop G_{c2} is out of service or is placed in manual with the controller output constant at its nominal value, then $M_2(s) = 0$. For this situation, the transfer

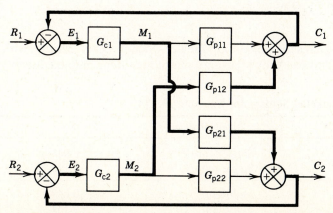

Figure 19.4. The hidden feedback control loop (in dark lines) for a 1-1/2-2 controller pairing.

function between C_1 and M_1 is merely $G_{p11}(s)$:

$$\frac{C_1(s)}{M_1(s)} = G_{p11}(s) \qquad (C_2 - M_2 \text{ loop open}) \qquad (19\text{-}7)$$

However, if the second feedback controller is in the automatic mode, then, in general, $M_2(s) \neq 0$. By using block diagram algebra, we can derive the following relation:

$$\frac{C_1(s)}{M_1(s)} = G_{p11} - \underbrace{\frac{G_{p12} G_{p21} G_{c2}}{1 + G_{c2} G_{p22}}}_{\text{interaction term}} \qquad (C_2 - M_2 \text{ loop closed}) \qquad (19\text{-}8)$$

Thus, the transfer function between C_1 and M_1 depends on the controller for the second loop G_{c2} via the interaction term. Similarly, transfer function $C_2(s)/M_2(s)$ depends on G_{c1} when the first loop is closed (see Exercise 19.1). These results have important implications for controller tuning since they indicate that the two controllers should not be tuned independently. Balchen and Mummé [2] have derived analogous results for general $n \times n$ processes which indicate the effect of closing all but one of the n feedback loops.

EXAMPLE 19.1

Wood and Berry [3] developed the following empirical model of a pilot-scale distillation column:

$$\begin{bmatrix} X_D(s) \\[2ex] X_B(s) \end{bmatrix} = \begin{bmatrix} \dfrac{12.8e^{-s}}{16.7s + 1} & \dfrac{-18.9e^{-3s}}{21s + 1} \\[2ex] \dfrac{6.6e^{-7s}}{10.9s + 1} & \dfrac{-19.4e^{-3s}}{14.4s + 1} \end{bmatrix} \begin{bmatrix} R(s) \\[2ex] S(s) \end{bmatrix}$$

where the notation is defined in Fig. 19.1b. Suppose that a multiloop control system consisting of two PI controllers is used. Compare the closed-loop set-point changes that result if the $X_D - R/X_B - S$ pairing is selected and:

(a) A set-point change is made in each loop with the other loop in manual.
(b) The set-point changes are made with both controllers in automatic.

Assume that the controller settings are based on the Ziegler–Nichols continuous cycling method in Table 13.1.

Solution
Table 19.1 compares the single-loop Ziegler–Nichols (Z–N) settings with alternative values obtained by McAvoy [4] using a multiloop tuning method due to Niederlinski

Table 19.1   Controller Settings for Example 19.1 [4]

Controller Pairing	Tuning Method	K_c	τ_I (min)
$x_D - R$	Single loop/Z-N	0.945	3.26
$x_B - S$	Single loop/Z-N	-0.196	9.00
$x_D - R$	Multiloop	0.647	10.20
$x_B - S$	Multiloop	-0.134	10.20

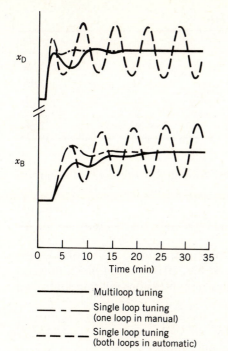

_____ Multiloop tuning

_ _ _ _ Single loop tuning
(one loop in manual)

_ _ _ _ Single loop tuning
(both loops in automatic)

Figure 19.5. The set-point responses for Example 19.1 [4].

[5]. McAvoy [4] has also evaluated the performance of these two multiloop control schemes. Some of his simulation results are shown in Fig. 19.5.

The Z–N settings provide satisfactory set-point responses for either control loop when the other controller is in manual. But when both controllers are in automatic, the control loop interactions produce unstable responses. The multiloop tuning method results in reduced interactions due to the more conservative controller settings.

EXAMPLE 19.2

Repeat the analysis of Example 19.1 for the following distillation column model reported by Luyben and Vinante [6]:

$$
\begin{bmatrix} T_{17}(s) \\ T_4(s) \end{bmatrix} = \begin{bmatrix} \dfrac{-2.16e^{-s}}{8.5s + 1} & \dfrac{1.26e^{-0.3s}}{7.05s + 1} \\ \dfrac{-2.75e^{-1.8s}}{8.25s + 1} & \dfrac{4.28e^{-0.35s}}{9.0s + 1} \end{bmatrix} \begin{bmatrix} R(s) \\ S(s) \end{bmatrix}
$$

where T_4 and T_{17} denote temperatures on the 4th and 17th trays from the bottom of the column, respectively.

Solution

McAvoy [4] has reported the PI controller settings shown in Table 19.2 and the set-point responses of Fig. 19.6. When both controllers are in automatic with Z–N settings, undesirable damped oscillations result due to the control loop interactions. The multiloop tuning method results in more conservative settings and more sluggish responses.

Control loop interactions are less of a problem for this example than they were

Table 19.2   Controller Settings for Example 19.2 [4]

Controller Pairing	Tuning Method	K_c	τ_I (min)
$T_{17} - R$	Single loop/Z-N	-2.92	3.18
$T_4 - S$	Single loop/Z-N	4.31	1.15
$T_{17} - R$	Multiloop	-2.59	2.58
$T_4 - S$	Multiloop	4.39	2.58

for Example 19.1. Section 19.2 discusses a methodology for predicting the importance of such interactions.

To further evaluate the effects of control loop interactions, again consider the block diagram for the 1-1/2-2 control scheme in Fig. 19.3a. Using block diagram algebra (see Chapter 10), we can derive the following expressions relating controlled variables and set points [7,8]:

$$C_1(s) = T_{11}(s)R_1(s) + T_{12}(s)R_2(s) \tag{19-9}$$

$$C_2(s) = T_{21}(s)R_1(s) + T_{22}(s)R_2(s) \tag{19-10}$$

where the closed-loop transfer functions are

$$T_{11}(s) = \frac{G_{c1}G_{p11} + G_{c1}G_{c2}\,(G_{p11}G_{p22} - G_{p12}G_{p21})}{\Delta(s)} \tag{19-11}$$

$$T_{12}(s) = \frac{G_{c2}G_{p12}}{\Delta(s)} \tag{19-12}$$

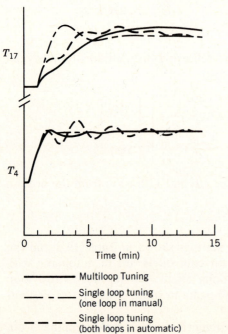

——————	Multiloop Tuning		
— · — · —	Single loop tuning (one loop in manual)		
— — — —	Single loop tuning (both loops in automatic)		

Figure 19.6   The set-point responses for Example 19.2 [4].

$$T_{21}(s) = \frac{G_{c1}G_{p21}}{\Delta(s)} \tag{19-13}$$

$$T_{22}(s) = \frac{G_{c2}G_{p22} + G_{c1}G_{c2}(G_{p11}G_{p22} - G_{p12}G_{p21})}{\Delta(s)} \tag{19-14}$$

and $\Delta(s)$ is defined as

$$\Delta(s) = (1 + G_{c1}G_{p11})(1 + G_{c2}G_{p22}) - G_{c1}G_{c2}G_{p12}G_{p21} \tag{19-15}$$

Two important conclusions can be drawn from these closed-loop relations. First, a set-point change in one loop causes both controlled variables to change since $T_{12}(s)$ and $T_{21}(s)$ are not zero, in general. The second conclusion concerns the stability of the closed-loop system. Since each of the four closed-loop transfer functions in Eqs. 19-11 to 19-14 has the same denominator, the characteristic equation is $\Delta(s) = 0$ or

$$(1 + G_{c1}G_{p11})(1 + G_{c2}G_{p22}) - G_{c1}G_{c2}G_{p12}G_{p21} = 0 \tag{19-16}$$

Thus, the stability of the closed-loop system depends on both controllers, G_{c1} and G_{c2}, and all four process transfer functions. An analogous characteristic equation can be derived for the 1-2/2-1 control scheme in Fig. 19.3b.

For the special case where either $G_{p12} = 0$ or $G_{p21} = 0$, the characteristic equation in Eq. 19-16 reduces to

$$(1 + G_{c1}G_{p11})(1 + G_{c2}G_{p22}) = 0 \tag{19-17}$$

Thus, for this situation, the stability of the overall system merely depends on the stability of the two individual feedback control loops and their characteristic equations,

$$1 + G_{c1}G_{p11} = 0 \quad \text{and} \quad 1 + G_{c2}G_{p22} = 0 \tag{19-18}$$

Note that if either $G_{p12} = 0$ or $G_{p21} = 0$, the third feedback control loop in Fig. 19.4 is broken. For example, if $G_{p12} = 0$, then the second control loop has no effect on C_1 while the first control loop serves as a source of disturbances for the second loop via transfer function G_{p21}.

The above analysis has been based on the 1-1/2-2 control configuration in Fig. 19.3a. A similar analysis and conclusions can be derived for the 1-2/2-1 configuration (see Exercise 19.2). The results in Eqs. 19-9 to 19-18 can be easily extended to block diagrams which include the transfer functions for the transmitters and control valves [8].

EXAMPLE 19.3

Consider a process that can be described by the transfer function matrix,

$$\mathbf{G_p}(s) = \begin{bmatrix} \dfrac{2}{10s + 1} & \dfrac{1.5}{s + 1} \\ \dfrac{1.5}{s + 1} & \dfrac{2}{10s + 1} \end{bmatrix}$$

Assume that the two proportional feedback controllers are to be used so that $G_{c1} = K_{c1}$ and $G_{c2} = K_{c2}$. Determine the values of K_{c1} and K_{c2} that result in closed-loop stability for both the 1-1/2-2 and 1-2/2-1 configurations.

Solution

The characteristic equation for the closed-loop system is obtained by substitution into Eq. 19-16:

$$a_4 s^4 + a_3 s^3 + a_2 s^2 + a_1 s + a_0 = 0 \qquad (19\text{-}19)$$

where $a_4 = 100$

$$a_3 = 20 K_{c1} + 20 K_{c2} + 220$$

$$a_2 = 42 K_{c1} + 42 K_{c2} - 221 K_{c1} K_{c2} + 141$$

$$a_1 = 24 K_{c1} + 24 K_{c2} + 8 K_{c1} K_{c2} + 22$$

$$a_0 = 2 K_{c1} + 2 K_{c2} + 1.75 K_{c1} K_{c2} + 1$$

Note that the characteristic equation in (19-19) is fourth order even though each individual transfer function in $\mathbf{G_p}(s)$ is first order!

The controller gains that result in a stable closed-loop system can be determined by applying the Routh criterion (Chapter 11) for specified values of K_{c1} and K_{c2}. The resulting stability regions, calculated by Gagnepain [7], are shown in Fig. 19.7. If either K_{c1} or K_{c2} is close to zero, the other controller gain can be an arbitrarily large, positive value and still have a stable closed-loop system. This result is a consequence of having process transfer functions that are first-order without time delay.

A similar stability analysis was performed for the 1-2/2-1 control configuration [7]. The calculated stability regions are shown in Fig. 19.8. A comparison of Figs. 19.7 and 19.8 indicates that the 1-2/2-1 control scheme results in a larger stability region since a wider range of controller gains can be used. For example, suppose that $K_{c1} = 2$. Then Fig. 19.7 indicates that the 1-1/2-2 configuration will be stable if $-0.9 < K_{c2} < 0.6$. By contrast, Fig. 19.8 shows that the corresponding stability limits for the 1-2/2-1 configuration are $-0.8 < K_{c2} < 2.2$.

This example illustrates that closed-loop stability depends on the control configuration as well as the numerical values of the controller settings. If PI control had been considered instead of proportional-only control, the stability analysis would have been much more complicated due to the larger number of controller settings and the higher dimension characteristic equation.

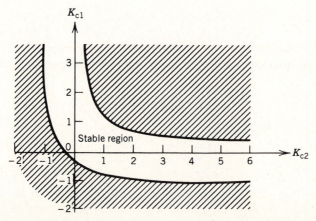

Figure 19.7. Stability region for Example 19.3 with 1-1/2-2 controller pairing [7].

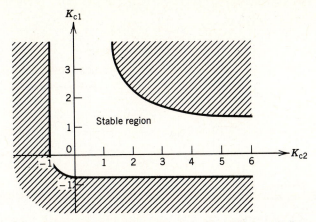

Figure 19.8. Stability region for Example 19.3 with 1-2/2-1 controller pairing [7].

Tuning Multiloop Control Systems

Several typical tuning procedures for multiloop control systems will be briefly described. For purposes of illustration, 2×2 processes will be considered, but the tuning methods are also applicable to higher dimension processes.

One widely used approach is to determine satisfactory controller settings for each loop with the other loop in manual. If these settings are also satisfactory when both loops are placed in automatic, then no further adjustment is required. However, if undesirable control loop interactions occur, then further tuning is necessary. Typically, this is done by making the controller settings more conservative, that is, by reducing the gains and increasing the integral times in one or more loops. For example, in a 2×2 control problem, one could choose to detune the control loop for the less important controlled variable.

An alternative approach is available for situations where one of the two controlled variables is more important than the other. In this situation, the controller for the more important controlled variable could be tuned with the other controller in manual. After satisfactory controller settings were obtained, the second controller would then be tuned with the first controller left in automatic. The controller gain of the less important loop would be kept low enough so it would not affect the operation of the more important loop. This approach tends to give controller settings that favor the more important controlled variable at the expense of the other.

Other tuning methods for multiloop control systems have been proposed but have not been widely applied. For example, the Niederlinski method [5] used in Examples 19.1 and 19.2 is based on a generalization of the continuous cycling method discussed in Chapter 13.

19.2 PAIRING OF CONTROLLED AND MANIPULATED VARIABLES

In this section we consider the important practical problem of how the controlled variables and the manipulated variables should be paired in a multiloop control scheme. An incorrect pairing can result in poor control system performance and reduced stability margins as was the case for the 1-1/2-2 pairing in Example 19.3. As an illustrative example, we consider the distillation column shown in Fig. 19.9. A typical distillation column has five controlled variables and five manipulated variables [1]. The controlled variables in Fig. 19.9 are product compositions x_D

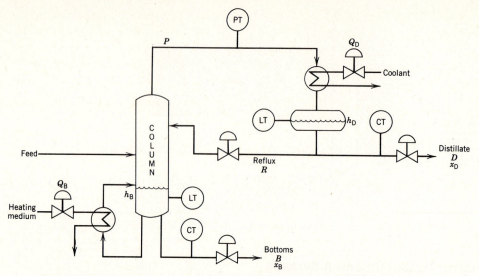

Figure 19.9. Controlled and manipulated variables for a typical distillation column.

and x_B, column pressure P, and the liquid levels in the reflux drum h_D and column base h_B. The five manipulated variables are product flows D and B, reflux flow R, and the heat duties for the condenser and reboiler, Q_D and Q_B. The heat duties are adjusted via the control valves on the steam and coolant lines. Note that if a multiloop control scheme consisting of five feedback controllers is used, there are $5! = 120$ different ways of pairing the controlled and manipulated variables. Some of these configurations would be immediately rejected as being impractical or unworkable, for example, any scheme that attempts to control base level h_B by adjusting distillate flow D or condenser heat duty Q_D. However, there may be a number of alternative pairings that seem promising; the question then facing the control system designer is how to determine the most effective pairing.

Next, we consider a systematic approach for determining the best pairing of controlled and manipulated variables, the Relative Gain Array Method [9]. A promising new approach based on singular value analysis [10,11] is described in Chapter 28.

Bristol's Relative Gain Array Method

Bristol [9] developed the first systematic approach for the analysis of multivariable process control problems. His approach requires only steady-state information and provides two important items of information:

1. A measure of process interactions.
2. A recommendation concerning the most effective pairing of controlled and manipulated variables.

Bristol's approach is based on the concept of a *relative gain*. Consider a process with n controlled variables and n manipulated variables. The relative gain λ_{ij} between a controlled variable, C_i, and a manipulated variable, M_j, is defined to be the dimensionless ratio of two steady-state gains:

$$\lambda_{ij} = \frac{(\partial C_i/\partial M_j)_M}{(\partial C_i/\partial M_j)_C} = \frac{\text{open-loop gain}}{\text{closed-loop gain}} \qquad (19\text{-}20)$$

for $i = 1, 2, \ldots, n$ and $j = 1, 2, \ldots, n$.

In Eq. 19-20 the symbol, $(\partial C_i / \partial M_j)_M$, denotes a partial derivative that is evaluated with all of the manipulated variables except M_j held constant. Thus, this term is the *open-loop gain* between C_i and M_j. Similarly, $(\partial C_i / \partial M_j)_C$ is evaluated with all of the controlled variables except C_i held constant. This situation could be achieved in practice by adjusting the other manipulated variables using controllers with integral action. Thus $(\partial C_i / \partial M_j)_C$ can be interpreted as a *closed-loop gain* that indicates the effect of M_j on C_i when all of the other feedback control loops are closed.

It is convenient to arrange the relative gains in a *relative gain array* (*RGA*), denoted by $\mathbf{\Lambda}$:

$$\mathbf{\Lambda} = \begin{array}{c} \\ C_1 \\ C_2 \\ \cdots \\ C_n \end{array} \overset{\begin{array}{cccc} M_1\ M_2 & \cdots & M_n \end{array}}{\begin{bmatrix} \lambda_{11}\ \lambda_{12} & \cdots & \lambda_{1n} \\ \lambda_{21}\ \lambda_{22} & \cdots & \lambda_{2n} \\ \cdots\cdots\cdots & & \cdots \\ \lambda_{n1}\ \lambda_{n2} & \cdots & \lambda_{nn} \end{bmatrix}} \tag{19-21}$$

The RGA has two important properties [9]:

1. It is normalized since the sum of the elements in each row or column is one.
2. The relative gains are dimensionless and thus not affected by choice of units or scaling of variables.

Calculation of the Relative Gain Array

The relative gains can easily be calculated from either steady-state data or a process model. For example, consider a 2×2 process for which a steady-state model is available. Suppose that the model has been linearized and expressed in terms of deviation variables as follows:

$$C_1 = K_{11}M_1 + K_{12}M_2 \tag{19-22}$$

$$C_2 = K_{21}M_1 + K_{22}M_2 \tag{19-23}$$

where K_{ij} denotes the steady-state gain between C_i and M_j. This model can be expressed more compactly in matrix notation as

$$\mathbf{C} = \mathbf{K}\,\mathbf{M} \tag{19-24}$$

Note that the steady-state (gain) model in Eq. 19-24 is related to the dynamic model in Eq. 19-4 by

$$\mathbf{K} = \mathbf{G_p}(0) = \lim_{s \to 0} \mathbf{G_p}(s) \tag{19-25}$$

Next, we consider how to calculate λ_{11}. It follows from Eq. 19-22 that

$$\left(\frac{\partial C_1}{\partial M_1} \right)_{M_2} = K_{11} \tag{19-26}$$

Before calculating $(\partial C_1 / \partial M_1)_{C_2}$ from Eq. 19-22, we first must eliminate M_2. This is done by solving Eq. 19-23 for M_2 with C_2 held constant at its nominal value, $C_2 = 0$:

$$M_2 = -\frac{K_{21}}{K_{22}} M_1 \tag{19-27}$$

Then substituting into Eq. 19-22 gives

$$C_1 = K_{11} \left(1 - \frac{K_{12} K_{21}}{K_{11} K_{22}} \right) M_1 \tag{19-28}$$

It follows that

$$\left(\frac{\partial C_1}{\partial M_1} \right)_{C_2} = K_{11} \left(1 - \frac{K_{12} K_{21}}{K_{11} K_{22}} \right) \tag{19-29}$$

Substituting Eqs. 19-26 and 19-29 into Eq. 19-20 gives an expression for the relative gain,

$$\lambda_{11} = \frac{1}{1 - \dfrac{K_{12} K_{21}}{K_{11} K_{22}}} \tag{19-30}$$

Since each row and each column of Λ in (19-21) sums to one, the other relative gains are easily calculated from λ_{11}:

$$\lambda_{12} = \lambda_{21} = 1 - \lambda_{11} \qquad \lambda_{22} = \lambda_{11} \tag{19-31}$$

Thus, the RGA for a 2×2 system can be expressed as

$$\Lambda = \begin{bmatrix} \lambda & 1 - \lambda \\ 1 - \lambda & \lambda \end{bmatrix} \tag{19-32}$$

where the symbol λ is now used to denote λ_{11}. Note that the RGA for a 2×2 process is always symmetric. However, this will not necessarily be the case for a higher-dimension process ($n > 2$).

For higher-dimension processes, the RGA can be calculated from the expression

$$\lambda_{ij} = K_{ij} H_{ij} \tag{19-33}$$

where K_{ij} is the (i, j) element of $\mathbf{K}$ in Eq. 19-24 and H_{ij} is the (i, j) element of $\mathbf{H} = (\mathbf{K}^{-1})^T$, that is, H_{ij} is an element of the transpose of the matrix inverse of $\mathbf{K}$. Since computer programs are readily available to perform matrix algebra, Eq. 19-33 can be easily evaluated on a digital computer.

Note that Eq. 19-33 does *not* imply that $\Lambda = \mathbf{K} (\mathbf{K}^{-1})^T$. This incorrect expression has been reported in several books and articles.

The derivation in Eqs. 19-22 through 19-33 has demonstrated that the RGA can be calculated from a linearized steady-state process model. An alternative approach is to calculate the RGA from process gains that are evaluated numerically either from a nonlinear process model or directly from experimental data. For example, K_{11} can be evaluated as $K_{11} = \Delta C_1/\Delta M_1$ where ΔC_1 is the steady-state change in C_1 that is produced by a steady-state change of ΔM_1 in M_1. However, the perturbations in the manipulated variables (ΔM_1, ΔM_2, etc.) must be chosen carefully so that the calculated gains and RGA do not change significantly with the size of the perturbations. This approach has been considered in detail elsewhere [12,13].

Measure of Process Interactions

The relative gain array can be used to provide a measure of the process interactions. Recall that the sum of the RGA elements is unity for each row and for each column.

For the 2×2 process in Eq. 19-32, five cases are possible:

1. **λ = 1.** In this situation, it follows from (19-20) that the open-loop and closed-loop gains between C_1 and M_1 are identical. Thus, opening or closing loop 2 has no effect on loop 1. It follows that C_1 should be paired with M_1 (i.e., a 1-1/2-2 configuration should be employed).

2. **λ = 0.** From Eq. 19-20 it follows that the open-loop gain between C_1 and M_1 is zero and thus M_1 has no direct effect on C_1. Consequently, M_1 should be paired with C_2 rather than C_1 (i.e., the 1-2/2-1 configuration should be utilized).

3. **0 < λ < 1.** From Eq. 19-20, we conclude that the closed-loop gain between C_1 and M_1 is larger than the open-loop gain. Thus, the control loops interact and the interactions are most severe when $\lambda = 0.5$ [1,12].

4. **λ > 1.** For this situation, closing the second loop reduces the gain between C_1 and M_1. Thus, the control loops interact. As λ increases, the degree of interaction becomes more severe.

5. **λ < 0.** When λ is negative, the open-loop and closed-loop gains between C_1 and M_1 have different signs. Thus, opening or closing loop 2 has a serious and undesirable effect on loop 1. It follows that C_1 should *not* be paired with M_1. For $\lambda < 0$ the control loops interact by trying to "fight each other" [1,12]. The degree of interaction becomes more severe as $\lambda \rightarrow -\infty$.

Based on these considerations, the RGA analysis for a 2×2 process leads to the conclusion that C_1 should be paired with M_1 only if $\lambda \geq 0.5$. Otherwise C_1 should be paired with M_2. This reasoning can be extended to $n \times n$ processes and leads to Bristol's original recommendaton for controller pairing [9]:

Recommendation: *Pair the controlled and manipulated variables so that the corresponding relative gains are positive and as close to one as possible.*

At this point, it is appropriate to make several remarks about the RGA approach.

1. The above recommendation is based solely on steady-state information. But process dynamics should also be considered in choosing a controller pairing. In particular, closed-loop stability should be checked using a theorem that is presented in the next section.

2. If $\lambda = 0$ or $\lambda = 1$, the two control loops for a 2×2 process either do not interact at all or exhibit only a *one-way interaction,* based on this steady-state analysis.[1] Furthermore, at least one of the four process gains must be zero according to Eq. 19-30.

A one-way interaction occurs when one loop affects the other loop but not vice versa. For example, suppose that **K** has the structure,

$$\mathbf{K} = \begin{bmatrix} K_{11} & K_{12} \\ 0 & K_{22} \end{bmatrix}$$

Then loop 1 does not affect loop 2 because $K_{21} = 0$ and thus M_1 has no effect on C_2. However, loop 2 does affect loop 1 via M_2 if $K_{12} \neq 0$. This one-way interaction does not affect the closed-loop stability since $G_{p21} = 0$ and, consequently, the characteristic equation in Eq. 19-16 reduces to the two equations in (19-18).

---

[1]However, Friedly [15] has shown that in unusual situations, significant steady-state interactions can occur even if $\lambda \approx 1$.

Thus, for this one way interaction, loop 2 tends to act as a source of disturbances for loop 1.

To illustrate how the RGA can be used to determine controller pairing, we consider four examples. Additional examples have been reported by McAvoy [13].

EXAMPLE 19.4

Consider the in-line blending system of Fig. 19.1a. It is proposed that w and x be controlled using a conventional multiloop control scheme with w_A and w_B as the manipulated variables. Derive an expression for the RGA and recommend the best controller pairing for the following conditions: $w = 4$ lb/min and $x = 0.4$.

Solution

Assuming perfect mixing, a process model can be derived from the following steady-state mass balances:

$$\text{Total mass:} \quad w = w_A + w_B \tag{19-34}$$

$$\text{Component A:} \quad xw = w_A \tag{19-35}$$

Substituting (19-34) into (19-35) and rearranging gives

$$x = \frac{w_A}{w_A + w_B} \tag{19-36}$$

The RGA for the blending system can be expressed as

$$
\begin{array}{c c}
 & \begin{array}{c c} w_A & w_B \end{array} \\
\begin{array}{c} w \\ x \end{array} & \left[\begin{array}{c c} \lambda & 1 - \lambda \\ 1 - \lambda & \lambda \end{array} \right]
\end{array}
$$

Relative gain λ can be calculated from Eq. 19-30 after the four steady-state gains are calculated:

$$K_{11} = \left(\frac{\partial w}{\partial w_A} \right)_{w_B} = 1 \tag{19-37}$$

$$K_{12} = \left(\frac{\partial w}{\partial w_B} \right)_{w_A} = 1 \tag{19-38}$$

$$K_{21} = \left(\frac{\partial x}{\partial w_A} \right)_{w_B} = \frac{w_B}{(w_A + w_B)^2} = \frac{1 - x}{w} \tag{19-39}$$

$$K_{22} = \left(\frac{\partial x}{\partial w_B} \right)_{w_A} = \frac{-w_A}{(w_A + w_B)^2} = -\frac{x}{w} \tag{19-40}$$

Substituting into Eq. 19-30 gives $\lambda = x$. Thus, the RGA is

$$
\begin{array}{c c}
 & \begin{array}{c c} w_A & w_B \end{array} \\
\begin{array}{c} w \\ x \end{array} & \left[\begin{array}{c c} x & 1 - x \\ 1 - x & x \end{array} \right]
\end{array}
$$

Note that the recommended pairing depends on the desired product composition x. For $x = 0.4$, w should be paired with w_B and x with w_A. Because all four relative gains are close to 0.5, control loop interactions will be a serious problem. On the other hand, if $x = 0.9$, w should be paired with w_A and x with w_B. In this case, the control loop interactions will be small. Note that for both cases, total flow rate w is controlled by the larger component flow rate, w_A or w_B.

EXAMPLE 19.5

The relative gain array for a refinery distillation column associated with a hydrocracker [14] is shown in Eq. 19-41:

$$
\Lambda = \begin{array}{c} \\ C_1 \\ C_2 \\ C_3 \\ C_4 \end{array}
\begin{array}{cccc} M_1 & M_2 & M_3 & M_4 \end{array}
\begin{bmatrix} 0.931 & 0.150 & 0.080 & -0.164 \\ -0.011 & -0.429 & 0.286 & 1.154 \\ -0.135 & 3.314 & -0.270 & -1.910 \\ 0.215 & -2.030 & 0.900 & 1.919 \end{bmatrix} \qquad (19\text{-}41)
$$

The four controlled variables are the compositions of the top and bottom product streams and the two side streams. The manipulated variables are the four corresponding flow rates.

Solution

To determine the recommended controller pairs, we identify the positive relative gains that are closest to one in each row and column. From the rows, it is apparent that the recommended pairings are: $C_1 - M_1$, $C_2 - M_4$, $C_3 - M_2$ and $C_4 - M_3$. Note that this pairing assigns M_2 to C_3 rather than to C_1 even though its relative gain of 3.314 is farther from 1.0. This choice is required because pairing any other manipulated variable with C_3 corresponds to a negative relative gain, which is undesirable.

The two previous examples have shown how the RGA can be calculated from steady-state gain information. For integrating processes such as the liquid storage system considered in Section 5.3, one or more steady-state gains do not exist. Consequently, the standard RGA analysis must be modified for such systems. Woolverton [16] has proposed that the RGA analysis proceed in the usual manner except that any controlled variable that is the output of an integrating element should be replaced by its rate of change. Thus, if a liquid level h is both a controlled variable and the output of an integrating element, then h would be replaced by dh/dt in the RGA analysis. This procedure is illustrated in Exercise 19.8.

Dynamic Considerations

An important disadvantage of the standard RGA approach is that it ignores process dynamics that can be an important factor in the pairing decision. For example, if the transfer function between C_1 and M_1 contains a very large time delay or time constant (relative to the other transfer functions), C_1 will respond very slowly to changes in M_1. Thus, in this situation, a $C_1 - M_1$ pairing is not desirable due to dynamic considerations. McAvoy [13, p. 214] has noted that dynamic interactions tend to be more important for 2×2 processes when $\lambda > 1$ rather than when $0 < \lambda < 1$. However, dynamic considerations can still affect the pairing decision even when $0 < \lambda < 1$, as illustrated in the following example.

EXAMPLE 19.6

Consider the transfer function model of Example 19.3 but with a gain of -2 in $G_{p11}(s)$ and a time delay of unity in each transfer function:

$$\mathbf{G_p}(s) = \begin{bmatrix} \dfrac{-2e^{-s}}{10s + 1} & \dfrac{1.5e^{-s}}{s + 1} \\[3mm] \dfrac{1.5e^{-s}}{s + 1} & \dfrac{2e^{-s}}{10s + 1} \end{bmatrix} \tag{19-42}$$

Use the RGA approach to determine the recommended controller pairing based on steady-state considerations. Do dynamic considerations suggest the same pairing?

Solution

The corresponding steady-state gain matrix is

$$\mathbf{K} = \begin{bmatrix} -2 & 1.5 \\ 1.5 & 2 \end{bmatrix} \tag{19-43}$$

Using the formula in Eq. 19-30, we obtain $\lambda_{11} = 0.64$. Thus, the RGA analysis indicates that the 1-1/2-2 pairing should be used. However, the off-diagonal time constants in Eq. 19-42 are only one-tenth of the diagonal time constants. Thus C_1 responds ten times faster to M_2 than to M_1; similarly, C_2 responds ten times faster to M_1 than to M_2. Consequently, the 1-2/2-1 pairing is favored based on dynamic considerations and a conflict exists between steady-state and dynamic considerations. A computer simulation of the two alternative control configurations for this example has shown that the 1-2/2-1 configuration provides better control [7,17]. Thus, for this example the RGA analysis provides an incorrect recommendation concerning the more effective controller pairing. Extensions of the RGA to treat dynamics are discussed later in this section.

Useful information about the stability of a proposed multiloop control system can be obtained using a theorem originally reported by Niederlinski [5] and later corrected by Grosdidier et al. [18]. Like the RGA analysis, the theorem is based solely on steady-state information. It is assumed that the steady-state gain matrix **K** has been arranged so that the diagonal elements correspond to the proposed pairing; that is, it is assumed that C_1 is paired with M_1, C_2 with M_2, and so on. This arrangement can always be obtained by reordering the elements of the **C** and **M** vectors if necessary.

The following theorem is based on three assumptions similar to those stated by Grosdidier et al. [18]:

1. Let $G_{pij}(s)$ denote the (i, j) element of process transfer function matrix, $\mathbf{G_p}(s)$. Each $G_{pij}(s)$ must be stable, rational, and proper; that is, the order of the denominator must be at least as great as the order of the numerator.
2. Each of the n feedback controllers in the multiloop control system contains integral action.
3. Each individual control loop is stable when any of the other $n - 1$ loops are opened.

Stability Theorem [18] *Suppose that a multiloop control system is used with the pairing* $C_1 - M_1$, $C_2 - M_2$, . . . , $C_n - M_n$. *If the closed-loop system satisfies Assumptions 1–3, then the closed-loop system is unstable if*

$$\frac{|\mathbf{K}|}{\prod\limits_{i=1}^{n} K_{ii}} < 0 \qquad (19\text{-}44)$$

where $|\mathbf{K}|$ *denotes the determinant of* $\mathbf{K}$.

Note that this theorem provides a sufficient (but not necessary) condition for instability. Thus, if the inequality is satisfied, the closed-loop system will be unstable. But if the inequality is not satisfied, the closed-loop system may or may not be unstable depending on the numerical values of the controller settings. For a 2 × 2 process, the inequality in Eq. 19-44 is both necessary and sufficient [18]. The inequality is also satisfied if the proposed pairing for a 2 × 2 system corresponds to a negative value of a relative gain [17]. McAvoy [13, p. 84] reports several examples where apparently reasonable RGA pairings result in unstable closed-loop systems. Thus, it is important to consider the process dynamics and also check to ensure that a proposed pairing does not satisfy the inequality in Eq. 19-44.

Since Assumption 1 requires that each $G_{pij}(s)$ be a rational function, the theorem does not strictly apply to processes that contain time delays. However, because time delays do not affect the steady-state matrix $\mathbf{K}$, the theorem still provides useful insight into the stability of such systems even though the analysis is no longer rigorous [18].

In recent years, several researchers have suggested alternatives to, or extensions of, Bristol's RGA method that consider the process dynamics as well as steady-state gains [13,15,17,19–21]. However, these newer methods tend to be more complicated than the standard RGA approach and have not yet gained widespread acceptance. Other recent papers [22,23] have extended the RGA approach to consider the effect of disturbances on multiloop control systems.

Additional information about RGA analysis and a wide variety of applications have been reported by Shinskey [1] and McAvoy [13].

19.3 STRATEGIES FOR REDUCING CONTROL LOOP INTERACTIONS

In Section 19.1 we described how process interactions between manipulated and controlled variables can result in undesirable control loop interactions. When control loop interactions are a problem, a number of alternative strategies are available:

1. "Detune" one or more feedback controllers.
2. Select different manipulated or controlled variables.
3. Consider a decoupling controller.
4. Consider a multivariable control scheme.

Alternatives 1 and 2 will be considered in this section. Decoupling control is the subject of Section 19.4, while more general multivariable control systems are introduced in Section 19.5.

The term *detuning* refers to using a conservative choice of controller settings that results in more sluggish closed-loop responses. For most control loops, this means using a smaller value of K_c and a larger value of τ_I than would be used for a SISO control loop. Detuning a controller tends to reduce control loop interactions but at the cost of sacrificing the performance of the detuned loop.

Selection of Different Controlled and Manipulated Variables

For some control problems, loop interactions can be significantly reduced by choosing alternative controlled and manipulated variables. For example, the new controlled or manipulated variable could be a simple function of the original variables such as a sum, or difference, or ratio [13,24,25]. Weber and Gaitonde [24] successfully controlled an industrial distillation column by using simple, nonlinear functions of x_D and x_B as the controlled variables rather than x_D and x_B, themselves. To illustrate this type of approach, we again consider the blending system of Example 19.4.

EXAMPLE 19.7 [13]

For the blending system of Example 19.4, choose a new set of manipulated variables that will reduce control loop interactions by making $\lambda = 1$.

Solution

From the expression for the relative gain in Eq. 19-30, it is clear that $\lambda = 1$ if K_{12} and/or $K_{21} = 0$. Thus, we want to choose manipulated variables so that the steady-state gain matrix has a zero for at least one of the off-diagonal elements. Inspection of the process model in Eqs. 19-34 and 19-36 suggests that suitable choices for the manipulated variables are $M_1 = w_A + w_B$ and $M_2 = w_A$. Substitution into the process model gives an equivalent model in terms of the new variables:

$$w = M_1 \tag{19-45}$$

$$x = \frac{M_2}{M_1} \tag{19-46}$$

Linearizing (19-46) gives the gain matrix $\mathbf{K}$ in Eq. 19-24 where

$$\mathbf{K} = \begin{bmatrix} 1 & 0 \\ -\dfrac{M_2}{M_1^2} & \dfrac{1}{M_1} \end{bmatrix} \tag{19-47}$$

vectors $\mathbf{C}$ and $\mathbf{M}$ are defined as $\mathbf{C} = [w, x]$ and $\mathbf{M} = [M_1, M_2]$. Because w depends on M_1 but not M_2, the only feasible controller pairing is $w - M_1$ and $x - M_2$. From Eqs. 19-47 and 19-30 it follows that $\lambda = 1$ since $K_{12} = 0$. Because $K_{21} \neq 0$, there will be a one-way interaction with the $w - M_1$ control loop generating disturbances that affect the $x - M_2$ loop but not vice versa. A practical implementation of this control scheme is shown in Fig. 19.10.

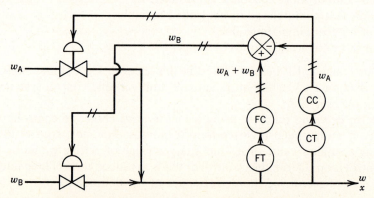

Figure 19.10.   A control scheme for Example 19.6 [15].

McAvoy [13, p. 136] suggests alternative manipulated variables, namely, $M_1 = w_A + w_B$ and $M_2 = w_A/(w_A + w_B)$. This choice is motivated by the process model in Eqs. 19-34 and 19-36 and means that the controlled variables are identical to the manipulated variables! Thus, **K** is the identity matrix, $\lambda = 1$, and the two control loops do not interact at all. This situation is fortuitous but also unusual because it is seldom possible to choose manipulated variables that are, in fact, the controlled variables.

The selection of appropriate manipulated and controlled variables that reduce control loop interactions tends to be an art rather than a science. However, Example 19.7 has indicated a general guideline: Select new manipulated or controlled variables so that one or more zeros appear in the steady-state gain matrix.

If neither controller detuning nor selection of new manipulated or controlled variables provides a satisfactory solution, then decoupling control or a multivariable control scheme should be considered. These topics are covered in the next two sections.

19.4 DECOUPLING CONTROL SYSTEMS

In decoupling control, the design objective is to reduce control loop interactions by adding additional controllers called *decouplers* to a conventional multiloop configuration. In principle, decoupling control schemes can provide two important benefits:

1. Control loop interactions are eliminated and consequently the stability of the closed-loop system is determined by the stability characteristics of the individual feedback control loops.
2. A set-point change for one controlled variable has no effect on the other controlled variables.

In practice, these theoretical benefits may not be fully realized due to imperfect process models. Typically, decouplers are designed using a simple process model that can be either a steady-state or dynamic model.

One type of decoupling control system for a process with two inputs and two outputs is shown in Fig. 19.11. Note that four controllers are used: two conventional feedback controllers, G_{c1} and G_{c2}, plus two decouplers, D_{12} and D_{21}. The input signal to each decoupler is the output signal from a feedback controller. In Fig. 19.11, the transfer functions for the transmitters and final control elements have been omitted for the sake of simplicity. Gould [8] has discussed the more general case where these transfer functions are included.

The decouplers are designed to compensate for undesirable process interactions. For example, decoupler D_{21} can be designed so as to cancel C_{21}, which arises from the process interaction between M_1 and C_2. This cancellation will occur at the C_2 summer if the decoupler output M_{21} satisfies

$$G_{p21}M_{11} + G_{p22}M_{21} = 0 \tag{19-48}$$

Substituting $M_{21} = D_{21}M_{11}$ and factoring gives

$$(G_{p21} + G_{p22}D_{21})M_{11} = 0 \tag{19-49}$$

But $M_{11}(s) \neq 0$ since M_{11} is a controller output that is time dependent. Thus, to satisfy Eq. 19-49, it follows that

$$G_{p21} + G_{p22}D_{21} = 0 \tag{19-50}$$

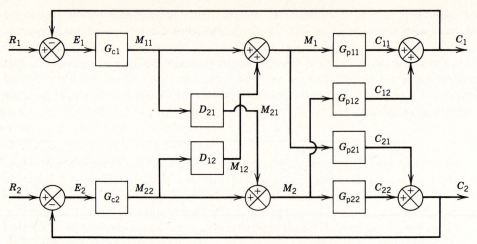

Figure 19.11. A decoupling control system.

Solving for D_{21} gives an expression for the ideal decoupler,

$$D_{21}(s) = -\frac{G_{p21}(s)}{G_{p22}(s)} \qquad (19\text{-}51)$$

In an analogous fashion, we can derive a design equation for $D_{12}(s)$ by imposing the requirement that M_{22} have no net effect on C_1. Thus, the compensating signal M_{12} and the process interaction due to G_{p12} should cancel at the C_1 summer,

$$G_{p12}M_{22} + G_{p11}D_{12}M_{22} = 0 \qquad (19\text{-}52)$$

The ideal decoupler is given by

$$D_{12}(s) = -\frac{G_{p12}(s)}{G_{p11}(s)} \qquad (19\text{-}53)$$

The ideal decouplers in Eqs. 19-51 and 19-53 are very similar to the ideal feedforward controller in Eq. 17-27 with $G_t = G_v = 1$. In fact, one can interpret the decoupler as a type of feedforward controller with an input signal that is a manipulated variable rather than a disturbance variable. We saw in Chapter 17 that the ideal feedforward controller may not be physically realizable. Similarly, ideal decouplers are not always physically realizable as demonstrated in the following example.

EXAMPLE 19.8

A process has the transfer function matrix:

$$\mathbf{G_p}(s) = \begin{bmatrix} \dfrac{5e^{-5s}}{4s + 1} & \dfrac{2e^{-4s}}{8s + 1} \\[3mm] \dfrac{3e^{-3s}}{12s + 1} & \dfrac{6e^{-3s}}{10s + 1} \end{bmatrix} \qquad (19\text{-}54)$$

Design ideal decouplers and indicate how they can be simplified based on practical considerations.

Solution

Substitution into Eq. 19-51 gives the ideal decoupler:

$$D_{21}(s) = -\frac{0.5(10s + 1)}{12s + 1} \tag{19-55}$$

Thus, this decoupler is a lead–lag unit. Since the lead and lag time constants are almost the same, they almost cancel and the decoupler can be approximated as a simple gain, $D_{21}(s) = -0.5$. In practice, the exact cancellations implied by Eqs. 19-50 and 19-52 can seldom be realized because of the inexact nature of process models. Fortunately, even approximate cancellations can be very beneficial in reducing control loop interactions while simplifying the controllers. Thus, replacing (19-55) by $D_{21}(s) = -0.5$ is a reasonable simplification.

Substitution into Eq. 19-53 gives the other decoupler,

$$D_{12}(s) = -\frac{0.25(4s + 1)e^s}{8s + 1} \tag{19-56}$$

which is physically unrealizable due to the prediction term e^s.

The reason this unfortunate situation occurs is apparent from inspection of $G_{p11}(s)$ and $G_{p12}(s)$. Since $G_{p11}(s)$ contains the larger time delay, M_2 affects C_1 sooner than M_1 does. Thus, it is impossible to completely eliminate the interaction between M_2 and C_1. Since the argument of the prediction term in Eq. 19-56 is relatively small in comparison with the process time constants, a reasonable approximation is given by

$$D_{12}(s) = -\frac{0.25(4s + 1)}{8s + 1} \tag{19-57}$$

If the time delay in the prediction term had been significant, then this type of approximation could result in poor control.

One alternative approach would be to use four decouplers as proposed by Waller [26]. In his approach, the decouplers are always physically realizable for all time delays but additional time delays may be introduced into the feedback control paths. A second option is to use *static decoupling,* that is, the steady-state version of Eq. 19-57 obtained by setting $s = 0$.

Alternative Decoupling Control System

An alternative to the decoupling control scheme of Fig. 19.11 is shown in Fig. 19.12 where the input signals to the decouplers are the manipulated variables M_1 and M_2. This decoupling configuration has the advantage that the decouplers are aware when a manipulated variable reaches a constraint [1]. In contrast, the decoupling scheme in Fig. 19.11 ignores saturation of the manipulated variables and may tend to "wind up" if saturation occurs. Thus, it may be necessary to monitor the manipulated variables and take appropriate corrective action to prevent windup.

Unfortunately, the alternative decoupling scheme in Fig. 19.12 has a significant disadvantage, namely, it tends to be more sensitive to modeling errors [26–28].[2] Consequently, the decoupling scheme in Fig. 19.11 or the variations proposed by Waller [26] are preferred.

---

[2]It can be shown that the "ideal decoupling" scheme analyzed by Luyben [27] is equivalent to the alternative decoupling scheme in Fig. 19.12, where the decouplers are given by Eqs. 19-51 and 19-53.

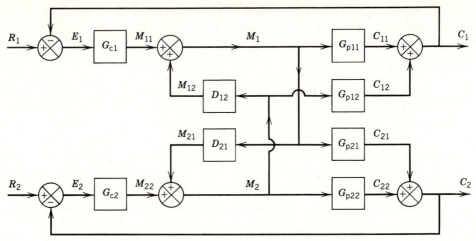

Figure 19.12.   An alternative decoupling control scheme.

Wood and Berry [3] reported an experimental application of decoupling control to a pilot scale distillation column. The 9-in. diameter, eight-tray column was used to fractionate a binary mixture of methanol and water. The amount of methanol (wt%) in each product stream, x_D and x_B, was controlled by adjusting the reflux flow R and the steam flow rate S to the reboiler. Decouplers were derived using the control configuration in Fig. 19.11, the design equations of (19-51) and (19-53), and the empirical process model of Example 19.1. The experimental results in Fig. 19.13 demonstrate that the decoupling (or noninteracting) control scheme outperformed a conventional dual loop PI control system for both increases and decreases in feed flow rate, a load variable. (In Fig. 19.13 the initial steady states were slightly different for the two control schemes.) Wood and Berry [3] also

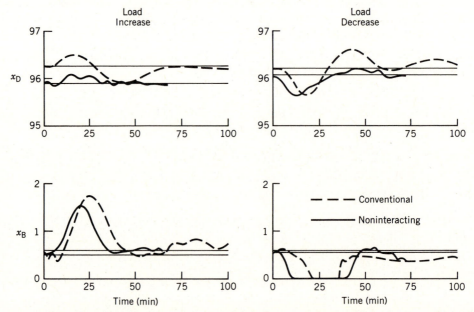

Figure 19.13.   An experimental application of decoupling (noninteracting) control to a distillation column [3].

reported that the IAE values for the decoupling control strategy were about one-half those for the conventional multi-loop control scheme.

Partial Decoupling

In some 2×2 control situations it is desirable to use only one of the two decouplers shown in Fig. 19.11, that is, D_{12} or D_{21}, and to set the other decoupler equal to zero. This approach is referred to as *partial* or *one-way* decoupling in contrast to the *complete* decoupling approach discussed above. Partial decoupling is an attractive approach for control problems where one of the controlled variables is more important than the other or where one of the process interactions is weak or absent. For example, if M_1 has no effect on C_2, then $G_{p21} = 0$ and D_{21} will be zero, according to Eq. 19-51. However, partial decoupling can also be advantageous even for highly interacting processes because it tends to be more robust, that is, less sensitive to modeling errors, than complete decoupling [1,13]. Partial decoupling can also provide better control than complete decoupling in some situations [17,28,29].

Static Decoupling

We have seen that for complete decoupling and partial decoupling, the design objective is to eliminate control loop interactions. A less ambitious approach is to design the decouplers so that only steady-state interactions between control loops are eliminated. This approach is referred to as *static* decoupling or *steady-state* decoupling [30]. The design equations for ideal, static decouplers can be obtained from Eqs. 19-51 and 19-53 by setting $s = 0$. Equivalently, the process transfer functions can be replaced with the corresponding steady-state gains:

$$D_{21} = -\frac{K_{p21}}{K_{p22}} \tag{19-58}$$

$$D_{12} = -\frac{K_{p12}}{K_{p11}} \tag{19-59}$$

Since static decouplers are merely constants, they are always physically realizable and easily implemented.

The advantage of static decoupling is that less process information is required, namely, steady-state gains rather than complete dynamic models. A disadvantage is that control loop interactions still exist during transient conditions. For example, a set-point change for C_1 will tend to upset C_2. However, if the dynamics of the two loops are similar, static decoupling can produce excellent transient responses.

Nonlinear Decouplers

So far, we have assumed that the decouplers can be represented by transfer functions with constant parameters. But if the process to be controlled is highly nonlinear or has time-varying dynamics, standard decouplers with constant parameters may not perform very well. In fact, their performance may be inferior to that obtained using conventional multiloop PID controllers [28]. For a highly nonlinear process, nonlinear decoupling is a reasonable alternative. Sometimes nonlinear decoupling can be achieved by selecting different controlled or manipulated variables, as discussed earlier. Alternatively, the decoupler gains can be allowed to vary in order to compensate for nonlinear process interactions. The books by Shinskey [1] and McAvoy [13] provide additional information on nonlinear decoupling.

Critique of Decoupling Control

The theoretical analyses of the previous sections and the illustrative example have demonstrated that decoupling control schemes can provide improvements over conventional multiloop control schemes. However, the performance of decoupling controllers depends strongly on the accuracy of the process model and how well the decouplers are tuned. For some processes, decoupling may provide worse results than conventional multiloop control even when a perfect model is available. For example, McAvoy and coworkers [13,28] have indicated that high-purity distillation columns with large relative gains are poor candidates for decoupling if the ideal decouplers in Eqs. 19-51 and 19-53 are used. McAvoy [13] reports a few guidelines for evaluating decoupling control based on an RGA analysis for 2×2 processes. A robustness analysis based on a singular value analysis (SVA) [10,11,31] can provide considerable insight concerning potential sensitivity problems associated with the implementation of decouplers. A brief introduction to SVA is presented in Chapter 28.

Although the previous discussion has emphasized decoupling 2×2 processes, decouplers can also be effectively used for $n \times n$ processes where $n > 2$. Here the design strategy is based on the requirement that the closed-loop transfer function matrix for set-point changes be diagonal [32]. This design approach is mathematically equivalent to the previously described strategy of cancelling undesired process interactions.

19.5 MULTIVARIABLE CONTROL TECHNIQUES

At the beginning of this chapter we listed four different strategies for controlling processes that are highly interacting. The last category, multivariable control, is a generic term that refers to the class of control strategies in which each manipulated variable is adjusted on the basis of errors in all of the controlled variables, rather than the error in a single variable, as is the case for conventional multiloop control. For example, a simple multivariable proportional control strategy for a 2×2 process (with transducer and valve gains of unity) could have the following form with four controller gains:

$$M_1(t) = K_{c11}E_1(t) + K_{c12}E_2(t) \tag{19-60}$$

$$M_2(t) = K_{c21}E_1(t) + K_{c22}E_2(t) \tag{19-61}$$

If $K_{c12} = K_{c21} = 0$, then the multivariable control system reduces to a 1-1/2-2 multiloop control system since each manipulated variable is adjusted based on a single error signal. Similarly, a 1-2/2-1 multiloop control system results if $K_{c11} = K_{c22} = 0$. Thus, multiloop control is a special case of the more general multivariable control. Note that the decoupling control schemes shown in Figs. 19.11 and 19.12 are also multivariable control strategies because each manipulated variable depends on both error signals.

Equations 19-60 and 19-61 illustrate multivariable proportional control for 2×2 processes. Multivariable control strategies have also been developed that include integral, derivative, and feedforward control action. Multivariable predictive control strategies are considered in Chapter 27. These methods are applicable to $n \times n$ processes where $n \geq 2$.

For additional information, see the texts by Ray [32], Douglas [33], and Friedland [34] or the survey articles by Ray [35] and Edgar [36].

SUMMARY

In this chapter we have considered control problems with multiple inputs (manipulated variables) and multiple outputs (controlled variables). Such MIMO control problems are more difficult to solve than single-input, single-output control problems due to the presence of process interactions. The process interactions can produce undesirable control loop interactions during closed-loop operation. The relative gain array (RGA) approach of Bristol provides a convenient way of determining the degree of process interaction and the best pairing of controlled and manipulated variables for a multiloop control scheme. Because the RGA approach is based only on steady-state information, the process dynamics should also be considered while selecting the best pairing.

If the control loop interactions are unacceptable for multiloop control, then a number of alternatives are available. One approach is to reduce the interactions by detuning one or more of the control loops. Alternatively, it may be possible to reduce the interactions by a different choice of manipulated or controlled variables. If this is not possible, decoupling control techniques or more general multivariable control techniques provide other options. Although these model-based control strategies can provide significant improvements over conventional multiloop control, their sensitivity to modeling errors has to be carefully considered.

REFERENCES

1. Shinskey, F. G., *Process Control Systems,* 3d ed. McGraw-Hill, New York, 1988, Chapters 8, 11.
2. Balchen, J. G., and K. I. Mummé, *Process Control: Structures and Applications,* Van Nostrand Reinhold, New York, 1988.
3. Wood, R. K., and M. W. Berry, Terminal Composition Control of a Binary Distillation Column, *Chem. Eng. Sci.* **28,** 1707 (1973).
4. McAvoy, T. J., Connection Between Relative Gain and Control Loop Stability and Design, *AIChE J.* **27,** 613 (1981).
5. Niederlinski, A., A Heuristic Approach to the Design of Linear Multivariable Interacting Control Systems, *Automatica* **7,** 691 (1971).
6. Luyben, W. L., and C. Vinante, Experimental Studies of Distillation Decoupling, *Kem. Teollisuus* **29,** 499 (1972).
7. Gagnepain, J.-P., M.Sc. Thesis, University of California, Santa Barbara (1979).
8. Gould, L. A., *Chemical Process Control: Theory and Applications,* Addison-Wesley, Reading, MA, 1969, Chapter 3.
9. Bristol, E. H., On a New Measure of Interactions for Multivariable Process Control, *IEEE Trans. Auto. Control* **AC-11,** 133 (1966).
10. Smith, C. R., C. F. Moore, and D. D. Bruns, A Structural Framework for Multivariable Control Applications, Paper TA-7 presented at the Joint Auto. Control Conf., Charlottesville, VA, 1981.
11. Moore, C. F., Application of Singular Value Decomposition to the Design, Analysis and Control of Industrial Processes, Proc. Am. Control Conf., Seattle, 1986, p. 643.
12. Smith, C. A., and A. B. Corripio, *Principles and Practice of Automatic Process Control,* Wiley, New York, 1985, Chapter 8.
13. McAvoy, T. J., *Interaction Analysis Theory and Application,* Instrum. Soc. of America, Research Triangle Park, NC, 1983.
14. Nisenfeld, A. E., and H. M. Schultz, Interaction Analysis in Control System Design, Paper 70-562, *Advances in Instrum.* Vol. 25, Pt. 1, Instrum. Soc. America, Pittsburgh, PA 1971.
15. Friedly, J. C., Use of the Bristol Array in Designing Noninteracting Control Loops. A Limitation and Extension, *IEC Process Des. Dev.* **23,** 469 (1984).
16. Woolverton, P. F., How to Use Relative Gain Analysis in Systems with Integrating Variables, *InTech* **27** (9), 63 (1980).

17. Gagnepain, J.-P., and D. E. Seborg, Analysis of Process Interactions with Applications to Multiloop Control System Design, *IEC Process Des. Dev.* **21,** 5 (1982).
18. Grosdidier, P., M. Morari, and B. R. Holt, Closed-Loop Properties from Steady-State Information, *IEC Fund.* **24,** 221 (1985).
19. Tung, L. S., and T. F. Edgar, Analysis of Control-Output Interactions in Dynamic Systems, *AIChE J.* **27,** 690 (1981).
20. Grosdidier, P., and M. Morari, Analysis of Interactions Using Structured Singular Values, Proc. Am. Control Conf., Seattle, 1986, p. 658.
21. Witcher, M. F., and T. J. McAvoy, Interacting Control Systems: Steady-State and Dynamic Measurement of Interaction, *ISA Trans.* **16** (3), 35 (1977).
22. Stanley, G., M. Marino-Galarraga, and T. J. McAvoy, Shortcut Operability Analysis 1. The Relative Disturbance Gain, *IEC Process Des. Dev.* **24,** 1181 (1985).
23. Shinskey, F. G., Disturbance-Rejection Capabilities of Distillation Column Control Systems, Proc. Am. Control Conf., Boston 1985, p. 1072.
24. Weber, R., and N. Y. Gaitonde, Non-Interactive Distillation Tower Analyzer Control, Proc. Am. Control Conf., Arlington, VA 1982, p. 87.
25. Waller, K. V., and D. H. Finnerman, On Using Sums and Differences to Control Distillation, *Chem. Eng. Commun.* **56,** 253 (1987).
26. Waller, K. V., Decoupling in Distillation, *AIChE J.* **20,** 592 (1974).
27. Luyben, W. L., Distillation Decoupling, *AIChE J.* **16,** 198 (1970).
28. Weischedel, K., and T. J. McAvoy, Feasibility of Decoupling in Conventionally-Controlled Distillation Columns, *IEC Fund.* **19,** 379 (1980).
29. Fagervik, K. C., K. V. Waller, and L. G. Hammarström, Two-Way or One-Way Decoupling in Distillation?, *Chem. Eng. Commun.* **21,** 235 (1983).
30. McAvoy, T. J., Steady-State Decoupling of Distillation Columns, *IEC Fund.* **18,** 269 (1979).
31. Arkun, Y., V. Manousiouthakis, and A. Palozoglu, Robustness Analysis of Process Control Systems: A Case Study of Decoupling Control in Distillation, *IEC Process Des. Dev.* **23,** 93 (1984).
32. Ray, W. H., *Advanced Process Control,* McGraw-Hill, New York, 1981, Chapters 3 and 6.
33. Douglas, J. M., *Process Dynamics and Control,* Vol. 2, Prentice-Hall, Englewood Cliffs, NJ, 1972, Chapters 8 and 9.
34. Friedland, B., *Control System Design,* McGraw-Hill, New York, 1986.
35. Ray, W. H., Multivariable Process Control—A Survey, *Computers & Chem. Eng.* **7,** 367 (1983).
36. Edgar, T. F., Advanced Control Strategies for Chemical Processes—A Review, in *Advances in Chemistry,* Vol. 124, (R. L. Squires and G. V. Reklaitis, Eds.), American Chemical Society, Washington, DC 1980.

EXERCISES

19.1. Derive an expression for the transfer function, $C_2(s)/M_2(s)$, in the 1-1/2-2 control configuration of Fig. 19.3a. Assume that the C_2 controller is in the manual mode with $M_2 \neq 0$ and that the C_1 controller is in automatic. If both controllers are placed in the manual mode, how does this transfer function change?

19.2. Derive an expression for the characteristic equation for the 1-2/2-1 configuration in Fig. 19.3b. Simplify and interpret this equation for the special situation where either G_{p11} or G_{p22} is zero.

19.3. For the stirred-tank heating system in Fig. 6.17, it is desired to control tank temperature T by adjusting flow rate w_h, and liquid level h by adjusting flow rate w_c. The primary load variable is inlet temperature T_h. Draw a block diagram for the multiloop control system assuming that each transmitter and control valve can be represented by a first-order transfer function and that PID controllers are employed.

19.4. Consider the stirred-tank heating system of Exercise 19.3 but now assume that the manipulated inputs are w_h and w. Suggest a reasonable pairing for a multiloop control scheme and justify your answer.

19.5. For the in-line blending system of Example 19.7, draw block diagrams for two multiloop control schemes:

(a) The standard scheme for $x = 0.4$.
(b) The less interacting scheme where $M_1 = w_A + w_B$ and $M_2 = w_A$.

You may assume that each transmitter and control valve can be represented by a first-order transfer function and that PI controllers are utilized.

19.6. A conventional multiloop control scheme consisting of two PI controllers is to be used to control the product compositions x_D and x_B of the distillation column shown in Fig. 19.1b. The manipulated variables are the reflux flow rate R and the steam flow rate to the reboiler S. Experimental data for a number of steady-state conditions are summarized below. Use this information to do the following:

(a) Calculate the RGA to determine the recommended pairing between controlled and manipulated variables.

(b) Does this pairing seem appropriate from dynamic considerations? Justify your answer.

Run	R (lb/min)	S (lb/min)	x_D	x_B
1	125	22	0.97	0.04
2	150	22	0.95	0.05
3	175	22	0.93	0.06
4	150	20	0.94	0.06
5	150	24	0.96	0.04

19.7. Derive an expression for the transfer function, $C_2(s)/M_1(s)$, in the 1-2/2-1 control configuration of Fig. 19.3b. Assume that the C_2 controller (with transfer function G_{c2}) is in the manual mode; M_1 is adjusted manually so that $M_1(s) \neq 0$. Controller G_{c1} is in the automatic mode. If both controllers are placed in manual, what is the corresponding expression for $C_2(s)/M_1(s)$?

19.8. A dynamic model of the stirred-tank heating system in Fig. 6.17 was derived in Chapter 6. Use this model to do the following:

(a) Derive an expression for the relative gain array.

(b) Design an ideal decoupling control system assuming that the transmitters and control valves have negligible dynamics.

(c) Are these decouplers physically realizable? If not, suggest appropriate modifications.

19.9. Repeat Exercise 19.8 for the situation where the pump on the exit line is replaced by a hand valve with a flow-head relation, $w = C_v\sqrt{h}$.

19.10. For the liquid storage system shown in the drawing, it is desired to control liquid levels h_1 and h_2 by adjusting volumetric flow rates q_1 and q_2. Flow rate q_6 is the major load variable. The flow–head relations are given by

$$q_3 = C_{v1}\sqrt{h_1} \qquad q_5 = C_{v2}\sqrt{h_2} \qquad q_4 = K(h_1 - h_2)$$

where C_{v1}, C_{v2} and K are constants.

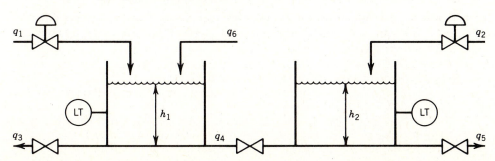

(a) Derive an expression for the relative gain array for this system.

(b) Use the RGA to determine the recommended pairing of controlled and manipulated variables for the following conditions:

Parameter Values

$K = 3$ gal/min ft
$C_{v1} = 3$ gal/min ft$^{0.5}$
$C_{v2} = 3.46$ gal/min ft$^{0.5}$
$D_1 = D_2 = 3.5$ ft (tank diameters)

Nominal Steady-State Values

$\bar{h}_1 = 4$ ft, $\quad \bar{h}_2 = 3$ ft

19.11. For the liquid-level storage system in Exercise 19.10:

(a) Derive a transfer function model of the form,

$$\mathbf{C}(s) = \mathbf{G_p}(s)\mathbf{M}(s) + \mathbf{G_L}(s)\, L(s)$$

where L is the load variable and $\mathbf{G_L}$ is a 2×1 matrix of load transfer functions.

(b) Draw a block diagram for a multiloop control system based on the following pairing: $h_1 - q_1/h_2 - q_2$. Do not attempt to derive transfer functions for the transmitters, control valves, or controllers.

19.12. For the flow–pressure process shown in the drawing, it is desired to control both pressure P_1 and flow rate F. The manipulated variables are the stem positions of the control valves, M_1 and M_2. For simplicity assume that the flow–head relations for the two valves are given by

$$F = 20M_1(P_0 - P_1)$$
$$F = 30M_2(P_1 - P_2)$$

The nominal steady-state conditions are $F = 100$ gal/min, $P_0 = 20$ psi, $P_1 = 10$ psi, and $P_2 = 5$ psi. Use the RGA approach to determine the best controller pairing.

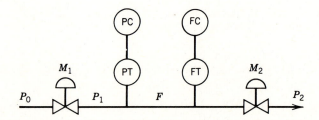

19.13. A blending system is shown in the drawing. Liquid level h and exit composition c_3 are to be controlled by adjusting flow rates q_1 and q_3. Based on the information below, do the following:

(a) Derive the process transfer function matrix, $\mathbf{G_p}(s)$.

(b) If a conventional multiloop control system is used, which controller pairing should be used? Justify your answer.

(c) Obtain expressions for the ideal decouplers $D_{21}(s)$ and $D_{12}(s)$ in the configuration of Fig. 19.11.

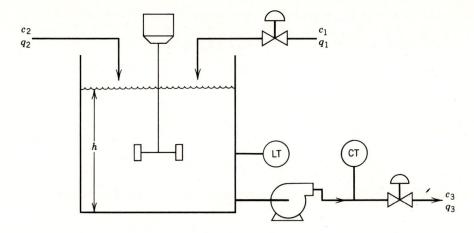

Available Information

    i. The tank is 3 ft in diameter and is perfectly mixed.

   ii. Nominal steady-state values are:

$$\bar{h} = 3 \text{ ft} \qquad\qquad \bar{q}_3 = 20 \text{ ft}^3/\text{min}$$

$$\bar{c}_1 = 0.4 \text{ mole/ft}^3 \qquad \bar{c}_2 = 0.1 \text{ mole/ft}^3$$

$$\bar{q}_1 = 10 \text{ ft}^3/\text{min}$$

 iii. The density of each process stream remains constant at $\rho = 60 \text{ lb/ft}^3$.

 iv. The primary load variable is flow rate q_2.

  v. Inlet compositions c_1 and c_2 are constant.

 vi. The transmitter characteristics are approximated by the following transfer functions with time constants in minutes:

$$G_{m11}(s) = \frac{4}{0.1s + 1} \quad (\text{mA/ft})$$

$$G_{m22}(s) = \frac{100}{0.2s + 1} \quad (\text{mA ft}^3/\text{mole})$$

 vii. Each control valve has a gain of 0.15 ft^3/min mA and a time constant of 10 s.

19.14. Repeat Exercise 19.13 for the situation where the manipulated variables are q_1 and q_2, and q_3 is the primary load variable.

— CHAPTER 20 —

Supervisory Control

In previous chapters we have considered the development of process models and the design of process controllers from an *unsteady-state* point of view. Such an approach focuses on obtaining reasonable dynamic responses of process outputs for set-point and load changes. Up to this point we have only peripherally mentioned how set points should be specified for an operating unit. The on-line calculation of set points, also called *supervisory control,* should allow the unit or plant to achieve maximum profits while satisfying operating constraints. The optimization algorithms employed to maximize profits (or minimize costs) are implemented with a digital computer, which is directly connected to the process controllers themselves. Set-point selection, when automated, has traditionally been considered as part of the field of steady-state process optimization and thus has not been covered in most process control textbooks. In supervisory control steady-state models are used, hence dynamic models and associated control theory are not normally required to perform the necessary calculations.

In this chapter we first discuss the basic ideas involved in supervisory control and then review typical applications. We also present guidelines for determining when supervisory control can be potentially advantageous. A procedure to formulate set-point selection as an optimization problem, involving development of the process economic model and the operating (steady-state) model for a plant or specific process, is outlined. Various optimization techniques that are popular in the process industries are described. Because a detailed review of optimization techniques is not possible in this book, the interested reader is referred to several texts on optimization methodology [1–3].

20.1 BASIC REQUIREMENTS IN SUPERVISORY CONTROL

In supervisory control process and economic models of the plant are used to optimize plant operations by maximizing daily profits, yields, or production rates. The computer program reviews operating conditions periodically, computes the new conditions that optimize a chosen objective function (e.g., profit), and adjusts plant controller set points, thus implementing the new, improved conditions.

The operating model discussed in the introduction typically is a steady-state model obtained either from fundamental knowledge of the process or from experimental data. It relates the plant operating conditions such as unit temperatures, pressure, and feed flow rates to product properties such as product yields (or distributions), production rates, and measurable characteristics, for example, purity, viscosity, and molecular weight. The economic model involves the costs of raw materials, values of products, and costs of production as functions of operating conditions, projected sales figures, and so on. An objective function can be expressed in terms of these quantities; for example, *operating profit* over some specified period of time might be expressed as

$$P = \sum_s F_s V_s - \sum_r F_r C_r - OC \tag{20-1}$$

where P = profit/time

$\sum_s F_s V_s$ = sum of product flow rates times respective product values

$\sum_r F_r C_r$ = sum of feed flow rate times respective unit cost

OC = operating costs/time

Both the operating and economic models typically will include constraints on:

1. *Operating Conditions*: Temperatures and pressures must be within certain limits.
2. *Feed and Production Rates*: A feed pump has a fixed capacity; sales are limited by market projections.
3. *Storage and Warehousing Capacities*: Storage tanks cannot be overflowed during periods of low demand.
4. *Product Impurities*: A product may contain no more than the maximum amount of some contaminant or impurity.

Implementation of Supervisory Control

The computer hardware that is used for process optimization is often connected to or comprises a part of the actual control system. In fact, the cost savings from supervisory control can often justify the implementation of a complete computer control system. Figure 20.1 illustrates the type of application of a *supervisory*

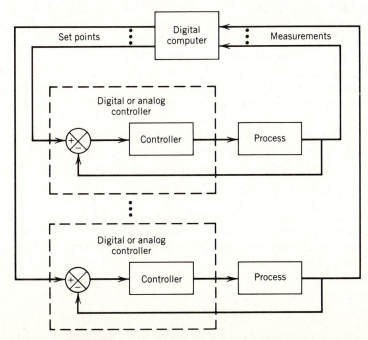

Figure 20.1. A supervisory digital computer system.

computer, which adjusts the set points of analog or digital controllers. A second type of application employs a supervisory computer connected to control hardware that is constructed of digital instrumentation modules, built around microcomputers. Alternatively, a single digital computer can be used to implement both the individual loop controllers (direct digital control) as well as the supervisory control function (see Chapter 21).

The supervisory computer is usually provided with a set of operating procedures that it must follow in changing set points. As examples, the rate at which a process temperature may be changed may have an upper limit (e.g., ≤3°F/min), or two pumps in parallel may be required to obtain flows in a certain range of feed rates. In control of batch processes, such procedures can be quite intricate (see Chapter 21).

20.2 APPLICATIONS FOR SUPERVISORY CONTROL

There are two valuable sources of plant data for analyzing which units or processes in a plant are potentially most attractive for application of supervisory control. These include the profit and loss statement and the operating records. The profit and loss statement for the plant contains information on sales, prices, manufacturing costs, and profits, while the operating records present information on material and energy balances, unit efficiencies, production levels, and feedstock usage.

Below we list some process attributes that are relevant to maximizing operating profits:

1. *Sales limited by production.*   In this type of market increased sales are desirable, and throughput should be increased. Increased production can be achieved by improved operating conditions and by optimal scheduling.
2. *Sales limited by market.*   This situation is susceptible to optimization only if improvements in efficiency at current production rates can be obtained. An increase in thermal efficiency, for example, usually leads to a reduction in manufacturing costs (e.g. utilities or feedstocks).
3. *Large throughputs.*   Units with large production rates offer great potential for increased profits. Small savings in production costs per unit throughput or incremental improvements in yield, which may be achievable by supervisory control, are greatly magnified in this type of facility.
4. *High raw material or energy consumption.*   These are major cost factors in a typical plant and they offer potential savings. For example, the optimal allocation of fuel supplies and steam in a plant can reduce costs by minimizing fuel consumption.
5. *Product quality better than specification.*   If the product quality is significantly better than that required by the customer, this can cause excessive production costs and wasted capacity. By operating closer to the maximum specification for the impurity (a constraint), cost savings can be obtained, but this also requires lower process variability (see Chapter 1, Figs. 1.6 and 1.7).
6. *Losses of valuable or hazardous components through waste streams.*   The chemical analysis of various plant exit streams, both to the air and water, should indicate if valuable materials are being lost. In addition, pollutant emissions should be minimized. Adjustment of air/fuel ratios in furnaces to minimize unburned hydrocarbon losses and to reduce nitrogen oxide emissions is one such example.

Latour [4,5] has discussed opportunities for the application of on-line optimization or supervisory control in refinery operations. He points out that there are three general types of optimization problems commonly encountered in industrial process operations (examples given in parentheses):

- Operating conditions (tower reflux ratio, reactor temperature)
- Allocation (fuel use, feedstock selection)
- Scheduling (batch processing, cleaning, maintenance)

The first two items are quite amenable to on-line optimization, while the third has not been treated as frequently. However, the increasing importance of batch processes may provide more opportunities for optimal scheduling (see Chapter 21).

As an example of optimizing operating conditions, recent interest in energy conservation has led to reducing the reflux ratios used in distillation columns, which relates directly to utilities consumption. In the past, large reflux ratios were used to ensure high product quality; that is, the process was over-refluxed. The reflux ratio was normally not changed in response to changing feed conditions, thus remaining in a "worst-case" operating mode. In recent years the relation among reflux ratio, feed rate, and product losses has been examined more closely. For example, a 5% decrease in recovery might involve a 50% decrease in utilities consumption and thus would yield higher profits. Other examples where optimization can be used in distillation include tower pressure optimization, treatment of multiple impurities, maximization of capacity versus separation quality, and adjustment of feed conditions [4].

Another refinery unit with considerable profit potential for supervisory control is the fluidized catalytic cracker (FCC) [4]. The FCC reaction temperature largely determines the conversion of a crude feedstock to lighter components. The product distribution (gasoline, middle distillate, fuel oil, light gases) changes as the degree of conversion is increased. Accurate process models of the product distribution as a function of FCC operating conditions and catalyst type is a necessity for on-line or off-line optimization. Feedstock composition, downstream unit capacities (e.g., distillation columns), individual product prices, product demand, feed preheat, gas oil recycle, and utilities requirements must be analyzed in optimizing an FCC unit. The large throughput of the FCC implies that a small improvement in yield translates to a significant increase in profits.

The second category of supervisory control problems discussed by Latour, allocation problems, involves the optimal distribution of a limited resource among several parallel (alternative) process units. Typical examples include [4]:

- *Steam Generators.* Optimum load distribution among several boilers of varying size and efficiency.
- *Refrigeration Units.* Optimum distribution of a fixed refrigeration capacity among several low temperature condensers associated with distillation columns.
- *Parallel Distillation Columns.* Minimization of "off-spec" products and utilities consumption while maximizing overall capacity.

Examples of scheduling problems encountered in continuous plants include catalyst regeneration, furnace decoking, and heat exchanger cleaning [4]. While these maintenance operations may or may not be performed under computer control, the computer can monitor the process of interest and notify the operator when maintenance operations ought to be carried out, consistent with maximizing profits (a trade-off between operating efficiency and lost production due to maintenance). In batch processing, optimal scheduling is crucial to avoid equipment under-utilization and excessive down-time [6].

20.3 THE FORMULATION AND SOLUTION OF OPTIMIZATION PROBLEMS

Once a process has been selected for application of supervisory control, an appropriate problem statement must be formulated and then solved. As mentioned earlier (Section 20.1), the optimization of set points requires that we select

1. An objective function to be maximized or minimized, incorporating the so-called economic model.
2. The operating or process model, which includes all constraints on the process variables.

Edgar and Himmelblau [3] have listed six steps that should be used in solving any practical optimization problem. A summary of the procedure with comments relevant to supervisory control is given below.

Step 1: *Identify the process variables.*   The important input and output variables for the process and/or unit must be identified. These variables are employed in developing the economic and operating models (see Steps 2 and 3 below). Here the input and output variables are expressed in absolute terms (rather than in deviation variable form as done in other chapters in this book). While we wish to optimize all variables listed, usually some of these variables are interrelated through the steady-state process model. In this step it is not necessary to specify (in an algebraic sense) which variables are independent and which ones are dependent; this issue is treated in Steps 3 and 4 below.

Step 2. *Select performance criteria and develop a mathematical expression for the objective function.*   To convert a verbal statement of the process optimization goals into a meaningful yet quantitative, objective function is sometimes rather difficult. Often such a verbal statement contains multiple objectives and implied constraints. As an example, we may wish to minimize the amount of off-spec product from a distillation column while at the same time prevent the "flooding" of the column. To arrive at a single objective based on profit, the amount of off-spec product can be related to the costs of utilities and feedstock and included with the objective function. The second requirement, namely that flooding be avoided, is a constraint on the vapor and/or feed flow rates (which also affect utilities consumption). This information is not normally incorporated in the objective function but is introduced as a constraint (see step 3 below). However, as discussed later, the constraint restricts the allowable operating conditions, which in turn may influence the objective function.

The specific objectives may vary depending upon plant configuration as well as supply/demand. Tables 20.1 and 20.2 show different operating objec-

Table 20.1   Alternative Operating Objectives for a Fluidized Catalytic Cracker [4]

1. Maximize gasoline yield subject to a fixed feed rate.
2. Minimize feed rate subject to fixed gasoline production.
3. Maximize conversion to light products subject to load compressor/regenerator constraints.
4. Optimize yields subject to fixed feed conditions.
5. Maximize gasoline production with fixed cycle oil production.
6. Optimize yields and feed rate, where product values vary stepwise with production.
7. Maximize feed with fixed product distribution.
8. Maximize fuel oil destruction for refinery.
9. Maximize FCC gasoline plus olefins for alkylate.

Table 20.2 Alternative Operating Objectives for Distillation Columns [4]

1. Maximize yield of more valuable components from a given feed, with purity specifications.
2. Maximize product purity at a given production rate from a given feed.
3. Minimize energy consumption; maintain reboiler and condenser products within purity specifications.
4. Optimize energy consumption versus product recovery value.
5. Maximize distillate production within specifications.
6. Optimize feed rate where capacity is balanced against recovery.

tives which may arise for a fluidized catalytic cracker and for a distillation column. Operation based on a quantified objective invariably is superior to arbitrary selection of set points by the operator. If profit is the ultimate objective function, Eq. 20-1 can be used to develop the appropriate expression.

Step 3. *Develop the models for the process and the constraints.* Basically what is required here is to develop steady-state input–output models for the process and to identify operating limits for the process variables. The input–output model of the process can be based upon the physics and chemistry of the process (see Chapter 2), or it can be based on empirical relations obtained from experimental process data (see Chapter 7).

The optimization problem statement involves both equality and inequality constraints. While the equality constraints relate the principal variables of the process to one another (the process steady-state model), the inequality constraints represent limits and allowable ranges for these variables. Inequality constraints arise often in supervisory control, partly due to the fact that many physical variables, such as compositions or pressures, can only be positive quantities. In addition, there may be maximum temperature or maximum pressure restrictions. These inequality constraints are a key part of the mathematical model and can have a profound effect on the optimum operating point. Later in this chapter we illustrate several cases where the optimum actually lies on a constraint.

Step 4. *Simplify the model and objective function.* Before undertaking any computation, it is very important that the mathematical statement developed in steps (1)–(3) be simplified or reduced in scope as much as possible without losing its essence. As a first step, we might decide to ignore variables that have a negligible effect on the objective function. This can be done based on engineering judgment, or by going through a more detailed mathematical or numerical analysis. Second, we may choose to eliminate some variables through algebraic substitution (by employing the process model equations). If optimization is to be performed on-line, the size of the optimization problem (in other words, the number of variables and the number of constraints) is a key consideration, due to computer time and storage limitations.

Step 5. *Compute the optimum.* This step involves the choice of technique or techniques to obtain the "best" answer for the problem. Most of the literature on the subject of optimization deals with this step. In general, the solution of most practical optimization problems requires the use of the digital computer. Over the past 15 years, much progress has been made in developing efficient and robust numerical methods for optimization calculations. Much is known about which methods are successful, although comparisons of candidate methods often are of an *ad hoc* nature, based on simple test cases. [2,3,7]. Virtually all optimization methods are iterative, and a good initial estimate of the optimum can reduce computer time. If there are a large number of inequality constraints, information about which constraints actually influence the optimum is also helpful.

Step 6.   *Perform sensitivity studies.*   It is useful to know which parameters in an optimization problem are the most important in determining the optimum. By varying model and cost parameters individually and recalculating the optimum, the most sensitive parameters can be identified. A sensitivity analysis is an important ingredient in any design calculation, not just one involving optimization, since the values of many parameters may be uncertain.

EXAMPLE 20.1

A section of a chemical plant makes two specialty products (E, F) and utilizes two raw materials (A, B) in limited supply. Each product is formed in a separate process, a schematic of which is shown in Fig. 20.2. The available materials A and B do not have to be totally consumed. The reactions involving A and B are as follows:

Process Data

$$\text{process 1:}\quad A + B \longrightarrow E$$

$$\text{process 2:}\quad A + 2B \longrightarrow F$$

Raw Material	Maximum Available (lb/day)	Cost (¢/lb)
A	40,000	15
B	30,000	20

Process	Product	Reactant Requirements (lb) per lb Product	Processing Cost	Selling Price of Product	Maximum Production Level (lb/day)
1	E	2/3 A, 1/3 B	15¢/lb E	40¢/lb E	30,000
2	F	1/2 A, 1/2 B	5¢/lb F	33¢/lb F	30,000

The processing cost includes cost of utilities and supplies. Labor and other costs are $200/day for process 1 and $350/day for process 2. These costs occur even if production of E or F is zero. Formulate the objective function as the total operating profit per day. List the equality and inequality constraints. (Steps 1–3)

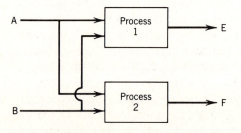

Figure 20.2.   A schematic of a chemical plant for Example 20.1.

Solution
We formulate the optimization problem using the first three steps delineated above.

Step 1. Define as variables the mass flow rates of reactants and products (see Fig. 20.2):

x_1 = lb/day A consumed
x_2 = lb/day B consumed
x_4 = lb/day F produced

Step 2. Using Eq. 20-1 to compute the operating profit per day, we need to identify product sales income, feedstock costs, and operating costs:

$$\text{Sales income (\$/day)} = \sum_s F_s V_s = 0.40x_3 + 0.33x_4$$

$$\text{Feedstock costs (\$/day)} = \sum_r F_r C_r = 0.15x_1 + 0.2x_2$$

$$\text{Operating costs (\$/day)} = OC = 0.15x_3 + 0.05x_4 + 350 + 200$$

Substituting into (20-1) yields the daily profit:

$$P = 0.4x_3 + 0.33x_4 - 0.15x_1 - 0.2x_2 - 0.15x_3 - 0.05x_4 - 350 - 200$$
$$= 0.25x_3 + 0.28x_4 - 0.15x_1 - 0.2x_2 - 550 \qquad (20\text{-}2)$$

Step 3. Not all variables in this problem are unconstrained. First consider the material balance equations, which in this case comprise the process model:

$$x_1 = 0.667x_3 + 0.5x_4 \qquad (20\text{-}3a)$$
$$x_2 = 0.333x_3 + 0.5x_4 \qquad (20\text{-}3b)$$

Note that $x_1 + x_2 = x_3 + x_4$ to satisfy the overall material balance. There are also limits on the amount of feedstock available and production levels:

$$0 \le x_1 \le 40{,}000 \qquad (20\text{-}4a)$$
$$0 \le x_2 \le 30{,}000 \qquad (20\text{-}4b)$$
$$0 \le x_3 \le 30{,}000 \qquad (20\text{-}4c)$$
$$0 \le x_4 \le 30{,}000 \qquad (20\text{-}4d)$$

Equation sets (20-2) through (20-4) constitute the optimization problem to be solved. Since the variables appear linearly in both the objective function and constraints, this formulation is referred to as a linear programming problem, which is discussed in Section 20.5.

20.4 UNCONSTRAINED OPTIMIZATION

Unconstrained optimization refers to the case where no inequality constraints are present and all equality constraints can be eliminated by variable substitution in the objective function. First we cover single-variable optimization, and then extend those concepts to optimization problems with multiple variables. Since the computational techniques are iterative in nature, we focus mainly on methods which are efficient in terms of computer time and thus can be applied on-line.

Single-Variable Optimization

Many process optimization problems involve the variation of a single quantity so as to maximize profit or some other objective function. Some examples of single-variable optimization, such as optimizing the reflux ratio in a distillation column or the air/fuel ratio in a furnace, have already been mentioned. While most processes actually are multivariable with several variables that could be optimized, often we choose to optimize only the most important variable. This keeps the supervisory strategy uncomplicated. One characteristic normally required in a single-variable optimization problem is that a single maximum (or minimum) occur in the region of search, that is, the objective function $f(x)$ be *unimodal* in the variable that is optimized, x. It is very difficult to predict whether a general function is unimodal prior to investigation, but in most physically-based single-variable situations the objective function does exhibit unimodality over the operating range of interest.

The selection of a method for one-dimensional search is based on the trade-off to minimize computer time between number of function evaluations and algorithm complexity. We can approach the optimum by evaluating the objective function at many values of x using a small grid spacing (Δx) over the allowable range of x values, but this method generally requires many iterations to reach the optimum. More efficient methods which minimize the distance from the optimum for a fixed number of iterations are preferred. Such a technique is an *equal-interval search*, which can rapidly converge to the optimum [1].

One type of equal-interval search is based on comparing the function values at three points that are equally spaced. We first identify the region where the optimum lies, $a \leq x^{\text{opt}} \leq b$. Reasonable values of a and b can be found from experience or by a quick search to bracket the optimum. The region $[a, b]$ is called the *interval of uncertainty* or *bracket* for the optimum. The initial interval of uncertainty is thus $L_0 = b - a$. Next consider two cases for optimization, as shown in Fig. 20.3. In case 1 the maximum has to lie between x_2 and b and we can eliminate the region $[a, x_2]$; hence the new bracket is $[x_2, b]$. In case 2, for a unimodal function, the maximum must lie between x_1 and x_3, and we eliminate the regions, $[a, x_1]$ and $[x_3, b]$. Hence the new bracket is $[x_1, x_3]$.

Thus, with three points, one-half of the original interval of uncertainty can be eliminated and the other half retained in either case. The intermediate point (and function value) bisecting the new interval of uncertainty is saved. At each subsequent iteration, we add two new points spaced so that the new interval of uncertainty is divided into four equal regions. A general formula for the interval of uncertainty after m iterations, L_m, is

$$L_m = (\tfrac{1}{2})^m L_0 \tag{20-5}$$

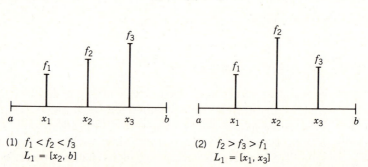

(1) $f_1 < f_2 < f_3$
$L_1 = [x_2, b]$

(2) $f_2 > f_3 > f_1$
$L_1 = [x_1, x_3]$

Figure 20.3. Two cases arising in a three-point equal-interval search.

One feature of the three-point equal interval search is that at each iteration we retain one function value from the previous iteration, requiring only two new function evaluations per iteration.

Equal-interval search is classified as a *region elimination* method, since the size of the interval of uncertainty after m iterations is predictable regardless of the shape of the function, as long as it is unimodal. Other one-dimensional search methods that are based on the region elimination concept include dichotomous search, Fibonacci search, and golden section search [1]. The latter two methods are generally more efficient than the equal-interval search since they converge faster to the optimum, but they are more complicated to program.

The region elimination methods mentioned above tend to be conservative in that they guarantee that after a certain number of iterations the optimum will be reached within a specified tolerance [3]. Faster techniques, such as interpolation methods, also utilize the notion of region elimination but the interval of uncertainty after m iterations is not predictable. However, they usually achieve a faster rate of convergence to the optimum [3]. Quadratic interpolation utilizes three points in the interval of uncertainty and is based on quadratic curve fitting. If x_1, x_2, and x_3 are three points in the interval and f_1, f_2, and f_3 are the corresponding function values, then we can fit a quadratic expression ($f = a_0 + a_1 x + a_2 x^2$) to the three data points (see Chapter 7). The resulting equation can be differentiated, set equal to zero, and solved for the optimum. Suppose three data points, (x_1, f_1), (x_2, f_2), and (x_3, f_3), have been recorded. The predicted optimum (x_4) based on these points [3] is

$$x_4 = \frac{1}{2} \frac{(x_2^2 - x_3^2)f_1 + (x_3^2 - x_1^2)f_2 + (x_1^2 - x_2^2)f_3}{(x_2 - x_3)f_1 + (x_3 - x_1)f_2 + (x_1 - x_2)f_3} \tag{20-6}$$

After only one iteration x_4 usually is not equal to x^{opt}, since the true function is not necessarily quadratic. However, x_4 is expected to be an improvement over x_1, x_2, and x_3. Saving the best two of the three previous points and finding the actual objective function at x_4, the search can be continued until convergence is indicated.

EXAMPLE 20.2

A free radical reaction involving nitration of decane is carried out in two sequential reactor stages, each of which operates like a continuous stirred tank reactor (CSTR). Decane and nitrate (as nitric acid) in varying amounts are added to each reactor stage, as shown in Fig. 20.4. The reaction of nitrate with decane is very fast and

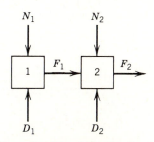

N_i = mol/s nitric acid (to stage i)
D_i = mol/s decane (to stage i)
F_i = mol/s reactor product (from stage i)   **Figure 20.4.** A schematic of a two-stage nitration reactor.

forms the following products by successive nitration: DNO_3, $D(NO_3)_2$, $D(NO_3)_3$ $D(NO_3)_4$, and so on. The desired product is DNO_3, while dinitrate, trinitrate, and so on, are undesirable products.

The flows of D_1 and D_2 are chosen to satisfy temperature requirements in the reactors, while N_1 and N_2 are optimized to maximize the amount of DNO_3 produced from stage 2, while satisfying an overall level of nitration, in this case $(N_1 + N_2)/(D_1 + D_2) = 0.4$. There is an excess of D in each stage, and $D_1 = D_2 = 0.5$ mol/s. A steady-state reactor model has been developed to maximize selectivity. Define $r_1 = N_1/D_1$ and $r_2 = N_2/(D_1 + D_2)$. The amount of DNO_3 leaving stage 2 (as mol/s in F_2) is given by

$$f_{DNO_3} = \frac{r_1 D_1}{(1 + r_1)^2 (1 + r_2)} + \frac{r_2 D_1}{(1 + r_1)(1 + r_2)^2} \tag{20-7}$$

This equation can be derived from the steady-state equations for a continuous stirred reactor with the assumption that all reaction rate constants are equal. Formulate a one-dimensional search problem that will permit the optimum values of r_1 and r_2 to be found. Employ equal interval search using a bracket of $0 \leq r_1 \leq 0.8$. Use enough iterations so that the final value of f_{DNO_3} is within $\pm$ 0.0001 of the maximum. Then compare your results with the use of quadratic interpolation to find the optimum.

Solution

Use the six steps described earlier to formulate mathematically the optimization problem.

Step 1. *Identify the process variables.* The process variables to be optimized are N_1 and N_2, the nitric acid molar flow rates for each stage. However, because D_1 and D_2 are specified, we can just as well use r_1 and r_2, since the conversion model is stated in terms of r_1 and r_2.

Step 2. *Select the objective function.* The objective is to maximize conversion of DNO_3. DNO_3 can be made into useful final products while other nitrates cannot. In terms of costs, we assume the unwanted by-products have a value of zero. The objective function f is given above in (20-7). We do not need to state it explicitly as a profit function, such as in Eq. 20-1, since the value (selling price) of DNO_3 is merely a multiplicative constant.

Step 3. *Develop models for the process/constraints.* The values of N_1 and N_2 are constrained through the overall nitration level:

$$\frac{N_1 + N_2}{D_1 + D_2} = 0.4 \tag{20-8}$$

This can be expressed in terms of r_1 and r_2 as

$$\frac{r_1 D_1 + r_2 D_1 + r_2 D_2}{D_1 + D_2} = 0.4 \tag{20-9}$$

Inequality constraints on r_1 and r_2 do exist, namely, $r_1 \geq 0$ and $r_2 \geq 0$, since all N_i and D_i are positive. We can ignore these constraints except when the search method incorrectly leads to negative values of r_1 or r_2.

Step 4. *Simplify the model.* Since $D_1 = D_2 = 0.5$, then from (20-9)

$$r_2 = 0.4 - 0.5r_1 \tag{20-10}$$

Table 20.3 One-Dimensional Search Iterations, Example 20.2 (Three-Point, Equal-Interval Search)

Iteration	x_1	f_1	x_2	f_2	x_3	f_3	Bracket	L_i
1	0.2	0.1273	0.4	0.1346	0.6	0.1324	[0.0, 0.8]	0.8
2	0.3	0.1325	0.4	0.1346	0.5	0.1344	[0.2, 0.6]	0.4
3	0.35	0.1339	0.4	0.1346	0.45	0.1348	[0.3, 0.5]	0.2
4	0.425	0.1348	0.45	0.1348	0.475	0.1347	[0.4, 0.5]	0.1

$$0.425 \le r_1^{opt} \le 0.475$$
9 function evaluations

We select r_1 to be the independent variable for one-dimensional search ($x = r_1$), and r_2 is therefore a dependent variable. Equation 20-10 implies that $r_1 \le 0.8$ and $r_2 \le 0.4$. Restricting r_1 to be between 0 and 0.8 ensures that the physical bounds are satisfied. After variable substitution there is only one independent variable. However, the objective function is too complicated to find the optimum analytically. In fact, differentiation may not be advantageous at all since a one-dimensional (iterative) search will still be required to solve the nonlinear equation, $df/dr = 0$.

Step 5. *Compute the optimum.* Since r_1 is bracketed between 0 and 0.8 (the interval of uncertainty), select the three interior points for equal interval search to be $r_1 = 0.2$, 0.4, and 0.6. The corresponding values of r_2 are 0.3, 0.2, and 0.1. Table 20.3 shows the sequence of steps for this method, along with objective function values and the interval of uncertainty. The tolerance on the objective function change is satisfied after only four iterations, with the value of r_1 that maximizes f located between 0.425 and 0.475. The converted mononitrate is 0.1348 mol/s from stage 2; the remainder of the nitrate is consumed to make higher molecular weight by-products. By repeating the calculation using quadratic interpolation (Table 20.4), the tolerance on the optimum r_1 is satisfied after three iterations; fewer function evaluations are required with this method (five vs. nine).

Step 6. *Sensitivity studies.* Based on the results in Tables 20.3 and 20.4, the yield is seen to be not significantly different from the optimum as long as $0.3 \le r_1 \le 0.6$. Practically speaking, this situation is beneficial because it allows a reasonable range of decane flows to achieve temperature control. If D_1 and D_2 change by more than 10%, we should recalculate the optimum. There also might be a need to reoptimize r_1 and r_2 if ambient conditions change (e.g., summer vs. winter operation). If the reaction is operated outside of the above range, then the economic loss could become important. Even a 1% change in

Table 20.4 One-Dimensional Search Iterations, Example 20.2 (Quadratic Interpolation)

Iteration	x_1	f_1	x_2	f_2	x_3	f_3	x_4
1	0.2	0.1273	0.4	0.1346	0.6	0.1324	0.4536
2	0.4	0.1346	0.6	0.1324	0.4536	0.1348	0.4439
3	0.4	0.1346	0.4536	0.1348	0.4439	0.1348	(Not needed)

$r_1^{opt} = 0.4439$
5 function evaluations

yield can be economically significant if production levels and the selling price of the product are sufficiently high.

Multivariable Optimization

In multivariable optimization problems, there is no guarantee that a given technique will reach the optimum in a reasonable amount of computer time. The numerical optimization of general nonlinear multivariable objective functions requires that efficient and robust techniques be employed. Efficiency is important since the solution requires an iterative approach. Trial and error solutions are usually out of the question for more than two or three variables. For example, consider a four-variable *grid* search where an equally spaced grid for each variable is prescribed. For 10 values of each of the four variables, there are 10^4 total function evaluations required to find the best answer out of the grid intersections. This computational effort still may not yield a result close enough to the true optimum. Grid search is a very inefficient method for optimization.

In multivariable optimization, the difficulty of dealing with multivariable functions often is resolved by treating the problem as a series of one-dimensional searches. From a given starting point, a search direction is specified, and the optimum is found by searching along that direction. Then a new search direction is determined, followed by another one-dimensional search. In choosing an algorithm to specify the search direction, we can draw upon extensive numerical experience with various optimization methods [2,3,7].

There are two basic types of unconstrained multivariable optimization algorithms, those requiring function derivatives and those that do not. The nonderivative methods have received great interest in supervisory control applications because these methods can be readily adapted to the case where experiments are carried out directly on the process. In such cases, an actual process measurement (such as yield) can be the objective function, and no mathematical model for the process is required. Two of these methods used in industrial control applications are the EVOP method [8] and the simplex or pattern search method [9].

EVOP is an acronym for *evolutionary operation*. A base point of operation is selected along with regularly spaced points around the base point (i.e., the process is operated and monitored at several new steady-state operating conditions or set points). In two-variable optimization, the new operating points form a rectangle with the base point at the center; for three dimensions, a rectangular polyhedron would be formed. After checking all of the new test points (2^N, where N is the number of independent variables), the process set points are changed to the operating conditions that optimize the objective function. Subsequent testing would use that point as the new base point. This procedure is repeated until no improvement from the base point is achieved. The major disadvantage of the EVOP method is the potentially large number of new operating points that must be checked at each iteration, particularly for three or more variables.

The sequential simplex method [9] uses a regular geometric shape (a simplex) to generate search directions. In two dimensions, the simplest shape is an equilateral triangle, and in three dimensions it is a regular tetrahedron. The objective function is evaluated at the vertices of the simplex. The general direction of search is projected away from the worst vertex through the centroid of the remaining vertices. A new simplex is formed by replacing the worst point by a reflected point. Thus, the direction of search can change, and only one new operating point must be evaluated at each iteration, as the simplex moves in space. This procedure is

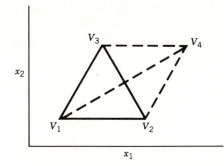

Figure 20.5. A reflection of a two-dimensional simplex to new point V_4.

sometimes called a pattern search. The sequential simplex method can readily be used for evolutionary improvement of a real process.

As an example of pattern search, consider the two-variable case. An equilateral triangle is formulated with three points (V_1, V_2, V_3); the objective function is evaluated at each of the vertices (f_1, f_2, f_3). The method of reflection to a new operating point is indicated by Fig. 20.5. Suppose point V_1 has the lowest value of f (out of f_1, f_2, f_3). The new point (V_4) is found by reflecting V_1 through the midpoint of the opposite side of the triangle (line connecting V_2 and V_3). Note that point V_4 can be found by vector addition, forming a new triangle, that is, $V_4 = V_2 + V_3 - V_1$. This algorithm normally leads to a zigzag pattern as the simplex moves to the optimum; see Fig. 20.6. The contours indicate constant values of the objective function, which in this case is to be maximized. The simplex points are (V_1, V_2, V_3) initially, then V_4 replaces V_1 because V_1 has the inferior value of f. Successive simplex coordinates are (V_2, V_3, V_4), (V_3, V_4, V_5), and (V_4, V_5, V_6). When the simplex is close to the optimum, there may be some oscillation, requiring that the simplex size be reduced. More details on the simplex search method can be found in Refs. 7 and 9. May et al. [10] utilized an improved version of this algorithm based on a flexible polyhedron in the supervisory control of a refrigeration system.

The optimization of multivariable nonlinear objective functions using function derivatives can also be employed in supervisory control. Such techniques as the conjugate gradient and variable metric methods [2,3] are extremely effective in solving such problems, but their application to on-line optimization has been previously hindered by large computing time and storage requirements. However, the availability of more powerful computers for plant applications will certainly increase the use of such methods in the near future.

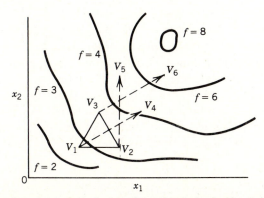

Figure 20.6. The pattern movement of a two-dimensional simplex for a maximization problem.

Occasionally, the optimum operating point can be computed analytically (e.g., using a quadratic objective function), but this situation is rare. In addition, most realistic multivariable problems include constraints, which must be treated using enhancements of various unconstrained optimization algorithms. In the next section we discuss one constrained optimization technique that is used extensively in the process industries.

20.5 CONSTRAINED OPTIMIZATION

When constraints are an important part of an optimization problem, more general methods must be employed, since the unconstrained optimum may correspond to unrealistic values of the operating variables. The general form of a nonlinear optimization problem allows for a nonlinear objective function and nonlinear constraints, that is,

$$\text{maximize } f(x_1, x_2, ..., x_{N_V}) \tag{20-11}$$

$$\text{subject to } h_i(x_1, x_2, ..., x_{N_V}) = 0 \quad (i = 1, ..., N_E) \tag{20-12}$$

$$g_i(x_1, x_2, ..., x_{N_V}) \leq 0 \quad (i = 1, ..., N_I) \tag{20-13}$$

In this case there are N_V process variables, with N_E equality constraints and N_I inequality constraints. Such problems pose a serious challenge to performing optimization calculations in real time. Yet there is a wide variety of powerful techniques that can be chosen [2,3,7].

An important class of constrained optimization problems has an objective function and constraints that are linear in form. The solution of these problems is highly structured and can be obtained rapidly. The accepted procedure, *linear programming* (LP), has become quite popular in the past twenty years, solving a wide range of industrial problems. It is increasingly being used for on-line optimization.

For processing plants there are several different kinds of *linear* constraints that often arise, which makes the LP method of great interest.

1. **Production Limitation.** This results from equipment throughput restrictions, storage limits, or market constraints (no additional product can be sold beyond some specific level). These expressions would have the form of $x_i \leq c_i$ or $g_i = x_i - c_i \leq 0$ (cf. Eq. 20-13).
2. **Raw Material Limitation.** Limits on feedstock supplies are quite common and often determined by production levels of other plants within the same company.
3. **Safety Restriction.** This type of constraint involves, for example, limitations on allowable operating temperature and pressure.
4. **Physical Property Specifications.** Constraints placed on the characteristics or composition of the final product fall into this category. For blends of various products, we usually assume that a composite property can be calculated through the averaging of pure component physical properties. For N_c components with physical property values ψ_k and volume fraction y_k, this corresponds to the linear inequality constraint ($\alpha = $ upper limit),

$$\sum_{k=1}^{N_c} \psi_k y_k \leq \alpha \tag{20-14}$$

5. Material and Energy Balances.   Whereas items 1–4 above generally are considered to be inequality constraints, the material and energy balances of the steady-state model can take the form of linear equality constraints.

Note that all of these constraints can change on a daily, even hourly basis, so analysis may need to be performed regularly.

The number of independent variables in a constrained optimization problem can be found by a procedure analogous to the degrees of freedom analysis in Chapter 2. If there are N_V process variables and the process model consists of N_E equations (all equality constraints), then the number of independent variables is $N = N_V - N_E$. This means N set points can be specified independently (to maximize profits) and the remaining $(N_V - N)$ set points can be found from the process model. However, the presence of inequality constraints changes the situation, since the N set points cannot be selected arbitrarily but must satisfy all of the inequality constraints.

The solution of optimization problems is profoundly influenced by the presence of inequality constraints, as can be illustrated by the following example. Examine two objective functions, namely the quadratic and linear objective functions, and see how an inequality constraint affects the location of the optimum. Figures 20.7a and b show a quadratic function, $f(x)$. The unconstrained optimization of a quadratic function is straightforward. By differentiating $f(x)$ with respect to x and setting

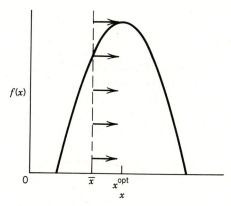

(a) Constrained case $(x \geq \bar{x})$, $x^{opt} = \dfrac{-a_1}{2a_2}$

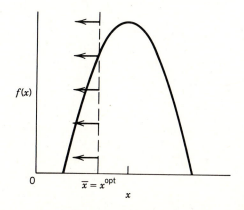

(b) Constrained case $(x \leq \bar{x})$, $x^{opt} = \bar{x}$

Figure 20.7.   The effect of an inequality constraint on the maximum of quadratic function, $f(x) = a_0 + a_1 + a_2 x^2$. (The arrows indicate the allowable values of x.)

the derivative equal to 0, a linear equation results that can be solved analytically:

$$\text{maximize } f = a_0 + a_1 x + a_2 x^2 \tag{20-15}$$

$$\frac{df}{dx} = 0 = a_1 + 2a_2 x \tag{20-16}$$

$$x^{\text{opt}} = -\frac{a_1}{2a_2} \tag{20-17}$$

Now impose a linear inequality constraint; this constraint may or may not affect the solution of the problem. We consider two possible cases: (1) $x \geq \bar{x}$ and (2) $x \leq \bar{x}$. In case 1, the constraint has no effect, that is, the unconstrained optimum (Eq. 20-17) satisfies the constraint (Fig. 20.7a). However, in the second case (Fig. 20.7b), the constraint becomes active and the optimum now lies on the constraint boundary, that is $x^{\text{opt}} = \bar{x}$.

Next consider a linear objective function, as shown in Figs. 20.8a and b. The first derivative of f can never be zero, as occurs for the unconstrained quadratic objective function. For the unconstrained optimization problem with linear objective function, there is no finite maximum, since $f(x)$ is unbounded (see Fig. 20.8a). Now impose a linear inequality constraint $x \leq \bar{x}$ on the solution to the problem. Figure 20.8b demonstrates that for a linear objective function, a finite optimum always occurs on a constraint ($x^{\text{opt}} = \bar{x}$).

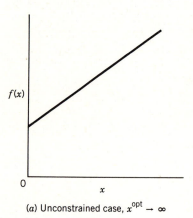

(a) Unconstrained case, $x^{\text{opt}} \to \infty$

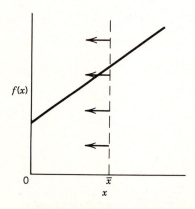

(b) Constrained case ($x \leq \bar{x}$), $x^{\text{opt}} = \bar{x}$

Figure 20.8.   The effect of a linear constraint on the maximum of linear objective function, $f(x) = a_0 + a_1 x$.

This fact was recognized many years ago and serves as the basis for linear programming (LP), which is probably the most powerful mathematical technique in constrained optimization for problems with large numbers of variables. The general linear programming problem can be stated as follows:

$$\text{minimize } f = \sum_{i=1}^{N_V} c_i x_i \tag{20-18}$$

subject to

$$x_i \geq 0 \qquad i = 1, 2, \ldots, N_V$$

$$\sum_{j=1}^{N_V} a_{ij} x_j \geq b_i \qquad i = 1, 2, \ldots, N_I \tag{20-19}$$

$$\sum_{j=1}^{N_V} a_{ij} x_j = d_i \qquad i = 1, 2, \ldots, N_E \tag{20-20}$$

The LP solution can be obtained by a method called the Simplex algorithm (not to be confused with the unconstrained method discussed in the previous section). The Simplex method, which is detailed in Refs. 1–3, is essentially an analytical procedure, although it must be implemented on a digital computer for most realistic problems. The algorithm can handle virtually any number of inequality constraints and any number of variables in the objective function (subject to computer time limitations of course). This technique utilizes the feature illustrated in Fig. 20.8*b* that only the constraint boundaries need to be examined to find the optimum. Because there are a limited number of potential intersections of constraint boundaries, the amount of computer time required to search for the optimum is reduced considerably compared to more general nonlinear optimization problems. Hence, many nonlinear optimization problems even with nonlinear constraints are often linearized so that the LP algorithm can be employed [2,3].

For problems involving two optimization variables, the Simplex procedure can be illustrated using a graphical technique, as shown below. For more than two optimization variables, we must resort to a numerical technique.

EXAMPLE 20.3

Consider the problem of minimizing fuel costs in a boilerhouse. The boilerhouse contains two turbine generators, each of which can be simultaneously operated with two fuels: fuel oil and medium Btu gas (MBG); see Fig. 20.9. The latter fuel

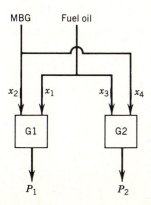

Figure 20.9. The allocation of two fuels in a boilerhouse with two turbine generators (G_1, G_2).

is produced as a waste off-gas from another part of the plant. It must be flared if it cannot be used on site. The goal of the supervisory control scheme is to find the optimum flow rates of fuel oil and MBG. The set points for the flow controllers are the search variables.

First consider a simplified version of this problem, which is amenable to graphical solution. The operating objective in the boilerhouse is to provide 50 MW of power while minimizing costs; this can be done by using as much of the MBG as possible while minimizing consumption of expensive fuel oil. The two turbine generators (G1, G2) have different operating characteristics; the efficiency of G1 is higher than that for G2.

Data collected on the fuel requirements for the two generators yield the following empirical relations:

$$P_1 = 4.5x_1 + 4x_2 \tag{20-21}$$

$$P_2 = 3.2x_3 + 2x_4 \tag{20-22}$$

where P_1 = power output (MW) from G1
P_2 = power output (MW) from G2
x_1 = fuel oil to G1 (tons/h)
x_2 = MBG to G1 (fuel units/h)
x_3 = fuel oil to G2 (tons/h)
x_4 = MBG to G2 (fuel units/h)

The total amount of MBG available is 5 fuel units/h. Each generator is also constrained by minimum and maximum power outputs: generator 1 output must lie between 18 and 30 MW, while generator 2 can operate between 14 and 25 MW.

Formulate the optimization problem by applying the methodology described in Section 20.2. Then solve for the optimum operating conditions (x_1, x_2, x_3, x_4) using a graphical LP analysis.

Solution

Step 1. *Identify the variables.*   Use P_1, P_2, and x_1 through x_4 as the six process variables (the set points). Not all of the variables are independent, since there exist several equality constraints (see Steps 3 and 4).

Step 2. *Select the objective function.*   The way to minimize the cost of operation is to minimize the amount of fuel oil consumed. This implies that we should use as much MBG as possible, since it has zero cost. The objective function can be stated in terms of variables defined above, that is, we wish to minimize

$$f = x_1 + x_3 \tag{20-23}$$

Conversely, we could maximize $-f = -x_1 - x_3$, but this distinction is not crucial here.

Step 3. *Develop models for process and constraints.*   The constraints given in the problem statement are as follows:

(1)   Power relations $P_1 = 4.5x_1 + 4x_2$ (20-21)

$P_2 = 3.2x_3 + 2x_4$ (20-22)

(2)   Power range $18 \leq P_1 \leq 30$ (20-24)

$14 \leq P_2 \leq 25$ (20-25)

(3)   Total power $50 = P_1 + P_2$ (20-26)

(4)   MBG supply $5 = x_2 + x_4$ (20-27)

Note that all variables defined above are nonnegative.

Step 4. *Simplification of the model and objective function.*   In Step 2 we identified the six variables ($N_V = 6$) involved in the objective function and specified the equality and inequality constraints. Since there are four equality constraints ($N_E = 4$), four of the six variables can be eliminated. Thus using $N = N_V - N_E$, we select two independent variables (x_1, x_3). Combining Eqs. (20-21), (20-22), (20-26), and (20-27) allows us to eliminate x_2 and x_4. Solving for P_1 and P_2 and applying the upper and lower limits in (20-24) and (20-25) gives

(a) $4.5x_1 + 6.4x_3 \geq 50$ (20-28)

(b) $4.5x_1 + 6.4x_3 \leq 62$ (20-29)

(c) $4.5x_1 + 6.4x_3 \geq 44$ (20-30)

(d) $4.5x_1 + 6.4x_3 \leq 55$ (20-31)

Similarly, combine the same four equations and eliminate P_1 and P_2. Then applying the constraints that x_1 and x_2 must be nonnegative gives:

(e) $2.25x_1 + 1.6x_3 \leq 20$ ($x_2 \geq 0$)   (20-32)

(f) $2.25x_1 + 1.6x_3 \geq 15$ ($x_4 \geq 0$)   (20-33)

Step 5. *Computation of the optimum.*   Here we demonstrate the graphical solution of the LP problem, since there are two variables which are independent. The graphical solution consists of three steps [1,3]:

1. Plot the constraint lines on the x_1–x_3 plane.
2. Determine the feasible region for x_1 and x_3 (those values of x_1 and x_3 that satisfy all of the inequality constraints).
3. Find the point or points in the feasible region that minimize the objective function f; this can be done by plotting the line $f(x_1, x_3) = C$ where C is a parameter indicating the different cost levels. The optimum will normally occur at an intersection of the constraints that form the feasible region.

Plotting Eqs. 20-28 through 20-33 in Fig. 20.10 defines the feasible region in the x_1–x_3 plane, that is, the region in which the optimum must lie. The four parallel lines arise from (20-28) to (20-31). Similarly, (20-32) and (20-33) are parallel lines. Note that some of the constraints are redundant and do not influence the feasible region. The polygon defined by the limiting inequality constraints is shown in Fig. 20.10 as a cross-hatched area. The line, $f = x_1 + x_3$, is also plotted for two arbitrary values of f ($f = 5$ and $f = 7$). These constant cost lines correspond to different operating conditions, that is, to different fuel oil distributions. To minimize f, the fuel oil consumption, the dashed line must be moved to the right to the edge of the feasible region where the inequality constraints are first satisfied. Using Fig. 20.10, the minimum cost corresponds to point A, which is a vertex (corner) of the feasible region. This is the intersection of constraints (a) and (f); moving away from vertex A into the interior of the feasible region corresponds to larger amounts of fuel oil consumption. At A, $f = 8.45$, $x_1 = 2.2$, and $x_3 = 6.25$, meaning that 2.2 tons/h of fuel oil are delivered to generator G1 while 6.25 tons/h are used in G2. G1 operates with all of the MBG, while G2 uses none, due to the low efficiency of G2. In terms of power production, $P_1 = 30$ and $P_2 = 20$.

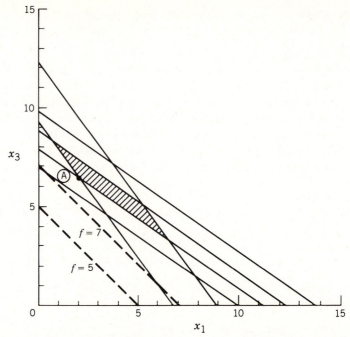

Figure 20.10.   The constraint lines for Example 20.3, with the feasible region identified as the cross-hatched area. Point A is the optimum. The objective function lines are dashed.

Step 6.   *Sensitivity analysis.*   There are many operating strategies that would be satisfactory although not optimal for the above problem. Any point in the feasible region satisfies the problem constraints, and the additional fuel oil consumption can be calculated for other possible operating points. The procedure discussed above can also be repeated easily if parameters in the original constraint equations are changed. For example, suppose the total power requirement is changed to 55 MW. Using algebraic substitution, a new two-variable optimization problem could be developed and solved to compute the new optimum.

The graphical analysis presented above is somewhat impractical if it is to be done by a process control computer or is to be performed fairly often. In addition, graphical analysis can only be performed for two independent variables, which restricts its applicability. The principle from graphical analysis, namely, that the optimum occurs at the intersection of two or more inequality constraints, forms the basis of the Simplex algorithm. This numerical method can be implemented by a digital computer and thus automate set-point selection. The Simplex algorithm also allows more detailed problem statements for boiler optimization to be considered, since it does not require algebraic elimination of variables. Equality constraints are treated directly by the Simplex algorithm.

In the process industries, the Simplex algorithm has been applied to a wide range of problems [3], including the optimization of a total plant utility system as well as to optimal allocation of boiler fuel. A general steam utility configuration typically involving as many as 100 variables and 100 constraints can be optimized using linear programming [11]. The updating of the process variables may be carried out on an hourly basis, since steam demands in process units may be changing. In addition, it may be economical to generate more electricity locally during times of peak demand, due to variable time-of-day electricity pricing by utilities.

SUMMARY

Although the economic benefits from feedback control are not always readily quantifiable, supervisory control, that is, set-point selection, offers a direct method of optimizing the steady-state profitability of a process or group of processes. The optimization of the set points is performed as frequently as necessary, depending on changes in operating conditions or constraints.

It is important to formulate the optimization problem carefully; a methodology for formulation and solution of optimization problems is presented in this chapter. There is a wide range of optimization techniques that can be used, depending on (1) the number of variables, (2) the nature of the equality and inequality constraints, (3) the nature of the objective function. Since we have presented only introductory concepts in optimization here, the reader is advised to consult other references on optimization before choosing a particular method for supervisory control.

REFERENCES

1. Beveridge, G. S., and R. S. Schechter, *Optimization: Theory and Practice,* McGraw-Hill, New York, 1968.
2. Reklaitis, G. V., *Engineering Optimization,* Wiley-Interscience, New York, 1984.
3. Edgar, T. F., and D. M. Himmelblau, *Optimization of Chemical Processes,* McGraw-Hill, New York, 1988.
4. Latour, P. R., On Line Computer Optimization, 1. What It Is and Where to Do It, *Hydro. Proc.* **58**(6), 73 (1979).
5. Latour, P. R., On Line Computer Optimization, 2. Benefits and Implementation, *Hydro. Proc.* **58**(7), 219 (1979).
6. Rosenof, H. P., and A. Ghosh, *Batch Process Automation,* Van Nostrand Reinhold, New York, 1987.
7. Himmelblau, D. M., *Applied Nonlinear Programming,* McGraw-Hill, New York, 1972.
8. Box, G. E. P., Evolutionary Operation: A Method of Increasing Industrial Productivity, *Appl. Statistics* **6,** 3 (1957).
9. Spendley, W., G. R. Hext, and F. R. Himsworth, Sequential Application of Simplex Designs in Optimization and Evolutionary Operation, *Technometrics* **4,** 441 (1962).
10. May, D. L., B. N. Norden, C. C. Andreasen, and C. H. Cho, Optimizing Plant Refrigeration Costs, *ISA Trans.* **18**(1), 71 (1979).
11. Bouilloud, P., Compute Steam Balance by LP, *Hydro. Proc.* **48**(8), p. 127 (1969).

EXERCISES

20.1. A laboratory filtration study has been carried out at constant rate. The filtration time (t_f, in hours) required to build up a specific cake thickness has been correlated as

$$t_f = 5.3x_c(e^{-3.6x_c} + 2.7)$$

where x_c = mass fraction solids in the cake. Find the value of x_c that maximizes t_f using quadratic interpolation.

20.2. You plan to use a one-dimensional search to find the operating pressure of the chemical reactor that gives the maximum yield of the reactor. A possible range of pressures has been identified by laboratory work to be between 1 and 5 atm. You would like to operate within 0.03 atm of the optimum pressure. How many tests would be required with a three-point equal-interval search? With quadratic interpolation?

20.3. Minimize the function $f = (x - 1)^4$. Use quadratic interpolation but no more than a maximum of ten function evaluations. The initial three points selected are $x_1 = 0$, $x_2 = 0.5$, and $x_3 = 2.0$. Repeat using three-point equal-interval search with an initial interval of uncertainty of $[-1, 2]$.

20.4. The thermal efficiency of a natural gas boiler versus air/fuel ratio is plotted in the drawing. Explain using physical arguments why a maximum occurs in the figure.

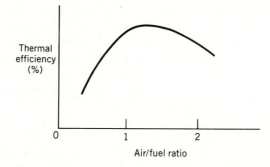

20.5. A specialty chemical is produced in a batch reactor. The time required to successfully complete one batch of product depends on the amount charged to (and produced from) the reactor. Using reactor data, a correlation is $t = 2.0P^{0.4}$, where P is the amount of product in pounds per batch and t is given in hours. A certain amount of nonproduction time is associated with each batch for charging, discharging, and minor maintenance, namely 14 h/batch. The operating cost for the batch system is \$50/h while operating. Other costs including storage depend on the size of each batch and have been estimated to be $C_I = \$800P^{0.7}$ (\$/yr). The required annual production is 300,000 lb/yr, and the process can be operated 320 days/yr (24 h/day). Total raw material cost at this production level is \$400,000/yr.

(a) Formulate an objective function using P as the only variable. (Show algebraic substitutions).
(b) Are there any constraints on P? (Give relations if applicable.)
(c) Solve for the optimum value of P analytically. Check that it is a minimum. Also check applicable constraints.

20.6. A refinery processes two crude oils that have the yields shown in the following table. Because of equipment and storage limitations, production of gasoline, kerosene, and fuel oil must be limited as shown below. There are no plant limitations on the production of other products such as gas oils. The profit on processing crude No. 1 is \$2.00/bbl and on crude No. 2 it is \$1.40/bbl. Find the optimum daily feed rates of the two crudes to this plant via linear programming (graphical method).

| | *Volume % Yields* | | *Maximum Allowable Production* |
	Crude No. 1	*Crude No. 2*	*Rate (bbl/day)*
Gasoline	70	31	6,000
Kerosene	6	9	2,400
Fuel oil	24	60	12,000

20.7. Linear programming is to be used to optimize the operation of the solvent splitter column shown in the drawing. The feed is naphtha which has a value of \$40/bbl in its alternate use as a gasoline blending stock. The light ends sell at \$50/bbl while the bottoms are passed through a second distillation column to yield two solvents. A medium solvent comprising 50 to 70% of the bottoms can be sold for \$70/bbl, while the remaining heavy solvent (30 to 50% of the bottoms) can be sold for \$40/bbl.

Another part of the plant requires 200 bbl/day of medium solvent; an additional 200 bbl/day can be sold to an external market. The maximum feed that can be processed in column 1 is 2000 bbl/day. The operational cost (i.e., utilities) associated with each distillation column is $2.00/bbl feed. The operating range for column 2 is given as the percentage split of medium and heavy solvent. Simplify the objective functions and constaints so that they are expressed in terms of the two independent variables, x_4 and x_5. Then solve the linear programming problem to get the maximum revenue and percentages of output streams in column 2.

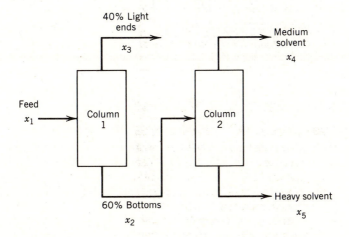

20.8. Reconciliation of inaccurate process measurements is an important problem in process control that can be solved using optimization techniques. The flow rates of streams B and C have been measured three times during the current shift (shown in the drawing). Some errors in the measurement devices exist. Assuming steady-state operation (M_A = constant), find the optimal value of M_A (flow rate in kg/h) that minimizes the sum of the squares of the errors for the material balance, shown.

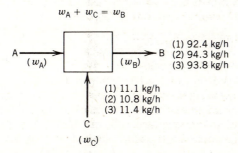

20.9. A reactor converts reactant BC to product CB by heating the material in the presence of an additive A (mole fraction = x_A). The additive can be injected into the reactor, while steam can be injected into a heating coil inside the reactor to provide heat. Some conversion can be obtained by heating without addition of A, and vice versa. The product CB can be sold for $50 per lb-mole. For 1 lb-mole of feed, the cost of the additive (in dollars per lb-mole feed) as a function of x_A is given by the formula, $2.0 + 10x_A + 20x_A^2$. The cost of the steam (in dollars per lb-mole feed) as a function of S is $1.0 + 0.003S + 2.0 \times 10^{-6} S^2$. ($S$ = lb steam/lb-mole feed). The yield equation is $y_{CB} = 0.1 + 0.3x_A + 0.0001S - 0.0001x_A S$.

$$y_{CB} = \frac{\text{lb-mole product CB}}{\text{lb-mole feed}}$$

(a) Formulate the profit function (basis of 1.0 lb-mole feed) in terms of x_A and S.

$$f = \text{income} - \text{costs}$$

(b) Maximize f subject to the constraints

$$0 \leq x_A \leq 1 \qquad S \geq 0$$

by any method you choose.

20.10. Carry out five steps of the simplex method to minimize the function $f(x_1, x_2) = x_1^2 + x_2^2$ starting with $(1, 1)$ and $(1, 1.5)$ as two corners of the simplex. Show each step on a contour plot.

20.11. Optimization methods can be used to fit equations to data. Parameter estimation involves the computation of unknown parameters that minimize the squared error between data and the proposed mathematical model. The step response of an overdamped second-order dynamic process can be described using the equation

$$\frac{y(t)}{K} = \left(1 - \frac{\tau_1 e^{-t/\tau_1} - \tau_2 e^{-t/\tau_2}}{\tau_1 - \tau_2} \right)$$

where τ_1, τ_2 are process time constants and K is the process gain.

The following normalized data have been obtained from a unit step test (K is equal to $y(\infty)$):

time, t	0	1	2	3	4	5
y_i/K	0.0	0.0583	0.2167	0.360	0.488	0.600
t	6	7	8	9	10	
y_i/K	0.692	0.772	0.833	0.888	0.925	

Use the Simplex method with starting points $(1, 0)$ and $(0.5, 0)$ to find values of τ_1 and τ_2 that minimize the sum of squares of the erors. Show the iterations on a contour plot.

20.12. A brewery has the capability of producing a range of beers by blending existing stocks. Two beers (suds and premium) are currently available, with alcohol concentrations of 3.5% for suds and 5.0% for premium. The manufacturing cost for suds is $0.25/gal and for premium it is $0.40/gal. In making blends, water can be added at no cost. An order for 10,000 gal of beer at 4.0% has been received for this week. There is a limited amount of suds available (9000 gal), and due to aging problems, the brewery must use at least 2000 gal of suds this week. What amounts of suds, premium, and water must be blended to fill the order at minimum cost?

20.13. A specialty chemicals facility manufactures two products A and B in barrels. Products A and B utilize the same raw material; A uses 120 kg/barrel while B requires 100 kg/bbl. There is an upper limit on the raw material supply of 9000 kg/day. Another constraint is warehouse storage space (40 m² total; both A and B require 0.5 m²/bbl). In addition, production time is limited, to 7 h per day. A and B can be produced at 20 bbl/h and 10 bbl/h, respectively. If the profit per bbl is $10 for A and $14 for B, find the production levels that maximize profit.

PART SIX

DIGITAL CONTROL TECHNIQUES

— CHAPTER 21 —————

Digital Computer Control

Since the early 1960s the number of applications of digital computers to process control has increased dramatically. During this period there have been a number of important changes: Where early computer control projects involved large, expensive, and heavily customized facilities, more recent projects make significant use of inexpensive off-the-shelf hardware components and require little or no programming expertise. If special-purpose optimization and control functions have to be included, generalized control languages permit the programming of user applications software in a minimum of time.

These changes have gone far beyond those expected to occur in the normal maturing of a new technological field. Changes in the computer control field have been revolutionary, resulting from an orders-of-magnitude increase in computer capabilities. This increase in hardware capability has itself been accompanied by a precipitous drop in prices. There has been a strong and developing incentive to use computers, particularly microcomputers, in totally new applications areas as a result of this radical change in economics. When digital computers first appeared, the few computer control projects that were undertaken had to use relatively large, *mainframe* computers. The projects cost on the order of a million dollars (in today's prices) and they could only be justified economically in terms of controlling around 100 loops or in very special circumstances. Today a distributed microcomputer-based instrumentation system is economically feasible for as few as 4–8 loops, and it can be expanded easily by adding modular units to include several hundred loops. Off-the-shelf communication equipment permits the interconnection of the primary control units with interactive operator terminals, with memory units for archival storage of plant operating data, and with other computers that can be used for advanced control algorithms such as supervisory or adaptive control. The problem today is not so much economic justification of a digital system as it is the choice of the correct system—single or multiple computer, microprocessor-based system, and so on.

In this chapter we first present a brief introduction to the field of digital process control and then discuss digital instrumentation systems from a functional point of view. Because an understanding of this applications area requires some knowledge of hardware, we briefly discuss this topic, focusing on the interface (or equipment) between the process and digital control units. Many digital control applications now require no user programming; however, nonstandard and more complex process facilities invariably do. Hence, a brief overview of programming considerations for this field is given. Implementation of the three-mode digital control algorithms developed in Chapter 8 is described as an example. Finally, the use of microcomputer-based programmable logic controllers and their application to the control of batch processes is briefly discussed.

21.1 ROLE FOR DIGITAL COMPUTER SYSTEMS IN PROCESS CONTROL

Two major areas for digital computer/digital control applications exist in the process industries [1, Chapter 4]. The first of these, which might be labeled *passive* applications, involves the acquisition and perhaps manipulation of process data by digital systems. The second, which we might label *active* applications, involves manipulation of the process by the computer as well.

The data acquisition area would appear to have little or nothing to do with process control. However, many computer control projects, particularly those associated with new processes, have been split into two phases. The first phase emphasizes characterization of the process and the second phase is concerned more directly with design and implementation of the computer control system. The first phase then deals predominantly with monitoring, alarming, and data reduction systems of the type shown schematically in Fig. 21.1. Note that even in system installations that are primarily intended for control, 70–80% of the computer inputs typically are *indicating-only*, that is, not used for control.

Figure 21.1 illustrates that some sort of interface between the process and the computer is necessary. One of the most troublesome features of any digital systems application is the measurement and conversion interface. We consider this component in more detail in Section 21.3.

Most data logging is done *on-line* with some type of digital equipment, perhaps not the final data reduction or control computer but rather one used to operate the measurement and conversion interface. The use of a microcomputer to supervise the interface is a common one and has given rise to the term *smart (or intelligent) front end.* A small computer used in this way can buffer or temporarily store information, sending it to a larger computer when convenient. The larger computer might serve as *host* to many front-end processors as shown in Fig. 21.2. This figure

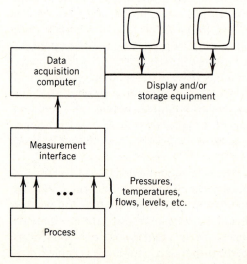

Figure 21.1. A digital data acquisition facility.

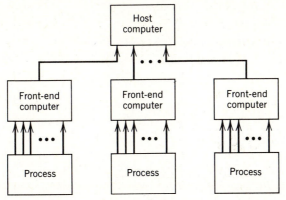

Figure 21.2. A host computer that is used to analyze results from several front-end data acquisition computers.

also illustrates a type of *distributed data processing,* the use of multiple computers (processors) distributed within a network. In this case, the network is a hierarchical one, a generally pyramidal form in which low-level functions are performed by less sophisticated digital equipment at the bottom. Results at this first level are passed upward to the larger data processors that are responsible for more complex operations. In Fig. 21.2 the front-end computers typically would acquire process data from the measuring instruments and convert the results to standard engineering units. The host computer typically would be responsible for averaging results, making data consistency checks, calculating material and energy balances, communicating summaries of results to the operator or to engineering personnel, and so on.

A related development that is of at least indirect interest for process control applications is the so-called *smart instrument.* Many measurement instruments presently are supplied with an internal microcomputer that can relieve the operator of the tedious chores associated with measurements. One example is the use of *autoranging* equipment to deal with, for example, voltage-level signals that may vary over a wide range of values, say millivolts up to volts. In this case a microcomputer may be used to control an internal amplifier between the input(s) from the process and the actual measurement circuitry. The gain of the amplifier is electronically switched by powers of 2 until the full precision of the measurement equipment is utilized. Such an approach can be used to maintain a reasonably uniform measurement precision over the entire input range. A second example that commonly is encountered is automatic calibration of the instrument.

There are numerous examples of instruments that must interact intelligently with the process they are monitoring, for example, in sampling process streams. A chromatograph controlled internally by a microcomputer falls into the category of smart instruments. A modern chromatograph ordinarily can be programmed to log a chromatogram, to perform baseline correction, to select peaks representing individual chemical species, to integrate the area under each peak, to convert the areas to concentrations, and to print out the results. If, in addition to the above duties, the chromatograph microcomputer is given the responsibilities of purging the chromatograph sample line, drawing in a sample from the appropriate process stream, and metering the proper amount into the chromatograph column, it obviously interacts with the process. Thus, sample lines and valves are manipulated, even though it is substantially still a monitoring unit.

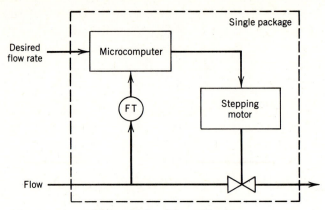

Figure 21.3. A "smart" valve.

Within the smart instruments category we can include the group of devices called *smart actuators*. In general such an actuator (final control element) contains some intelligent device, usually a microcomputer, which ensures that the actuator functions according to design even in the face of a changing environment. In addition, some desirable operating characteristics may be incorporated. For example, designers often choose an automatic valve with an equal percentage characteristic to compensate partially for the nonlinearity of flow measuring devices and for variations in the upstream or supply pressure (Chapter 9). If flow rate information is available from a flow transmitter, then a cascade control loop can be used to compensate exactly for these effects. A control valve driven by a stepping motor[1] can be constructed as a smart actuator by incorporating a microcomputer and flow transmitter, as shown schematically in Fig. 21.3. The microcomputer can be programmed to operate the valve so as to maintain any desired flow rate. This cascade control approach involves a feedback loop to provide a linear characteristic between flow set point and actual flow. Assuming tight control action is attained (a relatively easy task, in this case), the valve will supply exactly the desired flow rate and do this regardless of variations in upstream and downstream pressures. This unit could be used as a direct replacement for the secondary loop that regulates the flow of a process stream in a normal cascade control system.

The major *active* application of digital computers is in process control and plant optimization. At the present time these applications often utilize equipment that is designed around a microcomputer but that emulates the functions of the analog control equipment discussed in Chapter 8. When such equipment is connected together using digital communication techniques, it represents a distributed digital control system of the type described in the next section.

21.2 DISTRIBUTED INSTRUMENTATION AND CONTROL SYSTEMS

The traditional approach to digital computer control beginning around 1960 was to start with a single large control computer and design the control system around the characteristics of the computer rather than those of the process. The process control functions that could be accommodated easily by the computer were included, and those that could not were left out. Basing the control system so com-

---

[1] A stepping motor advances a fraction of a rotation for each pulse applied to its input line. Two lines can be used to obtain both forward and reverse operation.

pletely on the characteristics of the computer often led to compromises and to cumbersome systems that were unreliable, difficult to maintain, and virtually impossible to modify or expand. The development of microcomputer-based subsystems (smart instruments, controllers, logic sequencers, packaged operator interfaces, etc.) beginning in the mid 1970s led to an alternative approach. This configuration begins by replacing the bottom-level analog instruments and control units with more flexible digital ones on a one-for-one functional basis. The digital units then are interconnected, as required, using some form of digital communications network. Although the capability to incorporate general purpose computers in such systems for purposes of carrying out high-level data analysis and/or more sophisticated control existed earlier, large-scale conversions of analog to digital systems have only been made since about 1980. Whether or not high-level computational capability is present, such systems are referred to generically as *distributed digital instrumentation and control systems*. This is because of the network approach used to monitor and control the process.

The digital equipment can be distributed both geographically and functionally. *Geographical distribution* permits the individual equipment components that interact directly with the process to be placed adjacent to the pertinent process equipment. Other components such as the operator's interface can be placed wherever they are most convenient. With data concentration and serial digital data transmission techniques, one or more inexpensive coaxial cables (data highways) can replace the extensive amounts of analog wiring and/or pneumatic signal lines that ordinarily are required with classical analog control systems. *Functional distribution* allows the equipment to be partitioned in a logical way at the component level. Hence, separate units can be used that are designed specifically for data acquisition, single-loop control, batch process sequencing, data transmission, operator display, and so on. The major advantages of functional hardware distribution are flexibility in system design, ease of expansion, reliability, and ease of maintenance.

Many of the functional distribution advantages listed above have been associated historically with analog systems. Since the purchase costs of distributed digital hardware continue to be higher than analog hardware at the present time, why choose the digital system approach? The answer to this question is multifaceted. Essentially the same advantages that motivated early single-computer projects hold also for distributed systems:

- *Digital systems are more precise.*
- *Digital systems are more flexible.* Control algorithms can be changed and controller configurations can be modified without having to rewire the system.
- *Digital systems cost less to install and maintain.* Even though the equipment cost is higher, the installed cost typically is lower because the labor-intensive wiring required to connect a central control room to the analog instrumentation is eliminated.
- *Digital data in electronic data files are easier to deal with.* Operating results can be printed out in reports, displayed on color graphics terminals, stored in highly compressed form on disk or magnetic tape, and recalled almost instantaneously.

A representative distributed digital control system is depicted in Fig. 21.4. Such a system typically consists of one or more of the following elements:

1. **Local Control Unit (LCU).** Typically, this unit can implement 8 to 16 individual PID control loops, with 16 to 32 analog input lines, 8 to 16 output signals, and a limited amount of digital input, output, and internal logic capability.

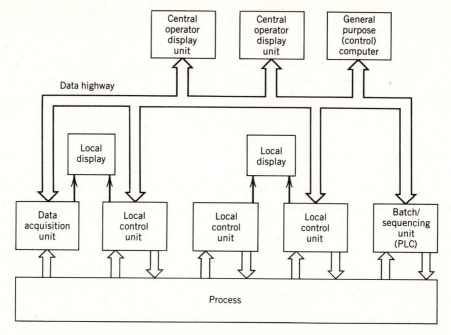

Figure 21.4.   A distributed digital instrumentation and control system.

2. **Data Acquisition Unit.**   Generally, this unit contains 2 to 16 times as many analog input and output channels as the LCU. This device is an intelligent remote unit of the type discussed in Section 21.2. Digital (discrete) as well as analog I/O can be handled. Typically, no control functions are available, and all output is specified either by the operator or by a higher-level (general purpose) control computer in the network.

3. **Batch Sequencing Unit.**   Typically, this unit contains a number of external event or timing counters, arbitrary function generators (e.g., for preprogrammed manipulation of process elements), and considerable digital input/output and internal logic capability. A more complete discussion of this unit is given in Section 21.6.

4. **Local Display.**   This device generally will contain only analog display stations and, perhaps, analog trend recorders. Alternatively, a video display may be used for readout of selected variables.

5. **Bulk Memory Unit.**   Process operating data covering extended periods of time can be stored in such units and recalled quickly. These usually consist of on-line mass storage disks or magnetic tape units.

6. **General Purpose Computer.**   A complex control system needs capabilities not usually supplied by the instrumentation system manufacturer. These might include supervisory (optimizing) control, advanced (e.g., adaptive) control, expert systems inventory checking, finished product storage allocation, recipe look-up for batch processes, and so on. A general purpose computer programmed by the customer or a third party ordinarily is required to obtain these functions.

7. **Central Operator Display.**   This unit typically will contain one or more consoles for operator communication with the system, multiple video color graphics display units, and hard-copy output devices such as printers.

8. **Data Highway.**   A serial digital data transmission link connecting all other components in the system may consist of just two wires or a coaxial cable. Most

commercial systems allow for redundant data highways to reduce the possibility of a loss of data from the monitoring and control units to the display, storage, and operational elements in the network.

9. Local Area Network (LAN). Many manufacturers now supply a port (gateway) device that allows their proprietary data highway to be connected to other instrumentation systems, to general purpose computers, or to other computational elements through a standard local area network. This approach permits the development of extended *information systems,* of the type discussed below.

An instrumentation system might be purchased that contains as little as one or two monitoring and/or control units interfaced to the process along with a local display unit. A big advantage of digital instrumentation systems compared to earlier single-computer digital systems is that the user can start out at a low level of investment. However, the true benefits of the distributed hardware accrue when a data highway or LAN is used to link the local process units with a central operator display and a high-level computer. Another obvious advantage of this type of distributed architecture is that complete loss of the data highway will not cause a complete loss of system capabilities. Often local units can continue operation with no significant loss of function over moderate or extended periods of time. Note that the high-level computer shown in Fig. 21.4 accesses the process only through lower-level units, not directly. Consequently, batch sequences, for example, would be modified by the higher-level computer. Also, most control units will include manual/auto/supervisory/computer operation for each local control loop. In the supervisory mode the set point can be supplied by the remote higher-level computer. In the computer mode the higher-level computer manipulates the final control element directly, with an automatic switch to a local backup PID or cascade controller if the higher-level computer or the data highway fails.

In designing the components of distributed instrumentation systems, manufacturers make extensive use of microprocessors, generally at least one in each separate unit or module. Each module is designed to perform a small subset of all system functions and to deal with a small subset of process inputs and outputs. Therefore, a table-driven programming approach can be used to implement system program functions supplied by the manufacturer in the lower-level microprocessor-based modules (see Section 21.5). These programs typically are not accessible to the user. They represent *firmware* code stored permanently in read-only memory that is used only by the local microprocessor.

Several trends in the development and application of digital instrumentation systems have now become apparent:

1. The control system easily generalizes into an information system. Key operating data, both present and historical (archived) data, are available within the system in a relatively accessible form. Terminals permit access to the distributed data bases, usually through one or more higher-level networks. By providing adequate protection for the data bases, managers and engineers can have access to required information in a timely and convenient way. This capability can be exploited at many levels in the typical large corporate structure; corporate, division, plant, area, unit, loop, and measurement point would be one example. Hierarchical information systems can be structured to give information in the form and amount necessary to make decisions. For example, three-level management information systems—corporate, plant, and unit—are commonplace [2]. When such a system is properly designed, information moves upward and decisions move downward in the network, easily and naturally.

2. Control system hardware costs, meaning microprocessor-based elements, for all practical purposes have become negligible relative to installed system costs. Payout times as low as three to four months have been reported [2]. The major costs now are associated with specialized firmware and software, with installation, with configuring the information flow structure (the choice of controlled and manipulated variables, control strategies, etc.), and with maintenance. The central operator display unit for a distributed system can be supplied as a workstation to help with installation of the system and with ongoing maintenance. Typically, it might take on a number of personalities—one for developing the system configuration, one for operator communication, one for dealing with maintenance personnel, and so on. In the configuration mode, development of the tables (internal data bases) needed to implement routine operations traditionally has been done interactively, for example, by query and answer or form fill-in (see Section 21.5). An alternative approach is for the user to assemble a set of icons (schematic elements) representing ISA (Instrument Society of America) standard control system components into the usual instrument diagram on the operator graphics display terminal. The workstation can then develop the data base needed by the operating digital control program from the graphics display file. Hence, the instrument diagram becomes the complete specification and documentation for both the control system and its operational program (its computer implementation) with a considerable reduction in development and documentation costs.

3. A very large problem is the lack of standardization—for a universally applicable programming language for control computers, for standardized data bases, for interconnection of devices, highways, and networks, and for standardized user program interfaces [2,3]. The distributed instrumentation system minimizes these problems, particularly if equipment from a single vendor is used.

4. More and more fault detection is being built into these systems, as is considerable redundancy of control functions [2]. The application of artificial intelligence techniques (see Chapter 18) within the decision-making hierarchy is a natural outgrowth of the computational power that already exists there [4].

The philosophy employed in designing and marketing digital instrumentation systems is to supply users with many of the benefits of digitally based hardware without requiring knowledge of the details of how digital systems operate. Their commercial success is in fact a function of how transparent the internal workings can be made to the operator and to the process engineer. Nevertheless, the most successful use of digital instrumentation systems requires the user to understand basic principles of digital hardware and programming. The next two sections deal with these topics.

21.3 GENERAL PURPOSE DIGITAL DATA ACQUISITION AND CONTROL HARDWARE

Almost any modern digital computer can be used in a process control application. Minicomputer- and microcomputer-based systems presently on the market are adequate for even very large applications with up to several thousand I/O points. From a utilitarian point of view, any general purpose, stored program computer with sufficient input/output capability, including interrupt handling ability discussed below, can be considered as a candidate for process control. Virtually any computer that fits this description will be composed of four subunits as shown in Fig. 21.5 [1, Chapter 8].

The *arithmetic unit* contains all hardware necessary to carry out arithmetic and

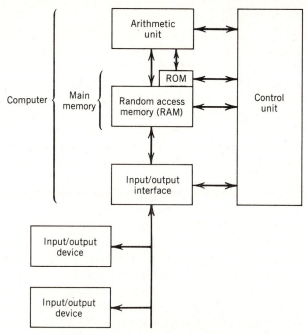

Figure 21.5. A general purpose digital computer.

logic commands, for example, to add two numbers, subtract two numbers, and check to see if one number is larger than a second number. All parts of the computer are constantly under the supervision of the *control unit*. This part of the computer is responsible for reading program statements from memory, interpreting them, and causing the appropriate actions to take place (e.g., adding two numbers in the arithmetic unit). The *memory unit* in a digital computer is used for the storage of data and of the computer program itself. Many industrial computer systems now utilize both random-access memory (RAM) and read-only memory (ROM) as part of main memory. RAM can be used for storage of both programs and data, whereas ROM is used to store system program elements (routines). ROM sometimes is referred to as *firmware*. The main memory unit, composed of solid-state components to store digital information, is often referred to as *fast memory* to distinguish it from *bulk memory* or *slow memory* which is usually a separate (peripheral) device.

The final unit in this simplified visualization of the digital computer is the *input/ output interface*. The I/O interface is necessary for the computer to communicate with the "outside world," that is, with all of its peripheral equipment or, if a data highway is used, with other devices in the network.

This interface contains the hardware logic necessary to detect and respond to external *events*. These events usually take the form of a request for some kind of action on the part of the computer which then must interrupt its normal processing sequence. The ability to respond to an external *interrupt* is a requirement for any computer, but it is in fact the very basis of design of a process control computer. Process control computer applications represent a special class of *real-time computing* applications. *Real time* implies that a task be completed within a specified time constraint. Certain operations, perhaps the computation of control signals to the process, must be programmed to take place in real time ("clock-on-the-wall" time). In addition, the computer must respond to external events, perhaps a process alarm situation, in real time.

Representation of Information

Control of processes almost always involves working with process information in analog format, for example, as shown in Table 21.1. The unifying feature of all of these variables, and the reason we refer to them as analog variables, is that they can take on any numerical value within their range, subject to the precision of the instrument that is used to measure them. For example, a reactor temperature may be measured as 251, 250.9, 250.92 °C, . . . , depending on the precision of the measurement device.

Digital information is expressed in a finite number of elements or states; for example, any integer number from 0 to 99 requires just two decimal numbers for representation. In applying a digital computer we are restricted to the use of the discrete memory elements available there. Since the modern digital computer is a binary machine, all internal data must be represented as binary numbers, and all arithmetic and logic must be reduced to binary operations. The important point here is that all process information that is not already in digital (binary) form must be converted to that form. Some process information is inherently digital. For example, a pump may be running (on) or shut down (off), and we could represent the on state by one, the off state by zero. Hence, only a single *bit* (binary element) in computer memory would be required to store the pump's operating state.

Most process measurements used for control will be in the form of analog signals that are changing in time. Two types of discretization are necessary for this information to be used within the digital computer: discretization in *magnitude* and discretization in *time*. As indicated above, the magnitude of an analog variable must be stored in one or several computer memory locations, each made up of a collection of bits called a word. If we consider a 0 to 10 V analog signal and assume that only a three-bit computer word is available, there would be only eight (2^3) different digital states that could be used to represent the full range of the analog variable. This feature is shown in Table 21.2. In each case the analog representation is obtained by linear interpolation.

It is easy to see that the degree of approximation is poor at this level of discretization. The obvious way to get more precision is to use more bits. In general, the degree of resolution is given by

$$\text{resolution} = [\text{full scale range}] \times \frac{1}{2^m - 1} \tag{21-1}$$

where m = number of bits in the representation. Note that the maximum error is one-half the resolution or $(\frac{1}{2})(\frac{1}{7}) = \frac{1}{14}$ for the example in Table 21.2 ($m = 3$).

In any digital data acquisition and control system, analog information is first converted to voltage or current form. Then it is converted to digital form by an electronic device called an *analog to digital converter* (A/D converter or ADC). The reverse process by which digital information is converted to analog form

Table 21.1   Some Examples of Analog Process Information

Variable	Measurement Range and Units
Pressure	100 to 200 psig
Pneumatic instrument	3 to 15 psig
Temperature	200 to 300 °C
Level	0 to 50 in.
Electronic instrument	4 to 20 mA

Table 21.2 Representation of a 0 to 10 V Analog Variable Using a Three-Bit Word

Contents of Word (Binary Representation)	Digital Equivalent	Analog Representation
0 0 0	0	0 to $\frac{1}{14}$
0 0 1	1	$\frac{1}{14}$ to $\frac{3}{14}$
0 1 0	2	$\frac{3}{14}$ to $\frac{5}{14}$
0 1 1	3	$\frac{5}{14}$ to $\frac{7}{14}$
1 0 0	4	$\frac{7}{14}$ to $\frac{9}{14}$
1 0 1	5	$\frac{9}{14}$ to $\frac{11}{14}$
1 1 0	6	$\frac{11}{14}$ to $\frac{13}{14}$
1 1 1	7	$\frac{13}{14}$ to 1

(usually a voltage or current) is carried out by a digital to analog converter (D/A converter or DAC). Most process control-oriented ADCs and DACs utilize a 10- to 12-bit representation (resolution better than 0.1%). Since most micro- and minicomputers utilize at least a 16-bit word, the value of an analog variable can be stored in one memory word.

We mentioned above that two types of discretization are required, the second being necessary to convert a signal that varies continuously in time to one that has values at only discrete instants in time. This sampling procedure is necessary so that an approximate representation of any signal can be maintained in computer memory (in successive words, perhaps). Chapter 22 contains more details on process considerations in signal sampling.

Process Interface

The computer input/output interface is too specialized to be used for general process information transfer. Further, it is not desirable to design a completely different interface for each process to be connected to the computer. The approach that traditionally has been taken is to reduce the total number of all process variables to several classes. Then a specialized device can be used to transfer all information of that class into or out of the computer. Most process information can be grouped into the four major categories in Table 21.3—digital, generalized digital, pulse, and analog [4, Chapter 3], [1, Chapter 10].

With many applications, Type 1 and 2 information can be handled by single devices, one for computer inputs and one for outputs. For a smaller facility it may be possible to dispense with one or another of the input/output units. For example, only analog inputs and digital outputs might be utilized.

The *digital signal* (*binary*) *interface* ordinarily is designed to match the computer word size. The input and output interfaces would be designed to have multiple registers, each with the same number of bits as the basic computer word. In this way a full word of process information (say 16 bits, representing 16 separate process binary variables such as pump on or off) can be transmitted to the computer at one time and stored. To determine the state of any particular process input line (say line 3, representing a motor drive), the computer would test the appropriate bit (Bit 3) to determine its state—0 or 1.

Since electronic components employed in computer interfaces and peripheral devices utilize voltage levels (usually 0 and +5 V) to represent the binary 0 and 1 states, one common design for the digital input device accepts process outputs

Table 21.3 Categories of Process Information

Type	Examples
1. Digital	Relay (open or closed) Switch (open or closed) Solenoid valve (open or closed) Motor drive (on or off) TTL circuit (0 or $+5$ V)
2. Generalized digital	Laboratory instrument output (e.g., binary coded decimal) Alphanumeric displays (e.g., binary coded decimal)
3. Pulse or pulse train	Turbine flowmeter (frequency of output pulse train proportional to flow rate) Stepping motor (pulse train opens or closes valve)
4. Analog	Thermocouple or strain gauge (millivolt level output) Operational amplifier (-10 to $+10$ V) Process instrumentation (4 to 20 mA)

of the same form. For example, a *contact-closure* device sends a 0 or 1 to the computer for each line depending on whether the corresponding process output relay is open or closed (the device detects a small current flow).

The *digital* (*or binary*) *output device* is a similar application of these ideas. This device is usually nothing more than a register or set of registers that hold or latch binary information (as it is transferred from the computer on command). The registers typically operate a set of relays or electronic switching circuits capable of driving process elements: solenoid valves, motor drives, display lights, and so forth.

It should be noted that process information of the second type, generalized digital information, can be processed into and out of the computer using the same type of devices. For example, one digit (0–9) of *binary coded decimal* (BCD) information can be represented by four bits of binary information. Hence, one 16-bit register in a digital input device could be used to transmit 4 digits of results (accuracy of 1 part in 10,000) from a particular instrument with BCD output that ranges from 0000 minimum to 9999 maximum.

The *input pulse interface* ordinarily consists of pulse counters in the form of a single register for each process line. Any counter can be reset to zero under program control and, after a specified length of time, the accumulated results (usually a binary or BCD count) can be transferred to the computer using techniques described above.

The *output pulse interface* consists of a device to generate a continuous train of pulses (of proper magnitude and period) followed by a gate. The gate can be turned on and off by the computer. Hence, by turning it on for a precomputed period of time, the desired number of pulses can be sent to the process.

The analog to digital converter is the device that is used for digitizing analog input information, usually in the form of voltage-level signals. For high-level signals, say in the range of 0 to 10 V, the conversion is carried out electronically and at great speed (on the order of 10 to 50 μs per conversion). Typically, many analog measurements are sent to the computer. To avoid the necessity of using a single ADC for each line, a *multiplexer* is used to switch selectively a number of analog signals (from 16 to 1024 or more) into a single ADC. All analog signals must be transduced to a single voltage range (normally on the order of 5 or 10 V) for conversion by a standard ADC. At these levels the switching device can be electronic (usually field effect transistor switches) rather than mechanical. Because many process analog signals to be input to the computer may be low-level (millivolt) signals from thermocouples or strain gauges, the use of low-level multiplexers is widespread. These devices must be mechanical in nature, usually consist of an array of mercury-wetted reed relays, and ordinarily feed a *programmable gain amplifier* to boost the signal level up to the range of the *sample and hold* and ADC. With the system shown in Fig. 21.6 the computer can convert any of a number of high- or low-level signals, amplify the signal if necessary, and convert to digital representation using only a single ADC.

A *digital to analog converter* (DAC) is much simpler and cheaper than an ADC. As a consequence, each analog output line from the computer to the process usually has its own dedicated DAC. Most DACs are designed to hold (that is, maintain) a previous analog output until commanded to perform another conversion. Conversion times for digital to analog converters are on the order of 5 to 20 μs, depending on the resolution of the converter which typically is 10 bits or 0.1% of full scale. Many industrial control computers generate analog outputs by sending a stream of pulses to a digital counter (register), whose digital output is converted to an analog signal via a type of DAC. By using two separate pulse lines to an up/down counter, both positive and negative changes can be effected. Such pulse-driven analog output devices have the advantage that a failure of the processor will not affect the analog output values. The stepper motor discussed previously has similar characteristics.

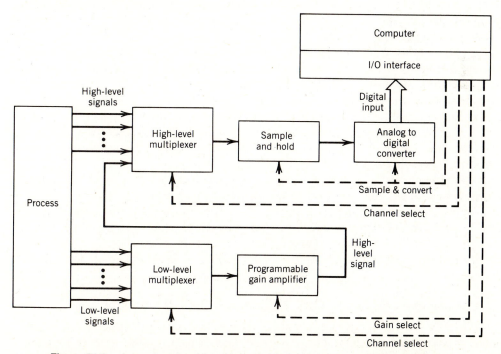

Figure 21.6. An analog signal interface suitable for both low- and high-level inputs.

Timing

The control computer must be able to keep track of time (real time or clock-on-the-wall time) in order that it be able to initiate data acquisition operations and calculate control outputs or to initiate a supervisory optimization on a desired schedule. It also may be programmed to log key process variables from time to time into a bulk storage unit or onto an output device, for example, a printer, to keep operations personnel informed. Hence all control computers will contain at least one hardware timing device. The so-called *real-time clock* represents one technique. This device actually is nothing more than a pulse generator that interrupts the computer on a periodic basis such as every 0.1 s and identifies itself as the interrupting device. System programs within the computer are required to update memory registers containing the time of day and to initiate operations that the user has programmed to execute at particular times or at particular time intervals. Sometimes a slightly more sophisticated hardware device is used that maintains the actual time and date, but it still must interrupt the computer periodically and the operating methods are substantially the same.

Operator Interface

Any general purpose supervisory or control computer system contains at least one terminal through which the system and the operator can communicate with each other [1, Chapter 9]. Usually the terminal permits graphical information to be displayed, such as variable trends or distributions (around set points) in the form of bar graphs. The use of *windowing* can be particularly helpful to the operator in displaying information of interest on a single terminal. Multiple displays are placed side by side or successively overlaying the screen as the operator interrogates the system for more details. Often these display consoles are color terminals for better visibility and recognition of key information. The operator ordinarily will communicate to the computer through the keyboard portion of a terminal—typing in requests for information or displays, changing controller parameters or set points, adding new control loops, and so on.

One or more printers invariably will be available to log process data in *hard copy*. Ordinarily these logs contain periodic updates of all key process variables and a running record of all important computer messages, such as alarms, and operator communications (e.g., set-point changes). One of the most important by-products of any digital computer-based control system is the vastly increased level of documentation of process operations. For some processes with strict environmental or health regulations such documentation may be the primary justification for a computer control system.

A general purpose computer system will include bulk storage units such as moving-head disks that can be used to store system and user programs. A disk-based bulk storage unit is most important in supporting key system programs, that is, the operating system and executive system, both of which are described below. Magnetic tape or disk may be used for long-term storage of programs and system records. There are numerous other I/O possibilities that are relatively less important.

Data Measurement, Signal Conditioning, and the Control Loop

In Fig. 21.7 we show a single direct digital control loop, an application where a computer has replaced an analog controller such as is given in Fig. 8.1. The loop shown in Fig. 21.7 is typical in the sense that all digital measurement/control loops require most of the same elements.

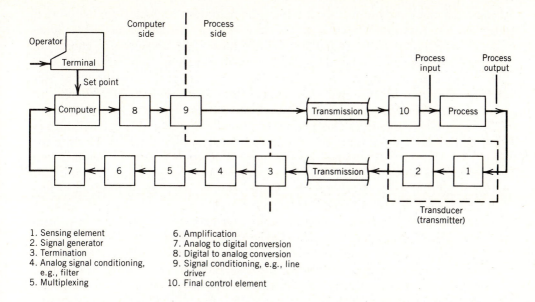

Figure 21.7. The components of a digital control loop.

1. Sensing element
2. Signal generator
3. Termination
4. Analog signal conditioning, e.g., filter
5. Multiplexing
6. Amplification
7. Analog to digital conversion
8. Digital to analog conversion
9. Signal conditioning, e.g., line driver
10. Final control element

We have discussed all of the elements in this loop except for some of the operations of the computer—in particular, digital signal conditioning (discussed in Chapter 22) and general digital control algorithms (covered in Chapter 26). After a brief consideration of computer software and programming in the next section, we outline the implementation of PID control algorithms that were introduced in Chapter 8.

21.4 DIGITAL CONTROL SOFTWARE

All computer programs must ultimately be reduced to the machine language[2] or, equivalently, the assembly language level [1,5]. Many of the earliest computer control projects required that all user programs be written in assembly language, an extraordinarily tedious procedure. Assembly language programs have the further disadvantages of being hard to document effectively, hard to debug and eliminate programming errors, and difficult to modify later (when inevitable process changes occur). Most user software is now written in higher-level languages such as BASIC, FORTRAN, PASCAL, or C, which have been extended to permit real-time operations. Other software is in the form of control-oriented programming languages supplied by vendors of process control computers. In this brief overview we concentrate on higher-level languages, discussing first the systems programs that are necessary for operation of user programs.

Operating Systems and Executive Programs

As discussed above, any real-time computer system must be able to respond to interrupts from external devices. These interrupts may come from the real-time

---

[2]Machine language consists of the elemental machine instructions written in binary code (zeros and ones). At this level instructions are stored in the computer memory unit, one instruction per memory word, and are executed sequentially via the computer's control unit.

clock, from input and output devices, from the operator terminal, from the process interface, or from other devices on the data highway or control network. Some system program has to be available to handle these interrupt servicing chores. In addition, with a general purpose computer there must be some supervisory software that oversees the loading and execution of user programs and that responds to user program requests, for example, to schedule other user programs at particular instants in time. The program that is responsible for handling these responsibilities is the *operating system* [1, Chapter 13]. It usually consists of a set of routines furnished by the vendor that can be utilized in programming and in executing the user's own programs. Even small general purpose computers now are supplied with extraordinarily versatile and complex operating systems. These can help the user develop and run applications software. Some vendors, particularly those supplying specialized process control computer systems, tend to supply a program that has only the necessary features for the loading and execution of the user's control programs. Such programs often are referred to as *executive programs*.

The main purpose of the operating system is to provide efficient use of the hardware resources and to coordinate the execution of multiple user programs. Through the interrupt handling capabilities, the operating system also must ensure that external events are dealt with in a real-time manner. Finally, the operating system provides many convenience features to the user, including a significant measure of security. This prevents one user program, for example, from overwriting the memory area used by another program in multiprogrammed or multiple-user systems.

Higher-Level Programming Languages

Most user-developed real-time programs now are written in the BASIC, FORTRAN, PASCAL, or C languages [1,5]. In many cases the user is able to utilize template routines supplied by the vendor, and is required only to duplicate these routines and to interconnect them to suit his or her own application purposes. Another approach involves the use of *table-controlled* programs, which are supplied by the vendor and require only that the user fill in the tables (a memory-resident data base organized as a table). The user supplies the necessary information specific to the application, for example, input and output channel numbers for each control loop, sampling period, and so on.

Real-time versions of both BASIC and FORTRAN are available from many computer vendors. BASIC usually is an *interpreted* language. The user program is stored in memory, virtually intact, and a program known as the interpreter scans it, one line at a time. Library subroutines are called to perform arithmetic and logical operations on program variables, to read in information from the process, and so on. Since many BASIC programs are interpreted rather than executed directly, they tend to run slower than other directly executable code such as a FORTRAN program unless a *compiled* version can be produced. However, BASIC programs are easy to write, and interpreted BASIC programs are easy to debug and to modify. Hence, the language lends itself to new environments, for example, in learning situations or with pilot plants. Unfortunately, there is no standardized form of BASIC for control applications.

FORTRAN is a *compiled* language, where the user program is reduced to machine language code by the FORTRAN compiler. This code is then loaded and executed directly, so it will always run faster than an equivalent interpreted program. However, FORTRAN errors are not easy to find, and to remove them means recompiling and reloading. Thus, FORTRAN is better suited for a stable, production environment where changes are seldom made and where run-time efficiency is important. A *standardized* real-time version of FORTRAN has been designed under the auspices of the ISA and accepted by the International Standards Organization (ISO). This version possesses a significant potential advantage: User programs generally are portable from one computer to another when written in standardized ISA/ISO FORTRAN [6].

A real-time version of any high-level language must be furnished with subroutines that can perform functions not supplied as part of data processing versions of the language. For example, routines to allow scheduling of user programs under supervision of the executive or operating system must be provided. The statement

Call TRNON(IPROG, IDAYTM, MODE)

might begin execution of (i.e., turn on) a user program designated by IPROG at a particular time of day specified by the variable IDAYTM. MODE is a variable that yields information concerning the success or failure of the call. A second case would be routines to allow access to process I/O equipment, for example

Call AIRD(NCHAN, NINCHN, NINVAL, MODE)

This call might access the ADC (analog input from a random set of channels) to obtain process information. NCHAN would specify the number of channels to convert. NINCHN would supply terminal connection data (channel number) for each desired variable. NINVAL would designate storage locations where the converted results would be put. MODE would again yield information concerning successful completion of the call. Other classes of subroutine calls necessary would be (1) for computer word manipulation at the bit level, (2) for building and using data files on the bulk storage device, and (3) for intercommunication between the executive system and user program elements or *tasks*. Most real-time executives permit the subdivision of the entire application program into program elements or tasks that execute independently of each other at specified times or in response to specific external events.

A simpler approach for the user is to utilize vendor-supplied firmware (ROM) or software to avoid writing programs. Because most process control applications reduce to a reasonably small and largely known set of operations, the process control computer vendor can develop control software that will handle 95 to 99% of the routine chores. For example, a standard PID control algorithm can be viewed as a simple sequence of events: read an analog input channel, compute controller output, write to an analog output channel (or send pulse train to a valve drive). Vendor-supplied routines contain instructions on how to specify the sampling period, the input and output channel numbers (or, equivalently, variable tag names), the set point, controller parameters, and so forth. After placing these data in established locations in a two-dimensional table in computer memory, as indicated in Fig. 21.8, the vendor routines can easily iterate through the table, thus handling any desired number of loops. Other operations, such as reporting alarms, can be dealt with in the same way [7, Section 2.2]. This approach is invariably used for dedicated local control units in a distributed instrumentation and control system.

The use of table-driven software or dedicated firmware-driven equipment can eliminate much of the drudgery of digital computer control programming [1, Chapter 16]. Most vendor-supplied software permits user-written programs to be attached to handle the 5 to 10% of a control project that might not be included in the vendor's complement of functions, for example, to implement a multiple-input/multiple-output control system for a complex process unit.

Control Loop 1	$\cdots$	*Control Loop n*
Sampling period	$\cdots$	Sampling period
Input channel	$\cdots$	Input channel
Output channel	$\cdots$	Output channel
Set point	$\cdots$	Set point
Controller settings · · ·	$\cdots$	Controller settings · · ·

Figure 21.8.   Process data table for table-driven software.

21.5 A TABLE-DRIVEN PID CONTROLLER

We now show, by means of a flow chart for the computer program, how the PID velocity algorithm derived in Chapter 8 can be implemented in a computer. Here it is assumed that some other program elements communicate with the operator (for example, in changing the controller mode or accepting new controller parameters).

Recall that in Chapter 8 we derived a velocity form for the PID control algorithm as Eq. 8-19 in which the incremental change in the controller output at the nth sampling instant, Δp_n, was obtained. By a slight rearrangement of (8-19) a form of the ideal PID controller velocity algorithm was obtained (8-20) that can be used to calculate the controller output p_n directly.

$$p_n = p_{n-1} + K_c\left[(e_n - e_{n-1}) + \frac{\Delta t}{\tau_I}e_n + \frac{\tau_D}{\Delta t}(e_n - 2e_{n-1} + e_{n-2})\right] \quad (21\text{-}2)$$

Note that in (21-2) the computer must at each sampling time store in memory (save) p_n, e_n, and e_{n-1}. At the next sampling time, these quantities represent p_{n-1}, e_{n-1}, and e_{n-2}, respectively.

Table 21.4 shows how the controller data and parameters might be stored in

Table 21.4   Controller Data and Parameters

Element Number	Parameter	Description
1	Δt	Sampling period for controller
2	N	No. of clock interrupts before next calculation is made
3	MA	0 = Manual mode 1 = Automatic mode
4	p_n	Controller output
5	T_R	Automatic set point
6	T_m	Most recent value of T_m
7	e_n	Most recent value of e
8	e_{n-1}	Next most recent value of e
9	K_c	Controller gain
10	τ_I	Integral time
11	τ_D	Derivative time

computer memory. Figure 21.9 illustrates how the three-mode algorithm is implemented. This program element is executed each time the control computer real-time clock causes an interrupt. If it is not yet time to implement control ($j > 0$), the routine exits. If in manual mode, the program would bypass calculation of p_n, leaving the stored value to be output again (a slight waste of time). However, it would continue to calculate and retain past values of e so that the transfer to automatic mode (MA $\rightarrow$ 1 by the operator) would be *bumpless* at any future time, if the process is operating at the set point.

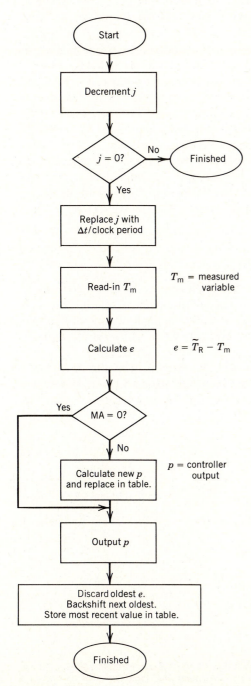

T_m = measured variable

$e = \tilde{T}_R - T_m$

p = controller output

Figure 21.9. A flowchart for the table-driven PID controller.

Finally, we see that extending the controller to multiple loops is quite easy. Simply augment the table and access the controller program once for each loop.

21.6 PROGRAMMABLE LOGIC CONTROLLERS AND BATCH PROCESS CONTROL

Most books on process control deal exclusively with control applications involving processes that operate continuously, as we have to this point. In implementing practical control systems, however, a large effort must be placed on operation of processes that are not continuous themselves, such as batch processes, or that contain equipment or control elements that operate discontinuously. A simple example of the latter category is the use of override control discussed in Section 18.3. In this case, alternative sensors (or final control elements) in a control loop are switched in/out by discrete logic based on the operating point of the process. Similarly, there are many instances where interlocks are required; for example, a flow control loop cannot be actuated unless a pump has been turned on. During start-up and shutdown of continuous processes many elements must be correctly sequenced; that is, upstream flows and levels must be established before downstream pumps can be turned on. Finally, batch processes represent a special area of rapidly growing importance that requires extensive sequencing of operations and activation/deactivation of control loops. Such processes are intentionally designed to operate much of the time in an unsteady-state manner, which may require special control techniques.

Programmable Logic Controllers

All of the features discussed above involve the application of logical decisions in implementing control. Such logic historically was built into control systems through the extensive use of hard-wired relay networks. In handling the flow control example above, an operator would turn on a pump via a manual switch, which simultaneously closes an attached relay and activates the associated flow controller.

By clever construction of parallel and series relay circuits, designers have been able to implement relatively sophisticated combinational logic and, with somewhat more difficulty, sequential logic as well. As an example, a set of three relays wired in series can be used to implement the AND condition shown in Fig. 21.10a. Only

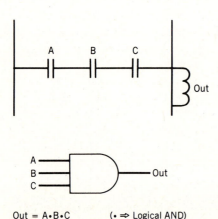

$$\text{Out} = A \cdot B \cdot C \qquad (\cdot \Rightarrow \text{Logical AND})$$

Figure 21.10a. Use of relays wired in series to implement AND logic.

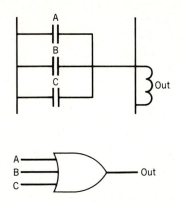

Figure 21.10b. Use of relays wired in parallel to implement OR logic.

Out = A + B + C   (+ ⇒ Logical OR)

when all three input relays are actuated will the output relay be actuated. Similarly, a set of relays wired in parallel could be used to implement the OR condition shown in Fig. 21.10b. Here actuation of any one or more of these input relays will cause the output relay to be actuated. Sequencing operations typically have required the use of latching relays (which hold a state indefinitely once actuated, much like a solid-state flip-flop) and delay relays (which delay a preprogrammed time interval before operating, once actuated).

The implementation of such logic requirements using a computer (usually a microprocessor) is a relatively recent development. Such special purpose computers are called *programmable logic controllers* (PLCs) because the computer is programmed to execute the desired Boolean logic and to implement the desired sequencing. In this case, the inputs to the computer are a set of relay contacts (outputs) representing the state of various process elements (e.g., two limit switches indicate whether a valve is fully open or fully closed). Various operator inputs (e.g., start/stop buttons) are also provided. The outputs from the computer are a set of relays energized (activated) by the computer that can turn pumps on or off, activate lights on a display panel, operate solenoid or motor-driven valves, and so on.

PLCs have substantially replaced hard-wired relay networks for a number of reasons:

- *Cost*. Both the design and construction of relay logic networks is expensive compared to the computer-based approach. Once input and output relays have been wired into the computer, the internal logic can be implemented quickly via a stored program.
- *Computational capability*. Virtually any function of the inputs can be calculated, for example, square root or products.
- *Flexibility*. Any minor change in process control logic requires rewiring of a relay network. With a PLC, a simple change in the stored program suffices.
- *Documentation*. Maintaining wiring and sequencing diagrams for hard-wired networks is troublesome. The PLC approach minimizes these difficulties, since a diagram of the internal logic can be obtained from the PLC at any time to indicate how it is designed to operate.

Control engineers often use a schematic way of representing PLC internal logic that is derived from the *ladder diagram* of hard-wired relay logic. As an example of the use of ladder diagrams, we consider a simple type of interlock system that

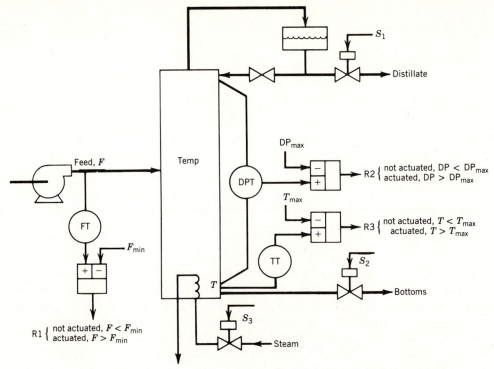

Figure 21.11. Distillation column instrumentation required for simple interlock protection.

might be used with a distillation column. Figure 21.11 shows a column that has been equipped to indicate a low feed condition (e.g., feed pump failure), high bottoms temperature, and high differential pressure. For the sake of simplicity, three streams—distillate, bottoms, and steam to the reboiler—have been provided with separate on/off (solenoid or motor-driven) valves[3] that can be used to shut off the product flows or heat to the column. The following interlocks are required for safe operation:

1. If the feed flow drops below a preset value, F_{min}, then both the distillate and bottoms streams should be shut off by operating valves S1 and S2.
2. If the feed flow is below F_{min},
     or
   If the column differential pressure DP exceeds the preset value DP_{max},
     or
   If the bottoms temperature exceeds the preset value T_{max},
   then the steam flow to the column should be shut off by operating valve S3.

Figure 21.12a shows the ladder diagram that represents the required interlocks. The first two *rungs* of the ladder represent interlock condition 1. Since R1 is actuated during *normal* operation, the diagram indicates that S1 should be actuated when R1 is *not* actuated. Figure 21.12b shows the traditional Boolean representation of the logical complement operation implied in the R1 relay symbol. The third rung in the diagram illustrates the OR-logic of interlock condition 2. Here, if R1 is not actuated or if R2 (or R3) is actuated, S3 will be actuated. Figure 21.12c shows the

---

[3]By suitable use of overrides, existing control valves could be used for this purpose.

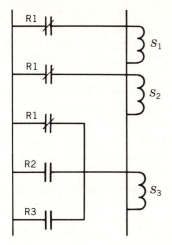

Figure 21.12a. A ladder diagram for distillation column interlocks.

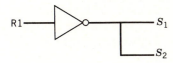

Figure 21.12b. The logic for interlock condition 1.

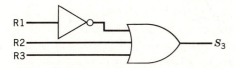

Figure 21.12c. The logic for interlock condition 2.

Boolean symbol for this OR-logic condition. Note that this simplified example does not deal explicitly with fail-safe operational modes. Nor does it illustrate the use of sequential logic that can be difficult to represent in traditional ladder diagrams. It may be necessary to actuate the steam valve immediately because of thermal lags in the reboiler whereas the distillate and bottoms flow valves could be actuated after some predetermined delay time. This approach may be useful because a certain amount of surge capacity is available in the distillate receiver and reboiler sump. Several examples of the application of PLCs and the use of ladder diagrams for sequential control are given in references [8] and [9].

A summary of the general characteristics of PLCs is given below in closing this discussion. Details are available elsewhere [9,10].

- *Inputs/outputs.* Up to several thousand discrete (binary) inputs and outputs can be accommodated. More sophisticated PLCs also can handle up to several hundred analog inputs and outputs for data logging and/or continuous (PID) control.
- *Logic handling capability.* All PLCs are designed to handle combinational binary logic operations efficiently. Since the logical functions that must be processed are stored in main memory, one measure of the PLC's capability is its memory scan rate. Another measure is the average time required to scan each step in a logic or ladder diagram; typical values range from 50 ms down to 10 μs. At the

faster speeds thousands of steps can be dealt with by a single unit. Most PLCs also handle sequential logic and are equipped with internal timing capability to delay an action by a prescribed amount of time, to execute an action for a prescribed time, and so on. Sequencing capability ordinarily can be mixed with combinational logic.

- *Continuous control capability.*   PLCs with analog I/O features usually include PID algorithms that can be used to control from several up to several hundred analog loops. More sophisticated PLCs incorporate virtually all commonly used control functions including PID, PID with on/off outputs, integral action only, ratio and cascade control, low- or high-signal select, lead–lag elements, and so forth. Such units are quite efficient because internal logic signals are available to switch controller functions. Examples include significant machine and process sequencing, override control, or where controllers must be turned on and off frequently (as in batch process control).

- *Operator communication.*   Smaller PLCs provide virtually no operator interface other than simple signal lamps to indicate the state of discrete inputs and outputs. Such systems often are networked as one element in a distributed instrumentation system with operator I/O provided by a separate unit in the network. Large PLC units often are provided with quite sophisticated CRT-based operator displays and one or more keyboards for operator interrogation. These units are intended to be self-sufficient, and they may provide most of the control functions typically provided by a general-purpose computer control system.

- *Programming the PLC.*   Control engineers make a distinction between *configurable* and *programmable*. The former term implies that logical operations (performed on inputs to yield desired output functions) can be put into the PLC memory, perhaps in the form of ladder diagrams by selecting from a menu supplied by the PLC itself, via direct interrogation by the PLC, and so on. The latter term implies that the logical operations are put into PLC memory in the form of a higher-level language such as BASIC. Control engineers prefer the simplicity of configuring the PLC to the alternative of programming it. However, some applications, particularly those involving complex sequencing, are best handled by a programming approach.

Batch Process Control

Batch (and semibatch) processes, that is, those with discontinuous feed and product stream flows, require an unusual amount of logic and sequencing in their control. Thus, they are well suited to the characteristics of PLCs. Batch processes—including reactors, distillation units, slurry mills, crystallizers and other separation units—often are used in preference to continuous flow units, especially when relatively small amounts of specialty products are required. In emulsion polymerization or crystallization, batch units provide better control of size distribution. For biochemical reactions, batch or semibatch processes are preferred because the living organisms involved—bacteria, mammalian cells, and so on—are quite difficult to keep uncontaminated for long periods of time in continuous flow systems.

Figure 21.13 illustrates a batch reactor schematically. Figure 21.14 shows a characteristic *batch cycle* consisting of (1) the charging (sequentially) of each of three reactants, (2) a heat-up, operation, and cool-down sequence (3) discharge of the final product mix for separation and subsequent processing. Additional details are given in [8,11].

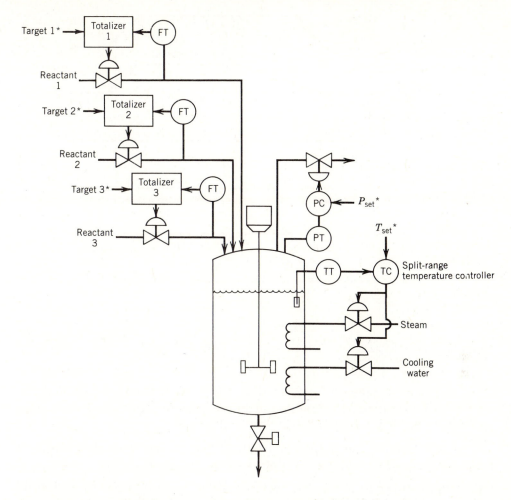

Figure 21.13 A schematic diagram of a batch reactor (* denotes an externally set parameter that is a function of time).

The important characteristics of batch processing procedures can be summarized as follows:

- Intermittent operations are primarily involved rather than continuous operations. Typically, a cycle will consist of start-up, operate, shutdown, cleanup, and changeover phases.
- As a consequence of the above, much sequencing of batch steps is required. Furthermore, widespread use of two-state actuators (pump on/off, valve open/closed) and two-state measurement elements (reactor filled or not) requires incorporating a considerable amount of logic, particularly in equipment interlocks.
- Conventional control theory, for continuous process, may not be applicable without some of the modifications discussed below.
- A large opportunity exists for the application of optimization techniques. Here the emphasis is not on choosing the best set of operating conditions (set points) periodically, as in supervisory control of continuous processes (Chapter 20), but in optimizing important characteristics of the batch cycle. For example, the cycle

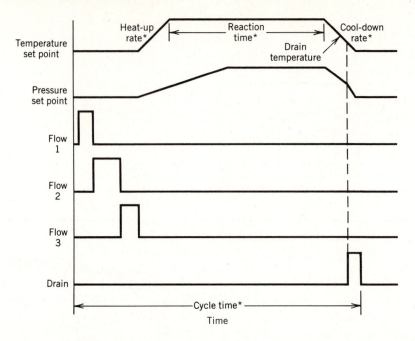

Figure 21.14. Set points and flow sequences of a one-batch reactor cycle (* denotes externally set function of time).

time itself can be minimized so as to maximize production rate through the unit, or reaction conditions (temperature and pressure) may be determined as functions of time to maximize the profitability of the products.

In implementing batch process control, several important differences in approach compared to continuous control should be noted:

- In charging materials to the unit, *totalizers* are often used to determine the end point of a charge (Fig. 21.13). Hence, the ability to control flow rate accurately is not as important as the ability to measure and integrate flow rate accurately. If a weigh mechanism is used, for example, a load cell on the batch unit, even measurement of flow rate is not particularly important.
- The start-up of a batch process often is carried out under operator control and with all process controllers turned off (placed in the manual mode).
- When feedback controllers are used during start-up, proportional-only control is often used to follow set points that are preprogrammed as functions of time. The use of proportional-only control, particularly during the start-up and shutdown periods, usually is desirable because (1) some small offset can be tolerated during the set-point tracking periods of operation; (2) reset windup might be a problem during these periods if PI control is used (however, the "batch switch" option can be employed to suspend the integral mode until the set point is approached); (3) gain settings for proportional-only controllers can be readily preprogrammed so the process response will be fast enough without danger of instability.
- On reaching the production part of the cycle (defined by generally constant operating conditions), controllers typically are switched to the PI operating mode. The preferred method is to calculate the initial integral mode contribution so as to avoid overshoot of the new (constant) set points [8].
- During the production period, prediction of end-point composition for batch reactors can be implemented. Often some type of indirect (inferential) method

is used, such as measuring off-gas evolution or agitator drive current demand (torque requirements). Similarly, for separation processes, cut determinations may have to be made indirectly, for example, based on temperature instead of composition.

- During shutdown of a process, the objective is to reach a state as quickly as possible where the unit can be discharged, cleaned, and readied for another cycle or another product (changeover). Hence, temperature set points are decreased as rapidly as constraints on equipment and product permit. Again, proportional-only control typically is used, or actuators are manipulated manually.

From the above descriptive information, it should be clear that PLCs can play an important role in controlling batch processes. Both logical and continuous variables can be handled within the same unit. Both binary combinational logic and sequencing operations can be carried out easily. Furthermore, more sophisticated PLCs easily implement all of the variations of P and PI control. Consequently, it is relatively easy to program a single unit to deal with all of the special requirements of batch processing with perhaps one or two exceptions. A single-unit PLC seldom is used to optimize batch cycle operations or to implement inferential control of reactor end point or separator cut points. These functions are better handled by a general purpose computer with its greater computational and algorithmic capabilities or by an expanded distributed control system. Hence, where these more sophisticated control requirements are found, the PLC should be networked to a separate optimizing computer such as in the digital instrumentation system depicted in Fig. 21.4.

SUMMARY

In this chapter we have briefly introduced the subject of digital computer control, emphasizing the key roles that hardware and software play in data acquisition and control systems. The digital instrumentation approach can replace analog control components and retains the most important advantages of classical analog control—the use of "bottom-up" design to match process requirements closely, acceptability by both operators and engineers, expandability, and reliability. The specific advantages of digital implementation are its better precision, ease of maintenance, and greater versatility. The addition of one or more general purpose computers in such a system permits the implementation of advanced control algorithms and supervisory optimizing programs, yielding general information and control systems suitable for plant-level and even corporate-wide management. Other capabilities are provided by programmable logic controllers in sequencing and batch process control.

Digitally-based control instrumentation represents a revolutionary change in the way process control systems are implemented. With digital systems the control engineer has the opportunity to go beyond the narrow limitations of standard analog control components to construct a system that is optimum for the information processing and control requirements of large processes or even of entire plants.

REFERENCES

1. Mellichamp, D. A. (Ed.), *Real-Time Computing with Applications to Data Acquisition and Control,* Van Nostrand-Reinhold, New York, 1983.
2. Haggin, J., Process Control on Way to Becoming Process Management, *Chem. Eng. News* **62,** 7 (April 1984).
3. Mellichamp, D., D. Bedworth, O. Pettersen, P. Rony, L. Bezanson, W. Higgins, and G. Korn, Real-Time Computing and the Engineering Support System, *IEEE Micro* **5,** 27 (October 1985).

4. Moore, R. L., and M. A. Kramer, Expert Systems in On-Line Process Control, in *Chemical Process Control* CPC-III (Morari, M., and McAvoy, T. J., Eds.), Elsevier, New York, 1986.

5. Harrison, T. J. (Ed.), *Minicomputers in Industrial Control,* ISA Press, Research Triangle Park, NC, 1978.

6. ANSI/ISA: S61.1 (1976) Standard. Industrial Computer System FORTRAN Procedures for Executive Functions, Process Input-Output, and Bit Manipulation, Instrum. Soc. of America, 1976. S61.2 (1978) Standard. Industrial Computer System FORTRAN Procedures for File Access and the Control of File Contention, Instrum. Soc. of America, 1978.

7. Deshpande, P. B., and R. H. Ash, *Computer Process Control,* ISA Press, Research Triangle Park, NC, 1981.

8. Cohen, E. M., and W. Fehervari, Sequential Control, *Chem. Eng.* **92,** 61 (April 1985).

9. Blickley, G. J., New Developments in Programmable Controllers and Peripherals, *Control Eng.* **34,** 76 (January 1987).

10. Flynn, W. R., Programmable Controller Update, *Control Eng.* **34,** 70 (January 1987).

11. Rosenof, H. P., and A. Ghosh, *Batch Process Automation: Theory and Practice,* Van Nostrand Reinhold, New York, 1987.

EXERCISES

21.1. Briefly answer the following questions:

(a) What minimum set of components should be included in a distributed instrumentation system if a relatively large number of control loops in a plant are to be handled with operator functions performed in a centralized control room?

(b) What are the advantages/disadvantages of constructing such a system out of components as compared to a single-computer system?

(c) If batch process control and/or equipment sequencing control will be incorporated in the digital computer systems, how would you modify your answers to parts (a) and (b)?

21.2. A computer uses an internal 16-bit data representation. If analog to digital converters yielding 10-, 12-, 14-, and 16-bit results are available, what degree of resolution (%) could be obtained with each? What is the smallest change in a 0 to 10 V analog signal that would be detectable in the digital representation for each case?

21.3. A process output (measured) variable is available from a transmitter with a 4–20 mA output. This signal is converted to a 1–5 V signal that is input to a microcomputer by an analog to digital converter. If the signal-to-noise ratio is 2000 over the entire range (1–5 V), what resolution should the ADC have? Explain your reasoning.

21.4. A microcomputer-based digital controller is used to implement 10 individual feedback control loops. The controller output in each case is obtained from a DAC that has a 4–20 mA output. Once the microcomputer has computed all the output values (e.g., via Eq. 21-2), the results are sent sequentially to the ten DACs. If it takes 10 μs for the microcomputer to execute the routine that passes data from computer to one DAC and if the settling time for each DAC is 50 μs, how much "skewing" will occur between the first and last output? Would this difference be significant if the sampling period for each loop is one minute? One second? *Assume that the microcomputer does not have to wait for one DAC to settle before sending an output value to the next one.*

21.5. A stepping motor used to drive a control valve rotates the valve stem 3° for each pulse that arrives at the drive electronics. If two complete rotations of the valve are required to go from fully closed to fully open, what is the degree of resolution (%) for the valve? If Eq. 8-18 is used to compute the change in valve position required at a particular sampling time and the control loop output change is given in terms of % open (closed), write an expression that gives the appropriate number of pulses required to drive this valve. Draw a flow chart that illustrates how the valve driver subroutine operates. *Assume that a positive Δp indicates that the valve should be opened and vice versa.*

21.6. A 0 to 10 V ADC converter is used in conjunction with a programmable gain amplifier that yields a selectable gain of 1, 2, 4, . . . , 1024. What gain should be selected for process output variables that range from 0 to

(a) 1.2 mV   (d) 800 mV
(b) 8 mV   (e) 1.2 V
(c) 80 mV   (f) 8 V

to obtain maximum precision (resolution) from the ADC? What would be the reading obtained by the ADC in each case if it has a 10-bit resolution?

21.7. The real-time clock in a control computer operates at a frequency of 100 Hz, that is, it interrupts the computer operations 100 times per second. Show, by means of a flow chart, how an internal routine can be programmed to furnish integer variables representing the actual time in hours, minutes, seconds, and tenths of seconds. Assume that a 24-h day is utilized.

21.8. What are the major differences between a compiled language such as FORTRAN and an interpreted language such as BASIC in terms of their abilities to operate in a control application environment?

21.9. A microcomputer program required to initiate all operations and to make all calculations for a single control loop requires 200 programming steps when running an interpreted language. Each step requires 1 ms. When running an equivalent compiled program 4000 programming steps are required; however, each of those requires only 1 μs. If 8 control loops are to be executed, what is the minimum sampling period that can be implemented in each case, assuming that the microcomputer has no other responsibilities?

21.10. Show, by altering Fig. 21.9, how the PID velocity algorithm can be modified to include:

(a) High and low limit checking on both input and output variables. How should the process data table (Tabel 21.4) be modified?
(b) Repeat part (a) for antireset windup on the controller output.

CHAPTER 22

Sampling and Filtering of Continuous Measurements

The specifications for a computer-based system to perform data acquisition and control must address several questions:

1. How often should data be acquired from a given measurement point; that is, what sampling rate should be employed?
2. Do the measurements contain a significant amount of *noise?* If so, can the data be conditioned (*filtered*) to reduce the effects of noise?
3. What digital control law should be employed?

In this chapter we will primarily be concerned with questions 1 and 2. Question 3 will be addressed in Chapter 26. The modeling and analysis of digital systems are covered in Chapters 23–25.

22.1 SAMPLING AND SIGNAL RECONSTRUCTION

As indicated in Chapter 21, when a digital computer is used for control, continuous measurements are converted into digital form by an analog to digital converter (ADC). This operation is necessary because the digital computer cannot directly process an analog signal; first the signal must be sampled at discrete points in time and then the samples must be digitized. The time interval between successive samples is referred to as the sampling period Δt. Two related terms are also used: the sampling rate $f_s = 1/\Delta t$ and the sampling frequency $\omega_s = 2\pi/\Delta t$. If Δt has units of minutes, then f_s has units of cycles per minute and ω_s has units of radians per minute.

Figure 22.1 shows an idealized periodic sampling operation in which the sampled signal $y^*(t)$ is a series of impulses that represents the measurements y_0, y_1, y_2, . . . at the sampling instants, t_0, t_1, t_2, ... The representation in Fig. 22.1 is also referred to as *impulse modulation* [1] and is used in the analysis of sampled-data systems. It is based on the assumption that the sampling operation occurs instantaneously.

In digital control applications the controller output signal must be converted from digital to analog form before being sent to the final control element. This operation is referred to as *signal reconstruction* and is performed by digital to analog converters (DACs), which were described in Chapter 21. The DAC operates as a zero-order hold as is shown schematically in Fig. 22.2. Note that the output signal

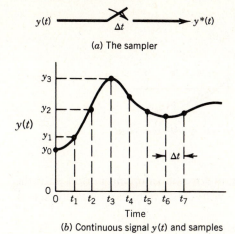

(a) The sampler

(b) Continuous signal $y(t)$ and samples

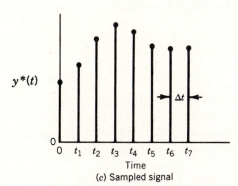

(c) Sampled signal

Figure 22.1.   Idealized, periodic sampling.

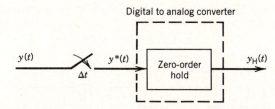

(a) Zero-order hold as a digital-to-analog converter

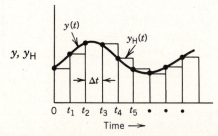

(b) Comparison of original signal, $y(t)$, and reconstructed signal, $y_H(t)$

Figure 22.2.   Digital-to-analog conversion using a zero-order hold.

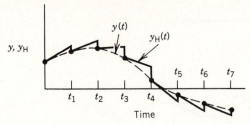

Figure 22.3.   Signal reconstruction with a first-order hold.

from the zero-order hold $y_H(t)$ is held constant for one sampling period until the next sample is received. The operation of the zero-order hold can be expressed as

$$y_H(t) = y_{n-1} \qquad \text{for } t_{n-1} \leq t < t_n \tag{22-1}$$

Other types of hold devices can be employed; for example, a *first-order hold* extrapolates the digital signal linearly during the time interval from t_{n-1} to t_n based on the change during the previous interval:

$$y_H(t) = y_{n-1} + \left(\frac{t - t_{n-1}}{\Delta t}\right)(y_{n-1} - y_{n-2}) \qquad \text{for } t_{n-1} \leq t < t_n \tag{22-2}$$

Figure 22.3 illustrates the operation of a first-order hold. Although second-order and other higher-order holds can be designed and implemented as special purpose DACs [1–3], these more complicated approaches do not offer significant advantages for most process control problems. Consequently, we will emphasize the zero-order hold since it is the most widely used hold device for process control.

Figure 22.4 shows the block diagram for a typical feedback control loop with a digital controller. Note that both continuous (analog) and sampled (digital) signals appear in the block diagram. The two samplers typically operate synchronously and have the same sampling period. However, *multirate sampling* in which one

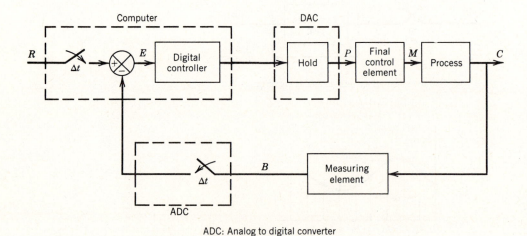

ADC: Analog to digital converter
DAC: Digital to analog converter

Figure 22.4.   Simplified block diagram for computer control.

sampler operates at a faster rate than the other is sometimes used [1–3]. For example, we may wish to sample a process variable and filter the measurements quite frequently while performing the control calculations less often in order to avoid excessive wear in the actuator or control valve.

The block diagram in Fig. 22.4 is symbolic in that the mathematical relations between the various signals (e.g., transfer functions) are not shown. Transfer functions for sampled-data systems and the analysis of block diagrams containing samplers will be considered in Chapters 24 and 25.

22.2 SELECTION OF THE SAMPLING PERIOD

In selecting a sampling period, two questions must be considered:

1. How many measurement points does the computer monitor?
2. What is the best sampling period from a process control point of view?

If a dedicated digital control system is connected to a single measurement point, then that measurement can be sampled as often as desired, or as rapidly as the computer can perform the sampling. However, rapid sampling of a large number of measurement points may unnecessarily load the computer and restrict its ability to perform other tasks. Before introducing a number of guidelines for choosing a sampling period, we consider an important practical problem that is referred to as *aliasing*.

Aliasing

The sampling rate must be large enough so that significant process information is not lost. The loss of information that can occur during sampling is illustrated in Fig. 22.5. Suppose that a sinusoidal signal is sampled at a rate of 4/3 samples per cycle (i.e., 4/3 samples per period). This sampling rate causes the reconstructed signal to appear as a sinusoid with a much longer period than the original signal, as shown in Fig. 22.5a. This phenomenon is known as *aliasing*. Note that if the original sinusoidal signal were sampled only twice per period, then a constant

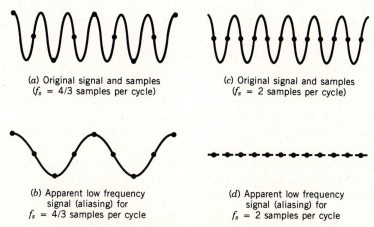

(a) Original signal and samples
(f_s = 4/3 samples per cycle)

(c) Original signal and samples
(f_s = 2 samples per cycle)

(b) Apparent low frequency
signal (aliasing) for
f_s = 4/3 samples per cycle

(d) Apparent low frequency
signal (aliasing) for
f_s = 2 samples per cycle

Figure 22.5. Aliasing error due to sampling too slowly.

sampled signal would result, as shown in Fig. 22.5*d*. According to Shannon's sampling theorem [1–3], a sinusoidal signal must be sampled *more* than twice each period to recover the original signal; that is, the sampling frequency must be more than twice the frequency of the sine wave.

Aliasing also occurs when a process variable that is *not* varying sinusoidally is sampled. In general, if a process measurement is sampled with a sampling frequency ω_s, high-frequency components of the process variable with a frequency greater than $\omega_s/2$ appear as low-frequency components ($\omega < \omega_s/2$) in the sampled signal. Such low-frequency components can cause control problems if they appear in the same frequency range as the normal process variations (e.g., frequencies close to the critical frequency ω_c, as discussed in Chapter 16). Aliasing can be eliminated by using an *anti-aliasing* filter, as discussed in Section 22.3.

Large Values Versus Small Values of Δ*t*

Sampling too slowly can reduce the effectiveness of the feedback control system, especially its ability to cope with disturbances. In an extreme case, if the sampling period is longer than the process response time, then a disturbance can affect the process and the influence of the disturbance will disappear before the controller takes corrective action. In this situation the control system cannot handle transient disturbances and is capable only of steady-state control. Thus, it is important to consider the process dynamics (including disturbance characteristics) in selecting the sampling period. For composition control, the time required to complete the composition analysis (e.g., using a gas chromatograph) sets a lower limit on the sampling period.

On the other hand, there is an economic penalty associated with sampling too frequently, namely that the number of measurement points the computer can handle decreases as the sampling period Δ*t* decreases. Since the optimum sampling period is application-specific, it is difficult to make any generalizations on this subject. However, a reasonable approach is to select a sampling period that is small enough to ensure that significant dynamic information is not lost, and then to examine whether the computer can handle the data acquisition requirements. If it cannot, then additional computing power should be considered. Commercial digital controllers which handle a small number of control loops (e.g., 8–16) typically employ a fixed sampling period of a fraction of a second. Thus, the performance of these digital controllers closely approximates continuous (analog) control.

If process conditions change significantly, then it may be necessary to change the sampling period. For example, if a feed flow rate to a processing unit is significantly increased, the residence time and hence the time constant for the unit are reduced. Consequently, it may be necessary to use a smaller sampling period in order to achieve satisfactory control. A simpler, more conservative approach would be to select the sampling period that corresponds to the worst possible conditions, that is, the smallest sampling period.

The signal-to-noise ratio (S/N) also influences the sampling period selection.[1] For low signal-to-noise ratios, rapid sampling should be avoided because changes in the measured variable from one sampling time to the next will be mainly due to high frequency noise rather than to the slower process changes. For low S/N values, a filter or filters should be used to condition the measurements in order to

---

[1]The signal-to-noise ratio (S/N) is usually defined as $S/N = \sigma_S^2/\sigma_N^2$, where σ_S^2 denotes the variance of the signal (or output) and σ_N^2 is the variance of a random input disturbance (i. e., "noise").

prevent the controller from acting on an *apparent* process excursion, as will be discussed in Section 22.3.

Guidelines for Selecting the Sampling Period

The selection of the sampling period remains more of an art than a science. A number of guidelines and rules of thumb have been reported for both PID controllers and model-based controllers such as the direct synthesis approach of Chapter 12 [1–4]. Representative results for PID controllers are summarized in Table 22.1.

In the early days of digital control, the suggestion was made that Δt should be selected according to the process variable being controlled (see Category 1 in

Table 22.1   Guidelines for the Selection of Sampling Periods for PID Controllers

Approach and Recommendation	*Comments*	*Reference*
1. Type of Physical Variable		
(a) Flow: $\Delta t = 1$ s		
(b) Level and pressure: $\Delta t = 5$ s	Ignore process dynamics	Williams [5]
(c) Temperature: $\Delta t = 20$ s		
2. Open-Loop System		
(a) $\Delta t < 0.1\tau_{max}$	$\tau_{max} \triangleq$ dominant time constant	Kalman and Bertram [6]
(b) $0.2 < \dfrac{\Delta t}{\theta} < 1.0$	For process model, $G(s) = Ke^{-\theta s}/(\tau s + 1)$	
(c) $0.01 < \dfrac{\Delta t}{\tau} < 0.05$	Based on (3b) and Ziegler-Nichols tuning (cf. Table 13.1)	Åström and Wittenmark [2, p. 187]
(d) $\dfrac{t_s}{15} < \Delta t < \dfrac{t_s}{6}$	t_s = settling time (95% complete)	Isermann [4]
(e) $0.25 < \dfrac{\Delta t}{t_r} < 0.5$	t_r = rise time for open-loop system[a]	Åström and Wittenmark [2, p. 61]
(f) $0.15\,(\Delta t)\omega_c < 0.50$	ω_c = critical frequency for continuous system (rad/s)	Åström and Wittenmark [2, p. 178]
(g) $0.050 < (\Delta t)\omega_c < 0.107$		Shinskey [7]
3. Miscellaneous		
(a) $\Delta t > \dfrac{\tau_I}{100}$	τ_I = integral time	Fertik [8]
(b) $0.1 < \dfrac{\Delta t}{\tau_D} < 0.5$	τ_D = derivative time	Åström and Wittenmark [2, p. 187]
(c) $0.05 < \dfrac{\Delta t}{\tau_D} < 0.1$		Shinskey [7]

[a] Åström and Wittenmark [2] use a rise time that differs from the definition in Section 5.4. They calculate the rise time by drawing a tangent through the inflection point of the step response; thus, it is equal to the dominant time constant. This procedure is illustrated in Fig. 7.4.

Table 22.1). According to this guideline, process variables that respond rapidly such as flow rates should be sampled more frequently than slower variables such as liquid level and temperature. However, the rules of thumb for Category 1 should be used with considerable caution since they ignore the dynamic characteristics of the individual elements in the feedback control loop. For example, if a distillation column has an open-loop response time of 4 h, a sampling period of 20 s for a temperature control loop would be much too small.

Guidelines 2a, 2b, and 2c in Table 22.1 are based on a simple transfer function model $G(s)$ which represents all of the components of the feedback control loop except the controller. The dominant time constant τ_{max} can be determined as follows. It is set equal to the largest time constant if one time constant is much larger than the others. Alternatively, τ_{max} can be set equal to the time constant of a first-order plus time-delay model of the process.

Guidelines 2d and 2e are based on the rise time t_r and settling time t_s of the open-loop step response of the process. Guidelines 2f and 2g relate the sampling period to the critical frequency ω_c which was defined in Section 16.2. These guidelines indicate that the recommended sampling periods are proportional to the rise time and the settling time but inversely proportional to the critical frequency.

Fertik [8] has proposed guidelines for the selection of Δt for noisy processes. He noted that the inequality in Guideline 3a should be satisfied to avoid a deadband for the integral control action, resulting from the accuracy limits of fixed-point computer calculations. Guidelines 3b and 3c relate the sampling period to the derivative time τ_D.

Next we consider a numerical example that illustrates the use of the guidelines in Table 22.1.

EXAMPLE 22.1

Consider a process with the transfer function,

$$G(s) = \frac{(2s + 1)e^{-4s}}{(10s + 1)(7s + 1)(3s + 1)}$$

where $G(s) = G_v(s)G_p(s)G_m(s)$ and the time constants and the time delay have units of minutes. Calculate recommended sampling periods for a PID controller based on Categories 2 and 3 in Table 22.1. For Category 3 assume that $\tau_I = 15$ min and $\tau_D = 4$ min.

Solution

To use the guidelines in Table 22.1, we need to calculate τ, θ, t_r, τ_{max}, t_s, and ω_c. The first five of these parameters can be calculated from the open-loop response to a unit step change as shown in Fig. 22.6. The critical frequency ω_c can be determined from the following phase angle expression:

$$-180° = -\tan^{-1}(10\omega_c) - \tan^{-1}(7\omega_c) - \tan^{-1}(3\omega_c)$$
$$+ \tan^{-1}(2\omega_c) - 4\omega_c (180/\pi)$$

A trial-and error solution gives $\omega_c \cong 0.21$ rad/min. The recommended sampling periods are shown in Table 22.2.

The results in Table 22.2 indicate that the recommended values of Δt vary over nearly two orders of magnitude, from 0.16 to 12 min. Guidelines 2b, 2d, and 2e produce the largest Δt values while Guidelines 2c, 2g, and 3c result in the smallest values.

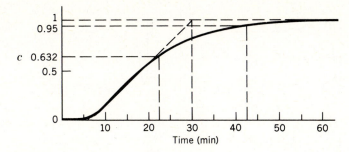

Thus:

$\theta = 6$ min $\qquad\qquad t_r = 30 - 6 = 24$ min

$t_s = 43$ min (95% settling time) $\tau = \tau_{max} = 22 - 6 = 16$ min

Figure 22.6. An open-loop step response and a graphical determination of rise time t_r and settling time t_s.

The effect of sampling period on control system performance for this example has been evaluated by Isermann [4]. His simulation results for set-point changes are shown in Fig. 22.7. He selected four sampling periods and used a digital PID controller with the controller settings chosen to minimize an Integral Squared Error performance index, as described in Chapter 12. Figure 22.7 indicates that the digital controller provides a good approximation to a continuous PID controller when $\Delta t = 1$ min because the manipulated variable changes in almost a continuous fashion. The response times and periods of oscillation increase as Δt increases until finally the closed-loop response is quite poor for $\Delta t = 16$ min. Based on these and other considerations, Isermann [4] concludes that an appropriate Δt for this example is in the range of 4 to 8 min.

Table 22.2   Recommended Sampling Periods for Example 22.1

	Guideline	*Numerical Value of Δt (min)*
(2a)	$\Delta t < 0.1\tau_{max}$	$\Delta t < 1.6$
(2b)	$0.2 < \dfrac{\Delta t}{\theta} < 1.0$	$1.2 < \Delta t < 6.0$
(2c)	$0.01 < \dfrac{\Delta t}{\tau} < 0.05$	$0.16 < \Delta t < 0.8$
(2d)	$\dfrac{t_s}{15} < \Delta t < \dfrac{t_s}{6}$	$2.9 \le \Delta t \le 7.2$
(2e)	$0.25 < \dfrac{\Delta t}{t_r} < 0.5$	$6 \le \Delta t \le 12$
(2f)	$0.15 < (\Delta t)\omega_c < 0.5$	$0.71 \le \Delta t \le 2.4$
(2g)	$0.05 < (\Delta t)\omega_c < 0.107$	$0.24 \le \Delta t \le 0.51$
(3a)	$\Delta t > \dfrac{\tau_I}{100}$	$\Delta t > 0.15$ min
(3b)	$0.1 < \dfrac{\Delta t}{\tau_D} < 0.5$	$0.4 < \Delta t < 2.0$
(3c)	$0.05 < \dfrac{\Delta t}{\tau_D} < 0.1$	$0.2 < \Delta t < 0.4$

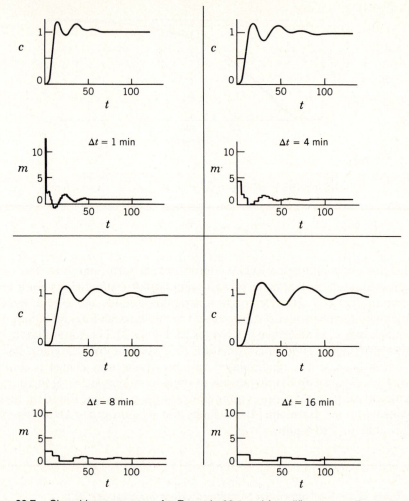

Figure 22.7. Closed-loop responses for Example 22.1 and four different sampling periods [4].

This example illustrates that published guidelines for selection of the sampling period can result in a very wide range of recommended values. Thus, although the guidelines provide useful information, a specific one should not be used blindly without some comparisons (see also Example 26.2).

22.3 SIGNAL PROCESSING AND DATA FILTERING

In process control, the noise associated with analog signals can arise from a number of sources: the measurement device, electrical equipment, or the process itself. The effects of electrically generated noise can be minimized by following established procedures concerning shielding of cables, grounding, and so forth [9]. Process-induced noise can arise from variations due to mixing, turbulence, and nonuniform multiphase flows. The effects of both process noise and measurement noise can be reduced by signal conditioning or filtering. In electrical engineering parlance, the term "filter" is synonymous with "transfer function," since a filter transforms input signals to yield output signals.

Analog Filters

Analog filters have been used for many years to smooth noisy experimental data. For example, an *exponential filter* can be used to damp out high-frequency fluctuations due to electrical noise; hence it is called a low-pass filter. Its operation can be described by a first-order transfer function or equivalently a first-order differential equation,

$$\tau_F \frac{dy(t)}{dt} + y(t) = x(t) \qquad (22\text{-}3)$$

where x is the measured value (the filter input), y is the filtered value (the filter output), and τ_F is the time constant of the filter. Note that the filter has a steady-state gain of one. The exponential filter is also called an *RC filter* since it can be constructed from a simple RC electrical circuit.

As shown in Fig. 22.5, relatively slow sampling of a high-frequency signal can produce an artificial low-frequency signal. Therefore it is desirable to use an analog filter to *prefilter* process data before sampling in order to remove high-frequency noise as much as possible. For these applications, the analog filter is often referred to as an *anti-aliasing filter*. This allows the sampling period to be selected independent of signal conditioning considerations. For applications where τ_F is less than three seconds, passive analog filters constructed from resistance–capacitance components are suitable. For slowly varying dynamic signals such as *drifts* where τ_F must be greater than three seconds, active analog filters are constructed using amplifiers. However, amplifier-based filters are more expensive than equivalent digital filters implemented via computer software. Consequently, for very slowly varying signals, it may be better to perform digital filtering since the required sampling period will not be very small and the extra computational burden on the computer will be quite modest.

The filter time constant τ_F should be much smaller than the dominant time constant of the process τ_{max} to avoid introducing a significant dynamic lag in the feedback control loop. For example, choosing $\tau_F < 0.1\tau_{max}$ satisfies this requirement. On the other hand, if the noise amplitude is high, then a larger value of τ_F may be required to *smooth* the noisy measurements. The frequency range of the noise is another important consideration. Suppose that the lowest noise frequency expected is denoted by ω_N. Then τ_F should be selected so that $\omega_F < \omega_N$ where $\omega_F = 1/\tau_F$. For example, suppose we specify $\omega_F = 0.1\omega_N$ which corresponds to $\tau_F = 10/\omega_N$. Then noise at frequency ω_N will be attenuated by a factor of 10 according to Eqs 14-20a and the Bode diagram of Fig. 14.2. In summary, τ_F should be selected so that $\omega_{max} < \omega_F < \omega_N$ where $\omega_F = 1/\tau_F$ and $\omega_{max} = 1/\tau_{max}$.

Digital Filters

In this section we consider several popular digital filters. A more comprehensive treatment of digital filtering and signal processing techniques is available elsewhere [10].

Exponential Filter. First we consider a digital version of the exponential filter. We will denote the samples of the measured variable as $x_{n-1}, x_n \ldots$ and the corresponding filtered values as $y_{n-1}, y_n \ldots$ where n refers to the current sampling instant. The derivative in (22-3) at time step n can be approximated by the backward difference:

$$\frac{dy}{dt} \cong \frac{y_n - y_{n-1}}{\Delta t} \tag{22-4}$$

Substituting in (22-3) and replacing $y(t)$ by y_n and $x(t)$ by x_n yields

$$\tau_F \frac{y_n - y_{n-1}}{\Delta t} + y_n = x_n \tag{22-5}$$

Rearranging gives

$$y_n = \frac{\Delta t}{\tau_F + \Delta t} x_n + \frac{\tau_F}{\tau_F + \Delta t} y_{n-1} \tag{22-6}$$

We define

$$\alpha \overset{\Delta}{=} \frac{1}{\tau_F/\Delta t + 1} \tag{22-7}$$

where $0 < \alpha \leq 1$. Then

$$1 - \alpha = 1 - \frac{1}{\tau_F/\Delta t + 1} = \frac{\tau_F}{\tau_F + \Delta t} \tag{22-8}$$

so that

$$y_n = \alpha x_n + (1 - \alpha)y_{n-1} \tag{22-9}$$

Equation 22-9 indicates that the filtered measurement is a weighted sum of the current measurement x_n and the filtered value at the previous sampling instant y_{n-1}. This operation is also called *single exponential smoothing*. Limiting cases for α are

$\alpha = 1$:   No filtering (the filter output is the raw measurement x_n).

$\alpha \to 0$:   The measurement is ignored.

In the above limits, note that $\tau_F = \Delta t(1 - \alpha)/\alpha$ by solving (22-7); hence, $\alpha = 1$ corresponds to a filter time constant of zero (no filtering).

Alternative expressions for α in (22-9) can be derived if the forward difference or other integration schemes for dy/dt are utilized [1].

Double Exponential Filter.   Another popular digital filter is the double exponential or second-order filter, which offers some advantages for eliminating high-frequency noise. The second-order filter is equivalent to two first-order filters in series where the second filter treats the output signal from the exponential filter in Eq. 22-9. The second filter can be expressed as

$$\bar{y}_n = \gamma y_n + (1 - \gamma)\bar{y}_{n-1} \tag{22-10}$$

$$\bar{y}_n = \gamma\alpha x_n + \gamma(1 - \alpha)y_{n-1} + (1 - \gamma)\bar{y}_{n-1} \tag{22-11}$$

Writing the filter equation in Eq. 22-10 for the previous sampling instant gives

$$\bar{y}_{n-1} = \gamma y_{n-1} + (1 - \gamma)\bar{y}_{n-2} \tag{22-12}$$

Solve for y_{n-1}:

$$y_{n-1} = \frac{1}{\gamma} \bar{y}_{n-1} - \frac{1-\gamma}{\gamma} \bar{y}_{n-2} \qquad (22\text{-}13)$$

Substituting (22-13) into (22-11) and rearranging gives the following expression for the double exponential filter:

$$\bar{y}_n = \gamma \alpha x_n + (2 - \gamma - \alpha)\bar{y}_{n-1} - (1 - \alpha)(1 - \gamma)\bar{y}_{n-2} \qquad (22\text{-}14)$$

A common simplification is to select $\gamma = \alpha$, yielding

$$\bar{y}_n = \alpha^2 x_n + 2(1 - \alpha)\bar{y}_{n-1} - (1 - \alpha)^2 \bar{y}_{n-2} \qquad (22\text{-}15)$$

The advantage of the double exponential filter over the exponential filter of Eq. 22-9 is that it provides better filtering of high-frequency noise, especially if $\gamma = \alpha$. The Bode diagrams in Figs. 14.2 and 14.3 provide a frequency response interpretation of this result. Note that second-order transfer function in Fig. 14.3 provides greater attenuation of high-frequency signals than the first-order system in Fig. 14.2. Although these Bode diagrams are for continuous systems (or filters), analogous results occur for digital systems (or filters).

A disadvantage of the double exponential filter is that it is more complicated than the exponential filter. Consequently, the single exponential filter has been more widely used in process control applications.

Moving Average Filter. A third type of digital filter is the moving-average filter which averages a specified number of past data points, by giving equal weight to each data point. The moving-average filter is usually less effective than the exponential filter, which gives more weight to the most recent data.

The moving-average filter can be expressed as

$$y_n = \frac{1}{J} \sum_{i=n-J+1}^{n} x_i \qquad (22\text{-}16)$$

where J is the number of past data points that are being averaged. Equation 22-16 implies that the previous filtered value, y_{n-1}, can be expressed as

$$y_{n-1} = \frac{1}{J} \sum_{i=n-J}^{n-1} x_i \qquad (22\text{-}17)$$

Subtracting (22-17) from (22-16) gives the recursive form of the moving-average filter:

$$y_n = y_{n-1} + \frac{1}{J}(x_n - x_{n-J}) \qquad (22\text{-}18)$$

The exponential and moving-average filters are examples of low-pass filters which are used to smooth noisy data by eliminating high-frequency noise.

Noise-Spike Filter. If a noisy measurement changes suddenly by a large amount and then returns to the original value (or close to it) at the next sampling instant, a *noise spike* is said to occur. Figure 22.8 shows two noise spikes appearing in the experimental temperature data for a fluidized sand bath. In general, noise spikes can be caused by spurious electrical signals in the environment of the sensor. If noise spikes are not removed by filtering before the noisy measurement is sent to

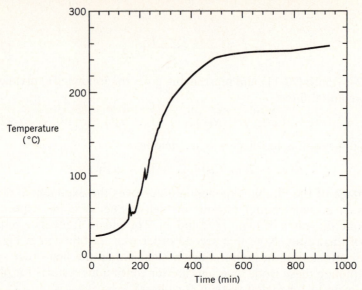

Figure 22.8. Temperature response data [11] from a fluidized sand bath contains two noise spikes.

the controller, the controller will cause large, sudden changes in the manipulated variable.

Noise-spike filters (or *rate of change* filters) are used to limit how much the filtered output is permitted to change from one sampling instant to the next. If Δx denotes the maximum allowable change, the noise-spike filter can be written as

$$y_n = \begin{cases} x_n & \text{if } |x_n - y_{n-1}| \le \Delta x \\ y_{n-1} - \Delta x & \text{if } y_{n-1} - x_n > \Delta x \\ y_{n-1} + \Delta x & \text{if } y_{n-1} - x_n < -\Delta x \end{cases} \tag{22-19}$$

If a large change in the measurement occurs, the filter replaces the measurement by the previous filter output plus (or minus) the maximum allowable change. This filter can also be used to detect instrument malfunctions such as a power failure, a break in a thermocouple or instrument line, or an ADC "glitch."

Other types of more sophisticated digital filters are available but have not been commonly used in process control applications. These include high-pass filters and bandpass filters [4,9,10].

22.4 COMPARISON OF ANALOG AND DIGITAL FILTERS

Analog and digital filters can be compared as follows:

1. Digital filters can be easily tuned (programmed) to fit the process. They are also easily modified.
2. Digital filters require the choice of a sampling period while analog filters do not.
3. Because digital filters require computation time and computer storage, they sometimes can limit the effectiveness of the data acquisition and control system. Analog filters are separate hardware devices and do not interact with the other computational tasks of the computer.

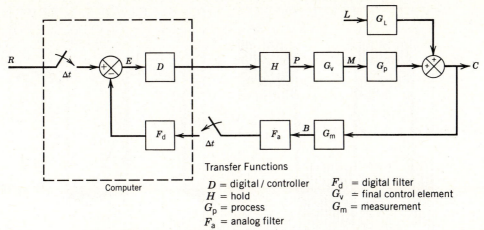

Transfer Functions

D = digital / controller F_d = digital filter
H = hold G_v = final control element
G_p = process G_m = measurement
F_a = analog filter

Figure 22.9. A block diagram with both analog and digital filters.

4. Analog filters are particularly effective for the elimination of high-frequency noise and aliasing.

It can be advantageous to use both analog and digital filters, as shown in the block diagram in Fig. 22.9. In this configuration high-frequency noise is filtered by the analog filter, while the digital filter suppresses lower-frequency (process) noise.

EXAMPLE 22.2

To compare the performance of alternative filters, consider a square wave signal with $f = 0.33$ cycles/min and an amplitude 0.5 corrupted by

(i) High-frequency sinusoidal noise (amplitude = 0.25, $f_N = 9$ cycles/min)
(ii) Random (Gaussian) noise with zero mean and a variance of 0.01.

Evaluate both analog and digital exponential filters as well as moving average filters.

Solution
(i) Sinusoidal noise
Representative results for high-frequency sinusoidal noise are shown in Fig. 22.10. The square wave with the additive noise is shown in Fig. 22.10a. The performance of two analog, exponential filters is shown in Fig. 22.10b. Choosing a relatively large filter time constant ($\tau_F = 0.4$ min) results in a filtered signal that contains less noise but is more sluggish, compared to the response for $\tau_F = 0.1$ min.

The effect of sampling period Δt on digital filter performance is illustrated in Fig. 22.10c. A larger sampling period ($\Delta t = 0.1$ min) results in serious aliasing because $f_s = 1/\Delta t = 10$ cycles/min, which is less than $2f_N = 18$ cycles/min. Reducing Δt by a factor of two results in much better performance. For each filter, a value of $\tau_F = 0.1$ min was chosen because this value was satisfactory for the analog filter of Fig. 22.10b. The smaller value of α (0.33 for $\Delta t = 0.05$ min vs. 0.5 for $\Delta t = 0.1$ min) provides more filtering.

The performance of two moving-average filters with $\Delta t = 0.05$ min is shown in Fig. 22.10d. Choosing $J = 7$ results in better filtering because this moving-average filter averages the sinusoidal noise over several cycles, while $J = 3$ gives a faster response but larger fluctuations.

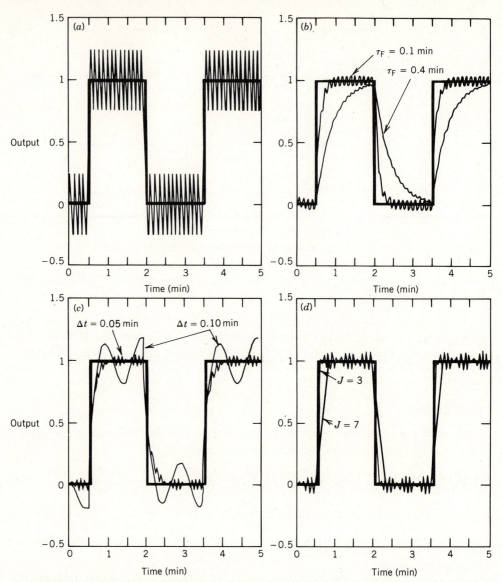

Figure 22.10. A comparison of filter performance for additive sinusoidal noise: (*a*) square wave plus noise; (*b*) analog exponential filters; (*c*) digital exponential filters; (*d*) moving-average filters.

(ii) Random noise

The filters considered in part (a) of this example were also evaluated for the situation where Gaussian noise was added to the same square wave signal. The simulations illustrating the effects of this noise level are shown in Fig. 22.11. Figure 22.11*a* shows the unfiltered signal after Gaussian noise with zero mean and a variance of 0.01 was added to the square wave signal. The analog, exponential filters in Fig. 22.11*b* provide effective filtering and again show the trade-off between degree of filtering and sluggish response that is inherent in the choice of τ_F. The digital filters in Fig. 22.11*c* and *d* are less effective even though different values of Δt and J were considered. Some aliasing occurs due to the high-frequency components of the random noise, which prevents the digital filter from performing as well as the analog filter.

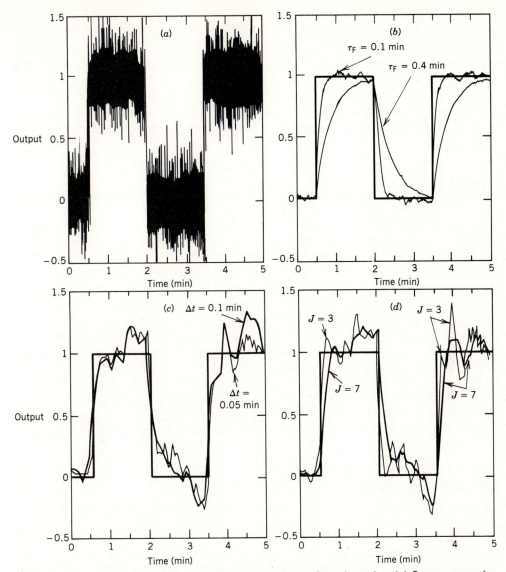

Figure 22.11. Comparison of filter performance for additive Gaussian noise: (*a*) Square wave plus noise; (*b*) analog exponential filters, (*c*) digital exponential filters; (*d*) moving-average filters.

In conclusion, both analog and digital filters can smooth noisy signals providing that the filter design parameters (including sampling period) are carefully selected.

22.5 EFFECT OF FILTER SELECTION ON CONTROL SYSTEM PERFORMANCE

Digital and analog filters are valuable for smoothing data and eliminating high-frequency noise, but they also affect control system performance. In particular a filter is an additional dynamic element in the feedback loop that causes a phase lag. Consequently, it reduces the stability margin for a feedback controller, compared to the situation where there is no filter [12]. Therefore, the controller may have to be retuned if the filter constant is changed. Derivative action can be included in the controller to provide phase lead which helps compensate for the phase lag due to the filter. However, when derivative action is used, it is important to filter noisy signals before the derivative control calculations are performed [2]. Because derivative action tends to amplify noise in the process measurement, filtering helps prevent controller saturation. Many electronic PID controllers in effect contain a high-frequency filter within their circuitry [13]. If the measurement signal is not filtered and process noise is significant, then derivative action should not be employed.

SUMMARY

When a digital computer is used for process control, measurements of the process variables are sampled and converted into digital form by an analog to digital converter (ADC). The sampling period Δt must be carefully selected. Sampling too slowly can produce aliasing and also reduce the effectiveness of the feedback control system. On the other hand, sampling too frequently tends to increase the data acquisition requirements of the computer. The choice of the sampling period should be based on the process dynamics, noise frequencies, signal-to-noise ratio, and the available computer control system.

Noisy measurements should be filtered before being sent to the controller. Analog filters are effective in removing high-frequency noise and avoiding aliasing. Digital filters are also widely used both for low-pass filters and other purposes such as the elimination of noise spikes. The choice of a filter and the filter parameters (e.g., τ_F) should be based on the process dynamics, the noise characteristics, and the sampling period. If a filter parameter is changed, it may be necessary to retune the controller since the filter is a dynamic element in the feedback control loop.

REFERENCES

1. Franklin, G. F., and J. D. Powell, *Digital Control of Dynamic Systems,* Addison-Wesley, Reading, MA, 1980.
2. Åström, K. J., and B. Wittenmark, *Computer Controlled Systems,* Prentice-Hall, Englewood Cliffs, NJ, 1984.
3. Ogata, K., *Discrete-Time Control Systems,* Prentice-Hall, Englewood Cliffs, NJ, 1987.
4. Isermann, R., *Digital Control Systems,* Springer-Verlag, New York, 1981, Chapter 27.
5. Williams, T. J., Economics and the Future of Process Control, *Automatica* **3,** 1 (1965).
6. Kalman, R. E., and J. E. Bertram, General Synthesis Procedure for Computer Control of Single-Loop and Multi-Loop Systems, *AIEE Trans.* **77,** Part 2, 602 (1958).
7. Shinskey, F. G., *Process Control Systems,* 3d ed., McGraw-Hill, New York, 1988.
8. Fertik, H. A., Tuning Controllers for Noisy Processes, *ISA Trans.* **14,** 4 (1975).
9. Wright, J. D., and T. F. Edgar, Digital Computer Control and Signal Processing Algorithms, in *Real-Time Computing,* (D. A. Mellichamp, Ed.), Van Nostrand Reinhold, New York, 1983, Chapter 22.
10. Oppenheim, A. V., and R. W. Shafer, *Digital Signal Processing,* Prentice-Hall, Englewood Cliffs, NJ, 1975.
11. Phillips, S. F., and D. E. Seborg, Adaptive Control Strategies for Achieving Desired Temperature Control Profiles During Process Startup, *IEC Res.* **27,** 1434 (1987).
12. Corripio, A. B., C. L. Smith, and P. W. Murrill, Filter Design for Digital Control Loops, *Instrum. Tech.,* **20**(1), 33 (1973).
13. Hougen, J. O., *Measurements and Control Applications,* 2d ed. ISA, Research Triangle Park, NC, 1979.

EXERCISES

22.1. A distillation column is subjected to a unit step change in feed flow rate F. Response data for the overhead product composition x_D are shown below. Previous experience has indicated that the transfer function,

$$\frac{X_D(s)}{F(s)} = \frac{5}{10s + 1}$$

provides an accurate dynamic model. Filter these data using an exponential filter with two different values of α, 0.5 and 0.8. Graphically compare the noisy data, the filtered data, and the analytical solution for the transfer function model.

Time (min)	x_D	Time (min)	x_D
0	0	11	3.336
1	0.495	12	3.564
2	0.815	13	3.419
3	1.374	14	3.917
4	1.681	15	3.884
5	1.889	16	3.871
6	2.078	17	3.924
7	2.668	18	4.300
8	2.533	19	4.252
9	2.908	20	4.409
10	3.351		

22.2. Show that the digital exponential filter output can be written as a function of previous measurements and the initial filter output y_0.

22.3. A signal given by

$$y(t) = t + 0.5 \sin{(t^2)}$$

is to be filtered with an exponential digital filter over the interval $0 \le t \le 20$. Using three different values of α (0.8, 0.5, 0.2), find the output of the filter at each sampling time. Do this for sampling periods of 1.0 and 0.1. Compare the three filters for each value of Δt.

22.4. The following product quality data were obtained from a reactor, based on a color evaluation of the product:

t (min)	x (color index)
0	0
1	1.5
2	0.3
3	1.6
4	0.4
5	1.7
6	1.5
7	2.0
8	1.5

(a) Filter the data using an exponential filter with $\Delta t = 1$ min. Use $\alpha = 0.2$ and $\alpha = 0.5$.
(b) Use a moving average filter with $J = 4$.
(c) Implement a noise-spike filter with $\Delta x = 0.5$.
(d) Plot the filtered data and the raw data for purposes of comparison.

22.5. The analog exponential filter in Eq. 22-3 is used to filter a measurement before it is sent to a proportional-only feedback controller with $K_c = 1$. The other transfer functions for the closed-loop system are $G_v = G_m = 1$, and $G_p = G_L = 1/(5s + 1)$. Compare the closed-loop responses to a sinusoidal load disturbance, $L(t) = \sin t$, for no filtering ($\tau_F = 0$) and for an exponential filter ($\tau_F = 3$ min).

22.6. Consider the first-order transfer function $Y(s)/X(s) = 1/(s + 1)$. Generate a set of data ($t = 1, 2, \ldots, 20$) by integrating this equation for $x = 1$ and randomly adding binary noise to the output, ± 0.05 units at each integer value of t. Design a digital filter for this system and compare the filtered and noise-free step responses for $\Delta t = 1$. Justify your choice of τ_F. Repeat for other noise levels, for example, ± 0.01 and ± 0.1.

CHAPTER 23

Development of Discrete-Time Models

Digital control systems inherently involve the processing of sampled signals. Thus, unless the sampling period is very small, it is more appropriate and convenient to perform the design and analysis of digital control systems using discrete-time models rather than continuous-time models. In this chapter we present several approaches for converting dynamic models based on differential equations to discrete-time models described by difference equations. We also consider a method for developing discrete-time models from experimental data. This method is analogous to the empirical approach for developing continuous-time models that was discussed in Chapter 7.

23.1 FINITE DIFFERENCE MODELS

A digital computer by its very nature deals internally with discrete data or numerical values of functions. To perform analytical operations such as differentiation and integration, numerical approximations must be utilized. Fortunately, formulas for such approximations are well established in the field of numerical analysis [1].

One way of converting continuous-time models to discrete-time form is to use finite difference techniques. In general, a nonlinear differential equation,

$$\frac{dy(t)}{dt} = f(y, x) \tag{23-1}$$

where y is the output variable and x is the input variable, can be numerically integrated (although with some error) by introducing a finite difference approximation for the derivative. For example, the first-order, backward difference approximation to the derivative at time $t = n\Delta t$ is [1]

$$\frac{dy}{dt} \cong \frac{y_n - y_{n-1}}{\Delta t} \tag{23-2}$$

where Δt is the integration interval that is specified by the user, y_n is the value of $y(t)$ at $t = n\Delta t$ and y_{n-1} denotes the value at the previous sampling instant $t = (n-1)\Delta t$. Substituting (23-2) into (23-1) and evaluating function $f(y, x)$ at the previous values of y and x (i.e., y_{n-1} and x_{n-1}) gives

$$\frac{y_n - y_{n-1}}{\Delta t} \cong f(y_{n-1}, x_{n-1}) \tag{23-3}$$

or

$$y_n = y_{n-1} + \Delta t \, f(y_{n-1}, x_{n-1}) \tag{23-4}$$

Equation 23-4 is a first-order difference equation that can be used to predict the value of y at time step n based on information at the previous time step $(n-1)$, namely, y_{n-1} and $f(y_{n-1}, x_{n-1})$. This type of expression is called a recurrence relation. It can be used to *numerically integrate* Eq. 23-1 by calculating y_n for $n = 0, 1, 2, 3, \ldots$ starting from known initial conditions, $y(0)$ and $x(0)$. In general, the resulting numerical solution, $\{y_n, n = 1, 2, 3, \ldots\}$, becomes more accurate and approaches the correct solution $y(t)$ as Δt decreases. However, for extremely small values of Δt, computer *roundoff errors* can be a significant source of error [1,2].

Equation 23-4 is the simplest numerical integration scheme and is referred to as Euler integration. The algorithm for Euler integration can also be expressed in the analogous form,

$$y_{n+1} = y_n + \Delta t \, f(y_n, x_n) \tag{23-5}$$

that results from using a forward difference approximation to the derivative rather than the backward difference in Eqn. (23-3):

$$\frac{dy}{dt} \cong \frac{y_{n+1} - y_n}{\Delta t} \tag{23-6}$$

Equation 23-4 can then be derived from Eq. 23-3 by evaluating $f(y, x)$ at the nth time step, followed by shifting all indices back one step. The forward difference derivation is the standard approach presented in numerical analysis books (e.g., [1]).

An important application of Eq. 23-4 in digital control is that it can be used as an approximate process model if the input and output signals are sampled. In this case, Δt is interpreted as the sampling period rather than the integration interval.

EXAMPLE 23.1

Develop a difference equation that approximates the differential equation,

$$\tau \frac{dy(t)}{dt} + y(t) = x(t) \tag{23-7}$$

or equivalently,

$$\frac{dy(t)}{dt} = -\frac{1}{\tau} y(t) + \frac{1}{\tau} x(t) \tag{23-8}$$

Solution

Note that if the initial condition is zero, that is, if $y(0) = 0$, these equations correspond to the first-order transfer function with unity gain:

$$\frac{Y(s)}{X(s)} = \frac{1}{\tau s + 1} \tag{23-9}$$

Using a backward difference approximation for the derivative and substituting y_{n-1} for $y(t)$ and x_{n-1} for $x(t)$ in Eq. 23-8, we obtain

$$\frac{\tau(y_n - y_{n-1})}{\Delta t} + y_{n-1} = x_{n-1}$$

Table 23.1   The Effect of Integration Interval on the Step Response of a First-Order Model ($\tau = 1$)

t	Numerical solution, y_n				Exact solution $y(t)$
	$\Delta t = 1.0$	$\Delta t = 0.25$	$\Delta t = 0.1$	$\Delta t = 0.01$	
0	0	0	0	0	0
0.5	—	0.438	0.410	0.395	0.393
1.0	1	0.684	0.651	0.634	0.632
2.0	1	0.900	0.879	0.866	0.865
3.0	1	0.968	0.958	0.951	0.951
4.0	1	0.990	0.985	0.982	0.982
5.0	1	0.997	0.995	0.993	0.993

Rearranging gives

$$y_n = \left(1 - \frac{\Delta t}{\tau}\right) y_{n-1} + \frac{\Delta t}{\tau} x_{n-1} \tag{23-10}$$

The new value y_n is a weighted sum of the previous value y_{n-1} and the previous input, x_{n-1}. Note that the digital filter derived in Chapter 22 differs from (23-10) because f is evaluated at the nth time step. This is because the filter is employed for smoothing the current measurement whereas (23-10) is used for prediction.

As Δt becomes infinitesimally small, the finite difference relation in Eq. 23-10 becomes a better approximation of the differential equation in (23-8). If Eq. 23-10 is used to numerically integrate (23-8) to $t = 5$ (nearly steady state) for a unit step change in $x(t)$, the integration error becomes substantial as Δt becomes quite large compared to τ, the time constant. Table 23.1 compares the analytical solution of (23-8) with the numerical solution obtained using (23-10) for $\tau = 1$ and various integration intervals.

The previous example has shown that a discrete-time version of a first-order differential equation can be derived using a backward difference approximation. Alternative discrete-time models can be obtained by using other finite difference approximations. Next we consider how discrete-time models can be derived for higher-order differential equations.

EXAMPLE 23.2

Develop a finite difference approximation for the second-order differential equation,

$$\frac{d^2 y}{dt^2} + a_1 \frac{dy}{dt} + a_0 y = x \tag{23-11}$$

Solution

The second-order derivative at the nth time step can be approximated by the finite difference approximation [1]:

$$\frac{d^2 y}{dt^2} \cong \frac{y_n - 2y_{n-1} + y_{n-2}}{(\Delta t)^2} \tag{23-12}$$

Substituting (23-12) and (23-2) into (23-11), replacing y and x by y_{n-1} and x_{n-1}, and rearranging yields

$$\left[\frac{1}{(\Delta t)^2} + \frac{a_1}{\Delta t}\right] y_n = \left[\frac{2}{(\Delta t)^2} + \frac{a_1}{\Delta t} - a_0\right] y_{n-1} - \frac{1}{(\Delta t)^2} y_{n-2} + x_{n-1} \quad (23\text{-}13)$$

All of the terms on the right side of (23-13) involve past information. Alternative discrete-time models can be obtained by other finite difference approximations, as discussed by Davis [1].

23.2 EXACT DISCRETIZATION FOR LINEAR SYSTEMS

For a process described by a linear differential equation, an alternative discrete-time model can be derived based on the analytical solution for a piecewise constant input. This approach yields an *exact* discrete-time model if the input variable is actually constant between sampling instants. Thus, this analytical approach eliminates the discretization error inherent in finite difference approximations for the important practical situation where the digital computer output (process input) is held constant between sampling instants. This signal is piecewise constant if the DAC acts as a zero-order hold, as discussed in Chapters 21 and 22.

As an illustrative example, consider the first-order system in Eq. 23-7. Assume that $x(t)$ is constant such that $x(t) = x(0)$, and that $y(0) \neq 0$. Taking the Laplace transform of (23-8) gives

$$sY(s) - y(0) = -\frac{1}{\tau} Y(s) + \frac{1}{\tau} \frac{x(0)}{s} \quad (23\text{-}14)$$

Solving for $Y(s)$,

$$Y(s) = \frac{1}{s + 1/\tau} \left[\frac{1}{\tau} \frac{x(0)}{s} + y(0)\right] \quad (23\text{-}15)$$

and taking inverse Laplace transforms gives

$$y(t) = x(0) (1 - e^{-t/\tau}) + y(0) e^{-t/\tau} \quad (23\text{-}16)$$

Equation 23-16 is valid for all values of t. Thus, after one sampling period, $t = \Delta t$, and we have

$$y(\Delta t) = x(0) (1 - e^{-\Delta t/\tau}) + y(0) e^{-\Delta t/\tau} \quad (23\text{-}17)$$

Next, we can generalize the analysis by considering the time interval, $(n-1) \Delta t$ to $n\Delta t$. For an initial condition $y[(n-1)\Delta t]$ and a constant input, $x(t) = x[(n-1) \Delta t]$ for $(n-1) \Delta t \le t < n\Delta t$, the analytical solution to (23-7) is

$$y(n\Delta t) = x[(n-1)\Delta t](1 - e^{-\Delta t/\tau}) + y[(n-1)\Delta t]e^{-\Delta t/\tau} \quad (23\text{-}18)$$

Note that the exponential terms are the same as for Eq. 23-17. Equation 23-18 can be written more compactly as

$$y_n = e^{-\Delta t/\tau} y_{n-1} + (1 - e^{-\Delta t/\tau}) x_{n-1} \quad (23\text{-}19)$$

Equation 23-19 is the exact solution to (23-7) providing that $x(t)$ is constant over each sampling period of length Δt. If Eq. 23-19 had been used in Example 23.1 to obtain the step response, the analytical solution $y(t)$ would have resulted for any sampling period (no integration error for the constant input).

23.3 HIGHER-ORDER SYSTEMS

To derive a discrete-time version of a higher-order differential equation, we can use the analytical solution and proceed as in the previous section. An alternative approach based on z-transforms will be presented in Chapter 24.

A linear differential equation of order p, when converted to discrete time, yields a linear difference equation also of order p. For example, consider the second-order system:

$$G(s) = \frac{Y(s)}{X(s)} = \frac{K(\tau_a s + 1)}{(\tau_1 s + 1)(\tau_2 s + 1)} \tag{23-20}$$

Using the analytical solution for a constant input provides the corresponding difference equation:

$$y_n + a_1 y_{n-1} + a_2 y_{n-2} = b_1 x_{n-1} + b_2 x_{n-2} \tag{23-21}$$

where $a_1 = -e^{-\Delta t/\tau_1} - e^{-\Delta t/\tau_2}$ (23-22)

$$a_2 = e^{-\Delta t/\tau_1} e^{-\Delta t/\tau_2} \tag{23-23}$$

$$b_1 = K\left(1 + \frac{\tau_a - \tau_1}{\tau_1 - \tau_2} e^{-\Delta t/\tau_1} + \frac{\tau_2 - \tau_a}{\tau_1 - \tau_2} e^{-\Delta t/\tau_2}\right) \tag{23-24}$$

$$b_2 = K\left(e^{-\Delta t(1/\tau_1 + 1/\tau_2)} + \frac{\tau_a - \tau_1}{\tau_1 - \tau_2} e^{-\Delta t/\tau_2} + \frac{\tau_2 - \tau_a}{\tau_1 - \tau_2} e^{-\Delta t/\tau_1}\right) \tag{23-25}$$

In Eq. 23-21 the new value of y depends on the values of y and x at the two previous sampling instants; hence it is a second-order difference equation. If $\tau_2 = \tau_a = 0$ and $K = 1$ in Eq. 23-21 through 23-25, the first-order difference equation in (23-19) results.

The steady-state gain of the second-order difference equation model can be found by considering steady-state conditions. Let $\bar{x}$ denote a step change in x and $\bar{y}$ the resulting steady-state change in y. At steady state, $x_{n-1} = x_{n-2} = \bar{x}$ and $y_n = y_{n-1} = y_{n-2} = \bar{y}$. Substituting these values into (23-21) gives

$$\bar{y} + a_1 \bar{y} + a_2 \bar{y} = b_1 \bar{x} + b_2 \bar{x} \tag{23-26}$$

Since y and x are deviation variables, the steady-state gain is simply $\bar{y}/\bar{x}$, the steady-state change in y divided by the steady-state change in x. Rearranging (23-26) gives

$$\text{Gain} = \frac{\bar{y}}{\bar{x}} = \frac{b_1 + b_2}{1 + a_1 + a_2} \tag{23-27}$$

Substitution of Eqs. (23-22) through (23-25) into (23-27) gives, Gain $= K$. This same result can be obtained by letting $s \to 0$ in (23-20). Thus the transfer function model in (23-20) and the corresponding discrete-time model in (23-21) have the same steady-state gains. This result also occurs for all other transfer function models.

Sets of differential equations can be converted to a discrete-time, difference equation model by using a technique referred to as the transition matrix approach [2]. General results for converting second-order and third-order differential equations (transfer functions) to difference equations have been derived by Neuman and Baradello [3] and are presented in Chapter 24.

23.4 FITTING DISCRETE-TIME EQUATIONS TO PROCESS DATA

In Chapter 7 we discussed methods for fitting continuous-time models (transfer functions) to process data. If a discrete-time model is desired, one approach is to fit a continuous-time model to experimental data and then to convert it to discrete form using the approach of Section 23.2. An alternative approach is to fit a discrete-time model to the data directly without first obtaining a continuous-time model. An important advantage of the latter approach is that the model parameters can be calculated using the well-known linear regression technique (see Chapter 7).

As a specific example, consider a rearranged version of the second-order difference equation (23-21). This model allows y_n to be predicted from data available at time $(n-1)$, or

$$\hat{y}_n = a_1' y_{n-1} + a_2' y_{n-2} + b_1 x_{n-1} + b_2 x_{n-2} \qquad (23\text{-}29)$$

where $\hat{y}_n$ denotes the prediction of y_n that is made at time $(n-1)$, $a_1' = -a_1$, and $a_2' = -a_2$. In developing a discrete-time model, model parameters a_1, a_2, b_1, and b_2 are treated as unknowns to be estimated by *fitting* the data. Note that the right side of (23-29) involves previous measurements of y rather than previous predictions. In general, $\hat{y}_n \neq y_n$, due to a number of practical considerations which include model inaccuracy, unmeasured disturbances, and measurement noise.

The objective of this parameter estimation problem is to determine numerical values for the unknown parameters so that the difference between the measurements and the predictions from Eq. 23-29 are as small as possible. The least-squares technique presented in Chapter 7 can be used here; the error criterion to be minimized is

$$\mathcal{E} = \sum_{n=1}^{r} (y_n - \hat{y}_n)^2 \qquad (23\text{-}30)$$

where y_n is the nth data point, $\hat{y}_n$ is the corresponding prediction made with measured values of x and y up to and including time $(n-1)$, and r is the number of data points.

EXAMPLE 23.3

Consider the step response data $\{y_n\}$ in Table 23.2 which were obtained from Example 7.3 and Fig. 7.8. At $t = 0$ a unit step change in x occurs, but the first output change is not observed until the next sampling instant. Fit the second-order difference equation (23-21) to the above input–output data using $\Delta t = 1$. Compare

Table 23.2 Step Response Data[a]

t	n	y_n
0	0	0
1	1	0.058
2	2	0.217
3	3	0.360
4	4	0.488
5	5	0.600
6	6	0.692
7	7	0.772
8	8	0.833
9	9	0.888
10	10	0.925

[a]$\Delta t = 1$; for $n < 0$, $y_n = 0$, and $x_n = 0$.

Table 23.3   Data Matrix for Estimating Parameters in Eq. 23-21

n	y_n	y_{n-1}	y_{n-2}	x_{n-1}	x_{n-2}
1	0.058	0	0	1	0
2	0.217	0.058	0	1	1
3	0.360	0.217	0.058	1	1
4	0.488	0.360	0.217	1	1
5	0.600	0.488	0.360	1	1
6	0.692	0.600	0.488	1	1
7	0.772	0.692	0.600	1	1
8	0.833	0.772	0.692	1	1
9	0.888	0.833	0.772	1	1
10	0.925	0.888	0.833	1	1

these results with models obtained in Example 7.3 using Harriott's method and nonlinear regression.

Solution

For the linear regression problem, there are four independent variables (y_{n-1}, y_{n-2}, x_{n-1}, x_{n-2}), one dependent variable (y_n), and four unknown parameters (a_1', a_2', b_1, b_2). We structure the data as shown in Table 23.3 and, using the least squares equations, solve four linear algebraic equations in four unknowns.

Table 23.4 compares the parameter values obtained by the three different approaches. The "Linear Regression" results were obtained by fitting a discrete-time model so as to minimize Eq. 23-30. The results labeled "Harriott's method" and "Nonlinear Regression" were obtained by fitting a continuous-time model (second-order, overdamped) to the data to obtain time constants τ_1 and τ_2 (see Example 7.3). In both cases, the model gain K was set equal to unity because the step responses were normalized. The continuous-time model was then converted to the corresponding discrete-time model via Eqs. 23-21 to 23-25.

The parameters obtained from linear regression in Table 23.4 are similar to the results for the other two methods but the differences are noteworthy. Using linear regression, four parameters were fit independently; with nonlinear regression only two parameters, τ_1 and τ_2, were estimated. Because the model gain was fixed at unity for nonlinear regression, this optimization method contained an implicit constraint. No such constraint was included in the linear regression approach. Consequently, the calculated model gain for linear regression, $K = 1.168$, is about 17% too large. This value was calculated from Eq. 23-27.

Linear regression would yield model parameters closer to those for continuous time, nonlinear regression if 10 to 20 additional data points at or near the steady state were added to the data set. Alternatively, the values of a_1', a_2', b_1, and b_2

Table 23.4   Comparison of Estimated Model Parameters for Example 23.2

	Linear Regression	*Harriott's Method*	*Nonlinear Regression*
a_1'	0.975	1.218	1.310
a_2'	−0.112	−0.348	−0.425
b_1	0.058	0.076	0.066
b_2	0.102	0.054	0.050
K	1.168	1.000	1.000

Table 23.5   Comparison of Simulated Responses for Various Difference Equation Models[a]

n	y	$\hat{y}_L$	$\hat{y}_H$	$\hat{y}_N$
1	0.058	0.058	0.076	0.066
2	0.217	0.217	0.201	0.192
3	0.360	0.365	0.374	0.375
4	0.488	0.487	0.493	0.496
5	0.600	0.596	0.599	0.602
6	0.692	0.690	0.691	0.695
7	0.772	0.767	0.764	0.768
8	0.833	0.835	0.830	0.833
9	0.888	0.886	0.876	0.880
10	0.925	0.932	0.921	0.924

[a]y, exact response for continuous system; $\hat{y}_L$, linear regression; $\hat{y}_H$, Harriott's method; $\hat{y}_N$, nonlinear regression.

could be optimized subject to the constraint that the process gain in Eq. 23-27 is unity; however, this would require a modified linear regression approach.

Table 23.5 compares the simulated responses for the three empirical models. Linear regression gives the best predictions because it fits the most parameters. Most of the differences among the models occur for the initial response ($n < 2$). In this particular example, it is difficult to distinguish graphically among the three models and the step response data for $n > 2$.

___

The above example has shown how we can fit a second-order difference equation model to data directly. The linear regression approach can also be used for higher-order models, provided that the parameters still appear linearly in the model. It is important to note that the calculated parameter values depend on the sampling period Δt that is selected, which in turn may be determined by the frequency of data collection. In general, changing Δt will result in different parameter estimates.

An attractive feature of linear regression is that it is not necessary to make a step change in x to estimate model parameters. An arbitrary input variation over a limited period of time would suffice. In particular, we need not force the system to a new steady state, a beneficial feature for industrial applications. However, as noted above, enough data should be taken to give an accurate steady-state gain.

SUMMARY

In this chapter we have presented several methods for developing discrete-time process models in the form of linear difference equations. If a continuous-time process model is available as one or more linear differential equations, then the corresponding discrete-time model can be derived analytically. This approach yields an *exact* discrete-time model provided that the process inputs are piecewise constant over the sampling period. An alternative approach is to use finite difference techniques to develop approximate discrete-time models. This approach is used when the continuous-time model consists of nonlinear differential equations and the analytical solution is not available.

An important advantage of discrete-time models is that they can be readily obtained by fitting experimental response data. Example 23.2 has illustrated that standard linear regression techniques can be employed to estimate the unknown model parameters in a linear difference equation model.

REFERENCES

1. Davis, M. E., *Numerical Methods and Modeling for Chemical Engineers*, Wiley, New York, 1984, Chapter 2.
2. Cadzow, J. A., and H. R. Martens, *Discrete-Time and Computer Control Systems*, Prentice-Hall, Englewood Cliffs, NJ, 1970.
3. Neuman, C. P., and C. S. Baradello, Digital Transfer Functions for Microcomputer Control, *IEEE Trans. Systems, Man, Cybernetics* **SMC-9** (12), 856 (1979).

EXERCISES

23.1. Consider the first-order differential equation

$$5\frac{dy}{dt} + y = 6x \qquad y(0) = 3$$

where $x(t)$ is piecewise constant and assumes the following values:

$$x(0) = 0 \qquad x(3) = 1$$
$$x(1) = 3 \qquad x(4) = 0$$
$$x(2) = 2 \qquad x(t) = 0 \qquad \text{for } t > 4$$

Develop a difference equation for this ordinary differential equation using $\Delta t = 1$ and

(a) exact discretization
(b) finite difference

Compare the integrated results for $0 \le t \le 10$. Examine whether $\Delta t = 0.1$ improves the finite difference model.

23.2. The following data were collected from a temperature sensor immersed in a gas stream. The input x is the flow rate deviation (in dimensionless units) and the sensor output y is given in millivolts. The flow (input) is piecewise constant over the intervals shown. The process is not at steady state initially, so y can change even though $x = 0$.

Time (s)	x	y
0	0	3.000
1	3	2.456
2	2	5.274
3	1	6.493
4	0	6.404
5	0	5.243
6	0	4.293
7	0	3.514
8	0	2.877
9	0	2.356
10	0	1.929

Fit a first-order model, $y_n = a_1' y_{n-1} + b_1 x_{n-1}$, to the data using the least-squares approach. Plot the responses of the fitted model and the actual data. Can you also find a first-order continuous transfer function to fit the data?

23.3. Fit a first-order discrete-time model to the step response data in Table 23.2. Compare your results with the graphical method for step response data, fitting the gain and time constant. Plot the two simulated step responses for comparison with the observed data.

23.4. A second-order differential equation is given as

$$\frac{d^2y}{dt^2} + 3\frac{dy}{dt} + 2y = x$$

For a unit step change in x (x changes from 0 to 1.0 at $t = 0$), calculate the responses ($0 \le t \le 5$) for $\Delta t = 0.1$ and $\Delta t = 0.5$. Perform the integration using the exact solution and a finite difference approximation (see Example 23.2). The initial conditions y and dy/dt are both zero at $t = 0$.

CHAPTER 24

Dynamic Response of Discrete-Time Systems

In earlier chapters we have seen that the Laplace transform provides a convenient way of analyzing the dynamic behavior of continuous-time systems. The Laplace transform is also applicable to discrete-time systems but is somewhat awkward to use. Thus we consider a related transform for discrete-time systems, the z-transform. The z-transform has the same utility as the Laplace transform in that its use leads to a compact mathematical description of a dynamic system and permits the use of algebraic operations for system analysis. By using z-transforms, transfer functions for discrete-time processes can be defined. This procedure facilitates subsequent analysis of sampled-data control systems, namely systems where a sampled signal appears. In this chapter we introduce z-transforms and pulse transfer functions and show how they can be used to calculate transient responses.

24.1 THE z-TRANSFORM

Consider the operation of an ideal, periodic sampler as shown in Fig. 24.1. The sampler converts a continuous signal $f(t)$ into a discrete signal $f^*(t)$ at equally spaced intervals of time. Mathematically it is convenient to consider *impulse sampling,* where $f^*(t)$ is the sampled signal formed by a sequence of *impulses* or Dirac delta functions:

$$f^*(t) = \sum_{n=0}^{\infty} f(n\Delta t)\, \delta(t - n\Delta t) \qquad (24\text{-}1)$$

Recall that the unit impulse $\delta(t)$ was defined in Chapter 3 as the limit of a rectangular pulse with infinitesimal width. The area under the pulse has a value of unity. Thus, it follows that if we integrate the sampled signal over a very small time period including the nth sampling instant,

$$\int_{n\Delta t^-}^{n\Delta t^+} f^*(t)\, dt = f(n\Delta t) \qquad (24\text{-}2)$$

In practice, impulse sampling is not attainable because the sampler remains closed for a small but finite amount of time. However, the time of closure is usually small (i.e., microseconds) compared to the sampling period and, consequently, impulse sampling provides a suitable idealization.

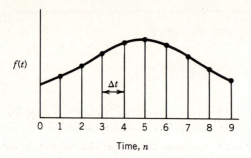

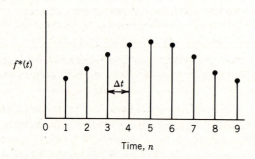

Figure 24.1 Sampled data impulse representation of continuous signal $f(t)$.

Next, consider the Laplace transform of Eq. 24-1, $F^*(s)$. The value of $f(n\Delta t)$ is considered to be a constant in each term of the summation and thus is invariant when transformed. Since $\mathcal{L}[\delta(t)] = 1$, it follows from the Real Translation Theorem (3-104) that the Laplace transform of a delayed unit impulse is $\mathcal{L}[\delta(t - n\Delta t] = e^{-n\Delta ts}$. Thus the Laplace transform of (24-1) is given by

$$F^*(s) = \sum_{n=0}^{\infty} f(n\Delta t)e^{-n\Delta ts} \qquad (24-3)$$

By introducing the change of variable, $z \triangleq e^{s\Delta t}$, we define $F(z)$, the z-transform of both $f^*(t)$ and $f(t)$, as

$$F(z) \triangleq Z[f^*(t)] = \sum_{n=0}^{\infty} f(n\Delta t)z^{-n} \qquad (24-4)$$

To simplify the notation, denote $f(n\Delta t)$ by f_n. Then (24-4) can be written as

$$F(z) = Z[f^*(t)] = \sum_{n=0}^{\infty} f_n z^{-n} \qquad (24-5)$$

In summary, a z-transform can be derived by taking the Laplace transform of a sampled signal and then making the change of variable, $z = e^{s\Delta t}$. Thus, the z-transform is a special case of the Laplace transform that is especially convenient for sampled-data systems. Although (24-5) is an infinite series, $F(z)$ can be written in closed form if the Laplace transform of $f(t)$ is a rational function [1].

Next we derive the z-transforms of several simple functions.

Step Function. To calculate the z-transform of a unit step input $S(t)$, set $f_n = 1$ for all $n \geq 0$. Note that $f_0 = 1$, which implies the sampled value is taken at $f(0^+)$. It follows from (24-5) that

$$F(z) = 1 + z^{-1} + z^{-2} + \cdots \qquad (24-6)$$

For $|z| > 1$ this infinite series converges, yielding

$$F(z) = \frac{1}{1 - z^{-1}} \tag{24-7}$$

Note that $|z| > 1$ corresponds to $e^{s\Delta t} > 1$ (or $s > 0$). This condition on s is the same as that used to derive the Laplace transform table in Chapter 3.

Exponential Function. For the exponential function $f(t) = Ce^{-at}$,

$$F(z) = \sum_{n=0}^{\infty} f(n\Delta t)z^{-n} = \sum_{n=0}^{\infty} Ce^{-an\Delta t}z^{-n} \tag{24-8}$$

Since Eq. 24-8 is a power series in $e^{-a\Delta t}z^{-1}$ that converges for $|e^{-a\Delta t}z^{-1}| < 1$ (which implies that $s > -a$), then

$$F(z) = \frac{C}{1 - e^{-a\Delta t}z^{-1}} \tag{24-9}$$

Properties of the z-transform. Some important properties of the z-transform are summarized below. Further information is available in Refs. 1 and 2.

 1. **Linearity.** The z-transform is a linear transformation, which implies that

$$Z[a_1 f_1(t) + a_2 f_2(t)] = a_1 Z[f_1(t)] + a_2 Z[f_2(t)] \tag{24-10}$$

where a_1 and a_2 are constants. This important property can be derived from the definition of the z-transform given in Eq. 24-1.

 2. **Real Translation Theorem.** The z-transform of a function delayed in time by an integer multiple of the sampling period is given by

$$Z[f(t - i\Delta t)] = z^{-i} F(z) \tag{24-11}$$

where i is a positive integer, provided that $f(t) = 0$ for $t < 0$. $F(z)$ is defined only for positive values of t.

The theorem can be easily proved. From (24-4),

$$Z[f(t - i\Delta t)] = \sum_{n=0}^{\infty} f(n\Delta t - i\Delta t)z^{-n} \tag{24-12}$$

Now substitute $j = n - i$

$$Z[f(t - i\Delta t)] = \sum_{j=-i}^{\infty} f(j\Delta t)z^{-j-i} \tag{24-13}$$

Since $f(j\Delta t) = 0$ for $j < 0$, we can write

$$Z[f(t - i\Delta t)] = z^{-i} \sum_{j=0}^{\infty} f(j\Delta t)z^{-j} \tag{24-14}$$

$$= z^{-i} F(z) \tag{24-15}$$

If the time delay is not an integer multiple of the sampling period, then the modified z-transform described below must be employed.

 3. **Complex Translation Theorem.** This theorem helps deal with z-transforms of functions containing exponential terms, which often arise with linear, continuous-time models.

$$Z[e^{-at}f(t)] = F(ze^{a\Delta t}) \tag{24-16}$$

To demonstrate the validity of (24-16), use (24-4) as the starting point:

$$Z[e^{-at}f(t)] = \sum_{n=0}^{\infty} e^{-an\Delta t}f(n\Delta t)z^{-n} \tag{24-17}$$

$$= \sum_{n=0}^{\infty} f(n\Delta t)(ze^{a\Delta t})^{-n} \tag{24-18}$$

$$= F(ze^{a\Delta t}) \tag{24-19}$$

We will illustrate the use of this theorem later in Example 24.2.

4.   Initial Value Theorem.   The initial value of a function can be obtained from its z-transform:

$$\lim_{n\to 0} f(n\Delta t) = \lim_{z\to\infty} F(z) \tag{24-20}$$

This result follows directly from Eq. 24-4, with the condition that $|z| > 1$.

5.   Final Value Theorem.   The final or large-time value of a function can be found from its z-transform, providing that a finite final value does exist:

$$\lim_{n\to\infty} f(n\Delta t) = \lim_{z\to 1} (1 - z^{-1})F(z) \tag{24-21}$$

Note that (24-21) is analogous to the final value theorem for Laplace transforms with stable poles (cf. Eq. 3-94).

To prove this theorem, substitute the definition of $F(z)$ into the right side of (24-21):

$$(1 - z^{-1})F(z) = (1 - z^{-1}) \sum_{n=0}^{\infty} f(n\Delta t)z^{-n} \tag{24-22}$$

$$= \{f(0) + [f(\Delta t) - f(0)]z^{-1}$$
$$+ [f(2\Delta t) - f(\Delta t)]z^{-2} + \cdots\} \tag{24-23}$$

Taking the limit as $z \to 1$, all of the terms in the infinite series except the last one cancel, yielding

$$\lim_{n\to\infty} f(n\Delta t) = \lim_{z\to 1} (1 - z^{-1})F(z) \tag{24-24}$$

6.   Modified z-Transform.   The modified z-transform is a special version of the z-transform that was developed to analyze continuous systems containing fractional time delays, namely those that are not an integer multiple of the sampling period. Suppose that a time delay θ is expressed as

$$\theta = (N + \sigma)\Delta t \tag{24-25}$$

where $0 < \sigma < 1$ and N is a positive integer. The sampled values of the delayed function are clearly not the same as those of the function with $\sigma = 0$. Using the real translation theorem, the z-transform of the delayed function $f(t - \theta)$ is

$$Z[f(t - \theta)] = \sum_{n=0}^{\infty} f(n\Delta t - N\Delta t - \sigma\Delta t)z^{-n} \tag{24-26}$$

Defining $m = 1 - \sigma$ and $k = n - N - 1$ yields

$$Z[f(t - \theta)] = \sum_{k=-N-1}^{\infty} f(k\Delta t + m\Delta t)z^{-k-N-1} \tag{24-27}$$

Since $f(n\Delta t) = 0$ for $n < 0$, the lower limit can be changed to zero. In addition z^{-N-1} can be factored out:

$$= z^{-N-1} \sum_{k=0}^{\infty} f(k\Delta t + m\Delta t)z^{-k} \tag{24-28}$$

Thus, it is convenient to define the modified z-transform by

$$F(z, m) \stackrel{\Delta}{=} z^{-N-1} \sum_{k=0}^{\infty} f(k\Delta t + m\Delta t)z^{-k} \tag{24-29}$$

where m is the modified z-transform variable. The modified z-transforms of some common functions have been tabulated by Ogata [1]. The theorems developed above (initial value, complex translation, etc.) can be extended to modified z-transforms. Unlike the z-transform, the modified z-transform contains information about the values of the function between samples $(0 < \sigma < 1)$. However, this information is available only if we know the continuous function $f(t)$. Applications of the modified z-transform have been discussed by Smith [3] and Deshpande and Ash [4].

Having introduced the properties of the z-transform, we now illustrate how they can be used to calculate z-transforms through a series of examples.

EXAMPLE 24.1

Derive the z-transform $F(z)$ for the function, $f(t) = t$.

Solution

Since $f(n\Delta t) = n\Delta t$, $F(z)$ is given by the formula,

$$F(z) = \sum_{n=0}^{\infty} f(n\Delta t)z^{-n} = \sum_{n=0}^{\infty} n\Delta t z^{-n} = \Delta t \sum_{n=0}^{\infty} nz^{-n} \tag{24-30}$$

To obtain a closed-form expression for the summation, let

$$S(z) = \sum_{n=0}^{\infty} nz^{-n} = z^{-1} + 2z^{-2} + 3z^{-3} + 4z^{-4} + \cdots \tag{24-31}$$

Multiplying by z^{-1} gives

$$z^{-1}S(z) = z^{-2} + 2z^{-3} + 3z^{-4} + 4z^{-5} + \cdots \tag{24-32}$$

Subtract (24-32) from (24-31):

$$S(z) - z^{-1}S(z) = z^{-1} + z^{-2} + z^{-3} + \cdots = \frac{1}{1 - z^{-1}} - 1 \tag{24-33}$$

Note that the left side is $(1 - z^{-1})S(z)$. Solving for $S(z)$ gives

$$S(z) = \frac{1}{(1 - z^{-1})^2} - \frac{1}{(1 - z^{-1})} = \frac{z^{-1}}{(1 - z^{-1})^2} \tag{24-34}$$

Hence, the z-transform of $f(t) = t$ is

$$F(z) = \Delta t\, S(z) = \frac{\Delta t z^{-1}}{(1 - z^{-1})^2} \tag{24-35}$$

EXAMPLE 24.2

Derive the z-transform of $\cos bt$. Then, using the complex translation theorem, find the z-transform of $f(t) = e^{-at} \cos bt$.

Solution
Applying the definition of the z-transform,

$$F(z) = Z(\cos bt) = \sum_{n=0}^{\infty} (\cos nb\Delta t)z^{-n} \tag{24-36}$$

To derive a compact formula for the power series, use Euler's relation for the cosine function:

$$\cos nb\Delta t = \tfrac{1}{2}(e^{jnb\Delta t} + e^{-jnb\Delta t}) \tag{24-37}$$

where $j = \sqrt{-1}$. This leads to

$$F(z) = \frac{1}{2}\left(\sum_{n=0}^{\infty} e^{jnb\Delta t}z^{-n} + \sum_{n=0}^{\infty} e^{-jnb\Delta t}z^{-n}\right) \tag{24-38}$$

We can then employ a previous result given in (24-8) and (24-9):

$$F(z) = \frac{1}{2}\left(\frac{1}{1 - e^{jb\Delta t}z^{-1}} + \frac{1}{1 - e^{-jb\Delta t}z^{-1}}\right)$$

$$= \frac{1}{2}\left(\frac{2 - (e^{jb\Delta t} + e^{-jb\Delta t})z^{-1}}{1 - (e^{jb\Delta t} + e^{-jb\Delta t})z^{-1} + z^{-2}}\right)$$

Using Euler's relation once more to return to trigonometric functions

$$F(z) = \frac{1 - z^{-1}\cos b\Delta t}{1 - 2z^{-1}\cos b\Delta t + z^{-2}} \tag{24-39}$$

Note that for $b = 2n\pi/\Delta t$, $F(z) = 1/(1 - z^{-1})$, which is the same expression as the z-transform of the unit step function. In other words, the sampled cosine has the identical appearance of the sampled unit step function ($f_n = 1$ for all n), an example of aliasing (see Fig. 22.5). Hence, the two z-transforms are identical in this special case.

To obtain the z-transform of the composite function of $f(t) = e^{-at}\cos bt$, apply the complex translation theorem:

$$Z(e^{-at}\cos bt) = F(ze^{a\Delta t}) \tag{24-40}$$

Equation 24-40 implies that we substitute $ze^{a\Delta t}$ every place z appears in (24-39). This step gives

$$Z(e^{-at}\cos bt) = \frac{1 - z^{-1}e^{-a\Delta t}\cos b\Delta t}{1 - 2z^{-1}e^{-a\Delta t}\cos b\Delta t + z^{-2}e^{-2a\Delta t}} \tag{24-41}$$

EXAMPLE 24.3

Find the modified z-transform of e^{-at} using Eq. 24-29. Show that the case $m = 0$ ($\sigma = 1$) corresponds to a pure time delay of one sampling period, that is, $\theta = \Delta t$.

Solution

Using (24-29) as the definition of the modified z-transform and substituting $(k + m)\Delta t$ and $N = 0$ for the value of t in e^{-at} yields

$$F(z, m) = z^{-1} \sum_{k=0}^{\infty} e^{-a(k+m)\Delta t} z^{-k} \tag{24-42}$$

$$= e^{-am\Delta t} z^{-1} \sum_{k=0}^{\infty} e^{-ak\Delta t} z^{-k} \tag{24-43}$$

Using (24-8) and (24-9) with k as the summation index gives

$$F(z, m) = \frac{e^{-am\Delta t} z^{-1}}{1 - e^{-a\Delta t} z^{-1}} \tag{24-44}$$

When $m = 0$ ($\sigma = 1$), the numerator of (24-44) becomes z^{-1}. This term indicates a one unit time delay ($N = 0$ and $\sigma = 1$ or $N = 1$ and $\sigma = 0$). Therefore it is consistent with Property 2 (the Real Translation Theorem) stated in Eq. 24-11.

As these examples illustrate, we can readily construct tables of z-transforms corresponding to functions $f(t)$ and $F(s)$. Table 24.1 provides a representative list of z-transforms; more extensive tables can be found in Ogata [1] and Deshpande and Ash [4].

EXAMPLE 24.4

Given the transform,

$$F(s) = \frac{1}{s(s + a)} \qquad (a > 0)$$

which might represent the step response of a first-order system, determine the final steady-state value, $\lim_{n \to \infty} f(n\Delta t)$.

Solution

From Table 24.1,

$$F(z) = \frac{1}{a} \left(\frac{1}{1 - z^{-1}} - \frac{1}{1 - e^{-a\Delta t} z^{-1}} \right)$$

Then from the Final Value Theorem,

$$\lim_{n \to \infty} f(n\Delta t) = \frac{1}{a} \lim_{z \to 1} \left[(1 - z^{-1}) \left(\frac{1}{1 - z^{-1}} - \frac{1}{1 - e^{-a\Delta t} z^{-1}} \right) \right]$$

$$\lim_{n \to \infty} f(n\Delta t) = \frac{1}{a} \lim_{z \to 1} \left[1 - \frac{z - 1}{z - e^{-a\Delta t}} \right] = \frac{1}{a}$$

This result agrees with the continuous-time limit of $f(t)$ as $t \to \infty$ obtained from the final value theorem for Laplace transforms. Note that if $a \leq 0$, $f(t)$ and $f(n\Delta t)$ are unbounded, and application of the Final Value Theorem would provide misleading results. Why?

Table 24.1 z-Transforms (Δt = Sampling Period)

Time Function $f(t)$	Laplace Transform $F(s)$	z-transform $F(z)$
$S(t)$, unit step	$\dfrac{1}{s}$	$\dfrac{1}{1 - z^{-1}}$
t	$\dfrac{1}{s^2}$	$\dfrac{\Delta t\, z^{-1}}{(1 - z^{-1})^2}$
t^{n-1}	$\dfrac{(n-1)!}{s^n}$	$\displaystyle\lim_{a\to 0} (-1)^{n-1} \dfrac{\partial^{n-1}}{\partial a^{n-1}} \left(\dfrac{1}{1 - e^{-a\Delta t}z^{-1}} \right)$
e^{-at}	$\dfrac{1}{s + a}$	$\dfrac{1}{1 - e^{-a\Delta t}z^{-1}}$
$\dfrac{1}{b - a}(e^{-at} - e^{-bt})$	$\dfrac{1}{(s + a)(s + b)}$	$\dfrac{1}{b - a}\left(\dfrac{1}{1 - e^{-a\Delta t}z^{-1}} - \dfrac{1}{1 - e^{-b\Delta t}z^{-1}} \right)$
$\dfrac{1}{ab}\left(S(t) + \dfrac{b}{a - b}e^{-at} - \dfrac{a}{a - b}e^{-bt} \right)$	$\dfrac{1}{s(s + a)(s + b)}$	$\dfrac{1}{ab}\left[\dfrac{1}{1 - z^{-1}} + \dfrac{b}{(a - b)(1 - e^{-a\Delta t}z^{-1})} - \dfrac{a}{(a - b)(1 - e^{-b\Delta t}z^{-1})} \right]$
te^{-at}	$\dfrac{1}{(s + a)^2}$	$\dfrac{\Delta t\, e^{-a\Delta t}z^{-1}}{(1 - e^{-a\Delta t}z^{-1})^2}$
$\dfrac{1}{b}e^{-at}\sin bt$	$\dfrac{1}{(s + a)^2 + b^2}$	$\dfrac{1}{b}\dfrac{z^{-1}e^{-a\Delta t}\sin b\Delta t}{1 - 2z^{-1}e^{-a\Delta t}\cos b\Delta t + e^{-2a\Delta t}z^{-2}}$
$e^{-at}\cos bt$	$\dfrac{s + a}{(s + a)^2 + b^2}$	$\dfrac{1 - z^{-1}e^{-a\Delta t}\cos b\Delta t}{1 - 2z^{-1}e^{-a\Delta t}\cos b\Delta t + e^{-2a\Delta t}z^{-2}}$
$\delta(t)$, unit impulse	1	1
$f(t - k\Delta t)$	$F(s)e^{-k\Delta ts}$	$F(z)z^{-k}$

24.2 INVERSION OF z-TRANSFORMS

Once a z-transform has been obtained (by whatever means), we need to be able to obtain the values of its corresponding time-domain function at the sampling instants. This is analogous to inverting Laplace transforms back to the time domain. The inversion of a z-transform $F(z)$ to its corresponding time domain function $f(t)$ is *not* unique because the inverse z-transform does not yield a continuous time function. Instead the values of the function are obtained only at the sampling instants. We know that a variety of continuous signals can be reconstructed from $f^*(t)$; that is, aliasing prevents the unique identification of the continuous function of time. On the other hand, the transformation from $F(z)$ to $f^*(t)$ (or, equivalently, from $F(z)$ to $f(n\Delta t)$) is unique. Consequently, we define the *inverse z-transform* operator, denoted by Z^{-1}, as follows:

$$f^*(t) = \{f(n\Delta t)\} = Z^{-1}[F(z)] \qquad (24\text{-}45)$$

The inverse z-transform consists of the sampled values $f^*(t)$, represented at the nth sampling instant as $f(n\Delta t)$.

To illustrate the inversion process, consider $F(z) = r_1/(1 - p_1 z^{-1})$. If $F(z)$

above is expanded as an infinite series,

$$\frac{r_1}{1 - p_1 z^{-1}} = r_1(1 + p_1 z^{-1} + p_1^2 z^{-2} + \cdots + p_1^n z^{-n} + \cdots) \quad (24\text{-}46)$$

Comparing this expression to (24-1), note that the inverse z-transform gives an expression for the value of the function at the nth sampling instant

$$f(n\Delta t) = f_n = r_1(p_1)^n \quad (24\text{-}47)$$

In most cases the z-transform to be inverted consists of a ratio of polynomials in z^{-1}. To invert such expressions, three methods are available:

(a) Partial fraction expansion
(b) Long division
(c) Contour integration

(a) Partial Fraction Expansion

This method is analogous to the procedure for expanding a complicated Laplace transform $F(s)$ into simpler functions prior to taking the inverse Laplace transform. Note that the z-transform table contains expressions that are functions of z^{-1} rather than z. Consequently, each term in the partial fraction expansion should be in this same form.

Suppose that $F(z)$ has the following form:

$$F(z) = \frac{V_1(z)}{V_2(z)} \quad (24\text{-}48)$$

where $V_1(z)$ = kth-order polynomial in z^{-1} (excluding a possible time-delay term z^{-N}, which can be factored)
$V_2(z)$ = mth-order polynomial in z^{-1} (the denominator is normalized so that the coefficient of z^0 is unity).

Assume that the denominator, $V_2(z)$, can be factored into m distinct real roots (i.e., poles of $F(z)$), denoted by $p_1, p_2, \ldots, p_m$. Thus,

$$F(z) = \frac{V_1(z)}{(1 - p_1 z^{-1})(1 - p_2 z^{-1}) \cdots (1 - p_m z^{-1})} \quad (24\text{-}49)$$

Then choose a partial fraction expansion of the form:

$$F(z) = \frac{r_1}{1 - p_1 z^{-1}} + \frac{r_2}{1 - p_2 z^{-1}} + \cdots + \frac{r_m}{1 - p_m z^{-1}} \quad (24\text{-}50)$$

Each numerator coefficient r_i can be calculated in a manner similar to that used for Laplace transforms (e.g., Heaviside's rule with $z = 1/p_i$). Taking the inverse z-transform of (24-50) term by term gives

$$f(n\Delta t) = Z^{-1}\left(\frac{r_1}{1 - p_1 z^{-1}}\right) + Z^{-1}\left(\frac{r_2}{1 - p_2 z^{-1}}\right)$$
$$+ \cdots + Z^{-1}\left(\frac{r_m}{1 - p_m z^{-1}}\right) \quad (24\text{-}51)$$

Since the inverse transform of $r_1/(1 - p_1 z^{-1})$ is $r_1(p_1)^n$, then

$$f(n\Delta t) = r_1(p_1)^n + r_2(p_2)^n + \cdots + r_m(p_m)^n \quad (24\text{-}52)$$

If there were only a single term in the z-transform, then $f(n\Delta t) = r_1 p_1{}^n$. For this simple case, we can examine how the sampled representation will vary as a function of the sign and magnitude of p_1. Figure 24.2 shows the discrete-time responses for different values of p_1 along with possible continuous-time interpretations for the case of a first-order z-transform.

In the special case when all roots are bounded by $0 \le p_i \le 1$ and Δt is known, then we can express the inverse z-transform for (24-46) as

$$f(n\Delta t) = r_1 e^{-q_1 n\Delta t} + r_2 e^{-q_2 n\Delta t} + \cdots + r_m e^{-q_m n\Delta t} \tag{24-53}$$

Relating (24-53) to (24-52), note that

$$q_1 = -\frac{1}{\Delta t} \ln p_1, \quad q_2 = -\frac{1}{\Delta t} \ln p_2, \cdots, \quad q_m = -\frac{1}{\Delta t} \ln p_m$$

If any $p_i < 0$, Eq. 24-52 should be used in place of (24-53).

If the denominator of $F(z)$ cannot be factored into real roots (i.e., complex roots appear), then the partial fraction expansion must contain a second-order polynomial in the denominator for each pair of complex roots. When inverting such a term, the appropriate quadratic form for damped sines and cosines in Table 24.1 should be used.

One other unusual case can arise in using the partial fraction expansion procedure. If the order of the numerator is greater than that of the denominator (i.e., $k > n$ in Eq. 24-48), then Eq. 24-50 is not strictly applicable. In this case, it is preferable to use other methods for generating the discrete-time sequence, such as the long division method discussed below.

EXAMPLE 24.5

Using a partial fraction expansion, find the inverse z-transform of $F(z) = 0.5 z^{-1}/[(1 - z^{-1})(1 - 0.5 z^{-1})]$ for a sampling period $\Delta t = 1$.

Solution

Expanding $F(z)$ into the sum of two fractions yields

$$F(z) = \frac{0.5 z^{-1}}{(1 - z^{-1})(1 - 0.5 z^{-1})} = \frac{r_1}{1 - z^{-1}} + \frac{r_2}{1 - 0.5 z^{-1}} \tag{24-54}$$

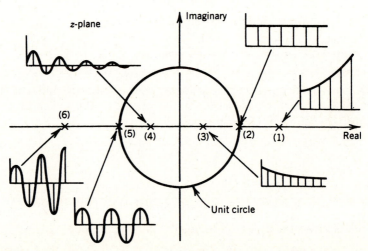

Figure 24.2 Time-domain responses for different locations of the root of $F(z)$.

Using Heaviside's rule (see Section 3.3) for finding r_1 and r_2, multiply $F(z)$ by $(1 - z^{-1})$ and set $z = 1$:

$$r_1 = \frac{0.5}{0.5} = 1$$

Next multiply $F(z)$ by $(1 - 0.5z^{-1})$ and set $z = 0.5$, that is, $z^{-1} = 2$:

$$r_2 = \frac{0.5(2)}{1 - 2} = -1$$

Substituting into (24-54) gives

$$F(z) = \frac{1}{1 - z^{-1}} - \frac{1}{1 - 0.5z^{-1}} \tag{24-55}$$

Equation 24-53 with $\Delta t = 1$ leads to

$$q_1 = -\frac{1}{\Delta t} \ln (1) = 0$$

$$q_2 = -\frac{1}{\Delta t} \ln (0.5) = 0.693$$

so that

$$f(n\Delta t) = 1 - e^{-0.693n\Delta t} \tag{24-56}$$

When Eq. 24-52 is used to check this result ($\Delta t = 1$, $r_1 = 1$, $p_1 = 1$, $r_2 = -1$, $p_2 = 0.5$), the same expression results since $e^{-0.693n} = (0.5)^n$.

(b) Long Division

Long division provides a second method for obtaining an inverse z-transform. In most cases it is considerably easier to use this method to obtain the inverse z-transform than to use partial fraction expansion. However, the result (an infinite series) may not be as useful as an analytical expression. Inversion via long division is an operation unique to discrete-time systems; no analogous method exists for continuous-time systems. From the definition of the z-transform as

$$F(z) = \sum_{n=0}^{\infty} f(n\Delta t) z^{-n} \tag{24-57}$$

the coefficients in the power series expansion of $F(z)$ are the values of $f(t)$ at the sampling instants:

$$F(z) = f(0) + f(\Delta t) z^{-1} + f(2\Delta t) z^{-2} + \cdots \tag{24-58}$$

Let $F(z)$ be a rational function represented by

$$F(z) = \frac{b_0 + b_1 z^{-1} + b_2 z^{-2} + \cdots + b_k z^{-k}}{a_0 + a_1 z^{-1} + a_2 z^{-2} + \cdots + a_m z^{-m}} \tag{24-59}$$

Dividing the denominator into the numerator by long division[1] gives

$$F(z) = c_0 + c_1 z^{-1} + c_2 z^{-2} + \cdots \tag{24-60}$$

---

[1] The order of division is based on dividing a_0 into the numerator and its remainders; see Example 24.6.

Referring back to (24-58), we can perform long division and equate the sequences $\{c_n\}$ and $\{f_n\}$ to obtain

$$f_0 = c_0 = \frac{b_0}{a_0} \tag{24-61}$$

$$f_1 = c_1 = \frac{b_1}{a_0} - \frac{b_0 a_1}{a_0^2} \tag{24-62}$$

$$f_2 = c_2 = b_2 - \frac{b_0 a_2}{a_0} - \frac{b_1 a_1}{a_0} + \frac{b_0 a_1^2}{a_0^2} \tag{24-63}$$

.
.
.

Some important properties of sampled-data systems can be obtained from long division of their z-transforms. For example, for a first-order z-transform,

$$F(z) = \frac{b_0}{1 - a_1 z^{-1}} \tag{24-64}$$

the equivalent sampled signal in the time domain can be found by long division, resulting in the infinite series (cf. (24–46)):

$$F(z) = b_0(1 + a_1 z^{-1} + a_1^2 z^{-2} + \cdots + a_1^n z^{-n} + \cdots) \tag{24-65}$$

Therefore, the sampled data response of this system is given by the sequence $\{b_0, b_0 a_1, b_0 a_1^2, \cdots\}$. This agrees with (24-61) through (24-63), taking into account the negative sign before a_1 in (24-64). Note that if $|a_1| > 1$, the magnitude of the signal grows steadily over time (see Fig. 24.2), but if $|a_1| < 1$, the signal will be attenuated over time. This coefficient in a first-order z-transform indicates if a system producing the signal is stable or unstable.

EXAMPLE 24.6

Repeat Example 24.5 using long division to generate the first five terms.

Solution
Dividing the denominator into the numerator,

$$
\begin{array}{r}
0.5z^{-1} + 0.75z^{-2} + 0.875z^{-3} + 0.9375z^{-4} + 0.9687z^{-5} + \cdots \\
\hline
1 - 1.5z^{-1} + 0.5z^{-2}\,\big|\,0.5z^{-1} \\
0.5z^{-1} - 0.75z^{-2} + 0.25z^{-3} \\
\hline
0.75z^{-2} - 0.25z^{-3} \\
0.75z^{-2} - 1.125z^{-3} + 0.375z^{-4} \\
\hline
0.875z^{-3} - 0.375z^{-4} \\
0.875z^{-3} - 1.3125z^{-4} + 0.4375z^{-5} \\
\hline
0.9375z^{-4} - 0.4375z^{-5} \\
0.9375z^{-4} - 1.4062z^{-5} - 0.4688z^{-6} \\
\hline
0.9687z^{-5} + 0.4688z^{-6}
\end{array}
$$

Note that $f(0)$ is zero in this expression. Long division does yield the same time sequence as does partial fraction expansion, with considerably less effort. However, the result is in the form of an infinite series.

(c) Contour Integration

A final method for inverting z-transforms utilizes a contour integral:

$$f(n\Delta t) = \frac{1}{2\pi j} \int_\Gamma F(z) z^{n-1}\, dz \qquad (24\text{-}66)$$

where the contour Γ must be appropriately specified. Although the integral can be evaluated using the residue theorem [1, p. 83], this method is seldom used in practice.

24.3 THE PULSE TRANSFER FUNCTION

In analogy with continuous systems, the analysis and design of sampled-data control systems is facilitated by the use of transfer functions based on z-transforms. The *pulse transfer function* represents a dynamic relationship and is defined as the ratio of the output and input z-transforms, assuming both output and input are initially at steady state, analogous to continuous-time systems. In addition, both the input and output signals are sampled at the same rate and synchronously.

Consider the sampled-data system in Fig. 24.3b where $X(z)$ and $Y(z)$ are z-transforms of the sampled input and output signals, $x(n\Delta t)$ and $y(n\Delta t)$, respectively. The response of a continuous linear process (Fig. 24.3a) is given by the convolution integral [1, p. 180],

$$y(t) = \int_0^t g(t - \tau) x^*(\tau)\, d\tau \qquad (24\text{-}67)$$

where $g(t)$ is the impulse response of the process (see Chapter 3), τ is the dummy variable of integration, and $x^*(\tau)$ is a series of impulses that can be expressed as

$$x^*(\tau) = \sum_{k=0}^{\infty} x(k\Delta t)\, \delta(\tau - k\Delta t) \qquad (24\text{-}68)$$

Substituting (24-68) into (24-67) gives

$$y(t) = \int_0^t g(t - \tau) \sum_{k=0}^{\infty} x(k\Delta t)\, \delta(\tau - k\Delta t)\, d\tau \qquad (24\text{-}69)$$

As shown by Ogata [1], an impulse $\delta(t)$ has the important property that, for an arbitrary function $h(\tau)$,

$$\int_0^t h(\tau)\, \delta(\tau - k\Delta t)\, d\tau = h(k\Delta t) \qquad \text{for } 0 \le \tau \le t \qquad (24\text{-}70)$$

(a) Continuous input

(b) Sampled input and output

Figure 24.3 Transfer function with (a) continuous input and (b) sampled input.

Thus, Eq. 24-69 reduces to

$$y(t) = \sum_{k=0}^{\infty} g(t - k\Delta t)x(k\Delta t) \tag{24-71}$$

In particular, for $t = n\Delta t$

$$y(n\Delta t) = \sum_{k=0}^{\infty} g(n\Delta t - k\Delta t)x(k\Delta t) \tag{24-72}$$

The z-transform of the output signal is defined by

$$Y(z) = \sum_{n=0}^{\infty} y(n\Delta t)z^{-n} \tag{24-73}$$

Substitution of $y(n\Delta t)$ from (24-72) gives

$$Y(z) = \sum_{n=0}^{\infty} \sum_{k=0}^{\infty} g(n\Delta t - k\Delta t)x(k\Delta t)z^{-n} \tag{24-74}$$

Let $i = n - k$:

$$Y(z) = \sum_{i=-k}^{\infty} \sum_{k=0}^{\infty} g(i\Delta t)z^{-(i+k)}x(k\Delta t) \tag{24-75}$$

Since $g(i\Delta t)$ is zero for $i < 0$, the lower limit on i can be changed to zero and the summations separated:

$$Y(z) = \left[\sum_{i=0}^{\infty} g(i\Delta t)z^{-i} \right]\left[\sum_{k=0}^{\infty} x(k\Delta t)z^{-k} \right] \tag{24-76}$$

or

$$Y(z) = G(z)X(z) \tag{24-77}$$

where $G(z)$ is defined as

$$G(z) \triangleq \sum_{i=0}^{\infty} g(i\Delta t)z^{-i} \tag{24-78}$$

We will refer to $G(z)$ as the *pulse transfer function* of the system. It relates the discrete-time input and output signals in the same manner as a transfer function in the s-domain relates continuous signals (see Fig. 24.4).

Other derivations of (24-77) are available [1]. Note that $G(z)$ can be calculated directly after $g(t)$, the impulse response function, has been determined (cf. Eq. 24-78).

EXAMPLE 24.7

Find the pulse transfer function for a first-order continuous process, $G(s) = K/(\tau s + 1)$.

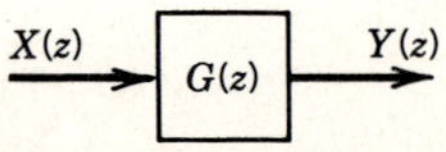

Figure 24.4   Pulse transfer function.

Solution

The continuous time impulse response $g(t)$ is found from

$$g(t) = \mathcal{L}^{-1}[G(s)] = \frac{K}{\tau} e^{-t/\tau} \tag{24-79}$$

The z-transform is

$$G(z) = \frac{K}{\tau} \sum_{n=0}^{\infty} e^{-n\Delta t/\tau} z^{-n} = \frac{K/\tau}{1 - e^{-\Delta t/\tau} z^{-1}} \tag{24-80}$$

This result can also be obtained from Table 24.1.

EXAMPLE 24.8

A second-order discrete process has the pulse transfer function

$$G(z) = \frac{-0.3225 z^{-1} + 0.5712 z^{-2}}{1 - 0.9744 z^{-1} + 0.2231 z^{-2}} \tag{24-81}$$

Determine its discrete-time response when forced by a sampled unit step input.

Solution

For a unit step input,

$$X(z) = 1 + z^{-1} + z^{-2} + z^{-3} + \cdots$$

$$= \frac{1}{1 - z^{-1}}$$

Here we use the closed-form expression to minimize the number of terms. Equation 24-77 is used to develop the expression for the step response $Y(z)$. We could apply the partial fraction expansion method to find $y(n\Delta t)$, but this would be relatively time-consuming due to the need to factor the denominator and use the Heaviside expansion. Therefore, long division is employed to determine the response. Substitution gives

$$Y(z) = G(z)X(z) = \frac{-0.3225 z^{-1} + 0.5712 z^{-2}}{1 - 0.9744 z^{-1} + 0.2231 z^{-2}} \frac{1}{1 - z^{-1}} \tag{24-82}$$

Applying the long division procedure yields

$$Y(z) = -0.3225 z^{-1} - 0.0665 z^{-2} + 0.2568 z^{-3} + 0.5136 z^{-4}$$
$$+ 0.6918 z^{-5} + 0.8082 z^{-6}$$
$$+ 0.8820 z^{-7} + 0.9277 z^{-8} + \cdots \tag{24-83}$$

We note in (24-83) that $y(n\Delta t)$ is steadily increasing and may be approaching a steady state value. Using the Final Value Theorem, we can calculate the value of $y(n\Delta t)$, as $n \rightarrow \infty$. Returning to (24-82), multiply by $(1 - z^{-1})$ and set $z = 1$. The ultimate (steady-state) value for the response is thus 1. Since at steady state both input and output values are 1, the gain of the pulse transfer function is also 1. This can be verified by evaluating $G(z)|_{z=1}$.

24.4 RELATING PULSE TRANSFER FUNCTIONS TO DIFFERENCE EQUATIONS

A pulse transfer function, representing a dynamic relation between an input and an output, has a unique correspondence with a difference equation. To demonstrate this, consider a general difference equation given by

$$a_0 y_n + a_1 y_{n-1} + \cdots + a_m y_{n-m} = b_0 x_n + b_1 x_{n-1} + \cdots + b_k x_{n-k} \quad (24\text{-}84)$$

where $\{a_i\}$ and $\{b_i\}$ are sets of constant coefficients and k and m are positive integers. Before taking the z-transform of (24-84), we recall the real translation theorem in Eq. 24-11:

$$Z(y_{n-i}) = Z[y(n\Delta t - i\Delta t)] = z^{-i} Y(z) \quad (24\text{-}85)$$

Taking the z-transform of both sides of Eq. 24-84 gives

$$a_0 Y(z) + a_1 z^{-1} Y(z) + \cdots + a_m z^{-m} Y(z)$$
$$= b_0 X(z) + b_1 z^{-1} X(z) + \cdots + b_k z^{-k} X(z) \quad (24\text{-}86)$$

Collecting terms, we solve for $Y(z)$:

$$Y(z) = \frac{b_0 + b_1 z^{-1} + \cdots + b_k z^{-k}}{a_0 + a_1 z^{-1} + \cdots + a_m z^{-m}} X(z) \quad (24\text{-}87)$$

The pulse transfer function of the discrete-time process is therefore given by

$$G(z) = \frac{Y(z)}{X(z)} = \frac{b_0 + b_1 z^{-1} + \cdots + b_k z^{-k}}{a_0 + a_1 z^{-1} + \cdots + a_m z^{-m}} \quad (24\text{-}88)$$

For most processes b_0 is zero, indicating that the input does not instantaneously affect the output. However, for proportional controllers and processes which are modeled simply by steady-state gains, all coefficients except a_0 and b_0 are zero. Also, the leading coefficient in the denominator can be set to unity by dividing both numerator and denominator by a_0.

Physical Realizability

In Chapter 4 we addressed the notion of physical realizability for continuous-time transfer functions. An analogous condition can be stated for a pulse transfer function, namely a discrete-time model cannot have an output signal that depends upon *future* inputs. Otherwise the model is not physically realizable. Consider the ratio of polynomials given in Eq. 24-88. The transfer function will be physically realizable as long as $a_0 \neq 0$, assuming that $G(z)$ has been reduced so that common factors in numerator and denominator have been cancelled. To show this property, examine Eq. 28-84. If $a_0 = 0$, the difference equation is

$$a_1 y_{n-1} = b_0 x_n + b_1 x_{n-1} + \cdots + b_k x_{n-k} - a_2 y_{n-2} - \cdots - a_m y_{n-m} \quad (24\text{-}89)$$

This equation requires a future input x_n to influence y_{n-1}, which is physically impossible (unrealizable). Another way to test physical realizability is to use long division. When the denominator in (24-88) is divided into the numerator, no terms with positive powers of z should occur for a physically realizable system.

EXAMPLE 24.9

Check the transfer function

$$\frac{Y(z)}{X(z)} = \frac{1 + 2z^{-1} + 3z^{-2}}{5z^{-1} + 2z^{-2}} \quad (24\text{-}90)$$

for physical realizability.

Solution

Note that the leading coefficient in the denominator is a power of z^{-1}, violating physical realizability ($a_0 = 0$ in (24-88)). The corresponding difference equation is

$$5y_{n-1} + 2y_{n-2} = x_n + 2x_{n-1} + 3x_{n-2} \tag{24-91}$$

An equivalent difference equation is

$$5y_n + 2y_{n-1} = x_{n+1} + 2x_n + 3x_{n-1} \tag{24-92}$$

This equation requires knowledge of the future input x_{n+1} to generate the current output value y_n. A similar conclusion can be deduced using long division, which yields positive powers of z. This exercise is left for the reader.

The Zero-Order Hold

Most sampled-data or computer control systems require a device to convert a digital output signal from the controller to an analog signal, which can then be utilized by a final control element such as a valve to manipulate the process. In many cases the final control element requires a continuous signal as input rather than a digital input to set its position (although in the specific case of a stepping motor, a continuous signal is generated from the digital input). The device usually employed for this purpose in process control is the digital-to-analog converter (DAC) which functions as a zero-order hold (ZOH), although there are other types of holds available (see Chapter 22). The zero-order hold converts the digital signal from the controller into a continuous staircase function. *This device ordinarily must appear in conjunction with a continuous process for digital control to be carried out.*

The process transfer function, when converted to a pulse transfer function, must incorporate the zero-order hold. The digital controller output signal may be thought of as an idealized sequence of impulses through the ZOH and then through the process. We know from Laplace transform theory that if $H(s)$ is the ZOH transfer function and $G(s)$ is the process transfer function (including the final control element), then the overall transfer function is $H(s)G(s)$. Although Table 24.1 gives the corresponding discrete-time form for a first- or second-order transfer function (see Example 24.7 for the first-order case), this table is based on an input that can be represented as a series of impulses. Thus, it cannot be used directly for the type of input that is characteristic of a ZOH (DAC) output, which is a staircase function. However, by taking the hold device into account, we can derive pulse transfer functions that are appropriate for process control calculations.

First we derive the appropriate expression for $H(s)$. Figure 24.5 shows the response of the hold device to an impulse input of unit strength. The hold yields a constant output value over the sampling period Δt, which is the same as the strength of the impulse. The impulse response of the ZOH over the interval

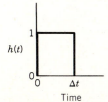

Figure 24.5 Response of a zero-order hold element to an impulse of unit magnitude.

$t = 0$ to $t = \Delta t$ can be written as the difference of two unit step functions, $\mathbf{S}(t)$ and $\mathbf{S}(t - \Delta t)$:

$$h(t) = \mathbf{S}(t) - \mathbf{S}(t - \Delta t) \tag{24-93}$$

From Eq. 3-22, the Laplace transform of $h(t)$ is

$$H(s) = \frac{1}{s} - \frac{e^{-s\Delta t}}{s} = \frac{1 - e^{-s\Delta t}}{s} \tag{24-94}$$

For processes with a piecewise constant input, the difference equation model must be based on the product $H(s)G(s)$ rather than $G(s)$ alone. Then Table 24.1 can be used to convert $H(s)G(s)$ to a z-transform.

EXAMPLE 24.10

For a first-order transfer function with gain equal to one,

$$G(s) = \frac{1}{\tau s + 1}$$

Show by transform techniques that a zero-order hold placed ahead of this process will yield the same difference equation model for the combined system (ZOH plus first-order process) as was derived in Eq. 23-19 for piecewise constant inputs to a first-order process.

Solution

First form the product $H(s)G(s)$:

$$H(s)G(s) = \frac{1 - e^{-s\Delta t}}{s} \frac{1}{\tau s + 1} \tag{24-95}$$

To convert this expression to its equivalent z-transform, we use a partial fraction expansion:

$$H(s)G(s) = \frac{1}{s} - \frac{1}{s + 1/\tau} - e^{-s\Delta t}\left(\frac{1}{s} - \frac{1}{s + 1/\tau}\right) \tag{24-96}$$

$HG(z)$ is defined as the z-transform of the combined ZOH plus process. $\mathcal{Z}$ is used as a shorthand way to denote the z-transform of the time-domain function given by the inverse Laplace transform,

$$HG(z) = \mathcal{Z}[H(s)G(s)] = Z\{\mathcal{L}^{-1}[H(s)G(s)]\}$$

$$= \mathcal{Z}\left(\frac{1}{s}\right) - \mathcal{Z}\left(\frac{1}{s + 1/\tau}\right) - \mathcal{Z}\left[e^{-s\Delta t}\left(\frac{1}{s} - \frac{1}{s + 1/\tau}\right)\right] \tag{24-97}$$

Table 24.1 is then used to convert each term in Eq. 24-97 into its equivalent z-transform. Since $\mathcal{Z}[e^{-s\Delta t}F(s)] = z^{-1}\mathcal{Z}[F(s)]$, Eq. 24-97 becomes

$$HG(z) = \left(\frac{1}{1 - z^{-1}} - \frac{1}{1 - e^{-\Delta t/\tau}z^{-1}}\right)$$

$$- z^{-1}\left(\frac{1}{1 - z^{-1}} - \frac{1}{1 - e^{-\Delta t/\tau}z^{-1}}\right) \tag{24-98}$$

or

$$HG(z) = (1 - z^{-1})\left[\frac{z^{-1}(1 - e^{-\Delta t/\tau})}{(1 - z^{-1})(1 - e^{-\Delta t/\tau}z^{-1})}\right]$$

$$= \frac{z^{-1}(1 - e^{-\Delta t/\tau})}{1 - e^{-\Delta t/\tau}z^{-1}} \tag{24-99}$$

Defining $a_1 \triangleq e^{-\Delta t/\tau}$, (24-99) can be written as

$$\frac{Y(z)}{X(z)} = HG(z) = \frac{(1 - a_1)z^{-1}}{1 - a_1 z^{-1}} \tag{24-100}$$

In difference equation form, (24-100) becomes

$$y_n - a_1 y_{n-1} = (1 - a_1) x_{n-1} \tag{24-101}$$

or

$$y_n = a_1 y_{n-1} + (1 - a_1) x_{n-1} \tag{24-102}$$

the same result as given in Eq. 23-19.

Several comments should be made about the procedure for transforming $H(s)G(s)$. First recognize that the combined transfer function $H(s)G(s)$ could have been converted to a series of impulse terms as in Example 24.7 by using the time-domain impulse response followed by transformation to the z-domain. The approach taken in Example 24.10, however, is more direct, because Table 24.1 provides the relation between s and z, thus bypassing the need to convert from the s-domain to the t-domain (and then to the z-domain). We can also generalize the results in (24-97) to (24-99) as

$$HG(z) = \mathcal{Z}(H(s)G(s)) = (1 - z^{-1})\mathcal{Z}\left(\frac{G(s)}{s}\right) \tag{24-103}$$

In the second term, the operator $\mathcal{Z}$ indicates that the z-transform equivalent of $G(s)/s$ can be determined using Table 24.1. Therefore, the calculation of $HG(z)$ first requires partial fraction expansion of $G(s)/s$, followed by transformation to its z-transform. Finally multiplication by $(1 - z^{-1})$ yields $HG(z)$.

Equation 24-103 can also be employed to illustrate an important property of the zero-order hold; in general, for dynamic systems,

$$HG(z) \neq H(z)G(z) \tag{24-104}$$

First calculate the z-transform of $H(s)$:

$$H(z) = \mathcal{Z}\left(\frac{1 - e^{-s\Delta t}}{s}\right) = (1 - z^{-1})\mathcal{Z}\left(\frac{1}{s}\right) \tag{24-105}$$

$$H(z) = \frac{1 - z^{-1}}{1 - z^{-1}} = 1 \tag{24-106}$$

This result is consistent with the fact that the gain of $H(s)$ is unity. Using $H(z) = 1$, Eq. 24-104 becomes

$$HG(z) \neq G(z) \tag{24-107}$$

Substituting (24-103) for $HG(z)$, the above expression is

$$(1 - z^{-1})\mathscr{Z}\left[\frac{G(s)}{s}\right] \neq \mathscr{Z}[G(s)] \qquad (24\text{-}108)$$

The inequality in 24-108 is valid, in general, except for the case $G(s) = K$. One other interesting characteristic of the zero-order hold is that

$$\lim_{\Delta t \to 0} HG(z) = G(s) \qquad (24\text{-}109)$$

This result seems intuitively correct in view of Fig. 22.2b. For example, if we substitute $z = e^{s\Delta t}$ in Eq. 24-99 and apply L'Hospital's rule for $\Delta t \to 0$, the z-transform reduces to the original first-order transfer function $G(s) = 1/(\tau s + 1)$. In contrast, the limit of $G(z)$ as $\Delta t \to 0$ does not equal $G(s)$ because of the discontinuous nature of the pulse transfer function $G(z)$ [1]; for example, see Eq. 24-80.

EXAMPLE 24.11

Derive the difference equation that corresponds to an integrating element, $G(s) = Y(s)/X(s) = 1/s$, using the ZOH and Eq. 24-103.

Solution

First determine $G(s)/s$, which is $1/s^2$. The z-transform is

$$\mathscr{Z}(1/s^2) = \frac{\Delta t\, z^{-1}}{(1 - z^{-1})^2} \qquad (24\text{-}110)$$

Multiply (24-110) by $(1 - z^{-1})$, yielding $HG(z)$:

$$HG(z) = \frac{\Delta t\, z^{-1}}{1 - z^{-1}} \qquad (24\text{-}111)$$

The corresponding difference equation is

$$y_n - y_{n-1} = \Delta t\, x_{n-1} \qquad (24\text{-}112)$$

The infinite series version of (24-112) can be obtained by long division of $HG(z)$ in (24-111) and conversion to discrete-time form:

$$y_n = \Delta t \sum_{k=1}^{n} x_{k-1} \qquad (24\text{-}113)$$

Equations 24-112 and 24-113 describe the two equivalent forms of the integrating element in discrete time.

Higher-Order Systems

As discussed in Chapters 6 and 7, many processes can be approximated by a second-order transfer function with time delay θ:

$$G(s) = \frac{Y(s)}{X(s)} = \frac{Ke^{-\theta s}}{(\tau_1 s + 1)(\tau_2 s + 1)} \qquad (24\text{-}114)$$

Assume that θ is an integer multiple of the sampling period ($\theta = N\Delta t$), and $x(t)$ is a piecewise constant input (i.e., $G(s)$ is placed in series with a ZOH). The following difference equation results:

$$y_n + a_1 y_{n-1} + a_2 y_{n-2} = b_1 x_{n-N-1} + b_2 x_{n-N-2} \tag{24-115}$$

The relations between (a_1, a_2, b_1, b_2) and (K, τ_1, τ_2) have been previously derived in Eqs. 23-21 through 23-25, based on the analytical solution of the differential equation. The pulse transfer function for (24-115) is

$$G(z) = \frac{Y(z)}{X(z)} = \frac{(b_1 z^{-1} + b_2 z^{-2}) z^{-N}}{1 + a_1 z^{-1} + a_2 z^{-2}} = \frac{(b_1 + b_2 z^{-1}) z^{-N-1}}{1 + a_1 z^{-1} + a_2 z^{-2}} \tag{24-116}$$

Note that (24-116) is a general expression for a second-order discrete-time model with a time delay of N sampling periods (the *apparent* time delay is one sampling period longer (i.e., $N + 1$) in (24-116) because the output cannot respond instantly). Neuman and Baradello [6] have derived difference equations incorporating the zero-order hold for a variety of linear process models. Table 24.2 gives pulse transfer functions for a number of transfer functions with the zero-order hold. Note that the transfer functions are given in pole–zero form (rather than using time constants). Only overdamped and integrating systems are considered.

24.5 EFFECT OF POLE AND ZERO LOCATIONS

For both continuous-time and discrete-time systems, the nature of the dynamic response is influenced by the location of the poles and zeros of the transfer function (see Section 6.1). In fact, a pole p_i of a continuous system maps into a pole in discrete time as $e^{p_i \Delta t}$ (recall the transformation $z = e^{s\Delta t}$). A process with two time constants (τ_1, τ_2) has two negative continuous-time poles $(-1/\tau_1, -1/\tau_2)$; therefore, after conversion to discrete-time, two positive poles in the z-plane $(e^{-\Delta t/\tau_1}, e^{-\Delta t/\tau_2})$ occur. This result is consistent with the expressions for a_1 and a_2 in a second-order difference equation (see Eqs. 23-22 and 23-23).

It is possible to categorize discrete-time responses for first- and second-order continuous-time processes with no zeros into eight different patterns [1], as shown in Fig. 24.6a–h. All responses exhibit an apparent one-unit time delay as noted above. For first-order systems, recall that in Fig. 24.2 we identified several possible responses depending on the pole location. The appropriate first-order difference equation is

$$y_n - p_1 y_{n-1} = x_{n-1} \tag{24-117}$$

Table 24.3 gives the response for the case of a unit impulse input at $n = 0$ ($x_0 = 1$ and $y_0 = 0$). The analytical solution is $y_n = (p_1)^n$. It is clear that a negative pole near the unit circle has a pronounced effect on the response. The alternation in sign of y is referred to as *ringing* of the output signal (see Chapter 26 for a discussion of ringing in control systems). Negative poles nearer the origin, although they produce a change in sign, are heavily damped and their results are not so noticeable. Positive poles do not cause the output to change in sign.

In Fig. 24.6 case c corresponds to a second-order overdamped system with two positive poles on the real axis. It has similar properties to case b. Cases d and e are noteworthy, because the two complex poles in continuous time map into a single pole in discrete time. In case e, the pole is located at $s = \pm \omega_s/2j$ in continuous time, indicating an undamped oscillation. Since ω_s is the sampling frequency, and $\omega_s = 2\pi/\Delta t$, the discrete-time pole $(e^{s\Delta t})$ is given by $e^{\pm \pi j}$. Via Euler's identity, the value of the single pole in discrete time is established as $e^{\pm \pi j} = \cos \pi \pm j \sin \pi = -1$. It is important to remember that a first-order discrete-time system

Table 24.2   Pulse Transfer Functions with Zero-Order Hold

Transfer Function $G(s)$	$HG(z) = \mathscr{Z}[H(s)G(s)]$	
$\dfrac{K}{s}$	$\dfrac{b_1 z^{-1}}{1 + a_1 z^{-1}}$	$a_1 = -1$ $b_1 = K\Delta t$
$\dfrac{K}{s + r}$	$\dfrac{b_1 z^{-1}}{1 + a_1 z^{-1}}$	$a_1 = -\exp(-r\Delta t)$ $b_1 = \dfrac{K}{r}[1 - \exp(-r\Delta t)]$
$\dfrac{K}{(s + r)(s + p)}$	$\dfrac{b_1 z^{-1} + b_2 z^{-2}}{1 + a_1 z^{-1} + a_2 z^{-2}}$	$a_1 = -\{\exp(-r\Delta t) + \exp(-p\Delta t)\}$ $a_2 = \exp[-(r + p)\Delta t]$ $b_1 = [K/rp(r - p)][(r - p) - r\exp(-p\Delta t) + p\exp(-r\Delta t)]$ $b_2 = [K/rp(r - p)][(r - p)\exp[-(r + p)\Delta t] + p\exp(-p\Delta t) - r\exp(-r\Delta t)]$
$\dfrac{K}{s(s + r)}$	$\dfrac{b_1 z^{-1} + b_2 z^{-2}}{1 + a_1 z^{-1} + a_2 z^{-2}}$	$a_1 = -\{1 + \exp(-r\Delta t)\}$ $a_2 = \exp(-r\Delta t)$ $b_1 = -(K/r^2)[1 - r\Delta t - \exp(-r\Delta t)]$ $b_2 = (K/r^2)[1 - \exp(-r\Delta t) - r\Delta t\exp(-r\Delta t)]$
$\dfrac{K(s + q)}{(s + r)(s + p)}$	$\dfrac{b_1 z^{-1} + b_2 z^{-2}}{1 + a_1 z^{-1} + a_2 z^{-2}}$	$a_1 = -\{\exp(-p\Delta t) + \exp(-r\Delta t)\}$ $a_2 = \exp[-(r + p)\Delta t]$ $b_1 = \dfrac{K}{p - r}\{\exp(-p\Delta t) - \exp(-r\Delta t) + (q/p)[1 - \exp(-p\Delta t)] - (q/r)[1 - \exp(-r\Delta t)]\}$ $b_2 = K\{(q/rp)\exp[-(r + p)\Delta t] + [(p - q)/rp]\exp(-r\Delta t) + [(q - r)/r(r - p)]\exp(-p\Delta t)\}$
$\dfrac{K}{(s + r)(s + p)(s + v)}$	$\dfrac{b_1 z^{-1} + b_2 z^{-2} + b_3 z^{-3}}{1 + a_1 z^{-1} + a_2 z^{-2} + a_3 z^{-3}}$	$a_1 = -\{\exp(-r\Delta t) + \exp(-p\Delta t) + \exp(-v\Delta t)\}$ $a_2 = \exp[-(r + p)\Delta t] + \exp[-(p + v)\Delta t] + \exp[-(r + v)\Delta t]$ $a_3 = -\exp[-(r + p + v)\Delta t]$ $b_1 = [K/(rpv)]\{(-q[\exp(-r\Delta t) + \exp(-p\Delta t) + \exp(-v\Delta t)]$ $+ \{[pv(q - r)]/[(p - r)(v - r)]\}[1 + \exp(-p\Delta t) + \exp(-v\Delta t)]$ $+ \{[rv(q - p)]/[(r - p)(v - p)]\}[1 + \exp(-r\Delta t) + \exp(-v\Delta t)]$ $+ \{[rp(q - v)]/[(r - v)(p - v)]\}[1 + \exp(-r\Delta t) + \exp(-p\Delta t)])$ $b_2 = [-K/(rpv)][(-q\exp[-(r + p)\Delta t] + \exp[-(p + v)\Delta t] + \exp[-(r + v)\Delta t]\}$ $+ \{[pv(q - r)]/[(p - r)]\}\{\exp(-p\Delta t) + \exp(-\omega\Delta t) + \exp[-(p + v)\Delta t]\}$

$$\frac{K(s+q)}{(s+r)(s+p)(s+v)} \qquad \frac{b_1 z^{-1} + b_2 z^{-2} + b_3 z^{-3}}{1 + a_1 z^{-1} + a_2 z^{-2} + a_3 z^{-3}}$$

$$+\ \{[rv(q-p)]/[(r-p)(v-p)]\}\{\exp(-r\Delta t) + \exp(-v\Delta t) + \exp[-(p+v)\Delta t] + \exp[-(r+v)\Delta t]\}$$
$$+\ \{[rp(q-v)]/[(r-v)(p-v)]\}\{\exp(-r\Delta t) + \exp(-p\Delta t) + \exp[-(r+v)\Delta t] + \exp[-(r+p)\Delta t]\}$$
$$b_3 = [K/(rpv)]\{-q\exp[-(r+p+v)\Delta t]$$
$$+\ \{[pv(q-r)]/[(p-r)(v-r)]\}\{\exp[-(p+v)\Delta t]$$
$$+\ \{[rv(q-p)]/[(r-p)(v-p)]\}\{\exp[-(r+v)\Delta t]$$
$$+\ \{[rp(q-v)]/[(r-v)(p-v)]\}\{\exp[-(r+p)\Delta t]\}$$

$$a_1 = -\{\exp(-r\Delta t) + \exp(-p\Delta t) + \exp(-v\Delta t)\}$$
$$a_2 = \exp[-(r+p)\Delta t] + \exp[-(p+v)\Delta t] + \exp[-(r+v)\Delta t]$$
$$a_3 = -\exp[-(r+p+v)\Delta t]$$
$$b_1 = [K/(rpv)]\{-q[\exp(-r\Delta t) + \exp(-p\Delta t) + \exp(-v\Delta t)]$$
$$+\ \{[pv(q-r)]/[(p-r)(v-r)]\}[1 + \exp(-p\Delta t) + \exp(-v\Delta t)]$$
$$+\ \{[rv(q-p)]/[(r-p)(v-p)]\}[1 + \exp(-r\Delta t) + \exp(-v\Delta t)]$$
$$+\ \{[rp(q-v)]/[(r-v)(p-v)]\}[1 + \exp(-r\Delta t) + \exp(-p\Delta t)]\}$$
$$b_2 = [-K/(rpv)]\{-q\exp[-(r+p)\Delta t] + \exp[-(p+v)\Delta t] + \exp[-(r+v)\Delta t]\}$$
$$+\ \{[pv(q-r)]/[(p-r)(v-r)]\}\{\exp(-p\Delta t) + \exp(-v\Delta t) + \exp[-(p+v)\Delta t] + \exp[-(r+v)\Delta t]\}$$
$$+\ \{[rv(q-p)]/[(r-p)(v-p)]\}\{\exp(-r\Delta t) + \exp(-v\Delta t) + \exp[-(r+v)\Delta t] + \exp[-(r+p)\Delta t]\}$$
$$+\ \{[rp(q-v)]/[(r-v)(p-v)]\}\{\exp(-r\Delta t) + \exp(-p\Delta t) + \exp[-(r+p)\Delta t]\}$$
$$b_3 = [K/(rpv)]\{-q\exp[-(r+p+v)\Delta t]$$
$$+\ \{[pv(q-r)]/[(p-r)(v-r)]\}\{\exp[-(p+v)\Delta t]$$
$$+\ \{[rv(q-p)]/[(r-p)(v-p)]\}\{\exp[-(r+v)\Delta t]$$
$$+\ \{[rp(q-v)]/[(r-v)(p-v)]\}\{\exp[-(r+p)\Delta t]\}$$

$$\frac{K(s+q)}{s(s+r)(s+p)} \qquad \frac{b_1 z^{-1} + b_2 z^{-2} + b_3 z^{-3}}{1 + a_1 z^{-1} + a_2 z^{-2} + a_3 z^{-3}}$$

$$a_1 = -\{1 + \exp(-r\Delta t) + \exp(-p\Delta t)\}$$
$$a_2 = \exp[-r\Delta t] + \exp[-p\Delta t] + \exp[-(r+p)\Delta t]$$
$$a_3 = -\exp[-(r+p)\Delta t]$$
$$b_1 = (K/rp)\{q\Delta t - (1 - q/p - q/r)[1 + \exp(-r\Delta t) + \exp(-p\Delta t)]$$
$$+\ \{[r(q-p)]/[p(p-r)]\}[2 + \exp(-r\Delta t)]$$
$$+\ \{[p(q-r)]/[r(r-p)]\}[2 + \exp(-p\Delta t)]\}$$
$$b_2 = (-K/rp)\{q\Delta t[\exp(-r\Delta t) + \exp(-p\Delta t)] + \exp[-(r+p)\Delta t]$$
$$-\ (1 - q/p - q/r)\{\exp(-r\Delta t) + \exp(-p\Delta t) + \exp[-(r+p)\Delta t]\}$$
$$+\ \{[p(q-r)]/[r(r-p)]\}[1 + 2\exp(-p\Delta t)]$$
$$+\ \{[r(q-p)]/[p(p-r)]\}[1 + 2\exp(-r\Delta t)]\}$$
$$b_3 = (K/rp)\{[[(q\Delta t - 1)rp + q(r+p)]/(rp)]\exp[-(r+p)\Delta t]$$
$$+\ \{[r(q-p)]/[p(p-r)]\}\exp(-r\Delta t)$$
$$+\ \{[p(q-r)]/[r(r-p)]\}\exp(-p\Delta t)\}$$

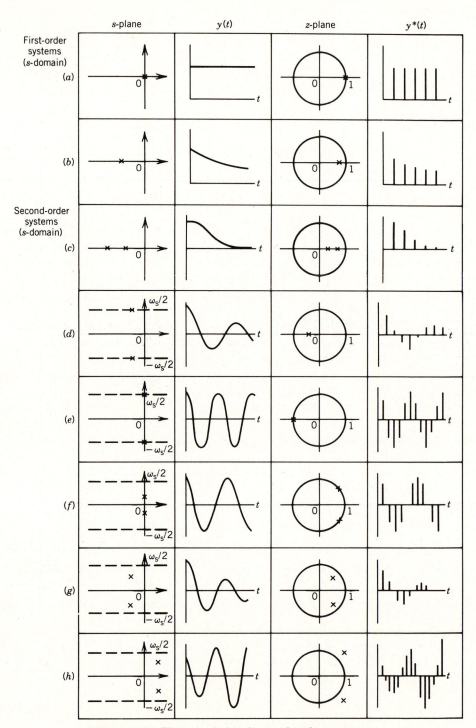

Figure 24.6 Effect of pole locations on impulse response.

Table 24.3 First-order Difference Output as a Function of Pole Location (unit impulse input)

		y_n		
n	x_n	$p_1 = -0.9$	$p_1 = -0.3$	$p_1 = 0.8$
0	1	0	0	0
1	0	1	1	1
2	0	-0.9	-0.3	0.8
3	0	$+0.81$	$+0.09$	0.64
4	0	-0.729	-0.027	0.512
5	0	$+0.656$	$+0.0081$	0.410
6	0	-0.590	-0.0024	0.328

can oscillate. Such behavior is not possible with a first-order continuous-time system.

Oscillation can also occur for second-order discrete-time systems, if the poles have imaginary (complex conjugate) values (see cases *f–h* in Fig. 24.6). When a positive or negative zero occurs in the discrete-time model, the degree of oscillation as well as its frequency can be affected. Unfortunately it is not possible to categorize this case easily, so we refer the reader to some examples presented by Franklin and Powell [5, pp. 32–35].

The mapping of zeros from continuous time to discrete time is unpredictable, mainly due to sampling effects. Consider the second-order difference equation resulting from a second-order continuous-time transfer function with no zero. In Eq. 24-116, the poles are found from factoring the denominator polynomial into two roots. The zero of the discrete-time transfer function is $-b_2/b_1$, which is fairly complicated when expressed in terms of τ_1 and τ_2 (see Eqs. 23-24 and 23-25). Therefore, there is no apparent simple relation between continuous- and discrete-time zeros. In addition, the sampling period can have a profound influence on the sampled response [7]. For example, an *inverse* response in continuous time (see Fig. 6.3) may not be observed at the sampling instants if the sampling rate is too slow.

24.6 CONVERSION BETWEEN LAPLACE AND z-TRANSFORMS

We have previously seen that Table 24.1 can be used to convert Laplace transforms to z-transforms and vice versa. However, implicit in this approach is the requirement that partial fraction expansion be performed to obtain the correct conversion. An alternative approach that avoids partial fraction expansion yields an approximate result merely by performing a variable substitution. No zero-order hold is explicitly considered in this approach.

As discussed earlier, the transform variable z was defined by $z = e^{s\Delta t}$ or $z^{-1} = e^{-s\Delta t}$. To obtain an approximate relation expressing s in terms of a ratio of polynomials in z, we can use the Padé approximation for $e^{-s\Delta t}$

$$e^{-s\Delta t} \cong \frac{2 - s\Delta t}{2 + s\Delta t} \tag{24-118}$$

Equating to z^{-1} gives

$$z^{-1} \cong \frac{2 - s\Delta t}{2 + s\Delta t} \tag{24-119}$$

or

$$s \cong \frac{2}{\Delta t} \frac{1 - z^{-1}}{1 + z^{-1}} \tag{24-120}$$

The approximation suggests that a Laplace transform can be converted to a z-transform by substituting (24-120) for s. Such an approach is known as *Tustin's method* [1]. A less accurate expression for s can be derived using the power series

$$e^{-s\Delta t} = 1 - s\Delta t + \frac{s^2 \Delta t^2}{2} - \cdots \tag{24-121}$$

Retaining only the first two terms, we have

$$z^{-1} = e^{-s\Delta t} \cong 1 - s\Delta t$$

or

$$s \cong \frac{1 - z^{-1}}{\Delta t} \tag{24-122}$$

which is equivalent to the backward difference formula we have used in Chapter 23. When the algebra involved in the substitution is not too complicated, (24-120) should be used instead of (24-122) to improve accuracy.

Ogata [1] has listed more accurate formulas for algebraic substitution into a transfer function $G(s)$. Approximate substitution is a procedure that should always be used with care. Exact conversion, especially of the process model (up to third order), is recommended. A bilinear transformation similar to (24-120) in form is sometimes used for stability analysis; its use will be discussed in Chapter 25.

EXAMPLE 24.12

Find an expression for the pulse transfer function of an ideal PID controller,

$$G_c(s) = K_c \left(1 + \frac{1}{\tau_I s} + \tau_D s \right) \tag{24-123}$$

using the approximation in (24-122). Compare your result with the velocity form of the PID algorithm given in Eq. 8-18.

Solution
Substituting (24-122) into (24-123) gives

$$G_c(z) = \frac{K_c(a_0 + a_1 z^{-1} + a_2 z^{-2})}{1 - z^{-1}} \tag{24-124}$$

where $a_0 = 1 + \dfrac{\Delta t}{\tau_I} + \dfrac{\tau_D}{\Delta t}$

$a_1 = -\left(1 + \dfrac{2\tau_D}{\Delta t} \right)$

$a_2 = \dfrac{\tau_D}{\Delta t} \tag{24-125}$

If e_n is the error signal and p_n is the output from the controller, then $G_c(z) = P(z)/E(z)$ and

$$(1 - z^{-1})P(z) = K_c(a_0 + a_1 z^{-1} + a_2 z^{-2})E(z) \tag{24-126}$$

Using the real translation theorem and converting the controller equation into a discrete-time form gives:

$$p_n - p_{n-1} = K_c a_0 e_n + K_c a_1 e_{n-1} + K_c a_2 e_{n-2} \qquad (24\text{-}127)$$

Substituting for a_0, a_1, and a_2 and collecting terms with respect to the controller settings K_c, τ_I, and τ_D gives

$$p_n - p_{n-1} = K_c \left[(e_n - e_{n-1}) + \frac{\Delta \tau}{\tau_I} e_n + \frac{\tau_D}{\Delta t}(e_n - 2e_{n-1} + e_{n-2}) \right] \qquad (24\text{-}128)$$

Note that this equation is identical to Eq. 8-18 which was derived using a finite-difference approximation.

SUMMARY

In this chapter we have introduced the z-transform and its properties, in much the same fashion as was done for Laplace transforms and continuous-time linear systems in Chapters 3 and 4. Operational use of the z-transform with linear process models has been emphasized here, since z-transforms are a convenient medium to analyze

Table 24.4   Discrete/Continuous Conversions for Linear Systems

Conversion	Method (section or equation)
(A) Impulse response $\rightarrow$ z-transform	Conversion table (§24.1)
(B) Laplace transform $\rightarrow$ z-transform	(1) Conversion table (§24.1, §24.2) (2) Approximate substitution (§24.6)
(C) z-transform $\rightarrow$ difference equation	Translation theorem (§24.2)
(D) Difference equation $\rightarrow$ z-transform	Translation theorem (§24.2)
(E) Laplace transform $\rightarrow$ difference equation (piecewise constant input)	Zero-order hold transfer function, Eq. 24-104 $\rightarrow$ partial fraction expansion $\rightarrow$ Table 24.1 $\rightarrow$ then use (C) (§24.2)
(F) Differential equation $\rightarrow$ difference equation	(1) Convert to Laplace transform; then use (E) (§24.4) (2) Analytical solution for piecewise constant input, Table 24.2 or Eq. 23-21 (3) Finite difference approximation (§23.1)
(G) Data $\rightarrow$ difference equation	(1) Linear regression (§23.3) (2) Nonlinear regression in continuous time (§7.1) $\rightarrow$ analytical solution (§23.2)
(H) z-transform $\rightarrow$ response	(1) Use (C) followed by integration of difference equation (§23.1) (2) Power series in z^{-k} by long division (§24.3) (3) Contour integration, Eq. 24-66

digital feedback control systems (Chapters 25 and 26). In Chapters 22–24 we have presented a rather diverse set of techniques for converting continuous models to discrete models and vice versa. Hence, a suitable epilog to this chapter would be to provide a summary of the possible avenues for interconversion. Table 24.4 gives a list of different approaches, the steps involved, and the pertinent sections or equations where the specifics are demonstrated.

REFERENCES

1. Ogata, K., *Discrete Time Control Systems,* Prentice-Hall, Englewood Cliffs, NJ, 1987.
2. Corripio, A. B., Module 3.3 in AIChemI Modular Instruction, Series A, Vol. 3, AIChE, New York (1983).
3. Smith, C. L., *Digital Computer Process Control,* InText, Scranton, PA, 1972.
4. Deshpande, P. B., and R. H. Ash, *Elements of Computer Process Control,* Instrum. Soc. of America, Research Triangle Park, NC, 1981.
5. Franklin, G. F., and J. D. Powell, *Digital Control of Dynamic Systems,* Addison-Wesley, Reading, MA, 1980.
6. Neuman, C. P., and C. S. Baradello, Digital Transfer Functions for Microcomputer Control, *IEEE Trans. Systems, Man, Cybernetics* **SMC-9** (12), 856 (1979).
7. Åström, K. J., and B. Wittenmark, *Computer-Controlled Systems,* Prentice-Hall, Englewood Cliffs, NJ, 1984.

EXERCISES

24.1. What is the z-transform $F(z)$ of the triangular pulse in the figure if the sampling period has the following values:

(a) $\Delta t = 5$ s
(b) $\Delta t = 10$ s

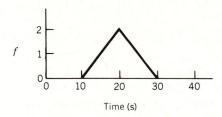

24.2. A temperature sensor has the transfer function,

$$\frac{T'_m(s)}{T'(s)} = \frac{1}{10s + 1}$$

where T'_m is the measured temperature and T' is the temperature (both in deviation variables). The temperature measurement is sampled every five seconds and sent to a digital controller. Suppose that the actual temperature changes in the following manner,

$$T(t) = \begin{cases} 350 \ °F & \text{for } 0 \le t < 4 \text{ s} \\ 370 \ °F & \text{for } 4 \le t < 12 \text{ s} \\ 350 \ °F & \text{for } t \ge 12 \text{ s} \end{cases}$$

(a) What is the z-transform of this signal, $T(z)$?
(b) Derive an expression for the z-transform of the measured temperature $T_m(z)$.
(c) The digital controller sounds an alarm if the sampled value of T_m exceeds 360 °F. Does the alarm sound?
(d) What is the maximum value of the measured temperature $T_m(t)$?

24.3. Suppose that

$$F(z) = \frac{1 - 0.2z^{-1}}{(1 + 0.6z^{-1})(1 - 0.3z^{-1})(1 - z^{-1})}$$

(a) Calculate the corresponding time-domain response $f^*(t)$.
(b) As a check, use the final value theorem to determine the steady-state value of $f^*(t)$.

24.4. Determine the inverse transform of

$$\frac{z(z + 1)}{(z - 1)(z^2 - z + 1)}$$

by the following methods:

(a) Partial fraction expansion.
(b) Long division.

24.5. Calculate the z-transform of the rectangular pulse shown in the drawing. Assume that the sampling period is $\Delta t = 2$ min. The pulse is $f = 3$ for $2 \le t < 6$.

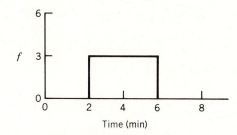

24.6. The pulse transfer function of a process is given by

$$\frac{Y(z)}{X(z)} = \frac{5(z + 0.6)}{z^2 - z + 0.41}$$

(a) Calculate the response $y(n\Delta t)$ to a unit step change in x using the partial fraction method.
(b) Check your answer in part (a) by using long division.
(c) What is the steady-state value of y?

24.7. The desired temperature trajectory $T(t)$ for a batch reactor is shown in the drawing.

(a) Derive an expression for the Laplace transform of the temperature trajectory, $T(s)$.
(b) Determine the corresponding z-transform $T(z)$ for sampling periods of $\Delta t = 4$ and 8 min.

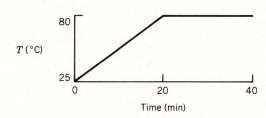

24.8. The dynamic behavior of a temperature sensor and transmitter can be described by the first-order transfer function,

$$\frac{T'_m(s)}{T'(s)} = \frac{e^{-2s}}{8s + 1}$$

where the time constant and time delay are in seconds

T = actual temperature

T_m = measured temperature

If the actual temperature changes as follows (t in seconds):

$$T = \begin{cases} 70\ ^\circ\text{C} & \text{for } t < 0 \\ 85\ ^\circ\text{C} & \text{for } 0 \le t < 10 \\ 70\ ^\circ\text{C} & \text{for } t \ge 10 \end{cases}$$

(a) What is the maximum value of the measured temperature T_m?

(b) If samples of the measured temperature are automatically logged in a digital computer every two minutes beginning at $t = 0$, what is the maximum value of the logged temperature?

24.9. The transfer function for a process model and a zero-order hold can be written as

$$H(s)G_p(s) = \left(\frac{1 - e^{-s\Delta t}}{s}\right) \frac{3.8e^{-2s}}{(10s + 1)(5s + 1)}$$

Derive an expression for the pulse transfer function of $H(s)G_p(s)$ when $\Delta t = 2$.

24.10. The pulse transfer function of a process is given by

$$\frac{Y(z)}{X(z)} = \frac{2.7z^{-1}(z + 3)}{z^2 - 0.5z + 0.06}$$

(a) Calculate the response $y(n\Delta t)$ to a unit step change in x using the partial fraction method.

(b) Check your answer in part (a) by using long division.

(c) What is the steady-state value of y?

24.11. A gas chromatograph is used to provide composition measurements in a feedback control loop. The open-loop transfer function is given by

$$G(s) = G_c H G_p G_m \qquad (G_v = 1)$$

and is

$$G(s) = \frac{B(s)}{E(s)} = 2\left(1 + \frac{1}{8s}\right)\left(\frac{1 - e^{-s\Delta t}}{s}\right)\left(\frac{10}{12s + 1}\right)e^{-2s}$$

(a) Suppose that a sampling period of $\Delta t = 1$ min is selected. Calculate $HG(z)$, the pulse transfer function of $G(s)$ with ZOH.

(b) If a unit step change in the controller error signal $e(t)$ is made, calculate the sampled open-loop response $b(n\Delta t)$ using $HG(z)$.

24.12. Determine the pulse transfer function with zero-order hold for the second-order process $G_p(s) = K/[(5s + 1)(3s + 1)]$ using partial fraction expansion in the s-domain. Check your results with those in Section 23.3. Note that Δt is unspecified here.

24.13. Find $HG(z)$ if $G(s) = (1 - 9s)/[(3s + 1)(15s + 1)]$ for $\Delta t = 4$ (use partial fraction expansion). What is the corresponding difference equation? Do you detect inverse response in the output y_n for a step change in the input at this sampling period?

24.14. Verify the z-transform in Table 24.1 for $f(t) = t^2$. What is the z-transform for $f(t) = 1 - e^{-at}$?

24.15. Find the response y_n for the difference equation

$$y_n - y_{n-1} + 0.21\, y_{n-2} = x_{n-2}$$

Let $x_0 = 1$, $x_n = 0$ for $n \ge 1$. Use long division as well as direct integration to check the results.

24.16. Use long division to calculate the first eight coefficients of the z-transform given by

$$F(z) = \frac{0.8z^{-1}}{(1 - 0.8z^{-1})^2}$$

24.17. Derive the pulse transfer function for an analog lead–lag device cascaded with a zero-order hold. The lead–lag device has the transfer function $(\tau_1 s + 1)/(\tau_2 s + 1)$. Check the steady-state gain of the pulse transfer function.

24.18. Determine the sampled function $f(n\Delta t)$ corresponding to the z-transform

$$F(z) = \frac{0.5z^{-1}}{1 - 1.5z^{-1} + 0.5z^{-2}}$$

Use partial fraction expansion ($\Delta t = 1$) and compare the results with the long division method for the first six sampled values ($n = 0, 1, \ldots, 5$).

24.19. For $G(s) = 1/[(s + 1)(s + 2)]$, obtain $G(z)$ for $\Delta t = 1$. Determine the response to a unit step change in the input. Repeat using Tustin's method (approximate z-transform) and compare the step responses for the first five samples.

24.20. To determine the effects of pole and zero locations, calculate and sketch the unit step responses of the pulse transfer functions shown below for the first six sampling instants, $n = 0$ to $n = 5$. What conclusions can you make concerning the effect of pole and zero locations?

(a) $\dfrac{1}{1 - z^{-1}}$

(b) $\dfrac{1}{1 + 0.7z^{-1}}$

(c) $\dfrac{1}{1 - 0.7z^{-1}}$

(d) $\dfrac{1}{(1 + 0.7z^{-1})(1 - 0.3z^{-1})}$

(e) $\dfrac{1 - 0.5z^{-1}}{(1 + 0.7z^{-1})(1 - 0.3z^{-1})}$

(f) $\dfrac{1 - 0.2z^{-1}}{(1 + 0.6z^{-1})(1 - 0.3z^{-1})}$

24.21. For the transfer functions shown below, determine the corresponding pulse transfer function $HG_p(z)$ for the system and a zero-order hold.

(a) $G_p(s) = \dfrac{1}{(s + 1)^3}$

(b) $G_p(s) = \dfrac{6(1 - s)}{(s + 2)(s + 3)}$

For sampling periods of $\Delta t = 1$ and $\Delta t = 2$, determine whether any poles or zeros of $HG_p(z)$ lie outside the unit circle for either process. Discuss the significance of these results.

— CHAPTER 25 —

Analysis of Sampled-Data Control Systems

The analysis of sampled-data control systems is complicated by the presence of both continuous and discontinuous (discrete) components in the control loop. Just as with continuous systems, a block diagram of the feedback control system can serve as the basis for analyzing the dynamic behavior of a digital system. The main focus of this chapter is the application of block diagram algebra to obtain the closed-loop transfer function and the analysis of the resulting characteristic equation to evaluate control system stability. However, care must be exercised in the development of the closed-loop transfer function. In particular, the rules for block multiplication used for continuous systems must be modified when sampling operations are included. Several techniques based on analysis of the characteristic equation of the closed-loop transfer function are presented which can be used efficiently to determine loop stability. Again, as with continuous systems, the stability of a sampled-data control system is a necessary condition for operability.

25.1 OPEN-LOOP BLOCK DIAGRAM ANALYSIS

If $G(s)$ is the transfer function between a continuous input and continuous output, the situation where both $x(t)$ and $y(t)$ are sampled is shown in Fig. 25.1. If the output is actually not physically sampled, a fictitious sampler can be inserted for purposes of analysis, as shown in Fig. 25.2. This procedure is employed because in z-transform analysis only values of the output at the sampling instants have any validity. It is usually assumed that both samplers act synchronously and have the same sampling period. The sampled input signal $x^*(t)$ (see Eq. 24-1) is a series of impulses. Then $X^*(s)$, the Laplace transform of $x^*(t)$, has been shown in Eq. 24-3 to be

$$X^*(s) = \sum_{n=0}^{\infty} x(n\Delta t)e^{-n\Delta ts} \tag{25-1}$$

or in z-transform notation,

$$X(z) = \sum_{n=0}^{\infty} x(n\Delta t)z^{-n} \tag{25-2}$$

Figure 25.1 Block diagram with sampled input and output signals.

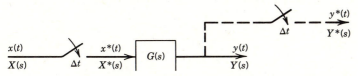

Figure 25.2 Addition of a fictitious sampler.

The sampled input signal passes through the continuous process $G(s)$, yielding a continuous output signal $y(t)$ (or $Y(s)$). Therefore, by the definition of the transfer function,

$$Y(s) = G(s)X^*(s) \tag{25-3}$$

Next consider the sampling of the output signal. As before, we define $Y^*(s)$ as the Laplace transform of $y^*(t)$, the impulse sequence obtained by sampling $y(t)$ (see Eq. 25-1). We will use an asterisk to denote the *star transform,* which indicates the impulse operator acting on the Laplace transform of a signal. It will be applicable to transfer functions as well as inputs and outputs. For example, to find the sampled signal $Y^*(s)$, the star transform is applied to both sides of Eq. 25-3 to obtain

$$Y^*(s) = [G(s)X^*(s)]^* = G^*(s)X^*(s) \tag{25-4}$$

Since $X^*(s)$ is a series of impulses, $G(s)X^*(s)$ represents a sequence of impulse responses for a continuous-time process. The star transform applied to this signal indicates the sampling of $G(s)X^*(s)$ to yield $Y^*(s)$. The proof of (25-4) using frequency-domain analysis has been given by Franklin and Powell [1, p. 86] and is a key result for block diagram analysis of sampled-data control systems. Note also that (25-4) is equivalent to

$$Y(z) = G(z)X(z) \tag{25-5}$$

where $G(z)$ is the pulse transfer function and $z = e^{s\Delta t}$. This relation was derived using a time-domain analysis in Chapter 24 (see Eq. 24-77 and Fig. 24.3).

Pulse Transfer Functions of Systems in Series

Consider the two systems shown in Figs. 25.3 and 25.4 and derive the pulse transfer function $Y(z)/X(z)$ for each system. For Fig. 25.3,

$$Y(s) = G_1(s)G_2(s)X^*(s) \tag{25-6}$$

Consequently, application of the star transform gives

$$Y^*(s) = [G_1(s)G_2(s)]^*X^*(s) \tag{25-7}$$

Since $v(t)$, the output from G_1, is not sampled, the product of G_1G_2 operates on $X^*(s)$ to form $Y^*(s)$. Then from the definition of the z-transform it follows that

$$\frac{Y(z)}{X(z)} = \mathcal{Z}[G_1(s)G_2(s)] = G_1G_2(z) \tag{25-8}$$

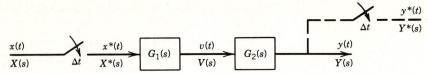

Figure 25.3 Two continuous systems in series with sampled input and output signals.

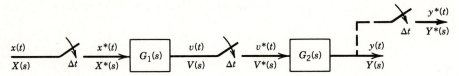

Figure 25.4 Two continuous systems in series with a sampler in-between.

Note that Eq. 25-8 continues use of the notation introduced on pg. 576 that $G_1 G_2(z) \triangleq \mathcal{Z}[G_1(s)G_2(s)]$.[1] It is important to note that in general

$$G_1(z)G_2(z) \neq G_1 G_2(z)$$

A special case of this equation was derived in Eq. 24-107, where G_1 was the zero-order hold.

Now consider the series of blocks shown in Fig. 25.4. Here

$$Y(s) = G_2(s)V^*(s) \tag{25-9}$$

Using Eq. 25-4,

$$V^*(s) = G_1^*(s)X^*(s) \tag{25-10}$$

Combining gives

$$Y(s) = G_2(s)G_1^*(s)X^*(s) \tag{25-11}$$

Next apply the star transform to both sides of the equation:

$$Y^*(s) = G_2^*(s)G_1^*(s)X^*(s) \tag{25-12}$$

or in terms of z-transforms

$$Y(z) = G_2(z)G_1(z)X(z) \tag{25-13}$$

Thus, in this case

$$\frac{Y(z)}{X(z)} = G_2(z)G_1(z) \tag{25-14}$$

The term $G_2(z)G_1(z)$ indicates that the z-transforms of G_1 and G_2 are obtained separately, then multiplied. The result is independent of the order of multiplication.

EXAMPLE 25.1

Show that if $G_1(s)$ and $G_2(s)$ are both first-order models, $G_2(z)G_1(z) \neq G_2 G_1(z)$.

---

[1]More completely, $G_1 G_2(z) \triangleq Z\{\mathcal{L}^{-1}[G_1(s)G_2(s)]\}$

Solution
Referring to Fig. 25.4, let

$$G_1(s) = \frac{K_1}{\tau_1 s + 1} \quad \text{and} \quad G_2(s) = \frac{K_2}{\tau_2 s + 1} \quad (\tau_1 \neq \tau_2) \quad (25\text{-}15)$$

Then by use of Table 24.1, we obtain

$$G_1(z) = \frac{K_1}{(1 - e^{-\Delta t/\tau_1} z^{-1})} \frac{1}{\tau_1} \quad \text{and} \quad G_2(z) = \frac{K_2}{(1 - e^{-\Delta t/\tau_2} z^{-1})} \frac{1}{\tau_2}$$

and

$$G_2(z)G_1(z) = \frac{K_2 K_1}{\tau_2 \tau_1 (1 - e^{-\Delta t/\tau_2} z^{-1})(1 - e^{-\Delta t/\tau_1} z^{-1})} \quad (25\text{-}16)$$

Referring to Fig. 25.3, we find

$$G_2(s)G_1(s) = \frac{K_2 K_1}{(\tau_2 s + 1)(\tau_1 s + 1)}$$

Using Table 24.1 again yields

$$G_2 G_1(z) = \mathcal{Z}\left[\frac{K_2 K_1}{(\tau_2 s + 1)(\tau_1 s + 1)} \right]$$

$$= \frac{K_2 K_1 (e^{-\Delta t/\tau_2} - e^{-\Delta t/\tau_1}) z^{-1}}{(\tau_2 - \tau_1)(1 - e^{-\Delta t/\tau_2} z^{-1})(1 - e^{-\Delta t/\tau_1} z^{-1})} \quad (25\text{-}17)$$

Therefore, by inspection of (25-16) and (25-17),

$$G_2(z)G_1(z) \neq G_2 G_1(z) \quad (25\text{-}18)$$

It should be clear on physical grounds that the two systems in Figs. 25.3 and 25.4 will have different pulse transfer functions betweeen X and Y. In Fig. 25.3 the input to the second transfer function is a continuous signal, $v(t)$, while in Fig. 25.4 the input is a sampled signal, $v^*(t)$, which is a series of pulses. Thus, we would expect the output signals $y(t)$ and probably their sampled counterparts $y^*(t)$ to be quite different.

EXAMPLE 25.2

Examine the influence of the zero-order hold on the continuous output responses $y(t)$ in Figs. 25.3 and 25.4 by letting $G_1(s) = H(s) = (1 - e^{-s\Delta t})/s$ and $G_2(s) = K/(\tau s + 1)$. The input signal $x(t)$ is a unit step input.

Solution
Figure 25.3 represents the normal usage of the ZOH cascaded with a process transfer function. The step input $x(t)$, after being sampled to form a series of unit impulses $x^*(t)$, is then sent through the zero-order hold, which reconstructs the unit step function again. In mathematical terms,

$$V(s) = G_1(s)X^*(s) \quad (25\text{-}19)$$

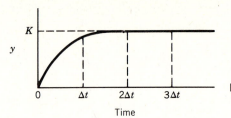

Figure 25.5 Response of a first-order system plus zero-order hold to a train of unit impulses (Example 25.2).

Because $X(z) = 1/(1 - z^{-1})$ and $z^{-1} = e^{-s\Delta t}$, then

$$V(s) = \frac{1 - e^{-s\Delta t}}{s} \frac{1}{1 - e^{-s\Delta t}} = \frac{1}{s} \tag{25-20}$$

Therefore,

$$Y(s) = G_2(s)V(s)$$

$$= \frac{K}{\tau s + 1} \frac{1}{s} \tag{25-21}$$

which is the transform for the step response of a first-order system. The corresponding transient response is shown in Fig. 25.5. In this case the unsampled output is independent of Δt because the input is a step function.

Next apply the same analysis to Fig. 25.4. This type of configuration would never be implemented in a process control application and is presented here only to illustrate the properties of the star transform. Equation 25-11 gives the appropriate expression for $Y(s)$; substituting for G_2, G_1^*, and X^* yields

$$Y(s) = \left(\frac{K}{\tau s + 1}\right)\left(\frac{1 - e^{-s\Delta t}}{s}\right)^* \left(\frac{1}{s}\right)^* \tag{25-22}$$

The sampled version of the hold circuit was calculated in Eq. 24-106, namely

$$H^*(s) = H(z) = 1 \tag{25-23}$$

The sampled step input $(1/s)^*$ is simply a series of unit impulses. Equation 25-22 therefore becomes

$$Y(s) = \frac{K}{\tau s + 1}(1 + e^{-s\Delta t} + e^{-2s\Delta t} + \cdots) \tag{25-24}$$

Inversion of $Y(s)$ to the time domain yields the unit impulse response for the output signal $y(t)$. At each sampling instant the impulse is repeated and causes a sudden increase (jump) in the value of $y(t)$, using the most recent value of $y(t)$ as the initial condition. The resulting time-domain response is shown in Fig. 25.6.

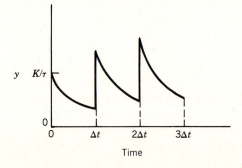

Figure 25.6 Response of a first-order system without hold device to a train of unit impulses (Example 25.2).

EXAMPLE 25.3

To further illustrate the mathematical effect of the sampling operation, compare the responses of the processes depicted in Figs. 25.3 and 25.4 for

$$G_1(s) = \frac{1}{s}$$

$$G_2(s) = \frac{1}{2s + 1}$$

Assume $X(s)$ is a unit step input ($= 1/s$) and ideal samplers are used. Determine the solution $y(t)$ for $0 \leq t \leq 6$.

Solution

In Fig. 25.3, $x^*(t)$ is a series of impulses of unity strength. When this signal passes through $G_1(s)$ (an integrator), the output is a series of steps as shown in Fig. 25.7. This signal $V(s)$ enters $G_2(s)$, a first-order transfer function, yielding $Y_1(s)$; $y_1(t)$ is a series of exponential rises. Figure 25.7 shows that the rises are cumulative, with a change in slope at each sampling instant. On the other hand, if a sampler is placed between $G_1(s)$ and $G_2(s)$, a different pattern emerges, as shown in Fig. 25.8. $V^*(s)$, the sampled signal based on $V(s)$, is a series of impulses in the time domain. When $V^*(s)$ forces $G_2(s)$ to yield $Y_2(s)$, an exponential decay after each impulse is obtained for $y_2(t)$ (see Fig. 25.8), although the initial value of each

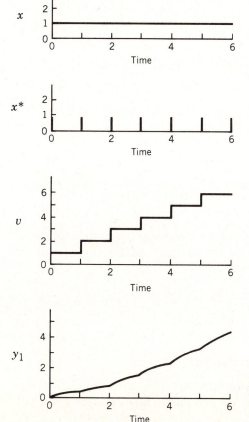

Figure 25.7 Responses for Example 25.3 based on Fig. 25.3.

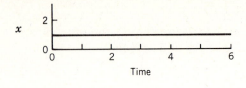

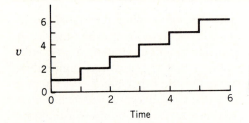

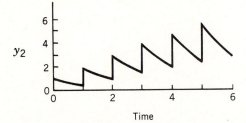

Figure 25.8 Responses for Example 25.3 based on Fig. 25.4.

decay transient grows in a monotonic fashion. The calculation of each response is left as an exercise for the reader. Note that the sampled values of $y(t)$ will be different for each case. For example, at $t = 5$,

$$y_1^*(5) = 5.91 \qquad \text{(Fig. 25.7)}$$

$$y_2^*(5) = 6.00 \qquad \text{(Fig. 25.8)}$$

Pulse Transfer Functions of Systems in Parallel

Consider the block diagrams in Figs. 25.9 and 25.10. Note that in Fig. 25.10 the sampler is not present in the lower branch. We will derive relations between $C(z)$

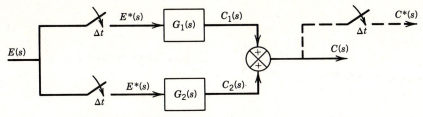

Figure 25.9 Addition in a block diagram with two samplers.

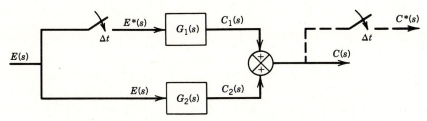

Figure 25.10 Addition in a block diagram with one sampler removed.

and $E(z)$ for each diagram. From Fig. 25.9,

$$C(s) = C_1(s) + C_2(s) \tag{25-25}$$

$$= G_1(s)E^*(s) + G_2(s)E^*(s) \tag{25-26}$$

$$= [G_1(s) + G_2(s)]E^*(s) \tag{25-27}$$

To find the sampled output signal, take the star transform:

$$C^*(s) = [G_1^*(s) + G_2^*(s)]E^*(s) \tag{25-28}$$

In terms of z-transforms,

$$C(z) = [G_1(z) + G_2(z)]E(z) \tag{25-29}$$

Hence, the overall pulse transfer function between C and E is

$$\frac{C(z)}{E(z)} = G_1(z) + G_2(z) \tag{25-30}$$

Now suppose that one sampler is removed, as shown in Fig. 25.10,

$$C(s) = C_1(s) + C_2(s) \tag{25-31}$$

$$= G_1(s)E^*(s) + G_2(s)E(s) \tag{25-32}$$

Thus,

$$C^*(s) = G_1^*(s)E^*(s) + [G_2(s)E(s)]^* \tag{25-33}$$

Expressing the relation in z-transforms yields

$$C(z) = G_1(z)E(z) + G_2E(z) \tag{25-34}$$

Note that Eq. 25-34 implies that a pulse transfer function of the form $C(z)/E(z)$ cannot be obtained for the system in Fig. 25.10. In contrast, the system in Fig. 25.9 does possess a pulse transfer function as defined in Equation 25-30. Because of the useful properties of transfer functions, the arrangement in Fig. 25.9 is preferred for analysis. However, Fig. 25.10 is applicable to one important case in digital process control, namely load changes, where the continuous load affects the process but is not sampled. This topic is discussed next.

25.2 DEVELOPMENT OF CLOSED-LOOP TRANSFER FUNCTIONS

Using the block multiplication and addition rules given above, we can analyze the block diagram of a sampled-data feedback control system, such as that shown in Fig. 25.11. This diagram is a simpler version of Fig. 21.7. In this diagram the Laplace transform variable, s, is used to denote continuous signals and transfer functions, and the z-transform variable z indicates the sampled representation of the signals as well as pulse transfer functions. Figure 25.11 is analogous to the continuous-time block diagram for feedback control, Fig. 10.8.

The discrete-time block labeled $D(z)$ represents the pulse transfer function for the digital controller. The controller output passes through the zero-order hold $H(s)$ before entering the process. This signal is the input to the final control element, possibly preceded by a transducer, which are combined in the block G_v in Fig. 25.11. Other arrangements are possible, such as the stepping motor-valve combination discussed in Chapter 21. The error signal $E(z)$ is the difference between the sampled set point, $\tilde{R}(z)$ and the sampled measurement $B(z)$. The closed-loop transfer function $C(z)/R(z)$ relates sampled values of the controlled variable $C(z)$ to the sampled values of the set point $R(z)$. Measurement gain K_m converts the physical set point R to an internal set point $\tilde{R}$, as discussed in Chapter 10. In other words, $\tilde{R}(z) = K_m R(z)$. Since $c(t)$ is a continuous signal that may not be actually sampled, a fictitious sampler is shown for the purpose of discrete-time analysis. The analysis of the closed-loop system given below mainly utilizes Laplace transforms, followed by conversion to z-transforms.

Set-Point Change

Consider first the servomechanism case where $L(s) = 0$. From Fig. 25.11 the measurement signal is related to P^* by

$$B(s) = H(s)G_v(s)G_p(s)G_m(s)P^*(s) \tag{25-35}$$

Therefore, $B^*(s)$ can be found by applying the star transform:

$$B^*(s) = [HG_vG_pG_m]^* P^*(s) \tag{25-36}$$

where $[HG_vG_pG_m]^*$ denotes the star transform of the product $H(s)G_v(s)G_p(s)G_m(s)$. The digital controller output $P^*(s)$ is related to $E^*(s)$ by

$$P^*(s) = D^*(s)E^*(s) \tag{25-37}$$

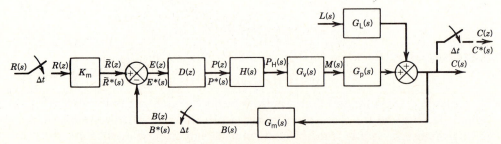

Figure 25.11 Block diagram of a general digital feedback control system.

At the comparator, the error is a sampled signal:

$$E^*(s) = \tilde{R}^*(s) - B^*(s) \tag{25-38}$$

Now substitute (25-38) for $E^*(s)$ in (25-37) and then replace $B^*(s)$ by (25-36):

$$P^*(s) = D^*(s)[\tilde{R}^*(s) - B^*(s)] \tag{25-39}$$

$$= D^*(s)[\tilde{R}^*(s) - [HG_vG_pG_m]^*P^*(s)] \tag{25-40}$$

Rearranging and solving for $P^*(s)$ gives

$$P^*(s) = \frac{D^*(s)\tilde{R}^*(s)}{1 + [HG_vG_pG_m]^*D^*(s)} \tag{25-41}$$

Recognize that $C(s)$ is related to $P^*(s)$ by

$$C(s) = H(s)G_v(s)G_p(s)P^*(s) \tag{25-42}$$

Using the star transform for the fictitious sampler,

$$C^*(s) = [HG_vG_p]^*P^*(s) \tag{25-43}$$

Therefore,

$$C^*(s) = \frac{[HG_vG_p]^*D^*(s)\tilde{R}^*(s)}{1 + [HG_vG_pG_m]^*D^*(s)} \tag{25-44}$$

Substituting for the setpoint, $\tilde{R}^*(s) = K_mR^*(s)$, we obtain the closed-loop transfer function:

$$\frac{C^*(s)}{R^*(s)} = \frac{[HG_vG_p]^*D^*(s)K_m}{1 + [HG_vG_pG_m]^*D^*(s)} \tag{25-45}$$

In z-transform notation,

$$\frac{C(z)}{R(z)} = \frac{HG_vG_p(z)K_mD(z)}{1 + HG_vG_pG_m(z)D(z)} \tag{25-46}$$

The characteristic equation for the closed-loop control system is obtained by setting the denominator of (25-46) equal to zero:

$$1 + HG(z)D(z) = 0 \tag{25-47}$$

where $HG(z)$ is defined as $HG_vG_pG_m(z)$. The roots of the characteristic equation determine the stability of the sampled-data control system, as discussed in Sect. 25.3.

A special case of (25-46) occurs when $G_m(s) = K_m$. Because the z-transform is a linear operator, the constant can be simply incorporated in it. Then

$$\frac{C(z)}{R(z)} = \frac{HG_vG_p(z)K_mD(z)}{1 + HG_vG_p(z)K_mD(z)} = \frac{HG(z)D(z)}{1 + HG(z)D(z)} \tag{25-48}$$

where $HG(z)$ is $HG_vG_pK_m(z)$. Note the similarity to the closed-loop transfer function defined for continuous-time systems in Chapter 10, Eq. 10-37.

Load Change

The regulator case (where $R(s) = 0$) can be derived using a procedure similar to the one above. For simplicity consider the case when $G_m(s) = K_m$. Equation

25-35 must be modified to include the effect of the load change:

$$B(s) = H(s)G_v(s)G_p(s)K_mP^*(s) + G_L(s)K_mL(s) \tag{25-49}$$

Applying the star transform yields

$$B^*(s) = K_m[HG_vG_p]^*P^*(s) + K_m[G_LL]^* \tag{25-50}$$

Because

$$C(s) = H(s)G_v(s)G_p(s)P^*(s) + G_L(s)L(s) \tag{25-51}$$

then

$$C^*(s) = [HG_vG_p]^*P^*(s) + [G_LL]^* \tag{25-52}$$

Since $P^*(s) = -D^*(s)B^*(s)$, we can multiply (25-50) by $-D^*(s)$ and solve for $P^*(s)$. Next, substitute that result for $P^*(s)$ in (25-52), which yields after rearrangement

$$C^*(s) = \frac{G_LL^*(s)}{1 + K_mHG_vG_p^*(s)D^*(s)} \tag{25-53}$$

In terms of z-transforms,

$$C(z) = \frac{G_LL(z)}{1 + K_mHG_vG_p(z)D(z)} = \frac{G_LL(z)}{1 + HG(z)D(z)} \tag{25-54}$$

Because the disturbance input is *not* a sampled signal, we are unable to define a closed-loop transfer function for the discrete-time case. The term $G_LL(z)$ represents the z-transform of the output from $G_L(s)$, caused by a continuous input $L(s)$ (note that $G_LL(z) \neq G_L(z)L(z)$). If $L(s)$ is specified to be a particular input (e.g., $L(s) = 1/s$), then $C(z)$ can be determined. Note, however, that the characteristic equation for (25-54) is the same as (25-48); thus the same stability analysis applies to both set-point and load changes.

EXAMPLE 25.4

For the system in Fig. 25.11, let

$$G_p(s) = \frac{K_pe^{-\theta s}}{\tau_ps + 1} \quad G_L(s) = \frac{K_L}{\tau_Ls + 1} \quad (\tau_L = \tau_p = \tau)$$

$$D(z) = K_c \quad H(s) = \frac{1 - e^{-s\Delta t}}{s} \quad G_m = 1 \quad G_v = 1 \tag{25-55}$$

where $\theta = N\Delta t$ and N is an integer. Derive $C(z)$ for a unit step change in load when the set point is held constant, that is, $L(s) = 1/s$ and $R(s) = 0$. Identify the characteristic equation.

Solution

To calculate the step response, derive the z-transforms indicated in Eq. 25-54. Thus,

$$\mathcal{Z}[G_L(s)L(s)] = \mathcal{Z}\left(\frac{K_L}{\tau s + 1}\frac{1}{s}\right) = \mathcal{Z}\left(\frac{K_L}{s} - \frac{K_L}{s + 1/\tau}\right) \tag{25-56}$$

Using Table 24.1,

$$\mathfrak{z}[G_L(s)L(s)] = K_L\left(\frac{1}{1 - z^{-1}} - \frac{1}{1 - az^{-1}}\right) = \frac{K_L(1 - a)z^{-1}}{(1 - z^{-1})(1 - az^{-1})} \quad (25\text{-}57)$$

where $a = e^{-\Delta t/\tau}$.

Similarly,

$$\mathfrak{z}[H(s)G_p(s)] = \mathfrak{z}\left(\frac{1 - e^{-s\Delta t}}{s} \frac{K_p}{\tau s + 1} e^{-\theta s}\right) \quad (25\text{-}58)$$

$$= K_p z^{-N}(1 - z^{-1})\mathfrak{z}\left[\frac{1}{s(\tau s + 1)}\right] \quad (25\text{-}59)$$

$$= K_p z^{-N}(1 - z^{-1})\frac{(1 - a)z^{-1}}{(1 - z^{-1})(1 - az^{-1})} = K_p z^{-N-1}\frac{1 - a}{1 - az^{-1}} \quad (25\text{-}60)$$

Substituting (25-57) and (25-60) into (25-54) with $D(z) = K_c$, and $K_m = 1$ gives

$$C(z) = \frac{\dfrac{K_L(1 - a)z^{-1}}{(1 - z^{-1})(1 - az^{-1})}}{1 + \dfrac{z^{-N-1}K_cK_p(1 - a)}{1 - az^{-1}}} \quad (25\text{-}61)$$

After rearrangement,

$$C(z) = \frac{K_L(1 - a)z^{-1}}{1 - (1 + a)z^{-1} + az^{-2} + K_cK_p(1 - a)z^{-N-1} - K_cK_p(1 - a)z^{-N-2}} \quad (25\text{-}62)$$

The characteristic equation is obtained by setting the denominator polynomial in (25-62) equal to zero. Note that the order of the characteristic equation depends on the magnitude of the time delay N. For example,

N	*Order of Polynomial*
0	2
1	3
2	4
etc.	

Because N is assumed to be an integer, the time-delay term merely increases the order of the polynomial. Contrast this situation to that for a continuous-time closed-loop system, where a time delay yields a transcendental function in the denominator (see Chapter 11). The implication here is that the number of roots of Eq. 25-62 is bounded (equal to the order of the polynomial), while such is not the case for a continuous-time system. In addition, special techniques exist for analyzing such polynomial equations, as is discussed in the subsequent section.

EXAMPLE 25.5

Assuming proportional control of a first-order process, determine the closed-loop transfer function for a set-point change, and evaluate the effect of K_c on the response. Let $K_m = 1$, $G_v = 1$, $G_p = K_p/(\tau_p s + 1)$, and $D = K_c$. What is the final value of $c(n\Delta t)$ for $R(z) = 1/(1 - z^{-1})$, a unit step change in set point?

Solution

The process transfer function is

$$HG(z) = \frac{K_p(1 - a)z^{-1}}{1 - az^{-1}} \qquad (25\text{-}63)$$

where $a = e^{-\Delta t/\tau_p}$. The closed-loop transfer function from (25-47) is

$$\frac{C(z)}{R(z)} = \frac{\dfrac{K_c K_p(1 - a)z^{-1}}{1 - az^{-1}}}{1 + \dfrac{K_c K_p(1 - a)z^{-1}}{1 - az^{-1}}} \qquad (25\text{-}64)$$

Rearranging yields

$$\frac{C(z)}{R(z)} = \frac{K_c K_p(1 - a)z^{-1}}{1 + [K_c K_p(1 - a) - a]z^{-1}} \qquad (25\text{-}65)$$

If the control system is to be stable, the root of the denominator polynomial must have an absolute value less than unity (see Fig. 24.2). Otherwise $C(z)$ will be unstable for a change in $R(z)$. If the term $[K_c K_p(1 - a) - a]$ is less than one, this absolute value condition is satisfied. Solving for $K_c K_p$ gives,

$$K_c K_p < \frac{1 + a}{1 - a} \qquad (25\text{-}66)$$

Therefore, for digital control, the maximum value of K_c is a function of K_p and a; hence, it is a function of the sampling period. As Δt approaches zero, $a \to 1$, and the ratio $(1 + a)/(1 - a)$ increases, thus increasing the maximum gain for stability. In contrast to this situation, recall that for a continuous first-order process under proportional feedback control, K_c can be infinite without causing instability. Thus for digital control the sampling period Δt must be selected judiciously so that the control loop remains stable, based on the inequality (25-66).

For a stable closed-loop system the final value $c(n\Delta t)$ for a unit step change in R can be calculated by the Final Value Theorem, Eq. 25-21. Substituting into Eq. 25-65 and simplifying, we obtain

$$\lim_{z \to 1} (1 - z^{-1})C(z) = \frac{K_c K_p(1 - a)}{1 + K_c K_p(1 - a) - a} = \frac{K_c K_p}{K_c K_p + 1} \qquad (25\text{-}67)$$

Thus for a proportional controller the same offset occurs for both continuous-time and digital systems.

25.3 STABILITY OF SAMPLED-DATA CONTROL SYSTEMS

For a satisfactory control system design, the closed-loop system must be stable. By analogy to Section 11.1, the following definition of stability is normally used for sampled-data systems [2]:

> **Definition.** *A linear sampled-data system is stable if the output sequence $\{y(n\Delta t)\}$ is bounded for any bounded input sequence $\{x(n\Delta t)\}$. Otherwise the system is said to be unstable.*

Note that this definition contains no mention of the process response during the intersample periods. Specifically, it is theoretically possible to have a system in which the continuous output is unbounded, even though a sampled sequence of the output remains bounded. An example of such an output is shown in Fig. 25.12. This type of *hidden oscillation* is rare in practice and can be readily detected by changing the sampling period or by using modified z-transforms [3].

The necessary and sufficient conditions for the stability of a linear sampled-data system can be expressed in two ways:

1. $\sum\limits_{n=0}^{\infty} |g(n\Delta t)| < \infty$, or

2. $G(z)$ has no poles on or outside the unit circle in the z-plane.

In the above conditions $G(z)$ is the pulse transfer function of the process and $\{g(n\Delta t)\}$ is the set of samples of the impulse response. The first condition implies that the impulse response $\{g(n\Delta t)\}$ will be bounded for the total time span ($n = 0$ to ∞). Long division can be used to calculate the value of $\sum_{n=0}^{\infty} g(n\Delta t)$ but this may be time-consuming. It is preferable to use the second condition, which is related to the partial fraction expansion procedure discussed in Chapter 24.

Recall in Eq. 24-49 we showed how the inverse z-transform depends on the poles of the transform ($p_1, p_2, \ldots, p_m$). For a pulse transfer function $G(z)$ and input $X(z)$, the output signal can be calculated:

$$Y(z) = G(z)X(z) = \frac{b_0 + b_1 z^{-1} + b_2 z^{-2} + \cdots + b_k z^{-k}}{(1 - p_1 z^{-1})(1 - p_2 z^{-1}) \cdots (1 - p_m z^{-1})} X(z) \quad (25\text{-}68)$$

Then, using partial fraction expansion (distinct roots are assumed),

$$Y(z) = \frac{r_1}{1 - p_1 z^{-1}} + \frac{r_2}{1 - p_2 z^{-1}} + \cdots + \frac{r_m}{1 - p_m z^{-1}}$$
$$+ [\text{contribution of input terms}] \quad (25\text{-}69)$$

where r_i = partial fraction coefficient (residue). In a closed-loop system, the terms shown in Eq. 25-69 include the influence of the poles of the closed-loop transfer function, plus the contribution of the set-point and/or disturbance inputs (in brackets). The input terms appear separately from the transfer function terms in the partial fraction expansion and are assumed to be bounded. Hence, the stability of the loop can be determined by the transfer function denominator terms. As discussed in Chapter 24, the sampled output response corresponding to (25-69) is

$$c(n\Delta t) = r_1(p_1)^n + r_2(p_2)^n + \cdots + r_m(p_m)^n + [\text{input terms}] \quad (25\text{-}70)$$

If the absolute value of any of the poles p_i is greater than unity, $c(n\Delta t) \to \infty$ as $n \to \infty$, thus violating the condition for stability. This is also true if the root is a complex number (in polar form, if $p = |p|e^{j\omega}$, then $p^n = |p|^n e^{jn\omega}$).

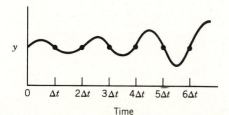

y

0   Δt $2\Delta t$ $3\Delta t$ $4\Delta t$ $5\Delta t$ $6\Delta t$

Time

Figure 25.12 A hidden oscillation.

Stability Regions in the s and z Planes

For both continuous-time and discrete-time systems, the condition of stability is determined by the roots of the characteristic equation (i.e., the system poles). For continuous systems the roots of the characteristic equation must have negative real parts. Graphically, the roots must lie in the left half of the s plane, shown in Fig. 25.13. Since

$$z = e^{s\Delta t} \tag{25-71}$$

we can map the stability region in s onto the complex z plane. For example, Fig. 24.6 showed the relation between poles in the s and z planes for several specific cases.

Because the s variable is a complex number we can write $s = \alpha + j\beta$. Substituting into Eq. 25-71 gives

$$z = e^{\Delta t(\alpha + j\beta)} = e^{\alpha\Delta t}e^{j\beta\Delta t} \tag{25-72}$$

$$= e^{\alpha\Delta t}(\cos \beta\Delta t + j \sin \beta\Delta t) \tag{25-73}$$

Thus a real value of s ($\beta = 0$) maps to a real value of z. Similarly a complex value of s maps to a complex point in the z plane, although sometimes two complex poles in the s plane map to a single value in the z plane due to aliasing (cf. Fig. 24.6). From (25-73) the magnitude of z is

$$|z| = e^{\alpha\Delta t} \tag{25-74}$$

$$\text{and} \quad \begin{array}{l} e^{\alpha\Delta t} > 1 \text{ for } \alpha > 0 \\ e^{\alpha\Delta t} = 1 \text{ for } \alpha = 0 \\ e^{\alpha\Delta t} < 1 \text{ for } \alpha < 0 \end{array}$$

Hence, if a root of a continuous-time system lies in the left half of the s plane, then $\alpha < 0$ and the corresponding root in the z plane is less than unity in magnitude, that is, $|z| < 1$. Graphically, the region of stability in the z plane is the interior of the unit circle, since the left half of the s plane maps into the inside of the unit circle of the z plane.

Stability Tests

The roots of the characteristic equation can be computed using a nonlinear equation solver to determine if a pulse transfer function is stable. Alternatively, the frequency

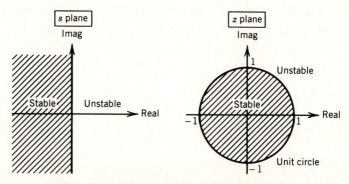

Figure 25.13 Stability regions in the s and z planes.

response of the pulse transfer function can be obtained ($z = e^{j\omega\Delta t}$), allowing Bode or Nyquist plots to be developed [2,3]. Such methods can be used for stability determination but do not readily lend themselves to traditional controller design (e.g., gain and phase margins). Either of these approaches can be time-consuming for high-order process models. Hence alternative techniques may be preferred for analyzing the stability of the characteristic equation. These techniques include

1. Modified Routh stability test
2. Jury's test
3. Schur-Cohn test

Modified Routh Stability Criterion.   In Chapter 11, the Routh stability test was used to analyze the characteristic equation of a continuous-time system for unstable roots, those roots that lie in the right half of the complex s plane. For sampled-data systems, this test can also be applied to determine whether any roots of the characteristic equation (in z) lie outside the unit circle. If a simple rational transformation from z to s can be found that maps the interior of the unit circle into the left half of the complex plane, then the Routh criterion can be used directly with z-transforms. Such a transformation (or mapping) is provided by the *bilinear transformation*

$$z = \frac{1 + w}{1 - w} \tag{25-75}$$

This complex mapping or transformation does not correspond to that for the original z-transform. It only approximates the original transformation $z = e^{s\Delta t}$, hence the use of w in place of s. However, the boundary separating the stable and unstable regions does map exactly [3,4] and consequently the stability determination is exact.

To apply this stability test, first determine the characteristic equation of the sampled-data system (written as positive powers of z):

$$\Gamma(z) = a_n z^n + a_{n-1} z^{n-1} + \cdots + a_1 z + a_0 = 0 \tag{25-76}$$

Note that if the time delay is an integer multiple of Δt, a polynomial in z will always result (see Example 25.4). Using the bilinear transformation, the characteristic equation is transformed to a function of w:

$$\Gamma(w) = \bar{a}_n w^n + \bar{a}_{n-1} w^{n-1} + \cdots + \bar{a}_1 w + \bar{a}_0 = 0 \tag{25-77}$$

also yielding a polynomial in w, where $\bar{a}_i$ = real constant coefficient ($i = 0, 1, \ldots, n$). Note that the $\bar{a}_i$ are not necessarily equal to the a_i.

The Routh test can then be applied directly to Eq. 25-77 to determine the number of roots that lie in the unstable right half of the w plane or, equivalently, how many roots of $\Gamma(z)$ lie outside the unit circle.

EXAMPLE 25.6

Consider the characteristic equation

$$\Gamma(z) = 2z^3 + z^2 + z + 1 = 0 \tag{25-78}$$

Use the bilinear transformation to check stability.

Solution

Substitution of the bilinear transformation (25-75) into (25-78) yields

$$\Gamma(w) = 2 \left(\frac{1+w}{1-w} \right)^3 + \left(\frac{1+w}{1-w} \right)^2 + \left(\frac{1+w}{1-w} \right) + 1 = 0 \qquad (25\text{-}79)$$

After some algebraic manipulation, Eq. 25-79 becomes

$$w^3 + 7w^2 + 3w + 5 = 0 \qquad (25\text{-}80)$$

Examination of (25-80) shows that the necessary condition for stability is satisfied because all coefficients are positive. The corresponding Routh array is

Row 1	1	3
Row 2	7	5
Row 3	16/7	0
Row 4	5	

Since no sign changes occur in column one, the system is stable. Equation 25-78 has actual roots of -0.739 and $0.119 \pm 0.814j$, all of which lie inside the unit circle.

Jury's Stability Criteria.

Jury's criteria provide a simple analytic test for stability of a closed-loop system. These criteria can be applied directly to a z polynomial and thus avoid application of the bilinear transform. Jury's method yields necessary and sufficient conditions for a polynomial in z with real coefficients to have all roots within the unit circle in the z plane. However, the number of unstable roots cannot be determined.

Consider the characteristic equation in (25-76). If a_n, a_{n-1}, a_{n-2}, . . . , a_1, a_0 are real, constant coefficients and a_n is positive, the Jury array is tabulated as shown in Table 25.1. The original coefficients $\{a_i\}$ are written in order of increasing subscript in the first row with the order reversed for the second row. Each succeeding pair of rows is calculated from the determinant relationships of Table 25.1, and has one less element than the previous pair. Additional rows are computed until $2n - 3$ rows are obtained. The last row contains three elements, s_0, s_1, s_2.

The Jury criteria state that the necessary and sufficient conditions for the roots of Eq. 25-76 to lie within the unit circle in the z plane are

$$\Gamma(z = 1) > 0$$
$$\Gamma(z = -1) > 0 \text{ for } n \text{ even} \qquad (25\text{-}81)$$
$$\Gamma(z = -1) < 0 \text{ for } n \text{ odd}$$

and

$$\left. \begin{array}{l} |a_0| < a_n \\ |b_0| > |b_{n-1}| \\ |c_0| > |c_{n-2}| \\ |d_0| > |d_{n-3}| \\ \qquad \cdot \\ \qquad \cdot \\ \qquad \cdot \\ |s_0| > |s_2| \end{array} \right\} \qquad n - 1 \text{ constraints} \qquad (25\text{-}82)$$

A proof has been given by Jury and Blanchard [5].

Note that the first set of conditions, Eq. 25-81, should be checked before the Jury array is constructed. If (25-81) is not satisfied, the characteristic equation has

Table 25.1   General Structure of the Jury Array

Row	z^0	z^1	z^2	$\cdots$	z^{n-2}	z^{n-1}	z^n
1	a_0	a_1	a_2	$\cdots$	a_{n-2}	a_{n-1}	a_n
2	a_n	a_{n-1}	a_{n-2}	$\cdots$	a_2	a_1	a_0
3	b_0	b_1	b_2	$\cdots$	$\cdots$	b_{n-1}	
4	b_{n-1}	b_{n-2}	b_{n-3}	$\cdots$	$\cdots$	b_0	
5	c_0	c_1	c_2	$\cdots$	c_{n-2}		
6	c_{n-2}	c_{n-3}	c_{n-4}	$\cdots$	c_0		
$\vdots$	$\vdots$	$\vdots$	$\vdots$	$\vdots$	$\vdots$		
$2n-5$	r_0	r_1	r_2	r_3			
$2n-4$	r_3	r_2	r_1	r_0			
$2n-3$	s_0	s_1	s_2				

$$b_k = \begin{vmatrix} a_0 & a_{n-k} \\ a_n & a_k \end{vmatrix} \qquad c_k = \begin{vmatrix} b_0 & b_{n-1-k} \\ b_{n-1} & b_k \end{vmatrix} \qquad d_k = \begin{vmatrix} c_0 & c_{n-2-k} \\ c_{n-2} & c_k \end{vmatrix}$$

$$s_0 = \begin{vmatrix} r_0 & r_3 \\ r_3 & r_0 \end{vmatrix} \qquad s_1 = \begin{vmatrix} r_0 & r_2 \\ r_3 & r_1 \end{vmatrix} \qquad s_2 = \begin{vmatrix} r_0 & r_1 \\ r_3 & r_2 \end{vmatrix}$$

roots outside the unit circle and the Jury array need not be calculated. If Eq. 25-81 is satisfied, the conditions in Eq. 25-82 must be investigated. If an inequality is violated, the rest of the array calculations are unnecessary.

EXAMPLE 25.7

Using the Jury stability criteria, determine if any roots of the characteristic equation

$$\Gamma(z) = 2z^4 - 3z^3 + 2z^2 - z + 1 = 0 \qquad (25\text{-}83)$$

lie in the unstable region.

Solution
The conditions of Eq. 25-81 are satisfied because

$$\Gamma(1) = 1 > 0$$
$$\Gamma(-1) = 9 > 0$$

Form the Jury array:

Row	z^0	z^1	z^2	z^3	z^4
1	1	-1	2	-3	2
2	2	-3	2	-1	1
3	-3	5	-2	-1	
4	-1	-2	5	-3	
5	8	-17	11		

Applicable conditions of (25-82) are

$$a_0 < a_4 \Rightarrow 1 < 2$$
$$b_0 > b_3 \Rightarrow 3 > 1$$
$$c_0 > c_2 \Rightarrow 8 \not> 11 \text{ (violation)}$$

Thus, the system is unstable. By factoring, the roots are $-0.152 \pm 0.661j$ and $0.902 \pm 0.523j$. Note that the second set of roots lies outside the unit circle.

Jury's stability test is subject to special cases similar to those that develop in the application of the Routh criteria. In the tabulation of the Jury array, a singular case can arise by either having the first and last elements of a row be zero or having a complete row of zeros. In either case, the tabulation terminates prematurely. Calculations for singular cases have been described by Ogata[2].

Schur-Cohn Criteria. While similar to Jury's criteria, the Schur-Cohn stability test is not attractive since roughly twice as many determinants must be calculated. Reference 2 contains more details on this method.

The Routh and Jury tests are attractive because they are analytical in nature, allowing one to calculate stability regions for an unspecified controller gain. However, if only the roots of the characteristic equation are needed (as in Examples 25.6 and 25.7), using a root-finding computer program would be more efficient.

SUMMARY

We have developed in this chapter the key equations for design of sampled-data control systems, namely the closed-loop transfer functions for both set-point and load changes and the characteristic equation. The characteristic equation can be used to perform a stability analysis of the digital control system, using several different methods. However, as with continuous-time systems, this result represents only a limit against which candidate controllers can be checked. The actual selection of controller settings is based on performance characteristics of the closed-loop system response, as discussed in the next chapter. Hence, stability is only one consideration, although it is an important one.

REFERENCES

1. Franklin, G. F., and J. D. Powell, *Digital Control of Dynamic Systems,* Addison-Wesley, Reading, MA, 1980.
2. Ogata, K., *Discrete-Time Control Systems,* Prentice-Hall, Englewood Cliffs, NJ, 1987.
3. Luyben, W. L., *Process Modeling, Simulation, and Control for Chemical Engineers,* McGraw-Hill, New York, 1973.
4. Saucedo, R., and E. E. Schiring, *Introduction to Continuous and Digital Control Systems,* Macmillan, New York, 1968.
5. Jury, E. I., and J. Blanchard, A Stability Test for Linear Discrete Systems in Table Form, *IRE Proc.* **49,** 1947 (1961).

EXERCISES

25.1. To determine the effect of pole and zero locations on the performance characteristics of digital controllers, calculate and sketch the responses of the following controller transfer functions to a unit step change in the error signal $e(n\Delta t)$. What conclusions can you make concerning the effect of pole–zero locations?

(a) $D(z) = \dfrac{1}{1 + z^{-1}}$

(b) $D(z) = \dfrac{1}{1 + 0.4z^{-1}}$

(c) $D(z) = \dfrac{1}{1 - 0.4z^{-1}}$

(d) $D(z) = \dfrac{1 + 0.4z^{-1}}{1 - 0.4z^{-1}}$

(e) $D(z) = \dfrac{1}{z^2 - 0.8z + 0.25}$

(f) $D(z) = \dfrac{1}{z^2 + 0.8z + 0.25}$

25.2. Derive an expression for the response $y(n\Delta t)$ of each system shown below to a unit step change in $x(t)$. Are the responses the same?

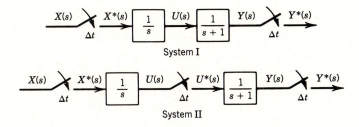

25.3. For the blending system shown in the drawing, it is desired to control exit concentration c_3 by adjusting flow rate q_2. Composition c_2 is the primary load variable. Using the information given below, do the following:

(a) Draw a block diagram for the controlled system. The symbols in the drawing should be used in the block diagram as much as possible.

(b) Derive (or state) the transfer function for each block in the block diagram.

(c) Derive an expression for the closed-loop transfer function, $C_3(z)/R(z)$. It is not necessary to substitute individual transfer functions.

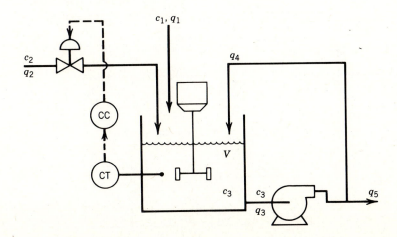

Available Information

i. The volume of liquid in the tank can be assumed to be constant since flow rates q_1, q_3, q_4, and q_5 are constant and q_2 is quite small.

ii. The density of all streams is constant at 90 lb/ft^3.

iii. The recycle line is 68.8 ft long and has an inside diameter of 4 in.

iv. The concentration transmitter (CT) has negligible dynamics. The transmitter output changes linearly from 3 to 15 psi as the concentration changes from 0 to 40 lb solute/ft^3.

v. The concentration controller (CC) is a digital PI controller with a gain of 0.5 and a reset time of 5 min. The controller output goes to a zero-order hold which outputs a signal to the control valve. The sampling period is 12 s.

vi. The steady-state characteristic of the control valve is given by

$$q_2 = 4 - 1.15 \sqrt{p - 3} \quad \text{ft}^3/\text{min}$$

where p is the controller output signal in psi. After a sudden change in p, the flow through the valve q_2 reaches a new steady-state value in 32 s (after five time constants).

vii. The tank is 6 ft in diameter and is perfectly mixed.

viii. The nominal steady-state values are:

$$\bar{q}_1 = 50 \text{ ft}^3/\text{min} \qquad \bar{q}_3 = 82 \text{ ft}^3/\text{min}$$

$$\bar{q}_2 = 2 \text{ ft}^3/\text{min} \qquad V = 100 \text{ ft}^3$$

$$\bar{c}_1 = 0.5 \text{ lb solute/ft}^3 \qquad \bar{c}_2 = 80 \text{ lb solute/ft}^3$$

ix. Feed composition c_1 is constant.

25.4. An irreversible, isothermal chemical reaction $2A \rightarrow B$ is carried out in the continuous stirred-tank reactor shown in the drawing. Since reactor temperature is maintained constant, only the product concentration c_3 need be controlled. This composition is sampled automatically every two minutes and the sample analyzed by a gas chromatograph (GC). The GC retention time of 2 min is essentially a pure time delay. The digital output signal from the GC is then sent to a digital PI controller which adjusts the feed composition c_1 via a transducer and a valve. All flow rates are kept constant including the recycle flow rate q_2.

(a) Draw a block diagram for the closed-loop system.

(b) Derive a transfer function for each block using the information given below.

(c) Derive an expression for the closed-loop transfer function, $C_3(z)/R(z)$.

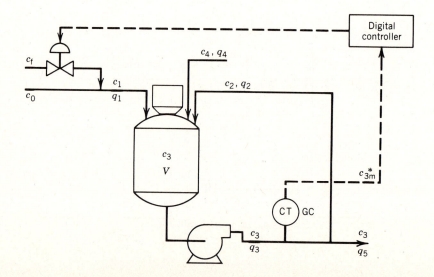

Available Information

i. Nominal steady-state values:

$$q_1 = 10 \text{ ft}^3/\text{min}, \, q_2 = 5 \text{ ft}^3/\text{min}, \, q_4 = 20 \text{ ft}^3/\text{min}$$

$$V = 150 \text{ ft}^3, \, \bar{c}_1 = 0.3 \text{ lb-mole A/ft}^3$$

$$\bar{c}_4 = 0.2 \text{ lb-mole A/ft}^3$$

ii. The rate of reaction is given by

$$r \, (\text{lb-mole A/ft}^3/\text{min}) = 15.4 \, c_3^2$$

where c_3 denotes the concentration of A in the reactor in lb-mole/ft^3.

iii. The transport-velocity lag (i.e., time delay) associated with the recycle line is 2 min.

iv. The GC sample valve is located a few feet from the reactor. The sample loop dynamics are negligible but the GC cycle time is 2 min.

v. The control valve–transducer combination provides the following steady-state relation between controller output p (mA) and feed composition c_1 (mol A/ft^3):

$$c_1 = 0.125 \sqrt{p - 4}$$

After a step change in p occurs, c_1 essentially reaches the new steady-state value in 32 s (five time constants).

vi. The output signal from the GC varies linearly from 4 to 20 mA as the measured composition changes from 0 to 0.3 lb-mole A/ft^3.

vii. The reactor can be assumed to be perfectly mixed.

viii. The density of each liquid stream is approximately the same.

25.5. A distillation column has an approximate transfer function between bottoms composition x_B and steam flow rate to the reboiler q_s of

$$G_p = \frac{0.1}{(3s + 1)^2}$$

A proportional digital controller is used to control x_B by adjusting q_s. A gas chromatograph with a time delay equal to the sampling period Δt is used to measure x_B. Find the ultimate gain of the system for values of $\Delta t = 1, 5, 10,$ and 20 min. Assume that $G_v = G_m = 1$. The process time constant is in minutes.

25.6. Product from a manufacturing process enters the warehouse shown in the drawing at a variable rate of q_i lb/h. Shipments from the warehouse are q_o lb/h, which can also vary with time. The production control strategy is to adjust q_i on the basis of the inventory W(lb) stored in the warehouse. Because it is expensive and inconvenient to monitor the warehouse contents on a continuous basis and to make continuous adjustments, a digital control strategy is used. In particular, a direct-acting sampled-data controller is used in conjunction with a zero-order hold device. Thus, the control law relating the manipulated variable and the sampled measurements of the inventory W^* for constant set point is given by

$$Q_i(s) = -K_c \, H(s) W^*(s)$$

where K_c is the controller gain, $H(s)$ is the transfer function for the zero-order hold device, W^* denotes samples of the inventory W, and all variables are in deviation form.

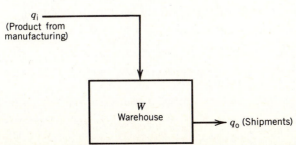

(a) Draw a block diagram for the controlled process assuming that measurement dynamics are negligible. A consistent set of units for the gain of each block can be assumed.

(b) Derive an expression for $W(z)$ that indicates how the inventory changes when a disturbance occurs in q_o.

(c) Use the Final Value Theorem to derive an expression for the steady-state value of W that results from a unit step disturbance in q_o.

(d) For what values of K_c will the closed-loop system be stable? (Assume the sampling period is $\Delta t = 1$ hr)

(e) For what values of K_c will the closed-loop response be oscillatory? (Assume $\Delta t = 1$ hr)

25.7. Determine whether the following two characteristic equations have any unstable roots:

$$\text{(a) } \Gamma_1(z) = 2 + 2z^2 - 3z^3 + 5z^4$$

$$\text{(b) } \Gamma_2(z) = 1 + 3.3z^{-1} + 3z^{-2} + 0.8z^{-3}$$

Use the modified Routh criterion.

25.8. A closed-loop control system under proportional-only digital control has

$$G_m = 1 \qquad HG_p(z) = \frac{0.2683z^{-1} + 0.4406z^{-2}}{(1 - 0.2636z^{-1})(1 - 0.7658z^{-1})}$$

Find the value of K_c that causes the loop to become unstable. Use the modified Routh criterion.

25.9. A process operating under proportional-only digital control has

$$G_p(s) = \frac{3}{(2s + 1)(5s + 1)} \qquad K_c = 4 \qquad G_m = 1 \quad \text{and} \quad \Delta t = 1$$

Using Jury's test, determine if the control system is stable. Check your results by calculation of the response to a set-point change.

25.10. Determine how the maximum allowable digital controller gain for stability varies as a function of Δt for the following system:

$$G_p(s) = \frac{1}{(s + 1)(5s + 1)}$$

$$G_c = K_c \qquad G_m = 1$$

Use the modified Routh criterion.

25.11. A temperature control loop includes a second-order overdamped model

$$HG(z) = \frac{(0.0826 + 0.0368z^{-1})z^{-1}}{(1 - 0.894z^{-1})(1 - 0.295z^{-1})}$$

and a digital PI controller

$$D(z) = K_c\left(1 + \frac{1}{8(1 - z^{-1})}\right)$$

Find the maximum controller gain K_{cm} for stability.

25.12. The following digital control algorithm has been proposed:

$$p_n = 2e_n + 1.3e_{n-1} + 0.7e_{n-2} - 0.8\,p_{n-1} - 0.4p_{n-2}$$

where p is the controller output and e is the error signal.

(a) Derive the corresponding pulse transfer function $P(z)/E(z)$.

(b) Is the controller stable?

(c) Suppose that a unit step change in $e(n\Delta t)$ occurs. What is the steady-state value of p?

25.13. A digital controller is used to control the liquid level of the storage tank shown in the drawing. The control valve has negligible dynamics and a steady-state gain of $K_v = 0.1$ ft^3/(min)(mA). The level transmitter has a time constant of 30 s and a steady-state gain of 4 mA/ft. The tank is 4 ft in diameter. The exit flow rate is not directly influenced by the liquid level; that is, if the control valve stem position is kept constant, $q_3 \neq f(h)$. Suppose that a proportional-only digital controller and a digital to analog converter with 4 to 20 mA output are used. If the sampling period for the analog to digital converter is $\Delta t = 1$ min, for what values of controller gain K_c is the closed-loop system stable? Will offset occur for the proportional controller after a change in set point?

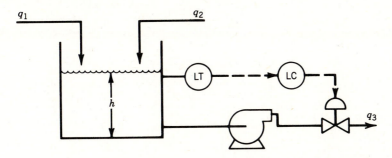

— CHAPTER 26 —————

Design of Digital Controllers

In the preceding chapter we described the principal components of a digital feedback control loop and presented several mathematical methods for analyzing the dynamic behavior of such a control system. In this chapter the main focus is on the digital controller: how it should be selected and tuned. We begin with the digital version of a PID controller and then show how other types of digital feedback controllers can be derived using the Direct Synthesis technique. Next, several digital feedforward control techniques are presented. The final topic considered in this chapter is combined load estimation and time-delay compensation, a special kind of digital controller which yields good performance for both load and set-point changes.

26.1 DIGITAL PID CONTROLLER

Digital versions of the PID controller in the form of difference equations were previously presented in Eqs. 8-16 and 8-18 without giving detailed derivations. As a starting point, consider the ideal continuous (analog) PID controller from Chapter 8:

$$p(t) = \bar{p} + K_c \left[e(t) + \frac{1}{\tau_I} \int_0^t e(t')\, dt' + \tau_D \frac{de(t)}{dt} \right] \quad (26\text{-}1)$$

To convert this expression to its digital equivalent, we use the following finite difference approximations:

$$\int_0^t e(t')\, dt' \approx \sum_{k=1}^n e_k \Delta t$$

$$\frac{de}{dt} \approx \frac{e_n - e_{n-1}}{\Delta t} \quad (26\text{-}2)$$

The digital PID controller equation can be written in two ways, the position form and velocity form. Substituting (26-2) into (26-1), we obtain the position form of the digital PID control algorithm.

614

Position Algorithm

$$p_n = \bar{p} + K_c \left[e_n + \frac{\Delta t}{\tau_I} \sum_{k=1}^{n} e_k + \frac{\tau_D}{\Delta t} (e_n - e_{n-1}) \right] \qquad (26\text{-}3)$$

Equation 26-3 can also be written as a z-transform expression. Let $p_n' = p_n - \bar{p}$ be defined as a deviation variable. Recall that the z-transform translation theorem yields

$$Z(e_n) = E(z)$$

$$Z(e_{n-1}) = z^{-1}E(z)$$

$$Z(e_1) = z^{-n+1}E(z)$$

Hence, the z-transform of Eq. 26-3 is

$$P'(z) = K_c \left[E(z) + \frac{\Delta t}{\tau_I} (z^{-n+1} + z^{-n+2} + \cdots + z^{-1} + 1)E(z) \right.$$

$$\left. + \frac{\tau_D}{\Delta t} (1 - z^{-1})E(z) \right] \qquad (26\text{-}4)$$

In the above equation, the summation (integral) term for large values of n approaches a limit of $1/(1 - z^{-1})$. Therefore, (26-4) can be simplified to

$$P'(z) = K_c \left[1 + \frac{\Delta t}{\tau_I} \left(\frac{1}{1 - z^{-1}} \right) + \frac{\tau_D}{\Delta t} (1 - z^{-1}) \right] E(z) \qquad (26\text{-}5)$$

The digital controller transfer function is

$$D(z) = \frac{P'(z)}{E(z)} = K_c \left[1 + \frac{\Delta t}{\tau_I} \left(\frac{1}{1 - z^{-1}} \right) + \frac{\tau_D}{\Delta t} (1 - z^{-1}) \right] \qquad (26\text{-}6)$$

Equation 26-6 is referred to as the *position* form of the control law because it yields the value of the controller output directly. It can also be derived using the backward difference approximation for s given in Eq. 24-122; see Example 24.12.

Velocity Algorithm

The *velocity* form of the PID controller is an attractive alternative to the position form because it avoids computing the summation in (26-3). Also, the velocity form does not require specification of the bias term $\bar{p}$ and is less prone to reset windup, as discussed below. Since the nominal steady-state value (or bias) $\bar{p}$ is a constant, the change in the controller output Δp_n is given by

$$\Delta p_n = p_n - p_{n-1} = p_n' - p_{n-1}' \qquad (26\text{-}7)$$

As shown in Chapter 8, the velocity form of the controller can be found by shifting (26-3) to obtain p_{n-1} and subtracting it from (26-3):

$$\Delta p_n = K_c \left[(e_n - e_{n-1}) + \frac{\Delta t}{\tau_I} e_n + \frac{\tau_D}{\Delta t} (e_n - 2e_{n-1} + e_{n-2}) \right] \qquad (26\text{-}8)$$

Taking the z-transform of (26-8) gives

$$\Delta P(z) = K_c \left[(1 - z^{-1})E(z) + \frac{\Delta t}{\tau_I} E(z) + \frac{\tau_D}{\Delta t} (1 - 2z^{-1} + z^{-2})E(z) \right] \qquad (26\text{-}9)$$

Note that Eq. 26-9 also can be obtained by multiplying both sides of (26-5) by $(1 - z^{-1})$.

Another variation of the digital PID controller is based on the more accurate *trapezoidal* approximation for the integral (cf. Eq. 26-1):

$$\int_0^t e(t') \, dt' \approx \sum_{k=1}^{n} \left(\frac{e_k + e_{k-1}}{2} \right) \Delta t \tag{26-10}$$

After this expression is substituted into (26-1), the velocity form of the control law becomes

$$\Delta p_n = K_c \left[(e_n - e_{n-1}) + \frac{\Delta t}{\tau_I} \left(\frac{e_n + e_{n-1}}{2} \right) + \frac{\tau_D}{\Delta t} (e_n - 2e_{n-1} + e_{n-2}) \right] \tag{26-11}$$

or as a *z*-transform (cf. Eq. 26-8),

$$\Delta P(z) = K_c \left[(1 - z^{-1})E(z) + \frac{\Delta t}{2\tau_I} (1 + z^{-1})E(z) \right.$$
$$\left. + \frac{\tau_D}{\Delta t} (1 - 2z^{-1} + z^{-2})E(z) \right] \tag{26-12}$$

Compared to Eq. 26-9, this more accurate approximation of the integral mode may not actually achieve a significant improvement in control loop performance. When each controller is tuned, slightly different values of the controller settings (K_c, τ_I, τ_D) may be obtained in tuning (26-11) versus (26-8). Note that both equations can be arranged to the same general form, $\Delta p_n = \alpha_1 e_n + \alpha_2 e_{n-1} + \alpha_3 e_{n-2}$.

Features of Digital PID Controllers

In Chapter 8 we discussed several modifications to the ideal PID controller that can improve the operation of the control system (e.g., antireset windup and derivative kick elimination). Similar modifications should be made to digital controllers, as discussed below.

1. *Elimination of Reset Windup.* In Eq. 26-3 reset windup can occur when the error summation grows to a very large value. Suppose the controller output is at an upper or lower limit, as the result of a large sustained error signal. Even though the measured variable eventually reaches its set point (where $e_n = 0$), the controller may be wound up because of the summation term. Until the error changes its sign for a period of time, thereby reducing the value of the summation, the controller will remain at its limit. Start-up situations performed under automatic control and other large set-point changes are particularly prone to this problem, leading to excessive overshoot by the controlled variable and saturation of the manipulated variable.

For the position algorithm, there are several modifications that can be made to reduce reset windup [1,2]:

a. Place an upper limit on the value of the summation. When the controller saturates, suspend the summation until the controller output moves away from the limit.

b. Back-calculate the value of e_n that just causes the controller to saturate. If saturation occurs, use this actual value as error term e_{n-1} in the next controller calculation.

Field testing [1] has indicated that approach (b) is superior to (a), although it is somewhat more complicated.

Note that in the velocity form of the algorithm, Eq. 26-8, no summation appears, avoiding the windup problem. The controller remains saturated until the error (e_n) decreases to a point where the control action returns to normal levels. However, the algorithm must be programmed to disregard Δp_n if p_n is at the limit, implying that p_n should be monitored at all times. In most situations, where the integral mode is present, the velocity algorithm is preferred over the position algorithm.

2. Elimination of Derivative Kick. When a set-point change is made, control algorithms (26-3) or (26-8) will produce a large immediate change in the output (an impulse) due to the derivative term. There are several methods available for eliminating derivative kick [3]:

a. As suggested in Chapter 8, most commercial controllers apply derivative action (or the discrete equivalent) to the measured variable rather than to the error signal. Thus, $e_n = r_n - b_n$, where b_n is the measured value of the controlled variable, is replaced by $-b_n$ in the derivative term, giving the position form:

$$p_n = \bar{p} + K_c \left[e_n + \frac{\Delta t}{\tau_I} \sum_{k=1}^{n} e_k - \frac{\tau_D}{\Delta t} (b_n - b_{n-1}) \right] \qquad (26\text{-}13)$$

Equation 26-8 for the velocity form can be changed in an analogous fashion.

b. Instead of a step change in the set point, ramp the set point to a new value, limiting the rate of change of r_n.

If measurement noise combined with a large ratio of derivative time to sampling period ($\tau_D/\Delta t$) is causing an overactive derivative mode, filter the error signal before calculating the derivative action (see Chapter 22).

3. Effect of Saturation on Controller Performance. One of the difficulties that arises in Eq. 26-3 is that a small change in the error can cause the controller output to saturate for certain values of the controller settings. Suppose that $K_c \tau_D/\Delta t = 100$ because of a small sampling period, and that e_n and p_n are both scaled from 0 to 100%. A small change in $\Delta e_n = e_n - e_{n-1}$ of 1% will yield a 100% change in p_n, exceeding its upper limit. Therefore, one must be careful to select controller settings and a value of Δt that do not cause scaling problems.

4. Comparison of Position and Velocity Algorithms. The position form of the PID algorithm (26-3) requires a value of $\bar{p}$, whereas the velocity form of the algorithm (26-8) does not explicitly require a steady-state value for the controller output. However, initialization of either algorithm is equally simple since manual operation of the control system usually precedes the transfer to automatic control. Hence $\bar{p}$ (or p_{n-1} for the velocity algorithm) is simply taken to be equal to the signal to the final control element at the time of transfer. As noted previously, the velocity form is less prone to reset windup problems.

To implement a velocity algorithm directly, a pulse up/down counter or stepping motor can be used in series with the controller to convert incremental changes to "position" (e.g., flow rate). Integral action is always recommended with the velocity algorithm to prevent *drift* of the process output from the set point. In (26-8), because $e_n = r_n - b_n$, note that r_n drops out when $\tau_I \to \infty$ (no dependence on set point).

5.   *Use of Dimensionless Controller Gain.*   In commercial digital control systems, the controller gain K_c is usually expressed as a dimensionless number. If the measured variable b_n and the signal to the final control element p_n are computed in terms of % full range, K_c will be dimensionless. This will also be true if the computer input and output have the same units (e.g., mA or V). However, where the input and/or output signals are converted internally into engineering units, the value of K_c will not be dimensionless.

6.   *Time-Delay Compensation.*   Many commercial distributed control systems offer the option of time-delay compensation with a digital PI or PID controller. The controller is implemented in difference equation form using a Smith predictor (see Fig. 18.10).

Physical Realizability of Digital Controllers

A digital controller cannot compute its output signal based on *future* process inputs; this *physical realizability* requirement was discussed in Chapter 24. Standard digital controller design methods incorporate the physical realizability requirement in the development of the controller.

EXAMPLE 26.1

Check the physical realizability of the PID controller transfer function in Eq. 26-6 using the standard form given in Eq. 24-88.

Solution
Equation 26-6 can be rearranged to give

$$D(z) = \frac{K_c\left[1 - z^{-1} + \dfrac{\Delta t}{\tau_I} + \dfrac{\tau_D}{\Delta t}(1 - 2z^{-1} + z^{-2})\right]}{1 - z^{-1}} \qquad (26\text{-}14)$$

or

$$D(z) = \frac{K_c\left[\left(1 + \dfrac{\Delta t}{\tau_I} + \dfrac{\tau_D}{\Delta t}\right) - \left(1 + \dfrac{2\tau_D}{\Delta t}\right)z^{-1} + \dfrac{\tau_D}{\Delta t}z^{-2}\right]}{1 - z^{-1}} \qquad (26\text{-}15)$$

A comparison of Eqs. 26-15 and 24-88 indicates that this controller is physically realizable because $a_0 \neq 0$.

26.2   SELECTION OF DIGITAL PID CONTROLLER SETTINGS

If the control algorithm is required to be of the PID form, there are several general approaches that can be chosen for specifying the controller settings:

1. Conversion of continuous controller settings
2. Integral error criteria based on digital simulation
3. Pole placement

Conversion of Continuous Controller Settings

For small values of Δt (relative to the process response time) the finite difference approximations for integral and derivative action discussed in Section 26.1 are reasonably accurate. Hence settings obtained for continuous (analog) controllers can be converted to discrete (digital) form using Eqs. 26-1 and 26-2. As noted by Bristol [4], the continuous and discrete PID controllers will have essentially the same behavior as long as $\Delta t/\tau_I \leq 0.1$. Åström and Wittenmark [5] have discussed the effect of sampling period for designing a wide range of digital controllers.

Use of the zero-order hold in digital process control systems requires a modification in the design procedure if Δt is not small. The zero-order hold causes an effective time delay in the signal to the final control element, thus narrowing the stability margin somewhat. The dynamic behavior of the sampler plus zero-order hold can be approximated by a time delay equal to one-half the sampling period [6,7]. Thus, it is a common practice in tuning PID controllers to add $\Delta t/2$ to the process time delay (θ) prior to computing K_c, τ_I, and τ_D [7,8].

While virtually any design method could be used to find K_c, τ_I, and τ_D, in practice simple techniques such as Cohen-Coon or minimum integral error criteria (see Chapter 12) have been employed to obtain digital PID controllers. This approach assumes that a first-order plus time-delay model approximation is applicable for the combined process, valve, and transmitter transfer function. For continuous second-order plus time-delay models, other design relations are available [9].

Digital Controllers Based on Integral Error Criteria

As the sampling period is increased from a very small value, some anomalous behavior may be encountered in the previous approach. Moore et al. [8] performed a study on the tuning of digital PI controllers based on the minimization of various integral criteria. For a first-order plus time-delay model, they compared the tuning parameters based on optimizing K_c and τ_I using digital simulation [10] with the controller settings obtained by using the continuous system correlations for ISE, IAE, and ITAE. In all cases the process time delay was increased by $\Delta t/2$, as discussed above. The comparison of K_c and τ_I for the two approaches is shown in Fig. 26.1, where θ/τ is the ratio of the time delay and the time constant in the process model. Note that the results for the two approaches are similar except when θ/τ is very small or $\Delta t/\tau$ is large. In these cases the controller gain must be modified (see the dashed lines in Fig. 26.1).

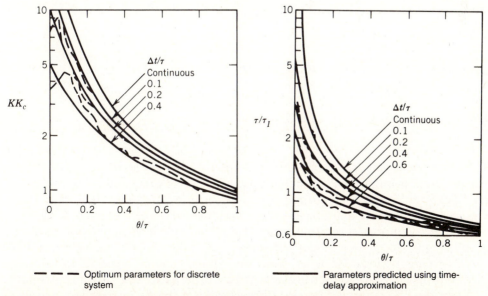

Figure 26.1 Comparison of tuning parameters for a PI controller predicted by the time-delay approximation to those of Lopez [10].

The sampling operation also influences the translation of stability characteristics from continuous to discrete time. Mosler et al. [11,12] have studied the design of PI digital controllers for first-order plus time-delay processes using the concepts of ultimate controller gain K_{cu} and ultimate period (see Chapter 16). They demonstrated that rather nonuniform results are obtained because of the sampling operation. Recall that a gain margin of 1.7 is typical for a well-tuned controller (see Section 16.4) and corresponds to a controller gain of $K_c = 0.6K_{cu}$, where K_{cu} is the ultimate gain. For a digital controller with rapid sampling, Mosler et al. obtained results similar to those for continuous control ($K_c = 0.6K_{cu}$). Larger values of Δt, however, in effect slow the response and change both the ratio of K_c/K_{cu} and the ultimate period.

EXAMPLE 26.2

A digital controller is used to control the pressure in a tank via a purge stream. The controller valve action is direct so that an increase in the tank pressure causes the pressure control valve opening to increase. Analysis of step response data gave a process model of

$$G = G_v G_p G_m = \frac{-20e^{-s}}{5s + 1}$$

The gain is expressed in kPa/% and the time constant is in minutes. The nominal sampling period is one minute.

Compare the closed-loop performance of discrete PI and PID controllers tuned for load changes ($G_L = G_p$) using design relations based on minimum ITAE. Check the effect of sampling period for each controller, with $\Delta t = 0.05, 0.25, 0.5,$ and 1 min.

Solution
First adjust the process time delay for each controller calculation by adding $\Delta t/2$. The appropriate controller settings are as follows (τ_I and τ_D in min):

	PI		PID		
Δt	K_c	τ_I	K_c	τ_I	τ_D
0.05	−0.21	2.48	−0.31	1.81	0.38
0.25	−0.18	2.69	−0.28	1.98	0.43
0.5	−0.17	2.89	−0.25	2.14	0.48
1.0	−0.14	3.27	−0.21	2.44	0.58

The controller settings are then substituted into Eq. 26-5 to obtain the digital form. Figure 26.2 shows that smaller sampling periods offer a faster closed-loop response for PI control. There is no change in performance for $\Delta t \leq 0.05$ min. Note that Fig. 26.1 for $\theta/\tau = 0.2$ and $\Delta t/\tau \leq 0.2$ predicts that very little correction in K_c and τ_I is necessary for digital control. Figure 26.3 indicates that when derivative action is added, the closed-loop response improves over the PI case, as expected, for small sampling periods. However, when the sampling period increases to 1.0, the PID controller actually causes instability. This is primarily due to a higher value of K_c for the PID controller (vs. PI), less accuracy in the $\Delta t/2$ correction, and the

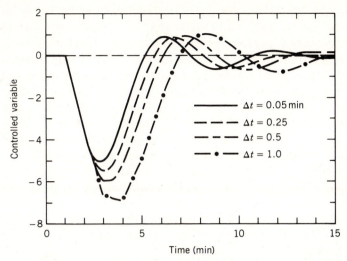

Figure 26.2   ITAE tuning for PI controllers with different sampling periods, Example 26.2.

larger discretization errors in the derivative mode for $\Delta t = 1$. Observe that kinks occur in the response in Fig. 26.3 as the sampling period grows larger, a result of using derivative action with first-order processes.

It is interesting to compare the results of this example with Table 22.1 and Example 22.1. The most reliable guidelines for choosing the sampling periods in this example are those based on closed-loop analysis, namely 2c, 3b, and 3c. In fact, constraints based on the derivative time τ_D are the most restrictive, since τ_D is fairly small for this process. Many of the other guidelines are actually misleading and would lead to instability for the PID controller.

Pole Placement

Digital feedback controllers can be designed using pole placement (root locus) in the z domain [13–15]. Such an approach is analogous to that given in Chapter 11, although the locations of desirable poles (roots) in the z domain differ from those

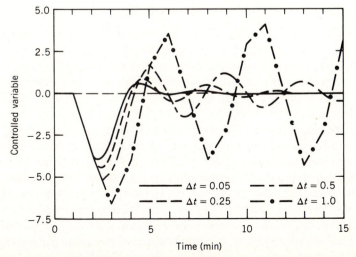

Figure 26.3   ITAE tuning for PID controllers with different sampling periods, Example 26.2.

for continuous time (the s domain), as discussed in Section 24.5 and shown in Fig. 24.6. Oscillation can be included in the response by specifying a pair of complex-valued poles or a single negative pole. However, the discrete root locus method is not recommended for designing PID controllers because the dynamic response is not uniquely determined by the closed-loop pole locations (the values of the transfer function zeros also are important). Therefore, all candidate designs obtained by this method should be checked by simulation to ensure that the closed-loop response is satisfactory.

In the next section we will consider several controller design techniques, called direct synthesis methods (cf. Section 12.2), which use information on poles and zeros of the transfer function. These methods differ from root locus in that the form of the controller is selected to cancel numerator and denominator terms in the process transfer function. Such an approach allows the designer to specify the closed-loop response with more certainty than with root locus. The resulting feedback controllers do not necessarily assume the familiar PID form, although often there are similarities.

26.3 DIRECT SYNTHESIS METHODS

In this section we extend the design methodology for direct synthesis of controllers first presented in Section 12.2 to the digital case. In Chapter 25, an equation was derived for the closed-loop transfer function for set-point changes, namely

$$\frac{C(z)}{R(z)} = \frac{HG(z)D(z)}{1 + HG(z)D(z)} \tag{25-48}$$

where $G_m(s) = K_m$ and $HG(z) = K_m HG_v G_p(z)$. Generally this equation can be used to find C/R once all components of the feedback loop (K_m, G_v, G_p, D) have been specified. The performance of the control system can then be evaluated by specifying the set-point change, for example, a step change, and observing the response. Suppose the problem were reversed: Given a desired input–output relation specified by the closed-loop transfer function $(C/R)_d$, what controller D will yield the desired performance? Solving for D gives

$$D = \frac{1}{HG} \frac{(C/R)_d}{1 - (C/R)_d} \tag{26-16}$$

Equation 26-16 is the digital equivalent of Eq. (12-3a), which was the basis for the Direct Synthesis design method discussed in Section 12.2 (continuous-time controllers). It is also the design equation for several digital controller design methods.

One important feature of Eq. 26-16 is that the resulting controller includes the reciprocal of the process transfer function. This feature can cause the poles of HG to become zeros of D, while the zeros of HG become poles of the controller, unless the poles and zeros of HG are canceled by terms in $(C/R)_d$. The inversion of HG in (26-16) can lead to operational difficulties. If HG contains a zero that lies outside the unit circle, then D will have an unstable pole that lies outside the unit circle. The product $HG(z)D(z)$ in (25-48) indicates that the unstable pole and zero will cancel, but only theoretically. In practice there will always be some model error that prevents exact cancelation. In this case, D is an *unstable* controller and thus would produce an unbounded output sequence for a step change in set point. However, the problems associated with unstable zeros can be successfully treated by judicious selection of $(C/R)_d$, as discussed below.

Digital controllers of the direct synthesis type also share one other characteristic, namely they contain time-delay compensation using a Smith predictor (see Chapter 18). Recall that N is the process time delay but the zero-order hold model (e.g., Eq. 24-116) exhibits an effective delay of $N+1$. Thus HG contains a time-delay term, z^{-N-1}, in the numerator. In Eq. 26-16, for D to be physically realizable, $(C/R)_d$ must also contain a term z^{-N-1}. In other words, if there is a term z^{-N-1} in the open-loop transfer function, the closed-loop process cannot respond before $N+1$ units of time (sampling intervals) have passed. Using $(C/R)_d$ of this form in Eq. 26-16 yields a D with the mathematical equivalence of time-delay compensation, because the time delay is eliminated from the characteristic equation.

Below we discuss three Direct Synthesis algorithms: minimal prototype control, Dahlin's method, and the Vogel-Edgar method. We also present a digital version of the Internal Model Control technique that was considered in Chapter 12.

Minimal Prototype Algorithm

The design criteria for this algorithm [7,15] are expressed in terms of the desired response to a change in set point:

1. The system must have zero steady-state error at the sampling instants for a specific set-point change (ramp, step, impulse, sinusoid, etc.). Note that this requirement implies that the controller will contain integral action.
2. The rise time should equal the minimum number of sampling periods.
3. The settling time should be finite.
4. Transfer functions D and $(C/R)_d$ must be physically realizable.

Including a time delay term in $(C/R)_d$ ensures that D will be physically realizable. Hence $(C/R)_d = z^{-N-1}$ for the minimal prototype design, implying that the closed-loop response will follow the set-point change exactly, except for the required time delay. The response for a unit step input in set point is

$$C(z) = (C/R)_d R(z) = z^{-N-1} \frac{1}{1 - z^{-1}} = z^{-N-1}(1 + z^{-1} + z^{-2} + \cdots) \quad (26\text{-}17)$$

This response implies a time delay of $N+1$ steps, that is, $c_n = 0$ for $n \le N$ and then $c_n = r_n = 1$ for $n \ge N+1$. This response corresponds to perfect control.

EXAMPLE 26.3

Develop the minimal prototype controller when HG is a general first-order plus time-delay model,

$$HG(z) = \frac{K(1 - a)z^{-N-1}}{1 - az^{-1}} \quad (26\text{-}18)$$

Plot the response for a ramp change up to the new set-point value ($r_n = 0.1n$, $n = 0$ to 9; $r_n = 1$ for $n \ge 10$) when $K = 1$, $a = 0.8187$ ($\tau = 1$, $\Delta t = 0.2$), and $N = 1$ ($\theta = 0.2$). In addition, evaluate the controller for a unit step change in the load variable, assuming $G_p(s) = G_L(s)$ and $K_m = G_v = 1$.

Solution

The desired closed-loop transfer function is $(C/R)_d = z^{-N-1}$, which implies that a unit step change in R is to produce a unit step change in C after $N+1$ sampling periods, with no steady-state error since $C/R|_{z=1} = 1$. Solving for D using (26-16) gives the minimal prototype controller for a general first-order plus time-delay model:

$$D(z) = \frac{1 - az^{-1}}{K(1 - a)(1 - z^{-N-1})} = \frac{1 - 0.8187z^{-1}}{0.1813(1 - z^{-2})} \tag{26-19}$$

Figure 26.4a shows the closed-loop response for a ramp set-point change. The set-point tracking appears to be satisfactory although the controlled variable consistently lags the set point during the ramping phase by $N+1$ ($=2$) time steps.

For a load change, Eq. 25-54 indicates that the term $LG_L(z)$ must be evaluated. If $L(s) = 1/s$ and $G_L(s) = e^{-\theta s}/(\tau s + 1)$, then by modifying Eq. 25-57,

$$\mathcal{Z}[G_L L(s)] = \frac{z^{-N-1}(1 - e^{-\Delta t/\tau})}{(1 - e^{-\Delta t/\tau}z^{-1})(1 - z^{-1})}$$

For $N = 1$, $\Delta t = 0.2$, and $a = e^{-\Delta t/\tau} = 0.8187$,

$$G_L L(z) = \frac{0.1813z^{-2}}{(1 - 0.8187z^{-1})(1 - z^{-1})}$$

Note that for a unit step change in load, $G_L L(z) = HG_L(z)L(z)$. Substituting the expressions for $G_L L$, HG, and D into Eq. 25-54, the z-transform of the closed-loop response is

$$C(z) = \frac{0.1813z^{-2}(1 + z^{-1})}{1 - 0.8187z^{-1}}$$

The load response in the time domain is shown in Fig. 26.4b.

For all values of N, the denominator of D includes $(1 - z^{-N-1})$, which contains the factor $(1 - z^{-1})$ for all values of N. This characteristic indicates that the controller provides integral action (see Eq. 26-6). Such a controller yields no offset for both load and set-point changes. However, for $N \geq 1$ $D(z)$ will also exhibit the undesirable feature of ringing, as discussed in the next example.

Note that the controller in (26-19) was derived analytically and thus is completely general in terms of the first-order process model parameters (K, a, N). This feature is also shared by the other types of Direct Synthesis controllers discussed in this section.

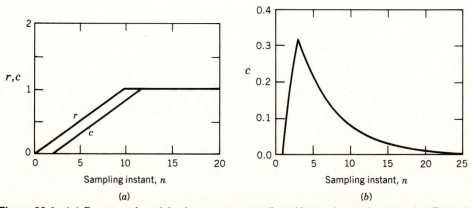

Figure 26.4 (a) Response for minimal prototype controller with ramping to the set point, Example 26.3. (b) Response of controlled variable for a unit step load change, Example 26.3.

EXAMPLE 26.4

If a continuous process model is a second-order plus time-delay transfer function, the discrete equivalent (with zero-order hold) is

$$HG(z) = \frac{(b_1 + b_2 z^{-1})z^{-N-1}}{1 + a_1 z^{-1} + a_2 z^{-2}} \qquad (26\text{-}20)$$

Derive the minimal prototype controller for $a_1 = -1.5353$, $a_2 = 0.5866$, $b_1 = 0.0280$, $b_2 = 0.0234$, and $N = 0$. These parameters correspond to a continuous second-order transfer function with $G(s) = 1/[(5s + 1)(3s + 1)]$ and $\Delta t = 1$; cf. Eqs. 23-22 through 23-25. Plot the response for a unit change in set point at $t = 5$ for $0 \le t \le 20$.

Solution

The desired minimal prototype closed-loop transfer function for $N=0$ is $(C/R)_d = z^{-1}$. Applying (26-16), the formula for the controller is

$$D(z) = \frac{1 + a_1 z^{-1} + a_2 z^{-2}}{b_1 z^{-1} + b_2 z^{-2}} \frac{z^{-1}}{1 - z^{-1}} \qquad (26\text{-}21)$$

$$= \frac{1 + a_1 z^{-1} + a_2 z^{-2}}{b_1 + (b_2 - b_1)z^{-1} - b_2 z^{-2}} \qquad (26\text{-}22)$$

Substituting the numerical values for a_1, a_2, b_1, and b_2, the controller is

$$D(z) = \frac{1 - 1.5353 z^{-1} + 0.5866 z^{-2}}{0.0280 - 0.0046 z^{-1} - 0.0234 z^{-2}}$$

From (26-22) it is clear that D is physically realizable. However, when this controller is implemented, an undesirable characteristic appears, namely *intersample ripple*. Figure 26.5 shows the response of the closed-loop system to a unit step change in set point at $t = 5$. Although the response does satisfy $c_n = 1$ for $n \ge 6$), the response is quite oscillatory; that is, intersample ripple occurs. This is caused by controller output (after passing through the zero-order hold) cycling back and forth between positive and negative deviations from the steady-state value. This behavior, called *ringing*, of course is unacceptable for a control system. Further discussion of ringing is provided later in this section.

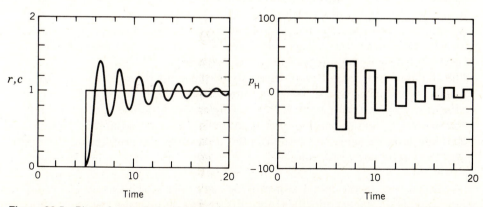

Figure 26.5 Plots of controlled variable c and controller output p_H (after zero-order hold) for minimal prototype control of a second-order process, Example 26.4.

A method that has been suggested to avoid intersample ripple is to use at least $m + N$ sampling periods to reach the final steady state, where m is the order of the denominator polynomial of the process transfer function and $N = \theta/\Delta t$ [14,15]. One way to accomplish this is to define $(C/R)_d$ as follows:

$$\left(\frac{C}{R}\right)_d = z^{-N}(\gamma_1 z^{-1} + \gamma_2 z^{-2} + \cdots + \gamma_m z^{-m}) \tag{26-23}$$

where $0 \leq \gamma_i \leq 1$. For $m = 2$ and $N = 0$, the closed-loop transfer function is

$$\left(\frac{C}{R}\right)_d = \gamma_1 z^{-1} + \gamma_2 z^{-2} \tag{26-24}$$

where γ_1 and γ_2 are determined below. For a unit step change in R, multiply the right side of (26-24) by $1/(1 - z^{-1})$; long division yields

$$C(z) = \gamma_1 z^{-1} + (\gamma_1 + \gamma_2)(z^{-2} + z^{-3} + \cdots) \tag{26-25}$$

To ensure no offset, $\gamma_1 + \gamma_2 = 1$, and for no overshoot, $\gamma_1 < 1$. The response requires two steps to reach the steady-state value of unity (assuming no time delay). For the model in (26-20) with b_1 and $b_2 > 0$, Luyben [14] has shown that if $\gamma_1 = b_1/(b_1 + b_2)$, no intersample ripple will occur. For Example 26.4 above, $\gamma_1 = 0.545$. This choice corresponds to retaining the open-loop zero in the closed-loop transfer function; that is,

$$\left(\frac{C}{R}\right)_d = \frac{b_1 z^{-1} + b_2 z^{-2}}{b_1 + b_2} \tag{26-26}$$

Substitution of (26-26) and (26-20) into (26-16) yields pole–zero cancellation. However, intersample ripple will be evident for other values of γ_1. Other algorithms that eliminate intersample ripple in second-order processes are due to Kalman (discussed in [7]) and Vogel and Edgar [16]. The Vogel-Edgar algorithm is discussed later in this section.

The minimal prototype controller provides a high-performance closed-loop response, although this last example illustrates there are several important disadvantages:

1. The design specifies only the response at the sampling instants. Thus, undesirable intersample rippling or large overshoots could be "hidden" in the closed-loop response. Vigorous control action is normally required, which may violate constraints on the controller output.
2. A minimal prototype response is highly tuned for the specific type of application for which it is designed. A design that yields a good set-point response may not yield a satisfactory load response and vice versa.
3. The minimal prototype design attempts to place all poles of the closed-loop transfer function as close as possible to the origin of the z plane so as to achieve a rapid response. Unfortunately, this configuration produces a response that is extremely sensitive to parameter changes in the process model [15]. Because of the undesirable sensitivity inherent in this controller, the usefulness of the minimal prototype design is limited.

A special case of the minimal prototype controller which has no intersample ripple is called a *deadbeat* controller (e.g., Eq. 26-26). In this case the specifications presented earlier apply to the continuous-time response as well as the discrete-time response. The controlled variable should equal the set point after a finite number of steps for all values of t. In addition, the output of the deadbeat controller (p) should exhibit a finite settling time. These additional design criteria are generally achievable only when the rise time is selected larger than for standard minimal prototype.

Dahlin's Algorithm

For most practical industrial applications, the minimal prototype response is difficult to achieve. The requirement that the controlled variable move from one set point to another over the span of just a few sampling periods is often physically too demanding, and inaccuracies in the process model may cause poor closed-loop performance. Dahlin's algorithm [17], also derived independently by Higham [18], is obtained from the same basic design equation of Eq. 26-16, but is less demanding in terms of closed-loop performance.

This algorithm specifies that the closed-loop performance of the system behave similarly to a continuous first-order process with time delay,

$$\left(\frac{C}{R}\right)_d = \frac{e^{-hs}}{\lambda s + 1} \tag{26-27}$$

where λ and h are the time constant and time delay of the closed-loop transfer function, respectively. Selecting $h = \theta = N\Delta t$ (the process time delay), the discrete form of Eq. 26-27 with a zero-order hold is

$$\left(\frac{C}{R}\right)_d = \frac{(1 - A)z^{-N-1}}{1 - Az^{-1}} \tag{26-28}$$

where

$$A = e^{-\Delta t/\lambda} \tag{26-29}$$

Substituting (26-28) into the controller synthesis formula (Eq. 26-16) yields the general form of Dahlin's control algorithm, which we call G_{DC}:

$$G_{DC} = \frac{(1 - A)z^{-N-1}}{1 - Az^{-1} - (1 - A)z^{-N-1}} \frac{1}{HG(z)} \tag{26-30}$$

As a special case, when HG is a first-order plus time-delay transfer function (Eq. 26-18), Dahlin's controller is

$$G_{DC} = \frac{1 - A}{1 - Az^{-1} - (1 - A)z^{-N-1}} \frac{1 - a_1 z^{-1}}{K(1 - a_1)} \tag{26-31}$$

For all values of N, $(1 - z^{-1})$ is a factor of the denominator, indicating the presence of integral action. This result is consistent with (26-27) and (26-28), which specify zero steady-state error for set-point changes.

The time constant λ for the closed-loop system serves as a convenient tuning parameter for the control algorithm. Small values of λ produce tight control while large values of λ give more sluggish control. This flexibility is especially useful in situations where the parameters of the process model, especially the time delay, are subject to error or are time-varying because of changes in the process. In a tightly controlled system, time-delay errors can cause poor control and an unstable response. By choosing a larger λ and "loosening" the control action, the controller can better accommodate the inaccurate model. As $\lambda \rightarrow 0$ (i.e., $A \rightarrow 0$), Dahlin's algorithm is equivalent to minimal prototype control in Eq. 26-19, but as discussed earlier such tight tuning is usually not desirable for process control applications.

EXAMPLE 26.5

Determine Dahlin's controller for the same process model used in Example 26.4. Set $\lambda = \Delta t = 1$. Plot the response and the controller output for a set-point change.

Solution

Applying Eq. 26-30, for $N = 0$, the controller transfer function is

$$G_{DC} = \left(\frac{0.632}{1 - z^{-1}} \right) \left(\frac{1 - 1.5353z^{-1} + 0.5866z^{-2}}{0.0280 + 0.0234z^{-1}} \right) \tag{26-32}$$

When this controller is implemented, the response $c(t)$ and the controller output $p(t)$ are shown in Fig. 26.6a. As in Example 26.4, intersample ripple occurs in the controlled variable, and the controller output after the zero-order hold alternates on either side of a constant value. The controller ringing is due to the presence of the term $(0.0280 + 0.0234z^{-1})$ in the denominator of (26-32), which corresponds to a controller pole at -0.836, quite close to the unit circle. As discussed below, this term, when transformed to the time domain, causes a change in sign at each sampling instant in the manipulated variable. Dahlin [17] suggested that ringing can be eliminated by setting $z = 1$ in the ringing term, in this case replacing $(0.0280 + 0.0234z^{-1})$ by a constant $(0.0280 + 0.0234 = 0.0514)$. Let the nonringing version of Dahlin's controller be called $\overline{G}_{DC}$. Figure 26.6b shows $c(t)$ and $p_H(t)$ for this case, indicating that the ringing behavior has disappeared. Interestingly, the closed-loop response now exhibits an overshoot, which contradicts the original design criterion of first-order approach to set point (Eq. 26-27). Therefore, the

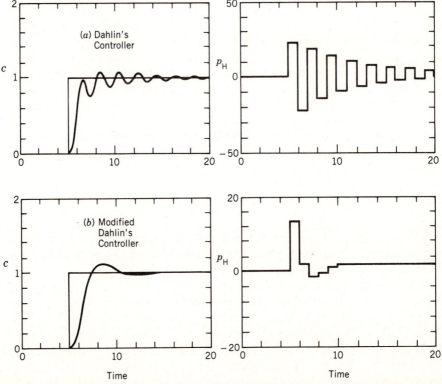

Figure 26.6 Comparison of ringing and non-ringing Dahlin's controllers for second-order process ($\lambda = 1$), Example 26.5 (c = controlled variable, p_H = controller output after zero-order hold).

closed-loop performance of Dahlin's controller modified for ringing is not always predictable. This lack of predictability represents a major disadvantage of the technique.

An Analysis of Ringing

In the previous examples the phenomenon of controller ringing was noted in conjunction with digital feedback control of a first or second-order process. Such behavior is unique to discrete-time direct synthesis methods and produces excessive actuator movement and wear. It is also unsettling to plant operators. To examine why ringing occurs, suppose D contains a stable pole p_1 that is located near $z = -1$ in the complex z plane. D can be factored as

$$D(z) = \frac{1}{1 - p_1 z^{-1}} D'(z) \tag{26-33}$$

The controller output based on an error signal E is therefore

$$P(z) = \left[\frac{1}{1 - p_1 z^{-1}} D'(z) \right] E(z) \tag{26-34}$$

If partial fraction expansion of P is carried out, we can isolate the effect of the pole p_1, assuming $D(z)$ and $E(z)$ are specified:

$$P(z) = \frac{r_1}{1 - p_1 z^{-1}} + [\text{other terms}] \tag{26-35}$$

When (26-35) is inverted to the time domain, the first term becomes $r_1(p_1)^n$, where n is the time step. As previously shown in Table 24.3, a negative pole near the unit circle has a pronounced effect on the response, causing the controller to oscillate or ring. On the other hand, negative poles near the origin are heavily damped and their results are not so noticeable. Positive poles do not cause the controller output to change in sign. Any digital control algorithm should contain some procedure for eliminating ringing pole(s) when they occur.

The most direct way to evaluate ringing with Dahlin's controller is to calculate P/R, since C/R may not exhibit oscillation at the sampling instants for the ringing case (see Fig. 26.5). By block diagram analysis;

$$\frac{P}{R} = \frac{D}{1 + HGD} \tag{26-36}$$

For Direct Synthesis algorithms with no model error, the formula for D in (26-16) can be substituted into (26-36) yielding the following equation for P/R:

$$\frac{P}{R} = \frac{1}{HG} \left(\frac{C}{R} \right)_d \tag{26-37}$$

For the special case of Dahlin's controller ($D = G_{DC}$; no ringing pole removed) and the second-order plus time-delay process model,

$$\frac{P}{R} = \frac{(1 + a_1 z^{-1} + a_2 z^{-2})}{b_1 z^{-1} + b_2 z^{-2}} \frac{(1 - A)z^{-1}}{1 - A z^{-1}} \tag{26-38}$$

In (26-38), for any input R, the term $b_1 + b_2 z^{-1}$ will cause ringing if b_1 and b_2

have the same sign, although the severity of ringing will depend on the relative sizes of b_1 and b_2. Two special cases of (26-38) can be considered:

1. $a_2 = b_2 = 0$ (first-order model):   There can be no ringing pole for any value of A.
2. $A = 0$ (minimal prototype): $b_1 + b_2z^{-1}$ may yield a ringing pole, depending on the signs of b_1 and b_2.

From the above, we conclude that increasing the order of the assumed process model has a major influence on the occurrence of ringing. Therefore, care must be taken when a higher-order model is chosen to represent the process. On the other hand, it can be shown that a controller designed using an inaccurate first-order model can also lead to ringing behavior.

Now consider the nonringing version of Dahlin's controller $\overline{G}_{DC}$. In this case, Eq 26-37 cannot be used because $C/R \neq (C/R)_d$ (no longer an exponential approach to set point). Hence we must analyze (26-36). After developing a nonringing version of Dahlin's controller ($\overline{G}_{DC}$) for a second-order plus time-delay model, and substituting it into Eq. 26-36, the resulting transfer function is

$$\frac{P}{R} = \frac{(1 - A)(1 + a_1z^{-1} + a_2z^{-2})}{(b_1 + b_2)[1 - Az^{-1} - (1 - A)z^{-N-1}] + (b_1 + b_2z^{-1})(1 - A)z^{-N-1}}$$

(26-39)

Recall that the controller $\overline{G}_{DC}$ should eliminate ringing. However, analysis of the roots of the denominator polynomial in Eq. 26-39 (left to the reader) indicates that an additional ringing pole could appear for $N = 1$. For $N \geq 2$ several new ringing poles can appear, although they may not be severe, depending on the tuning parameter A. If $A = 0$ (minimal prototype), the possibility of severe ringing exists even for $N = 1$. If there are errors in the model parameters (a_1, a_2, b_1, b_2), the extent of ringing may also increase depending on their specific values and the size of A.

Below we present an algorithm that is superior in performance to Dahlin's method, especially because of its ability to deal with the ringing phenomenon.

Vogel-Edgar Algorithm

Vogel and Edgar [16] have developed a controller that eliminates the ringing pole due to HG for processes that can be described by a second-order plus time-delay model (Eq. 26-20). The desired closed-loop transfer function is similar to that for Dahlin's controller (cf. (26-31))

$$\left(\frac{C}{R}\right)_d = \frac{(1 - A)}{1 - Az^{-1}} \frac{b_1 + b_2z^{-1}}{b_1 + b_2} z^{-N-1}$$

(26-40)

except that the zeros of the process model (HG) are also included as zeros of the closed-loop transfer function (but divided by $b_1 + b_2$ to preserve the steady state gain = 1). Although this choice of response characteristics may slow down the response somewhat, it makes the controller less sensitive to model errors and also reduces the possibility of ringing. The controller transfer function via (26-16) is

$$G_{VE}(z) = \frac{(1 + a_1z^{-1} + a_2z^{-2})(1 - A)}{(b_1 + b_2)(1 - Az^{-1}) - (1 - A)(b_1 + b_2z^{-1})z^{-N-1}}$$

(26-41)

Note that for $a_2 = b_2 = 0$ (a first-order process), Eq. 26-41 reverts to Dahlin's controller, Eq. 26-28. Because of the form of $(C/R)_d$ in (26-40), this controller does not attempt to cancel the numerator terms of the process transfer function and thus does not include the potential ringing pole. The controller output transfer function is

$$\frac{P(z)}{R(z)} = \frac{(1 - A)(1 + a_1 z^{-1} + a_2 z^{-2})}{(b_1 + b_2)(1 - A z^{-1})} \tag{26-42}$$

If A is chosen to be positive, no ringing will occur. Figure 26.7 compares the closed-loop responses for G_{DC} and $\overline{G}_{DC}$ (ringing pole removed) with the Vogel-Edgar controller (G_{VE}) for a second-order model with time delay. The tuning parameter A is selected in all cases to be 0.368 ($\lambda = \Delta t$). For this second-order system, G_{DC} is clearly unacceptable and G_{VE} is superior to $\overline{G}_{DC}$. For tighter tuning

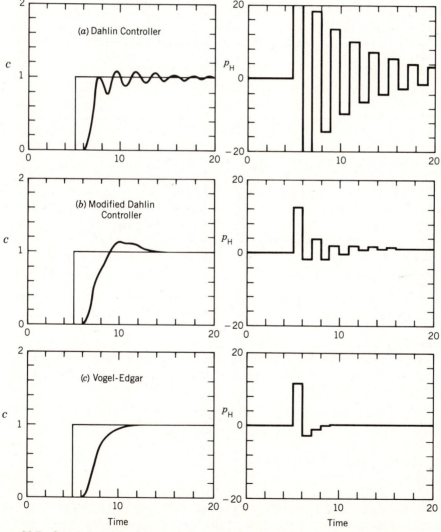

Figure 26.7 Comparison of responses for a step change in set point using (a) Dahlin's controller, (b) Dahlin's controller with ringing pole removed ($\overline{G}_{DC}$) and (c) Vogel-Edgar controller (G_{VE});

$$HG(z) = \frac{(0.0280 + 0.0234 z^{-1}) z^{-2}}{1 - 1.5353 z^{-1} + 0.5866 z^{-2}}$$

($\lambda = A = 0$), the controller output of G_{VE} will change only three times before reaching its steady-state value (see Eq. 26-42). Figure 26.8 makes a similar comparison but with a model that does not yield a ringing pole in G_{DC}. Note that setting $z = 1$ in $\overline{G}_{DC}$ makes the response worse for this process model.

Studies by Vogel and Edgar [16] have shown that their controller satisfactorily handles first-order or second-order process models with positive zeros (inverse response) or negative zeros as well as with simulated process and measurement noise. Many higher-order process models can be successfully controlled with G_{VE}. Neither G_{VE} nor G_{DC} are suitable for unstable process models, however. The robustness of the Vogel-Edgar controller is generally better than Dahlin's con-

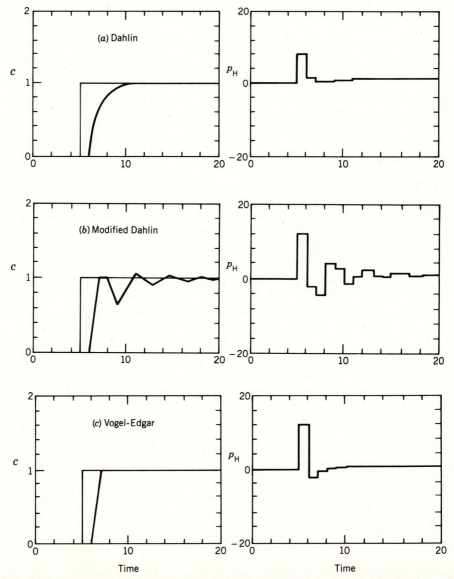

Figure 26.8 Responses for a step change in set point using (a) G_{DC} (Dahlin's controller), (b) $\overline{G}_{DC}$ (G_{DC} with $z = 1$ in numerator of $HG(z)$), and (c) G_{VE} (Vogel-Edgar controller);

$$HG(z) = \frac{(0.0791 - 0.0277z^{-1})z^{-2}}{1 - 1.5353z^{-1} + 0.5866z^{-2}}$$

troller, that is, when model errors occur. For processes with zeros outside the unit circle, Dahlin's controller can become unstable, while the stability of the Vogel-Edgar controller is unaffected.

Internal Model Control (IMC)

The general design methodology of Internal Model Control presented in Section 12.3 for continuous-time systems can be extended to sampled-data systems [19,20]. Figure 12.2 shows the block diagram used for IMC contrasted with that for conventional feedback control (G_v is included in G_p). Here we use the notation G_c^* instead of D for the controller transfer function because of the different block diagram structure and controller design methodology used with IMC. The perfect IMC controller is simply the inverse of the process model,

$$G_c^*(z) = 1/\widetilde{G} \tag{26-43}$$

However, usually a perfect controller is not physically realizable or may be impractical because of model error. The two key steps involved in the controller design are (cf. Section 12.3):

1. The process model is factored as

$$\widetilde{G}(z) = \widetilde{G}_+(z)\widetilde{G}_-(z) \tag{26-44}$$

where $\widetilde{G}_+$ contains the time-delay term z^{-N-1}, zeroes that lie outside the unit circle, and zeroes that lie inside the unit circle near $(-1, 0)$. $\widetilde{G}_+$ has a steady-state gain of unity.

2. The controller is obtained by inverting $\widetilde{G}_-$ (the invertible part of $\widetilde{G}$) and then multiplying by a first-order filter f to improve robustness of the controller as well as to ensure physical realizability of G_c^*:

$$G_c^*(z) = \frac{1}{\widetilde{G}_-(z)} \, f(z) \tag{26-45}$$

The filter f usually contains one or more tuning parameters. Zeroes of $\widetilde{G}$ that lie outside the unit circle (the so-called nonminimum phase zeroes) would yield unstable controller poles if such terms were included in $\widetilde{G}_-$ (instead of $\widetilde{G}_+$). Negative zeroes on the real axis near $z = -1$ cause a ringing controller if they are inverted; hence, they are included in $\widetilde{G}_+$ also. The closed-loop transfer function using the above design rules, assuming the process model is correct, is

$$\frac{C}{R} = \widetilde{G}_+(z)f(z) \tag{26-46}$$

EXAMPLE 26.6

Design an IMC controller for a first-order plus time-delay process given by $\widetilde{G}(s) = e^{-2s}/(5s + 1)$. For $\Delta t = 1$, this corresponds to the pulse transfer function

$$H\widetilde{G}(z) = \frac{0.1813z^{-3}}{1 - 0.8187z^{-1}} \tag{26-47}$$

Solution

For this example,

$$\widetilde{G}_+(z) = z^{-3} \tag{26-48}$$

$$\tilde{G}_-(z) = \frac{0.1813}{1 - 0.8187z^{-1}} \tag{26-49}$$

Let $f(z)$ be a first-order filter with tuning parameter α:

$$f(z) = \frac{1 - \alpha}{1 - \alpha z^{-1}} \tag{26-50}$$

Using Eq. 26-45, the IMC controller is

$$G_c^*(z) = \frac{1 - 0.8187z^{-1}}{0.1813} \frac{(1 - \alpha)}{1 - \alpha z^{-1}} \tag{26-51}$$

Note that this controller has a lead–lag structure. The resulting closed-loop transfer function is

$$\frac{C}{R} = z^{-3} \frac{(1 - \alpha)}{1 - \alpha z^{-1}} \tag{26-52}$$

This expression is the same as that for Dahlin's controller for a first-order system, Eq. 26-28, with $\alpha = e^{-\Delta t/\lambda}$. If $\alpha = 0$, the resulting deadbeat IMC controller yields the same performance as the minimal prototype controller discussed earlier.

The previous example does not give an indication of the flexibility and effectiveness of the IMC design procedure, since for low-order systems we obtained results similar to those derived by direct synthesis. However, the IMC design framework can readily be applied to higher-order systems [20], where direct synthesis is not as reliable. Details on treatment of process model zeroes and selection of the filter can be found in Refs. 19 and 20.

26.4 DIGITAL FEEDFORWARD CONTROL

Digitally based feedforward (FF) control involves the extension of theory for continuous transfer functions presented in Chapter 17. Here we return to the block diagram of a sampled-data, feedback control system (Fig. 25.8) and add the necessary components for feedforward control as shown in Fig. 26.9. It is assumed that the measured value of the disturbance is available as a sampled signal. As with continuous systems, we select G_f so that any disturbances are canceled (i.e., $C = R = 0$). This implies that $c(n\Delta t) = 0$ at the sampling instants, but not necessarily in between. Perfect control, such as is achieved in principle with con-

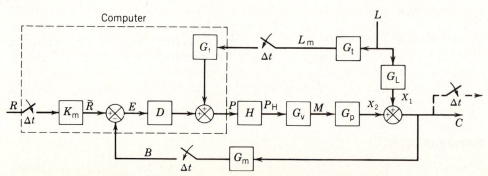

Figure 26.9 Block diagram for digital feedback/feedforward control.

tinuous feedforward control, may not be attainable. Considering only the feedforward path (and ignoring the feedback loop), the appropriate equation for perfect control (with $R = 0$, $C = 0$, and $L \neq 0$) is

$$LG_L(z) + LG_t(z)G_f(z)HG_vG_p(z) = 0 \tag{26-53}$$

Solving for G_f to obtain the FF controller gives

$$G_f(z) = \frac{-LG_L(z)}{LG_t(z)HG_vG_p(z)} \tag{26-54}$$

This result is not as attractive as for continuous control, because the load variable cannot be factored and cancelled, i.e., $LG_L(z)/LG_t(z) \neq G_L(z)/G_t(z)$ in general. However, the ratio HG_L/HG_t may be used in place of $LG_L(z)/LG_t(z)$ when L is a step input or a sequence of steps (piecewise constant input). This assumption removes the requirement that Eq. 26-54 be reevaluated for every different input $L(z)$. The resulting expression will be reasonably accurate except when the load change is not well approximated by a piecewise constant function.

If $G_t(s) = K_t e^{-\theta_t s}$ and $\theta_t = T \Delta t$, then $HG_t(z) = K_t z^{-T}$. Assuming a piecewise-constant load change, Eq. 26-54 becomes

$$G_f(z) = \frac{-HG_L(z)}{K_t z^{-T} HG_v G_p(z)} \tag{26-55}$$

The feedforward controller will not be realizable unless $\theta_L > (\theta_t + \theta_p)$, where θ_p is the process time delay. An unrealizable controller is indicated by a positive power of z premultiplying (26-55). For effective feedforward control, the manipulated variable must be able to act on the process output before the disturbance affects the response through the load transfer function, hence the time delay restriction.

Another way to handle the approximation of G_f is to tune G_f in the field for typical disturbances using adjustable parameters, such as is done for continuous control. In this way the mathematical intractability of Eq. 26-54 can be avoided. A lead–lag digital model with time delay is given by

$$G_f(z) = \frac{K_f(1 + b_f z^{-1})z^{-N_f}}{1 + a_f z^{-1}} \tag{26-56}$$

where K_f, a_f, b_f, and N_f can be tuned to give the desired compensation. This digital controller is roughly equivalent to the lead–lag compensator used in continuous feedforward control (see Eq. 17-33). Note that we can easily implement a pure time delay in G_f. This is an advantage for digital control because a time-delay term can only be approximated using lead-lag components for continuous feedforward control.

One other approach for designing digital feedforward control follows from Section 26.1. Suppose $G_f(s)$ can be obtained from Laplace transform models of the transmitter and process. Once $G_f(s)$ is computed, we can then discretize $G_f(s)$ to obtain $G_f(z)$, using any of the techniques presented in Chapter 24. Again, some on-line adjustment would need to be performed, since discretization errors will be present. However, G_f designed in this way should be a close approximation to the more rigorous version presented in (26-54).

EXAMPLE 26.7

A small distillation column separating methanol and water is controlled by the reflux flow rate; the controlled variable is the overhead composition of methanol.

The major disturbance variable is the composition of the feed stream to the column. By analysis of process dynamic data, the following transfer functions have been developed:

$$G_v G_p(s) = \frac{-5e^{-4s}}{(5s + 1)(3s + 1)} \qquad G_t = 0.2$$

$$G_L(s) = \frac{1.5e^{-4s}}{(7s + 1)(2s + 1)}$$

where the time delay and time constants are expressed in minutes. Develop a digital feedforward controller with $\Delta t = 1$ min, assuming that a piecewise-constant change in $L(s)$ occurs. Use the following strategies:

(a) Dynamic feedforward controller (check for intersample ripple)
(b) Steady-state feedforward controller
(c) A feedforward controller based on a tuned lead–lag unit, Eq. 26-56.

Compare the transient responses for a unit step change in load at $t = 10$ min.

Solution
We use the expression for the feedforward controller in (26-55) based on a piecewise-constant change in the load. For a different load disturbance, "perfect" control will not result, but this is a reasonable simplification for design purposes. For $G_t = K_t$, equation 26-55 becomes

$$G_f(z) = -\frac{HG_L(z)}{K_t HG_v G_p(z)} \qquad (26\text{-}57)$$

Using the discrete-time conversion formulas presented in Chapter 22 for a second-order plus time-delay model, $HG_v G_p(s)$ and $HG_L(s)$ can be calculated:

$$HG_v G_p(z) = \frac{(-0.1399 z^{-1} - 0.1171 z^{-2})z^{-4}}{1 - 1.5353 z^{-1} + 0.5866 z^{-2}}$$

$$HG_L(z) = \frac{(0.0435 z^{-1} + 0.0351 z^{-2})z^{-4}}{1 - 1.4734 z^{-1} + 0.5258 z^{-2}}$$

(a) For $K_t = 0.2$, the feedforward controller transfer function is

$$G_f(z) = \frac{-0.2174 z^{-1} + 0.1583 z^{-2} + 0.1419 z^{-3} - 0.1029 z^{-4}}{-0.1399 z^{-1} + 0.089 z^{-2} + 0.0990 z^{-3} - 0.0616 z^{-4}}$$

For a unit step load change, the response of the controlled variable is shown in Fig. 26.10. The deviations are quite small (note the scale of c has been multiplied by 1000). The intersample ripple (almost indetectable from the plot of p_H, the controller output) is due to a ringing pole in $G_f(z)$. Practically speaking, the small error in this case does not justify modifying $G_f(z)$ for ringing, although it may be necessary for larger disturbances. For comparison, see Fig. 26.11 discussed below.

(b) Steady-state feedforward control can be derived using $G_f(z = 1)$; in this case $G_f = 1.5$. This result is the same as would be obtained using a continuous-time approach. The response to a load change using this controller load change is shown in Fig. 26.11.

(c) A lead–lag expression for the feedforward controller allows four parameters to be optimized: K_f, a_f, b_f, and N_f. To achieve appropriate time-delay compensation, $N_f = 0$ ($\theta_L = \theta_p$). Also good steady-state behavior should be obtained,

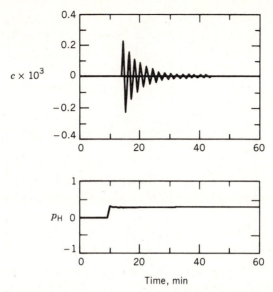

Figure 26.10 Responses of the controlled variable and controller output to a step change in the load using dynamic feedforward control, Example 26.6 (note expanded scale on c).

implying that $K_f(1 + b_f)/(1 + a_f) = 1.5$, the same value found in (b) above. This leaves two degreees of freedom, a_f and b_f. Suitable values found by visual tuning of the closed-loop response are $a_f = -0.90$ and $b_f = -0.89$. Figure 26.11 also shows the load response for this case.

26.5 COMBINED LOAD ESTIMATION AND TIME-DELAY COMPENSATION

In Section 26.3 we covered design principles for digital Internal Model Control, which is based on the block diagram presented in Fig. 12.2. The IMC block diagram can be expanded, as shown in Fig. 26.12, to include a block A^* in the feedback path as well as a load transfer function G_L. The block A^* can be used to predict the effect of the disturbance on the error signal to the controller. It must be chosen by the designer along with G_c^*, the feedback controller, given in Eq. 26-45. The use of G_c^* and A^* allows the designer to optimize closed-loop performance for both load and set-point changes (rather than obtain a compromise in performance

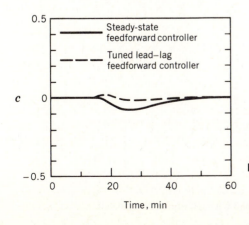

Figure 26.11 Comparison of responses for steady-state feedforward controller with a tuned lead–lag feedforward controller, Example 26.6.

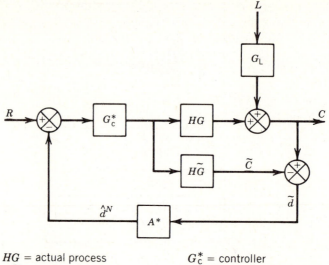

HG = actual process G_c^* = controller
$\widetilde{HG}$ = process model G_L = load transfer function
A^* = disturbance predictor/filter $\hat{d}^N = \tilde{d}$ predicted N steps ahead

Figure 26.12   Model-predictive block diagram for analytical predictor

between these two designs). In addition, A^* can provide for time-delay compensation, and when used for this purpose, it is known as an *analytical predictor* (AP).

The analytical predictor was originally developed for digital control systems as an alternative to the Smith predictor. Doss and Moore [21] proposed a discrete analytical predictor for second-order processes primarily to deal with unmeasured disturbances and their effects on the controlled variable. Suppose the load change is a step function (allowing the use of HG_L as discussed in Section 26.4), and $HG_L(z) = HG(z) = HG^*(z)z^{-N}$, where $HG^*(z)$ is that part of the process model that does not contain the time delay. Doss and Moore used Fig. 26.13 to configure the analytical predictor. Note that a prediction element z^N has been included in the feedback loop (i.e., $\hat{c}$ must be predicted N time steps ahead). Block diagram algebra then yields the closed-loop response for a set-point change R and/or a step load change L:

$$C(z) = \frac{D(z)HG(z)}{1 + D(z)HG^*(z)} R(z) + \frac{HG(z)L(z)}{1 + D(z)HG^*(z)} \qquad (26\text{-}58)$$

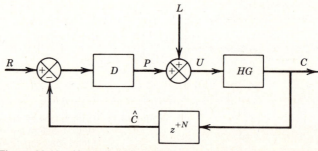

Figure 26.13   Alternative block diagram for the analytical predictor.

For either a set-point or load change, the denominator (characteristic equation) does not contain a time delay, which increases the stability margin of the controller $D(z)$. This derivation is based on a perfect model. The predictor element z^N cannot be implemented directly but there is a way to carry out such a prediction, as dicussed below. Meyer et al. [22] tested the AP algorithm on a pilot-scale distillation column, with some success. However there are some disadvantages of the AP, such as the assumption that $HG_L = HG_p$, which can be eliminated by a more general approach, called the Generalized Analytical Predictor or GAP [23,24].

Generalized Analytical Predictor

The derivation of the GAP algorithm can be performed for any selected transfer functions $\tilde{G}$ or G_L in Fig. 26.12; hence, the term *generalized*. Here we assume a first-order plus time-delay model for both $\tilde{G}$ and G_L and that no model error is present ($\tilde{G} = G$). Recall that $\tilde{G}$ includes the valve and transducer dynamics in addition to the process model.

Referring to Fig. 26.12, a current estimate of the effect of an entering distur-bance on the process, $\tilde{d}(z)$, can be obtained from the difference of the actual output C and the process model output $\tilde{C}$.

$$\tilde{d}(z) = C(z) - \tilde{C}(z) \tag{26-59}$$

The effect of the disturbance on the future output can be predicted by estimating the current load L and then predicting N steps into the future (N is the process model time delay). The prediction horizon can also be selected to be greater than N, for example, $N+1$ steps [23]. The load is assumed constant over the prediction horizon N. To estimate L from $\tilde{d}$, the load transfer function must be known or assumed. Assume that the load model is given by $H\tilde{G}_L$, which is a first-order transfer function. $H\tilde{G}_L$ is based on a piecewise constant input (recall that the load is assumed constant over the prediction horizon):

$$H\tilde{G}_L(z) = \frac{\bar{b}z^{-1}}{1 - \bar{a}z^{-1}} \tag{26-60}$$

The load model numerator and denominator coefficients are overlined to distinguish them from the process model coefficients a and b. Higher order models for $\tilde{G}_L$ can also be considered [23,24]. Note that no time delay has been assumed in Eq. 26-60. Since the disturbance is unmeasured and determined only by comparison of predicted and actual outputs, the true time delay of the load transfer function is unknown (and cannot be determined). The use of Eq. 26-60 differs from the AP, where it is assumed that $\tilde{G}_L = \tilde{G}$, that is, the load is additive with the control-ler output (see Fig. 26.13).

Assuming there is no process modeling error and that the load is a step input gives

$$\tilde{d}(z) = HG_L(z)L(z) \tag{26-61}$$

where $L(z)$ is the actual load and $HG_L(z)$ is the actual load transfer function derived for a piecewise constant input (based on the step change in L). Substituting the assumed load transfer function $H\tilde{G}_L(z)$ for $HG_L(z)$ and writing (26-60) and (26-61) as a difference equation yields

$$\tilde{d}_k = \bar{a}\tilde{d}_{k-1} + \bar{b}L_{k-1} \tag{26-62}$$

Assuming the step change in load occurred at time $k-1$, an estimate of the load based on measured values of $\tilde{d}_k$ and $\tilde{d}_{k-1}$ is given by rearranging (26-62):

$$\hat{L}_{k-1} = \frac{1}{\bar{b}}\,(\tilde{d}_k - \bar{a}\tilde{d}_{k-1}) \tag{26-63}$$

or in terms of z-transforms,

$$z^{-1}\hat{L}(z) = \frac{1 - \bar{a}z^{-1}}{\bar{b}}\,\tilde{d}(z) \tag{26-64}$$

The disturbance prediction one time delay ahead, $\hat{d}_{k+N}$, simulates the block z^N in Fig. 26.13. This prediction is based on the load estimate $\hat{L}$ held constant over N time steps. The predicted value $\hat{d}$ of the effect of the disturbance can be determined by solving Eq. 26-62 over a time horizon of N steps. Note that the term z^N cannot actually be implemented physically.

$$\hat{d}_{k+1} = \bar{a}\tilde{d}_k + \bar{b}\hat{L}_{k-1}$$

$$\hat{d}_{k+2} = \bar{a}\hat{d}_{k+1} + \bar{b}\hat{L}_{k-1}$$

$$\cdot$$
$$\cdot \tag{26-65}$$
$$\cdot$$

$$\hat{d}_{k+N} = \bar{a}\hat{d}_{k+N-1} + \bar{b}\hat{L}_{k-1}$$

The load estimate $\hat{L}_{k-1}$ is used throughout Eq. 26-65 because the load is assumed to be a step input so $\hat{L}_{k-1} = \hat{L}_k = \ldots = \hat{L}_{k+N-1}$. By successive substitution Eq. 26-65 can be simplified to a single closed-form prediction equation:

$$\hat{d}_{k+N} = \bar{a}^N\tilde{d}_k + \frac{1 - \bar{a}^N}{1 - \bar{a}}\,\bar{b}\hat{L}_{k-1} \tag{26-66}$$

If the load estimate is filtered before the prediction (because the $\tilde{d}_k$ data are noisy or subject to measurement errors), then Eq. 26-64 becomes

$$z^{-1}\hat{L}(z) = F_L(z)\frac{(1 - \bar{a}z^{-1})}{\bar{b}}\,\tilde{d}(z) \tag{26-67}$$

where F_L is a first-order filter with tuning parameter β (see Chapter 22):

$$F_L(z) = \frac{1 - \beta}{1 - \beta z^{-1}} \qquad 0 \le \beta < 1 \tag{26-68}$$

When the filter (26-68) is used ($\beta \ne 0$), the prediction equation in the z-domain is derived by taking the z-transform of (26-66) and combining it with (26-67):

$$\hat{d}^N(z) = A^*(z)[C(z) - \hat{C}(z)] \tag{26-69}$$

where $A^*(z) = \bar{a}^N + \dfrac{1 - \bar{a}^N}{1 - \bar{a}}\,F_L(z)(1 - \bar{a}z^{-1})$ \hfill (26-70)

and $\hat{d}^N = \tilde{d}$ predicted N steps ahead.

Because the GAP builds on the concept of internal model control, the IMC controller design procedure can be used for the block G_c^* in Fig. 26.12. The advantage of using the generalized analytical predictor compared to the AP is shown in the example below.

EXAMPLE 26.7

A first-order process with transfer function

$$G(s) = \frac{e^{-2s}}{5s + 1}$$

is controlled using an IMC controller ($\Delta t = 1$). Assume no modeling error ($G = \tilde{G}$). The load transfer function is known to be

$$G_L(s) = \frac{1}{s + 1}$$

A unit step change in the load occurs at time $t = 5$. Compare the performance of GAP, AP, and IMC with no load estimation ($A^* = 1$) for this system. Use a deadbeat IMC controller for G_c^* (filter time constant α of zero). Since it is assumed that there is no model error between G and $\tilde{G}$, use the deadbeat load estimator ($\beta = 0$ in Eq. 26-68).

Solution

The feedback IMC controller for this system was designed in Example 26.6. The deadbeat controller ($\alpha = 0$), using Eq. 26-52, is

$$G_c^*(z) = \frac{1 - 0.8183z^{-1}}{0.1817}$$

The disturbance prediction transfer function is given by

$$H\tilde{G}_L(z) = \frac{(1 - e^{-\Delta t/\tau_L})z^{-1}}{1 - e^{-\Delta t/\tau_L}z^{-1}}$$

where τ_L is the Laplace domain time constant of $\tilde{G}_L$, the load transfer function model. The simulation results for the following cases are shown in Figs. 26.14 to 26.16:

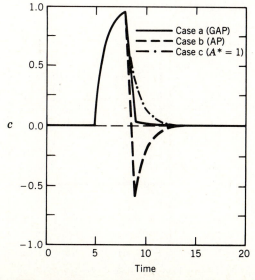

Figure 26.14 Controlled variable responses for Example 26.7.

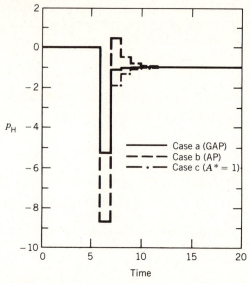

Figure 26.15 Controller outputs (after ZOH) for Example 26.7.

Case a: GAP with perfect prediction and no model errors ($\tau_L = 1$).

Case b: AP using $H\tilde{G}_L = H\tilde{G}$ (model error, $\tau_L = 5$).

Case c: IMC controller with no load estimation ($A^* = 1$, $\tau_L = 0$).

The output response for Case *b* exhibits severe overcorrection returning to the set point because $G_L \neq \tilde{G}$. Since the load estimator assumes perfect measurements ($\beta = 0$), the initial load estimates are much too large (Fig. 26.16) and the disturbance is overpredicted. This results in excessive controller action (Fig. 26.15) and output overshoot (Fig. 26.14). In contrast, the GAP in Case *a* yields an excellent output response with less control action. To illustrate the usefulness of the load estimation filter, compare Case *a* with Case *c*, where no load dynamics are assumed

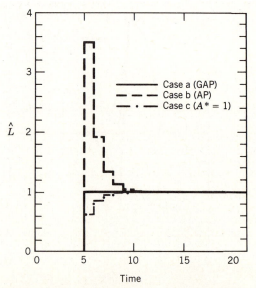

Figure 26.16 Load estimates for Example 26.7.

($A^* = 1$). The resulting response (Case c) is not as good as with $\tau_L = 1$ (Case a), but it shows improvement over the AP response. Thus, the GAP with the generalized disturbance predictor provides improved regulatory response compared to the AP when the load transfer function is not equal to the process transfer function. In addition, the GAP controller gives improved performance over the IMC controller with $A^* = 1$, which does not incorporate a load prediction scheme.

The GAP is relatively insensitive to unknown load transfer function dynamics, as shown by Wellons and Edgar [23]. If G_L is a first-order plus time-delay model, neither the gain nor the time delay need to be known for $\tilde{G}_L$. Only the time constant must be estimated (note that A^* in Eq. 26-70 depends solely on $\bar{a}$). Wellons and Edgar found that an approximate value for the time constant in $\tilde{G}_L$ was generally sufficient to achieve improved responses over AP, although it was necessary to use load filtering ($\beta \neq 0$) in those cases.

SUMMARY

In this chapter we have presented a number of different approaches for designing digital feedback controllers. Digital controllers that emulate electronic or pneumatic PID controllers can include a number of special features to improve operability. If the time delay is small (e.g., $\theta/\tau < 0.5$), a digital PID controller without time-delay compensation should provide satisfactory performance. Controllers based on Direct Synthesis techniques are most valuable for processes where the time delay is significant compared to the dominant time constant or the process is of high order (yielding an apparent time delay). Recently developed methods avoid ringing, eliminate offset, are very easy to tune, and provide a high level of performance for set-point changes. Load changes should be treated using either digital feed-forward control (when the load change is measurable) or an analytical predictor (for unmeasured load changes). Such predictive controllers represent very powerful approaches for achieving high-performance control systems when model parameters are reasonably well-known. These techniques can also be made adaptive when model parameters are changing or uncertain.

REFERENCES

1. Fertik, H. A., and C. W. Ross, Direct Digital Control Algorithm with Anti-Windup Feature. Preprint 10-1-ACOS-67, 2d Annual ISA Conference, Chicago, IL, Sept. 1967.
2. Corripio, A. B., Digital Control Techniques, Module 3.5 in AIChEMI Series, Series A, Vol. 3 (Process Control), AIChE, New York 1983.
3. Isermann, R., *Digital Control Systems,* Springer-Verlag, Berlin, 1981.
4. Bristol, E., Designing and Programming Control Algorithms for DDC Systems, *Cont. Eng.,* 24, (Jan. 1977).
5. Åström, K. J., and B. Wittenmark, *Computer Controlled Systems*, Prentice-Hall, Englewood Cliffs, NJ, 1984.
6. Mellichamp, D. A., D. R. Coughanowr, and L. B. Koppel, Identification and Adaptation in Control Loops with Time-Varying Gain, *AIChE J.* **12**(1), 83 (1966).
7. Smith, C. L., *Digital Computer Process Control,* International Textbook, Scranton, PA, 1972.
8. Moore, C. F., C. L. Smith, and P. M. Murrill, Simplifying Digital Control Dynamics for Controller Tuning and Hardware Lag Effects, *Instrum. Practice*, p. 45 (Jan. 1969).
9. Deshpande, P. B., and R. H. Ash, *Elements of Computer Process Control,* Instrum. Soc. of America, Research Triangle Park, NC, 1981.
10. Lopez, A. M., P. W. Murrill, and C. L. Smith, Tuning PI and PID Digital Controllers, *Inst. Cont. Syst.* **42**(2), 89 (1969).

11. Mosler, H. A., L. B. Koppel, and D. R. Coughanowr, Sampled Data Proportional Control of a Class of Stable Processes, *IEC Proc. Des. Dev.* **5**, 297 (1966).
12. Mosler, H. A., L. B. Koppel, and D. R. Coughanowr, Sampled Data Proportional-Integral Control of a Class of Stable Processes, *IEC Proc. Des. Dev.* **6**, 221 (1967).
13. Franklin, G. F., and J. D. Powell, *Digital Control of Dynamic Systems*, Addison-Wesley, Reading, MA, 1980.
14. Luyben, W. L., *Process Modeling, Simulation, and Control for Chemical Engineers*, McGraw-Hill, New York, 1973.
15. Ogata, K., *Discrete-Time Control Systems*, Prentice-Hall, Englewood Cliffs, NJ, 1987.
16. Vogel, E. F., and T. F. Edgar, A New Dead Time Compensator for Digital Control, *ISA/80 Proc.*, Houston, TX, Oct. 1980.
17. Dahlin, E. B., Designing and Tuning Digital Controllers, *Instrum. and Control Systems* **41**(6), 77 (1968).
18. Higham, J. D., Single-Term Control of First and Second Order Processes with Dead Time, *Control*, p. 136 (Feb. 1968).
19. Garcia, C. E., and M. Morari, Internal Model Control, 1. A Unifying Review and Some New Results, *IEC Proc. Des. Dev.* **21**, 308 (1982).
20. Zafiriou, E., and M. Morari, Digital Controllers for SISO Systems: A Review and a New Algorithm, *Int. J. Control* **42**, 885 (1985).
21. Doss, J. E., and C. F. Moore, The Discrete Analytical Predictor—A Generalized Dead-Time Compensation Technique, *ISA Trans.* **20**(4), 77 (1982).
22. Meyer, C., D. E. Seborg, and R. K. Wood, An Experimental Application of Time-Delay Compensation Techniques to Distillation Column Control, *IEC Proc. Des. Dev.* **17**, 1 (1978).
23. Wellons, M. C., and T. F. Edgar, The Generalized Analytical Predictor, *IEC Research* **26**, 1523 (1987).
24. Wong, K. P., and D. E. Seborg, A Theoretical Analysis of Smith and Analytical Predictors, *AIChE J* **32**, 1597 (1986).

EXERCISES

26.1. The block diagram of a sampled-data control system is shown in the drawing. The sampling period is $\Delta t = 1$ min.

(a) Design the digital controller $D(z)$ so the closed-loop system exhibits a minimal prototype response to a unit step change in the load variable L.

(b) Will this controller eliminate offset after a step change in the set point? Justify your answer.

(c) Is the controller physically realizable? Justify your answer.

(d) Design a digital PID controller based on the ITAE (set-point) criterion and examine its performance for a step change in set point.

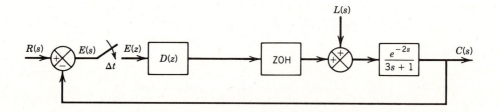

26.2. The exit composition c_3 of the blending system shown in the drawing is controlled using a digital feedback controller. The exit stream is automatically sampled every minute and the composition measurement is sent from the composition transmitter (CT) to the digital controller. The controller output is sent to a ZOH device before being transmitted to the control valve.

(a) Using the information given below, draw a block diagram for the feedback control system. (Use the symbols in the figure as much as possible.)

(b) Derive an expression for the pulse transfer function, C_3/Q_2, where C_3 and Q_2 are deviation variables.

(c) Suppose that the digital controller $D(z)$ is to be designed so that the closed-loop system exhibits a minimal prototype response to a unit step change in the load. Specify the form of the desired response, $(C/R)_d$. It is *not* necessary to derive an expression for $D(z)$ but you should justify your choice for $(C/R)_d$.

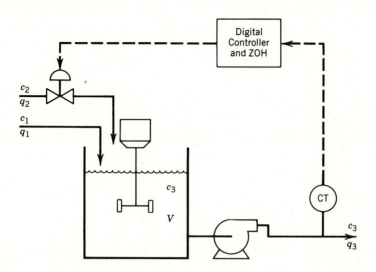

Available Information

i. Since flow rate q_2 is quite small, the liquid volume in the tank V remains essentially constant at 30 ft^3. The tank is perfectly mixed.

ii. The primary load variable is inlet composition c_2.

iii. The control valve has negligible dynamics and a steady-state gain of 0.1 ft^3/min mA.

iv. The composition transmitter (CT) has a steady-state gain of 2.5 mA/(lb-mole solute/ft^3). Composition samples are analyzed every minute, that is, the sampling period is $\Delta t = 1$ min. There is also a 1 min time delay associated with the composition analysis.

v. Nominal steady-state values (denoted by a bar) are

$$\bar{q}_2 = 0.1 \text{ ft}^3/\text{min} \qquad \bar{c}_2 = 1.5 \text{ lb-mole solute/ft}^3$$

$$\bar{q}_3 = 3 \text{ ft}^3/\text{min} \qquad \bar{c}_3 = 0.2 \text{ lb-mole solute/ft}^3$$

26.3. A digital controller has the following input/output relation:

$$p_n = a_0 e_n + a_1 e_{n-1} + a_2 e_{n-2} + \cdots$$
$$+ a_M e_{n-M} - b_1 p_{n-1} - b_2 p_{n-2} - \cdots - b_N p_{n-N}$$

where $p_n =$ controller output at $t = n\Delta t$ ($n =$ positive integer);
$\qquad e_n =$ error signal at $t = n\Delta t$;
$\qquad M, N$ are positive integers; and $\{a_i\}$ and $\{b_i\}$ are constants.

In answering the following questions, be as specific as possible:

(a) Suppose that this controller is to contain integral action; what restrictions (if any) does this place on the $\{a_i\}$ and $\{b_i\}$ coefficients?

(b) What conditions must be satisfied for the digital controller to have a steady-state gain?

(c) Derive an expression for the steady-state gain, assuming that it does exist.

26.4. The block diagram of a sampled-data control system is shown in the drawing. Design a deadbeat controller $D(z)$ that is physically realizable and based on a change in set point. The sampling period is $\Delta t = 1$ min. Calculate the closed-loop response when this controller is used and a unit step change in load occurs.

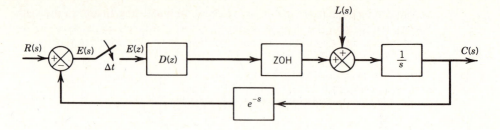

26.5. It is desired to control the exit temperature T_2 of the heat exchanger shown in the drawing by adjusting the steam flow rate w_s. Unmeasured disturbances occur in inlet temperature T_1. The dynamic behavior of the heat exchanger can be approximated by the transfer functions

$$\frac{T_2'(s)}{W_s'(s)} = \frac{2.5}{10s + 1} [=] \frac{°F}{lb_m/s}$$

$$\frac{T_2'(s)}{T_1'(s)} = \frac{0.9}{5s + 1} [=] \text{ dimensionless}$$

where the time constants have units of seconds and the primes denote deviation variables. The control valve and temperature transmitter have negligible dynamics and steady-state gains of $K_v = 0.2 \text{ lb}_m/s/mA$ and $K_m = 0.25 \text{ mA}/°F$. Design a minimal prototype controller that is physically realizable and based on a unit step change in the load variable T_1. Assume that a zero-order hold is used and that the sampling period is $\Delta t = 2$ s.

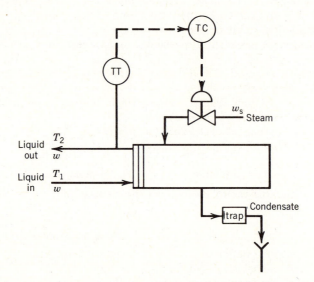

26.6. A second-order system with $K = 1$, $\tau_1 = 6$, and $\tau_2 = 4$ is to be controlled using the Vogel-Edgar controller with $\lambda = 5$ and $\Delta t = 1$. Assuming a step change in R,

(a) Write the difference equation for the output of the controller and compute p_n ($n = 0, 1, \ldots, 25$)

(b) Compute the controlled variable c_n over a similar time interval. Assume $G_v = G_m = 1$.

26.7. Compare minimal prototype, PID (minimum ITAE for set-point changes), and Dahlin controllers for $\Delta t = 1$, $\lambda = 1$, and $G_p(s) = 2e^{-s}/(10s + 1)$. Adjust for ringing if necessary. Plot the closed-loop responses for a set-point change as well as the controller output for each case.

26.8. Given a control system with $G_m = 1$ and $G(s) = 1.25e^{-5s}/(5s + 1)$, write the equation for Dahlin's controller with $\Delta t = 1$ and $\lambda = 1$. Will ringing occur?

26.9. Design feedback controllers for $G(s) = 1/[s(s + 1)]$, $G_m = 1$, and $\Delta t = 1$ using

(a) Minimal prototype response
(b) Vogel-Edgar algorithm

Check for physical realizability.

26.10. For $G(s) = 1/[(s + 1)(s + 2)]$ and $\Delta t = 1$, design a nonringing Dahlin controller with $\lambda = \Delta t$. Write the corresponding difference equation. Can you show the correspondence of this controller to a digital PID controller?

26.11. Compare Dahlin and Vogel-Edgar controllers for $G(s) = 1/[(2s + 1)(s + 1)]$ and $\lambda = \Delta t = 1$. Does either controller ring? Write the resulting difference equations for the closed-loop system (c_n related to r_n). Does overshoot occur in either case?

26.12. Design a digital controller for the liquid level in the storage system shown in the drawing. Each tank is 2.5 ft in diameter. The piping between the tanks acts as a linear resistance to flow with $R = 2$ min/ft^2. The liquid level is sampled every 30 s. The digital controller also acts as a zero-order hold device for the signal sent to the control valve. The control valve and level transmitter have negligible dynamics. Their gains are $K_v = 0.25$ ft^3/min/mA and $K_t = 8$ mA/ft, respectively. The nominal value of q_1 is 0.5 ft^3/min.

(a) Derive a deadbeat control algorithm based on a step change in inlet flow rate q_1. Assume that the load change occurs at the worst possible time with regard to sampling.
(b) Does the controller output exhibit any rippling?
(c) Derive an expression for Dahlin's algorithm. For what values of λ is the resulting controller physically realizable?
(d) If you were to tune this controller on line, what value of λ would you use as an initial guess? Justify your answer.

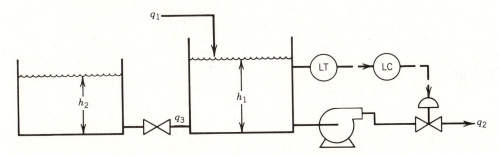

26.13. Feedforward control applications often utilize a controller that consists of a lead-lag unit:

$$G_f(s) = \frac{K(\tau_1 s + 1)}{\tau_2 s + 1} \qquad (\tau_1 \neq \tau_2)$$

Develop expressions for the controller output at the nth sampling instant p_n using the following methods:

(a) The pulse transfer function $G_f(z)$ corresponding to $G_f(s)$.
(b) Approximating s by Tustin's method.

Compare the unit step responses for the expressions in parts (a) and (b) when $K = 1$, $\tau_1 = 2$ min, $\tau_2 = 5$ min, and $\Delta t = 1$ min.

26.14. A second-order process with transfer function

$$G(s) = \frac{-5e^{-4s}}{(3s + 1)(5s + 1)} \qquad G_m = 1$$

is controlled with a digital IMC controller ($\Delta t = 1$) with no load estimation. There is no modeling error. The deadbeat controller filter constant ($\alpha = 0$ in Eq. 26-50) is used. A unit step change in load occurs at time $t = 10$, with a known load transfer function

$$G_L(s) = \frac{1}{7s + 1}$$

Compare IMC ($A^* = 1$) and GAP responses for a prediction horizon $N = 4$ and a deadbeat load estimation filter.

— CHAPTER 27 —

Predictive Control Techniques

Chapter 26 was devoted to design methods for digital controllers applied to SISO processes. These methods are based on z-transform analysis and often utilize a process model to predict the future values of the process outputs. For example, in time-delay compensation the prediction horizon is chosen to equal the time delay. In this chapter we consider a more general approach to *model predictive control*, where the process model is used to predict future outputs over a longer time period. Here, the prediction horizon is usually longer than the time delay; for example, it can be based on the open-loop response time.

Currently the most comprehensive and powerful model predictive control techniques are those based on optimization of a quadratic objective function involving the error between the set point and the predicted outputs. While these techniques are computationally intensive, they can be readily implemented on a process control computer. Design methods in this category are based on a particular type of discrete-time model (a convolution model) and include Dynamic Matrix Control (DMC) [1,2] and Model Algorithmic Control (MAC) [3,4]. Both of these techniques have been used successfully in commercial process control applications [2,3] involving MIMO processes. They also can handle inequality constraints on the controlled and manipulated variables.

In this chapter, we first introduce a modeling approach based on convolution models, which differs from the transfer function models used prior to this chapter. Then we show how the minimization of a quadratic objective function leads to a feedback controller, which can be designed by straightforward matrix operations. The controller can be tuned to achieve desirable performance. We show the effects of tuning parameters on closed-loop performance for several systems and compare predictive control results with those for PID control. Finally, the extension of the design method to MIMO processes is presented.

27.1 DISCRETE CONVOLUTION MODELS

In previous chapters, we have used transfer function models, usually first- or second-order models with time delay, to describe dynamic behavior. Such models are parametric in nature and require the model order to be specified. For processes that exhibit unusual dynamic behavior (see Fig. 27.1), it may be quite difficult to select appropriate first-order or second-order model structures. One approach that

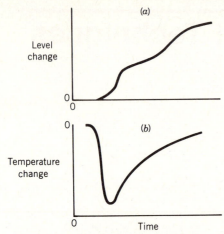

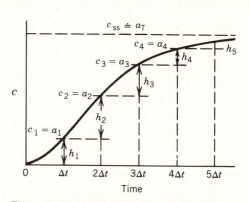

Figure 27.1 Two processes exhibiting unusual dynamic behavior [2]. (a) change in base level due to a step change in feed rate to a distillation column. (b) steam temperature change due to switching on soot blower in a boiler.

Figure 27.2 Identification of coefficients of step response (a_i) and convolution (h_i) models (for the situation where $c_0 = 0$).

avoids this problem is to employ a discrete impulse response (convolution) model. The advantage of the discrete convolution model is that the model coefficients, which may number as many as 50, can be obtained directly from the experimental step response without assuming a model structure. This flexibility is helpful in modeling unusual process behavior, such as that shown in Fig. 27.1. Theoretically, this modeling approach is applicable to any linear system. The convolution model also provides a convenient way to design a controller based on the use of optimization theory.

Step Response Model

To illustrate how a convolution model is developed, consider the typical open-loop step response shown in Fig. 27.2. The values of the unit step response are given by $a_0, a_1, a_2, \ldots, a_T$ using the sampling period Δt. We define $a_i = 0$ for $i \leq 0$. $T\Delta t$ may be taken to be the settling time of the process (the time for the open-loop step response to reach 99% completion) and the integer T is called the *model horizon*.

Now consider the step response resulting from a change Δm in the input. Let $\hat{c}_n$ be the predicted value of the output variable and m_n the value of the manipulated variable at the nth sampling instant. We also define c_n as the actual output; thus $\hat{c}_n = c_n$ if there is no modeling error and no disturbances. Both c and m are expressed as deviation variables. Denoting $\Delta m_i = m_i - m_{i-1}$, the convolution model is

$$\hat{c}_{n+1} = c_0 + \sum_{i=1}^{T} a_i \, \Delta m_{n+1-i} \qquad (27\text{-}1)$$

Equation 27-1 can be interpreted as the sum of a series of step changes Δm_i. Suppose the system is initially at an initial value c_0 and a step change in the input

Δm_0 is made with no subsequent input changes. Then $\hat{c}$ can be calculated as follows:

$$\hat{c}_1 = c_0 + a_1\,\Delta m_0$$

$$\hat{c}_2 = c_0 + a_2\,\Delta m_0$$

$$\cdot$$

$$\cdot \qquad\qquad (27\text{-}2)$$

$$\cdot$$

$$\hat{c}_T = c_0 + a_T\,\Delta m_0$$

Note that the predicted values of c simply follow the step response coefficients a_1, $a_2, \ldots, a_T$ (see Fig. 27.2), multiplied by the magnitude of the input change. If c_0 is the normal operating point, then $c_0 = 0$ because c is a deviation variable.

Suppose that two sequential input changes, Δm_0 and Δm_1, are made at times $i = 0$ and $i = 1$. Then using the principle of superposition,

$$\hat{c}_1 = c_0 + a_1\,\Delta m_0$$

$$\hat{c}_2 = c_0 + a_2\,\Delta m_0 + a_1\,\Delta m_1$$

$$\hat{c}_3 = c_0 + a_3\,\Delta m_0 + a_2\,\Delta m_1$$

$$\cdot$$

$$\cdot \qquad\qquad (27\text{-}3)$$

$$\cdot$$

$$\hat{c}_T = c_0 + a_T\,\Delta m_0 + a_{T-1}\,\Delta m_1$$

The generalization of Eq. 27-3 to T input changes is given by the convolution model (27-1).

In this chapter we express the model and predictive control law in terms of the manipulated variable m_n, rather than controller output p_n, as was done in Chapter 26. Typically the implementation of model predictive control requires sophisticated optimization calculations to calculate m_n, in contrast to the simple difference equations previously employed for p_n in Chapter 26. For model predictive control, it is usually assumed that values of m_n are transmitted as set points for the process input, for example, a flow rate. In this case, the control valve is arranged in a cascade loop as shown in Fig. 21.3, so that the manipulated variable closely tracks the set point. If the control valve has significant dynamics, however, it would be necessary to perform the predictive controller calculations using p rather than m, as done in Chapter 26. In this case, the predictive model would be based on $G_v G_p G_m$ rather than $G_p G_m$.

Impulse Response Model

Recall from Chapter 3 that the impulse response can be expressed as the first derivative of the step response. For a digital system with a zero-order hold the impulse response can be found by taking the first backward difference of the step response. The unit impulse response coefficients of the process, $h_1, h_2, \ldots, h_T$, then are given by

$$h_i = a_i - a_{i-1} \quad i = 1, 2, \ldots, T \tag{27-4}$$

$$h_0 = 0$$

and the discrete convolution model using the impulse response coefficients is

$$\hat{c}_{n+1} = c_0 + \sum_{i=1}^{T} h_i m_{n+1-i} \tag{27-5}$$

Equation 27-5 can be rearranged to Eq. 27-1 by substituting the expression for h_i in (27-4) and then grouping terms for each a_i.

Note that Fig. 27.2 illustrates the case where there is no time delay in the process model. Time delays are present when a_i (or h_i) = 0 for a number of terms. If there is a time delay of N units, then $a_0, a_1, \ldots, a_N$ are zero in (27-1). Similarly, $h_i = 0$ $(0 \le i \le N)$ in (27-5).

27.2 z-TRANSFORM ANALYSIS OF CONVOLUTION MODELS

The convolution models presented in the previous section are appealing because the step response can be employed directly without requiring detailed calculations. On the other hand, the impulse or step response coefficients can be computed by linear regression if the input is not a step function, using the methodology presented in Section 23.4. It is also straightforward to demonstrate the equivalence of convolution models and transfer function models, such as first or second-order with time delay. Taking the z-transform of (27-5), we have

$$\frac{\hat{C}(z)}{M(z)} = HG(z) = \sum_{i=1}^{T} h_i z^{-i} \tag{27-6}$$

which is the pulse transfer function based on a piecewise-constant input and a process model with T zeroes and no poles. Next we examine the properties of $HG(z)$ in the following example.

EXAMPLE 27.1

The process is a first-order continuous transfer function with zero-order hold, $\tau = 1$, $\Delta t = 0.2$, and $K = 1$.

(a) Derive a convolution model using long division of $HG(z)$
(b) Obtain the unit step response using the convolution model in the z-domain and compare with the result using continuous-time analysis.

Solution
(a) The zero-order hold transfer function for $\Delta t/\tau = 0.2$ (see Eq. 24-99) is

$$HG(z) = \frac{(1 - 0.8187)z^{-1}}{1 - 0.8187z^{-1}} = \frac{0.1813z^{-1}}{1 - 0.8187z^{-1}} \tag{27-7}$$

The convolution model equivalent to (27-7) can be obtained through the use of long division (see Section 24.2):

$$HG(z) = 0.1813z^{-1} + 0.1484z^{-2} + 0.1215z^{-3} + 0.0995z^{-4}$$
$$+ 0.0815z^{-5} + \cdots \tag{27-8}$$

(b) The unit step response is given by

$$\hat{C}(z) = HG(z)M(z) = HG(z)\frac{1}{1 - z^{-1}} \qquad (27\text{-}9)$$

Hence, by a second application of long division in (27-9),

$$\hat{C}(z) = 0.1813z^{-1} + 0.3297z^{-2} + 0.4512z^{-3} + 0.5507z^{-4} \qquad (27\text{-}10)$$
$$+ 0.6321z^{-5} + \cdots$$

We can compare the coefficients in (27-10) with the analytical step response, $c(t) = 1 - e^{-t/\tau}$, for a sampling interval of $\Delta t = 0.2$, and it agrees exactly. In addition, we can confirm that $h_i = a_i - a_{i-1}$ by comparing (27-8) and (27-10).

In the above example the response $\hat{C}(z)$ is based on the step response coefficients $\{a_i\}$ in Fig. 27.2. In other words,

$$\hat{C}(z) = \sum_{i=1}^{T} a_i z^{-1} \qquad (27\text{-}11)$$

By substituting (27-11) in (27-9) and multiplying both sides by $(1 - z^{-1})$, we get

$$(1 - z^{-1}) \sum_{i=1}^{T} a_i z^{-i} = HG(z) = \sum_{i=1}^{T} h_i z^{-i} \qquad (27\text{-}12)$$

Matching coefficients on both sides of (27-12) is consistent with Eq. 27-4, namely

$$h_i = a_i - a_{i-1}$$

The convolution model can also be written using incremental changes in the manipulated input,

$$\Delta m_n = m_n - m_{n-1} \qquad (27\text{-}13)$$

In transform notation,

$$\Delta M(z) = (1 - z^{-1})M(z) \qquad (27\text{-}14)$$

If we substitute for $M(z)$ in (27-6), the process response becomes

$$\hat{C}(z) = HG(z)\frac{\Delta M(z)}{1 - z^{-1}} = \left(\sum_{i=1}^{T} h_i z^{-1}\right)\frac{\Delta M(z)}{1 - z^{-1}} \qquad (27\text{-}15)$$

Using (27-12) gives

$$\hat{C}(z) = \left(\sum_{i=1}^{T} a_i z^{-1}\right)\Delta M(z) \qquad (27\text{-}16)$$

which can also be derived by taking the z-transform of Eq. 27-1.

EXAMPLE 27.2

For $c_0 = 0$, $\Delta m_0 = 0.2$, and $\Delta m_1 = 0.5$ and the first-order model given by (27-7), calculate the values of c_1, c_2, c_3, and c_4 using z-transforms. Compare the response obtained by use of the difference equations given by (27-3).

Solution

The original zero-order hold transfer function is

$$HG(z) = \frac{0.1813z^{-1}}{1 - 0.8187z^{-1}} \qquad (27\text{-}7)$$

We use Eq. 27-15 to calculate $C(z)$. For the specified input sequence, $\Delta M(z) = 0.2 + 0.5z^{-1}$, and

$$\hat{C}(z) = \frac{0.1813z^{-1}}{1 - 0.8187z^{-1}} \frac{0.2 + 0.5z^{-1}}{1 - z^{-2}}$$

$$= \frac{0.03626z^{-1} + 0.09065z^{-2}}{1 - 1.8187z^{-1} + 0.8187z^{-1}} \qquad (27\text{-}17)$$

By long division,

$$\hat{C}(z) = 0.03626z^{-1} + 0.1566z^{-2} + 0.2551z^{-3} + 0.3357z^{-4} + \cdots \qquad (27\text{-}18)$$

Next we use Eq. 27-5 to calculate $\hat{c}_1$, $\hat{c}_2$, $\hat{c}_3$, and $\hat{c}_4$, therefore employing a_1, a_2, a_3, and a_4 (given in the right side of Eq. 27-8):

$$\hat{c}_1 = 0 + 0.1813\,(0.2) = 0.03626$$

$$\hat{c}_2 = 0.1813\,(0.5) + 0.3297\,(0.2) = 0.1566$$

$$\hat{c}_3 = 0.3297\,(0.5) + 0.4512\,(0.2) = 0.2551$$

$$\hat{c}_4 = 0.4512\,(0.5) + 0.5507\,(0.2) = 0.3357$$

which agree with the values from long division.

27.3 MATRIX FORMS FOR PREDICTIVE MODELS

In this section we generalize the convolution model to include an arbitrary number of predictions. A central idea in predictive control is the use of *horizons*. We previously introduced the model horizon T as the time for the open-loop step response to reach 99% completion. We define two other horizons in this section, namely

1. the control horizon U
2. the prediction horizon V

The *control horizon U* is the number of control actions (or *control moves*) that are calculated in order to affect the predicted outputs over the *prediction horizon V*, i.e., over the next V sampling periods. Thus at time-step n the next U values of m are calculated $(m_n, m_{n+1}, \ldots, m_{n+U-1})$ as well as the next V output predictions $(\hat{c}_{n+1}, \hat{c}_{n+2}, \ldots, \hat{c}_{n+V})$. For example, we can specify a single input change $(U = 1)$ but evaluate the predicted response $(\hat{c}_{n+1}, \hat{c}_{n+2}, \ldots, \hat{c}_{n+V})$ over V future time steps.

Consider the general case of an arbitrary sequence of U input changes, Δm_0, $\Delta m_1, \ldots, \Delta m_{U-1}$, and an initial steady state $c_0 = 0$. The response can be calculated using the following matrix equation, which is based on Eq. 27-1 and a prediction horizon of V sampling periods:

$$
\begin{bmatrix} \hat{c}_1 \\ \hat{c}_2 \\ \hat{c}_3 \\ \cdot \\ \cdot \\ \cdot \\ \hat{c}_V \end{bmatrix}
=
\begin{bmatrix}
a_1 & 0 & 0 & \cdots & 0 \\
a_2 & a_1 & 0 & & 0 \\
a_3 & a_2 & a_1 & & 0 \\
\cdot & & & & \cdot \\
\cdot & & & & \cdot \\
\cdot & & & & \cdot \\
a_V & a_{V-1} & a_{V-2} & \cdots & a_{V-U+1}
\end{bmatrix}
\begin{bmatrix} \Delta m_0 \\ \Delta m_1 \\ \Delta m_2 \\ \cdot \\ \cdot \\ \cdot \\ \Delta m_{U-1} \end{bmatrix}
\qquad (27\text{-}19)
$$

The matrix in (27-20) has dimensions of $V \times U$. If $U = V$, the matrix is square. Both U and V are controller design parameters.

Single-Step Prediction

It will be useful below to express the discrete convolution model in recursive form. Using (27-5) and shifting the model back one time step, we can write

$$\hat{c}_n = \sum_{i=1}^{T} h_i m_{n-i} \tag{27-20}$$

Subtracting (27-20) from (27-5), we obtain a recursive form of the model expressed in incremental changes Δm:

$$\hat{c}_{n+1} = \hat{c}_n + \sum_{i=1}^{T} h_i \, \Delta m_{n+1-i} \tag{27-21}$$

Equation 27-21 is an open-loop prediction, in the sense that it does not provide any corrections for the influence of model errors or unmeasured load changes that may have occurred at any previous time step. To address this shortcoming, both DMC and MAC [1,5] utilize a corrected prediction of $\hat{c}_{n+1}$, denoted as c^*_{n+1}. The corrected value can be obtained by comparing the actual value of c_n with $\hat{c}_n$ and then shifting the correction forward, namely

$$c^*_{n+1} - \hat{c}_{n+1} = c_n - \hat{c}_n \tag{27-22}$$

This shift compensates for model errors as well as unmeasured load changes during the previous steps and acts like a feedback control correction. In recursive form,

$$c^*_{n+1} = \hat{c}_{n+1} + (c_n - \hat{c}_n) = c_n + \sum_{i=1}^{T} h_i \, \Delta m_{n+1-i} \tag{27-23}$$

Multistep Predictions

The major advantage of predictive control is that it incorporates predictions for a number of future time steps. This strategy enables the model-based control system to anticipate where the process is heading. The prediction horizon V is a design parameter that influences control system performance. A V-step predictor can be expressed in terms of incremental changes in the manipulated variable:

$$\hat{c}_{n+j} = \hat{c}_{n+j-1} + \sum_{i=1}^{T} h_i \, \Delta m_{n+j-i} \quad (j = 1, 2, \ldots, V) \tag{27-24}$$

Equation 27-24 can be applied sequentially V times to obtain $\hat{c}_{n+V}$. The recursive version of (27-24) which is analogous to (27-22) is

$$c^*_{n+j} = \hat{c}_{n+j} + (c^*_{n+j-1} - \hat{c}_{n+j-1}) \tag{27-25}$$

Equation 27-25 uses the difference between c^* and $\hat{c}$ (corrected value minus that predicted by the model) to update the new value of c^* using the model prediction at $t = (n+j)\Delta t$, assuming a constant future prediction error. To obtain the solution of (27-25) into the future we first set $c^*_n = c_n$, the current measured value.

Substituting (27-24) into (27-25) yields

$$c_{n+j}^* = c_{n+j-1}^* + \sum_{i=1}^{T} h_i \, \Delta m_{n+j-i} \tag{27-26}$$

for $j = 1, 2, \ldots, V$ and $c_n^* = c_n$.

As shown by Marchetti [6], Eq. 27-26 can be written in a more convenient vector–matrix form by taking the future incremental input changes Δm_{n+j} out of the summations and rearranging. For a prediction horizon V and control horizon U where $U \leq V$, Eq. 27-26 is equivalent to

$$\begin{bmatrix} c_{n+1}^* \\ c_{n+2}^* \\ c_{n+3}^* \\ \cdot \\ \cdot \\ \cdot \\ c_{n+V}^* \end{bmatrix} = \begin{bmatrix} a_1 & 0 & 0 & \cdots & 0 \\ a_2 & a_1 & 0 & & 0 \\ a_3 & a_2 & a_1 & & 0 \\ \cdot & & & & \cdot \\ \cdot & & & & \cdot \\ \cdot & & & & \cdot \\ a_V & a_{V-1} & a_{V-1} & \cdots & a_{V-U+1} \end{bmatrix} \begin{bmatrix} \Delta m_n \\ \Delta m_{n+1} \\ \Delta m_{n+2} \\ \cdot \\ \cdot \\ \cdot \\ \Delta m_{n+U-1} \end{bmatrix} + \begin{bmatrix} c_n + P_1 \\ c_n + P_2 \\ c_n + P_3 \\ \cdot \\ \cdot \\ \cdot \\ c_n + P_V \end{bmatrix} \tag{27-27}$$

where the $\{a_i\}$ are the same step response coefficients defined earlier, namely

$$a_i = \sum_{j=1}^{i} h_j$$

and

$$P_i = \sum_{j=1}^{i} S_j \quad \text{for} \quad i = 1, 2, \ldots, V \tag{27-28}$$

$$S_j = \sum_{i=j+1}^{T} h_i \, \Delta m_{n+j-i} \quad \text{for} \quad j = 1, 2, \ldots, V \tag{27-29}$$

The P_i terms are elements of the projection vector, which essentially includes future predictions of c based on previously implemented input changes. Note that both P_i and S_j depend only on h_i and past values of Δm. Observing Fig. 27.2, these input changes influence the future values of the response. In the next section, we show how the recursive model (27-26) is used to design the predictive controller.

27.4 CONTROLLER DESIGN METHOD

Controller design in model predictive control is based on the predicted behavior of the process over the prediction horizon. Values of the manipulated variables are computed to ensure that the predicted response has certain desirable characteristics. One sampling period after the application of the current control action, the predicted response is compared with the actual response. Using corrective feedback action for any errors between actual and predicted responses, the entire sequence of calculations is then repeated at each sampling instant.

The control objective is to have the corrected predictions c_{n+j}^* approach the set point as closely as possible. Denote the *set-point trajectory*, that is, the desired values of the set point V time steps into the future, as $r_{n+j}, j = 1, 2, \ldots, V$, and define the folowing vectors:

$$
\hat{\mathbf{E}} = \begin{bmatrix} r_{n+1} - c_{n+1}^* \\ r_{n+2} - c_{n+2}^* \\ \cdot \\ \cdot \\ \cdot \\ r_{n+V} - c_{n+V}^* \end{bmatrix} \qquad \hat{\mathbf{E}}' = \begin{bmatrix} E_n - P_1 \\ E_n - P_2 \\ \cdot \\ \cdot \\ \cdot \\ E_n - P_V \end{bmatrix}
$$

$\hat{\mathbf{E}}'$ is the predicted value of the process error, $r - c$, at V future sampling instants based on manipulated variable (input) changes up to time $(n - 1)\Delta t$ and the current error signal, $E_n = r_n - c_n$. P_i, an element of the projected (future) output vector, is defined in (27-28).

Note that both $\hat{\mathbf{E}}$ and $\hat{\mathbf{E}}'$ are vectors of predicted errors. $\hat{\mathbf{E}}'$ is an *open-loop* prediction because it is based only on past control actions. It does not include the current and future control actions (Δm_{n+j} for $j \geqq 0$). By contrast, $\hat{\mathbf{E}}$ is referred to as a *closed-loop prediction* because it is based on the current and future control actions.

With the above definitions Eq. 27-27 can be written as follows:

$$
\hat{\mathbf{E}} = -\mathbf{A}\,\Delta\mathbf{m} + \hat{\mathbf{E}}' \tag{27-30}
$$

where $\mathbf{A}$ is the $V \times U$ triangular matrix given in Eq. 27-20; $\Delta\mathbf{m}$ is the $U \times 1$ vector of future control moves. If a perfect match between the predicted output trajectory of the closed-loop system and the desired trajectory is required, then $\hat{\mathbf{E}} = \mathbf{0}$ and, from Eq. 27-30,

$$
\mathbf{0} = -\mathbf{A}\,\Delta\mathbf{m} + \hat{\mathbf{E}}' \tag{27-31}
$$

If the number of control moves and the number of predicted outputs are equal ($U = V$), then the solution to Eq. 27-31 is given by

$$
\Delta\mathbf{m} = (\mathbf{A})^{-1}\,\hat{\mathbf{E}}' \tag{27-32}
$$

Since in this case $\mathbf{A}$ is square, it is invertible if $a_1 \neq 0$ (no time delay). However, setting $U = V$ usually leads to an unsatisfactory controller compared to the case $U < V$. This is because $U = V$ results in a minimal prototype (deadbeat) controller, with excessive moves of the manipulated variable [8].

Dynamic Matrix Control (DMC) [1,2] requires that $U < V$ so that the resulting system of equations is overdetermined. Thus, only U future control actions ($\Delta\mathbf{m}$) are calculated and $\mathbf{A}$ is the $V \times U$ *dynamic matrix*. For $U < V$, the overdetermined system of Eq. 27-31 does not have an exact solution. It is possible, however, to obtain the best solution in the least-squares sense by minimizing the performance index

$$
J[\Delta\mathbf{m}] = \hat{\mathbf{E}}^T\hat{\mathbf{E}} \tag{27-33}
$$

The optimal solution is

$$
\Delta\mathbf{m} = (\mathbf{A}^T\mathbf{A})^{-1}\mathbf{A}^T\hat{\mathbf{E}}' = \mathbf{K}_c\hat{\mathbf{E}}' \tag{27-34}
$$

where $(\mathbf{A}^T\mathbf{A})^{-1}\mathbf{A}^T$ is the pseudoinverse matrix [7] and $\mathbf{K}_c$ is the matrix of feedback gains (with dimensions $U \times V$).

Generally, one would apply only the first control action Δm_n, observe c_{n+1}, correct the predictions, and apply Eq. 27-34 again. Thus, at each sampling instant, U future control actions are calculated but only the first one Δm_n is implemented.

The advantage of this procedure is that it gives early detection of modeling errors or disturbances and approximately corrects for them. Moreover, since only Δm_n (a scalar) needs to be calculated and $\mathbf{A}$ is a triangular matrix, it is not necessary to solve Eq. 27-34 completely. Only the first equation corresponding to the prediction c_{n+1}^* must be solved. Note that the error vector $\hat{\mathbf{E}}'$ must be updated after each new control move.

One difficulty with the control law of Eq. 27-33 is that it can result in excessively large changes in the manipulated variable. This undesirable phenomenon occurs when the $\mathbf{A}^T\mathbf{A}$ matrix is ill-conditioned or singular. Cutler and Ramaker [2] overcame this problem by multiplying the diagonal elements of $\mathbf{A}^T\mathbf{A}$ by a number greater than one before performing the matrix inversion. They called this modification *move suppression* since it moderated the changes Δm_n. An alternative approach [1,8], which is analogous to that used in optimal control methods, is to modify the performance index by penalizing movements of the manipulated variable:

$$J[\Delta\mathbf{m}] = \hat{\mathbf{E}}^T\mathbf{W}_1\hat{\mathbf{E}} + \Delta\mathbf{m}^T\mathbf{W}_2\Delta\mathbf{m} \tag{27-35}$$

where $\mathbf{W}_1$ and $\mathbf{W}_2$ are positive-definite weighting matrices (usually diagonal matrices with positive elements). The resulting control law that minimizes J is

$$\Delta\mathbf{m} = (\mathbf{A}^T\mathbf{W}_1\mathbf{A} + \mathbf{W}_2)^{-1}\mathbf{A}^T\mathbf{W}_1\hat{\mathbf{E}}' = \mathbf{K}_c\hat{\mathbf{E}}' \tag{27-36a}$$

Note that Eqs. 27-35 and 27-36 include the weighting matrices which allow the user to specify different penalties to be placed on the predicted errors. This provides a way to tune the controller.

As with Eq. 27-34, normally one applies only the first control action Δm_n, observes c_{n+1}, and then uses Eq. 27-36 to recalculate the control policy. In this updating scheme, only the first row of $\mathbf{K}_c$, containing V elements, is required in (27-36). We denote the first row of $\mathbf{K}_c$ as $\mathbf{K}_{c1}^T$. Thus from (27-36a) it follows that

$$\Delta m_n = \mathbf{K}_{c1}^T\hat{\mathbf{E}}' \tag{27-36b}$$

Using a standard z-transform analysis of this case, the controller transfer function $D(z)$, shown in Fig. 25.11, can be obtained from Eq. 27-36 [6]:

$$D(z) = \frac{M(z)}{E(z)} = \frac{g_0}{(1 + d_1z^{-1} + d_2z^{-2} + \cdots + d_{T-1}z^{-T+1})(1 - z^{-1})} \tag{27-37}$$

where the coefficients g_0 and d_j ($j = 1, 2, \ldots, T - 1$) are functions of the elements of $\mathbf{K}_{c1}^T$ and the impulse response coefficients. This version of the controller transfer function corresponds to a digital feedback controller that calculates the controller output from the current error signal and previous controller output signals. However, explicit predictions of the future process response are employed in implementing this control algorithm. Note that the digital controller in Eq. 27-37 contains integral action due to the pole at $z = 1$.

EXAMPLE 27.3

Consider a process model

$$HG(z) = h_{N+1}z^{-N-1} + h_{N+2}z^{-N-2} + \cdots + h_Tz^{-T} \tag{27-38}$$

Note that this model contains a time delay of N sampling periods because $h_i = 0$ for $i \leq N$. Suppose that the corrected prediction is to match exactly the set point r_n after $N + 1$ sampling periods, which takes into account the time delay; this requirement is equivalent to the closed-loop transfer function, z^{-N-1}. Show that the resulting predictive controller transfer function $D(z) = M(z)/E(z)$ is the same as the minimal prototype controller derived in Eq. 26-19. Use a modified form of (27-23) that takes into account the time delay, namely

$$c^*_{n+N+1} = \hat{c}_{n+N+1} + (c_n - \hat{c}_n) \qquad (27\text{-}39)$$

Solution

Since the corrected prediction does not match the set point r_n until time $t = (n+N+1)\Delta t$, we require that

$$c^*_{n+N+1} = r_n \qquad (27\text{-}40)$$

Thus, by substituting (27-40) and (27-20) into (27-39),

$$r_n = c_n + \sum_{i=N+1}^{T} h_i(m_{n+N+1-i} - m_{n-i}) \qquad (27\text{-}41)$$

The error is

$$e_n = r_n - c_n = \sum_{i=N+1}^{T} h_i(m_{n+N+1-i} - m_{n-i}) \qquad (27\text{-}42)$$

Taking the z-transform,

$$E(z) = (z^{N+1} - 1) \sum_{i=N+1}^{T} h_i z^{-i} M(z) \qquad (27\text{-}43)$$

Rearranging and denoting

$$HG(z) = \sum_{i=N+1}^{T} h_i z^{-i}$$

gives

$$\frac{M(z)}{E(z)} = \frac{1}{HG(z)} \frac{1}{z^{N+1} - 1} \qquad (27\text{-}44)$$

or

$$D(z) = \frac{M(z)}{E(z)} = \frac{1}{HG(z)} \frac{z^{-N-1}}{1 - z^{-N-1}} \qquad (27\text{-}45)$$

which is the same controller obtained by the direct synthesis approach (cf. Eq. 26-16). Note that time-delay compensation is contained in $D(z)$.

EXAMPLE 27.4

A process has the transfer function

$$G_p(s) = \frac{e^{-s}}{(10s + 1)(5s + 1)}$$

(a) Using a sampling period of $\Delta t = 1$, compute the discrete convolution model for $T = 70$.
(b) Using Eq. 27-33, design a predictive controller for two cases: $U = 2$, $V = 4$, and $U = 1$, $V = 3$, What is $\mathbf{K}_c$ for each case? Using the first row of $\mathbf{K}_c$, compare

Table 27.1 Step- and Impulse-Response Coefficients for $\Delta t = 1$

Sampling instant	Step-response coefficients	Impulse-response coefficients	Sampling instant	Step-response coefficients	Impulse-response coefficients
1	0.00000	0.00000	36	0.94052	0.00615
2	0.00906	0.00906	37	0.94610	0.00558
3	0.03286	0.02380	38	0.95116	0.00507
4	0.06718	0.03432	39	0.95576	0.00459
5	0.10869	0.04151	40	0.95993	0.00417
6	0.15482	0.04613	41	0.96370	0.00378
7	0.20357	0.04875	42	0.96713	0.00343
8	0.25343	0.04986	43	0.97023	0.00310
9	0.30324	0.04981	44	0.97305	0.00281
10	0.35216	0.04892	45	0.97560	0.00255
11	0.39958	0.04742	46	0.97791	0.00231
12	0.44506	0.04548	47	0.98000	0.00209
13	0.48833	0.04327	48	0.98189	0.00189
14	0.52921	0.04088	49	0.98361	0.00172
15	0.56762	0.03841	50	0.98516	0.00155
16	0.60353	0.03591	51	0.98657	0.00141
17	0.63697	0.03344	52	0.98784	0.00127
18	0.66801	0.03104	53	0.98900	0.00115
19	0.69673	0.02872	54	0.99044	0.00104
20	0.72323	0.02651	55	0.99099	0.00095
21	0.74765	0.02441	56	0.99184	0.00086
22	0.77008	0.02244	57	0.99262	0.00077
23	0.79067	0.02059	58	0.99332	0.00070
24	0.80953	0.01886	59	0.99395	0.00063
25	0.82679	0.01726	60	0.99453	0.00057
26	0.84257	0.01577	61	0.99505	0.00052
27	0.85697	0.01440	62	0.99552	0.00047
28	0.87011	0.01314	63	0.99595	0.00043
29	0.88208	0.01197	64	0.99633	0.00039
30	0.89298	0.01090	65	0.99668	0.00035
31	0.90290	0.00992	66	0.99700	0.00032
32	0.91193	0.00903	67	0.99728	0.00029
33	0.92014	0.00821	68	0.99754	0.00026
34	0.92759	0.00746	69	0.99777	0.00023
35	0.93437	0.00677	70	0.99799	0.00021

the closed-loop response (both controlled and manipulated variables) for a set-point change with that for a discrete PID controller with $K_c = 2.27$, $\tau_I = 16.6$, $\tau_D = 1.49$, and $\Delta t = 0.5$ (based on ITAE-tuning parameters). Repeat the comparison for a load change, using the ITAE load settings of $K_c = 3.52$, $\tau_I = 6.98$, $\tau_D = 1.73$.

Solution

(a) The unit step response of $G_p(s)$ can be generated analytically and is

$$c(t) = 1 - 2e^{-0.1(t-1)} \mathbf{S}(t - 1) + e^{-0.2(t-1)} \mathbf{S}(t - 1)$$
$$c(t) = 0 \quad \text{for } t \le 1$$

Convolution model coefficients (h_i) are shown in Table 27.1 for $0 \le t \le 70$

Table 27.2   Feedback Matrices K_c for Example 27.4

For $V = 3$ and $U = 1$:

$$K_c = [0 \quad 7.77 \quad 28.2]$$

For $V = 4$ and $U = 2$:

$$K_c = \begin{bmatrix} 33.1 & 48.6 & -13.4 \\ -71.3 & -97.0 & 57.1 \end{bmatrix}$$

and $\Delta t = 1$. For this model horizon, the step response is over ninety-nine percent complete.

(b) Using Eq. 27-33, the feedback matrix for each case is shown in Table 27.2. Figure 27.3 shows $c(t)$ and $m(t)$ for a set-point change, contrasting the performance of DMC with a PID controller. DMC gives a much better closed-loop response although it changes the manipulated variable more initially (note the expanded time scale for m). Using $U = 1$ limits the manipulated variable more than $U = 2$, while the controlled variable response is not degraded significantly. By increasing K_c, a PID controller can achieve settling times comparable to DMC, but it generally does so at the expense of greater overshoot and larger swings in the manipulated variable. Figure 27.4 shows the analogous comparison of DMC and PID for a load change. $U = 1$ gives a slightly slower response than $U = 2$ (but not appreciably). Compared to PID, the DMC controllers give a smaller peak error with the same settling time.

27.5   TUNING THE PREDICTIVE CONTROLLER

The predictive control technique presented in the previous section includes a number of design parameters which can be adjusted to give the desired response as

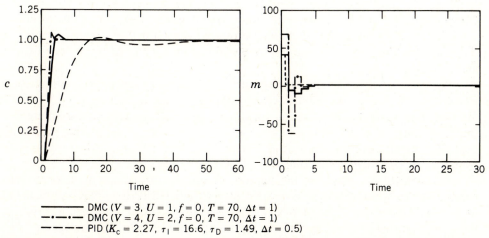

DMC ($V = 3$, $U = 1$, $f = 0$, $T = 70$, $\Delta t = 1$)
DMC ($V = 4$, $U = 2$, $f = 0$, $T = 70$, $\Delta t = 1$)
PID ($K_c = 2.27$, $\tau_I = 16.6$, $\tau_D = 1.49$, $\Delta t = 0.5$)

Figure 27.3 Comparison of controlled and manipulated variables for a set-point change, Example 27.4.

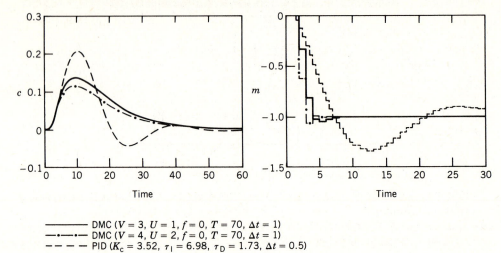

———— DMC ($V = 3$, $U = 1$, $f = 0$, $T = 70$, $\Delta t = 1$)
—·—·— DMC ($V = 4$, $U = 2$, $f = 0$, $T = 70$, $\Delta t = 1$)
— — — PID ($K_c = 3.52$, $\tau_I = 6.98$, $\tau_D = 1.73$, $\Delta t = 0.5$)

Figure 27.4   Comparison of controlled and manipulated variables for a load change, Example 27.4

well as an appropriate amount of controller effort. These parameters include

T =   model horizon

U =   control horizon

V =   prediction horizon

$\mathbf{W}_1$ =   weighting matrix for predicted errors

$\mathbf{W}_2$ =   weighting matrix for control moves

Δt =   sampling period

The model horizon $T\Delta t$ should usually be selected so that $T\Delta t \geq$ open-loop settling time, which is equal to the time for the open-loop step response to be 99% complete (in some cases, however, it is possible to select T corresponding to 95% complete). Values of T between 20 and 70 are typical of the recommendations in the literature [1,9]. T should be large enough so that no truncation problems arise in calculating the predicted values for the convolution model. This truncation problem is usually manifested as a "wiggle" in the actual response at $t = T\Delta t$, where the model error first becomes noticeable. A good rule of thumb is to choose T so that h_T is of the order of the measurement error for the output variable. In the case of integrating processes, some modifications in the approach need to be made [10,11].

Parameter V is the number of predictions that are used in the optimization calculations; it is also the dimension of the gain vector $\mathbf{K}_{c1}^T$. Increasing V results in more conservative control action which has a stabilizing effect but also increases the computational effort. When $V = N+1$ ($N =$ time delay), a minimal prototype controller results. Maurath et al. [9] recommend using V as a tuning parameter while Cutler [10] suggests setting $V = T + U$.

The control horizon U is the number of future control actions that are calculated in the optimization step to reduce the predicted errors. The parameter U is also the dimension of the matrix in Eq. 27-33 that must be inverted. Therefore, the computational effort increases as U is increased. A suitable first guess is to choose U so that $U\Delta t \cong t_{60}$, which is the time for the open-loop response to be 60% complete. For $U > 5$, the value of V is not crucial since only the first move is implemented in (27-36). However, too large a value of U results in excessive control action. For the special case where $U = V$, a minimal prototype controller [8] results. A smaller value of U leads to a robust controller that is relatively insensitive to model errors.

For unconstrained SISO control problems, Maurath et al. [9] recommend using Eq. 27-34 with $U = 1$ and V as a tuning parameter. To obtain a first guess for V when $U = 1$, $V\Delta t$ should be selected to be 50% of the open-loop response time. This approach results in a simple yet effective design procedure that requires only *one* tuning parameter, as demonstrated in Examples 27.4 and 27.5.

The weighting matrices $\mathbf{W}_1$ and $\mathbf{W}_2$ in Eq. 27-36 contain a potentially large number of design parameters. However, it is usually sufficient to select $\mathbf{W}_1 = \mathbf{I}$ and $\mathbf{W}_2 = f\mathbf{I}$ ($\mathbf{I}$ is the identity matrix and f is a scalar design parameter). Larger values of f penalize the magnitude of Δm more, thus giving less vigorous control. When $f = 0$, the controller gains are very sensitive to U, largely because of the ill-conditioning of $\mathbf{A}^T\mathbf{A}$ [8], and U must be made small, as discussed above. The approach of Maurath et al. [9] utilizes a singular value decomposition method to eliminate the ill-conditioning of $\mathbf{A}^T\mathbf{A}$ with $f = 0$, but this is not a serious problem with $U = 1$.

The sampling period Δt must also be selected as part of the design procedure. It should be small enough to ensure that important dynamic information is not lost (see Chapter 22). On the other hand, if Δt is too small, then, as discussed above, T must be made very large, which is undesirable. In Example 27.4, Δt is smaller than the time delay, which causes T to be rather large. The sampling period is not considered to be a tuning parameter because adjustment of U, V, and f is usually quite sufficient to obtain satisfactory closed-loop performance. However, it may be helpful to check the sensitivity of the response to Δt.

EXAMPLE 27.5

A process transfer function has a gain K, time delay θ, and dominant time constant τ_1 that vary with throughput,

$$G_p(s) = \frac{K(1 - 9s)e^{-\theta s}}{(3s + 1)(\tau_1 s + 1)}$$

(a) Tune a predictive controller using nominal values for the model of $K = 1$, $\theta = 0$, and $\tau_1 = 15$ and assuming that $G_p = \tilde{G}_p$. Use $\Delta t = 1$ and $T = 50$ in obtaining the convolution model. For $\mathbf{W}_1 = \mathbf{I}$, and $\mathbf{W}_2 = f\mathbf{I}$, vary U, V, and f to find a desirable closed-loop performance for a step change in set point and compare these results with $U = 1$ and $f = 0$ (tune V).

(b) Suppose that for another throughput the actual process parameters are $K = 1.5$, $\theta = 0.5$, and $\tau_1 = 10$ (two of the parameter changes are destabilizing). Evaluate the set-point change and retune the controller of part (a) for robustness, if necessary. Note that in this case the controller has been designed using an inaccurate process model and yet we want it to be satisfactory when the model errors are present.

Solution

(a) For no model error, the design parameters that gives the best response for the controlled variable are $V = 20$, $U = 5$, and $f = 2$. See Fig. 27.5 for the closed-loop response and controller output, respectively. The controller performance

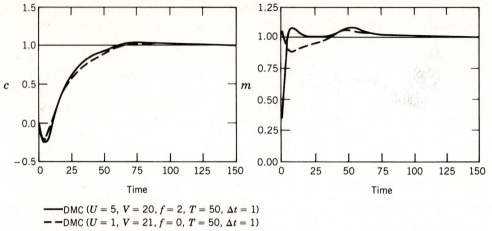

—— DMC ($U = 5$, $V = 20$, $f = 2$, $T = 50$, $\Delta t = 1$)
– – DMC ($U = 1$, $V = 21$, $f = 0$, $T = 50$, $\Delta t = 1$)

Figure 27.5 Comparison of controlled and manipulated variables for a set-point change with DMC controller, Example 27.5.

is not very sensitive to U; values of U as large as 10 are satisfactory. However, values of $f \leq 1$ ($5 \leq U \leq 10$) yielded too much controller effort. Also shown in Fig. 27.5 are the tuned responses for $U = 1$ and $f = 0$ ($V = 21$ via tuning). The latter case exhibits smaller swings in the manipulated variable, which is allowable without sacrificing performance for $c(t)$.

(b) Figure 27.6 shows that the controller with $U = 5$ and $f = 2$ becomes too oscillatory in the presence of model errors (an increased process gain and time delay destabilize the controller). The controller with $U = 1$ and $f = 0$ provides improved results for the inaccurate model, but the closed-loop response is too oscillatory. To increase the robustness of the DMC controller, the weighting on the controller effort f must be increased. For the $U = 1$ controller a value of $f = 10$ was found to be satisfactory. The closed-loop response for the case of model error in all three parameters is given in Fig. 27.7.

Treatment of Process Constraints

A key feature of many practical control problems is the presence of constraints on both controlled variables and manipulated variables. Inequality constraints arise commonly in process control problems due to physical limitations of plant equipment such as pumps, control valves, and heat exchangers. Constraints are often

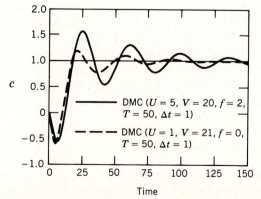

Figure 27.6 Comparison of set-point changes for DMC controllers with model errors, Example 27.5.

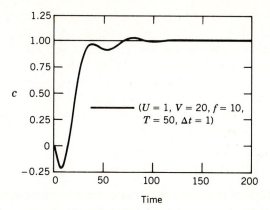

Figure 27.7 Set-point change for detuned DMC controller with model errors, Example 27.5 ($U = 1$, $V = 20$, $f = 10$, $T = 50$, $\Delta t = 1$).

imposed as part of the plant operating strategy or the control objectives. For example, the control objective may be to maximize a plant production rate while satisfying constraints on product quality and avoiding undesirable operating regimes (flooding in a distillation column, exceeding a metallurgical limit in a reactor, etc).

In Example 27.5, tuning parameters can be adjusted so as to satisfy constraints on the manipulated variable. Specifically, the weighting factor f and the control horizon U have the most impact on controller effort. Larger values of f penalize the manipulated variable more heavily, resulting in less control effort. However, adjustment of f is only an indirect way to satisfy upper and lower bounds on the input and output variables, that is, constraints of the form

$$\gamma^L \leq m_i \leq \gamma^U \tag{27-46}$$
$$i = 1, 2, \ldots$$

$$\delta^L \leq c_i \leq \delta^U \tag{27-47}$$

In other words, iterative selection of the tuning parameters must be carried out in the design procedure until the constraints are met, a procedure that is usually not very satisfactory.

A more direct method of satisfying constraints on current and future values of manipulated and controlled variables is to modify the unconstrained optimization problem posed in Eq. 27-35 by adding the constraints explicitly. This approach produces a mathematical programming problem, specifically a quadratic programming (QP) problem, because the objective function is quadratic while the process model and the inequality constraints are linear (upper and lower bounds). The size of the control moves can also be constrained using this method. If the objective function is changed to a linear form, then linear programming (LP) can be employed [12,13]. The reader should note that none of the feedback control techniques discussed in Chapter 26 explicitly treats constraints. Hence a method using a more general formulation (with constraints) should be advantageous, assuming that the additional computing requirements can be handled.

The choice between a LP or QP control strategy should be based on a number of considerations including control system performance and computational requirements. An advantage of QP control strategies is that they tend to be easier to tune since the control variables can be weighted in the quadratic performance index. By contrast, in the LP approach, weighting of the control variables in the performance index is not an effective tuning strategy [12]. Another advantage of the QP strategy is that it typically results in smoother responses than the LP approach.

Versions of both DMC and MAC use quadratic programming [1,3], while one version of DMC (LDMC) is based on linear programming [13].

A significant disadvantage of the QP approach is that the computational effort is typically four to five times larger than for LP. Consequently, there is considerable incentive to develop QP control strategies that are numerically efficient [14,15]. In some cases [16], it may be impossible to satisfy all constraints, especially on the controlled variables. Software needs to be able to treat such cases to arrive at an acceptable solution.

27.6 PREDICTIVE CONTROL OF MIMO SYSTEMS

The results developed in the previous section can be extended to multiple input–multiple output systems by using the principle of superposition. For a system with two inputs and two outputs, the predictive model becomes

$$\hat{c}_{1,n+1} = \sum_{i=1}^{T} h_{11i}m_{1,n+1-i} + \sum_{i=1}^{T} h_{12i}m_{2,n+1-i} \tag{27-48}$$

$$\hat{c}_{2,n+1} = \sum_{i=1}^{T} h_{21i}m_{1,n+1-i} + \sum_{i=1}^{T} h_{22i}m_{2,n+1-i} \tag{27-49}$$

where $\hat{c}_{1,n+1}$ denotes the predicted value of c_1 at the $(n+1)$ sampling instant. T is the model horizon for all four input–output models (select T to be the largest model horizon from among the four models).

In analogy with the derivations for SISO systems, this model can be transformed into the standard dynamic matrix form:

$$\hat{\mathbf{E}} = -\mathbf{A}\,\Delta\mathbf{m} + \hat{\mathbf{E}}'$$

where $\hat{\mathbf{E}}$ and $\hat{\mathbf{E}}'$ are vectors of length $2V$ and $\Delta\mathbf{m}$ is a vector of length $2U$. Thus,

$$\hat{\mathbf{E}} = \text{col}[\hat{E}_{1,n+1}, \hat{E}_{1,n+2}, \ldots, \hat{E}_{1,n+V}, \hat{E}_{2,n+1}, \hat{E}_{2,n+2}, \ldots, \hat{E}_{2,n+V}]$$

$$\Delta\mathbf{m} = \text{col}[\Delta m_{1,n}, \Delta m_{1,n+1}, \ldots, \Delta m_{1,n+U-1}, \Delta m_{2,n}, \Delta m_{2,n+1}, \ldots, \Delta m_{2,n+U-1}]$$

where "col" denotes a column vector form.

In the 2×2 case, the dynamix matrix $\mathbf{A}$ has the following structure:

$$\mathbf{A} = \begin{bmatrix} \mathbf{A}_{11} & \mathbf{A}_{12} \\ \mathbf{A}_{21} & \mathbf{A}_{22} \end{bmatrix}$$

where each partition is triangular:

$$\mathbf{A}_{ij} = \begin{bmatrix} a_{ij,1} & 0 & \cdots & 0 \\ a_{ij,2} & a_{ij,1} & & 0 \\ \cdot & & & \cdot \\ \cdot & & & \cdot \\ \cdot & & & \cdot \\ a_{ij,V} & a_{ij,V-1} & \cdots & a_{ij,V-U+1} \end{bmatrix}$$

Note that if each $\mathbf{A}_{ij}$ is $(V \times U)$, then $\mathbf{A}$ is $(2V \times 2U)$. The predictive control law has the same form as for the SISO case [6].

While we have presented a specific multivariable approach based on quadratic optimization, other methods could be employed. Internal Model Control (IMC) is an unconstrained predictive control technique that can be employed either in continuous or discrete time [17]. One advantage of using IMC is that rigorous analysis can be performed to tune the controller and to evaluate stability and robustness [18]. This approach allows one to draw parallels to existing multivariable control techniques such as linear-quadratic optimal control. For a more complete discussion of this subject, see Ref. 19.

Multivariable predictive control has been successfully applied to commercial processes such as a steam generator [4], distillation columns [3], fluid catalyic cracking units and hydrocrackers [20,21], and a multi-effect evaporation system [16].

SUMMARY

Predictive control techniques represent an alternative to the digital control design methods presented in Chapter 26. The techniques presented in this chapter are based on a specific type of model (convolution form), and optimization of a quadratic objective function is used to obtain the feedback law. Like Internal Model Control, the feedback signal is based on the difference between the actual controlled variable and its predicted value. Tuning can involve the adjustment of only one parameter, but additional design parameters can also be adjusted. A predictive control method can be advantageous for MIMO control problems when the process exhibits unusual dynamic characteristics or when it is crucial to meet constraints on the manipulated and/or controlled variables. For simple SISO control problems, the methods in Chapter 26 should be satisfactory.

REFERENCES

1. Prett, D. M., and C. E. Garcia, *Fundamental Process Control,* Butterworths, Boston, MA, 1988.
2. Cutler, C. R., and B. L. Ramaker, Dynamic Matrix Control—A Computer Control Algorithm, Proc. Joint Auto Control Conf, Paper WP5-B, San Francisco, 1980.
3. Richalet, J., A. Rault, J. L. Testud, and J. Papon, Model Predictive Heuristic Control: Applications to Industrial Processes, *Automatica* **14,** 413 (1978).
4. Mehra, R. K., R. Rouhani, and J. Eterno, Model Algorithmic Control: Review and Recent Developments, in *Chemical Process Control 2,* T. F. Edgar and D. E. Seborg (Eds.), Engineering Foundation, New York, 1982.
5. Rouhani, R., and R. K. Mehra, Model Algorithmic Control (MAC): Basic Theoretical Properties, *Automatica* **18,** 401 (1982).
6. Marchetti, J. L., Ph.D. Thesis, University of California, Santa Barbara, 1982.
7. Penrose, R., On Best Approximate Solutions of Linear Matrix Equations, *Cambridge Philos. Soc.* **52,** 17 (1956).
8. Marchetti, J. L., D. A. Mellichamp, and D. E. Seborg, Predictive Control Based on Discrete Convolution Models, *IEC Proc. Des. Dev.* **22,** 488 (1983).
9. Maurath, P. R., D. A. Mellichamp, and D. E. Seborg, Predictive Controller Design for SISO Systems, *IEC Research,* **27,** 956 (1988).
10. Cutler, C. R., Dynamic Matrix Control of Imbalanced Systems, *ISA Trans.* **21**(1), 1 (1982).
11. McDonald, K. A., and T. J. McAvoy, Optimal Averaging Level Control, *AIChE J.* **32,** 75 (1986).
12. Chang, T. S., and D. E. Seborg, A Linear Programming Approach for Multivariable Feedback Control with Inequality Constraints, *Int. J. Control* **37,** 583 (1983).
13. Morshedi, A. M., C. R. Cutler, and T. A. Skrovanek, Optimal Solution of Dynamic Matrix Control with Linear Programming Techniques (LDMC), Proc. Am. Control Conf., 1985, p. 199.
14. Ricker, N. L., Use of Quadratic Programming for Constrained Internal Model Control, *IEC Proc. Des. Dev.* **24,** 925 (1985).
15. Garcia, C. E., and A. M. Morshedi, Quadratic Programming Solution of Dynamic Matrix Control (QDMC), *Chem. Eng. Commun.* **46,** 73(1986).
16. Ricker, N. L., T. Sim, and C. M. Cheng, Predictive Control of a Multieffect Evaporation System, Proc. Am. Control Conf., 1986, p. 355.
17. Garcia, C. E., and M. Morari, Internal Model Control, 1. A Unifying Review and Some New Results, *IEC Proc. Des. Dev.* **21,** 308 (1982).
18. Garcia, C. E., D. M. Prett, and M. Morari, Model Predictive Control: Theory and Practice, *Automatica* (in press).
19. Morari, M., and E. Zafiriou, *Robust Process Control,* Prentice-Hall, Englewood Cliffs, NJ, 1988.

20. Garcia, C. E., and D. M. Prett, Advances in Industrial Model-Predictive Control, in *Chemical Process Control—CPC III,* M. Morari and T. J. McAvoy (Eds.), Elsevier, New York, 1986, p. 245.

21. Cutler, C. R., and R. B. Hawkins, Application of a Large Predictive Multivariable Controller to a Hydrocracker Second Stage Reactor, Proc. Am. Control Conf., 284 (1988).

EXERCISES

27.1. For the transfer function

$$G_p(s) = \frac{2e^{-s}}{(10s + 1)(5s + 1)} \qquad G_v = G_m = 1$$

(a) Find $HG(z)$ for $\Delta t = 1$. Using long division, calculate the corresponding convolution model as a z-transform.

(b) Derive the analytical unit step response for $G_p(s)$ and evaluate the values of a_i for $\Delta t = 1$. Compare your results with the h_i coefficients calculated in part (a).

(c) What value of T should be selected to give a response that is 95% complete?

(d) How could you find the impulse response coefficients if $\Delta t = 0.5$?

27.2. Use the convolution model in Exercise 27.1 to determine the change in c for $\Delta m_0 = 1$ and $\Delta m_1 = 2$. Calculate c_1, c_2, c_3, c_4, and c_5, assuming $c_0 = 0$. Use z-transforms and then check your answer using difference equations, such as were used in Example 27.2.

27.3. Calculations for a control horizon of $U = 1$ can be performed either analytically or numerically. Using the process model in Exercise 27.1, derive $\mathbf{K}_{c1}^T$ for $\Delta t = 1$, $T = 50$, and $V = 5$ using Eq. 27-33. Compare your answer with the analytical result stated by Maurath et al. [9], that is,

$$\mathbf{K}_{c1}^T = \frac{1}{\sum\limits_{i=1}^{V} a_i^2} [a_1 \, a_2 \, a_3 \cdots a_V]$$

27.4. In Example 27.3 a formula for the minimal prototype controller was derived. Suppose the desired response is an exponential approach to the set point, namely

$$c_{n+N+1}^* = \alpha c_{n+N}^* + (1 - \alpha)r_{n+N}$$

Derive the feedback controller $D(z)$ corresponding to this set-point requirement (cf. Eq. 27-45). Is this controller similar to a digital controller previously discussed in Chapter 26?

27.5. Repeat Example 27.4 but change the tuning parameters and sampling period in the following ways:

(a)	$\Delta t = 0.5$	$T = 50$	$U = 1$	$V = 3$	$f = 0$
(b)	$\Delta t = 1$	$T = 30$	$U = 1$	$V = 3$	$f = 0$
(c)	$\Delta t = 2$	$T = 40$	$U = 1$	$V = 5$	$f = 0$
(d)	$\Delta t = 2$	$T = 40$	$U = 5$	$V = 5$	$f = 0$
(e)	$\Delta t = 2$	$T = 40$	$U = 20$	$V = 20$	$f = 0$
(f)	$\Delta t = 2$	$T = 40$	$U = 3$	$V = 10$	$f = 0.01$
(g)	$\Delta t = 2$	$T = 40$	$U = 3$	$V = 10$	$f = 0.1$

In all of the above cases, examine the behavior of c_n and m_n. Compare your results with that for PID control (Figs. 27.3–27.4).

27.6. In Example 27.4, suppose we add a constraint on the manipulated variable, $m_n \leq 0.2$. Using $\Delta t = 2$ and $T = 40$, select values of U, V, and f so that this constraint is not violated for a load change of unity.

27.7. Often a controller tuned to give a good load response will not be satisfactory for a set-point change. Using Example 27.4, compare the following two sets of controller parameters for both load and set-point changes:

(a) $U = 7$ $V = 10$ $f = 0$
(b) $U = 3$ $V = 10$ $f = 0$

27.8. Compute the gain matrix $\mathbf{K}_c$ (Eq. 27-33) for Example 27.4 using the following two sets of parameters:

(a) $U = 5$ $V = 5$
(b) $U = 3$ $V = 5$

Note that in case (a) we have $\mathbf{K}_c = \mathbf{A}^{-1}$. Compare the two gain matrices.

27.9. In Exercise 27.8 evaluate the effect of using only the first row of $\mathbf{K}_c(\mathbf{K}_{c1}^T)$ versus the full matrix $\mathbf{K}_c$ for $U = 3$ and $V = 5$. Calculate the manipulated variable for the first three time steps. Do you think that the use of $\mathbf{K}_{c1}^T$ could lead to stability problems when $\mathbf{K}_c$ is stable?

27.10. Derive Eq. 27-37 and determine parameters g_0 and d_i for the case where $T = 10$.

27.11. Design a predictive controller for the process,

$$G_p(s) = \frac{e^{-6s}}{10s + 1} \qquad G_v = G_m = 1$$

Select T based on 95% completion of the step response and $\Delta t = 2$. Plot the closed-loop response for a set-point change using

(a) $U = 1$ $V = 7$ $f = 0$
(b) $U = 1$ $V = 5$ $f = 0$
(c) $U = 4$ $V = 30$ $f = 0$

27.12. A staged process has the transfer function

$$G_p(s) = \frac{e^{-6s}}{(5s + 1)^5}$$

Select T based on 95% completion of the step response and $\Delta t = 2$. Plot set-point changes for

(a) $U = 1$ $V = 12$ $f = 0$
(b) $U = 1$ $V = 20$ $f = 0$
(c) $U = 4$ $V = 30$ $f = 0$

PART SEVEN

PROCESS CONTROL STRATEGIES

— CHAPTER 28 —————————

The Art of Process Control

In the preceding chapters, we have concentrated on specific issues in the modeling and control of processes that are of interest to chemical engineers. The major emphasis has been on control problems with a single controlled variable and a single manipulated variable. These relatively simple examples have illustrated the development of dynamic models and the design and analysis of control systems. However, process control involves more than the straightforward application of physical principles and mathematical analysis. The practice of process control is also an art, relying on the insight, experience, and ingenuity of the practitioner.

In this chapter we illustrate the art of process control by considering a number of general issues:

- The influence of process design on process control
- The importance of degrees of freedom for process control
- Selection of control system configuration (manipulated and controlled variables)

We also describe a powerful analytical technique, singular value analysis, which can provide considerable insight into some of these issues. We conclude this chapter with an industrial case study that illustrates many of the control concepts presented earlier in this book.

28.1 THE INFLUENCE OF PROCESS DESIGN ON PROCESS CONTROL

Traditionally, process design and control system design have been separate activities. Control system design normally has not been initiated until after the plant design is well underway and major pieces of equipment have been ordered. This approach has serious limitations, since the plant design determines the process dynamic characteristics as well as the operability of the plant. In extreme situations, the plant may be uncontrollable even though the process design appears satisfactory from a steady-state point of view [1].

In recent years there has been a growing recognition of the importance of considering dynamics and control issues early in the plant design. This interplay between design and control has become especially important for modern processing plants, which tend to have a larger degree of material and energy integration and tighter performance specifications. As Hughart and Kominek [2] have noted: "The control system engineer can make a major contribution to a project by advising the project team on how process design will influence the process dynamics and the control structure."

Next we consider several specific examples of how process design affects dynamics and control.

EXAMPLE 28.1

A representative control configuration for the base of a distillation column is shown in Fig. 28.1. Bottoms composition is controlled by adjusting the product flow rate while liquid level is regulated by adjusting the steam control valve for the external reboiler. A typical olefins column might have the following specifications [2]:

Diameter	5 m
Liquid height	1 m
Specific gravity	0.51
Liquid holdup	10 000 kg
Bottoms product flow rate	200 kg/h

The time constant (residence time) for this oversized base section is 50 h. The rationale for having such a large liquid inventory in the column base is that the excess surge capacity will smooth out disturbances in the bottom stream, which is often used as the feed stream to the next column in a train. Hughart and Kominek [2] comment that a conventional composition controller with integral action will wind up in a few hours, causing the bottoms control valve to be full open or closed. They remark that "the results of this control action are: the operator is upset, the department manager is upset, and the composition controller gets placed on manual to stay there." They note that a redesigned column base section with a 5-min time constant when used with a proportional-only controller will seldom upset the feed flow to the next column, although proportional-only control will result in some offset.

EXAMPLE 28.2

Because of the rise in fuel prices since 1974, there has been considerable interest in reducing energy costs of distillation trains by heat integration or thermal coupling of two or more columns. Figure 28.2 compares a conventional distillation system

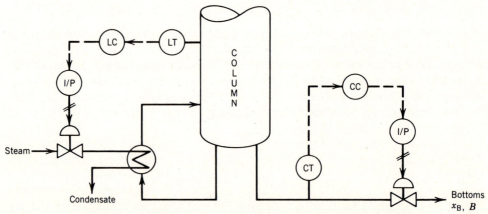

Figure 28.1   Control scheme for bottom section of a distillation column.

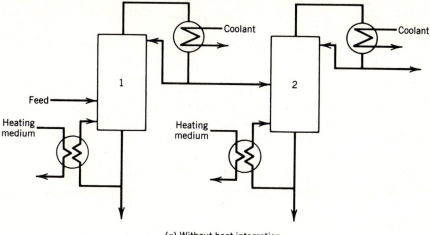

(a) Without heat integration

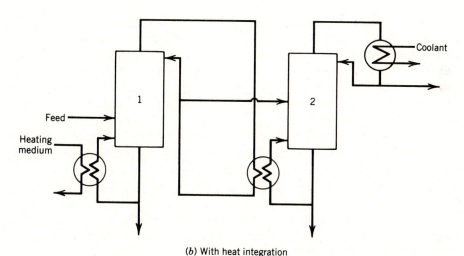

(b) With heat integration

Figure 28.2   Two distillation column configurations.

with a heat-integrated scheme. Heat integration reduces the energy costs by allowing the overhead stream from Column 1 to be used as the heating medium in the reboiler for Column 2. However, this column arrangement is more difficult to control for two reasons. First, the process is more highly interacting, because process upsets in one column affect the other column via the heat integration. Second, the heat integration scheme has one less manipulated variable available for process control, since the reboiler heat duty for Column 2 can no longer be independently manipulated. These disadvantages of heat integration can be overcome by utilizing more sophisticated control strategies [3].

EXAMPLE 28.3

A second type of heat integration is shown in Fig. 28.3b for a packed-bed reactor; the conventional design is shown in Fig. 28.3a. If the chemical reaction is exothermic, energy costs can be reduced by using the hot product stream to heat the

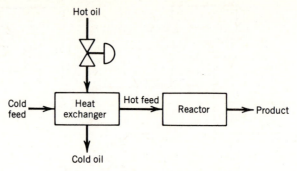

(a) Reactor with conventional feed preheater

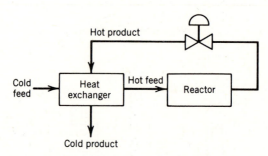

(b) Reactor with feed-effluent heat exchanger

Figure 28.3 Alternative reactor feed heating schemes.

cold feed in a heat exchanger. However, this reactor configuration has the same disadvantages as the heat-integrated distillation system in Fig. 28.2, namely, one less manipulated variable and unfavorable dynamic interactions. In particular, the feed-effluent heat exchanger introduces positive feedback and the possibility of a thermal runaway, since temperature fluctuations in the effluent are transmitted to the feed stream. Thus, they affect the rate of reaction and, consequently, the heat generation due to the reaction.

EXAMPLE 28.4

Two alternative temperature control schemes for a jacketed batch reactor are shown in Fig. 28.4. As indicated in Section 18.1, a common temperature control strategy for batch reactors consists of cascading the reactor temperature to the jacket coolant temperature. The configuration in Fig. 28.4a has the serious disadvantage that the coolant circulation rate and hence, the associated time delay in the coolant loop, also varies. As Shinskey [1] has noted, when the time delay varies with the manipulated variable, a nonlinear oscillation can develop. If the reactor temperature increases, the controller increases the coolant flow rate, which will result in a smaller time delay and a sharp temperature drop. On the other hand, when the reactor temperature is too low, the controller reduces coolant flow, which increases the time delay and results in a slow response. This nonlinear cycle tends to be repeated.

Shinskey [1] points out that this control problem can be solved by making a simple equipment design change, namely, by adding a recirculation pump as shown

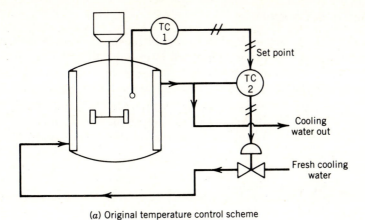

(a) Original temperature control scheme

(b) Temperature control scheme with coolant recirculation pump

Figure 28.4   Batch reactor with two temperature control strategies.

in Fig. 28.4b. Now the recirculation rate and process time delay are independent of the flow rate of fresh cooling water and the nonlinear oscillations are eliminated.

28.2 DEGREES OF FREEDOM FOR PROCESS CONTROL (REVISITED)

To formulate control objectives and to design a control system, it is necessary to select the appropriate number of manipulated variables. In particular, the number of process variables that can be manipulated cannot exceed the degrees of freedom. As indicated in Section 2.4, the degrees of freedom of a process are the number of process variables that must be specified to determine the remaining process variables. The degrees of freedom can be determined from a process model as follows:

$$N_F = N_V - N_E \tag{28-1}$$

where N_F is the degrees of freedom, N_V is the number of process variables, and N_E is the number of independent equations.

Effect of Feedback Control

Next we consider the effect of feedback control on the degrees of freedom of a process. In general, adding a single feedback loop (e.g., PI or PID control) introduces an additional equation, the control law, and thus uses up one degree of freedom, the manipulated variable. It can be argued that the control law also introduces a new process variable, the set point. However, the value of the set point is usually specified by the operator or is determined by a supervisory control system. Thus, the net result of controlling a process variable is to reduce the degrees of freedom by one.

Utilization of Degrees of Freedom

For an underspecified process ($N_F > 0$), the degrees of freedom are utilized in two ways:

1. Choice of manipulated variables.
2. Identification of process variables that are fixed (determined) by the process environment. For example, the ambient temperature or a feed stream that is an exit stream from an upstream unit would fit in this category.

Thus, we can write

$$N_F = N_M + N_S \tag{28-2}$$

where N_M = number of manipulated variables

N_S = number of process variables specified by the environment.

Note that Eq. 28-2 suggests the very important result that the number of manipulated variables is always less than or equal to the available degrees of freedom:

$$N_M \leq N_F \tag{28-3}$$

Now we consider three examples that illustrate how the number of manipulated variables can be determined using Eqs. 28-1 through 28-3.

EXAMPLE 28.5

Determine the degrees of freedom for the two liquid storage systems shown in Fig. 28.5. Assume that the tank cross-sectional area A and liquid density ρ are constant.

Solution

For System I, the unsteady-state mass balance is

$$A \frac{dh}{dt} = q_i - q \tag{28-4}$$

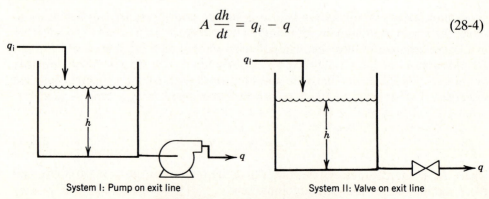

System I: Pump on exit line   System II: Valve on exit line

Figure 28.5 Liquid storage systems.

Thus, three process variables, h, q_i, and q, are related by a single equation. From Eq. 28-1 it follows that there are two degrees of freedom since $N_F = 3 - 1 = 2$. Thus, either q_i or q could be chosen as the manipulated variable with the other fixed by the environment, that is, the feed from or to an adjacent process unit.

For System II, Eq. 28-4 still applies but an additional equation is necessary to specify the flow–head relation associated with the valve. For example,

$$q = C\sqrt{h} \tag{28-5}$$

where C is a constant. Thus for System II, $N_V = 3$, $N_E = 2$ and consequently $N_F = 1$. In this case, only q_i is available to be manipulated.

By contrast, suppose that the degrees of freedom analysis is based on the steady-state mass balance rather than the unsteady-state balance. For System I, the steady-state version of Eq. 28-4 is

$$0 = q_i - q \tag{28-6}$$

Thus, it might be erroneously concluded that System I has $N_V = 2$, $N_E = 1$ and consequently $N_F = 1$ rather than the correct result, $N_F = 2$. This situation occurs because liquid level h appears only in the accumulation term of the unsteady-state balance, Eq. 28-4, but not in the steady-state balance, Eq. 28-6.

In general, the degrees of freedom analysis can be based on a steady-state analysis provided that the accumulation terms do not contain any additional process variables. The latter situation is characteristic of non-self-regulating (integrating) processes such as System I in Fig. 28.5.

EXAMPLE 28.6

Determine the degrees of freedom for the stirred-tank heater system in Fig. 2.1. Discuss an appropriate specification of manipulated variables, environmental variables, and controlled variables.

Solution
The dynamic model in Eqs. 2-11 and 2-12 contains six process variables (w_i, w, T_i, T, q, and V) in two equations. Thus, there are four degrees of freedom. Typically, $N_S = 2$ since the feed conditions (w_i and T_i) would be determined by upstream units. From Eq. 28-2 it follows that $N_M = 4 - 2 = 2$. For example, we could choose to manipulate q and w to control T and V, respectively.

EXAMPLE 28.7

Determine the available degrees of freedom for the blending system shown in Fig. 28.6. Discuss an appropriate specification of manipulated variables and environmental variables. Assume that the primary control objective is to control exit composition c_5 of a key component and that all of the streams have the same density. The makeup stream has a constant composition c_4, but its flow rate q_4 can be adjusted. Streams 1 and 2 have variable flow rates and compositions. In Fig. 28.6, the compositions of the key component have units of mols solute/m^3 while the q_i's denote volumetric flow rates, m^3/min.

Solution
Make the usual assumptions that each tank is perfectly mixed and has a constant cross-sectional area. Then the unsteady-state overall mass and component balances are

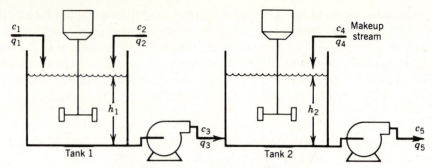

Figure 28.6 Two tank blending system.

$$\text{Tank 1} \qquad A_1 \frac{dh_1}{dt} = q_1 + q_2 - q_3 \qquad (28\text{-}7)$$

$$A_1 \frac{d(h_1 c_3)}{dt} = q_1 c_1 + q_2 c_2 - q_3 c_3 \qquad (28\text{-}8)$$

$$\text{Tank 2} \qquad A_2 \frac{dh_2}{dt} = q_3 + q_4 - q_5 \qquad (28\text{-}9)$$

$$A_2 \frac{d(h_2 c_5)}{dt} = q_3 c_3 + q_4 c_4 - q_5 c_5 \qquad (28\text{-}10)$$

There are four equations and twelve process variables: levels h_1 and h_2, compositions c_1 to c_5, and flow rates q_1 to q_5. From Eq. 28-1 it follows that there are $12 - 4 = 8$ degrees of freedom. There are four external disturbance variables (c_1, q_1, c_2, q_2); also c_4 is assumed to be constant. Thus, there are three remaining degrees of freedom that can be used to specify no more than three manipulated variables. For example, we could choose q_3, q_4 and q_5 as the manipulated variables. For controlled variables, we should select exit concentration c_5, because it is the primary controlled variable (see problem statement), and liquid levels h_1 and h_2, to maintain adequate liquid inventories without having the two tanks drain completely or overflow. Note that it is not necessary to control concentration c_3, since it is the composition of an intermediate stream rather than a product stream.

One reasonable multiloop control configuration consists of the following pairing of controlled and manipulated variables: $c_5 - q_4$, $h_1 - q_3$, and $h_2 - q_5$. However, further analysis is necessary to determine whether this proposed pairing is the best one from among the $3! = 6$ alternatives. In particular, both the steady-state and the dynamic characteristics of the process should be considered (see Section 19.2).

Required Number of Manipulated Variables

In the previous section, we have seen that the number of process variables that can be manipulated N_M is always less than or equal to the number of degrees of freedom N_F. The question then arises as to the number of manipulated variables necessary to control the N_C controlled variables. If offsets cannot be tolerated, then there must be at least as many manipulated variables as controlled variables, that is,

$$N_M \geq N_C \qquad (28\text{-}11)$$

For example, if a process has five controlled variables, then offsets can be avoided by using five PI or PID controllers. By contrast, if only four manipulated variables are available, then offset can be eliminated in only four of the five controlled variables. A similar situation exists for the selective control schemes of Chapter 18 where a single manipulated variable is shared by two controlled variables.

Occasionally, situations occur where there are more manipulated variables than controlled variables ($N_M > N_C$). For example, when split-range control valves are used for temperature control, two control valves, typically one for heating and one for cooling are employed to manipulate a single variable, (see Section 18.3). Often it is desirable to use extra manipulated variables to maximize profits (see Chapter 20).

28.3 CONTROL SYSTEM DESIGN CONSIDERATIONS

In Chapter 1 we described a general strategy for control system design (see Section 1.4 and Fig. 1.4). In this section we focus attention on two important steps: (1) specification of the control system objectives, and (2) selection of the control system configuration via the choice of controlled, manipulated, and measured variables.

The formulation of the control system objectives is strongly dependent on the objectives of the processing plant. The primary plant objective is to maximize profits by transforming raw materials into useful products while satisfying a number of important criteria:

1. **Safety.** It is essential that the plant be operated safely to protect the well-being of plant personnel and nearby communities. For example, safety considerations may limit the allowable temperature and pressure for a process vessel while explosion limits may restrict the hydrocarbon-to-oxygen ratio.
2. **Environmental Regulations.** Processing plants must be operated so as to comply with environmental regulations concerning air and water quality as well as waste disposal.
3. **Product Specifications.** For the plant to sell its products, there are usually product specifications that must be met concerning product quality and production rate.
4. **Operational Constraints.** In addition to the above three criteria, it is necessary that process variables satisfy certain other operating constraints. For example, distillation columns are operated so as to avoid conditions such as flooding and weeping; reactor temperatures are often limited so as to prevent degradation of the catalyst or the onset of undesirable side reactions.

The plant objectives, in turn, determine the objectives for the control system as well as the controller set points.

Classification of Process Variables

For purposes of control system design, it is convenient to classify process variables as either inputs or outputs, as is shown in Fig. 28.7. The output variables (y_1, y_2, . . . , y_N) are process variables that ordinarily are associated with exit streams of

Figure 28.7 Process with multiple inputs and multiple outputs.

a process or conditions inside a process vessel (e.g., compositions, temperatures, levels, and flow rates). A subset of the output variables is selected as variables to be controlled (i.e., *controlled variables*) in order to satisfy the plant and control objectives. The process inputs $(x_1, x_2, \ldots, x_M)$ are physical variables that affect the process outputs. Typically, the inputs are associated with inlet streams (e.g., feed composition or feed flow rate) or environmental conditions (e.g., ambient temperature). However, an exit flow rate from a process can also be an "input" from a control point of view if the flow rate is used as a manipulated variable or if the magnitude of the flow rate is determined by downstream units. For example, the flow rate of an exit stream from a storage tank may be adjusted by a downstream unit if the stream serves as a feed to that unit. Some of the process inputs are specified as manipulated variables, while the other inputs are considered to be disturbance variables, specified by the external environment.

In general, it is not feasible to control all of the output variables for a number of reasons [4]:

1. It may not be possible or economical to measure all of the outputs, especially compositions.
2. There may not be enough manipulated variables (refer to the degrees of freedom discussion in Section 28.2).
3. Potential control loops may be impractical because of slow dynamics, a low sensitivity to available manipulated variables, or interactions with other control loops.

In general, controlled variables are measured on-line and the measurements are used for feedback control. However, it is theoretically possible to control a process variable that is not measured by using a mathematical model of the process to calculate the value of the unmeasured controlled variable. This inferential control strategy (see Section 18.2) was proposed by Brosilow and co-workers [5] to control product composition of a distillation column by inferring mole fraction from flow rate and temperature measurements. Inferential control schemes have also been used for product dryness [6] and digester control in pulp mills [7].

Selection of Controlled Variables

The consideration of plant and control objectives has led to a number of suggested guidelines for the selection of controlled variables from the available output variables [4,8]:

Guideline 1.
Select variables that are not self-regulating. A common example is liquid level in a storage vessel with a pump in the exit line.

Guideline 2.
Choose output variables that may exceed equipment and operating constraints (e.g., temperatures, pressures, and compositions).

Guideline 3.
Select output variables that are a direct measure of product quality (e.g., composition, refractive index) or that strongly affect it (e.g., temperature, pressure).

Guideline 4.
Choose output variables that seriously interact with other controlled variables. The steam header pressure for a plant boiler that supplies several downstream units is an example of this type of output variable.

Guideline 5.

Choose output variables that have favorable dynamic and static characteristics. Ideally, there should be at least one manipulated variable that has a significant, direct, and rapid effect on each controlled variable.

These five guidelines should not be considered to be hard and fast rules. Also, for a particular application the guidelines may be inconsistent and thus result in a conflict. As an example of their use, an output variable such as temperature must be kept in limits (Guideline 2). Temperature could also affect other output variables (e.g., composition, pressure) and thus also should be selected according to Guideline 4. If there were a conflict, Guideline 2 would be the overriding concern in this situation. The use of these guidelines is illustrated by Example 28.8 later in this section.

Selection of Manipulated Variables

Based on the plant and control objectives, a number of guidelines have been proposed for the selection of manipulated variables from among the input variables [4,8]:

Guideline 6.

Select inputs that have large effects on the controlled variables.

Thus, for each control loop it is desirable that the steady-state gain between the manipulated variable and the controlled variable be as large as possible. Also, it is important that the manipulated variable have a large enough range. For example, if a distillation column operates at a steady-state reflux ratio of five, it will be much easier to control the level in the reflux drum by using the reflux flow rather than the distillate flow rate, since the reflux flow rate is five times larger. However, the effect of this choice on the product compositions must also be considered in making the final decision [3].

Guideline 7.

Choose inputs that rapidly affect the controlled variables.

Clearly, it is desirable that a manipulated variable affect the corresponding controlled variable as quickly as possible. Also, any time delays or time constants that are associated with the manipulated variable should be small relative to the dominant process time constant.

Guideline 8.

The manipulated variables should affect the controlled variables directly rather than indirectly [4].

Compliance with this guideline usually results in a control loop with favorable static and dynamic characteristics. For example, consider the problem of controlling the exit temperature of a process stream that is heated by steam in a shell and tube heat exchanger. It is preferable to throttle the steam flow to the heat exchanger rather than the condensate flow from the shell, since the steam flow rate has a more direct effect on the steam pressure inside the shell and thus on the steam saturation temperature and the rate of heat transfer.

Guideline 9.

Avoid recycling of disturbances.

As Newell and Lee [4] have noted, it is preferable *not* to manipulate an inlet stream or a recycle stream, because disturbances tend to be propagated forward or recycled back to the process. This problem can be avoided by manipulating a utility stream to absorb disturbances or an exit stream that allows the disturbances to be passed downstream, provided that the exit stream changes do not upset downstream process units.

Note that these guidelines may be in conflict. For example, a comparison of the effects of two inputs on a single controlled variable may indicate that one has a larger steady-state gain (Guideline 6) but slower dynamics (Guideline 7). In this situation a trade-off between static and dynamic considerations must be made in selecting the appropriate manipulated variable from the two input candidates.

Selection of Measured Variables

Safe, efficient operation of processing plants is made possible by the on-line measurement of key process variables. Clearly, output variables that are used as controlled variables should be measured. Other output variables are measured to provide additional information to the plant operators or for use in model-based control schemes such as supervisory control or inferential control. It is also desirable to measure selected input variables as well as output variables, since recorded measurements of manipulated inputs provide useful information for tuning controllers and troubleshooting control loops, as discussed in Chapter 13. Also, measurements of disturbance inputs can be used in feedforward control schemes (see Chapter 17).

In choosing which outputs to measure and in locating measurement points, both static and dynamic considerations are important.

Guideline 10.
Reliable, accurate measurements are essential for good control.

There is ample evidence in the literature [2,8] that inadequate measurements are a key contributor to poor control. Hughart and Kominek [2] cite common measurement problems that they observed in distillation column control problems: orifice runs without enough straight piping, sample lines with too much time delay, temperature probes located in insensitive regions, and flow measurements of liquids at saturation temperatures (e.g., distillate and bottoms streams) that involve liquid flashing in the orifice. They note that these measurement problems can be readily resolved during the process design stage but that improving a measurement location after the process is operating is extremely difficult.

Guideline 11.
Select measurement points that have an adequate degree of sensitivity.

For example, in distillation columns a product composition is often controlled indirectly by regulating a temperature near the end of the column if an analyzer is not available. However, for high-purity separations the location of the temperature measurement point can be important. If a tray near an end of the column is selected, the tray temperature tends to be insensitive, since the tray composition can vary significantly even though the tray temperature changes very little. By contrast, if the temperature measurement point is moved closer to the feed tray, the temperature sensitivity is improved, but disturbances entering the column at the ends (e.g., condenser, reboiler) are not sensed as quickly.

Guideline 12.
Select measurement points that minimize time delays and time constants.

Reducing dynamic lags and time delays associated with process measurements improves closed-loop stability and response characteristics. Hughart and Kominek [2] have observed distillation columns with the sample connection for the bottoms analyzer located 200 feet downstream from the column. This distance introduces a significant time delay and makes the column difficult to control, even more so because the time delay varies with the flow rate.

Next we consider an evaporator control problem that will be used to illustrate the guidelines developed in the previous sections.

EXAMPLE 28.8

An evaporator is used to concentrate a dilute solution of a single solute in a volatile solvent, as is shown in Fig. 28.8. Specify a control configuration for the evaporator for two situations:

a. The product composition x_B is measured on-line.
b. x_B is not measured on-line.

Solution

A dynamic model of the evaporator can be developed based on the following assumptions. For a more detailed discussion of evaporator models see Newell and Fisher [9].

1. Liquid is perfectly mixed as a result of the violent boiling that occurs.
2. The thermal capacitance of the vapor is negligible compared with the thermal capacitance of the liquid.
3. The dynamics of the steam chest are negligible.
4. The feed stream and bottoms stream have a constant molar density c and a constant heat capacity $C_{p\ell}$.
5. The vapor and liquid are in thermal equilibrium at all times.
6. Heat losses and heat of solution effects are negligible.

We can then write the following balances:

Total Material Balance

$$Ac \frac{dh}{dt} = F - B - D \qquad (28\text{-}12)$$

where F, B, and D are molar flow rates, A is the cross-sectional area, and c is the

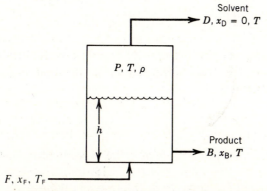

Figure 28.8 Schematic diagram of an evaporator.

molar density (mol/m^3). The average solution molecular weight is assumed to be constant, since changes in composition are small.

Solute Balance

$$Ac \frac{d(hx_B)}{dt} = Fx_F - Bx_B \tag{28-13}$$

where x_B and x_F denote mole fractions.

Energy Balance

$$AC_{p\ell} \frac{d(hT)}{dt} = C_{p\ell}FT_F - C_{p\ell}BT - C_{pv}DT + UA_s(T_s - T) - \Delta H_v E \tag{28-14}$$

where $C_{p\ell}$ = specific heat of the liquid
   C_{pv} = specific heat of the vapor
   A_s = heat transfer area
   U = overall heat transfer coefficient
   ΔH_v = latent heat of vaporization
   E = rate of evaporation (mol/h)
   T_s = steam temperature
   T_{ref} has been eliminated

In (28-14) the reference temperature associated with the energy flows has been eliminated algebraically (Chapter 2). For normal operation, the sensible heat changes in the liquid are small compared with the latent heats associated with the condensing steam and the evaporating solvent. Consequently, the heat transferred from the condensing steam is primarily used to evaporate solvent. Thus, the derivative in (28-14) is small, and this equation can be approximated by

$$0 = UA_s(T_s - T) - \Delta H_v E \tag{28-15}$$

Solving for the evaporation rate E gives

$$E = \frac{UA_s(T_s - T)}{\Delta H_v} \tag{28-16}$$

Material Balance on Vapor

$$\frac{1}{M} \frac{d(\rho V)}{dt} = E - D \tag{28-17}$$

where ρ is the vapor density, V is the volume of vapor, and M is the molecular weight of the solvent.

Equation of State

The density of the vapor ρ can be related to the pressure P and the temperature T by an equation of state:

$$\rho = \phi_1(P, T) \tag{28-18}$$

Vapor Pressure Relation

The pressure in the evaporator P is equal to the vapor pressure of the liquid solution P_{vap}, which depends on the temperature T. This temperature dependence can be expressed as $P_{\text{vap}} = \phi_2(T)$. Since $P = P_{\text{vap}}$, we can write

$$P = \phi_2(T) \tag{28-19}$$

Volume Relation

Since the liquid level can change, the vapor space volume V also can change. However, these two variables are related by

$$V_0 = V + Ah \qquad (28\text{-}20)$$

where V_0 is the fixed volume of the evaporator.

Thermodynamic Relation

If the steam is saturated, the relation between the steam pressure and temperature can be obtained from steam table data:

$$P_s = \phi_3(T_s) \qquad (28\text{-}21)$$

Thus, the simplified dynamic model of the evaporator consists of eight equations, (28-12), (28-13), and (28-16) through (28-21), and 14 variables, h, F, B, D, x_F, x_B, T, T_F, T_s, E, ρ, V, P, and P_s. Thus, there are six degrees of freedom. However, the feed conditions (F, T_F, and x_F) are normally fixed by operations in an upstream unit. Consequently, the maximum number of variables that can be manipulated is $6 - 3 = 3$.

Case (a): Product Composition x_B Is Measured On-Line

First, we consider the selection of the controlled variables. Because three degrees of freedom are available, three process variables can be controlled without offset by adjusting three manipulated variables. Since the primary objective is to obtain a product stream with a specified composition, mole fraction x_B is the primary controlled variable (Guideline 3). Liquid level h should also be controlled, since it is not self-regulating (Guideline 1). The evaporator pressure P should also be controlled, since it has a major influence on the evaporator operation (Guideline 2). Thus, three controlled variables are selected: x_B, h, and P.

Next we select the three manipulated variables. It has been assumed that the feed conditions cannot be adjusted; thus, the obvious manipulated variables are B, D, and P_s. Liquid flow rate B has a significant effect on h but a relatively small effect on P and x_B. Therefore, it is reasonable to control h by manipulating B (Guideline 6). Vapor flow rate D has a direct and rapid effect on P while having less direct effects on h and x_B. Thus, P should be paired with D (Guideline 6). This leaves a $P_s - x_B$ pairing for the third control loop, which is physically reasonable, since the most direct way of regulating x_B is by adjusting the amount of solvent that is evaporated via the steam pressure (Guideline 8).

Next, we consider which variables to measure. Clearly, the three controlled variables, x_B, h, and P, should be measured. It is also desirable to measure the three manipulated variables, B, D and P_s, since this information is useful for controller tuning and troubleshooting. If large and frequent disturbances in the feed stream occur, then measurements of load variables F and x_F could be used in a feedforward control strategy that would complement the feedback control scheme. It is probably not necessary to measure T_F, since sensible heat changes in the feedstream will typically be small compared with the heat fluxes in the evaporator.

A schematic diagram of the controlled evaporator is shown in Fig. 28.9.

Case (b): Product Composition Cannot Be Measured On Line

The controlled variables are the same as in Case (a) but, since the third controlled variable x_B cannot be measured on line, standard feedback control is not possible. However, a simple feedforward control strategy [10] can be developed based on a

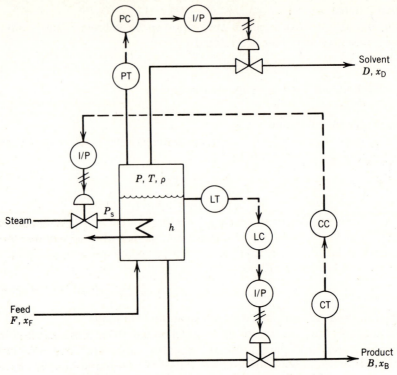

Figure 28.9   Evaporator control strategy for Case (a).

steady-state version of Eq. 28-13:

$$0 = \overline{F}\overline{x}_F - B\overline{x}_B \tag{28-22}$$

where the bars denote the nominal steady-state conditions. Rearranging gives us

$$\overline{B} = \overline{F}\frac{\overline{x}_F}{\overline{x}_B} \tag{28-23}$$

Equation 28-23 provides the basis for the feedforward control law. Replacing $\overline{B}$ and $\overline{F}$ by the actual flow rates, B and F, and replacing the nominal product composition $\overline{x}_B$ by the set-point value, $x_{B_{sp}}$, gives

$$B = F\frac{\overline{x}_F}{x_{B_{sp}}} \tag{28-24}$$

Thus, the manipulated variable B is adjusted based on the measured load variable F, the set point $x_{B_{sp}}$, and the nominal value of the feed composition $\overline{x}_F$.

The manipulated variables are the same as for Case (a): D, B, and P_s. B has already been used in the feedforward method of controlling x_B. Clearly, the P–D pairing is still desirable for the reasons given for Case (a). This leaves h to be controlled by adjusting the rate of evaporation via P_s. A schematic diagram of the controlled evaporator is shown in Fig. 28.10.

This control strategy has two disadvantages. First, it is based on the assumption that the unmeasured feed composition is constant at a known steady-state value. Second, the feedforward control technique was based on steady-state considerations. Thus, it may not perform well during transient conditions unless dynamic compensation is added as discussed in Chapter 17. Nevertheless, this scheme pro-

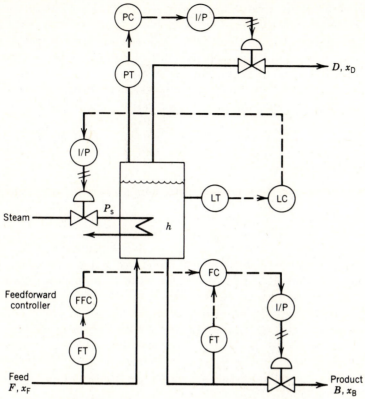

Figure 28.10 Evaporator control strategy for Case (b).

vides a simple, indirect method for controlling a product composition when it cannot be measured.

28.4 SINGULAR VALUE ANALYSIS

The previous section contained a number of qualitative guidelines for selecting controlled and manipulated variables. In this section we consider a powerful analytical technique, *singular value analysis (SVA)*, which can be used to solve the following important control problems:

1. Selection of controlled and manipulated variables.
2. Evaluation of the robustness of a proposed control strategy.
3. Determination of the best multiloop control configuration.

Singular value analysis and its extensions, including singular value decomposition (SVD), also have many uses in numerical analysis and the design of multivariable control systems which are beyond the scope of this book [11]. In this section we provide a brief introduction to SVA that is based on an analysis of steady-state process models.

Consider a process that has n controlled variables and n manipulated variables. We assume that a steady-state process model is available and that it has been linearized as in Section 19.2 to give

$$\mathbf{C} = \mathbf{K}\,\mathbf{M} \tag{28-25}$$

where $\mathbf{C}$ is the vector of n controlled variables, $\mathbf{M}$ is the vector of n manipulated variables, and $\mathbf{K}$ is the steady-state gain matrix. The elements of $\mathbf{C}$ and $\mathbf{M}$ are expressed as deviation variables. One desirable property of $\mathbf{K}$ is that the n linear equations in n unknowns represented by (28-25) be linearly independent. In contrast, if the equations are dependent, then not all of the n controlled variables can be independently regulated. This characteristic property of linear independence can be checked by several methods [12]. For example, if the determinant of $\mathbf{K}$ is zero, the matrix is singular and the n equations in (28-25) are not linearly independent.

Another way to check for linear independence is to calculate one of the most important properties of a matrix—its eigenvalues. The eigenvalues of matrix $\mathbf{K}$ are the roots of the equation:

$$|\mathbf{K} - \alpha\mathbf{I}| = 0 \tag{28-26}$$

where $|\mathbf{K} - \alpha\mathbf{I}|$ denotes the determinant of matrix $\mathbf{K} - \alpha\mathbf{I}$, and $\mathbf{I}$ is the $n \times n$ identity matrix. The n eigenvalues of $\mathbf{K}$ will be denoted by $\alpha_1, \alpha_2, \ldots, \alpha_n$. If any of the eigenvalues is zero, matrix $\mathbf{K}$ is singular, and difficulties will be encountered in controlling the process, as noted above. If one eigenvalue is very small compared to the others, then large changes in one or more manipulated variables will be required to control the process, as is shown at the end of this section.

Another important property of $\mathbf{K}$ is its *singular values*, $\sigma_1, \sigma_2, \ldots, \sigma_n$ [11,12]. The singular values are nonnegative numbers that are defined as the positive square roots of the eigenvalues of $\mathbf{K}^T\mathbf{K}$. The first r singular values are positive numbers where r is the rank of matrix $\mathbf{K}^T\mathbf{K}$. The remaining $n - r$ singular values are zero. Usually the nonzero singular values are ordered with σ_1 denoting the largest and σ_r denoting the smallest.

The final matrix property of interest here is the *condition number*, CN. Assume that $\mathbf{K}$ is nonsingular. Then the condition number of $\mathbf{K}$ is defined as the ratio of the largest and smallest nonzero singular values:

$$\text{CN} = \frac{\sigma_1}{\sigma_r} \tag{28-27}$$

If $\mathbf{K}$ is singular, then it is ill-conditioned and by convention CN $= \infty$. The concept of a condition number can also be extended to non-square matrices [11].

The condition number is a positive number that provides a measure of how ill-conditioned the gain matrix is. It also provides useful information on the sensitivity of the matrix properties to variations in the elements of the matrices. This important topic, which is related to control system robustness, will be considered later in this section. But first we consider a simple example.

EXAMPLE 28.9

A 2×2 process has the steady-state gain matrix shown below. Calculate the determinant, eigenvalues, and singular values of $\mathbf{K}$.

$$\mathbf{K} = \begin{pmatrix} 1 & K_{12} \\ 10 & 1 \end{pmatrix} \tag{28-28}$$

Use $K_{12} = 0$ as the base case; then recalculate the matrix properties for a small change to $K_{12} = 0.1$.

Solution

By inspection, the determinant for the base case is $|\mathbf{K}| = 1$. The eigenvalues can be calculated as follows:

$$|\mathbf{K} - \alpha\mathbf{I}| = \begin{vmatrix} 1-\alpha & 0 \\ 10 & 1-\alpha \end{vmatrix} = 0 \tag{28-29}$$

Thus, $(1 - \alpha)^2 = 0$ and the eigenvalues are $\alpha_1 = \alpha_2 = 1$.

Now calculate the singular values. From Eq. 28-28,

$$\mathbf{K}^T\mathbf{K} = \begin{pmatrix} 1 & 10 \\ 0 & 1 \end{pmatrix}\begin{pmatrix} 1 & 0 \\ 10 & 1 \end{pmatrix} = \begin{pmatrix} 101 & 10 \\ 10 & 1 \end{pmatrix} \tag{28-30}$$

The eigenvalues of $\mathbf{K}^T\mathbf{K}$, denoted by α', can be calculated from $|\mathbf{K}^T\mathbf{K} - \alpha'\mathbf{I}| = 0$, which again yields a second-order polynomial:

$$(101 - \alpha')(1 - \alpha') - 100 = 0 \tag{28-31}$$

Solving (28-31) gives $\alpha'_1 = 101.99$ and $\alpha'_2 = 0.01$. The singular values of $\mathbf{K}$ are then

$$\sigma_1 = \sqrt{101.99} = 10.1 \tag{28-32}$$

$$\sigma_2 = \sqrt{0.01} = 0.1 \tag{28-33}$$

and the condition number is

$$\text{CN} = \frac{\sigma_1}{\sigma_2} = \frac{10.1}{0.1} = 101 \tag{28-34}$$

This $\mathbf{K}$ matrix is considered to be poorly conditioned because of the large CN value.

Now consider the case where $K_{12} = 0.1$. The determinant of $\mathbf{K}$ is zero, which indicates that $\mathbf{K}$ is singular for this perturbation. The eigenvalues of $\mathbf{K}$ calculated from (28-26) are $\alpha_1 = 2$ and $\alpha_2 = 0$. The singular values of $\mathbf{K}$ are $\sigma_1 = 10.1$, $\sigma_2 = 0$, and the condition number is $\text{CN} = \infty$ because $\mathbf{K}$ is singular.

This example shows that the original $\mathbf{K}$ matrix (with $K_{12} = 0$) is poorly conditioned and very sensitive to small variations in the K_{12} element. The large condition number (CN = 101) indicates the poor conditioning. In contrast, the reasonable value for the determinant ($|\mathbf{K}| = 1$) gives no indication of poor conditioning. The example demonstrates that the condition number is superior to the determinant in providing a more reliable measure of ill-conditioning and potential sensitivity problems.

Processes with poorly conditioned $\mathbf{K}$ matrices tend to require large changes in the manipulated variables in order to influence the controlled variables. This assertion can be justified as follows. Solving Eq. 28-25 for $\mathbf{M}$

$$\mathbf{M} = \mathbf{K}^{-1}\mathbf{C} \tag{28-35}$$

and substituting set point $\mathbf{R}$ for $\mathbf{C}$ gives,

$$\mathbf{M} = \mathbf{K}^{-1}\mathbf{R} \tag{28-36}$$

The inverse of $\mathbf{K}$ in (28-36) can be calculated from the standard formula,

$$\mathbf{K}^{-1} = \frac{\text{adjoint of } \mathbf{K}}{|\mathbf{K}|} \tag{28-37}$$

The adjoint of K is formed from its cofactors [12].

If $|\mathbf{K}|$ is small, we conclude from (28-36) and (28-37) that the required adjustments in **M** will be very large, resulting in excessive control actions. For a poorly conditioned model, $|\mathbf{K}|$ is either small or can become small for slight variations in the elements of **K** as illustrated by Example 28.9.

Small values of $|\mathbf{K}|$ also lead to large values of the relative gain array (cf. Chapter 19), which is widely used as a measure of process interactions. For a 2×2 process the relative gain array is characterized by a single parameter λ. The following expression for λ can be obtained by rearranging Eq. (19-30):

$$\lambda = \frac{K_{11}K_{22}}{K_{11}K_{22} - K_{12}K_{21}} = \frac{K_{11}K_{22}}{|\mathbf{K}|} \tag{28-38}$$

Thus if $|\mathbf{K}|$ is small, λ becomes very large which indicates that the process interactions are extremely strong leading to control difficulties.

For the two cases in Example 28.9, the relative gains are $\lambda = 1$ ($K_{12} = 0$) and $\lambda = \infty$ ($K_{12} = 0.1$), respectively. The value of $\lambda = 1$ is quite misleading because it suggests that the process model in Example 28.9 has no interactions and that a 1-1/2-2 controller pairing will be suitable. However, the large condition number of 101 for this example implies that the process is poorly conditioned and thus will be difficult to control with any controller pairing or even with advanced control techniques [13]. These examples demonstrate that the RGA approach can provide misleading results for poorly conditioned systems.

Next we briefly describe other uses of SVA for process control problems. Moore [14] has suggested that the condition number can be used as an index of controllability. He cited the case of a tunnel dryer with four heaters and eight temperature sensors. Singular value analysis of the 8×4 gain matrix led to the conclusion that close regulation of even four of the eight temperatures was not possible. In fact, it appeared that only two sensors could be practically controlled using the heaters at their current locations.

Roat et al. [15] analyzed the choice of manipulated variables for a complex, four-component distillation column, depicted in Fig. 28.11. The four components

Table 28.1   Condition Numbers for the Gain Matrices Relating Column Controlled Variables to Various Sets of Manipulated Variables [15]

Controlled Variables

x_D = Mole fraction of propane in distillate D

x_{64} = Mole fraction of isobutane in tray 64 sidedraw

x_{15} = Mole fraction of n-butane in tray 15 sidedraw

x_B = Mole fraction of isopentane in bottoms B

Possible Manipulated Variables

L = Reflux flow rate	B = Bottoms flow rate
D = Distillate flow rate	S_{64} = Sidedraw flow rate at tray 64
V = Steam flow rate	S_{15} = Sidedraw flow rate at tray 15

Strategy Number[a]	Manipulated Variables	Condition Number
1	L/D, S_{64}, S_{15}, V	9 030
2	V/L, S_{64}, S_{15}, V	60 100
3	D/V, S_{64}, S_{15}, V	116 000
4	D, S_{64}, S_{15}, V	51.5
5	L, S_{64}, S_{15}, B	57.4
6	L, S_{64}, S_{15}, V	53.8

[a]*Note:* In each control strategy, the first controlled variable is paired with the first manipulated variable, and so on. Thus, for Strategy 1, x_D is paired with L/D, and x_B is paired with V.

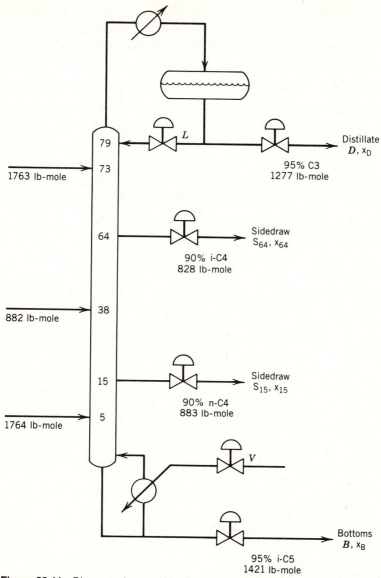

Distillate
D, x_D

95% C3
1277 lb-mole

Sidedraw
S_{64}, x_{64}

90% i-C4
828 lb-mole

Sidedraw
S_{15}, x_{15}

90% n-C4
883 lb-mole

V

Bottoms
B, x_B

95% i-C5
1421 lb-mole

1763 lb-mole

882 lb-mole

1764 lb-mole

L

Figure 28.11   Diagram of a complex, four-component distillation column [15].

were: propane, isobutane, n-butane, and isopentane. There were six possible manipulated variables, and ratios of these variables were also permissible. Table 28.1 shows the condition numbers for six schemes that were evaluated for the column. Note that three of the strategies have roughly the same low CN. Subsequently, these were subjected to further evaluation using dynamic simulation. Based on simulation results, the best control strategy was number 4 in Table 28.1 [15].

There are several other extensions of SVA, requiring further computational steps, that can be employed in control systems analysis:

1. Analysis of controller pairing to augment the RGA [14]
2. Choice of sensor location (e.g. distillation columns) [14,15]
3. Evaluation of the robustness of multivariable control schemes such as decoupling and internal model control [13]

Computer software packages are readily available to make SVA and SVD calculations for processes with any number of inputs or outputs [11]. As mentioned above, it is also helpful to carry out dynamic simulations to verify the SVA results.

28.5 INDUSTRIAL CASE STUDY: THREE-REACTOR SYSTEM

Cardner [16,17] has described a model-based control strategy for a three-reactor system at an industrial chemical plant. A flow diagram of the reactor system is shown in Fig. 28.12. Important process information can be summarized as follows.

1. **Reactor.** Each fixed-bed, catalytic reactor is operated adiabatically. Since the reaction is exothermic, the exit stream from each reactor must be cooled before entering the next reactor. The feed enters the first reactor at approximately 100°C.

2. **Chemistry.** Product C is formed from reactant B and reaction gas A in the liquid phase reaction:

$$A_{(g)} + B_{(\ell)} \rightarrow C_{(\ell)}$$

Reactant B can also be consumed in undesired side reactions. The kinetics of the desired reaction with respect to B can be expressed as follows:

For $3\% \leq c_B < 100\%$ zero-order reaction
For $c_B < 3\%$ first-order reaction

where c_B is the concentration of B in mol%. The nominal value of c_B in the exit stream from the third reaction is 0.15%; thus the third reactor must operate in both reaction rate regimes.

3. **Modes of Operation.** The reactor system operates in several different modes as discussed by Cardner. We will consider only the series mode of operation, shown in Fig. 28.12, in which all of reactant B is introduced via the feed stream to the first reactor.

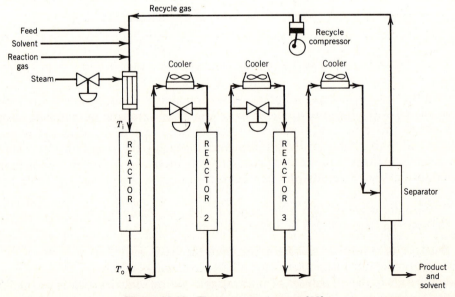

Figure 28.12 Three-reactor process [16].

4. **Conversion.** Approximately one-third of the feed is reacted in each of the three reactors. Virtually all of the feed is reacted by the time it leaves the third reactor. (Recall that the nominal value of the exit composition from the third reactor, the *leakage,* is $c_B = 0.15\%$.)
5. **Gas Separation.** The unreacted gas is recycled back to the first reactor from the gas–liquid separator.
6. **Process Gains.** The conversion across a reactor is proportional to the temperature rise, $\Delta T = T_o - T_i$, where T_o is the exit temperature and T_i is the inlet temperature. The process gain for the transfer function, $T_o(s)/T_i(s)$, is nonlinear and varies widely with reactor conditions. For the normal operating range of the reactor, the gain may change by a factor of ten. In the third reactor, the process gain for transfer function, $C_B(s)/T_i(s)$, can actually change by several orders of magnitude because of exponential dependence on reactor temperature and the importance of the first-order reaction rate.
7. **Reactor Dynamics.** The dynamic behavior of each reactor can be approximated by a time constant of 75 s and a time delay of 135 s. Thus, the time-delay-to-time-constant ratio is quite large, 1.8.
8. **Available Measurements.** The various flow rates and the inlet and exit temperatures for each of the three reactors are measured. The exit composition from the third reactor is measured with a continuous analyzer. However, this measurement is not used directly for control purposes because of the large process time delay associated with each reactor and the fact that the analyzer is not always in service.

Control Objectives

Since the feed material B is expensive, it is desirable to minimize both by-product formation and the amount of unreacted feed leaving the third reactor. By-product formation can be reduced by properly selecting the conversion for each reactor, and the amount of unreacted feed leaving the third reactor can be reduced by increasing the reaction temperature. But below a certain feed concentration, the rate of by-product formation increases rapidly and is further accelerated by elevated temperatures. Thus, control objectives can be summarized as follows:

1. Distribute conversion over the three reactors in a manner that minimizes by-product formation.
2. Maintain the leakage from the third reactor at a point (less than 0.1%, if possible) that minimizes the sum of the yield loss and unreacted feed.
3. Provide satisfactory control for all plant operating modes (not just the series mode discussed here) with a control scheme that is reliable and convenient to use.

Development of the Control Strategy

From the description of the process and the control objectives, it is evident that each reactor exhibits highly nonlinear behavior with the process dynamics dominated by the process time delay. Thus, it is unlikely that a conventional PI or PID controller would provide satisfactory control. Instead, some type of gain compensation and time-delay compensation should be employed (see Chapter 18). In addition, because the analyzer measurement is not suitable for control purposes, an inferential control strategy should be considered where the conversion in each reactor is calculated from a process model and available temperature and flow measurements.

The process model can also be used to maximize the overall yield by determining the desired conversion for each reactor. Each conversion can be controlled by manipulating the reactor inlet temperature via a heat exchanger on its inlet stream. On-line optimization will be required, because the optimum conversion for each reactor will change as the catalyst activity changes or the production rate changes. In general, the catalyst activity changes at different rates for each reactor. Also, small changes in feed impurities can produce significant changes in catalyst activity in just a few hours.

The various parts of the model-based, inferential control scheme will now be described. Additional information has been reported by Cardner [16].

On-Line Optimization

An overview of the resulting control scheme for the three-reactor system is shown in Fig. 28.13. Once each hour the optimizer reads the appropriate process data provided by the data acquisition system and performs supervisory control calculations. These calculations are based on a process model that consists of steady-state mass and energy balances. Before beginning the calculations, the process model is calibrated by calculating the current catalyst activity from 15-min moving averages of the analyzer and process variable measurements. The optimizer then determines the temperature (and consequently the conversion) for each reactor that will maximize the overall yield of desired product by minimizing the yield loss. By definition, yield loss = unreacted feed + by-products. This constrained, non-linear optimization problem is solved using a nonlinear programming method proposed by Newell and Himmelblau [18]. Typical variation of the yield loss with temperature for the on-line reactor optimization is shown in Fig. 28.14. The objective function to be minimized is the yield loss for a single reactor. Increasing the reactor exit temperature reduces the amount of unreacted feed but increases by-product formation. Thus, an optimum temperature T^{opt} exists. For the industrial application, the optimum values of all three reactor exit temperatures are determined simultaneously via nonlinear programming [16].

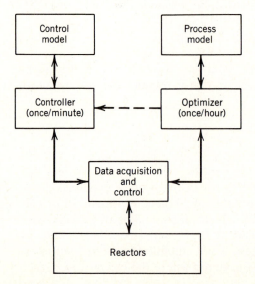

Figure 28.13 An overview of the optimizing computer control system.

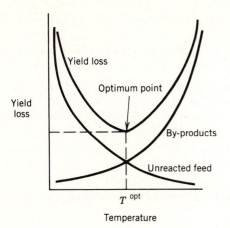

Figure 28.14 Optimization of yield loss by varying temperature (yield loss = by-products + unreacted feed).

After determining the optimum operating point, the optimizer then sends the corresponding set point for each reactor conversion and the leakage from the third reactor to the block in Fig. 28.13 labeled "controller." The calculated catalyst activity is also transmitted to the controller since this information is used in the control calculations.

Control Calculations

The control calculations are executed once per minute to provide dynamic regulation. The inferential controller is based on a process model that is used to calculate the conversion across each reactor and the leakage from the third reactor. The controller calculates the reactor inlet temperatures that are required to achieve the

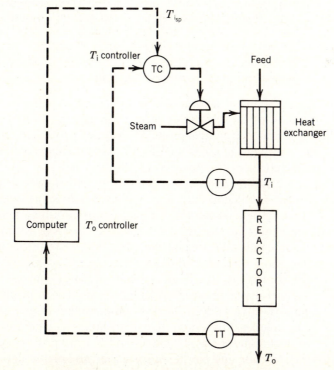

Figure 28.15 Inlet temperature control scheme.

conversion and leakage targets. These T_i targets are then used as set points for the T_i control loops, as is shown in Fig. 28.15 for the first reactor. For the second and third reactors, inlet temperature is controlled by manipulating the bypass flow rate around the cooler (Fig. 28.12).

The controller provides time-delay compensation via an analytical predictor (Chapter 26) and automatically adjusts the controller gains. The controller action is based on deviations in reactor temperature, conversion, and leakage [16].

The control model is also used to detect thermocouple failures or off-scale readings. If a single thermocouple is malfunctioning, the control model can detect this and provide an estimated value based on the energy balance so that control can be maintained. If more than one thermocouple is determined to have failed, the controller "disengages from all control" [17] and sends an appropriate message to the operator.

Control System Performance

Cardner [16] reported that the inferential (optimizing) control system has been very successful, having been in place for a number of years. Significant reductions in operating costs ("loss") were obtained with the advanced control scheme, as is shown in Fig. 28.16. An indication of the accuracy of the process model is shown in Fig. 28.17 for swings of 1°C in the inlet temperature to the first reactor. Although the model predictions of leakage were imperfect, the model-based control scheme provided significant improvement in yield loss. Cardner [16] has demonstrated that the inferential control scheme performed well during a series of extreme upsets in feed flow rate. He noted that an advantage of the inferential control scheme was that it reduced undesirable process interactions by controlling the three-reactor system as if it were a single reactor. A disadvantage was that the inferential control scheme could result in inappropriate responses during extreme upset (transient) conditions because only steady-state models were used in the optimizing and regulating controllers.

SUMMARY

The practice of process control is as much an art as it is a science. Although a thorough understanding of the principles of process dynamics and control is required, the solution of practical control problems depends very much on the insight, experience, and ingenuity of the control engineer. In particular, knowledge of the dynamic and static behavior of the process is indispensable.

In this chapter we have considered a number of broad issues that occur in process control problems. Difficult control problems can arise because of the limitations in the process design. Although a clever control strategy may provide a solution, in extreme situations the plant may not be operable without a design change. It is important to consider process dynamics and control issues *during* plant design, rather than waiting until after the design has been completed.

Selection of the controlled, manipulated, and measured variables represents a major decision in control system design. Guidelines have been presented for the selection of these important variables. The maximum number of manipulated variables is limited by the degrees of freedom that are available. Singular value analysis provides a powerful technique for analyzing a number of important control problems that include the selection and pairing of controlled and manipulated variables, and robustness analyses of alternative control schemes.

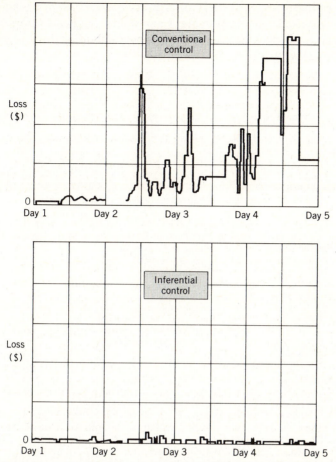

Figure 28.16   Comparison of losses due to nonoptimum operation for conventional control vs. inferential (optimizing) control.

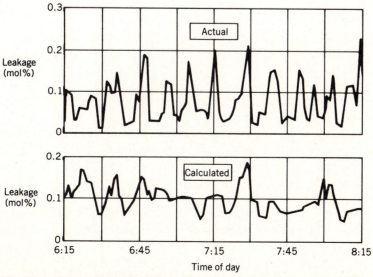

Figure 28.17   A comparison of actual and calculated reactant leakage exiting the third reactor for a cyclic disturbance.

Finally, an industrial case study of a three-reactor system has illustrated that advanced model-based control strategies can provide significant improvements over conventional control schemes. Such an approach embodies the underlying philosophy and message in this book.

REFERENCES

1. Shinskey, F. G., Uncontrollable processes and what to do about them. AIChE Annual Meeting, Los Angeles (Nov. 1982).
2. Hughart, C. L. and K. W. Kominek, Designing distillation units for controllability, *Instrum. Technol.* **24** (5), 71 (1977).
3. Shinskey, F. G., *Distillation Control,* 2d ed., p. 260–63, McGraw-Hill, New York, 1984.
4. Newell, R. B. and P. L. Lee, *Applied Process Control,* Prentice-Hall of Australia, Brookvale, NSW, Australia, 1988.
5. Brosilow, C. B. and M. Tong, The structure and dynamics of inferential control systems, *AIChE J.* **24,** 492 (1978).
6. Shinskey, F. G., *Process Control Systems,* 3d ed., McGraw-Hill, New York, 1988.
7. Cegrell T. and T. Hedqvist, A New Approach to Continuous Digester Control, Proc. 4th IFAC Sympos. on Digital Computer Applications to Process Control, Zurich, 1974, Pt. 1, p. 300.
8. Hougen, J. O., *Measurement and Control Applications,* 2d ed., ISA, Research Triangle Park, NC, 1979.
9. Newell, R. B. and D. G. Fisher, Model development, reduction and experimental evaluation for an evaporator, *IEC Process Des. Dev.* **11,** 213 (1972).
10. Findley, M. E., Selection of control measurements, in *AIChEMI Modular Instruction,* Series A, Vol. 4, (T. F. Edgar, Ed.), AIChE, New York, 1983, p. 55.
11. Klema, V. C. and A. J. Laub, The singular value decomposition: Its computation and some applications, *IEEE Trans. Auto. Control,* **AC-35,** 164 (1980).
12. Strang, G., *Linear Algebra and Its Applications,* Ch. 2, Academic Press, New York, 1976.
13. Skogestad, S., M. Morari, and J. C. Doyle, Robust control of ill-conditioned plants: High-purity distillation, *IEEE Trans. Auto. Control,* **33,** 1092 (1988).
14. Moore, C. F., Application of Singular Value Decomposition to the Design, Analysis, and Control of Industrial Processes, Proc. Am. Control Conf., 1986, p. 643.
15. Roat, S. D., J. J. Downs, E. F. Vogel, and J. E. Doss, The integration of rigorous dynamic modeling and control system synthesis for distillation columns: An industrial approach. *Chemical Process Control* CPC-III, (M. Morari and J. J. McAvoy, Eds.), CACHE–Elsevier, New York, 1986, p. 99.
16. Cardner, D. V., Model inferential optimizing computer control of three series reactors. ISA/84, Houston, TX (Oct. 1984).
17. Cardner, D. V., private communication (1984).
18. Newell, J. S. and D. M. Himmelblau, A new method for nonlinearly constrained optimization, *AIChE J.* **21,** 479 (1975).

— APPENDIX —

Professional Software For Process Control

Program	Source	Features	Price (*US $*)
ACS (Advanced Control System)	Prof. Gerry Sullivan, Dept. of Chem. Eng., University of Waterloo, Waterloo, Ont. Canada N2L 3G1	IBM mainframe, can be used for simulation or real-time operations, trend and faceplate displays, process schematics. Multiloop control custom strategies, application module for furnace, interface with SPEEDUP simulation package.	Ask
CC	Systems Technology, Inc., 13766 S. Hawthorne Blvd., Hawthorne, CA 90250-7083	PC, single and multiple loop, time and z-domain, frequency domain, root locus.	$3500 professional, $1100 academic (10 copies)
CLADP	Cambridge Control High Cross Madingby Road, Cambridge, England	PC or workstation. Controller design using pole/zero evaluation, matrix manipulation, Bode plots, Nichols charts, root locus, s-, z-, w-domains. Frequency response and state space methods for multivariable controllers. Icon-based version available for block diagrams (SIMBOL).	Ask
CONSYD	Prof. Harmon Ray Dept. of Chem. Eng. University of Wisconsin Madison, WI 53706	Mainframe, dynamic system analysis, multivariable control, all domains, modeling and identification.	$1500 educational
CTRL-C	Systems Control Technology 1801 Page Mill Road	Mainframe or workstation, matrix analysis, multivariable control,	Ask

Professional Software For Process Control (*Continued*)

Program	Source	Features	Price (US $)
	P.O. Box 10180 Palo Alto, CA 94303-0888	utilizes MATLAB, LINPAK, and EISPACK, state space and frequency domain analysis.	
CYPROS	CAMO A/S Jarleveien 4 N-7041 Trondheim, Norway	Mainframe, parameter estimation and Kalman filtering, simulation of continuous or discrete systems, optimal multivariable control, adaptive control, block-oriented simulation.	Ask
EASY5	Boeing Computer Services P.O. Box 24346 Mail Stop 7L-23 Seattle, WA 98124-0346	Mainframe or workstation, uses block diagram structure to solve nonlinear modeling and simulation problems, classical and modern control design tools, continuous or discrete time, all domains.	Ask
KEDDC	Ingenieurbuero Erbele Jahnstr. 73 D-7441 Grossbettlingen FR Germany	Mainframe, handles SISO or MIMO design using either classical or modern tools, time or frequency domain. Includes a variety of system identification methods, continuous or digital controller design. Performs block-oriented simulation. Also designs adaptive controllers.	$500–$1000 depending on system size.
MATLAB (w/control system toolbox)	Math Works Inc. Suite 250 20 N. Main St. Sherborn, MA 01770	PC, workstation or mainframe, classical and modern control tools, model conversions between domains.	Ask
MATRIX-X	Integrated Systems 151 University Ave. Palo Alto, CA 94301	Mainframe or workstation, based on MATLAB, provides classical or modern control tools, continuous or discrete time, adaptive control.	$8000
PC-PARSEL	Instrument Society of America, P.O. Box 12277, Research Triangle Park, NC 27709	PC, up to third-order transfer functions, PID controllers based on correlations, time and frequency domain analysis.	$450

Professional Software For Process Control (*Continued*)

Program	Source	Features	Price (*US $*)
Process Plus 3.0	Gerry Engineering Systems, 13310 West Red Coat Center Lockport, IL 60441	PC, PID loop simulator, provides 10 different industrial control algorithms, robustness plots, Bode plots.	$750
PROCOSP	D. R. Lewin Dept. of Chem. Eng. Technion-Israel Institute of Technology Haifa 32000 Israel	PC, interactive program for PID control.	Ask
ROBEX	Prof. Manfred Morari, Dept. of Chem. Eng. Cal Tech Pasadena, CA 91125	PC, SISO design, user-friendly for a wide range of skill levels, designs robust PID/IMC controllers.	Ask
SIM TUNE	CHC Systems P.O. Box 61114 Phoenix, AZ 85082-1114	PC, simple models, valve and transmitter characteristics, PID controllers, tuning diagnostics.	$850
TUTSIM	Applied i 200 California Ave. No. 214 Palo Alto, CA 94306	PC, suitable for analog, digital and nonlinear control. Uses block diagrams or bond graphs to implement system equations term by term. Operating characteristics of the model are displayed graphically.	$300–$600 professional, $30 student
UC-Online	Prof. Alan Foss Dept. of Chem. Eng. Univ. of California Berkeley, CA 94720	PC, can be used for simulation or real-time operations, trend displays, handles multiloop configurations and variable limits, PID, nonlinear elements, cascade, application modules for distillation and other operations.	$495

Index